TECHNICAL DESCRIPTIVE GEOMETRY

TECHNICAL DESCRIPTIVE GEOMETRY

B. LEIGHTON WELLMAN

Professor of Mechanical Engineering
Head of Division of Engineering Drawing
Worcester Polytechnic Institute

SECOND EDITION

McGRAW-HILL BOOK COMPANY

New York · Toronto · London · 1957

TECHNICAL DESCRIPTIVE GEOMETRY

Library of Congress Catalog Card Number 56-8872

15 16 17 18 19 20 – MAMM – 7 5 4 3 2 1 0

69234

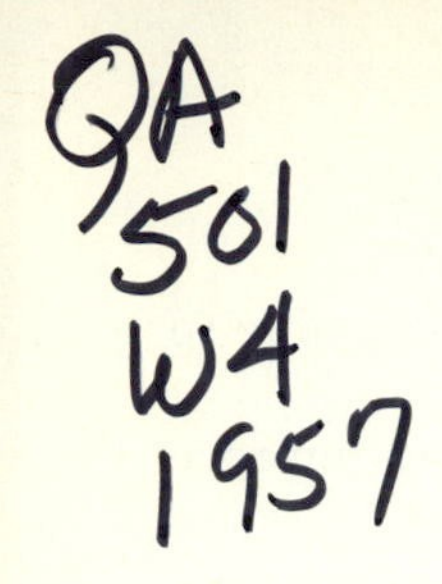

PREFACE

This book was written to provide students and industrial draftsmen with a complete up-to-date treatment of the important subject of descriptive geometry. With this purpose in mind the subject has been developed realistically from a natural, practical viewpoint, beginning with the most elementary concepts and progressing by easy stages to the complex intersection and development problems found in modern applications. Consistent with industrial practice the auxiliary view method is employed throughout the text. Views are classified simply as "adjacent" or "related," and attention is concentrated on the direction of sight for these views. Each view shows the object itself, for there are no imaginary planes and projections. The reference line has no spatial significance but serves only as a convenient device for constructing new views. Thus the draftsman thinks entirely in terms of the object and the logical relationships between the views. The emphasis is on *logic*, not on imagination.

Traditionally, descriptive geometry has been taught as a course in visualization. The emphasis has been placed on imagination, and therefore the unimaginative student has been foredoomed to failure. Years of experience have convinced the author that this approach is faulty. Descriptive geometry is a science based on sound facts and should be taught as a course in logical reasoning. Visualization must follow, not precede, reasoning. Imagination can aid, but must not determine, the solution.

As described in Chap. 1 of this text, the visualization of an object from a multiview drawing should be based on logical conclusions derived from observation and accurate analytical thinking. There must be no guesswork or flights of imagination. Students have frequently said to the author: "I can't *imagine* what this new view should look like, so how can I draw it?" The answer has always been: "Good! Since you have no preconceived idea, you can go ahead and draw the view without prejudice. Follow exactly the rules and principles that you know apply here. When the view is finished, you'll see what it looks like and you'll know that it is right." In this method of teaching, the student is taught to depend solely on established facts. He learns that he must take no action for which he cannot advance a sound reason. He attacks each problem by observing the hard, undeniable facts that are given; he recalls the proved principles that he knows he can rely upon; he then goes on by logical reasoning to the inescapable conclusion.

In this second edition every effort has been made to further improve the readability of the text. Each article has been reexamined, tested, and rewritten for maximum clarity. For easier reference boldface subheadings have been introduced, and key ideas are concisely expressed in italic type clearly set off from the adjacent text. Every article has been carefully planned to keep the text discussion and the corresponding illustration on the same or facing page. Whenever necessary the illustration has been repeated on the next page. Cardinal principles are stated in the form of rules—*not to be memorized* but to provide for quick review in as few words as possible. In similar manner the complete analysis of important problems has been followed by a concise summary, or "thumbnail analysis." Several new constructions have been added to the appendix.

Many illustrations have been added, others have been improved or simplified, and some have been completely redrawn. Complicated or difficult drawings that require several stages of construction are shown progressively in two to six separate drawings. Pictorial illustrations have been used freely to teach fundamental concepts. The scheme of notation is simple and easily remembered. A unique feature is that given points are labeled in boldface type to distinguish them from constructed and required points.

In response to many requests two new chapters have been added: Vector Applications, and Geology and Mining Applications. In these new chapters theory and basic principles are thoroughly explained and then illustrated with specific practical examples. In vector analysis several new graphical procedures have been introduced, and the principle of concurrency has been effectively applied to the general space-equilibrium problem. In geology problems the standard auxiliary-view methods are supplemented by the geologist's shorter and more compact one-view methods.

The arrangement of the material in twelve chapters with advanced topics at the end of each chapter will enable the instructor to plan a course of any desired length and yet preserve a logical and complete continuity. Chapter 1, although elementary, is a very important introduction to the succeeding chapters and should always be assigned. Chapters 2 to 12 may then be assigned in whole or in part, depending upon the scope of the course and the individual preference of the instructor. Chapters 9 and 10 can be assigned as given, in reverse order, or concurrently. Chapters 11 and 12 should be assigned only after adequate study of the related point, line, and plane topics in Chapters 2 through 5. Some of the advanced material may well be reserved for use in later courses. However, whether assigned or not, many of these topics will interest the more advanced students and should, of course, be available for reference.

All problems are arranged in groups at the back of the book. To simplify reference between the problem statement and the illustration, 65 new figures have been added. There are now 1,692 problems, of which 1,532 refer to specific fully dimensioned illustrations of the given data. Thus problems that are to be solved graphically are also presented graphically, a practical feature that saves time for both student and instructor. The problems have been carefully varied in each group to range from simple, straightforward problems to those requiring some original thinking. Enough problems have been provided to avoid repetition for four or five years. Numerical answers are given at the end of every problem that can be so answered, but, if desired, the instructor can invalidate any answer simply by changing one given dimension.

The author wishes to acknowledge the many valuable suggestions received from those who taught and studied the first edition. Their kind and generous criticisms are deeply appreciated. He is especially indebted to Prof. John M. Coke for material and suggestions on Geology and Mining Applications and to Prof. Frank A. Heacock for reviewing the chapter on Vector Applications. To his wife, Marjorie, he again expresses his appreciation for many hours of patient typing and proofreading.

B. LEIGHTON WELLMAN

CONTENTS

CODE OF LINES AND SYMBOLS

	Visible object lines and required lines.
	Hidden object lines.
	Revolved position of a line.
	Center line; axis of symmetry; axis of geometric solids.
	Construction lines; path of revolved points.
	Parallels connecting and aligning points in adjacent views (Art. 1·5).
F | R; a_F → a_R	Arrows indicate sequence of construction: point a_R located from point a_F.
F | R; c_F → ← c_R	Opposed arrows indicate independently located points that check by alignment.
T / F	Reference line between adjacent views (Art. 2·4).
Edge	Edge view of a plane, unlimited in extent (Art. 4·6).
C——P	Edge view of a cutting plane (Art. 4·25).
Axis; A | B; Axis – T. L.	Axis of revolution in adjacent views (Art. 5·1).
	Visible element of a surface (Art. 6·3).
	Hidden element of a surface.
5——5	Intersection of cutting planes with base planes (Art. 9·13).
	Bend line on a development (Art. 10·4).
A, B, C, D, E, F, . . etc.	Capital letters in text refer to points in space (Art. 1·6); lower-case letters refer to specific views of points.
$\mathbf{a_T}$, $\mathbf{b_T}$, $\mathbf{c_T}$, . . $\mathbf{a_F}$, $\mathbf{b_F}$, . . etc.	Boldface letters in figures indicate given points; subscripts indicate type of view (Art. 1·6).
e_T, f_F, g_R, . . x_A, y_B, . . etc.	Lightface letters in figures indicate required or constructed points.
a_T^R, c_F^R, e_R^R, . . m_A^R, p_B^R, . . etc.	Revolved positions of points.
1 2 3 . . 1′ 2′ 3′ . . 1″ 2″ 3″ . . 1R 2R 3R	A series of points each located by the same process; view subscripts omitted.
p_T	Line intersecting a surface.

CORRECTION SYMBOLS FOR INSTRUCTOR'S USE

✓	Correct within allowed limits.
✓+	Correct, but slightly outside allowed limits.
✓H	Correct, but slightly higher than correct value.
✓L	Correct, but slightly lower than correct value.
✓‡	Correct method, but considerably outside allowed limits.
X	Incorrect.
A	Error in alignment of points in adjacent views.
R	Error in reference-line measurement between related views.
M	Error due to incorrect measurement of answer.
LO	Error in laying out problem from given data.
S	Give scaled value — not distance on paper.
WS	Wrong scale used.
Q	Answer given does not agree with distance on paper.
WV	Wrong type of view.
PV	Poor choice or placement of views.
N	Inadequate notation. Points, lines, etc., not labeled.
V	Line should be visible.
H	Line should be hidden.
℄	Center line required.
//	Lines should be parallel.
⊥	Lines should be perpendicular.
⊗	Note obvious minor error within encircled area.
C	Curve poorly drawn.
T	Point of tangency inaccurate or incorrect.
D	Incorrect dimensioning.
I	Inaccurate work.
Inc	Incomplete solution.
P	Poor drafting; dull pencil, improper weight of lines.
U	Untidy work.

1. MULTIVIEW ENGINEERING DRAWINGS

1.1. The problem

We live in a world of three-dimensional objects and have become accustomed to a scheme of describing these objects by referring to their length, height, and depth. For countless centuries, ever since a cave man first drew a saber-toothed tiger on the wall of his cave, one great problem has perplexed every artist and draftsman: How can three-dimensional objects be accurately described on a two-dimensional surface? Length and height are easily shown, but depth, the third dimension, has always been elusive. The flat, depthless drawings of the ancient Egyptians show some progress, and by the subtle use of shades and shadows and a better understanding of perspective the painters of the Renaissance achieved a remarkable success. Photography has captured perspective perfectly, and the skilled cameraman can plan the lighting of the object to produce a very realistic picture.

But for many purposes, particularly in engineering work, a photograph may be unobtainable since the object to be pictured may exist only in the mind of the inventor or designer. Once the object has been made it can be photographed, but today the designer is seldom the maker; hence, before the object can be made, a complete and exact description must be conveyed from the mind of the designer to the mind of the maker. This is the function of the engineering drawing.

Many kinds of drawings have been evolved for this purpose of conveying exact size and shape description of an object from one mind to another, but none is completely satisfactory—not even the photograph. Each

Fig. 1·1. A photograph.

type of drawing has merit, but each is also deficient in some respect. In order to appreciate the superiority of the multiview engineering drawing it is desirable to consider the deficiencies of the other methods.

1·2. Pictorial drawings

Figure 1·1 is a photographic picture of a simple object. Evidently the camera was aimed toward the corner of the object and held slightly above it in order to show two sides and the top surfaces of the object. At a glance it is apparent that this is an L-shaped block with a circular hole in the lower leg of the L. One would also conclude that the rear, or far, side of the object is of the same size and L shape as the front, or near, side; that the depth of the object along the left side is the same as the depth along the right side; and that all the corners are square, or right-angled. These observations are true, but suppose that we attempt to verify these conclusions by actual measurement on the photograph.

Figure 1·2 shows the same photograph with true scales of measurement placed along the front and rear edges of the object. The width at the rear, which we believe to be the same as that at the front, is obviously shorter. Similar measurements of other apparently equal distances on the object produce the same result, and it is soon discovered that the shortest distances are always those farthest from the camera. This

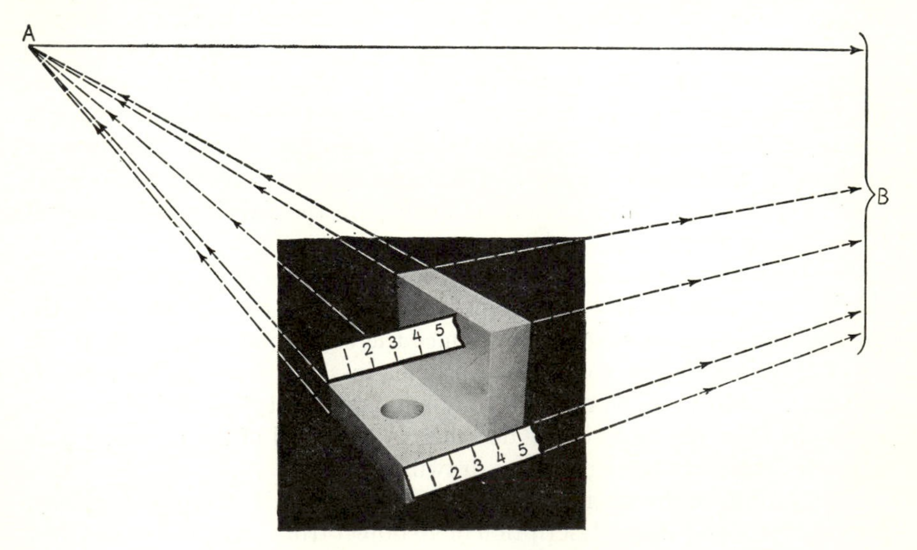

Fig. 1·2. Measurements on the photograph.

reduction of size with distance is caused by the fact that edges of the object which are actually parallel do not appear parallel in the photograph. This is convincingly demonstrated in Fig. 1·2 by the dash lines. When the edges of the object are extended in straight lines, these lines are not parallel but converge to a point. The horizontal lines of the object parallel to the depth dimension converge, when extended, at a point A on the left, while the horizontal lines parallel to the front side of the object converge at point B off the paper. This *perspective* effect is common to all true pictures or photographs and makes impossible the direct measurement of lengths and angles on the surfaces of the object. It should also be noted that accurate delineation may be obscured or confused by poorly arranged lighting.

If the object is not available for photographing, a line drawing can be made that will be an exact reproduction of the photograph. Such a *perspective drawing* is shown in Fig. 1·3. Shades and shadows can thus be eliminated, and the drawing can be made without having the actual object available, but the apparent convergence of parallel lines still prevents direct measurement of lengths and angles. This method of pictorial representation is very useful since, for example, it enables an architect to produce an exact picture of a proposed building. But for the purpose of actually constructing the building the perspective drawing is worthless since it does not show true distances.

The reduction of size with distance, which is inherent in the perspective drawing, can be eliminated by making all parallel edges of the object parallel in the drawing. Many variations of this scheme are possible, and

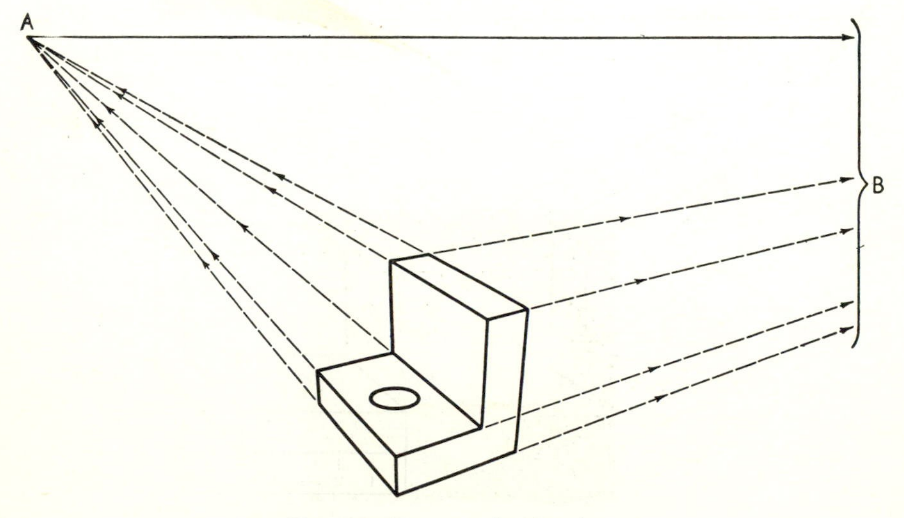

Fig. 1·3. A perspective drawing.

two of the most common are shown in Figs. 1·4 and 1·5.[1] The rulers placed on the drawings show that equal parallel lines on the object are now equal in the picture. This is a distinct advantage in spite of the unnatural appearance of the object, and such drawings are much easier to construct than perspective drawings. But not all equal distances appear equal: diagonals AB and CD, in each case, are actually equal on the object, but they are obviously unequal in the drawings. In the *isometric* drawing (Fig. 1·4) none of the right-angle corners appears as a right angle, and in the *oblique* drawing (Fig. 1·5) only those angles on the front surfaces appear in their true size. Thus these distorted pictures are only partly successful in showing true distances, but they do have the advantage of being easy to construct.

In each of the drawings thus far considered the object has been placed in front of the observer in a position that enabled him to see three sides of the object simultaneously. But it is now evident that in each case some or all of the true distances on the object have been distorted, and it might be concluded that a change in the position of the object might produce better results. Figure 1·6 shows the same object as it would appear in perspective from three other positions. In each case the object has been raised until the intermediate horizontal surface is now exactly at the same level as the observer's eye. In Fig. 1·6(b) the object has been raised to eye level from its previous position, but it has not been turned. A straight line drawn horizontally from the observer's eye to the object would strike the object at the corner marked A. This imaginary *line of sight* is pictured in Fig. 1·6(a) as an arrow.

This new position shown in Fig. 1·6(b) has produced an entirely different picture of the object. The top surface, which is now above eye level,

[1] Methods of construction will be found in any standard text on engineering drawing (see also Art. 3·21).

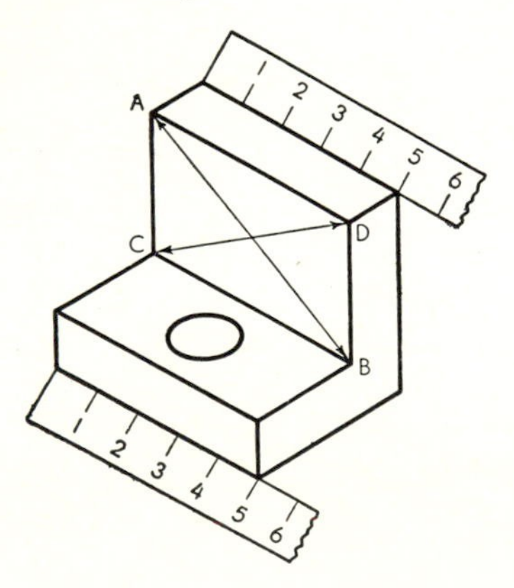

Fig. 1·4. An isometric drawing.

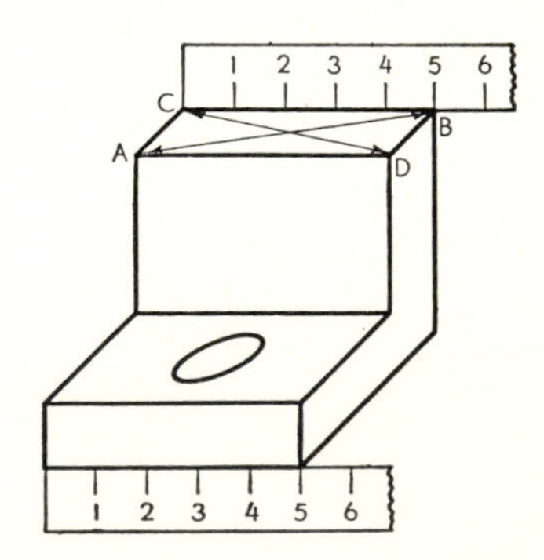

Fig. 1·5. An oblique drawing.

cannot be seen, and neither can the bottom surface, which is below eye level, but the intermediate surface, which is exactly at eye level, appears as a straight line because it is being viewed edgewise. The front and side surfaces are oblique to the line of sight, and true distances on these surfaces are still distorted. The existence of a hole is no longer evident.

Keeping the object at the same level, but turning it around to the right, the front surfaces can be brought directly before the eye as in Fig. 1·6(c). The line of sight is now perpendicular to the front surfaces and would strike the object at point B. Unfortunately the distinctive L shape of the object is no longer shown, and the hole is still invisible. In fact, with the object pictured thus, it is unrecognizable. But something has been accomplished: the two rectangular front surfaces of the object actually appear rectangular in the drawing, and many of the lines of equal length now appear equal. An obvious exception is that the width of the rear part of the object still appears smaller than the width of the nearer front part. This is attributable, as previously explained, to the fact that the rear part of the object is farther from the eye than the front portion, and hence appears reduced in size. The picture has a very flat appearance because the depth dimension is now entirely missing.

If the object is turned to the left instead of to the right, then the right-side surface can be brought directly before the eye so that the line of sight now strikes the object at point C. Figure 1·6(d) shows that the right side

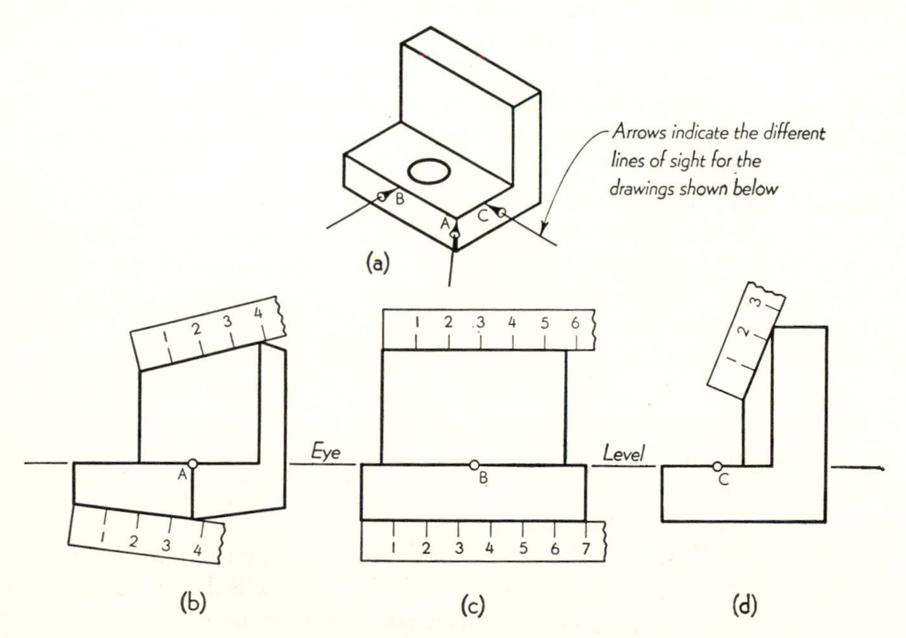

Fig. 1·6. Perspective drawings for other positions.

now appears in its true shape, and there is some evidence of depth although the depth dimension is very short.

To summarize the entire discussion of pictorial drawings we conclude:

1. Figure 1·3 gives the most realistic description of the object but also shows the greatest distortion of true distances.

2. Figures 1·4 and 1·5 give an adequate but unreal description of the object and show some true distances.

3. Figures 1·6(*c*) and (*d*) each give incomplete descriptions of the object, but they also show many true distances and true shapes.

It was stated in Art. 1·1 that the primary function of an engineering drawing is to convey an exact description of the size and shape of an object. It has now been demonstrated that no single picture of an object can show the three dimensions—length, height, and depth—in true size and shape. For this reason engineering drawings always consist of *two or more* separate views (or pictures) and hence are called *multiview drawings.*

1·3. The principal views

When the object shown in Fig. 1·6(*a*) was viewed from a position directly in front of point *B*, the picture obtained was that shown in Fig. 1·6(*c*). If we now imagine the observer to be an infinite distance away from the object, then all of the horizontal lines of sight would be exactly parallel, as shown in Fig. 1·7(*a*). Under these conditions the distance from the front of the object to the observer would be practically the same as the distance from the rear of the object to the observer, and as a result reduction of size with distance would be negligible. Thus viewed from an infinite distance the object would appear as in Fig. 1·7(*b*). The front surfaces of the object, which are perpendicular to the horizontal lines of sight, now appear in true shape and proportion, and the depth dimension has completely disappeared. This is the *front view* of the object.

In similar manner, if the observer looks horizontally in a direction perpendicular to the right side of the object and again views it from an infinite distance, the resulting view would be that shown in Fig. 1·7(*c*). Here the right side appears in true proportion, but the length dimension, which is parallel to the horizontal lines of sight, has disappeared. This is the *right-side view* of the same object.

A third view [Fig. 1·7(*d*)] can be obtained by looking straight down on the object from an infinite height. Again it is notable that one dimension, namely, the height, has disappeared. This is the *top view* of the object.

These three views obtained by viewing the object from three mutually perpendicular directions are called the *principal views*. The following important facts regarding these views should be carefully noted:

1. Surfaces that are perpendicular to the lines of sight for a given view appear in true size and shape in that view.
2. The lines of sight for each view are perpendicular to the lines of sight for each of the other two views.
3. Each view shows only two of the three dimensions of the object.
4. Taken separately no one of the three views can provide a complete description of the object.

The following considerations will indicate why the four facts stated above are important. Given an object that is to be described with the three principal views, the draftsman must first decide upon the direction of the line of sight for each view. Since he is primarily interested in showing as many of the various surfaces in their true size as possible, he selects the directions of sight perpendicular to the main surfaces of the object. He must not forget, however, that the three directions selected must be mutually perpendicular. This is an essential condition since each view should show only two of the three dimensions. For example,

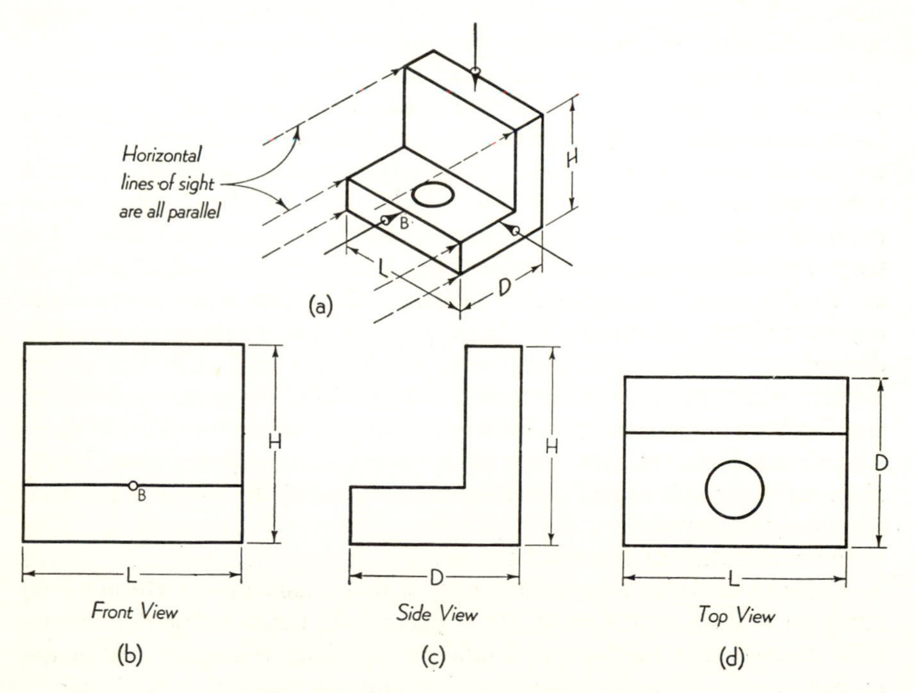

Fig. 1·7. Principal views of the object.

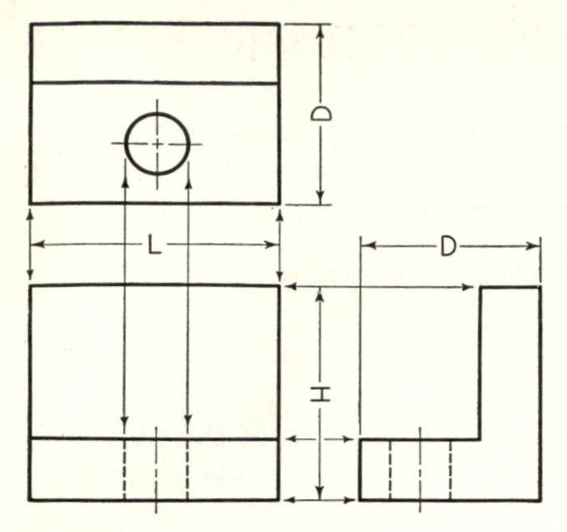

Fig. 1·8. American standard arrangement of views.

the front view shown in Fig. 1·7(*b*) shows height *H* and length *L*, but not depth *D*, because the lines of sight are parallel to the depth dimension. If the right-side view in Fig. 1·7(*c*) is to show only height *H* and depth *D*, then the length dimension must be parallel to the direction of sight for that view and hence perpendicular to the direction of sight for the front view. The fourth fact stated above is a logical sequel to the first three since no three-dimensional object is fully described by two dimensions only.

1·4. Arrangement of the views

It has been demonstrated that the principal views provide an accurate means of shape description, and also that two or more views are necessary for complete description of a given object. In Fig. 1·7 each of the three views has been labeled to indicate from what direction the view was taken. This labeling of the views would not be necessary if the views were always placed on the paper in a logical standard arrangement.

In Fig. 1·8 the three principal views have been arranged in accordance with standard practice in the United States and Canada. The top view is placed directly above the front view, and the right-side view is placed to the right of the front view. Not only is this a logical and natural placement of the views, but it also correlates those dimensions which are common to two views. The length *L* of the object appears equally in the front and top views; hence these equal lengths have been aligned one above the other in Fig. 1·8. The height *H* is the same whether the object is viewed from the front or from the side, and this equality is evidenced by the horizontal alignment of the front and right-side views. Finally, although direct alignment is not possible, the depth *D* shown in the top view must exactly equal the distance *D* shown in the side view. The distance between the views may be varied to suit the space available, but the *position* and *alignment* of the views just described must be rigorously observed.

In none of the previous drawings has the depth of the hole been shown. In order to indicate that the cylindrical hole goes entirely through the lower part of the object, dash lines representing the invisible sides of the hole are shown in the front and side views of Fig. 1·8. It is customary on engineering drawings to show all lines and surfaces in each view, and

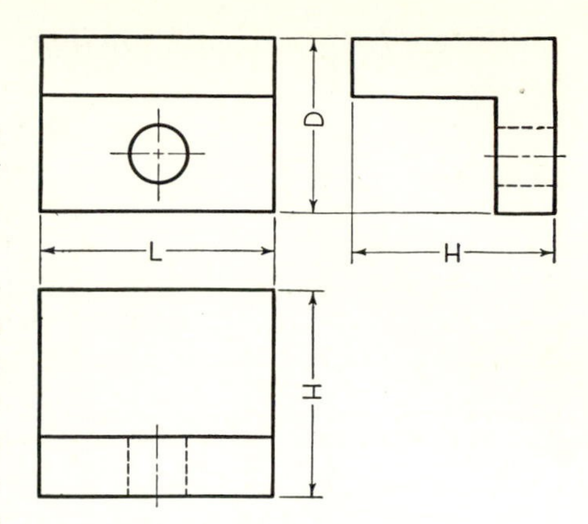

Fig. 1·9. Alternate standard arrangement of views.

those lines or contours which are hidden or invisible in any view are shown as dash lines (see Fig. 1·10).

An acceptable alternate arrangement of the principal views is shown in Fig. 1·9. Here the top and front views are placed exactly as in Fig. 1·8, but the right-side view has been aligned opposite the top view instead of opposite the front view. This scheme is entirely logical since the top view and the side view both show the same common dimension D. It was necessary, of course, to turn the side view to the position shown in order to achieve alignment with the top view. It should be noted that the height H in the side view must still equal the same height H appearing in the front view.

Although three mutually perpendicular views will completely describe any object, these views need not be confined to top, front, and right side. The object can also be viewed from the bottom, rear, or left side. There are therefore six possible principal views, and the standard arrangement for all six is shown in Fig. 1·10. It is apparent that this is simply an expansion of the arrangement shown in Fig. 1·8. Logically, the left-side view is placed to the left, and the bottom view is placed below the front view. Although the rear view could be aligned opposite the right-side, top, or bottom views, it has arbitrarily been placed as shown in Fig. 1·10 for standardization.

Since the purpose of a multiview drawing is to describe completely and clearly the size and shape of an object, then for clarity and simplicity the number of views employed should be a minimum. Examination of the six views in Fig. 1·10 shows that there is much repetition. The right- and left-side views are identical, though reversed, and one would be sufficient: traditionally, the right is preferable. The top and bottom views differ only in visibility and reversed position. This is also true of the front and rear views. Since draftsmen prefer views with a minimum of hidden lines, the bottom and rear views should be discarded. Thus, although six views can be drawn, three are obviously unnecessary in this case, and the three views originally drawn in Fig. 1·8 are retained.

Under certain circumstances two views may be sufficient, but care must be observed in selecting the two views to avoid ambiguity. For example, Fig. 1·11(a) shows only the top and front views of Fig. 1·8. These two views are not sufficient since they could represent the original object or

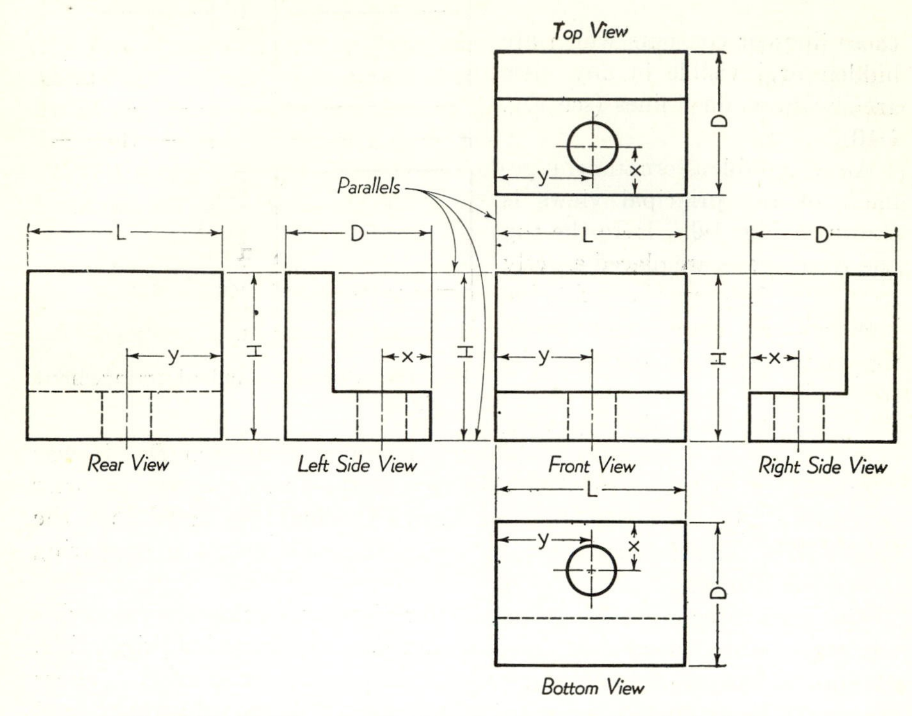

Fig. 1·10. American standard arrangement for six principal views.

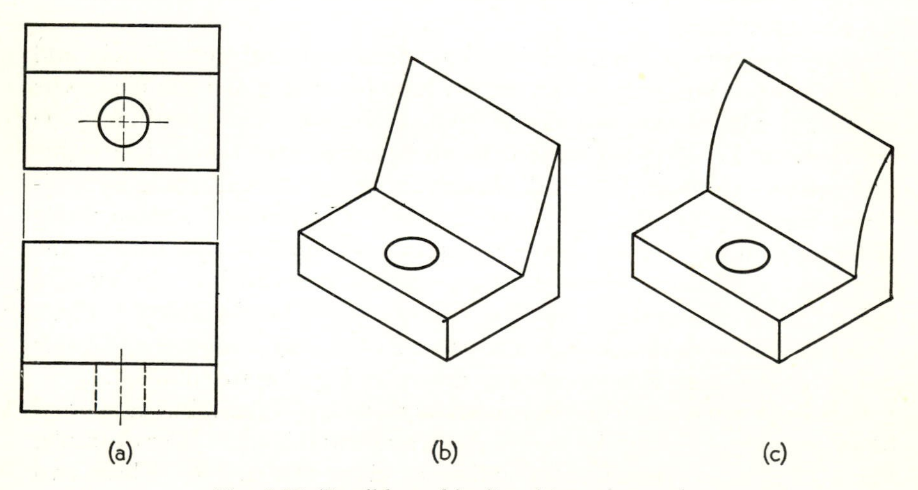

Fig. 1·11. Possible ambiguity of two views only.

either of the two objects shown at the right in Fig. 1·11 at (*b*) and (*c*).

If the object shown in Fig. 1·8 is turned around so that the left side is toward the front, then the top and front views shown in Fig. 1·12(*a*) would completely describe the object, and it could not be mistaken for either of the objects pictured in Figs. 1·11(*b*) and (*c*). Figures 1·12(*b*) and (*c*) show top and front views of these alternate objects for comparison.

1·5. Relationships between views

The general features of multiview drawings having been considered, the following three terms are now defined as they will be employed throughout this text:

1. Any two views placed side by side to align their common dimension shall be designated as ***adjacent views.***
2. The parallel lines connecting and aligning adjacent views shall be called ***parallels*** (see Fig. 1·10).
3. All views adjacent to the same view shall be designated as ***related views.***

In Fig. 1·8 the top and front views are adjacent views, and the front and side views are adjacent views, but the top and side views are related views. In Fig. 1·9 the top and side views are adjacent views, but the front and side views are related views. In Fig. 1·10 the top, bottom, and right- and left-side views are all related views since each is adjacent to the front view.

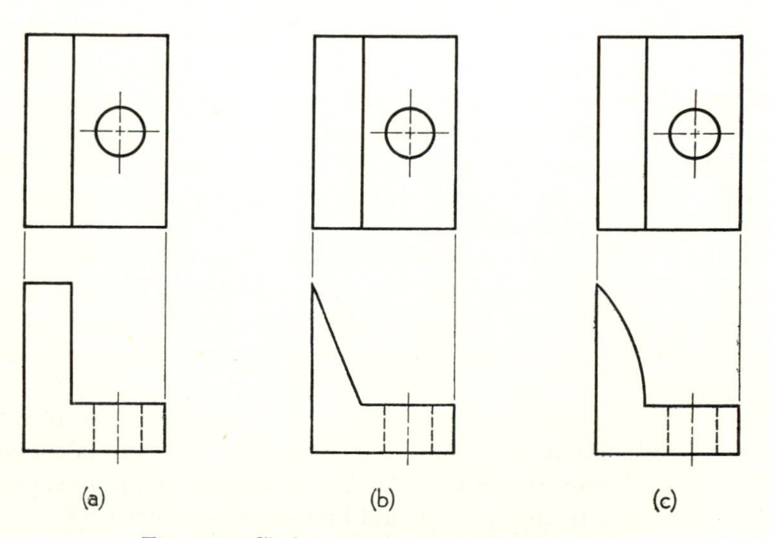

Fig. 1·12. Choice of views to avoid ambiguity.

The use of parallels should be noted in Fig. 1·10 (repeated on the next page). These are very light lines constructed by the draftsman to assist him in aligning the various adjacent views. A few or all of these lines are frequently left on the finished drawing to guide the eye of the reader in comparing the views. The parallels may also be thought of as representing the direction of the line of sight for any view.

Employing the terms defined above, we may now state the following *three basic rules* that govern the relationships of all views of a given object:

- ***Rule 1. Rule of perpendicularity.*** *The lines of sight for any two adjacent views must be perpendicular.*

- ***Rule 2. Rule of alignment.*** *Every point on the object in one view must be aligned on a parallel directly opposite the corresponding point in any adjacent view.*

- ***Rule 3. Rule of similarity.*** *The distance between any two points on the object measured along the parallels must be the same in all related views.*

That Rule 1 states an important and necessary condition has already been demonstrated in Art. 1·3, and the arrangement of views proposed in Art. 1·4 shows that Rule 2 states a very desirable condition. Reference to Fig. 1·10 will explain the application of Rule 3 to related views. Consider the position of the hole in the top, bottom, and right- and left-side views. The top view shows that the center of the hole is located a distance x from the front surface of the object. Since distance x is measured in the direction of the parallels connecting the top and front views, which are adjacent views, it must, according to Rule 3, also appear similarly in all views that are related to the top view. Comparison of the four related views which are adjacent to the front view shows that this is true. It should be noted that distances such as x must always be measured in the direction of the parallels.

The distance y shown in the top view of Fig. 1·10 must obviously agree with distance y in the front and bottom views because of the alignment of these views (Rule 2). However, the distance y shown in the front view must also appear similarly in the rear view since these are related views (Rule 3).

Figure 1·13 illustrates a common error in the construction of related views. Here the right-side view has been incorrectly drawn in a reversed position. This error can be avoided if we note that the *front* surface of the object in each related view should be *toward* the *front* view.

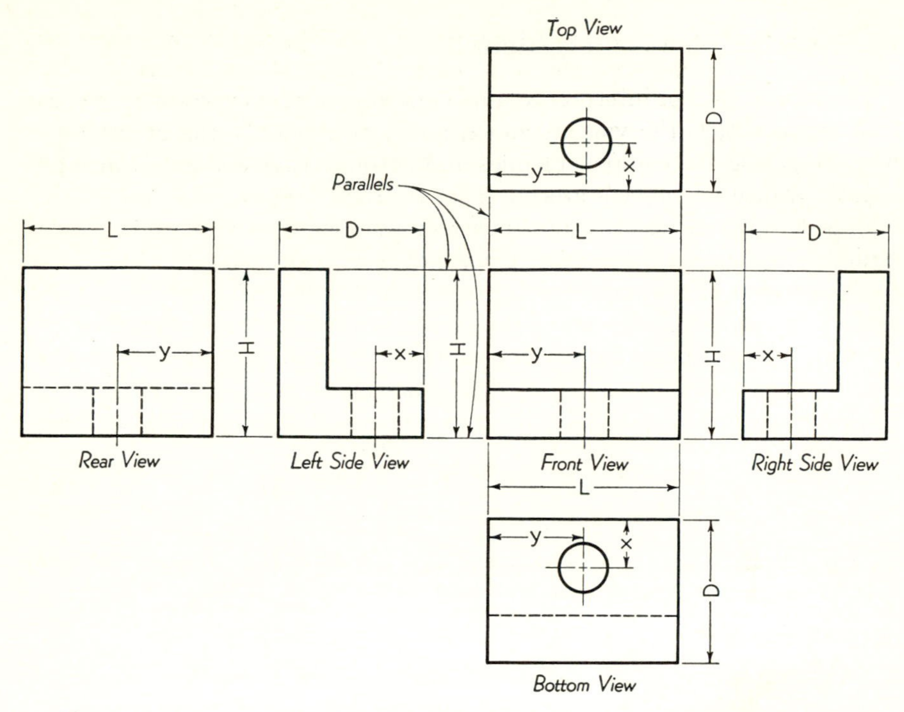

Fig. 1·10 (repeated). American standard arrangement for six principal views.

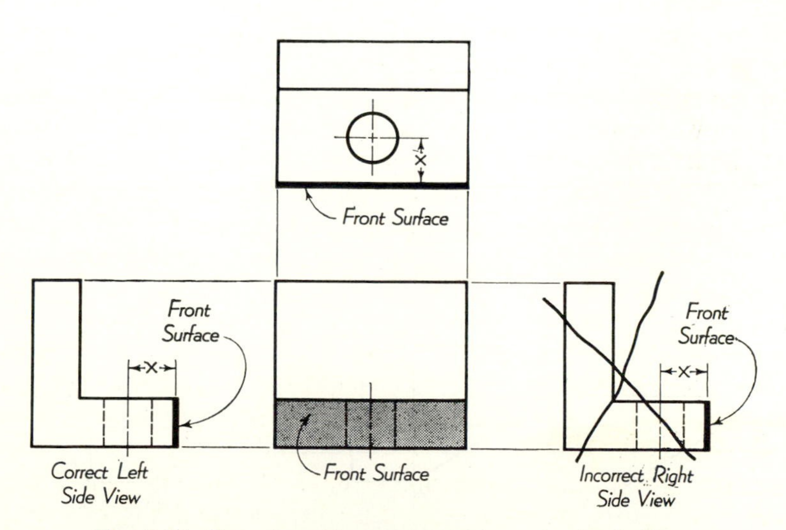

Fig. 1·13. Correct and incorrect positions of related views.

1·6. Principal views of a straight line

In order to read or interpret a multiview drawing it is necessary to analyze and compare the various views, studying not only the object as a whole but also the component lines and surfaces that compose the solid object. Since all objects are bounded by surfaces and all surfaces are bounded by lines, we shall now consider the principal views of a single straight line and later the principal views of a single plane surface.

The position of a straight line in space can be described by showing it in each of the principal views, and its length, or extent, can be indicated by labeling the two points at its extremities. In Fig. 1·14 are shown the *seven* typical positions of a line *AB*. The capital letters *A* and *B* refer to the actual line in *space;* the lower-case letters *a* and *b* are used to label

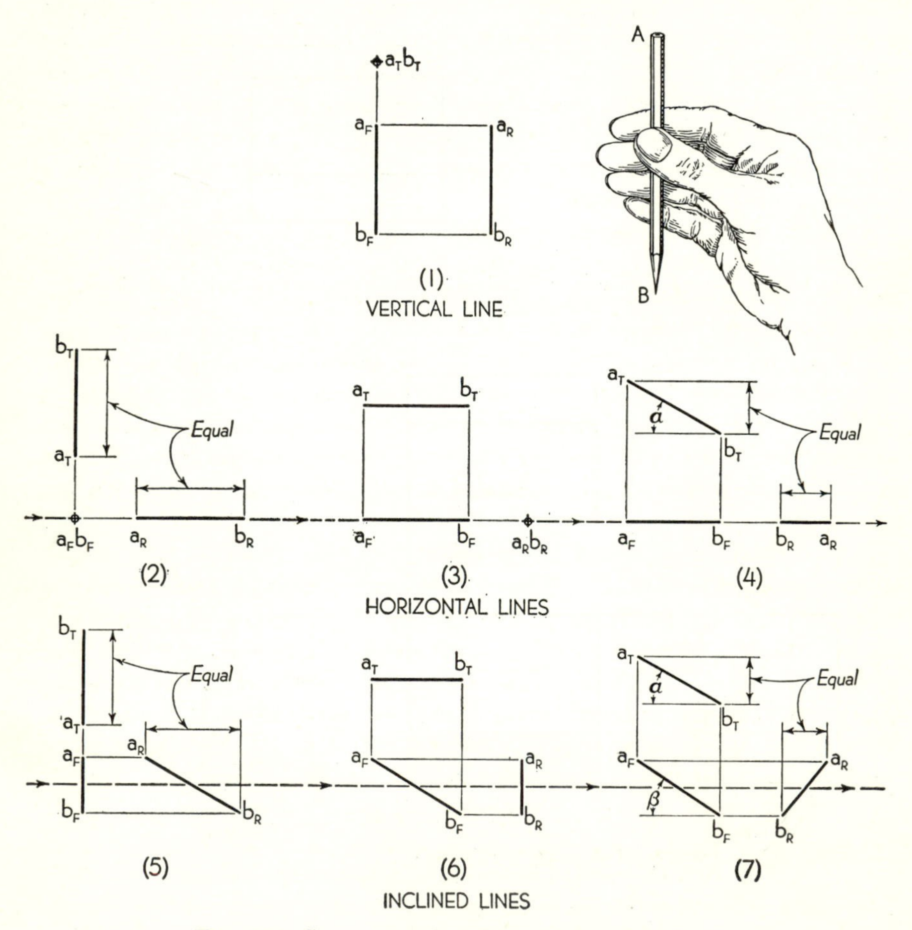

Fig. 1·14. Seven typical positions of a straight line.

these same points as they appear in each of the *views*. As a further distinction, the top view of point A is labeled a_T, the front view is labeled a_F, and the right-side view a_R.

The various positions of line AB in Fig. 1·14 have been classified as vertical, horizontal, or inclined. The vertical line is unique by definition, for the word "vertical" has only one meaning—"perpendicular to the surface of the earth." Since the upper end of a vertical line is directly above the lower end, the line will appear in the top view as a dot, or point. We must refer to the front (or side) view to see that in this case A is the upper end of the line.

A horizontal line has all points on the line at the same height, or elevation. In Fig. 1·14(2), (3), and (4) a light dash line has been drawn horizontally through the front and side views to indicate that points a_F and b_F and a_R and b_R are at the same level, or height, in these views. Thus a line may be identified as horizontal by this characteristic, but the line may appear in the top view in a variety of positions. The positions shown at (2) and (3) are special positions that cause the line to appear as a point in either the front or the side view. The position shown at (4) illustrates the general case of a horizontal line where the line appears in the top view at any angle α.

An inclined line has one end of the line higher than the other end. Here again, a light dash line has been drawn horizontally through the front and side views of each of the three inclined lines shown in Fig. 1·14, in order to show that in every case point a_F is higher than point b_F and a_R higher than b_R. Thus lines that are inclined can be recognized as inclined only in the front or side views, and the line may appear in the top view in various positions. The positions shown at (5) and (6) are special positions of an inclined line, but the position shown at (7) illustrates the most general case where the angles α and β may assume any values consistent with the rules of Art. 1·5.

It should be noted that the top, front, and side views of each of the seven lines shown in Fig. 1·14 have been drawn in accordance with the rules of Art. 1·5. Point a_F is always directly below a_T and is always aligned directly opposite a_R (Rule 2, the rule of alignment). The distance from a_T to b_T in the top view measured along the parallels always equals the distance from a_R to b_R along the parallels in the side view (Rule 3, the rule of similarity). These equal distances have been indicated in Fig. 1·14.

1·7. Principal views of a plane surface

In order to study the various positions that a plane surface might assume, imagine a square sheet of cardboard as the plane surface. The corners of the square can be labeled A, B, C, and D. If this plane is held in a horizontal position, then the principal views will appear as in Fig. 1·15(1). Since all points on a horizontal plane are at the same height, the front and side views of this plane will appear as horizontal lines. In other words, the plane appears in each of these two views as an *edge*. The top view shows the plane in its true size and shape. The horizontal plane is thus unique since it will always appear true size in the top view and as an edge in the front and side views.

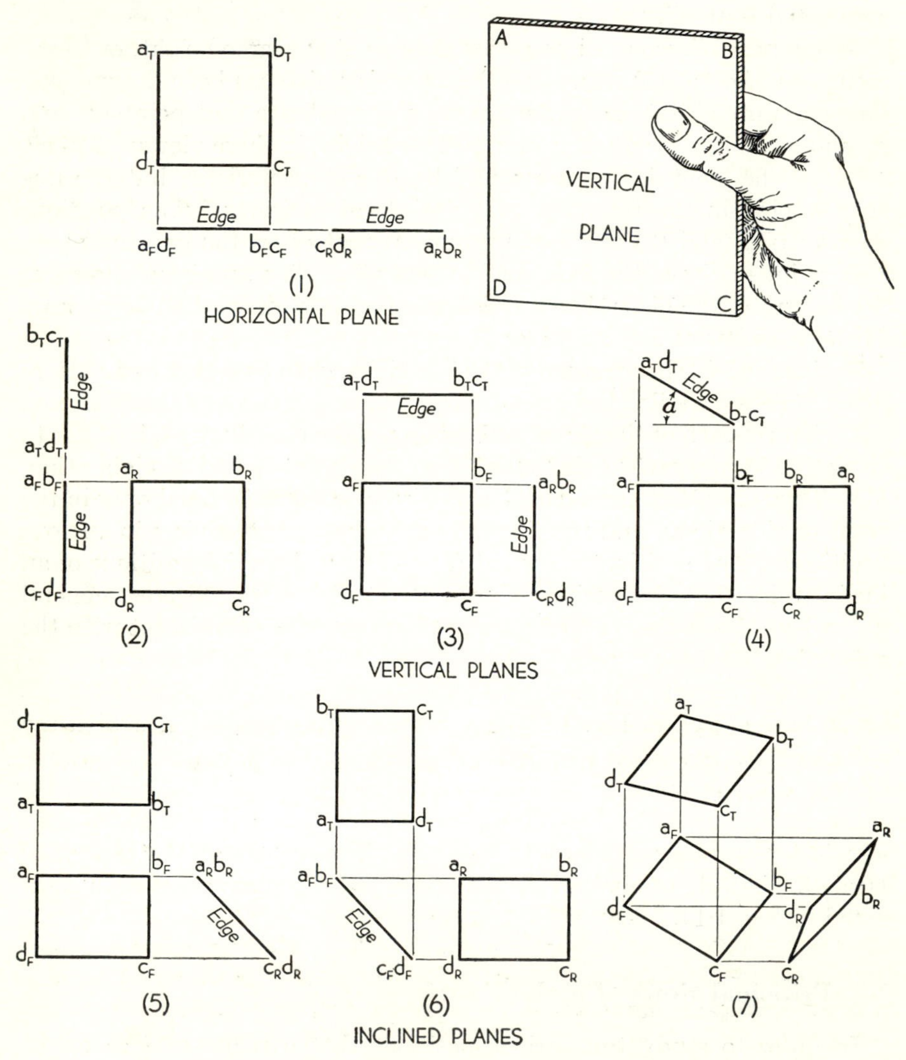

Fig. 1·15. Seven typical positions of a plane surface.

If the plane is held in a vertical position, then the top view will always show the plane appearing as an edge since the upper edge of the plane will always be directly above the lower edge. If the vertical plane is turned so that it also appears as an edge in the front view, as shown in Fig. 1·15(2), then the side view will show it in true size. If the vertical plane is held so that the front view shows true size, then the side view shows the plane as an edge as in Fig. 1·15(3). If, however, the vertical plane is turned at some angle α, then in the front and side views the plane will appear neither as a line or edge, nor as a square or true size. Thus a vertical plane may appear in a variety of positions, but it can always be identified as vertical by its appearance in the top view as an edge.

Inclined planes are those which are neither horizontal nor vertical. Therefore, an inclined plane cannot appear in the top view as a true size or as an edge. Examination of the top views of the three inclined planes shown in Fig. 1·15 reveals that in no case does the plane appear as a square, which is its true shape. An inclined plane can appear as an edge in either the front or the side view, as illustrated in positions (5) and (6), or it may be turned so that it does not appear as an edge in any view, as in position (7). This last case illustrates the most general position that a plane may assume. In summary, an inclined plane may appear as an edge in the front or side views, but it never appears as a true size in any of the principal views.

For the purpose of the above discussion the plane surface shown in Fig. 1·15 was assumed to be a square sheet of cardboard. In the seven positions shown, involving a total of 21 views, it is notable that the square plane surface always appeared either as a *line* or as a *four-sided figure.* When it appeared in any view as a four-sided figure, that figure was either a square, a rectangle, or a parallelogram, but obviously it did not appear as a triangle or three-sided figure. These statements may appear to be self-evident, but they form the basis for a very useful principle:

- ***Rule 4. Rule of configuration.*** *Every plane surface, regardless of shape, always appears either as an edge or as a figure of similar configuration.*

Figure 1·16 illustrates the above principle by showing how five different plane figures might appear in the principal views. Figure 1·16(1) shows an inclined triangular surface which appears as an edge in the front view, a position similar to that shown in Fig. 1·15(6). The triangle in the top view is different in size and shape from that seen in the side view, but both are triangles; hence these views are of *similar configuration.* The L-shaped surface shown in Fig. 1·16(2) is of different size and shape in the top and front views but still retains its characteristic outline. The plane figure shown in Fig. 1·16(3) does not appear as an edge in any view, and although the angles of the figure are different in each view, there is a similarity of configuration in all three of the principal views. A circle standing on edge is shown in Fig. 1·16(4). The front and side views both show the circle as an ellipse. If the reader now understands the application of this principle, he should have no difficulty in selecting the incorrect view in Fig. 1·16(5).

1·8. Reading the multiview drawing

The multiview drawing is the written language of the engineer, and it is the experience of most engineers and teachers that it is easier to write the language than it is to read it. If given a pictorial sketch similar to that shown in Fig. 1·3, 1·4, or 1·5, the average student has little difficulty in making the corresponding engineering drawing such as is shown in Fig. 1·8 or 1·9. But when the picture is lacking and only the multiview drawing is available, then the draftsman or workman must analyze the various views and evolve a mental picture of the object described by the drawing. The beginner views with awe the rapidity with which the experienced draftsman reads and interprets the views, and he frequently concludes that this is the result of some special gift of intuition. Actually, like the art of reading the printed page or a musical score, it is the result of logi-

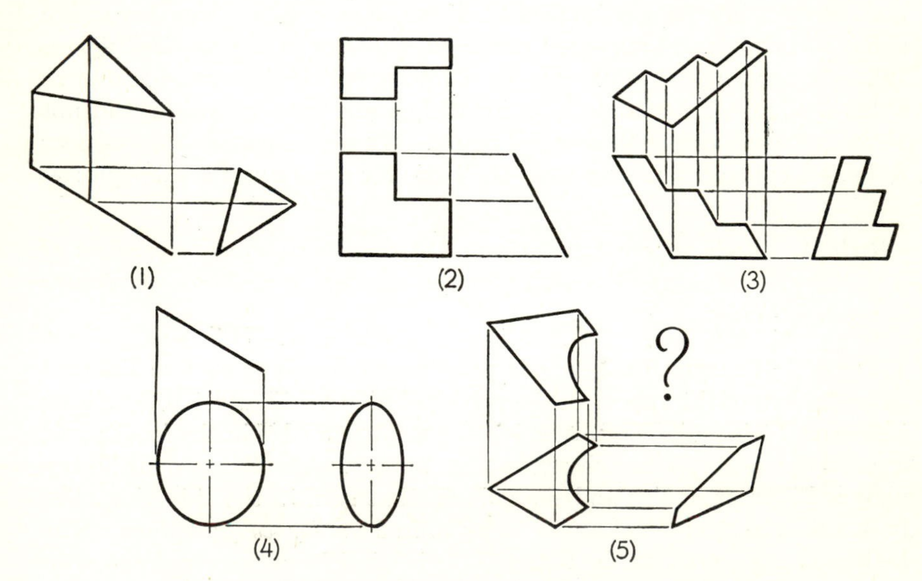

Fig. 1·16. Similar configuration of various plane surfaces.

cal reasoning, orderly thought, and much practice. The general principles and specific methods involved in reading a multiview drawing will now be discussed.

A full page of printed matter cannot be absorbed at a glance but must be read word by word, sentence by sentence in orderly sequence. Similarly, a complex drawing must be read by methodically examining the details of its component parts, the geometrical solids that compose the object, the plane or curved surfaces that bound the solids, and the lines that bound the surfaces. How minutely these details must be analyzed depends, of course, on the form of the object. Some of the simpler objects are the hardest to visualize correctly. The following examples will illustrate the methods of analysis.

1·9. Analysis by solids

Figure 1·17 shows the three principal views of an object, which can be analyzed easily by examination of the geometrical solid forms involved. A general survey of the views brings out these salient features:

1. Viewed from the top the object is rectangular. Visible on or above the rectangular surface *A* are three circular areas *B*, *C*, and *D*. On the right end toward the front the group of dash lines at *E* indicate something hidden below the rectangular surface *A*.

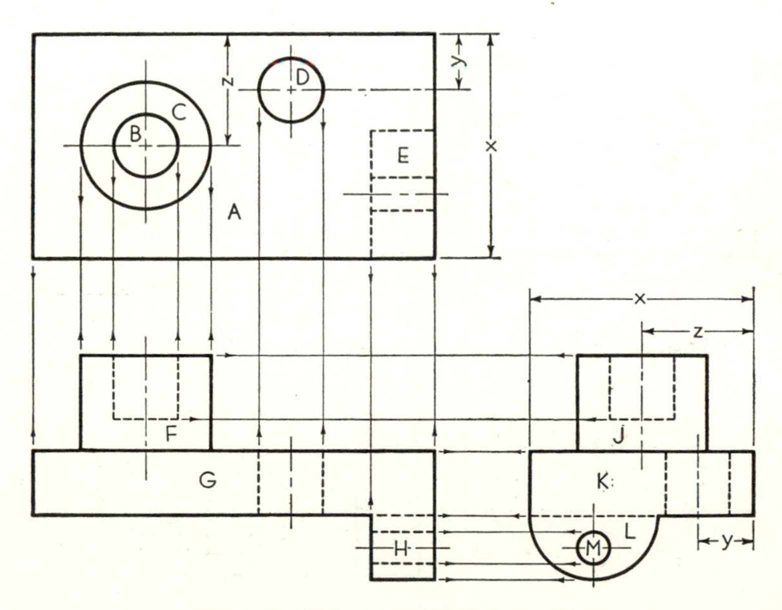

Fig. 1·17. Object for analysis by solids.

2. Viewed from the front the object shows three visible rectangular areas *F*, *G*, and *H*. Area *F* is above, and *H* is below, area *G*. The various dash lines indicate that there are other features hidden behind areas *F*, *G*, and *H*.

3. Viewed from the right side there are four visible areas *J*, *K*, *L*, and *M*. Area *L* is semicircular; *J* and *K* are rectangular. Area *M* is circular. Again, dash lines indicate details that are invisible when viewed from the right side.

Having inspected each view in a general way, we must now compare the views, noting particularly the alignment of the areas in one view with those in adjacent views. To facilitate this comparison, parallels have been drawn lightly between adjacent views. Correlating the various areas and features, we find:

1. Area *A* in the top view extends from extreme left to extreme right and hence can be aligned only with area *G* in the front view. But area *G* is directly opposite rectangular area *K*. Hence we conclude that the areas *A*, *G*, and *K* are the top, front, and right-side views of a rectangular prism which appears to be the main part of the object.

2. Area *F* in the front view extends above the main part of the object and hence should be visible in the top view. This is confirmed by the fact that the visible circular area *C* lies directly above area *F*. Thus this upward-projecting part of the object must be a cylinder. Area *J* represents the cylinder as seen from the right side.

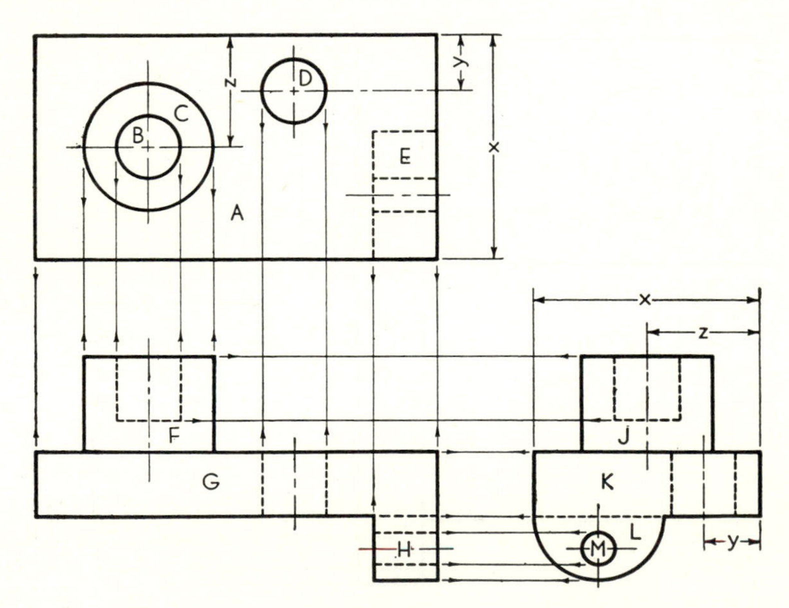

Fig. 1·17 (repeated). Object for analysis by solids.

3. Area H in the front view extends below the main part of the object and hence should be invisible in the top view. This is confirmed by the group of dash lines at E, which are directly above area H. The rectangular shape of areas E and H might be mistaken for a rectangular solid until we see that the semicircular area L directly opposite area H must be the side view of this same part of the object.

At this point the reader begins to assemble in his mind the general outlines of the object. He should see in imagination a rectangular block, with a cylinder standing on the left end of it, and extending downward from the right-front corner of the block a semicircular projection that faces to the right. To this mental picture can now be added the other details, one by one, as follows:

4. The small circle B in the top view is directly above the dash lines in area F in the front view; hence this must be a round hole in the center of the cylinder. Only the front and side views show the depth of this hole.

5. The dash lines in the front view directly below circle D show that this is a round hole in the rectangular prism and that the hole passes completely through the prism from top to bottom. Similar dash lines in area K show this same hole in the side view.

6. The circular area M in the side view is easily identified now in similar manner as a hole through the semicircular projection.

The reader should now have a clear mental picture of the entire object, and by referring to Fig. A·30 in the Appendix he can compare his mental picture with a pictorial drawing of this same object.

In the above analysis a complete mental picture has been achieved simply by observing the alignment between the top and front views and between the front and side views (the rule of alignment, Art. 1·5). Although it was not necessary to the analysis, the rule of similarity also applies. Notice that such distances as x, y, and z shown in the top view must exactly equal the corresponding distances in the related side view.

1·10. Analysis by surfaces

In Fig. 1·17 the various features of the object were very fortunately arranged, for at no time did an area in one view align with two possible areas in the adjacent view. When this condition occurs, it is frequently necessary to consider the separate surfaces that bound the solids. Figure 1·18 shows a simple object that will illustrate this method. Since there are no curved lines in any of the three views, the object must be bounded only by plane surfaces. Then the three areas A, B, and C in the top view represent three visible rectangular plane surfaces, and since each of these three areas is adjacent, or contiguous, to each of the other two, they must all lie in different planes. If areas B and C were in the same plane, then there would be no dividing line between them. This can be stated as a rule:

- ***Rule 5. Rule of contiguous areas.*** *No two contiguous areas can lie in the same plane.*

Therefore, by this rule, areas D and E in the front view must lie in different planes, and the same is true of areas F and G in the side view.

As an aid to discussion, Fig. 1·19 shows only the top and front views of Fig. 1·18, but with the visible areas in each view slightly separated to emphasize their individuality. Comparing the two views it is evident that area A matches either area D or area E by alignment. But, on the basis of Rule 4, the shape of area A is not at all like areas D or E; area A is four-sided, and area D is five-sided; area A has parallel sides, and area E is trapezoidal in shape. Therefore area A cannot be the top view of either surface D or surface E. Areas B and C do not align with D or E; nor are they of similar configuration. From these observations the following logical conclusions can be drawn:

1. Areas D and E must represent surfaces that appear *edgewise* in the top view. Hence they are *vertical* surfaces.

2. Areas A, B, and C must represent surfaces that appear *edgewise* in the front view.

But final acceptance of the above conclusions must be withheld until the side view has been considered, and it must be remembered that we have studied only the visible surfaces in each view.

In Fig. 1·20 the visible areas of the front and side views have been slightly separated without appreciably disturbing their correct alignment. Checking alignment only, area D is the same height as area G and directly opposite it. The same is true of areas E and F. But area D cannot be

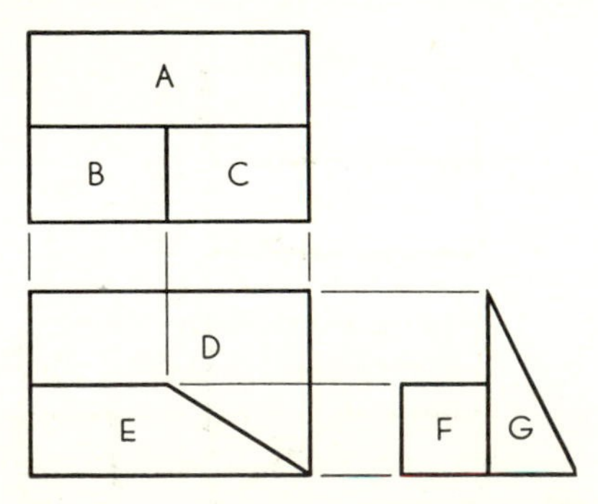

Fig. 1·18. Object for analysis by surfaces.

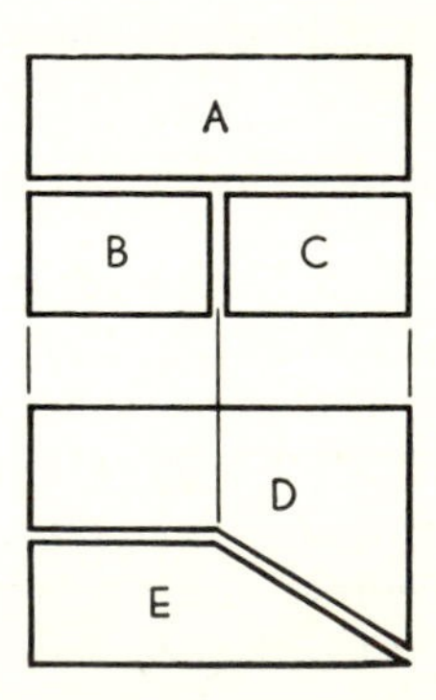

Fig. 1·19. Top and front views with areas separated.

the front view of surface G, nor area E the front view of surface F, because of the dissimilar configuration of these aligning areas. We can now add to our previous conclusions the following facts:

3. Areas D and E represent surfaces that appear *edgewise* in the side view.

4. Areas F and G represent surfaces that appear *edgewise* in the front view.

Since it has been established that most of the areas shown must appear as lines in the adjacent views, it only remains to identify the correct matching lines in each view. For example, since surfaces D and E are vertical surfaces (conclusion 1), visible in the front view, and appear as lines in the side view (conclusion 3), the thick vertical lines labeled D and E in the side view of Fig. 1·21 must correctly represent these surfaces. Thus we now know that surface E is on the front of the object, while surface D is set back a distance x behind it. The lines that represent surfaces D and E in the top view can now be identified.

The visible triangular surface G in the side view must appear as a line (conclusion 4), of the same height (alignment), on the right side (visible) of the front view. The line is labeled in Fig. 1·21. Surface F must be an inclined surface since it appears as the sloping line in the front view. But surface C also appears as a line in the front view (conclusion 2), and thus C and F are, respectively, the top and side views of the same inclined surface. By continuing this process of reasoning and verifying all conclusions by checking in each of the three views, the shape and position of each visible surface can be determined. Attention has been focused in the above discussion on only the visible surfaces; they should always be considered first, and then if necessary the same methods can be employed to identify the hidden surfaces.

The alert reader has probably noted already that conclusion 2 was not entirely correct. Considering only the top and front views, surface A

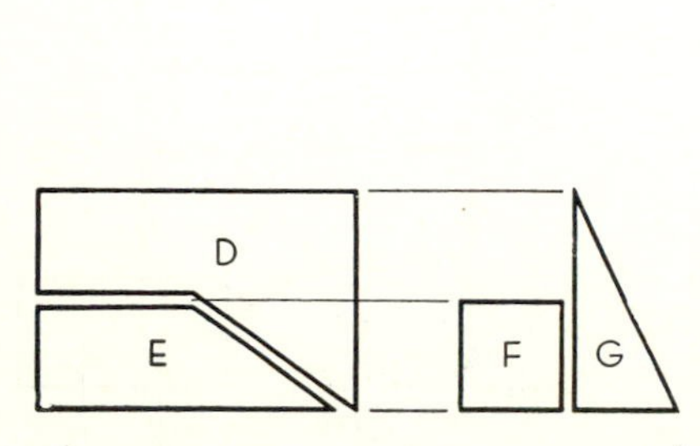

Fig. 1·20. Front and side views with areas separated.

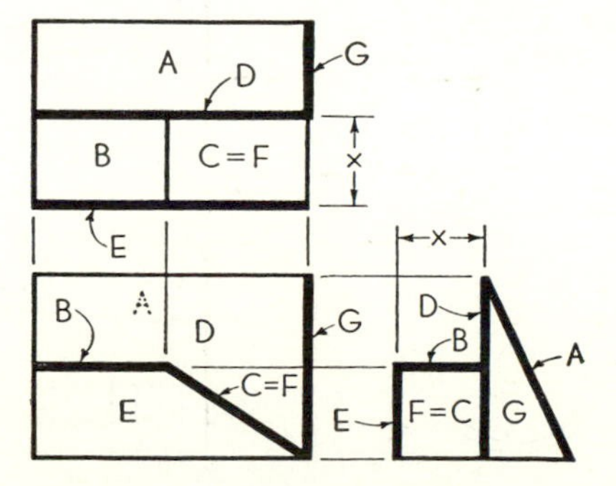

Fig. 1·21. Surfaces appearing edgewise shown with thick lines.

could be a horizontal plane surface on the top of the object and would thus appear as a line in the front view. But the side view (which was not considered in conclusion 2) shows that surface *A* is not horizontal but slopes down toward the lower-rear edge of the object. This apparent exception to the rule simply emphasizes again that two views of an object are not always sufficient (see Fig. 1·11). Figure A·30 in the Appendix shows a pictorial drawing of this object.

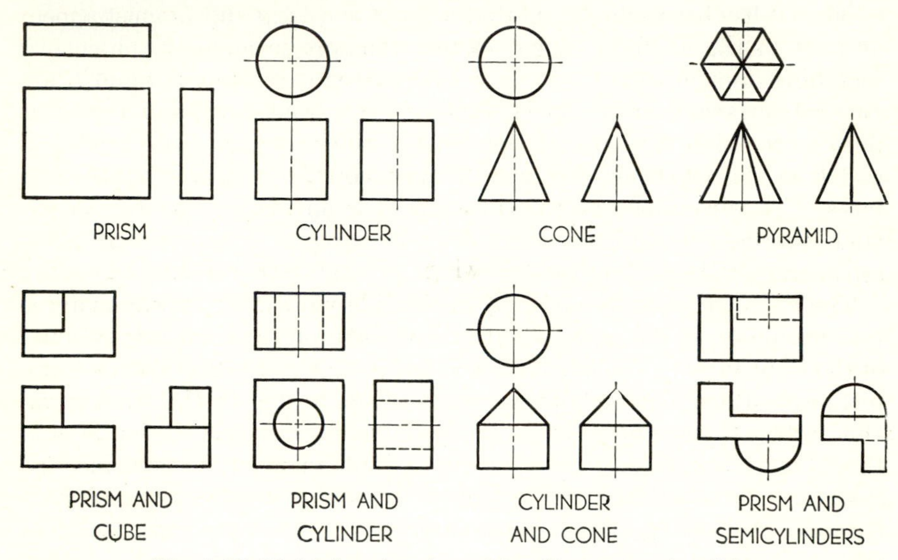

Fig. 1·22. Multiview drawings of familiar geometric solids.

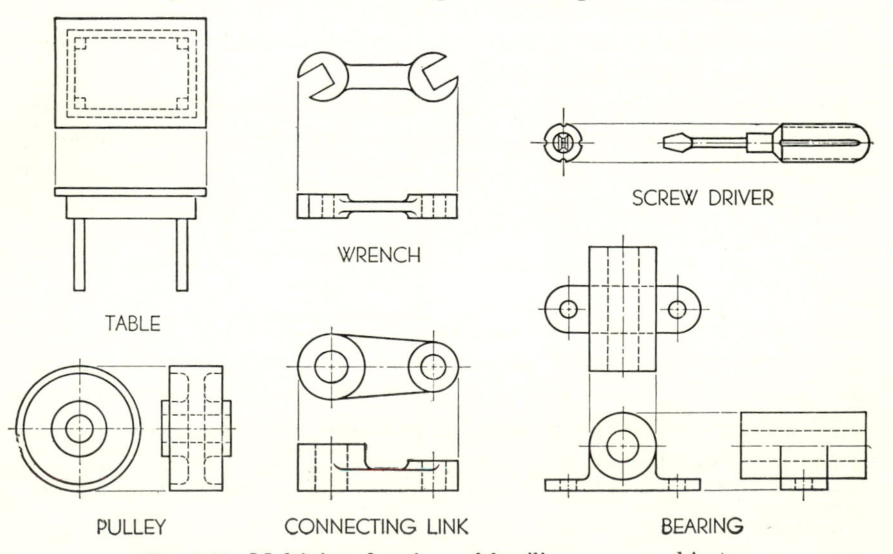

Fig. 1·23. Multiview drawings of familiar common objects.

1·11. General suggestions for reading a drawing

In the previous articles the reading of a multiview drawing has been reduced to an exacting and thorough process of analysis. The analysis by solids and the analysis by surfaces have purposely been explained in considerable detail in order to convince the unimaginative student that he too can learn to read a drawing. With a clear knowledge of principles and methods and by diligent application to many practice problems the ability to read engineering drawings can be acquired.

As skill and perception increase with practice, it will be discovered that such lengthy analysis is usually unnecessary. Three factors contribute to this increased facility: (1) the recognition of recurring geometrical shapes as component parts of less familiar objects; (2) the recognition of familiar objects; (3) the rapid, almost subconscious application of analytical methods.

Figure 1·22 shows multiview drawings of a few simple geometric solids, alone and in combination. Such objects assuredly require no analysis but are recognizable at a glance just as a pianist recognizes certain combinations of notes. Figure 1·23 shows multiview drawings of a few very familiar common objects. To the layman the views of the table, wrench, and screwdriver produce an immediate mental picture of the object. Similarly, to the engineer who has seen many machines the views of the pulley, connecting link, and bearing call forth instant recognition. But following general recognition both the layman and the engineer must then examine the views in detail to determine what features, if any, are peculiar to this particular table or bearing.

It should be remembered that each view of an object requires a different viewpoint. When the reader looks at the top view, he should imagine that he is actually directly above the object looking straight down. When the eye is shifted to the front view, the observer must mentally change his position and then imagine that he is directly in front of the object. Thus as the reader glances from one view to another, he is continually changing his viewpoint; the object itself is stationary.

In the next chapter we shall discuss the detailed construction of the principal views of any object and evolve a simple systematic procedure for this construction that will then enable us to draw not only the principal views but also a view of the object as it would appear from any conceivable direction. The application of this powerful technique to the solution of problems involving space distances and relationships is the *science of descriptive geometry.*

PROBLEMS. Group 1.

2. AUXILIARY VIEWS

2·1. What auxiliary views are

In Fig. 1·10 of the previous chapter the six principal views of a simple object were shown. Three of these views were so like the other three that they were discarded. The remaining three views shown in Fig. 1·8 (or in Fig. 1·9) were entirely adequate. But not all objects are so simple, and not all objects consist of only horizontal and vertical surfaces. If an object has inclined surfaces, then these surfaces will appear shortened or distorted in the principal views. Figure 1·15 shows that the square of cardboard never appeared as a square in any view when it was inclined.

In order to see an inclined surface in its true size the observer must look directly at it; that is, his direction of sight must be perpendicular to the surface. None of the principal views has an inclined direction of sight; hence the desired view would have to be a special view. Since such a view is usually provided in addition to the principal views, it is called an *auxiliary view*.

Auxiliary views can be drawn to show an object from any desired direction, and thus these views are a very valuable aid to the draftsman since almost all his problems can be solved easily if viewed from the proper direction.

2·2. Construction of a third principal view

Since the construction of an auxiliary view is very similar to the construction of a principal view, the latter problem will be discussed first. Assume that the front and right-side views of the object are given as shown in Fig. 2·1 and that it is required to draw the top view of this object. The planning, analysis, and construction of the new view should be performed in four stages, as illustrated in the figure and explained below.

Stage 1. *Plan the general position of the new view.*

Since the top view must lie directly above the front view, parallels are drawn upward from the ends of the front view. For a pleasing, well-balanced drawing the marginal distances D and D', E and E' should be equal. If possible, distances B and C should be approximately equal, but distances A and A' must be *exactly* equal since this is the depth dimension of the object as it appears in the top and side views.

Stage 2. *Analyze the given views, and label corresponding surfaces and points in the two given views.*

The visible surface T in the side view is an outstanding feature because of its odd irregular shape. There is no area of similar configuration in the front view; hence surface T must appear there as an edge, and the inclined line labeled T must be this surface. Surfaces R and S can be similarly identified in each view. As an additional aid in construction each corner of surface T has been labeled in the side view. By following the horizontal parallels to the front view the same points can be located there.

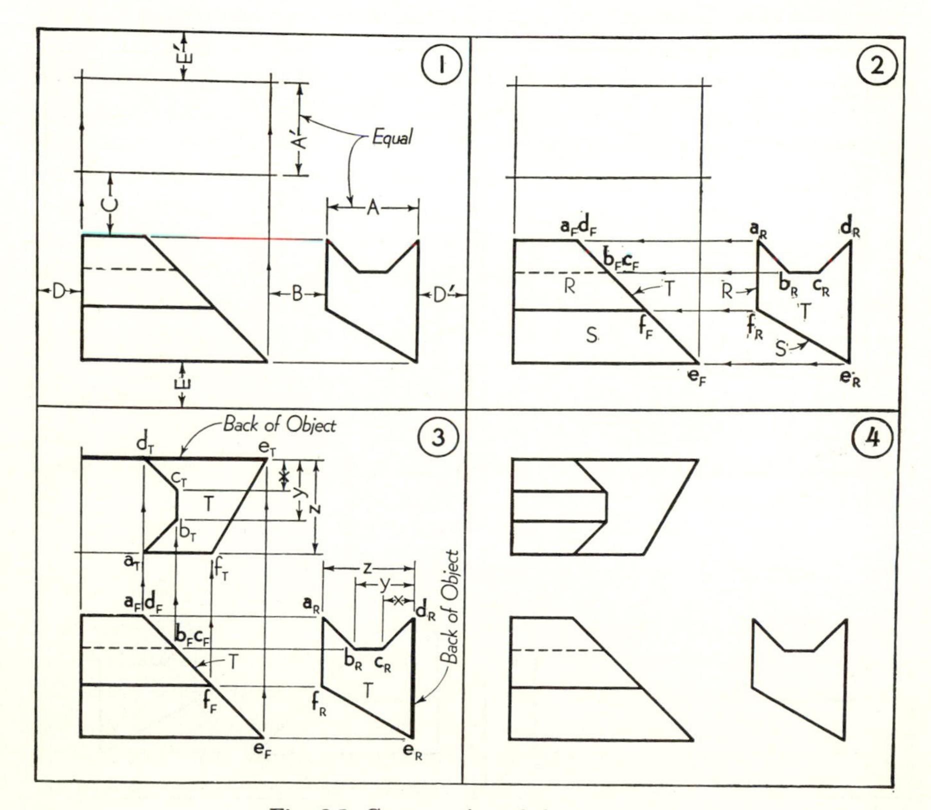

Fig. 2·1. Construction of the top view.

Stage 3. *By alignment and similarity locate the labeled points and surfaces in the new view.*

From each labeled point in the front view parallels are extended upward to the top view. Then a_T must lie directly above a_F, b_T above b_F, etc. (Rule 2). But the right-side view and the top view are *related views*, and since the side view shows points d_R and e_R on the back of the object (shown as a heavy line), the top view must also show these same points on the back of the object at d_T and e_T (Rule 3). Point c_R is distance x in front of the back of the object in the side view. This same distance x must be laid off in the top view to locate c_T directly above c_F. Point b_T is located similarly a distance y from the back of the object, and points a_T and f_T, which are on the front surface, are located distance z from the back. The top view of surface T is then completed by connecting the points in the top view in the same alphabetical sequence shown in the side view.

Stage 4. *Complete the view by inspection or by further analysis.*

After surface T has been drawn, the top view can be completed by inspection, but if the draftsman's mental picture of the object is still

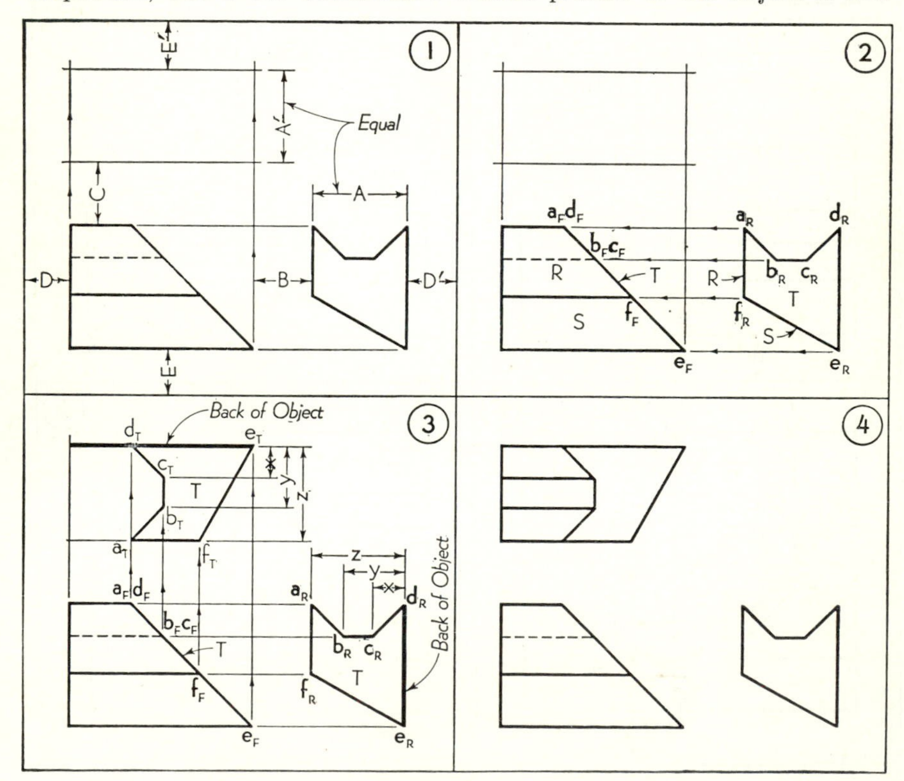

Fig. 2·1 (repeated). Construction of the top view.

hazy, other surfaces such as *R* and *S* can be identified, labeled, and constructed in the same manner until the view is completed.

2·3. Use of a reference line

In Fig. 2·1 distances along parallels were transferred from the side view to the top view by taking all such measurements from the back surface of the object. This was convenient because the back surface of the object appeared as an edge in both the top and side views. But the position of an object may be such that none of its surfaces appears as an edge. Such an object is shown in Fig. 2·2. Here the top and front views of a tilted oblique prism are given, and it is required to construct the right-side view.

Stage 1. *Selecting a reference line.*

Since distances along the parallels in the top view will have to be transferred to corresponding parallels in the related side view, provision must be made for a base of measurement, or *reference line.* Lacking a line or surface on the object that is perpendicular to the parallels, we can arbi-

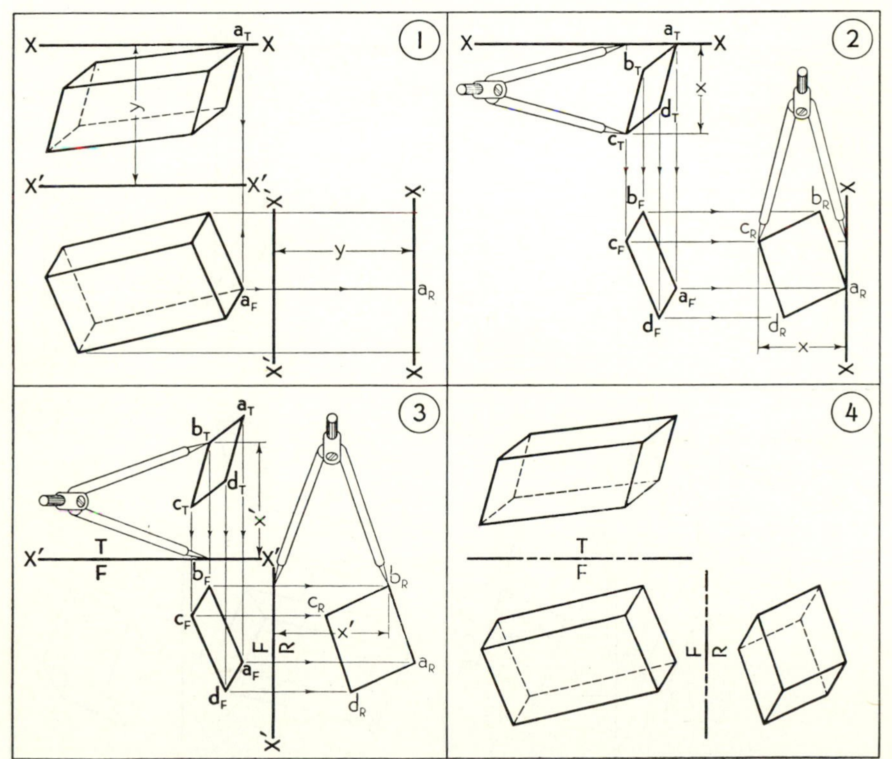

Fig. 2·2. Construction of a side view by using a reference line.

trarily select a line such as X–X in the top view. If this reference line X–X passes through point a_T in the top view, it must also pass through point a_R in the side view, and therefore line X–X is placed in the side view to give the desired distance between the front and side views.

Stage 2. *Construction by using reference line X–X.*

Only the right-end surface of the prism is shown, and the four corners of this surface have been labeled in the top and front views. To locate the four points in the side view parallels are drawn from each of the points in the front view to the side view. The distance x from c_T to line X–X in the top view is then transferred with dividers to the corresponding parallel in the side view to locate c_R. The respective distances for points b_T and d_T are similarly transferred to locate b_R and d_R. Since the right end of the prism is visible in the right-side view, solid lines connect the points.

Stage 3. *Construction by using alternate reference line X′–X′.*

The construction shown here is the same as that shown in stage 2 except that the reference line X'–X' was used. Referring back to stage 1, it is apparent that if the arbitrarily selected reference line X–X in the top

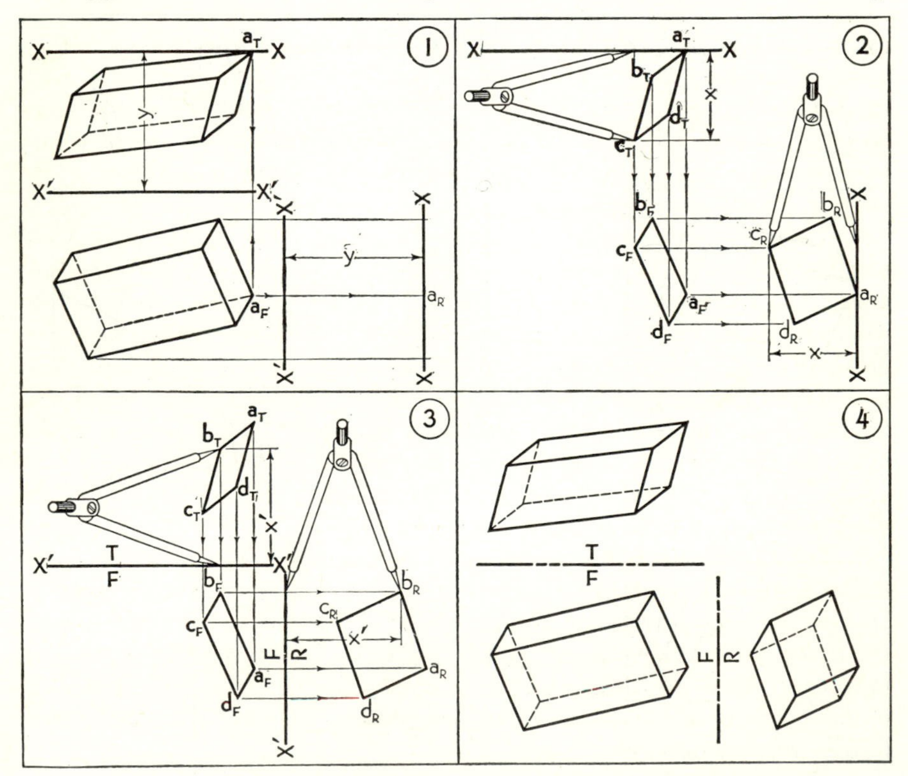

Fig. 2·2 (repeated). Construction of a side view by using a reference line.

view is moved some distance y to a new position X'–X' and the line X–X in the side view is changed by the same distance y, then the resulting side view will be the same as before. Thus the reference line can be placed anywhere in the two related views except that: (1) it *must be perpendicular* to the parallels connecting the adjacent views, and (2) measurements must be consistently *toward or away* from the common adjacent view. Note, for example, that in stage 2 distance x is measured from line X–X *toward* the front view in both the top and side views; in stage 3 distance x' is measured from line X'–X' *away* from the front view in both the top and side views. Failure to observe this condition would result in drawing a left-side view on the right side (see Fig. 1·13).

Stage 4. *Completing the view.*

The side view has been completed by locating the other four corners of the prism in the same manner that the first four were located and then connecting the points.

2·4. Reference-line notation

The reference line is simply an artificial device employed as an aid in the construction of additional views. In the construction of principal views it is frequently unnecessary, as was illustrated in Fig. 2·1 where measurements were made from the back of the object, but in the construction of most auxiliary views it is practically a necessity. It is therefore desirable that some systematic scheme of usage and notation be employed throughout this text.

Although, as previously explained, the reference line may be placed anywhere, there is considerable advantage in placing it between the adjacent views as shown in stage 4 of Fig. 2·2. Used thus, the reference line will always lie between the adjacent views, effectively separating the areas allotted to each view. Another advantage in this position is that measurements along parallels in related views will always be *away* from the common adjacent view. Drawings requiring only the two given views for the solution obviously do not need a reference line between the views. In subsequent text illustrations of such problems, however, we shall retain the line to separate the views clearly.

In order to distinguish the reference line it can be drawn with two short dashes and one long dash used alternately. In this text the reference line will also be labeled with letters such as T–F, F–R, etc., to associate it with the two adjacent views on either side of the line (see Fig. 2·2). It should be evident, however, that the basic method of construction just described is independent of the relative position of the reference line and of the letters or numerals used to label it.

PROBLEMS. Group 2.

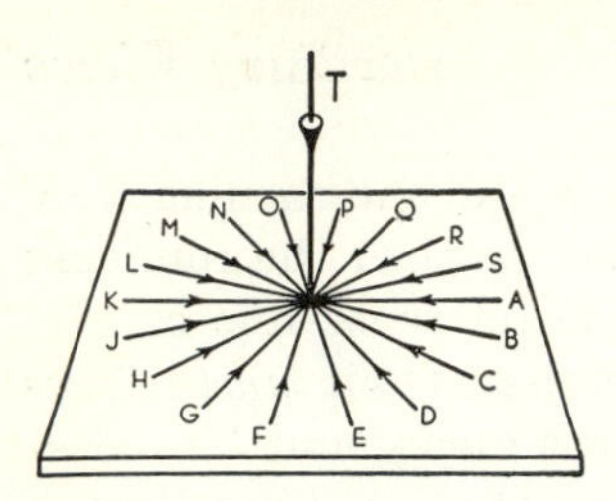

Fig. 2·3. Lines perpendicular to a given line T.

2·5. Top-adjacent auxiliary views

In Art. 1·5 it was stated that the lines of sight for any two adjacent views must be perpendicular (Rule 1). But there can be an infinite number of lines perpendicular to any given line. Given the vertical line T in Fig. 2·3, there is an infinite number of lines A, B, C, etc., which may be selected as shown, all of which are perpendicular to line T. Therefore, if the vertical line T represents the direction of sight for a top view, then the perpendicular lines shown in Fig. 2·3 represent the directions of sight for an infinite number of views that could be adjacent to the top view. But all these top-adjacent views would necessarily be views taken with *horizontal lines of sight* since 90° from vertical can only be horizontal. Because man is an animal who walks erect and habitually views his environment with horizontal lines of sight, such views are not only important in engineering work but also give a natural viewpoint. This type of view will be referred to as an *elevation*, or *top-adjacent*, view.

Figure 2·4 shows a top view and four top-adjacent views of a pyramid. In the upper-right corner of the illustration is a picture of the pyramid and arrows showing the directions of sight for each of the views. Looking down on the pyramid, the arrows A, B, C, and F would appear as shown in the top view. Then the front view, which is obtained by looking in the direction of arrow F, lies directly below the top view, and the parallels connecting the two views are parallel to the direction of sight F. Similarly view A (which is simply the right-side view in the alternate position as shown in Fig. 1·9) must be located by extending the parallels to the right, parallel to arrow A. The positions of views B and C are also determined by extending parallels in the direction of the desired line of sight.

Each of the five corners of the pyramid has been labeled in the given top and front views, and the reference line T–F has been assumed anywhere between the two views. From these two adjacent views any other top-adjacent view, such as view B, may be drawn. The step-by-step procedure, similar to that used for additional principal views, is as follows:

Step 1. Parallels from each point in the top view are extended parallel to the desired direction of sight B.

Step 2. Reference line T–B is drawn perpendicular to the parallels at any convenient distance from the top view.

Step 3. Views B and F are related views, and therefore distances along the parallels must be the same in each view. Point o_F, for example, is distance x from T–F, hence point o_B must also be distance x from T–B. Distance y (spanned by the dividers in Fig. 2·4) from the pyramid base to T–F must similarly be transferred to auxiliary view B as shown.

Step 4. The five points thus located are connected with straight lines to complete the view, observing that line o_Ba_B is hidden.

The construction of views A, B, and C differs only in the direction of sight because all top-adjacent views are necessarily equidistant from their respective reference lines and thus have a common feature:

All top-adjacent views show the true height of the object.

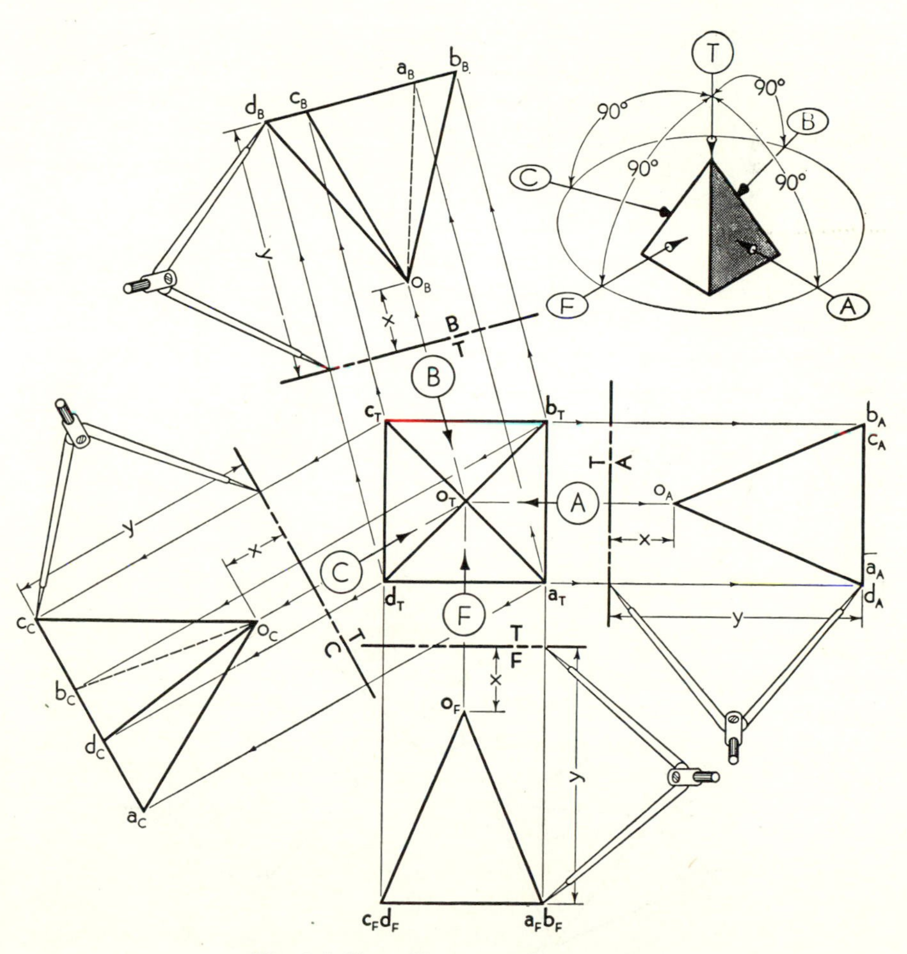

Fig. 2·4. Top-adjacent auxiliary views.

2·6. Front-adjacent auxiliary views

In the previous article the top view was the *common adjacent view* for a series of elevation, or top-adjacent, views. In similar manner any one of the other principal views may also serve as the nucleus for an infinite number of adjacent views. In conformity with Rule 1, however, each pair of adjacent views must again have lines of sight that are perpendicular.

Figure 2·5 illustrates the construction of auxiliary views adjacent to another principal view, in this case the front view. Shown are the same

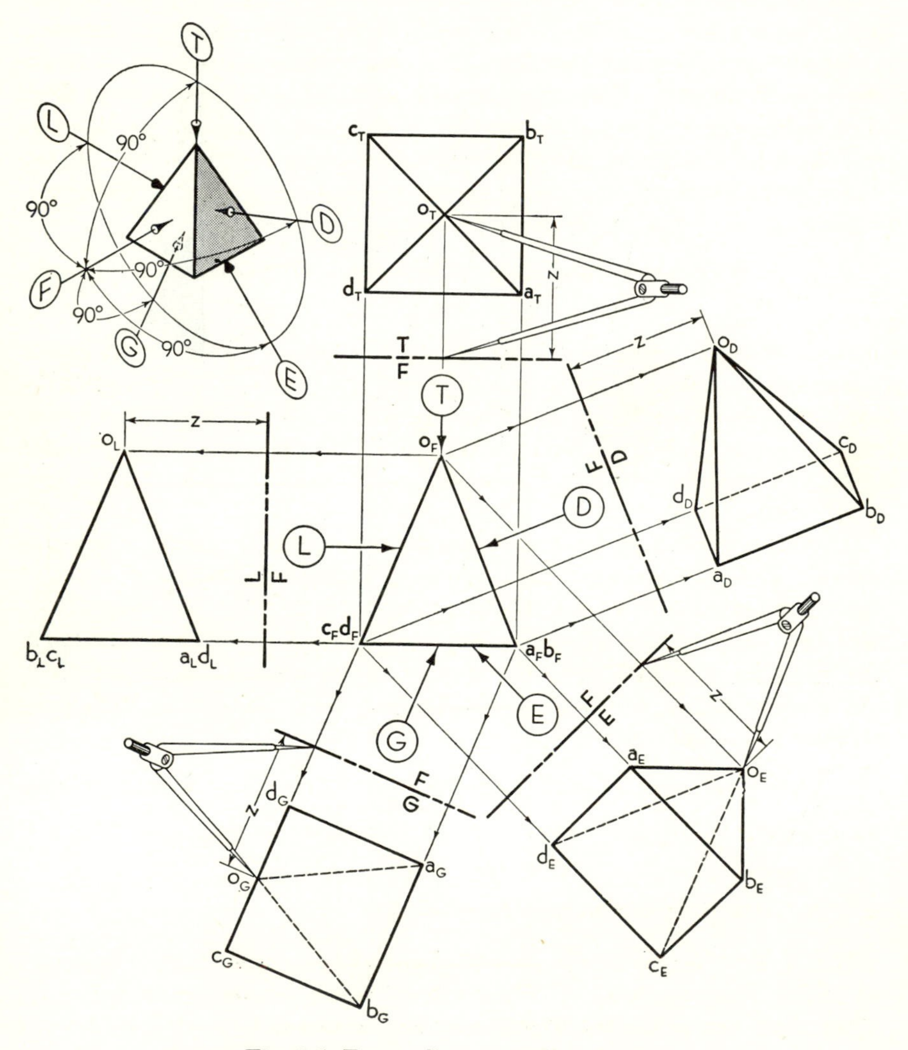

Fig. 2·5. Front-adjacent auxiliary views.

top and front views of a pyramid that were given in Fig. 2·4. Views D, E, G, and L are additional views drawn adjacent to the front view. The pictorial sketch emphasizes the fact that the direction of sight for each of the front-adjacent views must be at right angles to the direction of sight for the front view. The same directional arrows are also shown as they appear in the front view, and each front-adjacent view is located on the corresponding parallels.

The steps of construction for front-adjacent views are exactly analogous to those described for top-adjacent views: (1) parallels are extended from the front view in the desired direction of sight; (2) the reference line is placed at any desired distance from the front view; (3) distances along the parallels may be taken from any other related view to locate the various points. Thus, for example, point o_T is distance z from T–F; then points o_D, o_E, o_G, and o_L must all be distance z from their respective reference lines. The same is true for any other point on the object. Therefore, though front-adjacent views may appear to be very dissimilar, they all have one common characteristic:

> *All front-adjacent views show the true depth of the object from front to back.*

It has already been seen that *all* top-adjacent views have *horizontal* lines of sight. With four important exceptions front-adjacent views have *inclined* lines of sight. These four exceptions are the top and bottom views, which have vertical lines of sight, and the right- and left-side views, which have horizontal lines of sight. (View L in Fig. 2·5 is a left-side view.)

Auxiliary views can also be drawn adjacent to the right- or left-side views. These views have not been illustrated because the method of construction is identical with that for front-adjacent views. Views adjacent to a side view would, of course, have lines of sight perpendicular to the direction of sight for the side view, and therefore side-adjacent views can include a top, bottom, front, and rear view, as well as an infinite number of inclined views.

> *All side-adjacent views show the true length of the object from left to right.*

PROBLEMS. Group 3.

2·7. Auxiliary-adjacent auxiliary views

It has now been demonstrated that additional views may be drawn of an object showing it as it would appear when viewed from directions other than the principal directions. But views drawn adjacent to the top, front, or side views are limited to those having lines of sight perpendicular to the line of sight for one of these principal views. In order to see the object from a direction not perpendicular to any of the principal directions it is necessary to draw an auxiliary view adjacent to a previously drawn auxiliary view. This construction, which is quite similar to those previously described, is illustrated in Fig. 2·6.

As before, the top and front views of a pyramid are given in Fig. 2·6; it is required to construct the auxiliary views *A*, *B*, and *C*. The directions of sight for views *A*, *B*, and *C* have been selected entirely at random simply to illustrate the method of constructing each view; in other words, no effort has been made to view the object from a specific predetermined direction.

View *A* is a top-adjacent view taken from an arbitrarily selected direction. It is therefore an elevation view and has been constructed as explained in Art. 2·5. Note that distance *x* in view *A* equals distance *x* in the front view because views *A* and *F* are related views.

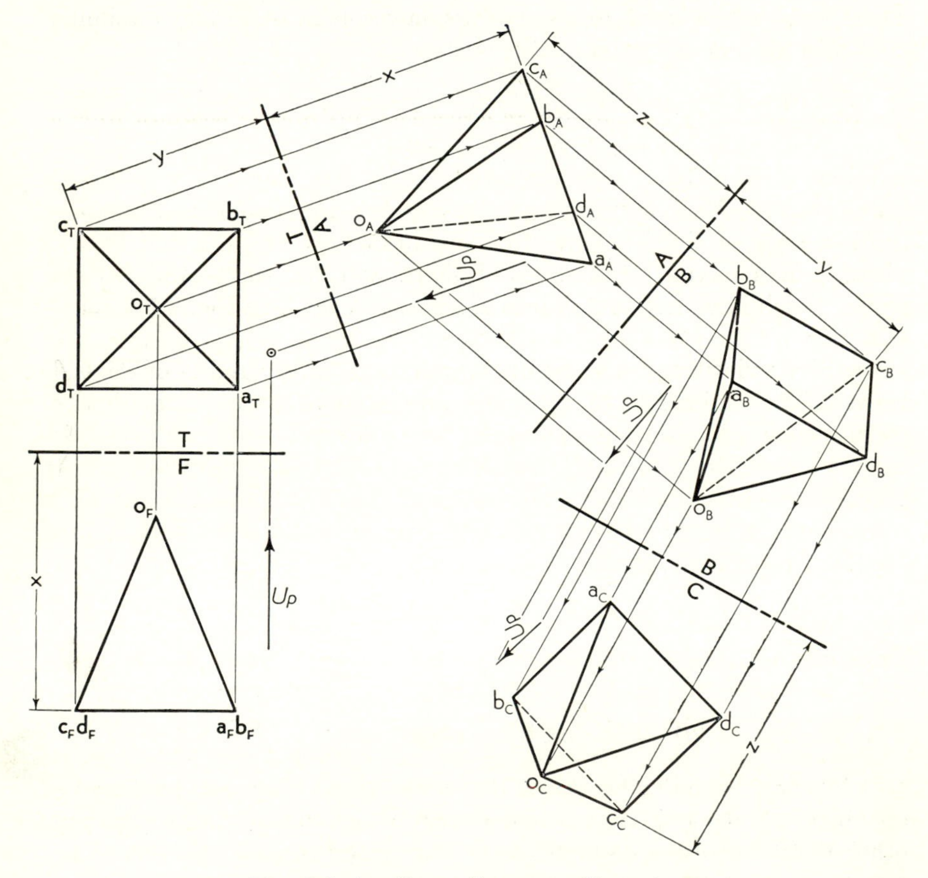

Fig. 2·6. Auxiliary-adjacent auxiliary views.

If from view A parallels are extended in any desired direction, a new auxiliary view, such as view B, may be constructed adjacent to view A. The construction of view B is identical in method with any other additional view: (1) parallels are extended from each corner of the object in view A; (2) the reference line A–B is drawn perpendicular to the parallels at any convenient distance from view A; (3) distances along the parallels in view B must be obtained from another related view, in this case the top view. (The top view and view B are related views because each is adjacent to view A.) As an example of this measurement, note that points c_T and c_B are each distance y from reference lines T–A and A–B, respectively.

View B is an auxiliary-adjacent view. In the same way that view B has been drawn adjacent to view A, another view, such as view C, can be drawn adjacent to view B; and, if desired, still another view can be drawn adjacent to view C, and so on, ad infinitum. But regardless of the number of views drawn, the third step of construction (applying Rule 3) is always the same:

Distances along the parallels in a new view are always obtained from a view related to the one being constructed.

The only view related to view C is view A; hence point c_C is located distance z from B–C because point c_A is distance z from A–B.

Although it is relatively easy for the draftsman to imagine the position of the observer for each of the principal views, this visualization becomes more difficult for views such as B and C. Fortunately the construction and use of the views are not dependent upon such imaginative effort since, as will be shown later, the direction of sight for any view can, if necessary, be determined by analytical methods.

PROBLEMS. Group 4.

2·8. The aspect of the views

Figures 2·4 to 2·6 show, in total, 12 different views of the same object. It is evident from the front view in each figure that the pyramid stands on its base in a natural upright position. In apparent contradiction of this fact, however, most of the auxiliary views seem to show the object in a tipped position—in some cases almost upside down. But the object is never moved from the position shown in the original top and front views; it is only the observer who changes position. Then must the observer stand on his head to see the object as it is shown in view B of Fig. 2·4? Not at all—it is only the paper that is upside down.

In Fig. 2·4, views *A*, *B*, *C*, and *F* are elevation views observed with horizontal lines of sight. They show the pyramid as it would appear to an observer who walked around it, pausing at each of the four positions to look toward the object. The reader can obtain the same effect by rotating the book and looking at each view when it appears to be upright.

To see each of the front-adjacent views in Fig. 2·5 in its natural aspect the reader should turn each view until point *O* is uppermost. Then, if the base of the pyramid is visible, the observer must be underneath the object looking upward. This is the approximate viewpoint for views *E* and *G*.

To see view *B* in Fig. 2·6 properly the whole figure should be turned almost upside down. The pyramid will then appear upright, and the base will be visible, hence the observer is lower than the object and is looking upward at an angle. With the page held in approximately this same position, view *C* also appears upright, but here the base is hidden; hence the observer must be looking downward at an angle from a position higher than the object. In general, the various auxiliary views can be approximately oriented to appear in their natural aspect by comparison with the front view: the higher points in the front view should be the higher points in the auxiliary view when that view is in its natural aspect.

For more exacting purposes a vertical line is an accurate index to the aspect of any view. In the top and front views of Fig. 2·6 a vertical arrow has been drawn pointing upward. In the top view this arrow appears as a point, and in the front view it appears upright in its full true length. This arrow can also be drawn in views *A*, *B*, and *C* as shown and will thus clearly indicate in each view the correct upward direction. Obviously, if the object itself contains a vertical line, that line can perform the same function as the arrow.

2·9. Visibility

To construct new views of an object that contains only straight lines it has been demonstrated that it is only necessary to locate the corners of the object and then to connect these points with straight lines. Care must be exercised to connect in the new view only those points which are connected in the given views. There remains, however, the question of *visibility*—which lines are visible, and which are hidden. This can be determined completely by inspection of the new view and its adjacent views.

Although the objects that a draftsman may be required to draw can be infinite in variety and complication, a few general rules can be stated as follows.

■ ***Rules of visibility for solids***

a. The outside lines of every view will be visible.

b. The corner or edge of the object nearest to the observer will be visible.

The nearest corner or edge is determined by reference to any adjacent view, where it will appear as the point or line nearest to the reference line. For example, in view B of Fig. 2·7 (views B and C of Fig. 2·6) point a_B must be the corner nearest to the observer since the adjacent view C shows point a_C nearest to reference line B–C.

c. The corner or edge farthest from the observer will usually be hidden if it lies within the outline of the view.

In view B of Fig. 2·7 line o_Bc_B is hidden because adjacent view C shows line o_Cc_C farthest from reference line B–C. In view C line b_Cc_C is hidden because adjacent view B shows that this line is farthest from the observer. Exceptions to Rule c may occur if the object contains holes or voids through which the line may appear visible.

d. Crossing edges that are approximately equidistant from the observer must be tested for visibility at the crossing point.

This rule determines the visibility of edges that are not obviously the nearest or farthest from the observer. In view D of Fig. 2·8 edges AC

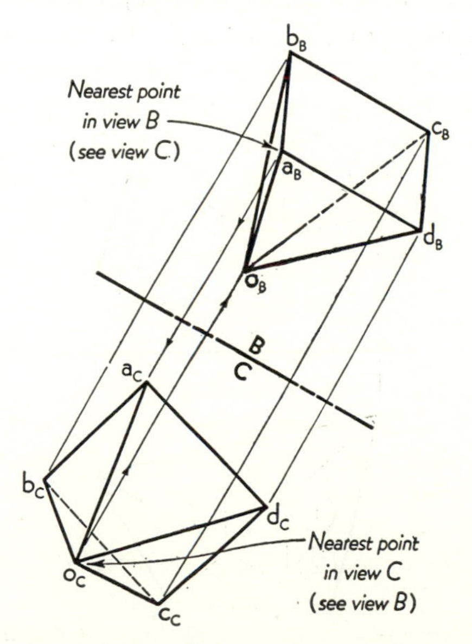

Fig. 2·7. Visibility in any two adjacent views.

and BD cross at the point labeled 1,2. If point 1 is on line AC and point 2 is on BD, then the adjacent view E shows that point 1 is nearer to reference line D–E and hence nearer to the observer in view D. Line a_Dc_D is therefore visible, and line b_Dd_D is hidden.

In similar manner, a parallel extended from crossing point 3,4 in view E to the same lines in view D shows that point 3 on BD is the nearer point; hence line b_Ed_E is visible, and a_Ec_E is hidden.

e. If a point in an adjacent view lies on a parallel that does not pass through any part of the adjacent view, then that point will be visible in the new view.

Points a_A, b_A, c_A, and d_A in Fig. 2·9 illustrate Rule *e*; they are all visible because the parallels from a_F, b_F, c_F, and d_F do not pass through any part of view F. The validity of Rule *e* is evident if it is remembered that the parallels actually represent lines of sight from the object to the observer; if the line of sight for any point is obviously not intercepted by any solid part of the object, then that point must be visible. Rule *e* also confirms the visibility of points x_A, e_A, and h_A, determined previously by Rules *a* and *b*. It should be noted that points to which Rule *e* does not apply are not necessarily hidden: Rule *e* does not apply to point j_A because the parallel from j_F to j_A passes through view F, but point j_A is visible by Rule *a*.

View A in Fig. 2·9 also illustrates the application of the first three rules. By rule *a*, points e_A, f_A, g_A, h_A, j_A, k_A and the consecutive lines

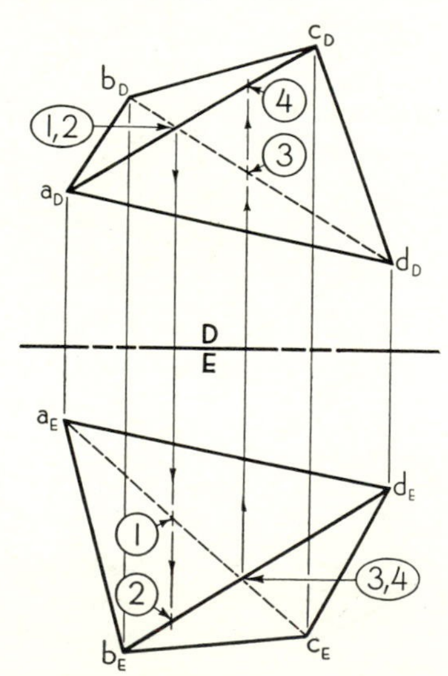

Fig. 2·8. Visibility of crossing lines in adjacent views.

that join them must all be visible since they lie on the outside perimeter of the view. By Rule *b*, point x_A must be visible because adjacent view *F* shows x_F nearest to reference line *A–F*. Since the lines x_Ah_A, x_Af_A, and x_Ak_A are external edges of the object connecting corners known to be visible, then the lines must be visible also. Note that point y_A is the exception to Rule *c* because it can be seen through the hole in the block.

The five rules given here are applicable to all solid objects and will correctly determine the visibility of many points and lines. They do not, however, cover all possible cases; hence it will be necessary to draw the remaining lines by inspection. If the rules are fully utilized, the remaining lines will be chiefly internal lines and those on the far side of the object; obviously such lines will usually be hidden. In view *A* of Fig. 2·9, for example, the internal lines that form the corners of the hole are all hidden except a_Ay_A, which can be seen through the hole.

In summary, it should be noted that all rules of visibility except Rule *a* are based on one cardinal principle:

> *Visibility of the inside lines in any view is primarily determined by reference to an adjacent view.*

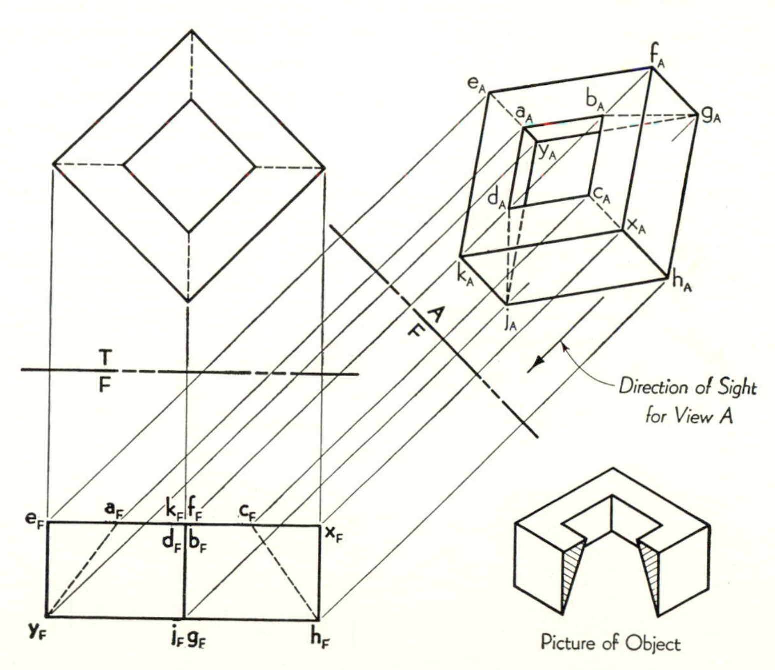

Fig. 2·9. Illustration of rules of visibility.

POINTS AND LINES

3·1. Location of a point

In order to describe the location of a single specific point, either verbally or graphically, its position must be explained by reference to some other point whose location is known. The known point then becomes a "reference point," or "origin of measurements," and all other points may be located from it by some system of three-dimensional measurements. The Cartesian rectangular coordinate system, shown in Fig. 3·1(*a*), is most generally employed in mathematics, particularly in solid analytic geometry. Through the origin *O* pass three mutually perpendicular axes *X*, *Y*, and *Z*, and any point, such as *A*, may be located by stating the three distances *x*, *y*, and *z*.

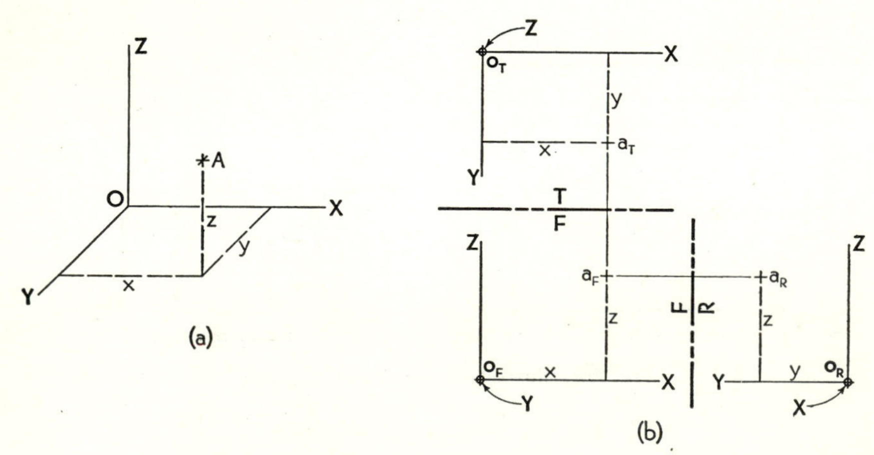

Fig. 3·1. Cartesian coordinates.

This system of measurement can be utilized in the multiview drawing as shown in Fig. 3·1(*b*). Here are shown the top, front, and right-side views of the axes and point in Fig. 3·1(*a*). It is immediately apparent that no one of the three views can completely describe the position of point *A* with respect to point *O*. The top view, in which the *Z* axis appears as a point, can show only the *x* and *y* distances, the front view can show only the *x* and *z* distances, and the side view shows only the *y* and *z* distances. But any two of the three views taken together will give *x*, *y*, and *z* and thus locate point *A*.

Figure 3·2 shows a practical application of coordinate axes to an airplane. The origin has been selected in the fuselage center plane (*YZ* plane) at the point where the leading edge of the wing intersects this plane of symmetry. Directions upward, to the right (left side of airplane), and to the rear are considered positive. By employing these rigged axes (*X* and *Y* axes level in flying position) any point on the airplane can be located by its coordinates. Point *A*, for example, on the wing tip has the *x*, *y*, and *z* coordinates shown in the figure.

It is sometimes convenient to describe a point as being to the left or right of, above or below, in front of or behind the point of reference. Here the terms "above" or "below" mean only "higher than" or "lower than" and *not* that the point is *directly* above or below the reference point. "Left" or "right" and "in front" or "behind" have the same broad meaning. For example, point *A* might be described as 2 in. to the right, 1½ in. above, and 1 in. in front of point *O*. If the position of

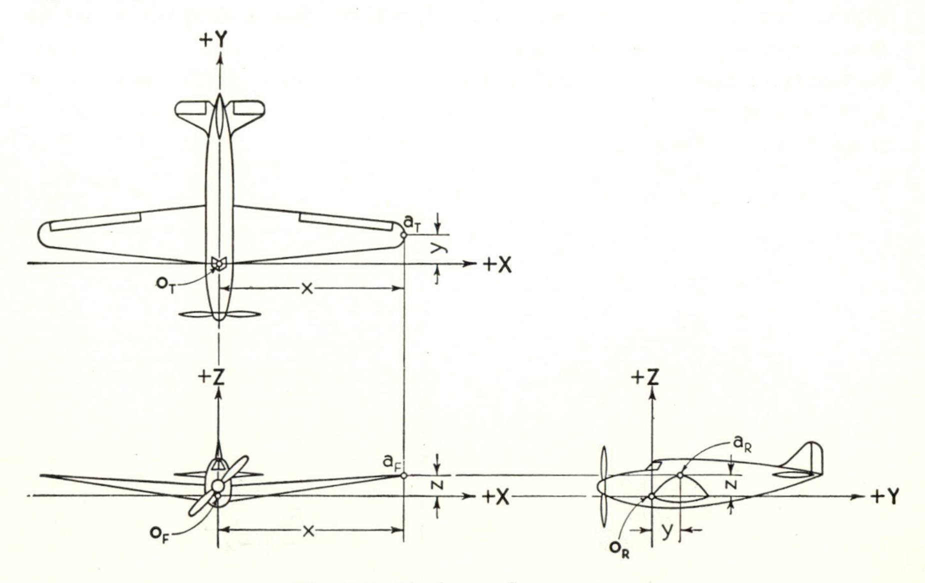

Fig. 3·2. Airplane reference axes.

point O is given as in Fig. 3·3, then point A can be located in each of the three views. Since the top view can show distances to the left or right, and to the front or back, but not *height*, the point a_T is located 2 in. to the right of point o_T and 1 in. in front of o_T. (*Toward* the front is always *toward* the front view.) Point a_F must lie directly below a_T, and hence it is unnecessary to remeasure the 2-in. distance; but the front view also shows height, and here the point a_F must appear 1½ in. higher than point o_F. (*Upward* is toward the *top* view.) With the top and front views of point A located the right-side view can be completed in the usual way. The labeled directional arrows in Fig. 3·3 are, of course, never shown on engineering drawings but are introduced here simply for emphasis.

3·2. Maps and bearing of a line

When a portion of the earth's surface must be represented, a map drawing is employed; the map is actually a one-view drawing showing only the top view. Although many types of maps are used, the topographic map, which shows to large scale the exact shape of the earth's surface, is of importance to the engineer. Such a map would appear, in its simplest form, as in the top view of Fig. 3·4. Here are shown a series of contour lines representing a hill. Each of the irregularly shaped contour lines is an imaginary line on the surface of the earth that connects points at the same elevation, or height. The numbers on the contours indicate the elevation in feet above sea level, and thus three dimensions are shown on a single view (see also Art. 12·1).

The reference point O is called a *bench mark*, and any other point, such

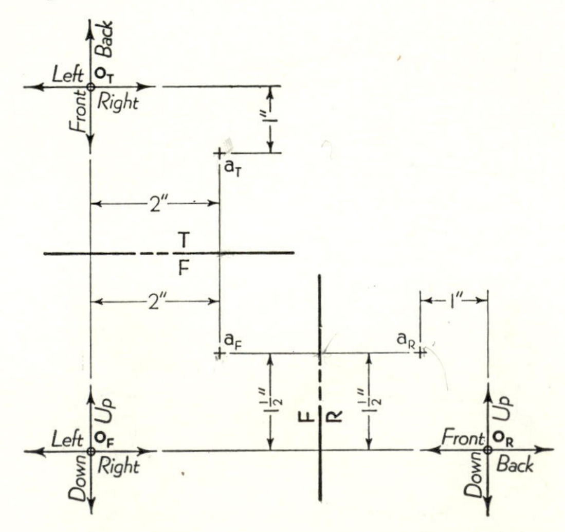

Fig. 3·3. Locating point A from point O.

as A, may be located from it. For example, point A is 80 ft due east and 46 ft north of the bench mark. If point A is on the surface of the ground, then it must be at an elevation of 920 ft above sea level because it lies on that contour line. Since point O is on the 890-ft contour line, point A is 30 ft higher than point O.

Bearing of a line is the angle by which the line deviates east or west from a north-south line as shown in the top, or map, view.

The direction of north is always understood to be toward the top of the drawing unless otherwise indicated by a north-pointing arrow. With respect to the compass directions given in Fig. 3·4, note the bearing of each of the four different lines shown there. Since the bearing of a line is solely a map direction, it can be seen and measured *only in the top view.* The bearing of a line is *entirely independent of its slope or inclination.*

Point A in Fig. 3·4 can also be located by stating the *bearing* of line o_Ta_T as N 60° E from o_T to a_T (it would be S 60° W if taken from a_T to o_T) and giving the horizontal *map distance* between the two points as 92.5 ft. It should be noted that the 92.5-ft distance is *not* the actual straight-line distance between points O and A but only the horizontal distance between the points that are shown in the top view.

To assist in visualizing the shape of the hill described by the contours in the top view, Fig. 3·4 also shows a front-elevation view. Such an

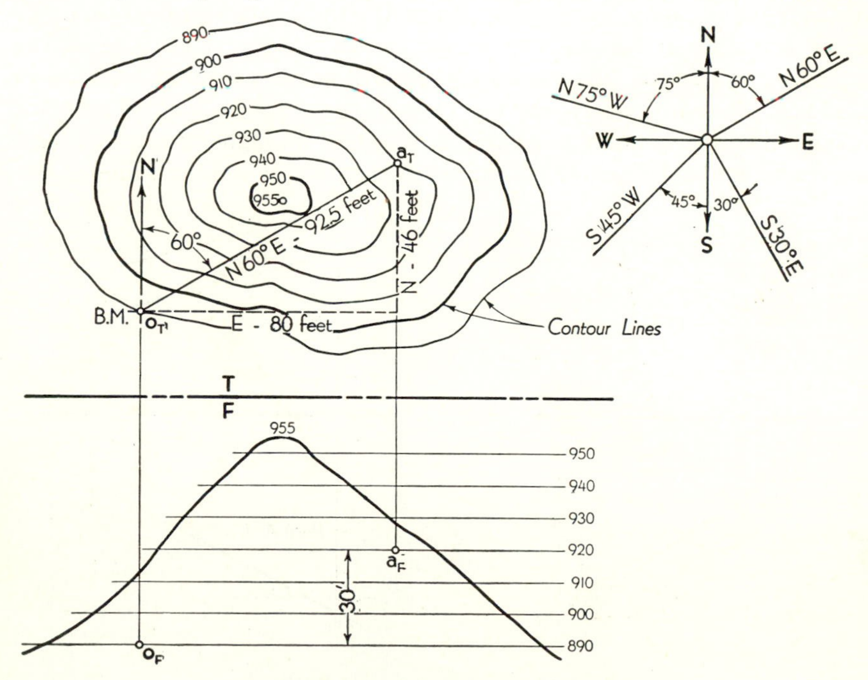

Fig. 3·4. Locating point A on a map.

elevation view is rarely used since the top view conveys all necessary information.

3·3. Location of a line

A straight line is usually designated by labeling its extreme points, but any two points on a straight line will fix the direction and position of that line. Therefore, to locate a line it is only necessary to locate any two points on it. For the sake of brevity, the single word "line" shall hereafter imply a straight line.

Suppose, for example, that the location of point A is given as shown in Fig. 3·5(a) and that the point B is known to be 1½ in. to the right of A, ½ in. behind A, and 1 in. lower than A. Then the top and front views of point B can be located with respect to point A as shown in the figure. Note again that height measurements—up or down, higher or lower—can be shown only in an elevation view, never in the top view. The line AB can now be constructed simply by drawing a straight line through points A and B in each view.

A straight line can also be located as shown in Fig. 3·5(b). Assume again that the location of point A is given as shown in the figure; assume also that point B is known to be S 75° E of point A a map distance of 2 in. and is 1 in. higher in elevation than point A. Through point a_T a line of indefinite length is drawn toward the right (eastward) at an angle of 75° with the south-pointing arrow (north is toward the top of the paper). On this line the map distance of 2 in. is measured from a_T to locate b_T. The line a_Tb_T is the top view of the required line. The point b_F must lie directly below b_T and must also be 1 in. higher than a_F. This locates b_F, and the line a_Fb_F is the front view of the required line.

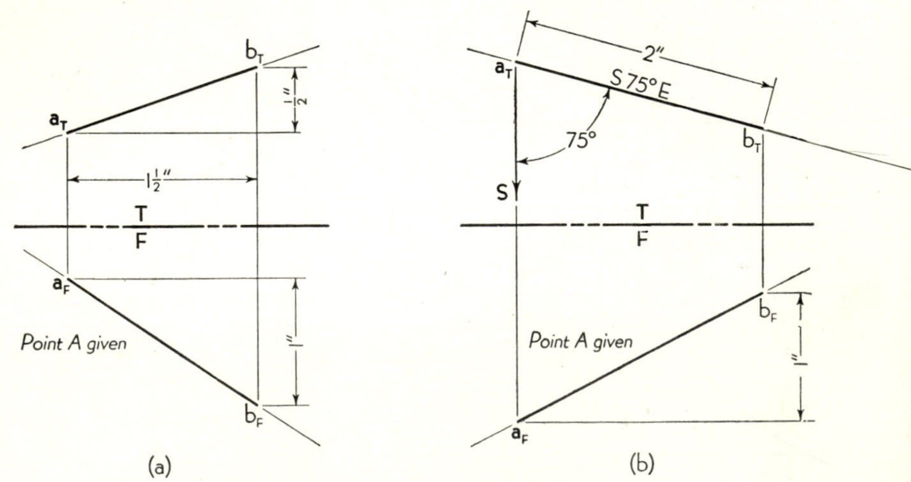

Fig. 3·5. Location of a straight line.

Depending upon the requirements of the problem, any line AB may be limited to the segment between points A and B, or it may be extended indefinitely at either end. In both cases the line may be correctly referred to as the line AB.

The location of a straight line may also be described in terms of its true length (actual straight-line distance between extremities), its bearing, and the angle it makes with a horizontal plane (see Art. 3·8).

The location of curved lines requires at least three points on the line plus additional information as to the shape of the curve. In graphical work it is common practice to locate a large number of points and then to draw through these points a smooth, or "fair," curve that satisfies the eye of the draftsman.

3·4. A point on a line

If a point actually lies on a line, then it must appear on that line in all views. Thus in Fig. 3·6(*a*) point X does lie on the line AB, but points Y and Z do not. The fact that y_F lies on line a_Fb_F means only that point Y *may* lie on line AB or it *may* lie directly before or behind the line. The position of y_T shows that the point Y is in front of line AB. Similarly, point Z lies directly below line AB but not on it.

When the top and front views of a line are both perpendicular to the reference line, the line is called a *profile line*. Such a line is shown in Fig. 3·6(*b*). In the top and front views points R and S both appear to lie on the line CD, but in the right-side view it is quite evident that point R does not lie on the line. Any new view (except a bottom or rear view), such as view A, will similarly demonstrate that point R does not lie on line CD.

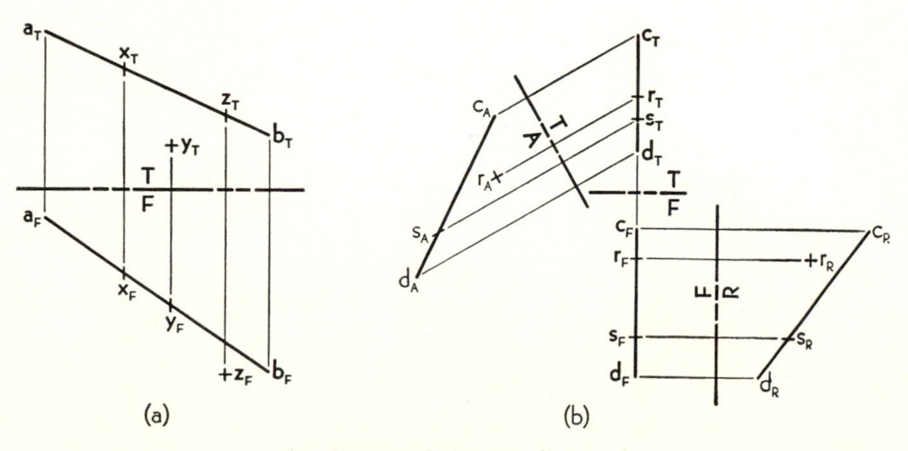

Fig. 3·6. Point on a line.

3·5. Location of a point on a line

If a point is known to lie on a given line, then one coordinate distance is sufficient to fix its position. In Fig. 3·7(*a*), for example, the line *AB* is given as shown, and the point *X* must lie on *AB*, 1 in. to the right of point *A*. A parallel drawn 1 in. to the right of point *A* locates x_T and x_F and solves the problem.

If the point *X* on *AB* must be ½ in. behind point *A*, then x_T must be located first as shown in Fig. 3·7(*b*). Point x_F is then located on a_Fb_F directly below x_T.

If the line *AB* is a profile line and the point *X* on *AB* must be ¾ in. below point *A*, then x_F must be located first as shown in Fig. 3·7(*c*). A new view, such as view *R*, must then be drawn to locate x_R and to determine the distance *d*. Point x_T can finally be located on a_Tb_T by transferring distance *d* to the top view.

The following theorem is a valuable adjunct to the previous discussion:

> *A point on a line divides the line, and all views of that line, into two segments whose ratio is always the same.*

Thus, if point *X* is the mid-point of line *AB*, then x_T will be the mid-point of line a_Tb_T, x_F the mid-point of a_Fb_F, etc. This theorem can be utilized to locate a point on a profile line without drawing a new view. If, for example, in Fig. 3·7(*c*), it were noted that the line segment x_Fb_F is exactly one-third of the length a_Fb_F, then x_T could be located simply by making x_Tb_T one-third of a_Tb_T. However, in general, unless the division of the line is simple and convenient, it is better to employ the additional view.

PROBLEMS. Group 5.

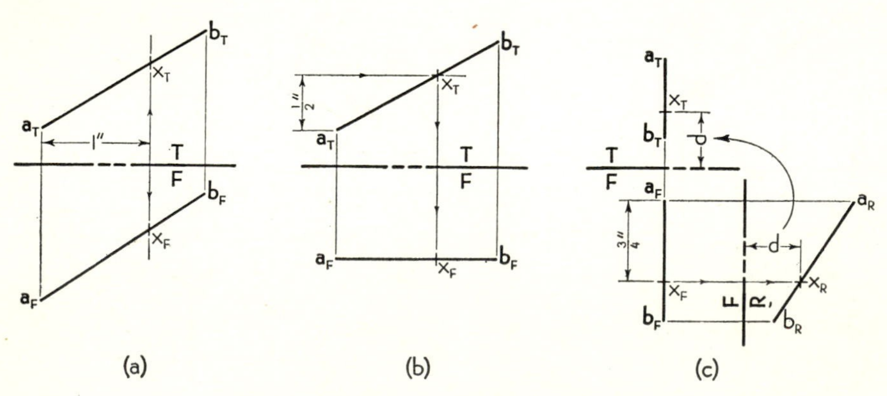

Fig. 3·7. Point on a line by coordinates.

3·6. True length of a line (basic view, type I)

The *true length* of a line is the actual straight-line distance between its two ends. If the reader will hold his pencil up before him to represent a line in an oblique position, he can then view the pencil from various directions. It will soon be apparent that in order to see the full true length of the pencil the observer must look in a direction perpendicular to the pencil: in other words, both ends of the pencil must be equidistant from the observer's eye. If one end of the pencil is farther away than the other, the pencil will necessarily appear shorter than it really is.

The lines shown in Fig. 3·8 are called *principal lines* because each of these lines appears true length in one of the principal views. The *vertical* line (*a*) is perpendicular to all horizontal directions of sight and therefore appears true length (*T.L.*) in the front view (and in all other elevation views). All *horizontal* lines (*b*) and (*c*) will appear true length in the top view because both ends of the line will always be equidistant from the downward-looking observer. An inclined line can never appear true length in the top view, but it may be a *frontal* line (true length in the front view), as in (*d*), or a *profile* line (true length in the side view), as in (*e*). An *oblique*, or generally inclined, line does not appear true length in any of the three principal views. In general, then, a line will appear true length in one view if any adjacent view shows that the ends of the line are equidistant from the observer.

The following rule, in two parts, summarizes the above observations:

- ***Rule 6. Rule of true lengths***
  - *a. If a line appears as a point in one view, it will appear true length and perpendicular to the reference line in any adjacent view.*
  - *b. If a line appears parallel to the reference line in one view, it will appear true length in the adjacent view.*

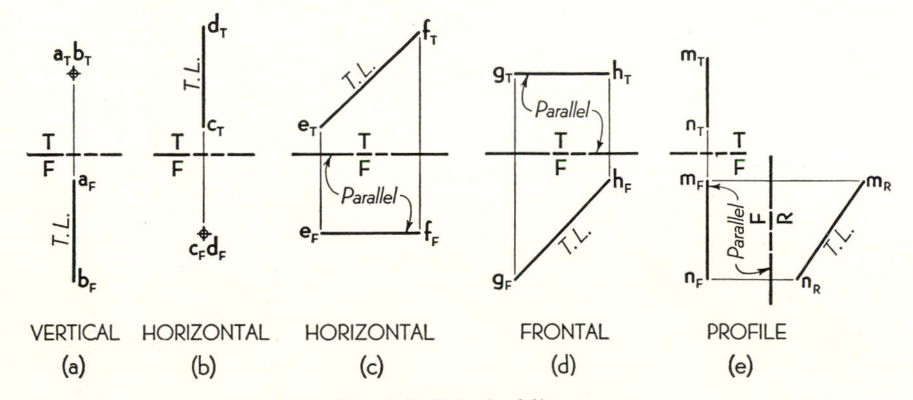

Fig. 3·8. Principal lines.

When a line is oblique, as shown in Fig. 3·9, and does not show its true length in any of the principal views, it is then necessary to draw a new view to show the true length. Line a_Tb_T is not a true length because a_Fb_F is not parallel to the reference line *T–F*, and a_Fb_F is not true length because a_Tb_T is not parallel to *T–F* (Rule 6*b*). But if a new view is drawn, such as top-adjacent view *A*, with the new reference line *T–A* parallel to a_Tb_T, then the line a_Ab_A must be true length. Or the new view can be drawn adjacent to the front view. Both views *B* and *C* have been so

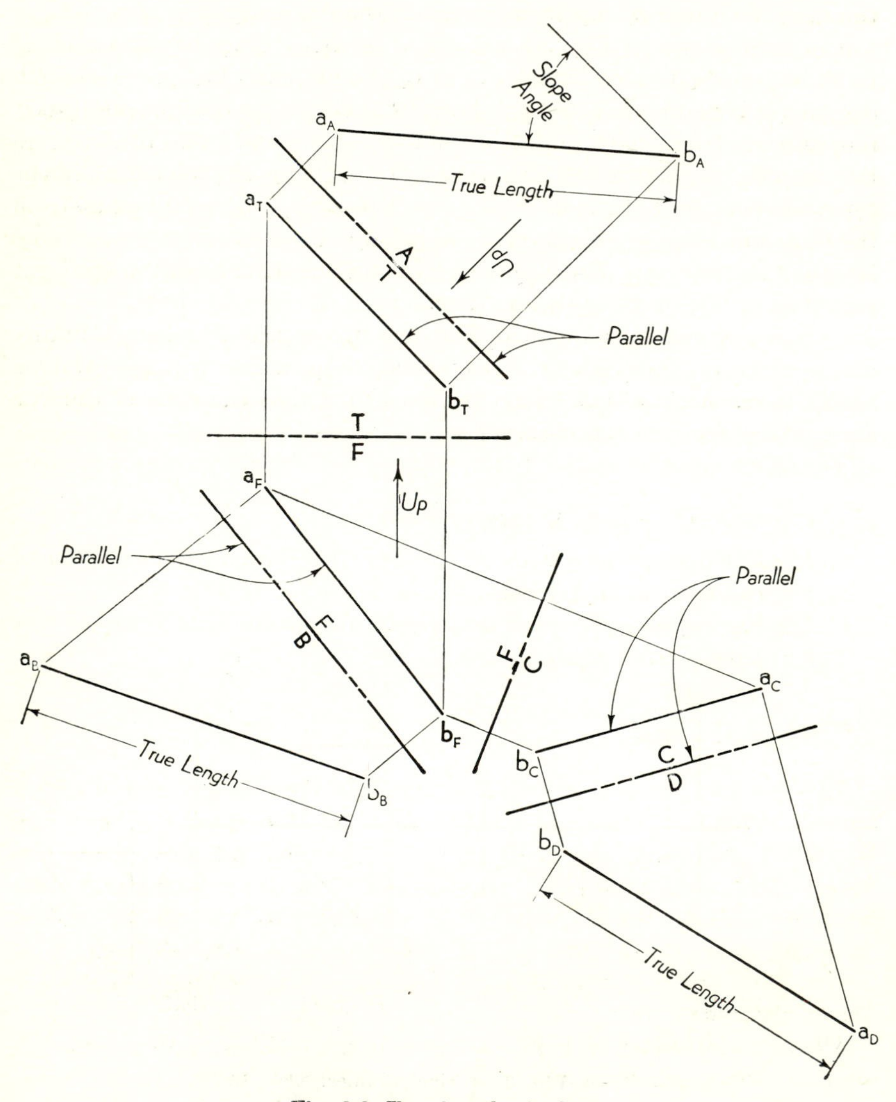

Fig. 3·9. True length of a line.

drawn, but only view B shows the line AB in its true length because only the reference line $F\text{–}B$ is parallel to a_Fb_F. The line a_Cb_C is not a true length and hence must be shorter, for no line ever appears longer than it really is. However, a new view D can be drawn adjacent to view C as shown, and a_Db_D will again be a true length. In summary:

> *A true-length view can be drawn adjacent to any other view by choosing the new reference line parallel to the line in the given view.*

In engineering design it is frequently necessary to determine the actual true length in feet or inches of oblique cables, guy wires, beams, center distances, etc. that occur in machines or structures. The graphical method described in this article is a rapid and simple means of obtaining such lengths whenever the required degree of accuracy is commensurate with the scale and precision of the drawing. On a design drawn to a scale of $\frac{1}{8}$ in. equals 1 ft a drafting inaccuracy of $\frac{1}{50}$ in. is equivalent to an error of approximately 2 in. in the actual structure. If an error of this magnitude is intolerable, then either the scale of the drawing must be increased or the distance must be calculated mathematically as described in Art. A·28 in the Appendix.

In subsequent articles it will be shown that there are only *four* basic and very important types of views. A view that shows a line in its true length is the first of these four, but since the four basic views must be studied and used in sequence, they are outlined here for preliminary reference.

The four basic types of views show:

I. *The true length of a line.*
II. *A line appearing as a point.*
III. *A plane appearing as an edge.*
IV. *The true size of a plane.*

3·7. Slope of a line

The *slope* of a line is the tangent of the angle that the line makes with a horizontal plane. The angle itself is called the *slope angle.* That a line is inclined, or sloping, can be determined from the front view or any elevation view by the simple fact that one end of the line will be higher than the other (see Art. 1·6 and Fig. 1·14). In Fig. 3·10 it is readily evident from the given views that each of the three lines OA, OB, and OC is inclined, but the fact that all three lines have exactly the same slope is not so apparent.

The pictorial sketch in Fig. 3·10 shows that the slope angle of each line lies in a vertical plane; therefore, if we would see the angle in its true magnitude, we must look directly at the plane in which it lies. The slope angle of line OA, for example, will be seen by looking in the direction of the horizontal arrow F, and thus the slope angle of line OA appears true size in the front view. But when the direction of sight is perpendicular to the plane of the slope angle, it is also perpendicular to the line, and consequently the line will always appear in its true length when the slope angle appears in its true size. This explains why the angles X and Y do not represent the true slope angle of lines OB and OC; these lines are not true length in the front view.

In order to determine the slope of line OB a new view must be drawn looking in the direction of arrow A. Since arrow A is horizontal, the new view must be an elevation view, and it must show line OB in its true length. All top-adjacent views are elevation views (Art. 2·5), and therefore we draw the reference line T–A parallel to $o_T b_T$ and construct the new view A. In this view the line is seen in its true length, the horizontal plane again appears as an edge parallel to the reference line, and the slope angle is the angle between the line and the horizontal plane. To see view A in its natural aspect we should, of course, turn the drawing around until the horizontal plane is actually horizontal (note upward-pointing arrows).

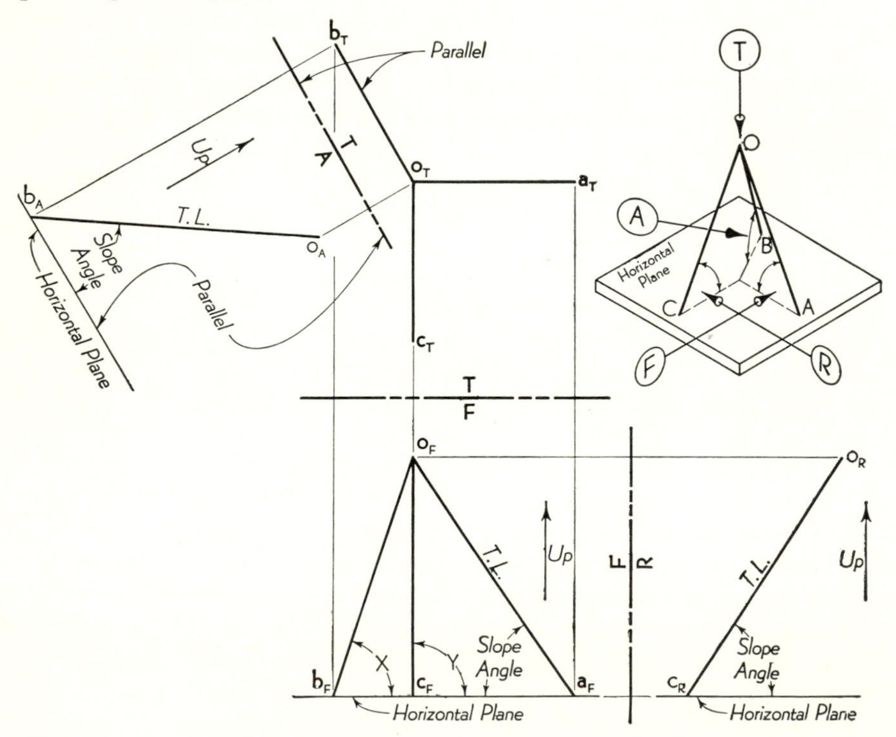

Fig. 3·10. Slope angle of a line.

The slope of a profile line such as *OC* can be found in the same manner except that in this case either a top-adjacent or a front-adjacent view can be used since both will be elevation views showing the line in true length. Note, however, that in a front-adjacent elevation such as view *R* the horizontal plane is perpendicular to the reference line *F–R*; in top-adjacent views the horizontal plane is always parallel to the reference line.

From the above discussion the following rule can be stated:

▪ ***Rule 7. Rule for slope of a line.*** *The slope angle of a line can be seen true size only in* ***that elevation view*** *which shows the line in its* ***true length.***

Attention is here directed to the double requirement in Rule 7. For a given line an infinite number of elevation views can be drawn (Art. 2·5); for that same line an infinite number of views can be drawn perpendicular to the line to show its true length; but there is only *one true-length elevation* view that will show the true size of the slope angle.

In Fig. 3·9, there are three different views that show the true length of line *AB*, but only one of these, namely, view *A*, is an elevation view that shows the slope angle.

The slope angle is usually measured in degrees, but it is also common engineering practice, particularly in railroad and highway work, to express the slope in per cent grade. The per cent grade of a line is the tangent of the slope angle multiplied by 100. Thus a highway that gains elevation at the rate of 5 ft vertically for each 100 ft horizontally has a 5 per cent grade. The slope angle would be the angle whose tangent is 5/100, or 2°52′.

The slope angle of an oblique line can be measured only in a true-length elevation view, but per cent grade is more convenient to use because it can be established directly in the top and front views, as shown in Fig. 3·11. Here a line *AB*, of indefinite length, begins at *A* and bears N 60° E with a rising grade of 25 per cent. The given bearing fixes the direction of $a_T b_T$; hence the top view is drawn first. Along $a_T b_T$ a horizontal distance of 100 units (any convenient scale divisions) is laid off to locate x_T. But x_F must be 25 units higher than a_F, and is so located to establish the front view of the required line.

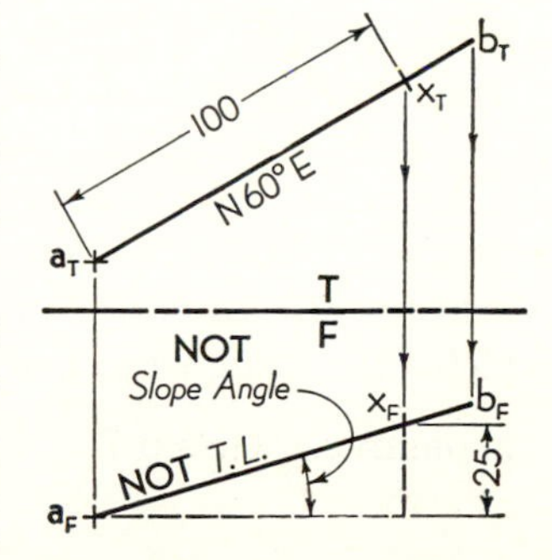

Fig. 3·11. Line of 25 per cent grade.

PROBLEMS. Group 6.

3·8. Location of a line of given bearing, slope, and true length

The solution of this problem is simply a reverse process since previously the top and front views of the line were given and it was required to determine its bearing, slope angle, and true length. Now these characteristics of the line are given, and it is required to construct the top and front views.

Problem. The required line *AB* is the center line of a tunnel which starts at a given point *A*, has a bearing of N 40° E, is 112 ft long, and slopes downward from *A* to *B* with a 35 per cent grade. Figure 3·12 shows the location of point *A* and four stages of construction.

Stage 1. *Lay off bearing and begin the construction of a true-length elevation view.*

The direction of north is taken as toward the top of the paper. Since the bearing of the line is independent of true length and slope (Art. 3·2), we may draw through a_T a line (indefinite in extent) at an angle of 40° to the right (eastward) of the north line. The reference line *T–A* is then drawn parallel to this bearing line in order to provide a true-length elevation view.

Stage 2. *Lay off per cent grade* (*or slope angle*) *in the true-length view.*

Point a_A is located first as shown. Parallel to *T–A* (horizontally) a distance of 100 units (the unit can be any convenient scale division) is measured from point a_A to locate point 1. From point 1 a distance of 35 units is measured downward (perpendicular to *T–A*) to locate point 2. This establishes the 35 per cent grade. If the slope angle is given instead of the per cent grade, the angle can be laid off with a protractor or by the more accurate tangent method described in Art. A·1 of the Appendix.

Stage 3. *Lay off true length in the true-length view.*

Through a_A and point 2 a line is drawn indefinite in extent. On this line a distance of 112 ft is measured to locate b_A. This completes the true-length view *A* of line *AB*.

Stage 4. *Complete the top and front views.*

A parallel drawn to the top view from b_A locates b_T on the bearing line, thus completing the top view of line *AB*. Point b_F is located directly below b_T, the same distance *y* from *T–F* that b_A is from *T–A*. This completes the front view, and the required views of line *AB* are as shown.

PROBLEMS. Group 7.

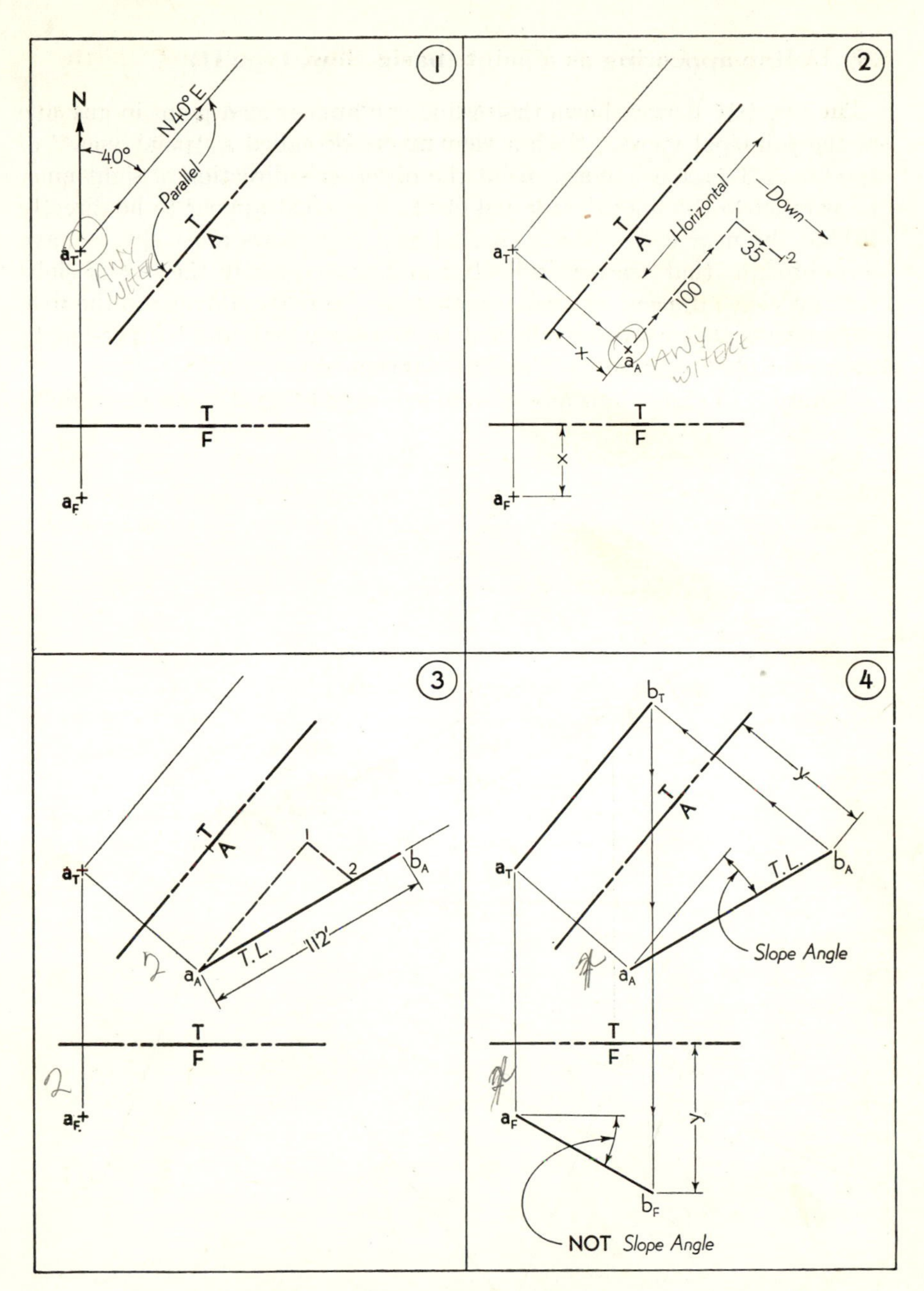

Fig. 3·12. Location of a line of given bearing, slope, and true length.

3·9. A line appearing as a point (basic view, type II)

In Fig. 1·14 it was shown that a line can appear as a point in any one of the principal views. Such a view might be called a "point view" of the line. To see a line as a point the observer's direction of sight must be parallel to the line; the far end of the line must appear to lie directly behind the near end. But by Rule 1 adjacent views must always have lines of sight that are perpendicular, and consequently the line of sight for any view adjacent to a point view must be perpendicular to the line. This means that all views adjacent to a point view must be true-length views. Rule 6*a* expresses this same relationship.

Figure 3·13 shows how any generally inclined line *AB* can be made to

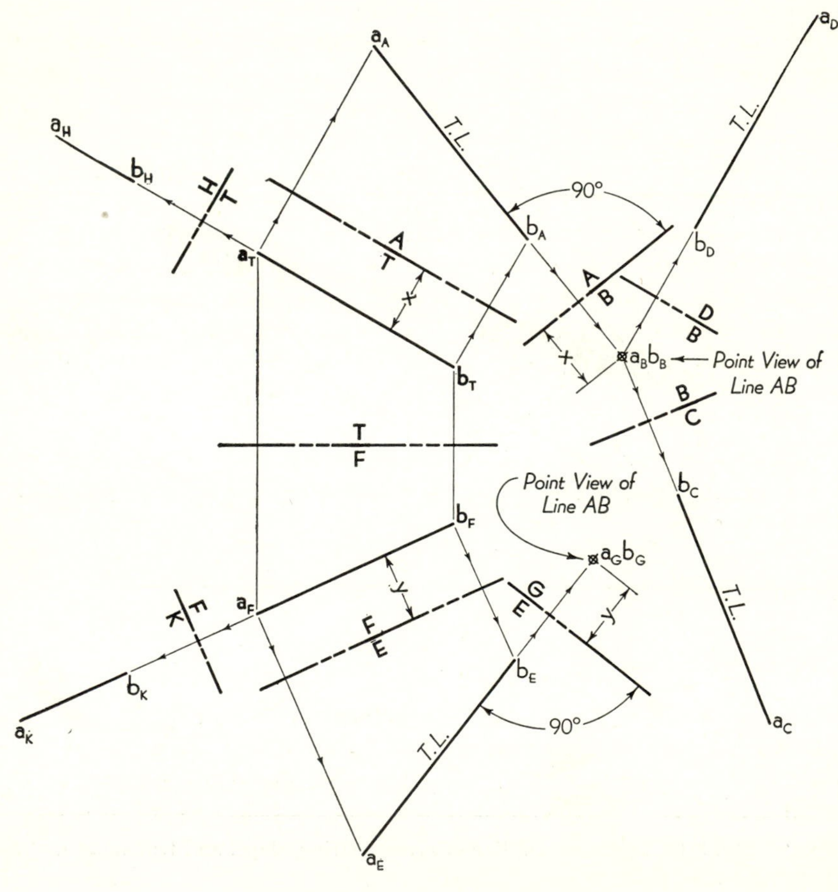

Fig. 3·13. Line appearing as a point.

appear as a point. Since neither the top nor the front view shows the line in its true length, it is impossible for any view adjacent to either of these views to show the line as a point. A true-length view, such as view A, must first be constructed. In order that points A and B shall appear superimposed the next reference line A–B must be taken perpendicular to the true-length line a_Ab_A. Then points a_B and b_B will both lie on the same parallel, and both will be distance x from reference line A–B (views B and T are related views), and the line AB will appear as a point. In similar manner view E also shows line AB in true length, and view G again shows it as a point (note distance y in related views G and F).

It is interesting to note that, although the point views B and G have each been derived from a different set of views and occupy different positions on the paper, they are actually identical views. Both show the line as it would appear to an observer looking along the line from B toward A. If reference line A–B or G–E had been placed at the opposite end of the true-length line, the line would still appear as a point, but the observer would then be looking at the line from A toward B. For the purpose of linear measurements either viewpoint is equally serviceable.

Views H and K have been constructed simply to emphasize that a point view cannot be obtained from an oblique view of the line.

Views A, C, and D demonstrate the general validity of Rule 6*a* that all views adjacent to a point view will show the line in its true length.

To be able to show any given line as a point is a very powerful and fundamental accession to knowledge. A view that shows a line as a point is the second of the four basic types of views. We shall call this *basic view, type* II.

- ***Rule 8. Rule of point views.*** *A point view of a line must be adjacent to a* ***true-length view,*** *and the direction of sight must be* ***parallel*** *to the line.*

PROBLEMS. Group 8.

3·10. Parallel lines

Having discussed the various characteristics of a single line, we shall consider next the relationships of two straight lines. Two straight lines must be parallel, intersecting, or nonparallel and nonintersecting (also called *skew* lines). Figure 3·14 shows seven different views of a pair of parallel lines AB and CD, and with only two exceptions the lines appear as two parallel lines in every view. In view D the lines coincide, and in view B they both appear as points. But coincidence is also parallelism, and the point view shows that both lines have the same direction. Therefore we can express this observation in a rule as follows:

▪ ***Rule 9. Rule of parallel lines.*** *Parallel lines will appear parallel in all views.*

This property of parallel lines is very useful to the draftsman as a means of detecting errors or inaccuracies in the various views. But it also means that he must be able to recognize which lines are actually parallel in the given views. With one exception, two lines are actually parallel if they appear parallel in each of *two* adjacent views. The exception occurs when the lines in the adjacent views are perpendicular to the reference line. In Fig. 3·15 the profile lines *AB* and *CD* appear parallel in both the top and the front view, but the right-side view shows that they are not parallel. Again, in Fig. 3·14, views *T* and *C* alone are not obvious proof of parallelism and must be supplemented by some additional view.

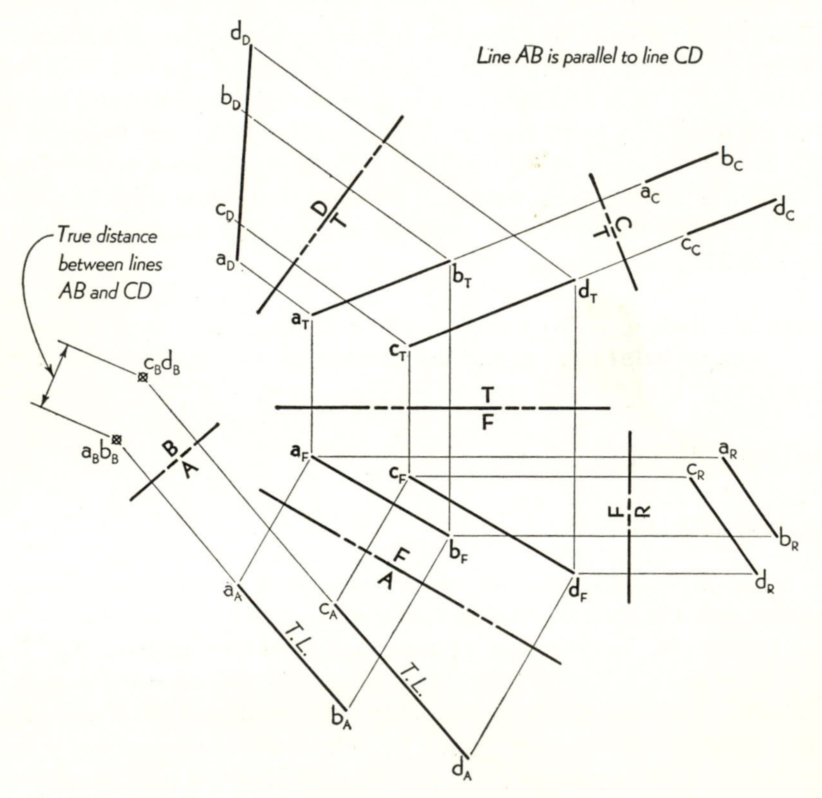

Fig. 3·14. Parallel lines.

3·11. Location of a line through a given point parallel to a given line

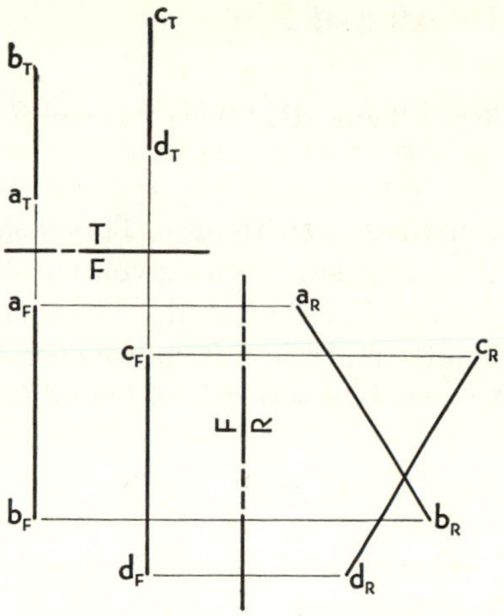

The required line must pass through the given point and appear parallel to the given line in all views. If the required line is so located in any two adjacent views, its position is established unless the given line is a profile line.

Fig. 3·15. Nonparallel profile lines.

In Fig. 3·16(a) the given line is AB, and the given point is C. In each view the required line is drawn through point C parallel to line AB and indefinite in extent. This establishes the position and direction of the required line, and any other points on the line such as E and F may now be assumed as desired.

In Fig. 3·16(b) the given line AB is a profile line. Again the required line can be drawn through point C in the top and front views parallel to line AB and indefinite in extent, but the line is not definitely established in these views until a second point, such as D, can be located on it. This requires some new view such as view R. In view R the required line must pass through c_R and be parallel to $a_R b_R$. Point d_R can now be selected anywhere on the required line, and points d_F and d_T can be located by alignment and measurement.

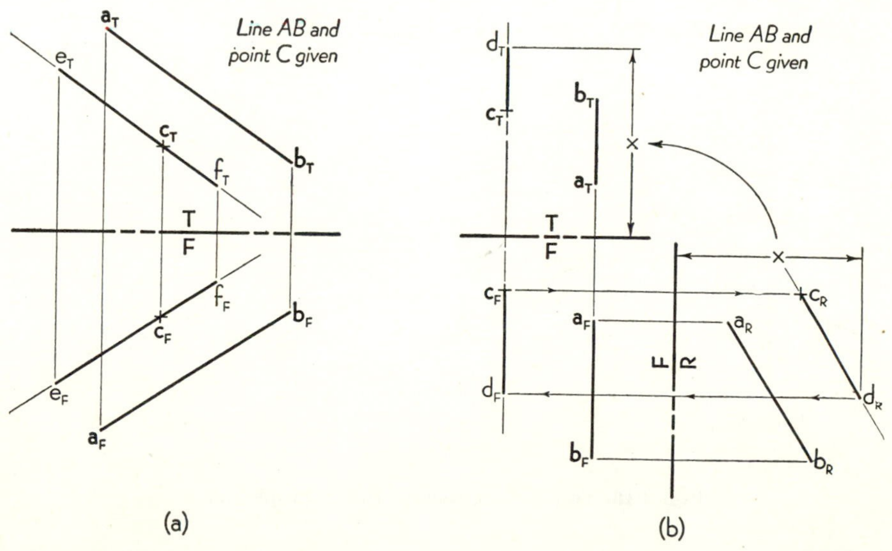

Fig. 3·16. Line through point C parallel to line AB.

3·12. True distance between two parallel lines

The true distance between two parallel lines is the perpendicular distance between them. This distance will appear in its true length in the view that shows the given parallel lines as points. In Fig. 3·14 a pair of parallel lines *AB* and *CD* are shown. View *A* shows these lines in true length, and view *B* shows each line as a point. The distance from a_Bb_B to c_Bd_B is the required true distance between the lines.

3·13. Intersecting lines

If two lines intersect, then the point of intersection must be a point that lies on both of the lines. Thus, in Fig. 3·17(*a*), lines *AB* and *CD*

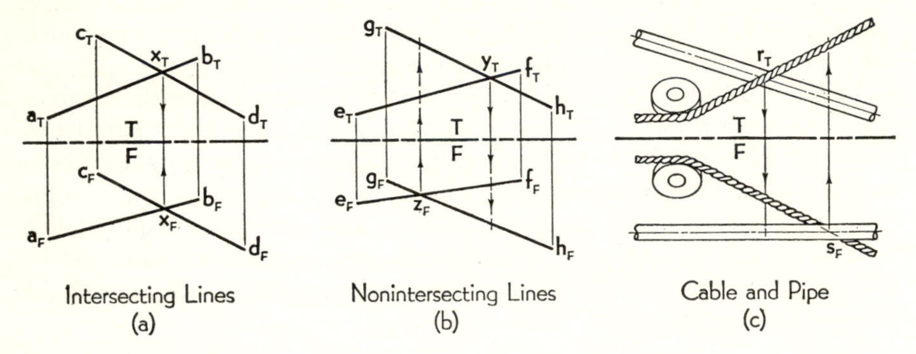

Fig. 3·17. Intersecting and nonintersecting lines.

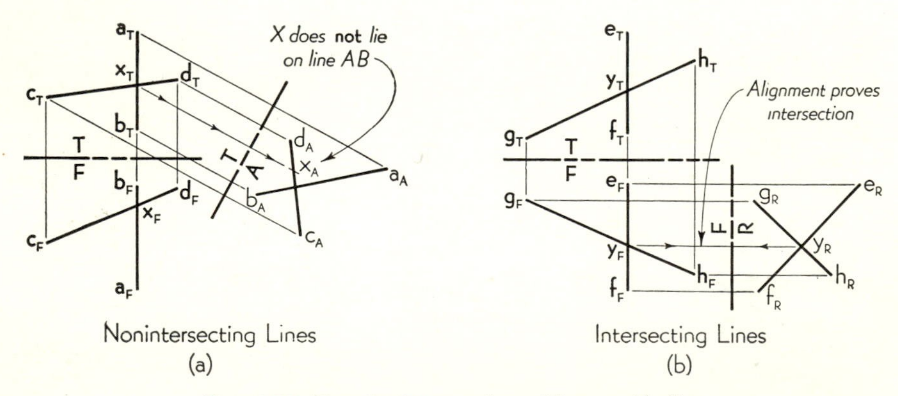

Fig. 3·18. Test for intersection with a profile line.

actually intersect at point X because the intersection of a_Tb_T and c_Td_T, which is x_T, lies directly above the intersection of a_Fb_F and c_Fd_F, which is x_F. Point X thus qualifies as a point on both AB and CD and proves that the lines intersect.

In Fig. 3·17(*b*) lines e_Tf_T and g_Th_T cross at the point y_T; but when a parallel is drawn to the front view, it is apparent that y_F cannot lie simultaneously on both e_Ff_F and g_Fh_F. In the front view these lines appear to intersect at z_F, but, extending a parallel to the top view, it is again seen that z_T cannot lie on both e_Tf_T and g_Th_T. Therefore lines EF and GH do not intersect since they have no common point. At the apparent intersection point y_T line EF is actually passing *over* line GH, and at z_F line EF passes in *front* of line GH.

By the same method of inspection it is obvious in Fig. 3·17(*c*) that the cable does not intersect the pipe but passes over it at point r_T and behind it at point s_F. The actual amount of clearance between cable and pipe can be determined by the method of Art. 3·17.

Figure 3·18 illustrates again that the profile line requires special treatment. In Fig. 3·18(*a*) lines a_Tb_T and c_Td_T cross at x_T, lines a_Fb_F and c_Fd_F cross at x_F, and point x_T lies directly above x_F. But because the line AB is a profile line, it is questionable whether point X is actually on the line AB. If lines AB and CD are drawn in some new view, such as view A, then it is readily apparent that point x_A lies on line c_Ad_A but not on a_Ab_A. Hence these two lines do not intersect.

In Fig. 3·18(*b*) the new-view test has been applied by adding view R. The intersection point y_R is in exact alignment with y_F, hence point Y is a point common to both lines, and lines EF and GH do intersect at Y.

Figure 3·19 shows two lines AB and CD whose point of intersection lies well beyond the left edge of the paper. Whether these lines extended would actually intersect or not can be determined within the limits of the paper by connecting any two points on one of the lines to any two points on the other line to form a pair of crossing lines. If the lines AD and BC intersect at a common point, such as X, then the original lines AB and CD will also intersect when extended.

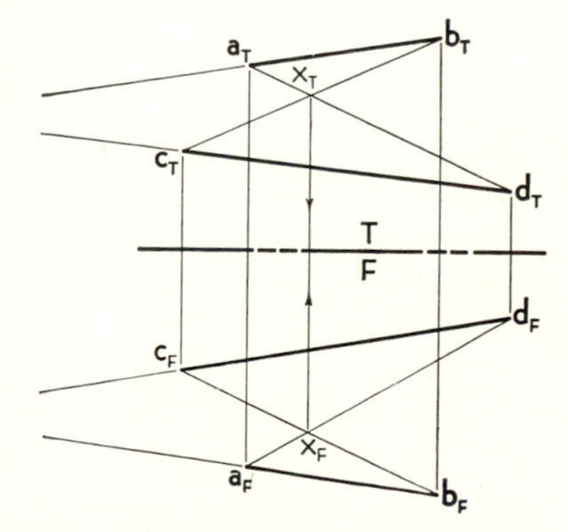

Fig. 3·19. Test for intersecting lines.

PROBLEMS. Group 9.

3·14. Perpendicular lines

Although two intersecting lines may make any angle with each other, perpendicular lines occur most frequently, and their special properties deserve careful study. For this purpose let us consider a simple example of perpendicular lines—a spoked wheel and axle. Each of the spokes is perpendicular to the axle at a common point, the center of the hub.

The pictorial sketch in Fig. 3·20 shows the axle CD in an inclined position, and in the top and front views only the center line of the axle is shown as line CD. If the wheel and axle are viewed in a direction parallel to the axle (along arrow B), then the center line of the axle will appear as a point, and each spoke will appear in its true length. But to obtain view B in the multiview drawing the axle must first be seen in its true length, as in view A (Rule 8).

With the axle center line shown in each of the four views, a few of the spokes can now be drawn. For simplicity the wheel is represented in view B as a single circle and each pair of spokes as a diameter. Since all spokes are true length in view B, then in view A they must all appear parallel to reference line A–B (Rule 6), hence perpendicular to the axle. Thus any pair of spokes taken at random in view B, such as diameter 1–2, must be drawn perpendicular to the axle in view A, and points 1 and 2 can then be located in the top and front views by alignment and similarity.[1] In similar manner the specially selected diameters 3–4, 5–6, and 7–8 have been located first in view A and then in the top and front views. Remembering that each of these four diameters is actually at right angles to axle CD, let us observe the appearance of these right angles in the top and front views where CD does not appear true length.

Diameter 1–2, selected at random, does *not* appear perpendicular to CD in either the top or front views.

Diameter 3–4 coincides with the axle center line in the top view and is *not* perpendicular to CD in the front view.

Diameters 1–2 and 3–4 have thus been used to demonstrate that a 90° angle may in some views appear as small as zero, as large as 180°, or of any intermediate value.

Diameter 5–6 appears as a point in view A and hence must appear true length in the adjacent top view (Rule 6*a*) and be perpendicular to the axle. In the front view, diameter 5–6 is horizontal because it appears true length in the top view. Note now that this diameter is the *only one* that appears *true length* in the top view and that it is also the only diameter that appears *perpendicular* to the axle in that view.

[1] In general, the author prefers subscripts in the text illustrations for ease of reference in the accompanying reading matter; but when no confusion will result, the view subscripts will be omitted. The student may exercise the same prerogative with discretion.

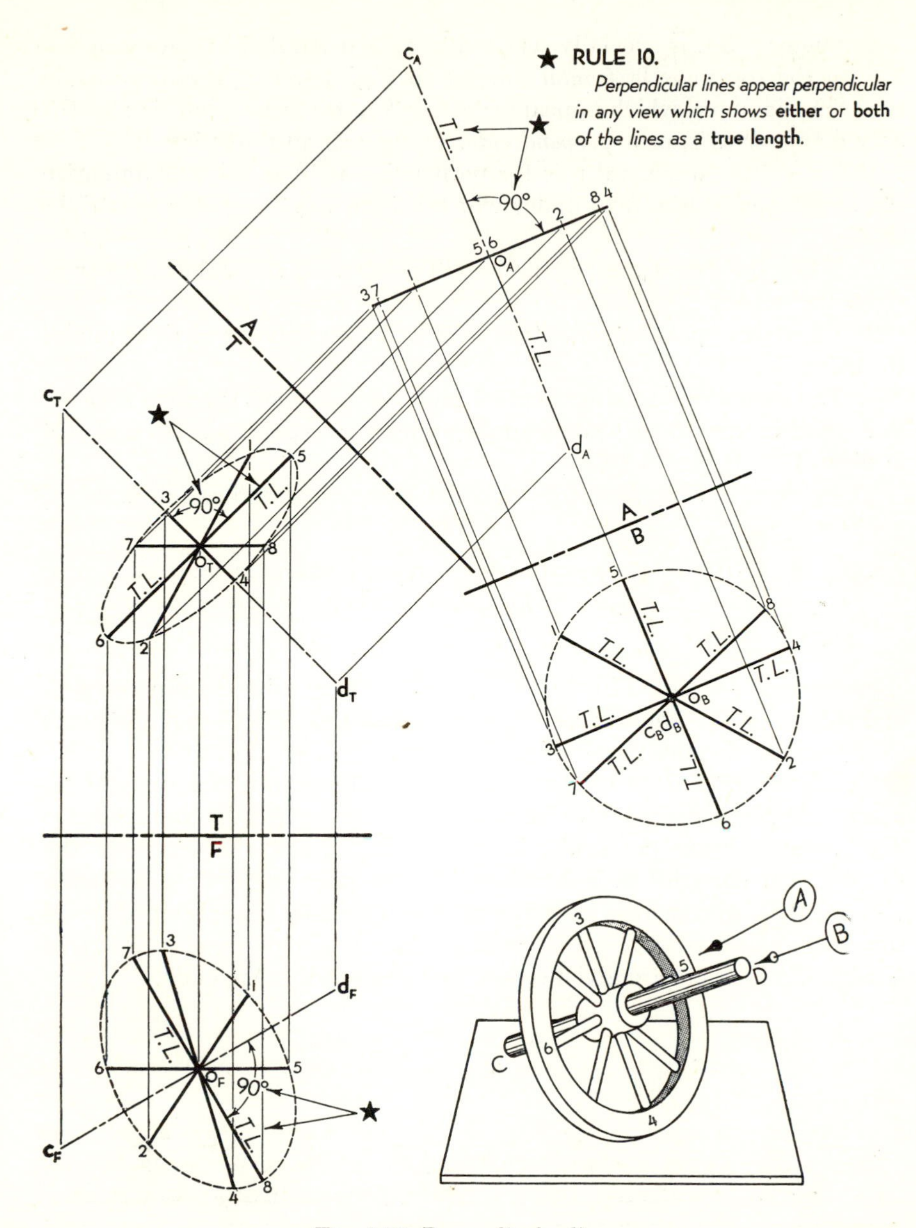

Fig. 3·20. Perpendicular lines.

Diameter 7–8 was specially selected to appear parallel to reference line *T–F* in the top view [a *frontal* line, as in Fig. 3·8(*d*)]. But note again that this diameter, which appears *true length* in the front view, is also the only diameter to appear *perpendicular* to the axle in that view.

Before stating a general rule for perpendicular lines, let us summarize the conditions in Fig. 3·20 under which a right angle appears as a right angle:

1. When the axle appears true length in view *A*, all spokes appear at right angles to it.
2. When any spoke appears true length, it also appears at right angles to the axle.
3. The exceptions to statements 1 and 2 occur when the axle appears as a point in view *B* or when a spoke appears as a point as *O*–5 and *O*–6 in view *A*.

The general rule for perpendicular lines may be stated as follows:

- ***Rule 10. Rule of perpendicular lines.*** *Perpendicular lines appear perpendicular in any view which shows* ***either*** *or* ***both*** *of the lines as a* ***true length.***

The exceptions noted in statement 3 above are omitted from the general rule since the appearance of one line as a true length and the other as a point is adequate proof of perpendicularity.

The converse of Rule 10 is also true: if two lines appear perpendicular in any view and one or both of the lines is a true length, then the two lines are actually perpendicular in space. In Fig. 3·21, for example, the five different triangles each show a right angle in one or both of the given views. By noting which lines appear true length and applying Rule 10 the reader can determine by inspection which triangles are actually right triangles. Answers are given on the last page of the Appendix.

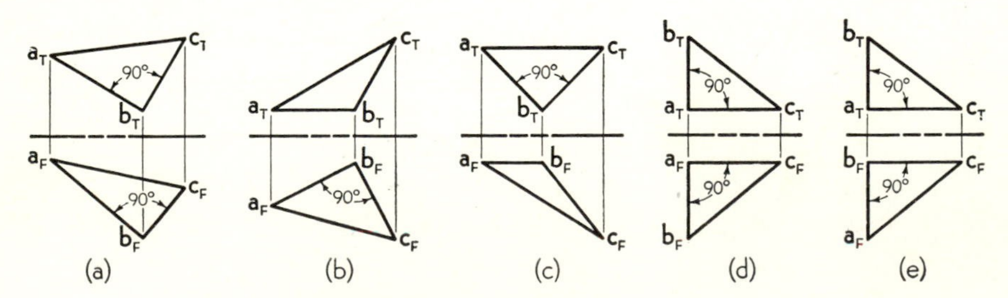

Fig. 3·21. Which triangles are right triangles in space?

3·15. Location of a perpendicular at a given point on a line

At a given point on a line an infinite number of lines can be drawn perpendicular to the given line; hence additional restrictions are necessary to obtain a unique solution. If the required perpendicular is given in only one view, as in Fig. 3·22(*a*), then its position is fixed, and it can be established in the adjacent view by applying Rule 10. Line a_Fc_F does not appear at right angles to a_Fb_F in the front view, and therefore a_Fc_F cannot be a true length. But the two lines *AC* and *AB* must appear perpendicular when one of them appears in true length. This reasoning suggests several methods of solution.

In Fig. 3·22(*b*) the front-adjacent view *A* has been drawn to show line *AB* in true length. In this view the lines must appear perpendicular; hence a_Ac_A is drawn at 90° to a_Ab_A and c_A is aligned with c_F. By transferring distance x to the related top view point c_T is established, and the top view can be completed.

In Fig. 3·22(*c*) the same solution has been obtained by drawing top-adjacent view *B* to show *AB* in true length. Again the two lines must appear perpendicular in this view, and point c_B must be distance y from reference line *T–B*. With c_B established, point c_T can now be located by alignment as shown.

The solution could also be made in a third but similar way (not shown) by drawing a front-adjacent view to show line *AC* in true length.

PROBLEMS. Group 10.

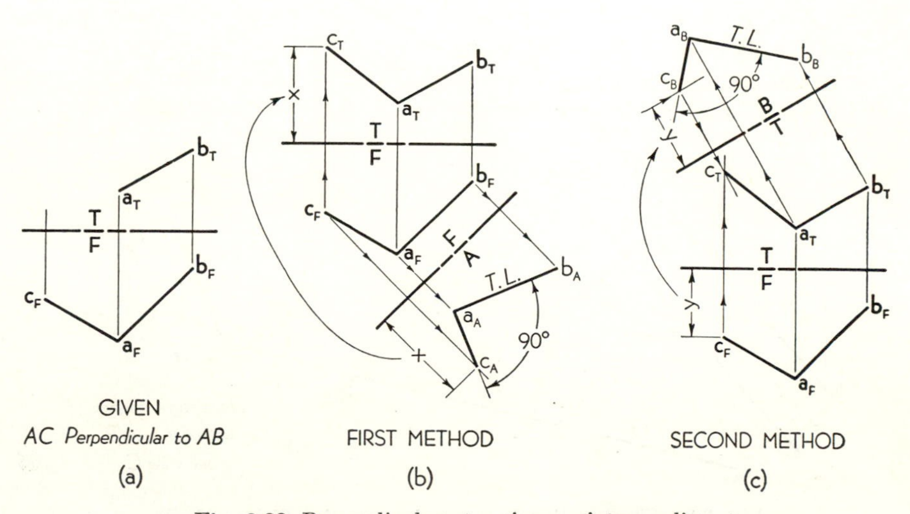

Fig. 3·22. Perpendicular at a given point on a line.

3·16. Shortest line from a point to a line (line method)[1]

Analysis. The shortest line from a given point to a given line is perpendicular to the line. It will appear perpendicular when the given line appears true length. Concisely expressed for ease of remembrance, the method is to:

Show the given line as a true length.

Problem. Figure 3·23 shows the top and front views of a pipeline *AB* and a tank outlet at point *C*. A connecting pipe from the outlet must join the given pipe with a standard 90°-tee fitting. Determine the length of the connecting pipe, and locate the point *X* where the pipes must be joined.

Construction. The required pipe is perpendicular to the given pipe and is therefore the shortest line from point *C* to line *AB*. A new view *A* has been drawn adjacent to the front view in order to show line *AB* in its true length (view *A* could also have been taken adjacent to the top

[1] See Art. 4·13 for another method.

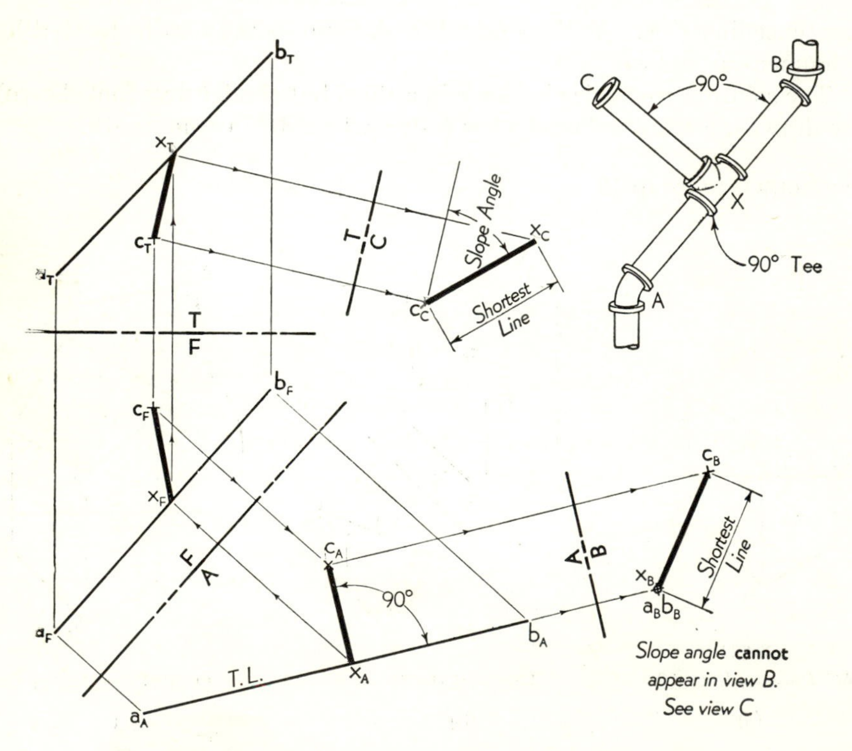

Fig. 3·23. Shortest line from a point to a line (line method).

view). From c_A the shortest line is drawn perpendicular to a_Ab_A to locate point x_A, the junction of the shortest line CX with the line AB. Since point X lies on line AB, points x_F and x_T can be located simply by alignment. The required shortest line from point C to line AB is therefore line CX, as shown in the given views.

To obtain the true length of line CX a new view must be drawn for the purpose since none of the present three views shows CX in true length. View B, which shows the given line as a point, also shows line CX in true length.

Caution. It should be noted that the direction of sight for view B is *inclined* (parallel to inclined line AB), and although this view shows the true length of CX it *cannot* also show the slope of CX. The slope angle of CX can be seen only in an *elevation* view, which shows the line in its true length (Rule 7); view C is such a view (see Art. 3·7).

PROBLEMS. Group 11.

3·17. Shortest line between two skew lines (line method)[1]

The shortest line connecting two skew lines (nonintersecting, nonparallel lines) must be perpendicular to both the given lines. To determine the length and position of this shortest line is a common engineering problem. When two oblique pipes must be connected by a third pipe, it is desirable to use only right-angled tees and elbows and to have the shortest possible pipe. Two tunnels can be connected most economically by the shortest tunnel. Clearance distance between guy wires, electrical conductors, control cables, braces, etc., must be checked by finding the shortest distance between the members.

Analysis. Since the required shortest distance must be perpendicular to each of the given skew lines, it will appear perpendicular to either line when that skew line appears in true length. But if either skew line appears as a point, then the shortest distance will appear true length, and it will therefore be perpendicular to the other skew line. Therefore the method of solution should be concisely remembered as:

Show one of the skew lines as a true length and then as a point.

Problem. Figure 3·24 shows a top and front view of an airplane-rudder control cable and a nearby fuselage brace. The center lines of brace and cable are represented here as lines AB and CD. The clearance distance and its location between the two center lines are to be determined.

[1] See Art. 4·21 for another method.

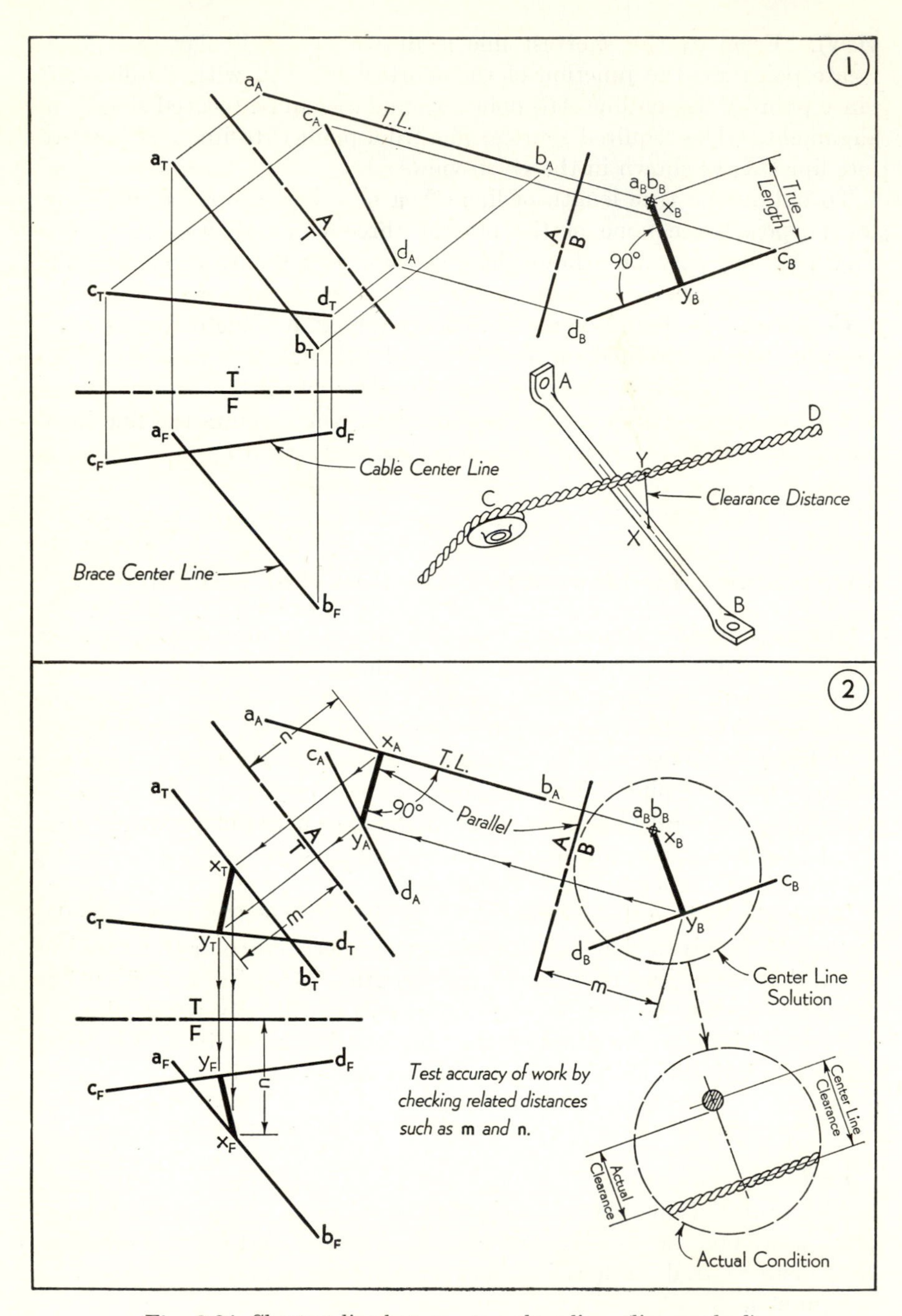

Fig. 3·24. Shortest line between two skew lines (line method).

Stage 1. *Construct the necessary views.*

One of the two lines must be shown as a true length and then as a point. Either line may be selected, and the true-length view may be drawn adjacent to either the top or the front view. In Fig. 3·24 AB has been drawn true length in view A. Line CD is also shown here. The required shortest distance must be perpendicular to a_Ab_A, but its exact location is still unknown; hence view B is drawn to show line AB as a point. Line CD is *not* true length in view B, but the shortest line will be true length and hence will appear perpendicular to c_Bd_B. Thus the line x_By_B is located in view B and is the required true clearance distance between center lines.

Stage 2. *Locate the shortest line in the given views.*

The amount of clearance has been found in view B; it remains to show the location of this shortest distance XY in the given views. Point y_A is located by alignment, and x_Ay_A is drawn perpendicular to a_Ab_A, which is a true length in view A. The position of x_Ty_T and x_Fy_F can now be determined by extending parallels from view A to view T and then to view F. Locating points on lines by successive alignment may result in accumulated error, and it is therefore wise also to check measurements from the reference line in related views. The distance m from y_T to T–A should equal that from y_B to A–B, and the distance n from x_F to T–F should equal that from x_A to T–A, etc.

PROBLEMS. Group 12.

3·18. Location of a line through a given point and intersecting two skew lines

This problem is simply a variation of the problem discussed in the previous article, except that the required line must pass through a specific point and therefore will not necessarily be perpendicular to the skew lines. This type of problem is encountered in engineering work, for example, when it is necessary to connect two skew braces with a third stiffening brace that must be anchored at a specific point.

Analysis. If a new view is drawn to show one of the given lines as a point, then in this view the required line must pass through the given point X and the point view of the line. The required line can be extended to intersect the other skew line at a definite point Y. With two points, X and Y, on the required line now known, its position is fixed. This solution thus requires that we:

Show either of the given skew lines as a point.

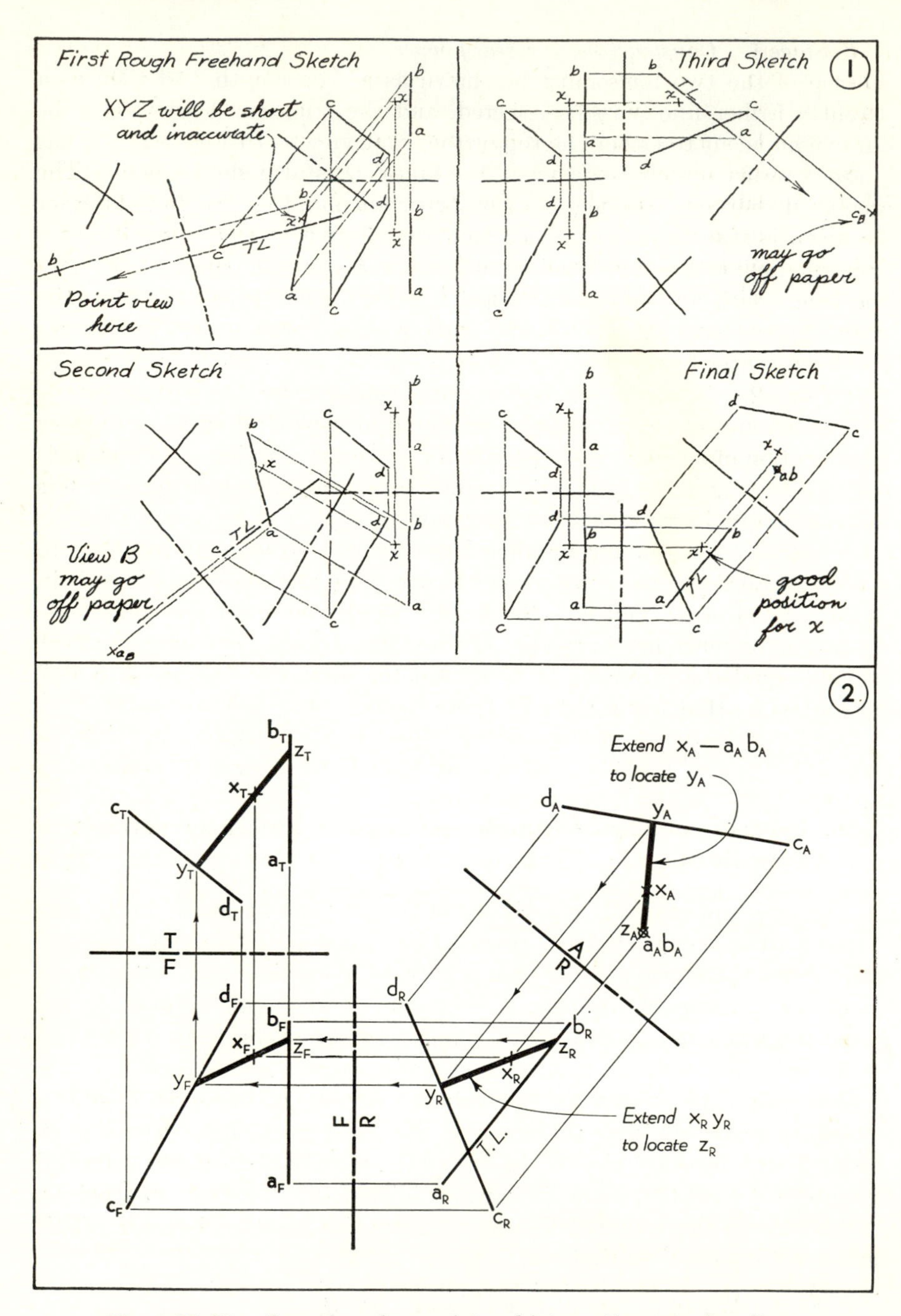

Fig. 3·25. Line through a given point and intersecting two skew lines.

Problem. Let the lines AB and CD in Fig. 3·25 represent the center lines of two braces that are to be connected by a third brace passing through point X. The direction of the required brace is unknown.

Stage 1. *Make rough freehand sketches of possible solutions.*

Since either line can be shown as a point, the draftsman has a choice of four possible solutions. The first reference line may be taken parallel to a_Tb_T, c_Td_T, a_Fb_F, or c_Fd_F. Although all four constructions will yield the same answer, the student should learn to examine each of the four possibilities before making a choice. In some cases this can be done with sufficient accuracy by mentally estimating the position of the new views. In more complex problems it may be necessary to make rough freehand sketches, as shown in Fig. 3·25. Such sketches are usually accurate enough to indicate space requirements and occasionally reveal potential sources of inaccuracy. In future problems involving two or more new views preliminary sketches are recommended to save time, improve accuracy, and avoid needless erasures.

Stage 2. *Locate the required line in accurately constructed new views.*

The solution shown in Fig. 3·25 has been made by showing line AB as a point in view A. View R was drawn first to show line AB as a true length. It should be noted that the direction of sight for the two new views is determined solely by the line AB. Line CD and point X are simply carried over into the new views by alignment and measurement.

In view A the required line is drawn from a_Ab_A through x_A to intersect c_Ad_A at point y_A. Point y_R can now be located on line c_Rd_R by alignment. In view R the required line is then $y_Rx_Rz_R$, where z_R has been found by extending the line y_Rx_R. With the required line now completely determined in two views it can be drawn in the front and top views by alignment and verifying measurements. Note that none of the four views shows the line YXZ in true length.

PROBLEMS. Group 13.

3·19. Principal views of objects with inclined axes

It is sometimes necessary to show in the top and front views an object of regular shape in an inclined, or tilted, position. In general, this is most easily done by first drawing auxiliary views that show the object in the simplest manner. Figure 3·26 illustrates the consecutive stages in the construction of the top and front views of a tilted hexagonal prism (an ordinary nut simplified by omitting threads and chamfered corners).

Stage 1. *Establish the necessary new views.*

It is required that the axis MN of the object shall coincide with the inclined line AB shown in the top and front views and also that one side

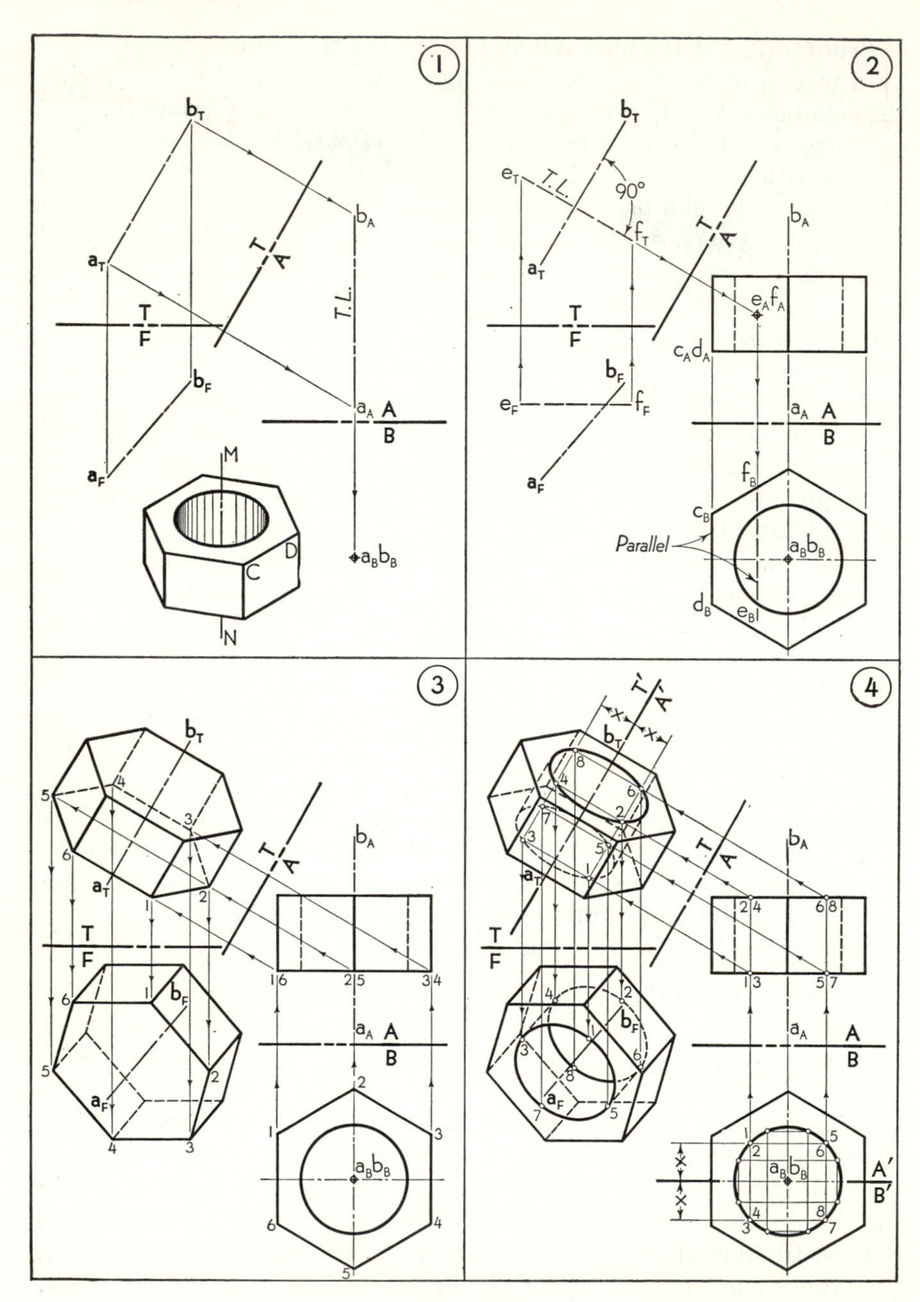

Fig. 3·26. Principal views of an object with inclined axis.

of the hexagon, such as line CD, shall be horizontal. The simplest views of this object will be those which show the axis AB as a point and as a true length, and therefore the first step is to construct the auxiliary views A and B.

Stage 2. *Position the object as required in the new views.*

In order correctly to orient the hexagon in view B it is necessary to determine the direction of line CD in that view. Since line CD must be horizontal, it will be a true length in the top view; and since it is also perpendicular to axis AB, it will appear in the top view perpendicular to a_Tb_T. But the exact location of line CD in the top and front views is as yet unknown; hence a substitute line EF, parallel to the proposed CD line, is assumed anywhere in these views. The line EF is then drawn in views A and B. Now when the hexagon is constructed in view B, it must be placed so that one side, c_Bd_B, is parallel to e_Bf_B. When the object is also drawn in view A, the two auxiliary views are complete and it is apparent that these two views are exactly like a top and front view of the untilted object.

Stage 3. *Locate object corners and edges in the given views.*

For illustration six corners of one hexagonal surface have been numbered in view B. The corresponding corners are also numbered in the adjacent view A. These same points are then located in the top view on parallels from view A and by measurements taken from related view B. Following the completion of the top view, points in the front view are located by alignment and by measurements from view A. As a check on accuracy, note that parallel lines on the object are parallel in all views.

Stage 4. *Locate curved lines in the given views.*

The circular ends of the hole will appear elliptical in the top and front views. To construct these ellipses a series of points are assumed on the circle in view B, and the corresponding points are located in view A. Each of these points can be located in the top and front views in exactly the same manner that the corners of the hexagon were located in stage 3. The points are then connected with a smooth curve to form the required ellipses. The number of points to be used varies with the size of the circle, ranging from 12 to 24 points on each circle. To reduce the labor of construction the points are usually equally spaced around the circle and symmetrically placed with respect to axes parallel and perpendicular to the reference line A–B. By assuming a new displaced reference line A'–B' on the center line in view B and a correspondingly displaced line T'–A' in the top view, the eight symmetrical numbered points in view B can all be transferred to the top view with just one setting of the dividers to distance x. In constructing the front view each setting of the dividers serves to locate two points.

The construction outlined here for drawing the principal views of inclined objects is obviously most pertinent to regular geometric solids, but the principle of first constructing simple auxiliary views and the method of orienting the object in these views can easily be applied to objects of any shape.

PROBLEMS. Group 14.

3·20. Auxiliary views in a prescribed direction

Analysis. The direction of sight for a required auxiliary view may be specified as parallel to a given line. Any new view that shows this given line as a point will also show the object in the prescribed direction. Therefore:

Show as a point the line representing the direction of sight.

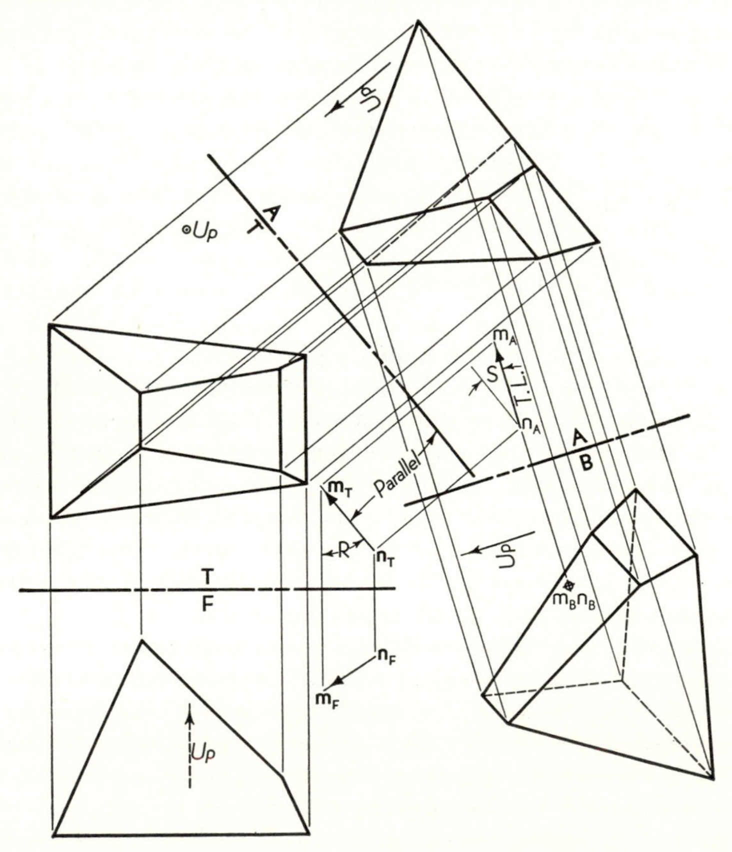

Fig 3·27. Auxiliary view in a prescribed direction.

Construction. In Fig. 3·27 it is required to show the given object as it would appear if viewed in a direction parallel to the arrow MN. The auxiliary elevation view A has been drawn first to show line MN as a true length. This view has been taken adjacent to the top view because this elevation view of the object again shows the base surface edgewise and hence is easy to construct. In view B, which shows the line MN as a point, note that the observer is looking along the arrow from N *toward* M, which is the direction indicated by the arrowhead. It should also be noted that the line MN may be assumed in any *position* relative to the object; only the *direction* of the line, fixed by angles such as R and S, determines the resulting auxiliary view.

In order to determine the natural aspect of view B a vertical arrow (labeled "Up") has been introduced in the top and front views and has then been drawn in views A and B also. If Fig. 3·27 is turned until the "Up" arrow in view B appears vertical, it will be seen that view B is an excellent pictorial representation of the object. By a proper choice of the angles R and S pictorial drawings may be produced that will show any object in its most advantageous position. Views such as view B, taken in an oblique direction, are called *axonometric* views. Another method of constructing such pictorial drawings is discussed in the next article.

PROBLEMS. Group 15.

3·21. Axonometric views and drawings

To illustrate the various axonometric views that may be obtained by varying angles R and S of the line of sight, consider the 1-in. cube shown in Fig. 3·28. Here the direction of sight has been taken parallel to the cube diagonal AB, and therefore angle R is 45° and angle S (the slope angle of the direction of sight) is 35°16′. In view B the diagonal AB appears as a point, and the view should be rotated until the vertical edge BC of the cube appears vertical. This view B is an *isometric* view of the cube, so called because each of the 12 edges of the cube has been foreshortened equally; each edge in this view is 0.8165 of the actual true length, and the angles X and Y are each exactly 30°.

Because of the equal foreshortening and the convenient 30° angles, it is customary to construct isometric views directly as shown at the right in Fig. 3·28. Since, however, the edges of the cube are all equal in length, these edges are drawn full size, and the resulting isometric *drawing* is exactly similar to the isometric *view* but is larger. For pictorial purposes this enlargement is not objectionable.

Although the isometric drawing is easy to construct, it sometimes produces a confusing and misleading picture of some objects. In such cases the angles R and S can be changed to shorten one or two of the cube edges in some convenient scale ratio. Figure 3·29 shows axonometric drawings of a cube for a variety of scale ratios. Each of the cubes shown is actually larger than a true axonometric view because they have been drawn to scale proportion without correction for foreshortening. When two sides of the cube are drawn to the same scale, the drawing is called *dimetric;* when all three sides are drawn to different scales, it is called *trimetric.*

The table on page 78 gives the values of angles R and S and angles X and Y for various ratios of the x, y, and z sides of the cube. As shown in view B of Fig. 3·28, X is the left-axis angle, Y is the right-axis angle; x is the left edge, y the right edge, and z the vertical edge of the cube.

As an illustration of method, Fig. 3·30 shows the construction of a trimetric drawing of the same object shown in Fig. 3·27. The type of drawing desired is determined by selecting one of the cubes in Fig. 3·29. In this case cube (9) will be used. Reference to the table shows that this type of view is obtained by taking the direction of sight at an angle

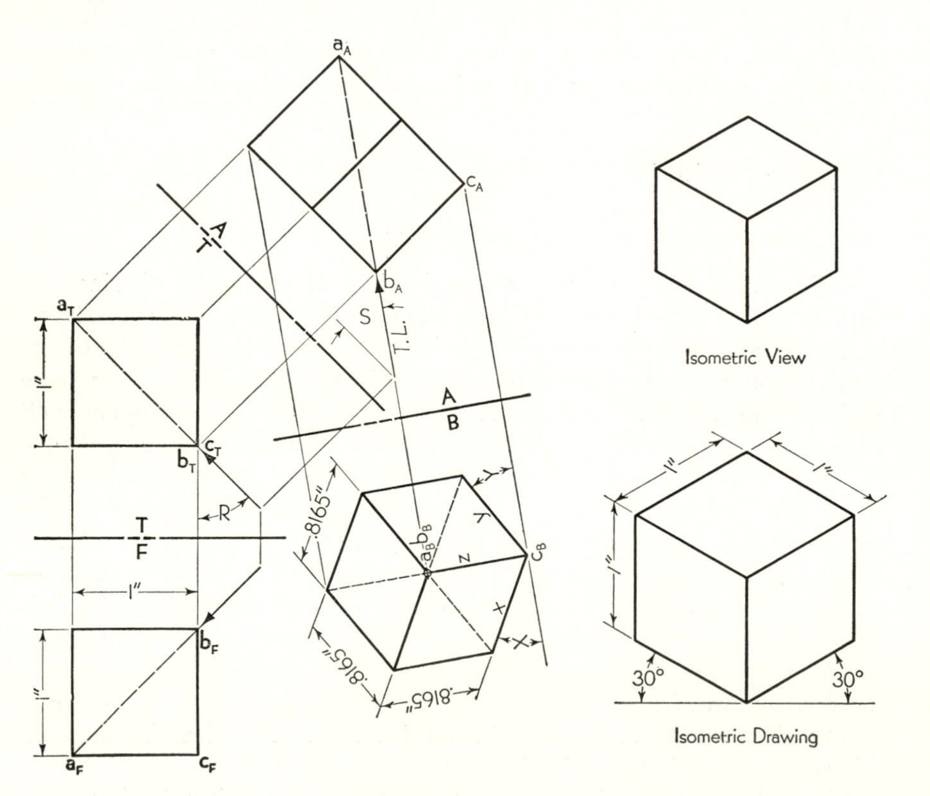

Fig. 3·28. Construction of an isometric view and drawing.

R of 39°8′ and an angle S of 22°3′. Viewed in this direction the three equal dimensions of a cube, x, y, and z, will all be foreshortened but will appear in the ratios 7/8, 3/4, and 1, respectively. The z edge of the cube will be vertical, but the x edge will be inclined 17°0′ with the horizontal and the y edge 24°46′ with the horizontal. With these values the axes for the pictorial drawing in Fig. 3·30 can now be constructed, starting at point O. Through this point the left axis OX is drawn at an angle of 17°0′ and the right axis OY at 24°46′.

The next step is to "box in" the given object as shown by the dash lines in the top and front views. This imaginary box, which exactly contains the object, should then be constructed in the pictorial drawing, where it is shown with dash lines. The length of the box along OX is

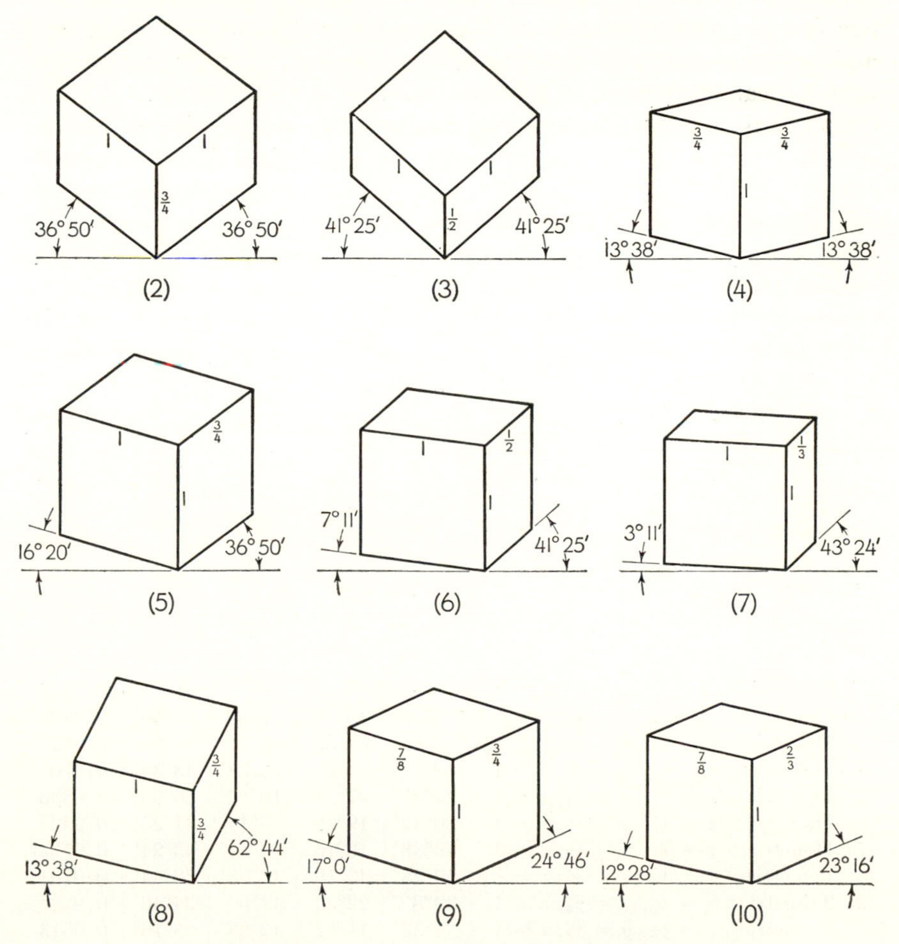

Fig. 3·29. Axonometric drawings of a cube.

then laid off equal to seven-eighths of the actual length $o_T x_T$; the width of the box OY is laid off equal to three-fourths of the actual width $o_T y_T$; the vertical height OZ is laid off equal to the actual height $o_F z_F$. The remaining edges of the box can then be drawn parallel to these three edges to complete it.

The work of reducing object dimensions to the required scale ratios can be minimized by using a triangular diagram such as that shown in Fig. 3·30. To construct the triangles, base AB is drawn to any length that exceeds the longest dimension of the box. Side BC, perpendicular to AB, is made seven-eighths of length AB. Side BD, equal to three-fourths of AB, is also laid off on this side of the triangle. Side AB is graduated in small divisions, and vertical ordinates are drawn as shown. Now if the dividers, set to some true distance such as AG, are placed along line AB, they can be pivoted about point G and reset to distance GH, which will thus be seven-eighths of the original AG distance. Similarly, true distance AE can be reduced to three-fourths size on ordinate EF.

Oblique lines on the object, such as line MN, must be located in the pictorial drawing by locating the points M and N individually by coordinates. Point M lies on the top surface of the box a distance a from the left end and a distance b from the front side of the box. To locate point M in the pictorial drawing, distance a is transferred to the diagram, reduced to seven-eighths size, and then laid off at a'. Distance b, reduced to three-fourths size, is then laid off at b', parallel to OY, to locate point M. Point N, lying wholly within the box, requires three

Axonometric Drawing Proportions

Type of drawing	Scale ratios	Direction of sight angles		Angle of drawing axes		True fore-shortening ratio
		R	S	X	Y	
(1) Isometric	$x = 1,\ y = 1,\ z = 1$	45°	35°16′	30°	30°	0.8165
(2) Dimetric	$x = 1,\ y = 1,\ z = 3/4$	45°	48°30′	36°50′	36°50′	0.8835
(3) Dimetric	$x = 1,\ y = 1,\ z = 1/2$	45°	61°52′	41°25′	41°25′	0.9428
(4) Dimetric	$x = 3/4,\ y = 3/4,\ z = 1$	45°	14°2′	13°38′	13°38′	0.9701
(5) Dimetric	$x = 1,\ y = 3/4,\ z = 1$	32°2′	27°56′	16°20′	36°50′	0.8835
(6) Dimetric	$x = 1,\ y = 1/2,\ z = 1$	20°42′	19°28′	7°11′	41°25′	0.9428
(7) Dimetric	$x = 1,\ y = 1/3,\ z = 1$	13°38′	13°16′	3°11′	43°24′	0.9733
(8) Dimetric	$x = 1,\ y = 3/4,\ z = 3/4$	19°28′	43°19′	13°38′	62°44′	0.9701
(9) Trimetric	$x = 7/8,\ y = 3/4,\ z = 1$	39°8′	22°3′	17°0′	24°46′	0.9269
(10) Trimetric	$x = 7/8,\ y = 2/3,\ z = 1$	35°38′	17°57′	12°28′	23°16′	0.9513

dimensions *r*, *s*, and *t* to locate it. Distance *r* is reduced to seven-eighths size and laid off along *OX* at *r'*; distance *s*, reduced to three-fourths size, is next laid off parallel to *OY*; finally distance *t*, full size, is drawn parallel to *OZ* to locate point *N*.

By repetition of the above procedure all corners of the object are located, and the drawing is completed. The reader should now compare this trimetric drawing with the axonometric view *B* in Fig. 3·27. Except for size they are the same because angles *R* and *S* in Fig. 3·27 were taken as 39°8′ and 22°3′ (see table). All linear distances in view *B* are 0.9269 of those shown on the trimetric drawing.

Although a trimetric drawing has been used here to illustrate method, the dimetric drawings are most frequently employed because only one scale reduction is necessary. For further details of construction the interested student is referred to the many excellent texts on engineering drawing.

PROBLEMS. Group 16.

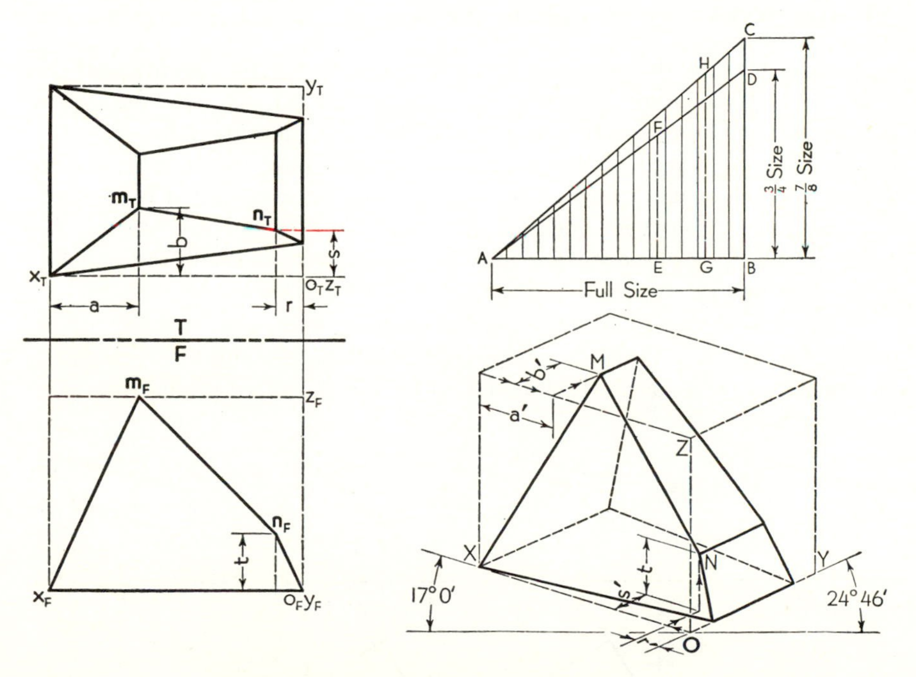

Fig. 3·30. Construction of a trimetric drawing.

4. PLANE SURFACES

4·1. Representation of a plane surface

A plane may be defined as a surface in which any two points may be connected by a straight line and the straight line will lie entirely within the surface. The position of any plane surface is definitely fixed by the location of any three points, not in a straight line, in the plane. In actual practice, a plane surface is limited to the area bounded by certain straight lines or curves; theoretically, and for the purpose of solving many problems, a plane surface may be considered to extend indefinitely beyond its bounding lines.

Figure 4·1 shows four ways of representing a generally inclined, or oblique, plane in a multiview drawing; points A, B, and C are located identically in each case. In Fig. 4·1(a) two of the three points, namely,

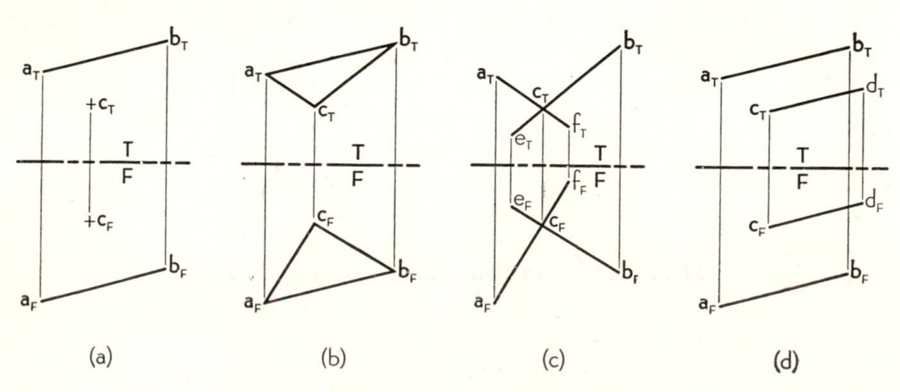

Fig. 4·1. Representation of a plane surface.

A and *B*, have been connected by a straight line, and thus the plane is defined by the point *C* and the line *AB*. In Fig. 4·1(*b*) all three points have been connected, and the triangle *ABC* defines the same plane in slightly different guise. In Fig. 4·1(*c*) lines *AC* and *BC* have been extended beyond point *C* to form a pair of intersecting lines that also define the plane—thus demonstrating that additional points, such as *E* and *F*, may be located in a plane simply by extending a straight line in that plane. In Fig. 4·1(*d*) a new line *CD* has been drawn through point *C* parallel to line *AB* and arbitrarily terminated at point *D*; thus two parallel lines also define a plane.

Whereas any three points, not in a straight line, establish a definite plane, it must be remembered that any additional points will have to be located by accurate construction to ensure that they will lie in the same plane as the first three (see Art. 4·5). If two lines are employed to define a plane, they must be either parallel or intersecting; if the lines are not parallel and the point of intersection does not lie on the paper, the lines should be tested for intersection by the construction shown in Fig. 3·19.

4·2. Location of a line in a plane

Given a plane, defined and represented by one of the methods described above, it is frequently necessary to locate additional lines in that plane. In Fig. 4·2, for example, the plane is defined by given points *A*, *B*, and *C*, connected to form a triangle. Line r_Ts_T is the top view of a line which must also lie in plane *ABC*, but the location of r_Fs_F is unknown.

The given line r_Ts_T crosses the lines a_Tb_T and b_Tc_T at points x_T and y_T, respectively. Since line *RS* must lie in the same plane as line *AB*, then line *RS* must actually intersect line *AB* at the point *X*. Point x_F must therefore lie directly below x_T on the line a_Fb_F. By the same reasoning, point *Y* must be the intersection of lines *RS* and *BC*, and y_F is located on b_Fc_F. Then r_Fs_F must pass through x_F and y_F, and the front view of line *RS* is thus established.

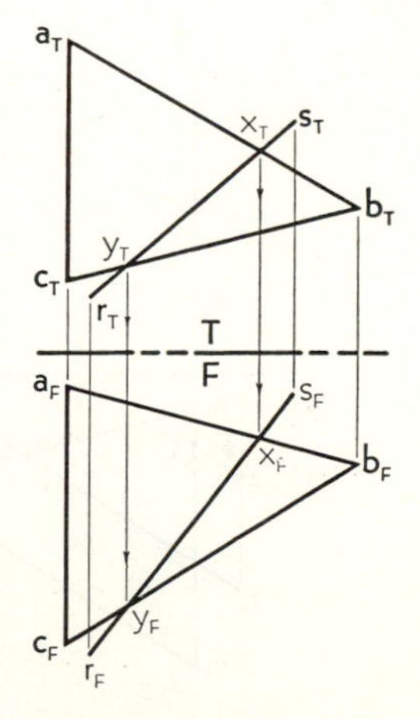

Fig. 4·2. Location of a line in a plane.

In Fig. 4·3 the given line r_Ts_T crosses line a_Tb_T and is parallel to b_Tc_T. The point of intersection *X* of lines *RS* and *AB* can be located as before. Since lines *RS* and *BC* appear parallel in the top view and are known to lie in the same plane, they must actually be parallel lines, in which case they will appear parallel in all views (Rule 9). Consequently, r_Fs_F is drawn through point x_F parallel to b_Fc_F.

In Fig. 4·4 the given line r_Ts_T crosses line a_Tc_T and b_Tc_T but is not parallel to any given line on the plane. The intersection of *RS* and *BC* at *X* is easily located and is shown in the front view at x_F. The intersection of lines *RS* and *AC* is evident in the top view; but line *AC* is a profile line, and to locate the intersection point in the front view would necessitate drawing an additional view. This could be done (as shown in Fig. 3·6), but easier solutions are available. For example, any number of new lines may be arbitrarily created in the plane *ABC* simply by connecting any point on one of the given lines to any point on another. By selecting point *E* anywhere on line *AB* and connecting this point to point *C* we have the new line *CE*, which also lies in plane *ABC*. Point *Y*, the intersection of lines *RS* and *CE*, is thus determined, and y_F is located on c_Fe_F. The line r_Fs_F can now be drawn through x_F and y_F. Another variation of this same idea is to select the new point, such as *F*, on the extension of one of the given lines; line *CF* is an example of this. Point *Z* is the intersection of lines *RS* and *CF* and provides another point, z_F, on the required line r_Fs_F.

In each of the three cases described here the line *RS* was assumed to be given in the top view, but it should be apparent that these methods are equally applicable when the line is given in the front view or any adjacent view. It is especially useful to note how easily a given plane such as *ABC* in Fig. 4·4 can be transformed into plane *CEF* or *ACF* without altering the position of the given plane.

Exceptions. The plane shown here is a generally inclined plane, and it is therefore possible, with one exception, to assume r_Ts_T (or r_Fs_F) in any desired position in the plane. The one exception occurs if line *RS*

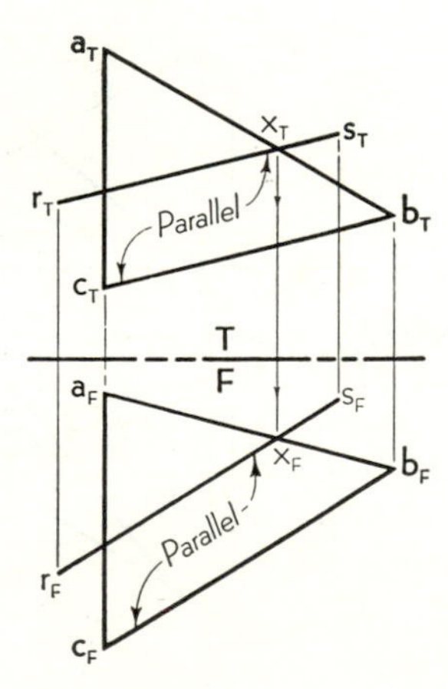

Fig. 4·3. Parallel line in a plane.

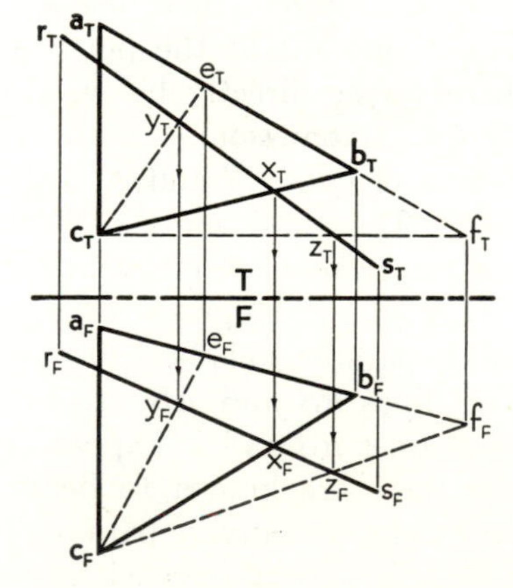

Fig. 4·4. Creating new lines in a plane.

is given as a point in one view because then its direction in space is completely fixed. It is obviously impossible, for example, for a vertical line to lie in an inclined plane. If the given plane appears as an edge in one of the principal views, then the new line cannot be selected at random in that view but must coincide with the edge view of the plane in which it must lie.

4·3. True-length lines in a plane

The previous article showed how various additional lines may be located in a given plane. How to locate such lines so that they not only will lie in the plane but also will appear true length in any desired view is a very basic construction in this scheme of descriptive geometry.

Horizontal lines. Assume that it is required to construct a line in plane *ABC* (Fig. 4·5) which will also be true length in the top view. To be true length in the top view the required line must be horizontal and must therefore be parallel to reference line *T–F* in the front view. Any horizontal line, such as c_Fm_F, may be assumed in the front view. Point *M* lies on line *AB* and can be located in the top view at m_T. The line c_Tm_T is now the true-length view of line *CM*, which lies in plane *ABC*. The horizontal line could also have been drawn in a higher or lower position in plane *ABC*, for example at *PQ*, but note that p_Tq_T would have the same direction as c_Tm_T. However, line *CM* is longer, hence more accurate, than *PQ*, and its location requires only one new point *M*.

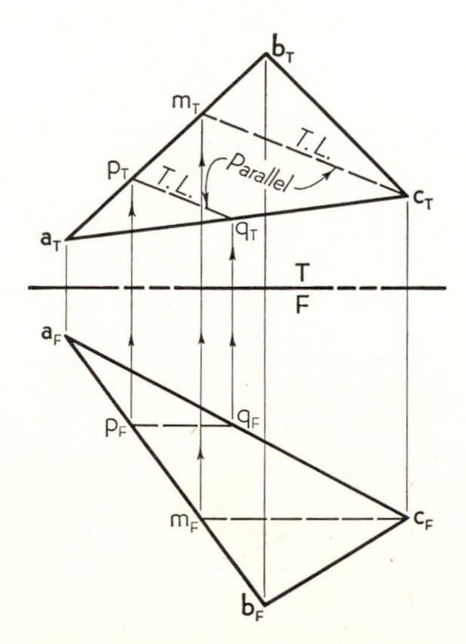

Fig. 4·5. Horizontal lines in a plane.

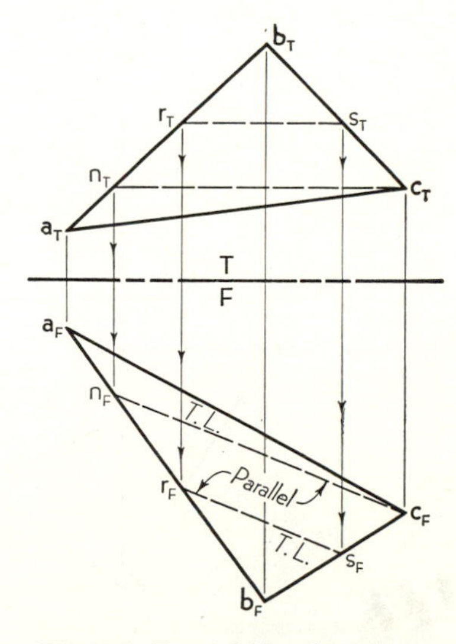

Fig. 4·6. Frontal lines in a plane.

Frontal lines. In similar manner, if the required line must be true length in the front view (a *frontal* line), then it must be assumed parallel to reference line *T–F* in the top view, as shown in Fig. 4·6. Line $c_T n_T$ has been so drawn, and $c_F n_F$ is the required true length in the front view. Any other frontal line in the plane, such as *RS*, has the same direction as line *CN* in the front view.

Summary. In general, to establish in a plane a line that will appear true length in any given view the line must be assumed in the plane parallel to the reference line in an adjacent view. In Fig. 4·7 the profile line *BT*, which is true length in view *R*, was first assumed in the adjacent front view parallel to reference line *F–R*. Line *BV*, which is true length in view *A*, was first assumed in the top view at $b_T v_T$ parallel to reference line *T–A*.

4·4. Strike of a plane

The *strike* of a plane is the bearing of a horizontal line in the plane. This term is employed by mining engineers and geologists to describe the direction of the various strata, or layers, of the earth's crust. The use of strike in mining applications is more fully discussed in Chap. 12, but strike can also be used to describe the direction of any plane surface.

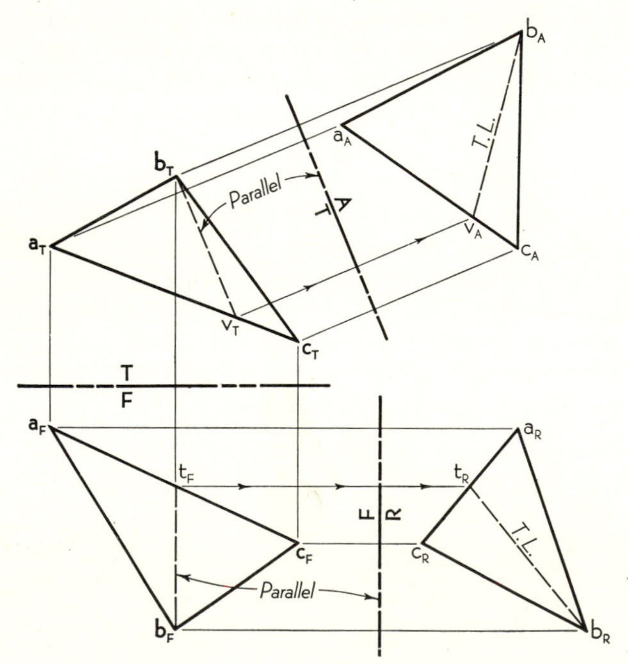

Fig. 4·7. Other true-length lines in a plane.

In Fig. 4·8 points A, B, and C represent three points in a given plane. To determine the strike of this plane assume any horizontal line in the plane, such as AD. Then a_Td_T is a *strike line* of the plane, and the strike is the bearing, N 65° E, of this line. The strike is usually lettered along the strike line as shown. The strike line may be taken at any level in the plane, but line AD, the longest line, can be most accurately located.

PROBLEMS. Group 17.

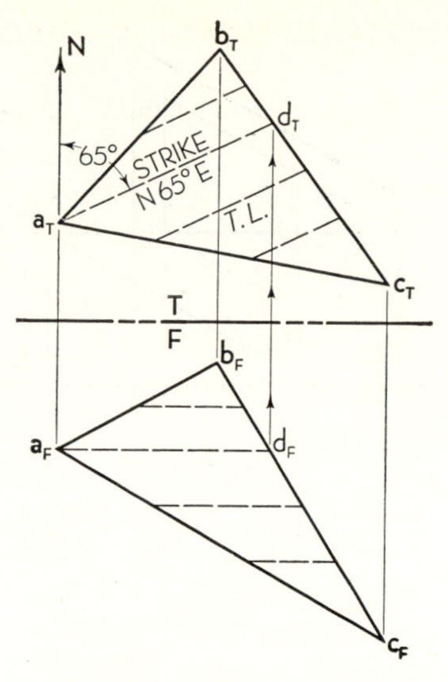

Fig. 4·8. Strike of a plane.

4·5. Location of a point in a plane

To locate a point in a given plane the point must first lie on some line in that plane. If the required point does not lie on one of the given lines of the plane, then some new line must be assumed in the plane to contain the given point. Since an infinite number of lines may be assumed in a plane passing through a given point, there are an infinite number of ways of locating the point.

Assume in Fig. 4·9 that the point x_F is given as shown and that point X must lie in the plane defined by the parallel lines AB and CD. It is required to locate point x_T. A line, such as m_Fn_F, may be drawn through the point x_F in any desired direction. Then MN is a new line in the plane and may be located in the top view because points M and N lie on lines AB and CD, respectively. Point x_T is now determined since it must lie on m_Tn_T directly above x_F.

If the required point is given in some position such as y_T, it may be necessary to assume several new lines in the plane or to extend the given lines. In Fig. 4·9, y_F has been found by adding the lines AC and BD to the given plane and then drawing any random line $y_Tr_Ts_T$ to intersect these lines. When points r_F and s_F have been located in the front view on the corresponding lines, point y_F is located on the extension of r_Fs_F.

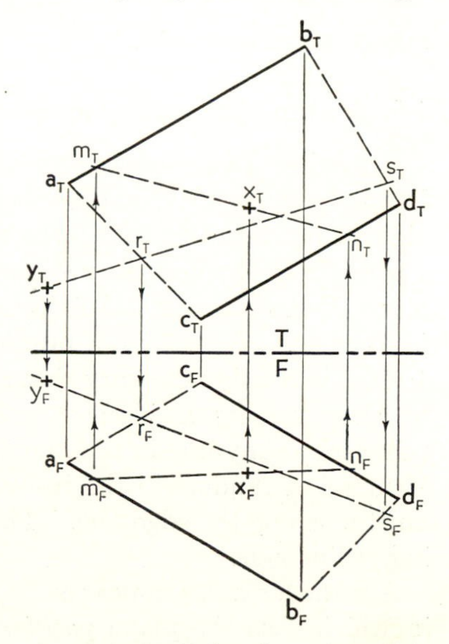

Fig. 4·9. Points in a plane.

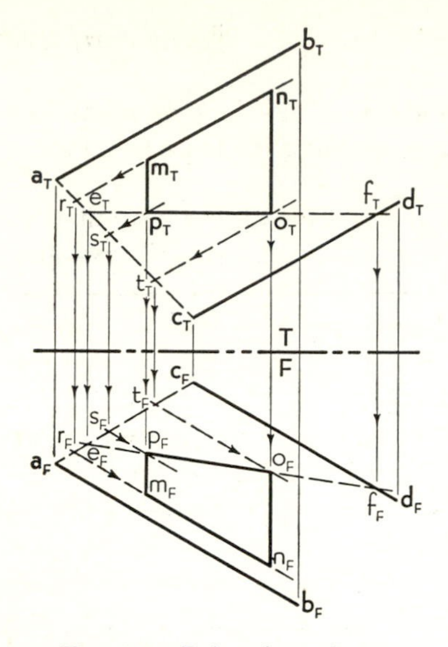

Fig. 4·10. Points in a plane.

If a series of points must be located on a given plane, it is frequently simpler to employ a series of parallel lines on the plane. In Fig. 4·10 the plane is again defined by the parallel lines *AB* and *CD*, and it is required to construct the front view of the figure *MNOP* which lies in the given plane and whose top view is given as shown. Points *A* and *C* are first connected to introduce a new line in the plane. Since line m_Tn_T of the figure is parallel to lines a_Tb_T and c_Td_T, then line m_Fn_F will also be parallel to a_Fb_F and c_Fd_F. Extending line m_Tn_T locates r_T on line a_Tc_T; from point r_F a line is drawn parallel to a_Fb_F, and points m_F and n_F are found on this line. By assuming a second parallel line through p_T, the point s_T is determined, and the line s_Fp_F is drawn parallel to a_Fb_F to locate point p_F. A third parallel line *OT* is used to find point o_F. As a check, the line *OP* can be extended to intersect lines *AC* and *CD* at *E* and *F*, respectively. It should be noted that profile lines *MP* and *NO* cannot be determined in the front view by extending them to intersect the given plane lines.

PROBLEMS. Group 18.

4·6. A plane appearing as an edge (basic view, type III)

A vertical plane always appears as an edge in the top view, and therefore its *position* may be described completely by a single line in the top view to represent the edge view of the plane, as in Fig. 4·11(*a*). In such a case the front view serves only to show the bounds, or limits, of the plane; if the plane is indefinite in extent, nothing need be shown in the front view. To assume a random point *X* in the vertical plane of Fig. 4·11(*a*) it is only necessary that x_T shall lie somewhere along the *edge line* which represents the plane as an edge; point x_F may be anywhere directly below x_T. Any line in the plane, such as the horizontal line *RS*, may be assumed at random in the front view, but the top view of all such lines must lie along the edge line. The strike of this vertical plane is the bearing of line r_Ts_T.

A plane may also appear as an edge in the front view, as in Fig. 4·11(*b*), in which case the plane *position* is fully established by the front view only. In order that any point *X* shall lie in this plane it is only necessary that x_F lie on the edge line. A horizontal line in this plane can appear only as a point in the front view, and thus the strike of all *edge-front* planes is due north.

Whenever a plane appears as an edge in any view and is unlimited in extent, we shall indicate its position in that view by a single solid line, indefinite in length but distinguished by three or four short dashes at its ends and labeled "Edge." Additional points and lines may always be assumed in such planes if necessary, but in general these planes need not be described or bounded in the adjacent views.

Oblique planes. A generally oblique plane does not appear as an edge in any of the principal views; but if a new view is drawn with the direction of sight parallel to the plane, that view will show the plane as an edge. But if the direction of sight is parallel to the plane, it is also parallel to some line in that plane, and thus it is only necessary to look in the direction of some line in the plane to see the plane as an edge. We may therefore choose *any* line in the plane and show that line as a point, and the plane will appear as an edge.

In Fig. 4·12 are shown the top and front views of a generally oblique plane *ABC*. Selecting the line *AB* as any line in the plane, a new view *A* is drawn to show this line in its true length. View *B* shows line *AB* as a point; and since lines *AC* and *BC* now obviously coincide, the plane appears only as a line or edge. Had any other line in the plane been selected to appear as a point, the plane would still appear as an edge although viewed from a different direction. Because, as will be demonstrated in subsequent articles, a view showing a plane as an edge so

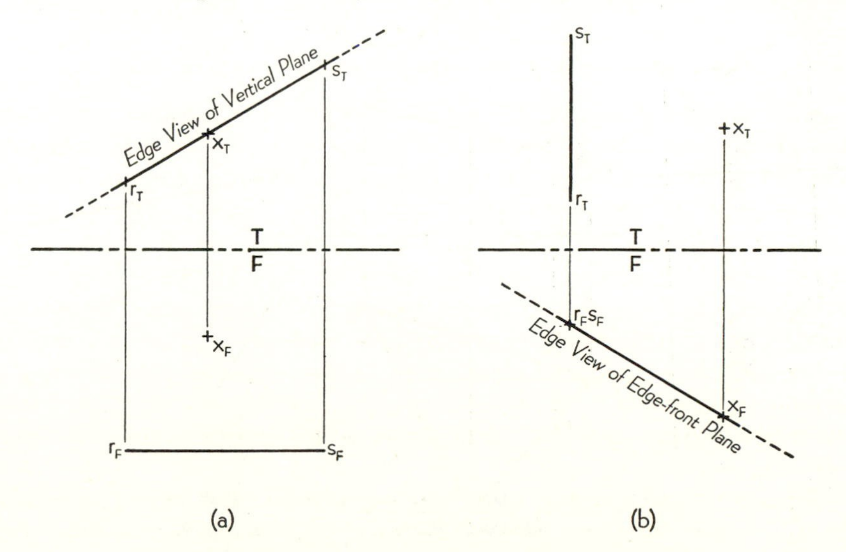

Fig. 4·11. Planes appearing edgewise in principal views.

greatly simplifies many problems, we shall call this *basic view, type* III. The following rule expresses the general method:

- ***Rule 11. Rule of edge views.*** *Any plane will appear as an edge in that view which shows **any** line in the plane as a point.*

Figure 4·12 also shows in views C and D a further important simplification in the method of showing a plane as an edge. To show line AB as a point required the construction of two views, A and B, because line AB had to be shown as a true length before it could be shown as a point. Rule 11 states that *any* line on the plane may be used; therefore, if the line selected in the plane were already true length in one of the given views, the necessity for a true-length view would be eliminated. The horizontal line AD in the plane appears true length in the top view (Art. 4·3), and therefore the elevation view C, which shows line AD as a point, also shows the plane as an edge. In similar manner the edge view D can

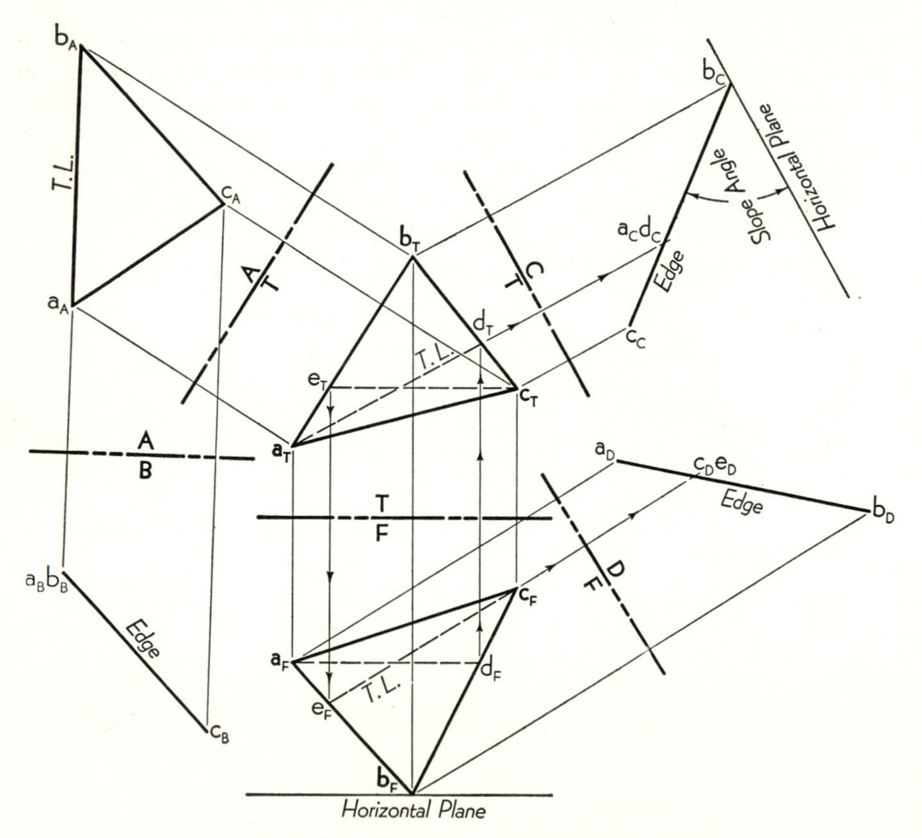

Fig. 4·12. An oblique plane appearing as an edge.

be drawn adjacent to the front view by employing the frontal line *CE*, which is true length in the front view.

This use of assumed true-length lines to produce a plane as an edge will be the recommended procedure in future problem solutions. Only rarely is it necessary to see a particular line as a point when the plane appears as an edge.

4·7. Slope of a plane

The slope angle of a plane is the angle that the plane makes with a horizontal plane. The slope angle of a plane may be expressed in degrees or in per cent grade just as for a line (Art. 3·7).

As a simple illustration see Fig. 4·13(*b*), which shows an easel resting on a horizontal surface. The plane of the easel, *ABCD*, is inclined to the horizontal plane, and the desired slope angle lies in the vertical plane *EFG*, which is perpendicular to plane *ABCD*. To see this angle in its true size the observer must look in a direction perpendicular to plane *EFG*, or parallel to arrow *A*. Looking in this *horizontal* direction, the easel would appear as shown in Fig. 4·13(*c*); plane *EFG* appears in true size, plane *ABCD* appears as an edge, and the horizontal plane also appears as an

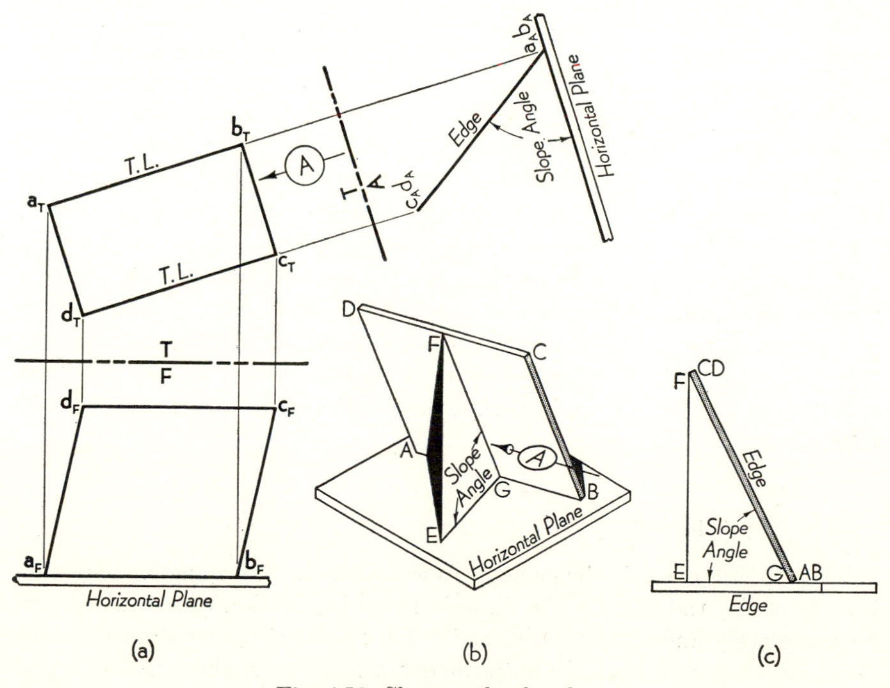

Fig. 4·13. Slope angle of a plane.

edge. This, then, is the type of view that must always be obtained to see the slope angle of a plane; the given plane and a *horizontal plane* must both appear as edges in the same view.

Figure 4·13(*a*) shows the easel surface *ABCD* and a horizontal plane. A new view must be drawn to show plane *ABCD* and the horizontal plane as edges. But a horizontal plane will appear edgewise only when viewed in a horizontal direction, in other words, in an *elevation* view. Therefore, if the new view is a top-adjacent view, it will also be an elevation view (Art. 2·5). To make plane *ABCD* appear as an edge in a top-adjacent view requires a true-length line in the plane in the top view (Art. 4·6). Following this line of reasoning the top-adjacent view *A* is drawn to show line *AB* as a point and plane *ABCD* as an edge. The horizontal plane appears parallel to the reference line *T–A*, and the required angle is measured as shown.

Summarizing the foregoing analysis in the form of a rule, we have the following:

▪ ***Rule 12. Rule for slope of a plane.*** *The slope angle of a plane can be seen only in* ***that elevation view*** *which shows the plane as an* ***edge.***

The reader should note the similarity between this rule and Rule 7 (Art. 3·7). It is *not sufficient* that the given plane appear as an edge; it must also appear in an *elevation* view in order that a horizontal plane may simultaneously appear as an edge. In Fig. 4·12 are shown three different edge views of a plane *ABC*, but only one of these views, view *C*, is an elevation view that can show the slope angle of the plane.

It is of interest to note here that view *A* in Fig. 4·13(*a*) also shows the slope of line *BC*, because line *BC* appears true length in elevation view *A*. Thus line *BC* has the same slope angle as the plane. Since no line in a plane can have a greater slope than the plane itself has, line *BC* (or any parallel line) must be the *steepest* line in the plane. But line *BC* is perpendicular to line *AB* because the two lines appear perpendicular in the top view, where line *AB* is true length (Rule 10). Thus we conclude that the *steepest* line in a plane is always *perpendicular* to a *horizontal* line in that plane.

4·8. Dip of a plane

The *dip* of a plane is the geologist's term for the slope angle. By stating the strike and the dip the direction and inclination of a stratum can be completely and simply described. By showing the location and elevation of the strike line on a map the exact location of the stratum is also fixed. Such a description does not, however, indicate the extent or limits of the plane surface. The use of dip in mining applications is more fully discussed in Chap. 12.

PROBLEMS. Group 19.

4·9. Shortest line from a point to a plane

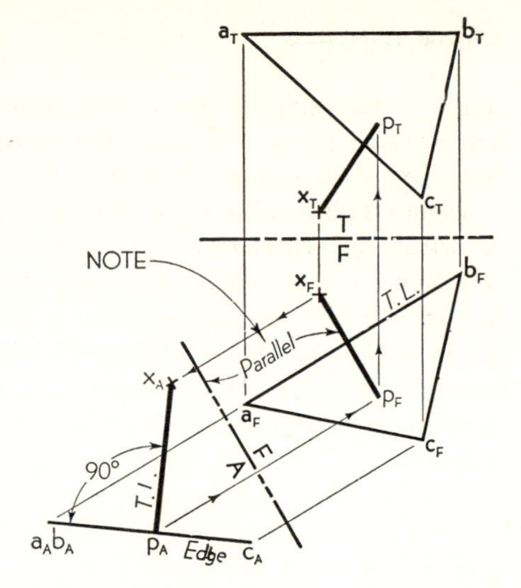

Fig. 4·14. Shortest line from a point to a plane.

Analysis. The shortest line from a point to a plane must be perpendicular to the plane. In any view that shows the plane as an edge the shortest line will appear perpendicular to the edge view. Since the direction of sight will then be parallel to the plane, it will also be perpendicular to the shortest line, and this line will therefore appear in true length. In brief, the method requires that we:

Show the given plane as an edge.

Construction. Figure 4·14 shows a plane *ABC* and a point *X*. *Any* edge view of the plane will show the desired shortest line, and the quickest way to obtain an edge view is directly from the top or front view, as was done in views *C* and *D* in Fig. 4·12. Inspection shows that line a_Fb_F is true length in the front view (a_Tb_T is parallel to *T–F*), and therefore reference line *F–A* is taken perpendicular to a_Fb_F.

In view *A* the shortest line x_Ap_A is drawn perpendicular to the plane to determine point p_A. In order to show line *XP* in the given views it is necessary to observe that x_Fp_F must be parallel to reference line *F–A* because x_Ap_A is true length (Rule 6*b*). Failure to note this simple condition leaves the draftsman with no means of locating either p_F or p_T. With point p_F thus located, p_T is found by alignment and measurement.

The shortest line may also be located directly in the given views without recourse to an edge view of the plane, but explanation of this method must necessarily be deferred until later (Art. 4·32).

PROBLEMS. Group 20.

4·10. Shortest grade line from a point to a plane

The shortest distance from a point to a plane is a perpendicular to the plane (Art. 4·9), but in some cases the slope angle of the perpendicular may be undesirably steep. In mining, for example, the entrance passages leading to a seam of coal or vein of ore should be short for economy, but they should also have the correct slope for drainage, ventilation, and transportation of the ores. Such passages are therefore designed as the shortest line of *specified grade* from the point of entrance to the plane of the stratum. Definitions of some of the special terms used in mining are listed in Art. 12·3.

Analysis. The given plane must appear as an edge, but it is also essential that the new edge view shall show the slope angle of the required grade line. Slope angle can be measured only in an elevation view, and therefore:

Show the plane as an edge in an elevation view.

Problem. In Fig. 4·15 points A, B, and C represent three known points on the upper surface of a layer of coal. Point X represents the mine opening on the surface of the ground. It is required to locate: (1) the shortest horizontal passage, (2) the shortest inclined passage having a slope angle of 20°, and (3) the shortest (perpendicular) passage from point X to the coal bed.

Construction. The assumed horizontal line CD is true length in the top view (a *strike line*), and view A is constructed to show CD as a point and the plane as an edge. From point x_A each of the required passages is drawn to meet the plane at points r_A, q_A, and p_A. Since each passage

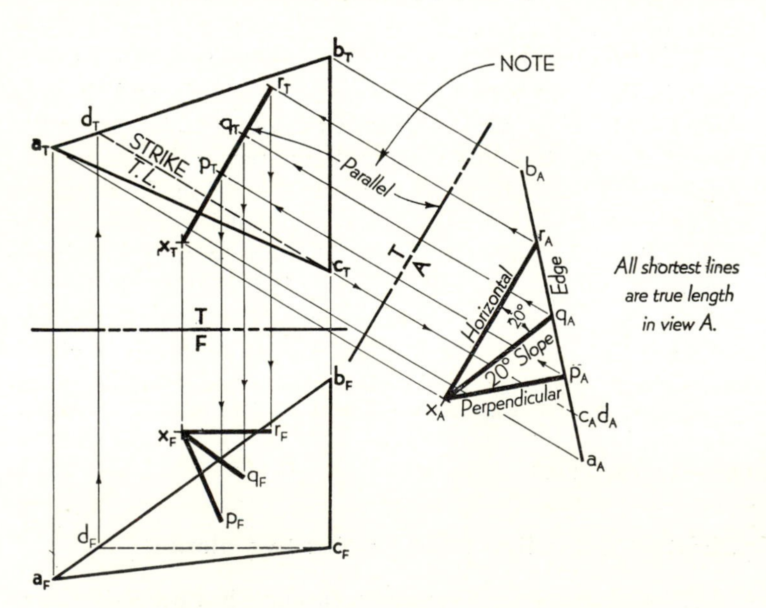

Fig. 4·15. Shortest grade line from a point to a plane.

appears in its *true length* in this *elevation view*, each one also shows its slope angle (Rule 7). Thus the horizontal passage is drawn parallel to reference line *T–A*, the 20° passage at 20° with line *T–A*, and the shortest passage perpendicular to the plane.

It is interesting to note that all shortest lines, of whatever grade, have the *same bearing* and are perpendicular to the strike line. It also should be noted again that points p_T, q_T, and r_T can be located correctly only by observing that x_Tp_T, x_Tq_T, and x_Tr_T must be parallel to reference line *T–A*.

PROBLEMS. Group 21.

4·11. True size of a plane (basic view, type IV)

At the end of Art. 3·6 it was stated that there are only *four* basic types of views. Three of these basic views have already been described and applied to the solution of numerous problems. With the acquisition of *basic view, type* IV, our kit of basic tools will be complete. It may also be desirable at this time to call attention to the interdependence and necessary sequence of the basic views: a plane cannot appear as an edge until some line on the plane appears as a point, and a line cannot appear as a point until it first appears as a true length. This order of construction must always be observed, and it will now be shown that basic view, type IV, must always follow basic view, type III.

To see a plane surface in its true size and shape the observer must look in a direction perpendicular to the plane. In this way every point on the plane will be equidistant from the eye of the observer, who, it must be remembered, is an infinite distance from the plane (Art. 1·3). A direction of sight that is perpendicular to a plane can easily be established in any view showing the plane as an edge. The true-size view can then be seen by looking in this direction.

Figure 4·16 gives the top and front views of a generally inclined plane *ABC*. It is required to show this plane in true size. The first step must be to show the plane as an edge; either view *A* or view *C* will serve. These edge views of the plane are obtained by showing the true-length lines *AD* and *AE* as points in views *A* and *C*, respectively. This is the same procedure as that used to construct views *C* and *D* in Fig. 4·12.

A true-size view of plane *ABC* can now be obtained by looking in the direction of either arrow *B* or arrow *D*. These arrows, which represent the direction of sight, are, of course, perpendicular to the edge views of the plane, and therefore the reference lines *A–B* or *C–D* must be taken parallel to the plane. Points *A*, *B*, and *C* can now be located in views *B* and *D* by the customary construction for auxiliary-adjacent views (Art. 2·7). Obviously, either view *B* or view *D* is a solution of the problem.

One would not ordinarily construct both; but since both views show the plane as seen from the same direction, the two triangles $a_Bb_Bc_B$ and $a_Db_Dc_D$ must be identical in size and shape. This constitutes an excellent check on the accuracy of the work.

This then is the procedure for obtaining basic view, type IV. The rule is as follows:

- ***Rule 13. Rule of true-size views.*** *A true-size view of a plane must be adjacent to an* ***edge view,*** *and the direction of sight must be* ***perpendicular*** *to the plane.*

PROBLEMS. Group 22.

4·12. Angle between two intersecting lines

Analysis. Two intersecting lines define a plane, and the angle between these two lines lies in this plane. To see this angle in its true size the observer must look in a direction perpendicular to the plane of the angle. In other words, the angle will appear true size in any view that shows the plane of the two lines in true size.

Construction. Assume that the two given intersecting lines are *AB* and *BC* in Fig. 4·16 and that it is required to determine the true size of the angle *ABC*. Then in the true-size views *B* and *D* the required angle is angle $a_Bb_Bc_B$ or angle $a_Db_Dc_D$. If it is also required to bisect angle *ABC*, then the bisector *BX* must be constructed in a true-size view (in view *B*, for example) as shown. Point *X*, which lies on line *AC*, may be carried back into the given views to show the bisector in the top and front views. It should be noted that the bisector of angle *ABC* does not necessarily bisect angles $a_Tb_Tc_T$ or $a_Fb_Fc_F$ and must therefore always be located first in a true-size view.

PROBLEMS. Group 23.

4·13. Shortest line from a point to a line (plane method)

Analysis. The given point and line define a plane. If this plane is shown in true size, then the shortest (perpendicular) line may be drawn in this view. This problem was previously solved in Art. 3·16 by the line method.

Construction. In Fig. 4·16 assume that line *AB* is the given line and point *C* the given point. Adding such lines as *AC* and *BC* does not affect

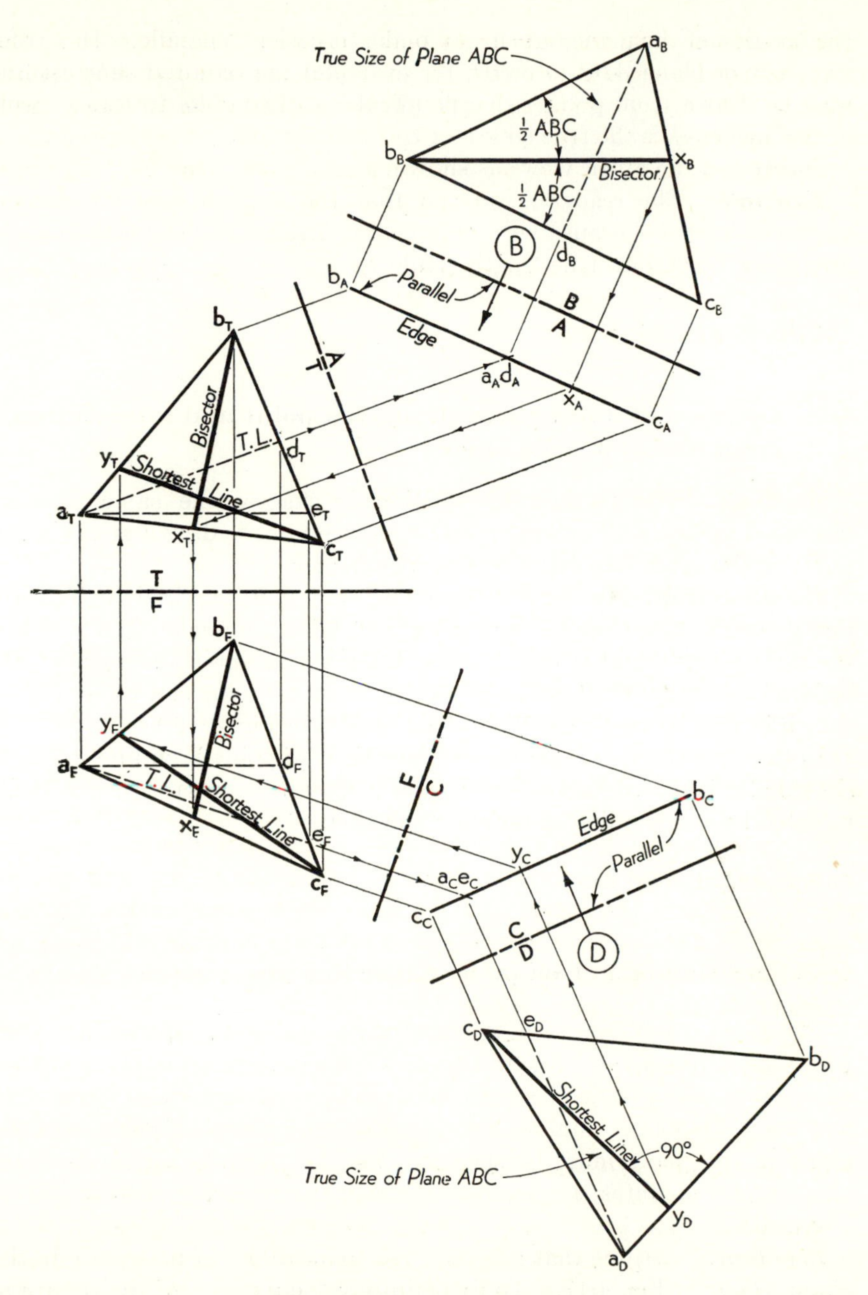

Fig. 4·16. True size of a plane.

the position of the plane but simply makes it easier to handle. In a true-size view of plane *ABC* (view *D*, for example) the required shortest line may be drawn from point c_D perpendicular to line a_Db_D to locate point y_D. Then c_Dy_D is the true length of the shortest line. This line can now be located in the given views as shown.

Caution. The reader is warned that the slope of line *CY* cannot appear in view *D* because view *D* is *not* an *elevation* view even though it does show line *CY* in true length (Rule 7).

PROBLEMS. Group 24.

4·14. Location of a line through a given point and intersecting a given line at a given angle

Analysis. The method of solution is exactly the same as that described in Art. 4·13 except that the required line is drawn at the given angle to the given line instead of perpendicular.

Construction. In Fig. 4·16 assume that it is required to locate a line through point *C* and intersecting line *AB* at an angle of 75°. Then the same views *C* and *D* would be used, but the required line c_Dy_D would be drawn in view *D* at an angle of 75° with a_Db_D instead of at 90° as shown in the figure. (This line is not shown in Fig. 4·16.)

In this case there are two possible positions of point *Y* on line *AD*. It should also be noted that in some cases it may be necessary (if practical) to extend the given line to obtain a solution.

PROBLEMS. Group 25.

4·15. Location of a given plane figure in a given plane

The given plane figure will appear in true size and shape in that view which shows the given plane in true size. This, then, is simply a further application of the principles outlined in the preceding articles (Arts. 4·11 to 4·14). To obtain a unique solution, however, the exact position of the figure in the plane must be specified, and the form of these specifications usually dictates the manner in which the true-size view should be obtained.

Problem. Assume that it is required to locate a 1¼-in. square in the plane *ABCD* in Fig. 4·17. To fix definitely the position of the square in the plane, let us further assume that the center of the square must be 1⅜ in. from point *A* in the plane, 1 in. lower than point *B*, and that two sides of the square must be horizontal.

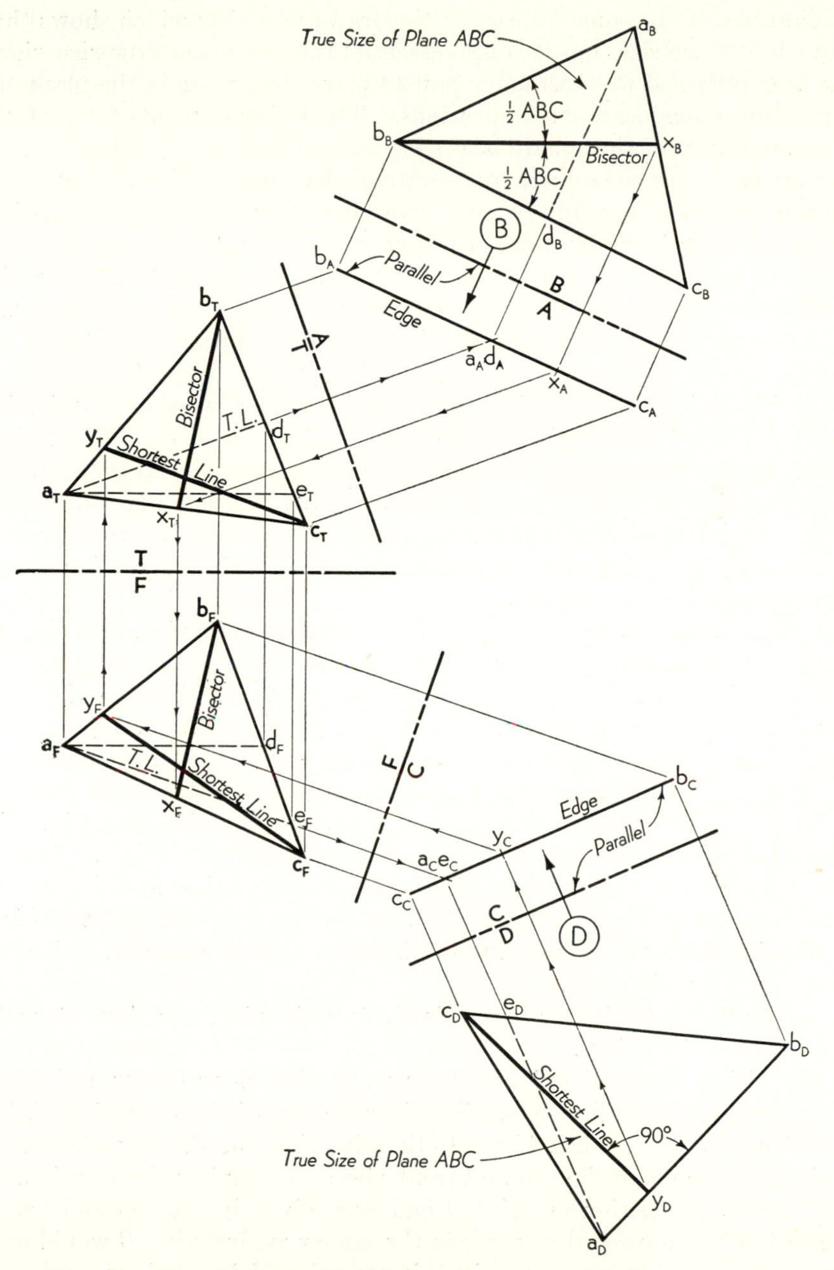

Fig. 4·16 (repeated). True size of a plane.

Analysis. The square must first be drawn in a view which shows the plane in its true size, but it is also essential that when the true-size view has been obtained we shall know how to place the square in the plane to satisfy the given location specifications. The 1⅜-in. distance from point A to the center of the square is a measurement that will lie in the plane and can therefore be laid out true length in the true-size view. The 1-in. distance will not lie in the plane because it is simply a difference of elevation; difference of elevation, or height, can be laid out true length only in elevation views. The two sides of the square that will appear horizontal in the front view will appear true length in the top view and will therefore be parallel to any line in the plane that appears true length in the top view.

Having reviewed the specifications in this manner, we should logically decide: (1) that the 1⅜-in. distance shall be measured in the true-size view, (2) that elevation views should preferably be employed to facilitate the measurement of the 1-in. distance, and (3) that some horizontal line should be installed in the plane as a guide for positioning the square. After this reasoning process is completed, the actual construction work can be started—never before.

Stage 1. *Construct an elevation view and a true-size view, and show a horizontal line in the plane in all views.*

The horizontal line EF has been drawn in the plane. Instead of drawing it at any random elevation it has been taken 1 in. below point b_F, and it is now known that the center of the square must lie somewhere along the line EF. The line EF may also be used to show the plane $ABCD$ as an edge in view A; and since view A is an elevation view, the 1-in. difference of height could also be laid off in this view as shown in the figure. View B is next drawn to show plane $ABCD$ in true size, and line EF is also shown in this view. Since the center of the square must be 1⅜ in. from point A, an arc of 1⅜ in. radius is struck with a_B as center to locate point o_B on the line e_Bf_B.

Stage 2. *Establish the square in the true-size view, and then return it to the given views.*

The 1¼-in. square has been constructed with o_B as center and with two sides parallel to line e_Bf_B. The four corners of the square, R, S, T, and V, can now be extended back to the edge view and then by alignment and measurement to the top and front views to complete the solution.

If the exact position of point O had been given in the top and front views, the solution could be made in the same way, but view B would not be necessary. The procedure in this case should be obvious, and the details are left to the reader.

PROBLEMS. Group 26.

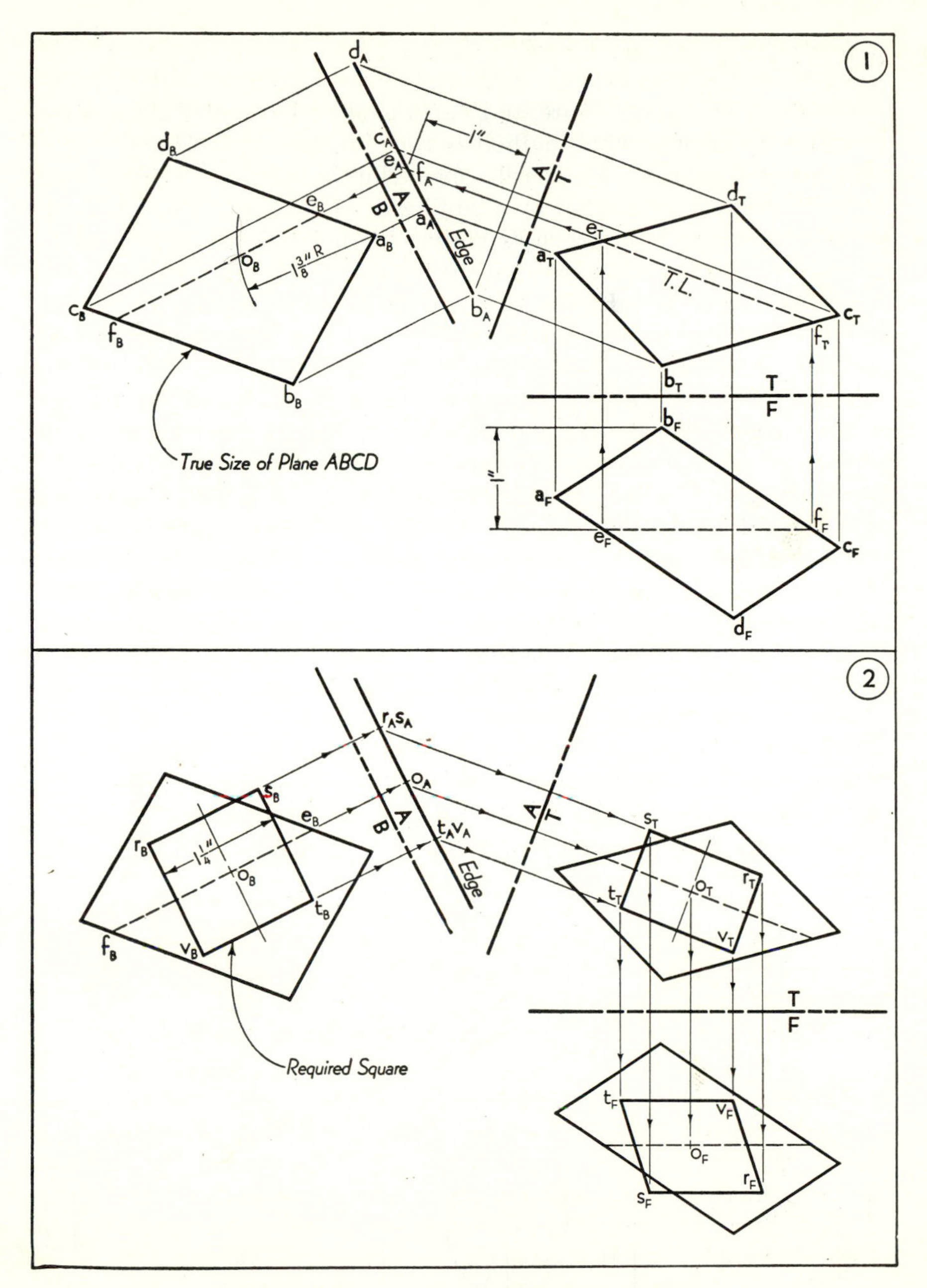

Fig. 4·17. Location of a plane figure in a plane.

4·16. Views of a circle

The circle is a plane figure and can therefore be located on a given plane by the same general method outlined in the previous article. If the plane is oblique, the circle will appear elliptical in the principal views. These ellipses can be plotted by assuming a series of points on the circle in the true-size view and then extending the points back into the given views by alignment and measurement. (This method was employed in stage 4 of Fig. 3·26.) However, this method is tedious and inaccurate, and it will now be shown that the ellipses can be constructed by shorter and simpler methods.

A circle of diameter D appears true size in the front view of Fig. 4·18(*a*). In this view *all* diameters appear true length. In the top view the circle appears as an edge of length D parallel to the reference line (Rule 13). If this circle is revolved about its vertical diameter, it will then appear as in Fig. 4·18(*b*). The length of the edge view still equals diameter D, and in the front view the circle now appears as an ellipse. The vertical diameter, which appears as a point in the top view, is now the *only diameter* appearing *true length* in the front view—and this diameter is also the *major axis* of the ellipse. The diameter most foreshortened is at right angles to the major diameter and appears true length in the edge view. This is the *minor axis* of the ellipse.

In Fig. 4·18(*c*) the same circle has been tipped backward so that it now appears elliptical in the top view (the front view has been omitted). In the top-adjacent view A the circle again appears as an edge of length D. In the top view all diameters but *one* are foreshortened, and that one is

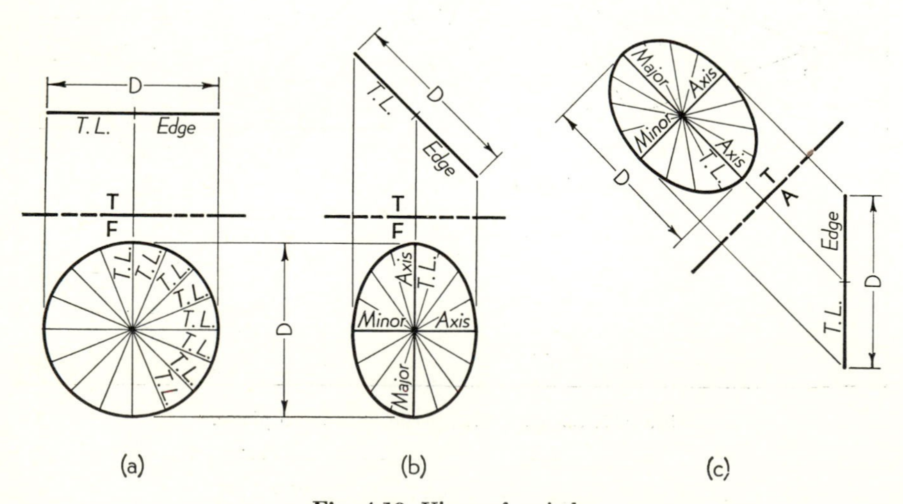

Fig. 4·18. Views of a circle.

true length because it appears as a point in the edge view. Thus the *one true-length diameter* is again the *major axis* of the ellipse. The minor axis again appears true length in the edge view.

From the above observations the following valuable conclusion can be drawn:

> *In every oblique view of a circle one diameter always appears true length, and that diameter is always the major axis of the ellipse. The minor axis always appears true length in an adjacent edge view.*

Whenever the major and minor axes of an ellipse are known, it can be constructed by any one of several methods (see Arts. A·7, A·9, and A·11 in the Appendix). How to establish the ellipse axes for an oblique circle will be shown in the next two articles.

4·17. Circle on a plane (edge-view method)

Problem. It is required to represent a circle of diameter D with center at O in the plane $ABCD$ of Fig. 4·19.

Stage 1. *Locate the major axis of the ellipse in the top view.*

Through point O draw a horizontal line EF in the plane. On the true-length line e_Tf_T lay off the true diameter D of the given circle to establish the major axis 1–2.

Stage 2. *Locate the minor axis of the ellipse in the top view.*

Construct a top-adjacent view A showing the plane as an edge. On the edge view lay off the true diameter D of the circle with center at o_A.

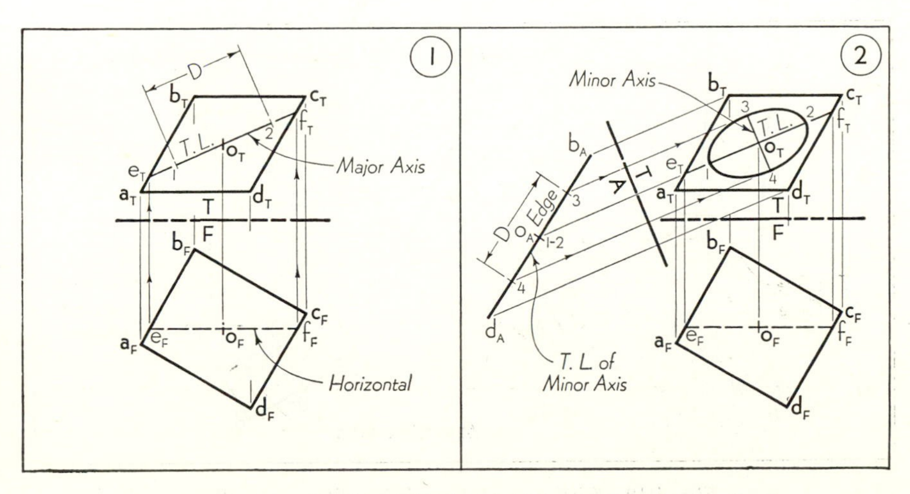

Fig. 4·19 (stages 1 and 2). Circle on a plane (edge-view method).

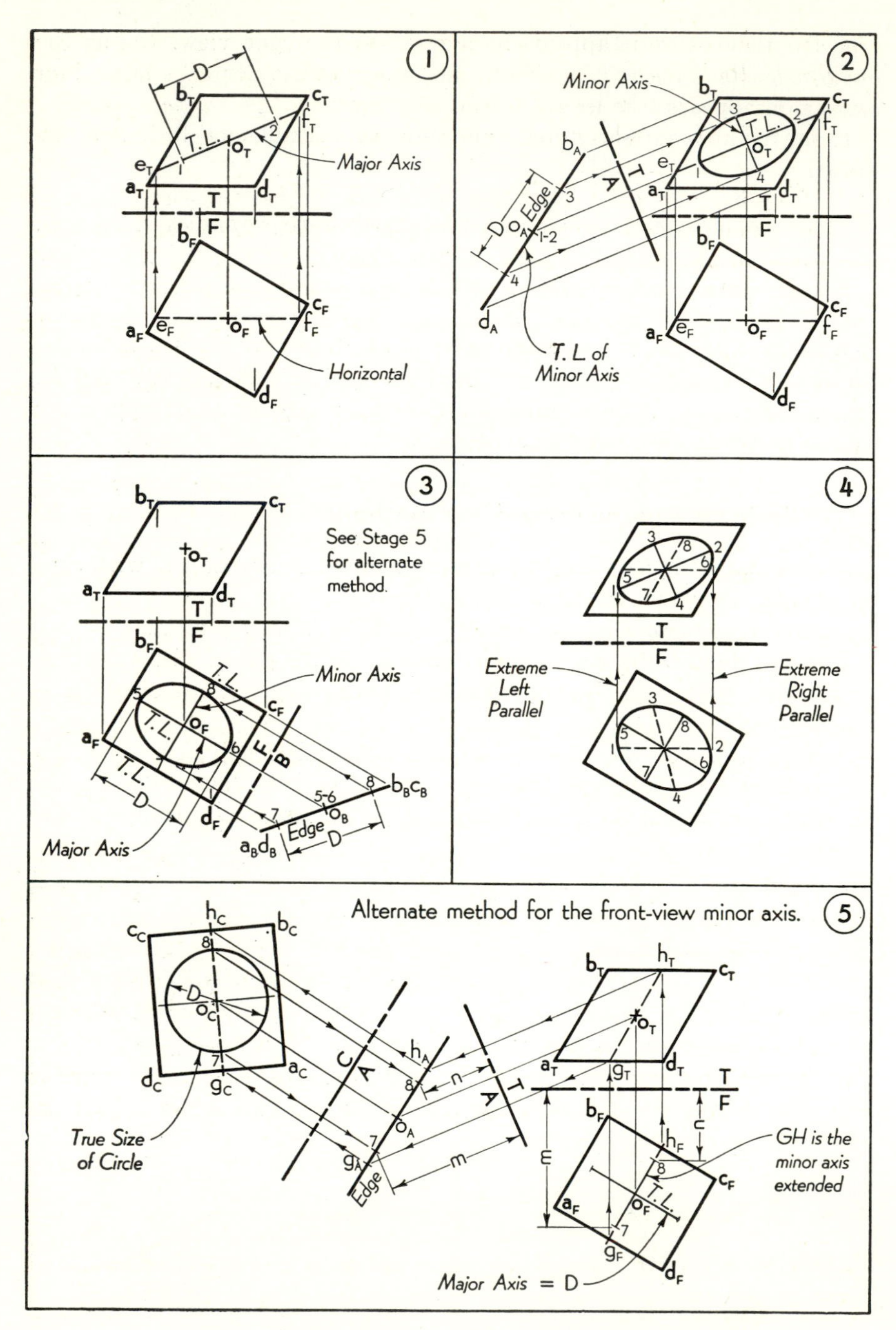

Fig. 4·19. Circle on a plane (edge-view method).

Extend points 3 and 4 to the top view to establish the minor axis at right angles to the major axis. With the major and minor axes thus determined the top-view ellipse can now be drawn by the *trammel method* or by any of the standard methods (see Appendix, Arts. A·7, A·9, and A·11).

Stage 3. *Locate the major and minor axes in the front view.*

Through o_F draw a true-length line (parallel to a_Fd_F in this case), and on this line lay off the major axis 5–6 equal to D. Construct a front-adjacent edge view B, and lay off diameter D in this view. Extend points 7 and 8 to the front view to establish the minor axis. Draw the ellipse on these axes. Note that the construction of the front-view ellipse is exactly analogous to that used for the top-view ellipse, but the two constructions are entirely independent. Error or inaccuracy in one view does not affect the other view.

Stage 4. *Check the alignment of the top- and front-view ellipses.*

As a final check on the accuracy of the work note that the extreme left and right points on the top ellipse should lie directly above the extreme left and right points on the front ellipse. In order to demonstrate that the ellipse axes in one view *do not necessarily correspond* to the axes in an adjacent view, each pair of major and minor axes has been shown in the adjacent view with dash lines. These dash-line axes are called *conjugate axes;* they will appear perpendicular in a true-size view but not in an oblique view. An ellipse can be constructed from a pair of conjugate axes (see Appendix, Art. A·9), but constructions based on the major and minor axes are usually simpler.

Stage 5. *Alternate method for the front-view minor axis.*

After the front-view major axis has been established (see stage 3), draw the minor axis perpendicular to the major axis and extend it to intersect the given plane lines at g_F and h_F. Construct a true-size view C of the plane, and locate the circle and line GH in this view. Points 7 and 8, where line g_Ch_C crosses the true-size circle, are then extended back to the edge view. Distances m and n can now be transferred to the front view to establish the ends of the minor axis at 7 and 8 on g_Fh_F.

This method is especially suitable when a true-size view must be drawn for other purposes, such as to determine the location of point O or the diameter of the circle.

4·18. Circle on a plane (two-view method)

An ellipse is fully determined when its *major diameter* and *any other diameter* are known, because the minor diameter can then be found by the construction given in the Appendix, Art. A·8. By employing this construction (which is simply an inversion of the trammel method) a cir-

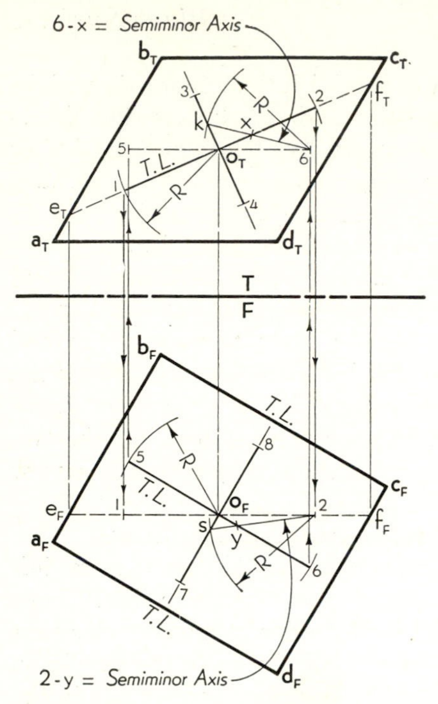

Fig. 4·20. Circle on a plane (two-view method).

cle can be drawn at a given point on a generally oblique plane without the aid of additional views.

Problem. It is required to represent a circle of diameter D (radius R) with center at O in the plane $ABCD$ of Fig. 4·20. This problem is the same as that shown in Fig. 4·19.

Construction. Through point O draw a horizontal line EF in the plane. On e_Tf_T lay off the true radius R of the given circle on each side of o_T to locate points 1 and 2, and thus establish the major axis in the top view. Locate points 1 and 2 on line e_Ff_F in the front view. In similar manner, through point O draw a frontal line (parallel to AD in this case). On each side of o_F lay off radius R on the true-length line to locate points 5 and 6, thus establishing the major axis in the front view. Locate points 5 and 6 on the corresponding line in the top view.

In the top view 1–2 is the major diameter and 5–6 is a second diameter. Through o_T draw a perpendicular to the major axis. With R as radius and center at 6 (or 5) strike an arc cutting the perpendicular at k. Draw line 6–k to intersect the major axis at x. Then 6–x is the length of the semiminor axes o_T–3 and o_T–4. In the front view 5–6 is the major diameter and 1–2 is a second diameter. The construction described above can be repeated here to obtain the distance 2–y which is the length of the semiminor axes o_F–7 and o_F–8. The required ellipses can now be drawn by any of the constructions based on the major and minor axes.

The method of this article is much shorter than the edge-view method (Art. 4·17), but it can be applied only to the *generally oblique plane* [Fig. 1·15(7)]. When the given plane appears as an edge in one of the principal views [as in (4), (5), or (6) of Fig. 1·15], then the major axes are the *same line* in adjacent views, and the edge-view method must be used. The accuracy of the two-view method diminishes as the angle between the axes becomes smaller, and in such cases the length of the minor axis should be verified with an edge view.

PROBLEMS. Group 27.

4·19. Location of a plane through one line and parallel to a second line

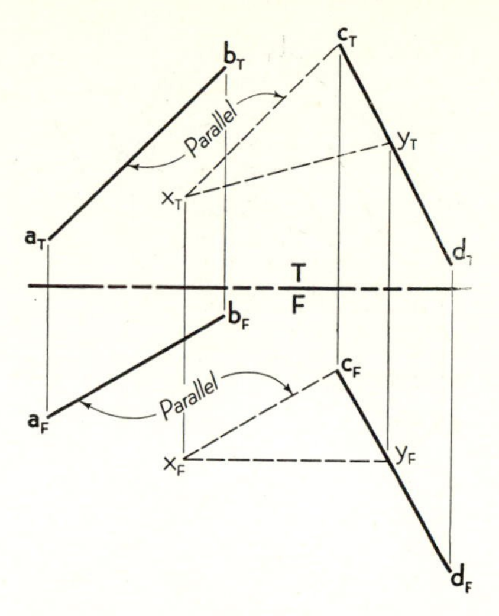

Fig. 4·21. Plane through one line and parallel to a second line.

Analysis. The required plane will be defined by two intersecting lines: the given line through which the plane must pass and a second line which is parallel to the given second line.

Problem. In Fig. 4·21 it is required to construct a plane that will pass through, or contain, line CD and also be parallel to line AB.

Construction. Line CD is one line in the required plane, and it is only necessary to draw another line that intersects CD and is parallel to line AB. The intersection point may be taken anywhere along line CD; in Fig. 4·21 the point C has been selected. Through point c_T line $c_T x_T$, of any desired length, is drawn parallel to $a_T b_T$; through point c_F line $c_F x_F$ is drawn parallel to $a_F b_F$. The plane DCX thus formed is the required plane, and any other lines may now be assumed in it, as, for example, the horizontal line XY. This solution may easily be verified by constructing any edge view of plane DCX and observing that line AB appears parallel to the plane.

If the two given lines are parallel to each other, the solution becomes indeterminate since *any* plane through one of these lines will always be parallel to the second line.

PROBLEMS. Group 28.

4·20. Location of a plane through a given point and parallel to two given lines

Analysis. The required plane will be defined by two lines intersecting at the given point, each line being parallel to one of the two given lines.

Problem. In Fig. 4·22 it is required to construct a plane through point O and parallel to lines AB and CD.

Construction. Through point O the line RS is drawn of any desired length and parallel to line AB; line TV is similarly drawn through point O and parallel to line CD. The plane $RSTV$, containing point O, is therefore the required plane. The solution may be verified, and checked for accuracy, by drawing a new view of the plane showing it as an edge; in

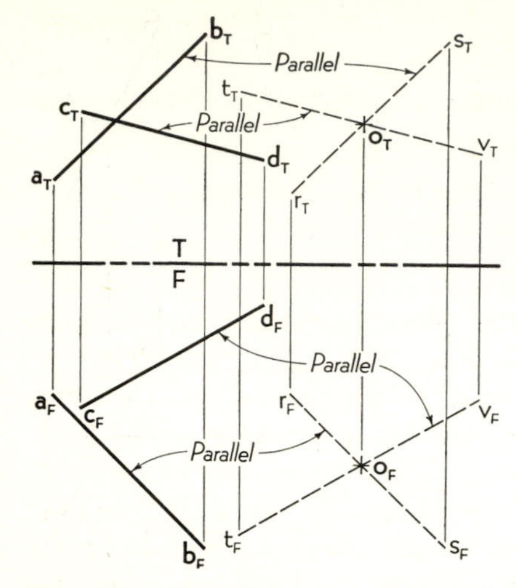

Fig. 4·22. Plane through a point and parallel to two given lines.

this edge view lines AB and CD should appear parallel to the plane.

The location of a plane through a given point and parallel to a given plane may be determined in the same manner as above by employing any two lines in the given plane.

PROBLEMS. Group 29.

4·21. Shortest line between two skew lines (plane method)

Analysis. A plane is passed through one of the given lines parallel to the second given line. Since the second line is then parallel to the newly formed plane, every point on this line must be equidistant from the plane, and the perpendicular distance from the line to the plane is the same as the shortest distance between the two skew lines. An edge view of the plane will show the perpendicular to the plane in its true length, and a true size view of the plane will show this same perpendicular as a

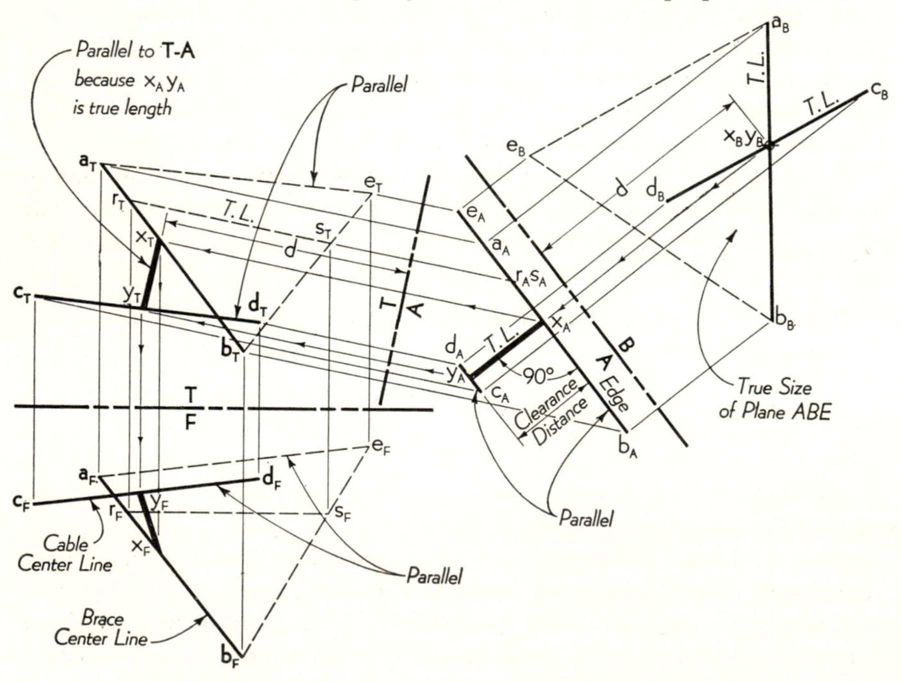

Fig. 4·23. Shortest line between two skew lines (plane method).

point and thus reveal its exact location. To summarize the *two steps* of the solution:

1. *Pass a plane through one line parallel to the second line.*
2. *Show the plane as an edge and then in true size.*

This problem was solved in Art. 3·17 by the line method.

Problem. In Fig. 4·23 determine the clearance distance and its location between the brace and cable center lines (same as in Fig. 3·24).

Construction. The line *AE* has been drawn parallel to given line *CD* through point *A* on the line *AB* to form the plane *ABE*. The horizontal line *RS* has been located in plane *ABE* and then shown as a point in view *A* to provide an edge view of the plane. When line *CD* is also located in view *A*, it should be found, of course, that $c_A d_A$ is exactly parallel to the plane. This is an excellent check on the accuracy of the work at this stage.

In view *A* the required shortest line must be perpendicular to the plane and must connect lines $a_A b_A$ and $c_A d_A$. Thus the *direction* and *true length* of the required line is shown in this view, but its exact *location* is still unknown. To locate the shortest line, view *B* is drawn to show plane *ABE* in its true size. Line *CD* will also appear true length in view *B*, and the shortest line—which must appear as a point in view *B*—is the point $x_B y_B$ at the intersection of lines $a_B b_B$ and $c_B d_B$. Extending a parallel from $x_B y_B$ in view *B* back to view *A* locates the exact position of $x_A y_A$, and the line *XY* can now be located in the top and front views by direct alignment but should be checked, of course, by reference-line measurements such as distance *d*.

Although this problem could have been solved just as easily by taking the edge view adjacent to the front view, it is notable that view *A* is an elevation view which shows line *XY* in true length; therefore, if the slope angle of line *XY* were also required, it could be measured directly in view *A*.

PROBLEMS. Group 30.

4·22. Shortest grade line between two skew lines

This problem is very similar to the one just described in Art. 4·21, except that the required connecting line is *not* perpendicular to the two given skew lines but instead must have a certain specified grade or slope angle.

Analysis. A plane passed through one of the given lines parallel to the second given line will establish *all* shortest lines. If this plane appears as an edge, the shortest line must be one that appears *true length.* If the edge view is also an *elevation* view, the required line can be drawn at the *specified grade.* A view showing the shortest grade line as a point will *locate* its intersection with the given lines. Summarizing the *three steps* of the solution:

1. *Pass a plane through one line parallel to the second line.*
2. *Show the plane as an edge in an elevation view.*
3. *Establish grade in the edge view, and then construct an adjacent view that will show the required line as a point.*

Caution. The line method of Art. 3·17 *cannot* be used because the view that shows the required line in true length also shows an *inclined* line as a point (a_Bb_B in Fig. 3·24), and such a view cannot be an elevation view.

Problem. In Fig. 4·24 lines *AB* and *CD* are the center lines of two mine tunnels that must be connected by the shortest possible incline having a 30 per cent grade upward from tunnel *AB* to tunnel *CD*.

Construction. The plane *ABE* is first constructed through line *AB* and parallel to line *CD* (line *AE* has been drawn parallel to line *CD*). This plane must next appear as an edge in an *elevation* view. Using the horizontal line *AF*, view *A* is constructed to show plane *ABE* as an edge and line *CD* parallel to the plane. Any shortest incline of whatever grade desired will appear true length in this view, but its exact location can be determined only by means of a second new view.

To locate the shortest incline of 30 per cent grade, this grade must first be established in view *A*. At any convenient spot in view *A* a right triangle is constructed that is 100 units in a horizontal direction (parallel to *T–A*) and 30 units in a vertical direction. The hypotenuse of this triangle is a line of 30 per cent grade upward from line *AB* toward line *CD* and will be parallel to the true length of the incline in this view. If reference line *A–B* is now taken at right angles to the 30 per cent grade line, the shortest incline of 30 per cent grade will appear in view *B* as a point m_Bn_B, which must lie at the intersection of lines a_Bb_B and c_Bd_B. This determines the exact location of the required incline *MN*, and by alignment and verifying measurements it can now be extended back into the given views.

Problem. In Fig. 4·24 it is required to locate the shortest horizontal passage connecting tunnels *AB* and *CD*.

Construction. The same method may be employed, but in this case the new view *C* is constructed by taking reference line *A–C* perpendicular to a horizontal line—in other words, at right angles to reference line *T–A*. In view *C* the shortest horizontal line appears as point x_Cy_C at the intersection of lines a_Cb_C and c_Cd_C.

Because view *C* shows a horizontal line *XY* as a point, it must be an elevation view and thus could have been obtained directly from the top view without using view A. Therefore, for the *particular* case of locating the shortest *horizontal* line, the solution can be shortened by drawing a new elevation view D that will show the true-length line AF as a *true length* again, instead of using views A and B. Thus, in Fig. 4·24, a new view taken in the direction of arrow D, using reference line T–D, would also show line XY as a point at the intersection of lines a_Db_D and c_Dd_D. This solution is not shown, and it is left to the reader to test its correctness on an actual problem.

Caution. This short solution is valid *only* for the shortest *horizontal* line and cannot be used if the required line has a slope.

PROBLEMS. Group 31.

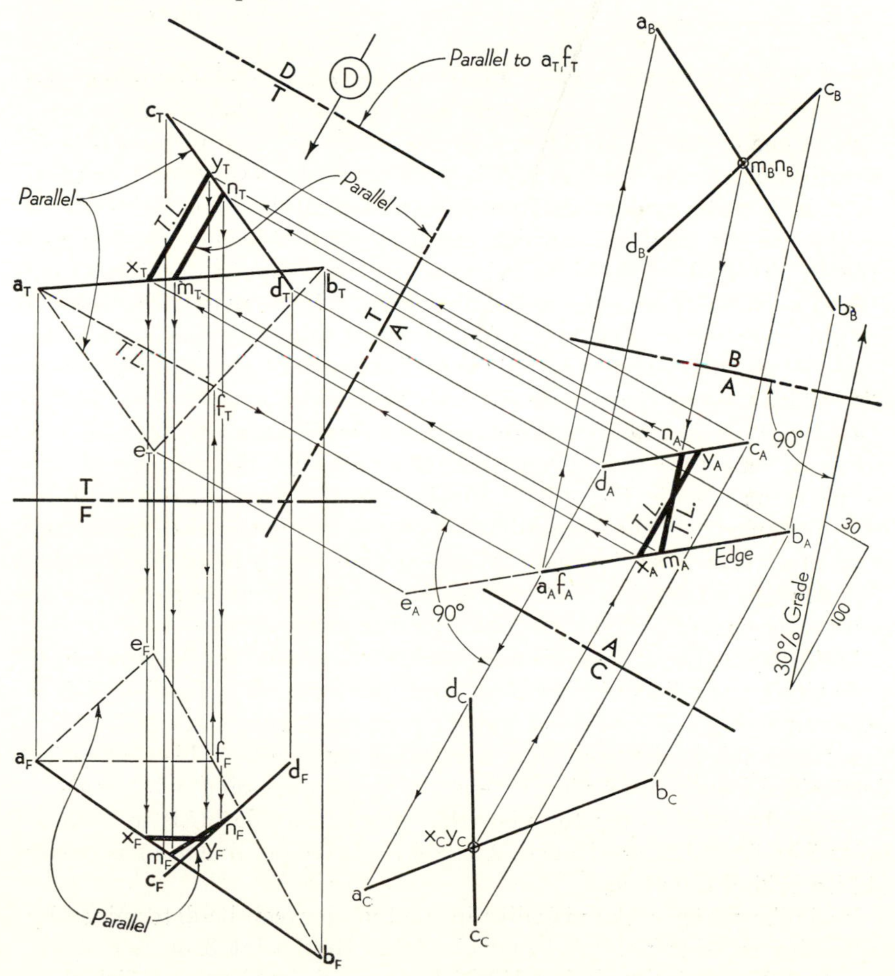

Fig. 4·24. Shortest grade line between two skew lines.

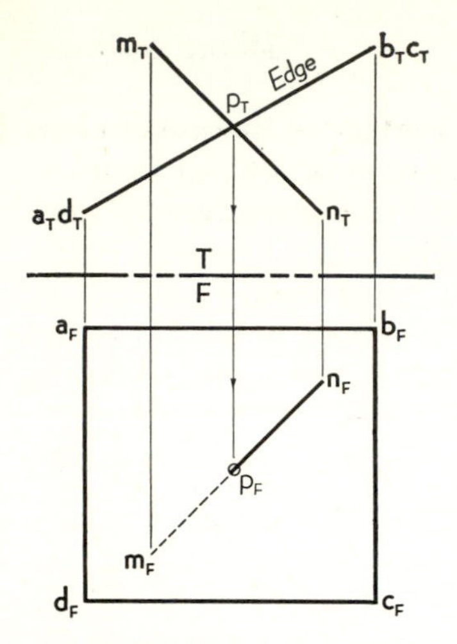

Fig. 4·25. Line intersecting a vertical plane.

4·23. Intersection of a line and a plane (edge-view method)

A straight line, if neither in the plane nor parallel to it, will intersect the plane at a single point, which must lie, of course, both in the plane and on the line. This problem of finding the point of intersection of a line and a plane surface is the basic element in so many problems in descriptive geometry that it might well be ranked in importance with the four basic types of views.

Plane as an edge. The simplest case occurs when the given plane appears as an edge in one of the given views. In Fig. 4·25 the given plane *ABCD* is a vertical plane, and the line *MN* is an oblique line. In the top, or edge, view, it is quite evident that point p_T at the intersection of m_Tn_T and $a_Td_Tb_Tc_T$ is the only point which can lie on line *MN* and also in plane *ABCD*. This, then, is the top view of the required point of intersection, *P*, and p_F must lie on m_Fn_F in the front view. If the plane is considered to be opaque (which is customary), then in the front view that part of the line, *NP*, which lies wholly in front of the plane will be visible and the part *MP*, which the top view shows is wholly behind the plane, will be hidden.

Oblique planes. When the given plane is oblique, as in Fig. 4·26, it is necessary only to construct an edge view to reduce the problem to the simple case shown in Fig. 4·25. In Fig. 4·26 the plane *ABCD* has been drawn as an edge in the front-adjacent view *A* in order to take advantage of the fact that lines *AB* and *CD* appear true length in the front view. The point p_A where line *MN* intersects plane *ABCD* appears in view *A*, and points p_F and p_T are then found by direct alignment.

Visibility. The correct visibility of line m_Fn_F can now be determined from either the adjacent view *A* or from the top view. In view *A* it is apparent that line *MP* is in front of the plane and hence is the visible portion of line *MN* in the front view (see Rule *b*, Art. 2·9). This can also be determined from the top view by examining some crossing point, such as 1,2, and observing in the top view that point 1 on line *BC* is in front of point 2 on line *MN* (see Rule *d*, Art. 2·9). Therefore line n_Fp_F is partly hidden behind the plane.

In similar manner, the visibility in the top view can be determined by examining crossing point 3,4 and observing that point 3 on line *AB* is directly above point 4 on line *MN*. In the top view line m_Tp_T is therefore below the plane.

The *edge-view method* described here is particularly suitable when it is necessary to find the intersection points of a number of lines all intersecting the same plane (see Fig. 4·35, for example). A shorter method, preferable for one or two lines, is given in Art. 4·25.

PROBLEMS. Group 32.

4·24. Intersection line between two planes (edge-view method)

Two plane surfaces, if not parallel, will intersect along a straight line that is common to both planes. Since the position of a straight line may be fixed by any two points on the line, it is only necessary to find two points that lie in both planes. By selecting any two lines in one of the planes and determining where these lines intersect the other plane the necessary two points on the line of intersection can be found.

One plane an edge. Figure 4·27 illustrates the case where one of the given planes appears as an edge in one of the given views. Here the edge-front plane *ABCD* intersects the oblique plane *RST*. In the front view it is readily apparent that line *RS* intersects plane *ABCD* at point *P* and line *ST* at point *Q*. Points *P* and *Q* lie in plane *RST* and also in plane *ABCD* and must therefore be two points on the required line of intersection. Points p_T and q_T are located in the top view on lines r_Ts_T and s_Tt_T,

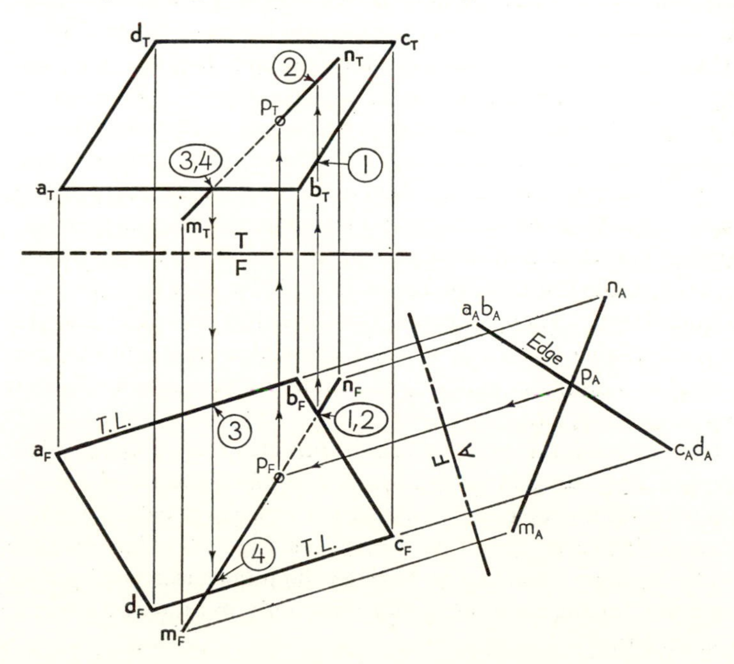

Fig. 4·26. Line intersecting an oblique plane (edge-view method).

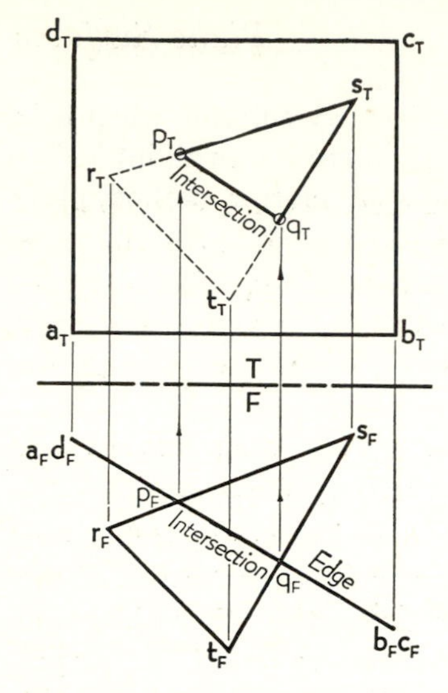

Fig. 4·27. Oblique plane intersecting an edge-front plane.

respectively, by alignment. The line PQ is the intersection line of the two planes if the plane surfaces are limited by their definitive lines; if the planes are considered to be indefinite in extent, then the line PQ may also be extended indefinitely. Note that line RT, as shown by the front view, lies wholly below plane $ABCD$, and therefore the area $RPQT$ is hidden in the top view.

Both planes oblique. If both the given planes are oblique, as shown in Fig. 4·28, the solution can be simplified by constructing a new view that will show one of the planes as an edge. Either plane may be selected, and the edge view may be taken as top-adjacent or front-adjacent. In Fig. 4·28 plane ABC has been shown as an edge in view A. When plane RST is also shown in view A, the intersection points of lines RS and RT can be located at p_A and q_A, respectively. These two points can now be extended back to the top view to locate p_T and q_T, remembering that they lie on lines RS and RT. Points p_F and q_F are similarly located by alignment. The fact that point P lies outside the defined limits of plane ABC does not affect the direction of the line of intersection; but if plane ABC is limited, the line PQ may be terminated at its intersection with line AB.

Visibility. Visibility may be determined in each view by examining any of the numerous points where a line of one plane crosses a line of the other plane (Rule d, Art. 2·9). When the correct visibility has been so established at any *one* crossing point in a view, the visibility at the rest of the crossing points in that view can be determined by inspection. Line ST, for example, is found to be below line BC by checking points 1 and 2 in the front view. Then ST must also be below line AB; line AB must change from visible to hidden when it crosses intersection PQ; line RS is therefore above lines AB and AC, and thus the process continues around the area of overlapped planes.

Shorter methods for finding the intersection of oblique planes are explained in Art. 4·26, but the *edge-view method* described here has the advantage of simplicity and clarity and is especially useful when a series of planes all intersect the same plane (see Fig. 4·35, for example). It should also be used when one or both of the planes is given by strike and dip, since the dip angle can be constructed only in an elevation edge view (see Art. 12·10).

PROBLEMS. Group 33.

4·25. Intersection of a line and a plane (cutting-plane method)

In Art. 4·23 the intersection of a line and an oblique plane was found by the edge-view method (Fig. 4·26). Although this is probably the most obvious type of solution, it has the disadvantage of requiring an extra view. It will now be shown that this problem can be solved, in the given views only, by the *cutting-plane method* (except in the case of a profile line).

Vertical cutting plane. Figure 4·29 shows the top and front views of the oblique plane *ABC* and the oblique line *MN*. At the right, in this same figure, is a pictorial representation of the same line and plane. In the pictorial view a *second* plane has been introduced. This is the *vertical cutting plane*, which has been placed so that the line *MN* lies wholly in it. The vertical plane and the given plane *ABC* must intersect in a straight line, and this line has been shown in the picture and labeled *RS*. The given line *MN* and the intersection line *RS* both lie in the vertical plane

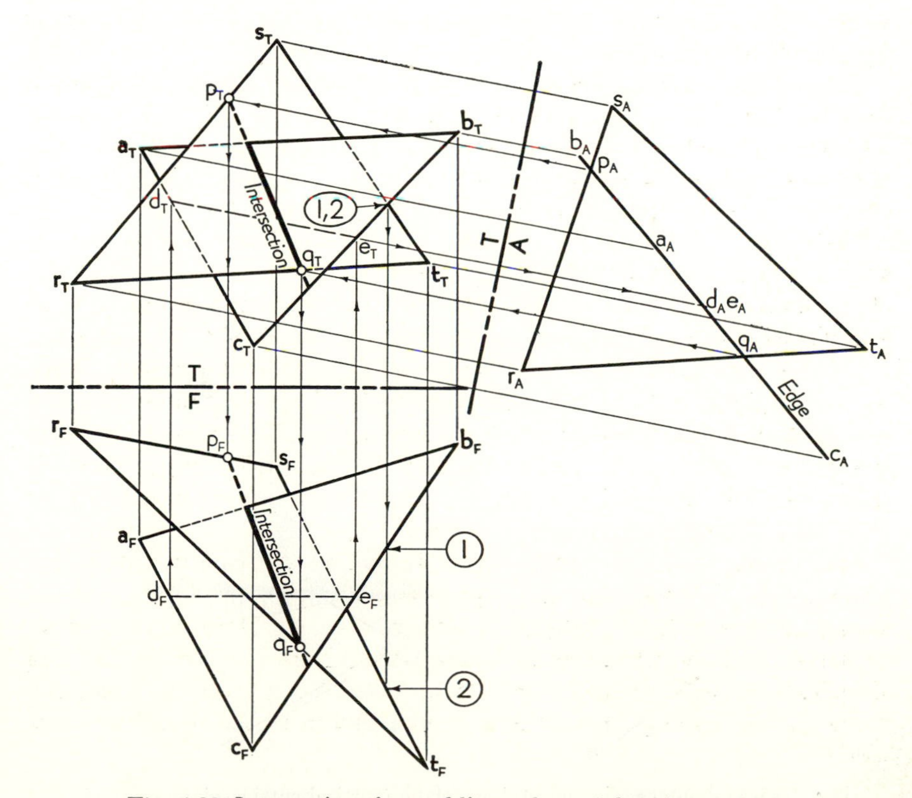

Fig. 4·28. Intersection of two oblique planes (edge-view method).

now and must either be parallel or intersecting. If they intersect as shown at some point P, then that point must be the desired intersection point of line MN and the plane ABC.

Let us now apply the same method to the multiview drawing in Fig. 4·29. First a vertical plane must be introduced to contain the line MN. Such a plane will appear as an edge in the top view and must coincide with the line $m_T n_T$. This is represented by the line labeled C-P (for cutting plane) in the top view. This vertical plane may be indefinite in extent, and this fact is indicated by the three short dashes at the extremities of the line C-P. The position of the cutting plane is fully described in the top view and therefore need not be bounded, or outlined, in the front view (as explained in Art. 4·6 and Fig. 4·11).

The next step is to determine the intersection line of the cutting plane and plane ABC by noting that line AB intersects the cutting plane at point R and line BC at point S. Line $r_F s_F$ is the front view of this line of intersection. (Compare this step with Fig. 4·27.) The top view now shows that lines MN and RS both lie in the cutting plane, and the front view shows that these lines intersect at point p_F. This is the required intersection point of line MN and plane ABC, and p_T can now be located by alignment on line $m_T n_T$ in the top view.

Edge-front cutting plane. The cutting plane may also be taken as an edge-front plane, as shown in Fig. 4·30. In the pictorial view the edge-front plane has been tipped at an angle to contain the line MN, and the intersection of the two planes has been located at TV. The point P is

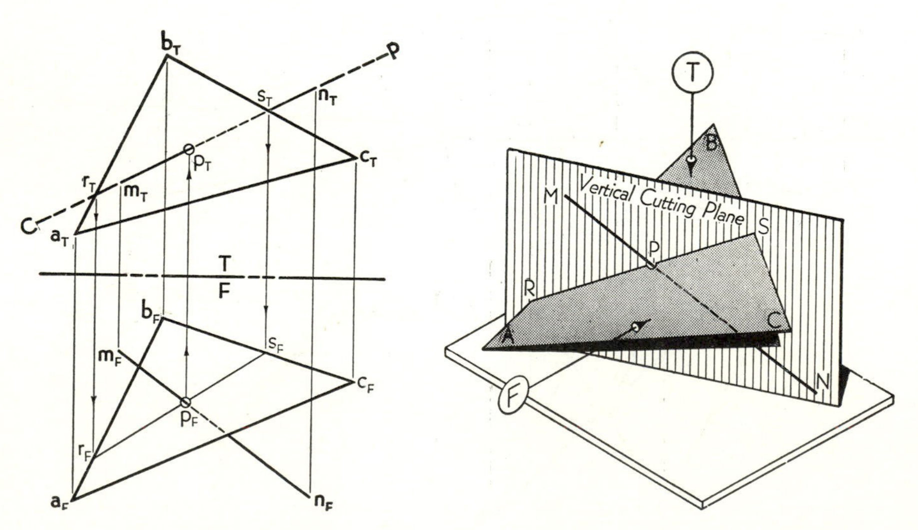

Fig. 4·29. Line intersecting an oblique plane (vertical cutting-plane method).

the required intersection point since it lies on the line MN and also in the plane ABC.

In the multiview drawing in Fig. 4·30 the cutting plane is made to appear as an edge in the front view and coinciding, of course, with line $m_F n_F$. The intersection of the two planes, TV, is next located in each view (compare with Fig. 4·27), and the required intersection point of line MN and plane ABC is found at p_T. Point p_F is found by alignment.

If the intersection lines RS or TV had appeared parallel to line MN in either Fig. 4·29 or 4·30, there would be no intersection point P, and we should correctly conclude that the line MN is parallel to the given plane.

If the given line is a profile line, the cutting-plane method cannot yield a solution in the top and front views only, because the cutting plane appears as an edge in both views. Some new view, such as a side view, would then be necessary. Since an extra view must be drawn, it is usually better to use the edge-view method in this case.

Summary. Because of the importance and frequent use of the cutting-plane method we shall summarize this principle in the following rule:

▪ ***Rule 14. Rule of a line intersecting a plane.*** *The intersection of a line and a plane must lie on the intersection line of the given plane and a cutting plane that contains the line.*

PROBLEMS. Group 34.

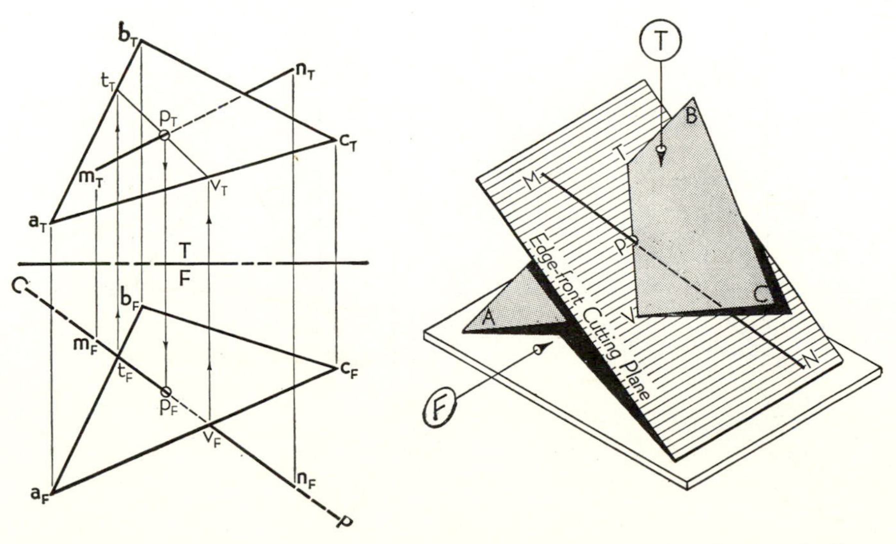

Fig. 4·30. Line intersecting an oblique plane (edge-front cutting-plane method).

4·26. Intersection of two planes (individual-line method)

Analysis. If any line on one of the planes is selected, its intersection with the other plane can be found by the cutting-plane method of Art. 4·25. This determines one point on the line of intersection of the two planes. A second point (and a third or fourth if desired) can be found in the same manner by using some other line in one of the planes. The required line of intersection passes through these individual-line intersection points.

Construction. In Fig. 4·31 the given planes are *ABC* and *RST*. Arbitrarily choosing line *RS* first, its intersection with plane *ABC* is found

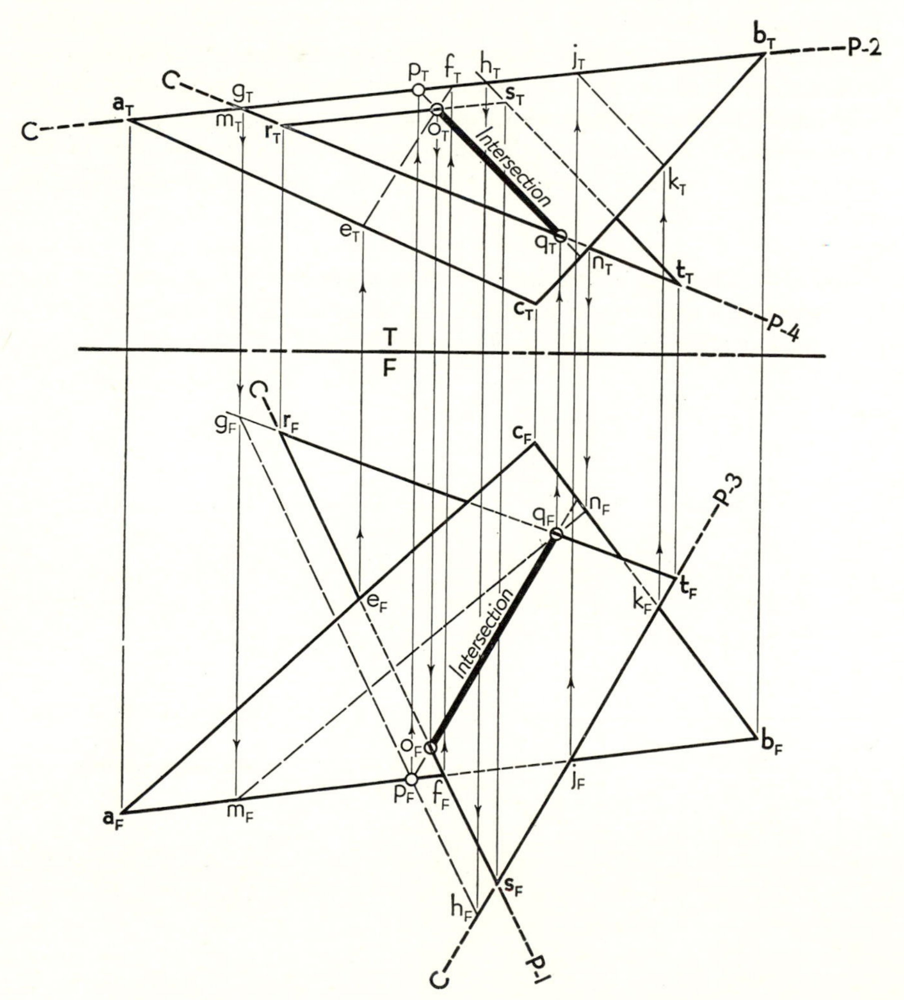

Fig. 4·31. Intersection of two oblique planes (individual-line method).

by using the edge-front cutting plane 1 (*C-P*-1). (A vertical cutting plane through r_Ts_T would work just as well.) This cutting plane intersects plane *ABC* along the line *EF*, and in the top view e_Tf_T is seen to intersect r_Ts_T at point o_T. Point o_F lies directly below o_T on the line r_Fs_F, and the point *O*, where line *RS* intersects plane *ABC*, is one point on the required line of intersection of the two planes. (This operation is identical with that shown in Fig. 4·30.)

To obtain a second point we may again choose a line at random. The point where line *AB* intersects plane *RST* has been found by using a vertical cutting plane *C-P*-2. In this case it was necessary to extend lines *RT* and *ST* in each view to locate the cutting-plane intersection *GH*. Since the line g_Fh_F intersects a_Fb_F at point p_F, this is the front view of a second point on the line of intersection of the two planes. The top view of this same point lies on a_Tb_T, of course.

We now have two points *O* and *P* on the required line of intersection; but unfortunately they are too close together to determine the line accurately, and we should try to find a third point. For this purpose cutting plane *C-P*-3 has been taken through line *ST*. This gives the intersection line *JK*, and it is found that j_Tk_T is parallel to s_Tt_T. This means that line *ST* is parallel to plane *ABC* and is therefore parallel to the required line of intersection of planes *ABC* and *RST*. Hence we may now draw the intersection line through points *O* and *P* parallel to line *ST*. As a further check a fourth cutting plane has been taken through line *RT* to intersect plane *ABC* along line *MN* and thus locate point *Q*. Visibility of the bounding lines of the planes should be determined after the intersection has been found.

In using the individual-line method it should be remembered that the choice of lines is not restricted to those which define the planes. Occasionally these given lines produce points that lie off the paper, and in this case the draftsman should assume other lines in the planes or employ the *auxiliary cutting-plane method* described in the next article.

4·27. Intersection of two planes (auxiliary cutting-plane method)

Analysis. If any third plane intersects the two given planes, then all three planes must meet at some common point *P* (except when the third plane is parallel to the line of intersection of the first two). This common point *P* is one point on the required line of intersection of the given planes. If the third plane appears as an edge in one of the given views, no additional views will be needed for the solution.

The pictorial sketch in Fig. 4·32 illustrates this analysis. The given planes, defined by the parallel lines *AB* and *CD* and by triangle *RST*, are shown extended beyond the defining lines. The edge-front cutting plane intersects plane RST along line RK and intersects plane $ABCD$ along line EF. These lines extended meet at P, one point on the required intersection line.

Construction. In the multiview drawing of Fig. 4·32 the edge-front cutting plane C-P-1 has been taken through point r_F and is horizontal. It need not pass through any particular point, and it need not be horizontal; but it should, for the sake of accuracy, cut each of the given planes at the widest part. The intersection of this cutting plane and plane RST is the line RK. This same cutting plane also intersects plane $ABCD$ along the line EF. In the top view, line r_Tk_T is extended to intersect the extension of line e_Tf_T at the point p_T. This point P is one point on the required line of intersection; and since P also lies in the cutting plane C-P-1, the point p_F is located on the edge view of C-P-1.

The necessary second point on the line of intersection has been found in Fig. 4·32 by using a vertical cutting plane C-P-2. This plane cuts the

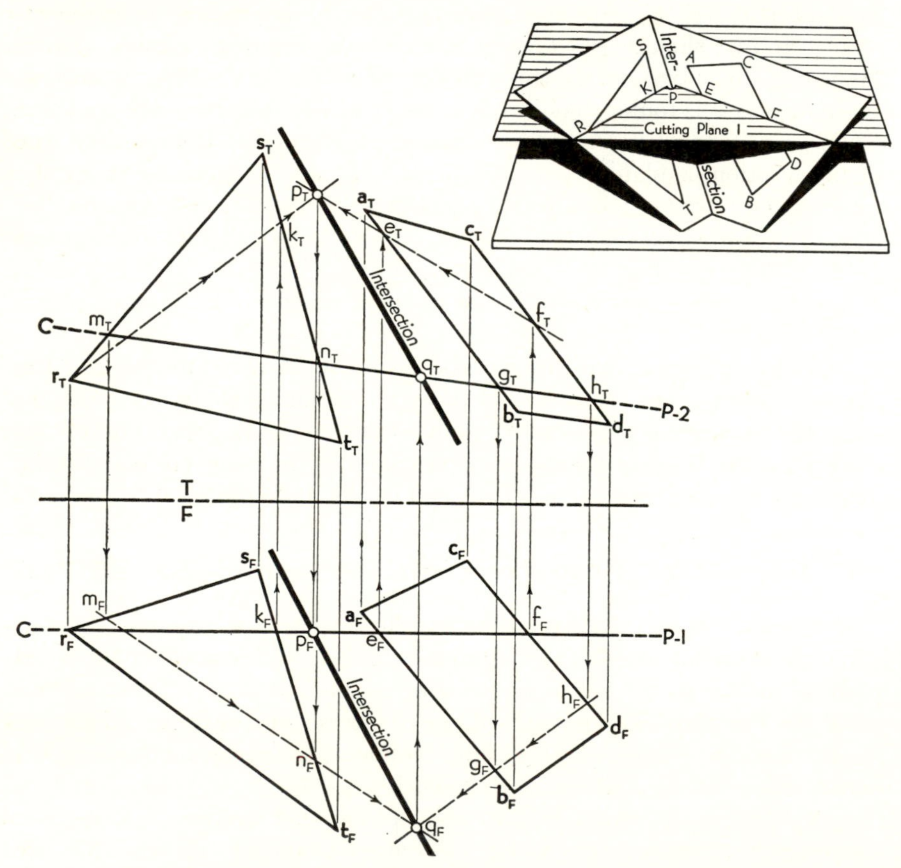

Fig. 4·32. Intersection of two oblique planes (auxiliary cutting-plane method).

line MN from plane RST and line GH from plane $ABCD$. In the front view these two lines $m_F n_F$ and $g_F h_F$ are extended to intersect at point q_F. Point q_T is located above q_F on the edge view of the cutting plane used to locate Q. The line of intersection of the two given planes may now be drawn through points P and Q, indefinite in extent because the planes are not limited in this case.

Although two cutting planes locating two points are sufficient to determine the line of intersection, it is usually advisable to locate one or two more points as a check. All such points should lie along the same straight line. To obtain additional points, new cutting planes may be assumed anywhere in either view at any desired angle with reference line T–F. It is sometimes convenient to make the entire solution with a series of *parallel* vertical (or edge-front) planes to take advantage of the resulting parallel intersection lines.

If in Fig. 4·32 the lines $r_T k_T$ and $e_T f_T$ had been parallel, it would then be known that the required line of intersection, if any, must be parallel to lines RK and EF, but its exact location would still be unknown. In this event the second cutting plane should be taken at a radically different angle with T–F or should be taken next in the top view (but not parallel to $r_T k_T$, of course). If this second cutting plane also produces parallel lines, we can only conclude that the two given planes are actually parallel to each other and that there is no intersection line.

PROBLEMS. Group 35.

4·28. Polyhedrons

Any solid, bounded entirely by plane surfaces, may properly be referred to as a polyhedron. The plane surfaces are called the *faces* of the solid, and the lines of intersection of the faces are called the *edges*. There are, however, five *regular* polyhedrons, known as the "five platonic solids," that are of historical interest. A regular polyhedron is a convex solid whose faces are all equal regular polygons. Each of the five solids, shown in Fig. 4·33, derives its name from the number of faces it possesses: the *tetrahedron* has 4 faces, each an equilateral triangle; the *cube* (*hexahedron*) has 6 square faces; the *octahedron* has 8 equilateral triangular faces; the *dodecahedron* has 12 pentagonal faces; the *icosahedron* has 20 equilateral triangular faces.

With the exception of the cube, the regular polyhedrons are of very little practical value in engineering. They are important, however, in that branch of mineralogy called crystallography. Most minerals occur in characteristic crystal forms, and the crystals assume the shape of regular polyhedral solids. The various minerals may thus be classified according to their degree of symmetry, axial ratios, and interfacial angles.

The interested student is advised to consult a text on crystallography for more information on this fascinating subject.

A *prism* is a polyhedron which has two congruent parallel faces called its *bases* and whose remaining lateral faces are parallelograms. The bases may be of any polygonal shape, as shown in Fig. 4·34. When the lateral faces of the prism are rectangular and perpendicular to the base, the solid is a *right* prism; otherwise, the prism is *oblique*. The *altitude* of a prism is the perpendicular distance between the planes of the bases. A *truncated* prism is a solid formed by cutting off one end of a prism by a plane not parallel to the base.

A *pyramid* is a polyhedron which has a polygon for its base and whose remaining lateral faces are triangles having a common vertex called the *vertex* of the pyramid (see Fig. 4·34). The *axis* of a pyramid is the straight line connecting the center of the base to the vertex. When the axis is perpendicular to the base the pyramid is a *right* pyramid; otherwise, the pyramid is *oblique*. The *altitude* of a pyramid is the perpendicular distance from the vertex to the plane of the base. A *truncated* pyramid (also called a *frustum*) is a solid formed by cutting off the vertex portion of a pyramid by a plane.

The representation of a prism or pyramid in the principal views, when the axis of the object is inclined, has already been discussed in Art. 3·19.

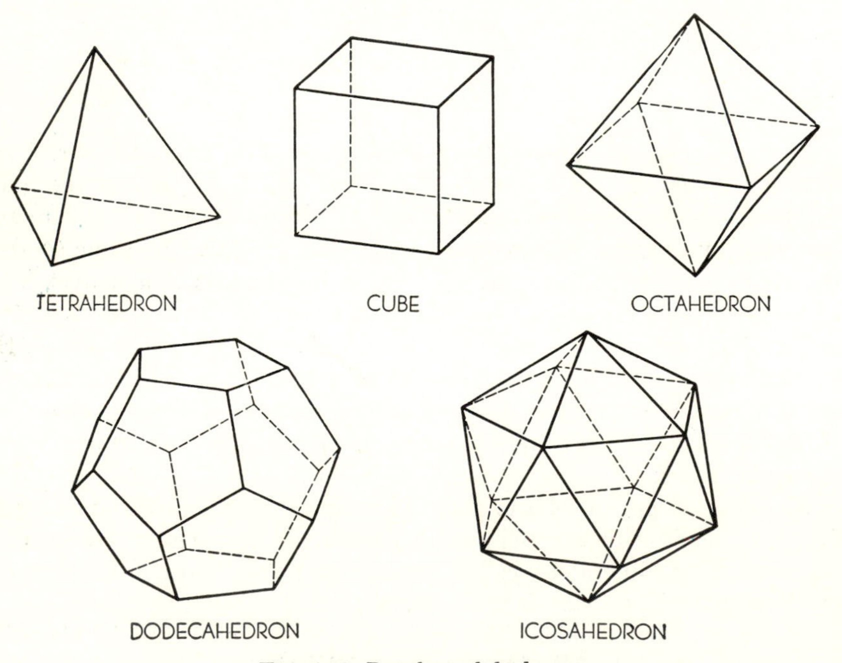

Fig. 4·33. Regular polyhedrons.

4·29. Intersection of a plane and a polyhedron (edge-view method)

If a plane intersects a polyhedron, it cuts a straight line on each of the intersected faces. These straight lines form a polygonal plane figure that is called a *plane section* of the polyhedron. When the intersecting plane is at right angles to the axis of the polyhedron, the plane section is called a *right section*.

Analysis. A view showing the given plane as an edge will show the point where each edge of the polyhedron intersects the plane. These points locate the vertices of the required plane section.

Problem. Figure 4·35 shows an oblique parallelepiped (a prism whose bases are parallelograms) in an oblique position. The line of intersection of the prism and plane *ABCD* is required.

Construction. The necessary edge view may be either top-adjacent or front-adjacent; but since lines *AB* and *CD* appear as true lengths in the top view, view *A* has been selected as shown. In view *A* it is now obvious that plane *ABCD* intersects the lateral faces of the polyhedron but does not intersect the bases. This is one distinct advantage of the edge-view method, for it is frequently not apparent in the given views which faces of the polyhedron will be cut by an oblique plane.

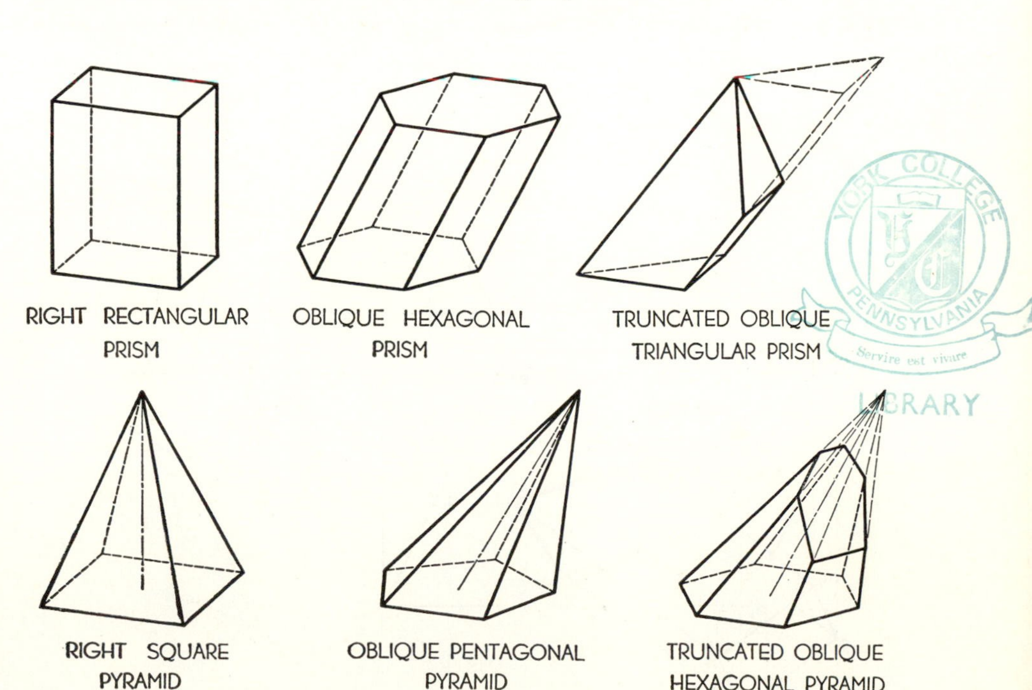

Fig. 4·34. Prisms and pyramids.

View A also shows that the lateral edge EM intersects plane $ABCD$ at point X, edge FN at point Y, etc. The line $x_Ay_Aw_Az_A$ is therefore an edge view of the required plane section, and points X, Y, Z, and W are the vertices of the plane figure. By extending parallels back to the top view, points x_T, y_T, etc., may be located on the corresponding edges of the prism. Points x_F, y_F, etc., may be similarly located in the front view but should also be checked by reference-line measurement.

The visibility of the plane section must agree with the visibility of the prism. Points x_F, y_F, and z_F, for example, must be visible points since they lie on visible edges of the prism in the front view, and therefore lines x_Fy_F and y_Fz_F are visible lines. Visibility of the edges of the prism and

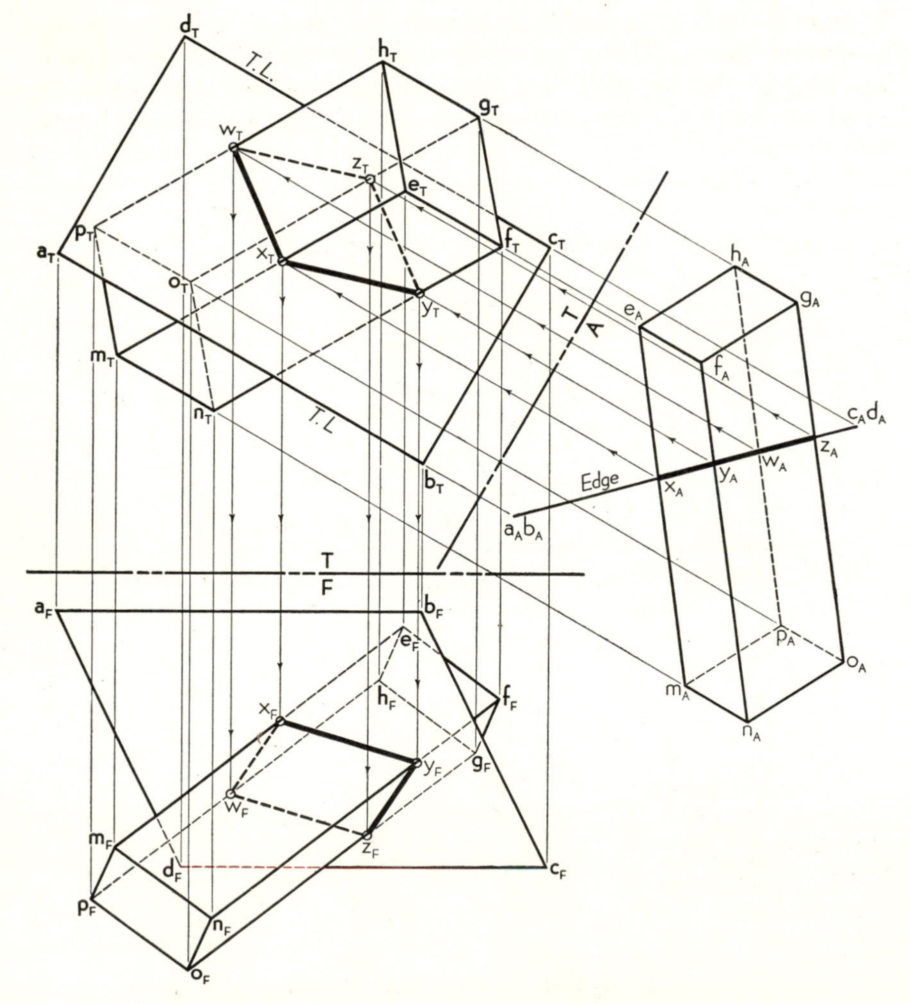

Fig. 4·35. Intersection of a plane and a polyhedron (edge-view method).

plane in each view can then be determined by the standard rules of visibility (Art. 2·9).

4·30. Intersection of a plane and a polyhedron (cutting-plane method)

Analysis. The required plane section can be determined by finding the intersection of each edge of the polyhedron with the plane. This can be done directly in the given views by the cutting-plane method of Art. 4·25.

Problem. Figure 4·36 shows an oblique triangular pyramid *MNOP*. The line of intersection of the pyramid and plane *ABC* is required.

Construction. The solution begins by selecting any one of the six edges of the pyramid and determining where this edge line intersects the plane. In Fig. 4·36 the lateral edge *MP* is selected first, and the edge-front cutting plane *C-P*-1 is taken through line *MP*. This first cutting plane

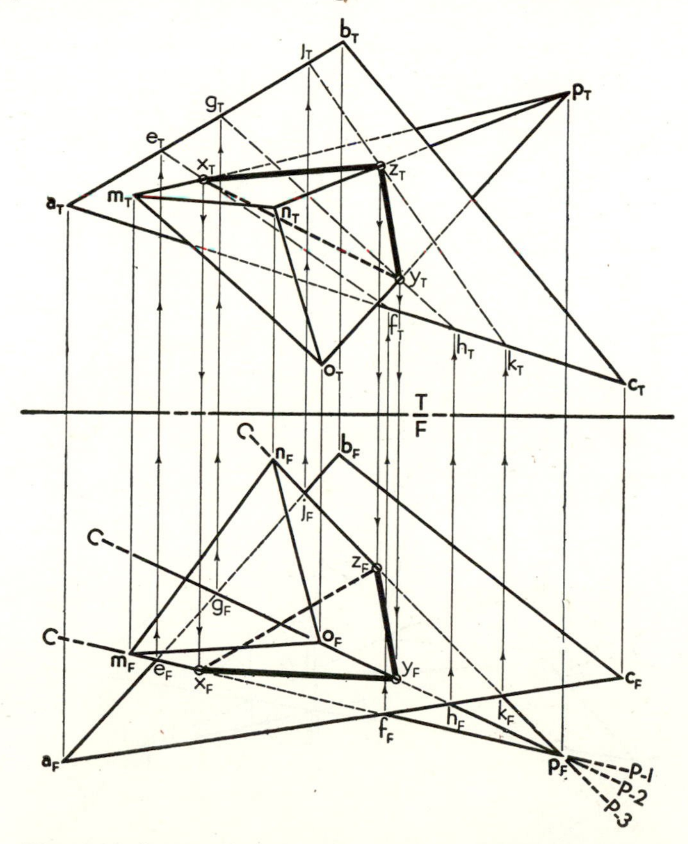

Fig. 4·36. Intersection of a plane and a polyhedron (cutting-plane method).

intersects plane *ABC* along the line *EF*. In the top view the line *EF* is seen to intersect the edge *MP* at point *X*, and thus one point on the line of intersection is located. In similar manner *C-P*-2 and *C-P*-3 are taken through edges *OP* and *NP*, respectively, to cut plane *ABC* along the lines *GH* and *JK* and locate points *Y* and *Z*. Connecting points *X*, *Y*, and *Z* in proper visibility gives the required plane section. Note that line $x_T y_T$, for example, is hidden in the top view because it lies on the hidden face $m_T p_T o_T$.

If, in Fig. 4·36, line $e_T f_T$ had fallen to the left of line $m_T p_T$, it would then have been necessary to extend line $m_T p_T$ to locate point x_T. But if point *X* lies on line *MP* extended, then line *XY* will intersect edge *MO*, and line *XZ* will intersect edge *MN*, and we should therefore know that the given plane intersects the base plane of the pyramid. Point *X* would not then be a vertex of the plane section, but it would provide an excellent check on the accuracy of the base-plane intersection.

PROBLEMS. Group 36.

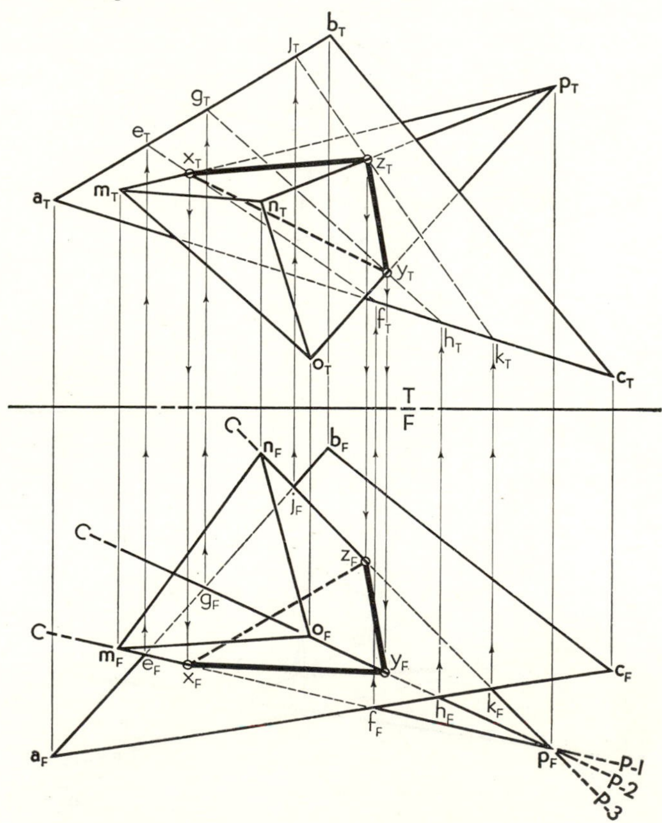

Fig. 4·36 (repeated). Intersection of a plane and a polyhedron (cutting-plane method).

4·31. Intersection of a line and a polyhedron

Analysis. If a line passes through a convex polyhedron, it must intersect the surface of the solid at two points. By assuming any plane containing the given line and determining the plane section cut from the polyhedron by this plane the two points of intersection of the line may be found. Although any oblique plane might be used, the solution will be shorter and simpler if a vertical or edge-front plane is selected.

Problem. Figure 4·37 shows an oblique prism and a line MN. The points of intersection of the line and prism are required.

Construction. Assume the edge-front cutting plane C-P through line MN. This plane intersects the lateral edges of the prism at points R, S, T, and V, and the plane section appears in the top view as the quadrilateral $r_Ts_Tt_Tv_T$. The intersection of line MN with the plane section locates points P and Q, which are the required intersection points. The correct visibility of line MN may now be determined. In the top view, line r_Ts_T lies on the visible face $a_Tb_Tf_Te_T$ and is therefore visible; point p_T lies on the visible line r_Ts_T and is therefore visible. By similar reasoning, q_T is hidden. Then the line m_Tn_T must be visible from m_T to p_T, hidden from p_T to q_T inside the prism, and hidden from q_T to edge c_Tg_T, where it

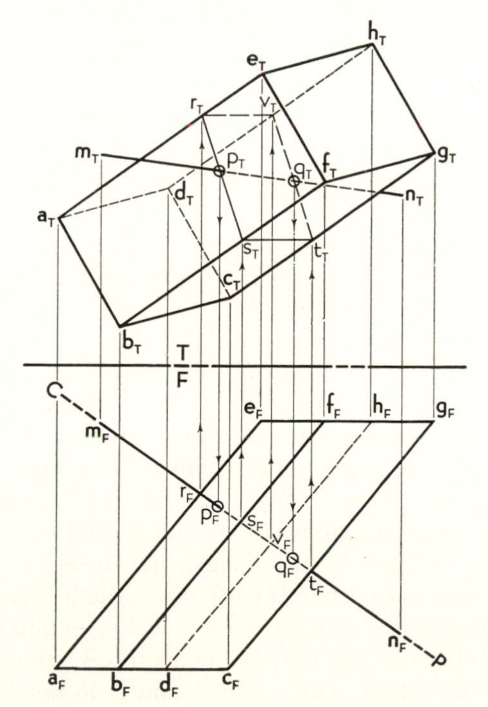

Fig. 4·37. Intersection of a line and a polyhedron.

emerges from under the prism. Similar analysis provides the correct visibility of $m_F n_F$.

PROBLEMS. Group 37.

4·32. Location of a line perpendicular to a plane

In Art. 4·9 a perpendicular to a plane was constructed by showing the plane as an edge. In the articles that follow it will be demonstrated that perpendicular lines and planes can be established by using only the given views. In some cases, however, this shorter method becomes inaccurate because of poor line intersections; the edge-view method should then be employed as a check.

Analysis. If a line is perpendicular to a plane, it is also perpendicular to all lines in that plane (whether these plane lines intersect the perpendicular or not). Those lines in the plane which are parallel to the reference line in one view will appear true length in the adjacent view (Art. 4·3). Therefore the required perpendicular will appear perpendicular in each view to any true-length line in the plane in that view (Rule 10, Art. 3·14).

Problem. In Fig. 4·38 the given plane is *ABC*, and the required perpendicular must pass through the given point *X*.

Construction. The first step is to determine the direction of a true-length line in the plane in each view. Line *EF* appears true length in the top view, and line *GH* is true length in the front view. The required perpendicular to the plane must be perpendicular to both these lines, and hence $x_T y_T$ must appear perpendicular to true length $e_T f_T$ and $x_F y_F$ perpendicular to true length $g_F h_F$ (Rule 10). This determines the direction of line *XY* in each view, and the point *Y* may, of course, be located anywhere along the required perpendicular.

Discussion. The horizontal line *EF* may be taken at *any desired level* on the plane because all such horizontal lines will have the *same direction* in the top view as $e_T f_T$, and it is only the *direction* of the true-length line that determines the *direction* of $x_T y_T$. Similarly, line *GH* could have been assumed in any parallel position on the plane without altering the *direction* of $x_F y_F$. Because these two true-length lines are assumed entirely at random, we can attach no significance to the labeled points 1 and 2 where the perpendicular crosses the true-length line in each view. The perpendicular does *not* actually intersect the true-length lines at these points, and—contrary to common student belief—points 1 and 2 do *not* represent the point where the perpendicular intersects the plane. Even casual inspection will show that points 1 and 2 do not lie on a common parallel. If it is necessary to determine also where the required perpendicular intersects the plane, then the method of Art. 4·25 should be employed *after* the direction of the perpendicular has been determined (Art. 4·36).

Summary. It has now been demonstrated how the *direction* of a perpendicular to a plane may be determined in the given views, and this principle is important enough to warrant stating it as a rule:

▪ ***Rule 15. Rule of perpendiculars to a plane.*** *A line perpendicular to a plane will appear perpendicular to any line in the plane which appears true length in that same view.*

PROBLEMS. Group 38.

4·33. Location of a plane perpendicular to a line

Analysis. The position of some one point in the required plane must be given. Then the plane may be defined by two lines intersecting at the given point, each line being perpendicular to the given line.

Problem. In Fig. 4·39 the given line is AB, and it is required to locate the perpendicular plane through the given point X.

Construction. In the top view, line t_Tv_T of any desired length is drawn through x_T perpendicular to a_Tb_T. But t_Tv_T must be true length,

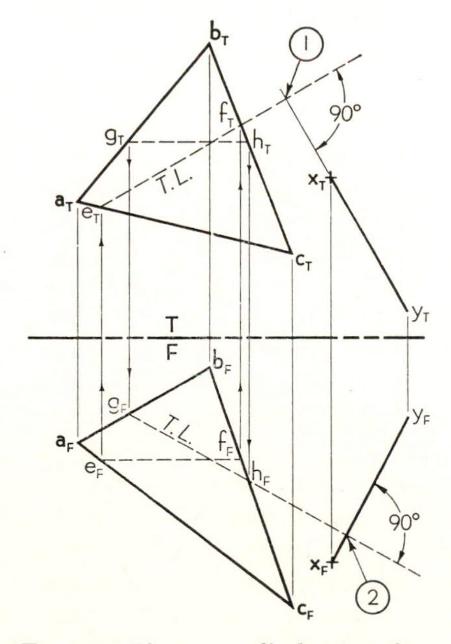

Fig. 4·38. Line perpendicular to a plane.

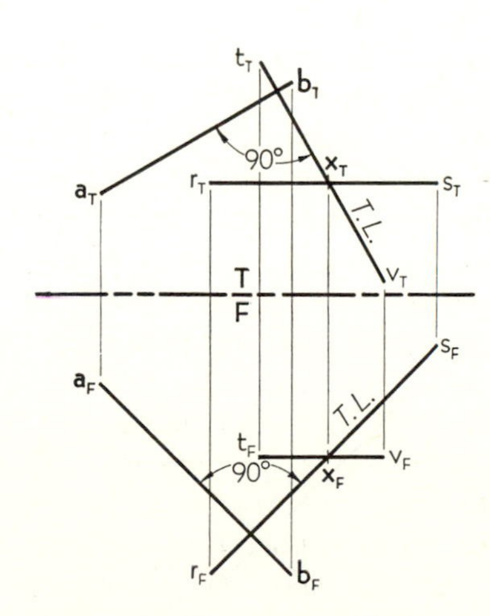

Fig. 4·39. Plane perpendicular to a line.

and therefore t_Fv_F must be parallel to the reference line T–F. This establishes one line in the required plane. The second line, RS, is constructed to appear perpendicular to AB in the front view; that is, r_Fs_F is drawn through x_F perpendicular to a_Fb_F, and r_Ts_T must be parallel to reference line T–F. The plane of intersecting lines RS and TV is therefore the required plane.

PROBLEMS. Group 39.

4·34. Location of a plane through a given line and perpendicular to a given plane

Analysis. Since two intersecting lines determine a plane and one line in the required plane is given, it only remains to locate a second line in the required plane. But if the required plane must be perpendicular to the given plane, then it must contain a line perpendicular to the given plane. Therefore the second line must be so constructed that it intersects the given line and is perpendicular to the given plane.

Problem. In Fig. 4·40 the given line is the profile line XY, and the given plane is ABC.

Construction. The necessary second line in the required plane may intersect line XY at any desired point, but to avoid the need for an additional view of this profile line (as in Fig. 3·6) we shall wisely choose the

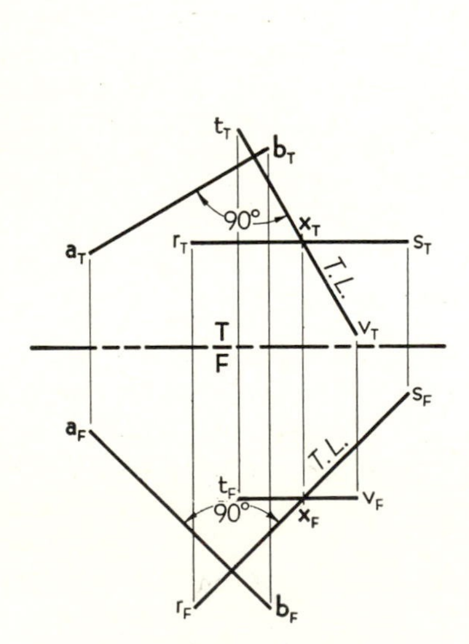

Fig. 4·39 (repeated). Plane perpendicular to a line.

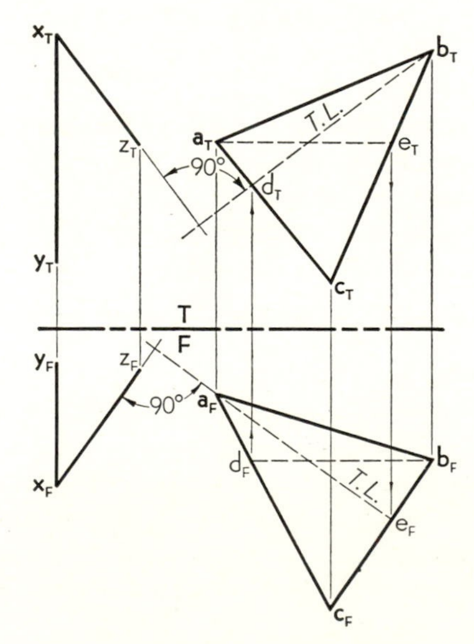

Fig. 4·40. Plane through a line and perpendicular to a given plane.

intersection point at either X or Y rather than at some intermediate point. Through point X the line XZ, of any desired length, is drawn perpendicular to plane ABC, using exactly the same construction as shown in Fig. 4·38. The plane YXZ is the required plane.

PROBLEMS. Group 40.

4·35. Location of a plane through a given point and perpendicular to two given planes

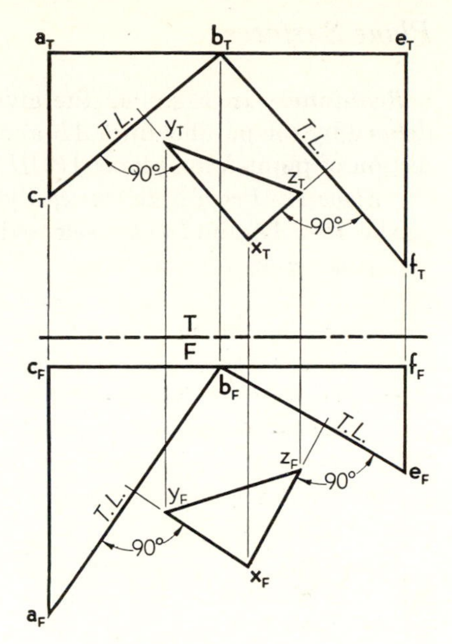

Fig. 4·41. Plane through a point and perpendicular to two given planes.

Analysis. The required plane may be defined by two lines intersecting at the given point, each line being perpendicular to one of the given planes. The solution may be made then by performing the construction shown in Fig. 4·38 twice, once for each line and plane.

Problem. In Fig. 4·41 the given point is X, and the two given planes are ABC and BEF.

Construction. Line XY has been drawn perpendicular to plane ABC by observing that line BC is true length in the top view and line AB is true length in the front view. Line XZ is similarly drawn perpendicular to plane BEF, and the required plane is then XYZ. Points Y and Z may, of course, be selected anywhere along the respective perpendiculars.

PROBLEMS. Group 41.

4·36. Projection of a point on a plane

The projection of a point on a plane is the point of intersection with the plane of a perpendicular line from the point to the plane. Thus the projection of a point on a plane may be thought of as the shadow of the point cast on the plane by light shining in a direction perpendicular to the plane.

Analysis. The projection of a point on a plane may be determined by using only the given views if we observe that the solution involves two operations, each of which may be performed in the given views only:

1. *Construct a perpendicular to the plane from the given point.*
2. *Locate the intersection of this perpendicular with the plane.*

Problem. In Fig. 4·42 the given point is X, and the given plane is defined by the parallel lines AB and CD. It is required to locate the projection of point X on plane $ABCD$.

Stage 1. *From X construct a perpendicular of indefinite length.*

The lines AE and BD are selected to provide true-length lines in the top and front views, respectively. From x_T a line of indefinite length is drawn at right angles to a_Te_T, and from x_F a line is drawn at right angles to b_Fd_F. This is the required perpendicular.

Stage 2. *Locate the intersection of perpendicular and plane.*

A vertical cutting plane C-P has been assumed through the perpendicular. This cutting plane intersects the given plane along the line RS, and the intersection of r_Fs_F and the perpendicular in the front view locates p_F, the front view of the required projection P. Point p_T may then be located in the top view.

Caution. The sequence and independence of the two operations should be observed. For example, the crossing of lines a_Te_T and x_Tp_T has no significance whatsoever and has nothing to do with the subsequent location of point P. The lines AE and BD are true-length lines selected

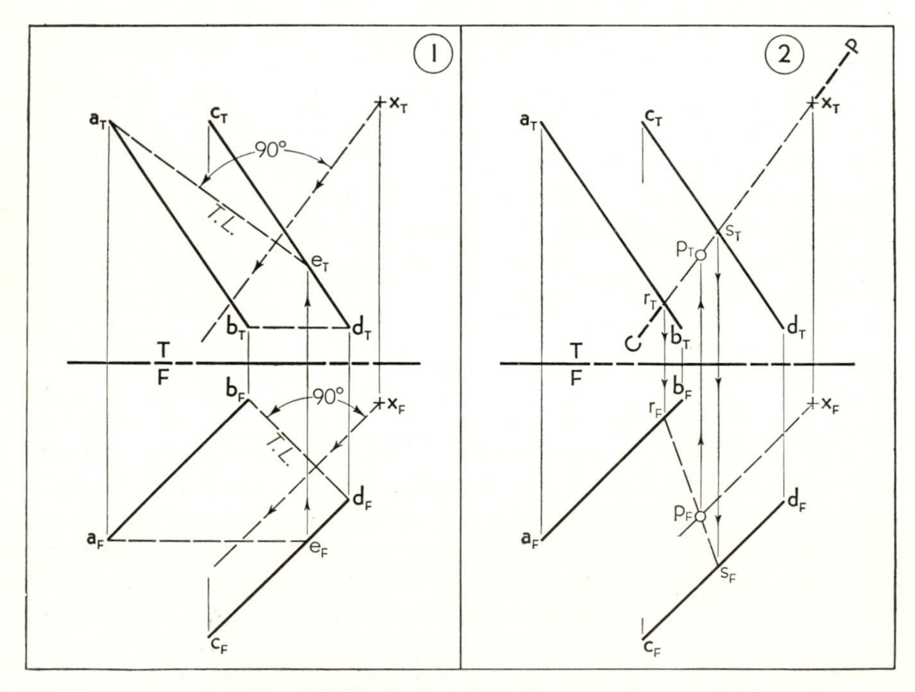

Fig. 4·42 Projection of a point on a plane.

at random and thus could not be expected to locate a specific point such as P.

PROBLEMS. Group 42.

4·37. Projection of a line on a plane

Analysis. This problem is simply an extension of that discussed in the previous article, since the projection of a line on a plane may be determined merely by projecting any two points of that line on the plane. A line passing through the two point projections is the required line projection.

Problem. In Fig. 4·43 stage 1 shows the given plane $ABCD$ and the line MN, which is to be projected on the plane.

Stage 2. *Draw two perpendiculars from the line to the plane.*

In the plane $ABCD$ line AB is true length in the top view, and line AK has been constructed to appear true length in the front view. These lines establish the *direction* of the perpendiculars drawn from M and N in each view.

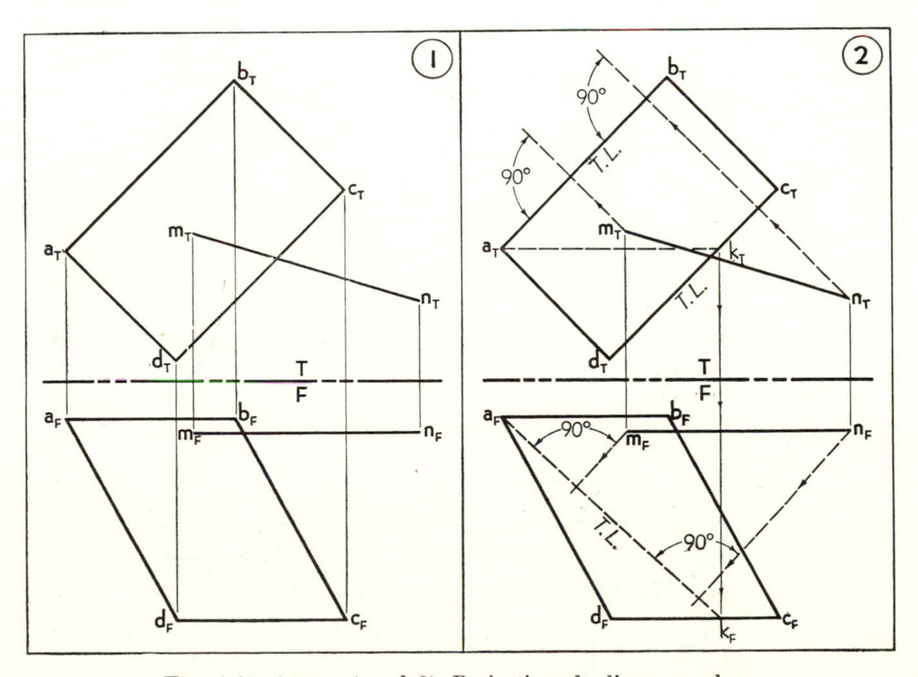

Fig. 4·43. (stages 1 and 2). Projection of a line on a plane.

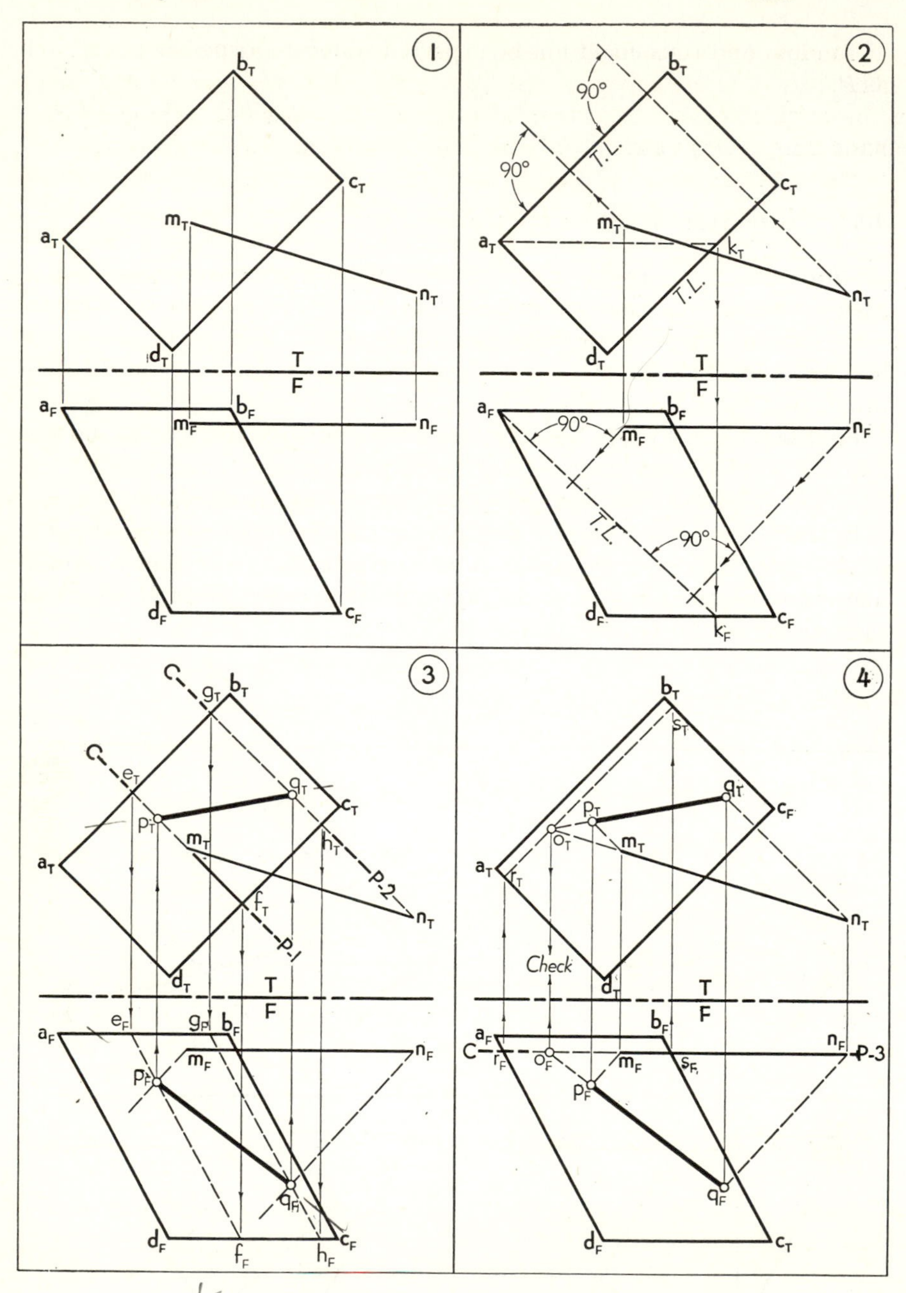

Fig. 4·43 Projection of a line on a plane.

Stage 3. *Find the intersections of perpendiculars and plane.*

The two vertical cutting planes *C-P*-1 and *C-P*-2, each containing one of the perpendiculars, have been assumed in the top view. (Edge-front cutting planes could also have been used.) Cutting planes 1 and 2 intersect plane *ABCD* along the lines *EF* and *GH*, respectively; and in the front view the intersection of e_Ff_F and the perpendicular from m_F locates point p_F and the intersection of g_Fh_F and the perpendicular from n_F locates point q_F. The line p_Fq_F is the front view of the required projection of line *MN* on plane *ABCD*, and p_Tq_T is located in the top view by alignment. The solution is thus completed in stage 3.

Stage 4. *Check the accuracy of the work.*

Note that the line m_Fn_F extended intersects the projection p_Fq_F extended at point o_F. But point *O* must be a point in plane *ABCD*, since it lies on *PQ*, which is in plane *ABCD*, and point *O* must also be a point on *MN*. Therefore point *O* must be the point of intersection of line *MN* and plane *ABCD*; and the point o_T, found at the intersection of lines m_Tn_T and p_Tq_T extended, must lie on the same parallel as o_F. That point *O* is the intersection point of line *MN* in plane *ABCD* is further verified by the cutting plane *C-P*-3, which cuts plane *ABCD* along the line *RS*.

PROBLEMS. Group 43.

4·38. Dihedral angle (line of intersection given)

The angle formed by two intersecting planes is called a *dihedral angle.* This dihedral angle must be measured in a plane that is perpendicular to each of the intersecting planes, in other words, perpendicular to the line of intersection of the two planes.

Analysis. The plane of the dihedral angle will appear true size when the line of intersection of the two given planes appears as a point. In the point view both given planes will appear as edges. The solution therefore requires that we:

Show the line of intersection as a point.

Problem. In Fig. 4·44 the given planes are *ABC* and *ABD*, intersecting along the common line *AB*.

Construction. View *B* has been drawn to show line *AB* as a point. Both planes appear as edges in this view, and the required dihedral angle appears as shown. It is evident that only one point in each plane outside the intersection line is necessary to determine the dihedral angle, but additional points are sometimes advisable as a check on accuracy.

4·39. Dihedral angle (line of intersection not given)

Analysis. The dihedral angle will appear true size in a view that shows both planes as edges. If *one* of the planes is shown first as an *edge* and then as a *true size*, the *second* plane will not necessarily appear as either an edge or as a true size in these views. But *any view* of a plane taken adjacent to a *true-size view* will always show that plane as an edge again. Therefore the direction of the third auxiliary view, adjacent to the second, is selected to show the *second* plane as an edge. With both planes now appearing as edges, the dihedral angle may be measured. In summary:

> *Show one plane in true size; then select the third view to show the second plane as an edge.*

Problem. In Fig. 4·45 the given planes are *ABC* and *RST*. The line of intersection is not given.

Construction. Arbitrarily choosing plane *ABC*, views *A* and *B* are drawn to show this plane first as an edge and then as a true size. Plane *RST* is simply carried along in these views, where it appears obliquely. Actually, plane *ABC* need not be outlined in view *B* because regardless of where reference line *B–C* is placed the plane *ABC* will always appear in view *C* as an edge parallel to the reference line. With the edge appearance of plane *ABC* in view *C* thus assured, view *C* must now be selected

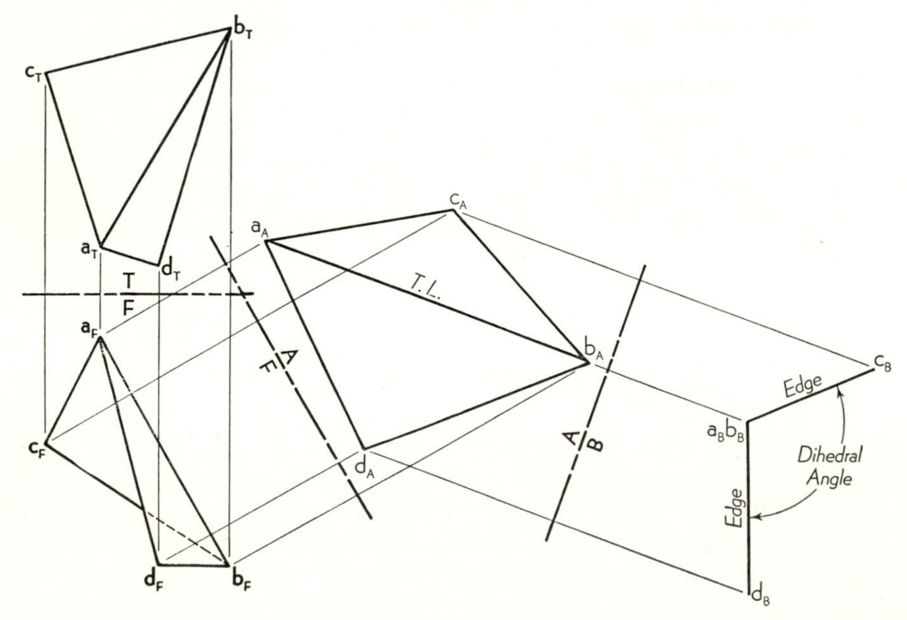

Fig. 4·44. Dihedral angle (line of intersection given)

to show plane *RST* as an edge. To make plane *RST* appear as an edge in view *C* some line in plane *RST* must appear as a point in that view, and therefore this line must be true length in view *B*, which in turn means that it must be parallel to reference line *A–B* in the adjacent view *A*. Line $f_A s_A$ is so drawn and line $f_B s_B$ located in view *B*. View *C* is then drawn to show line *FS* as a point at $f_C s_C$ and plane *RST* as an edge. Plane *ABC* will appear as an edge parallel to reference line *B–C*, and the dihedral angle may be measured as shown. It will be observed that the supplementary angle labeled "*D.A.*" is also an acceptable answer. Choice of these two angles necessarily depends upon the practical application of the result.

Alternate method. An indirect method that does not require the line of intersection of the given planes is based on the following idea: If two intersecting lines are drawn through any assumed point so that each line is perpendicular to one of the planes, then the angle between the two lines will be the *supplement* of the dihedral angle between the two planes. The angle between the two intersecting lines can be determined by using only two new views. The disadvantage of this indirect method is that either of two supplementary angles may be obtained depending upon the location of the assumed point, and consequently it is frequently difficult to know which angle is the one desired.

PROBLEMS. Group 44.

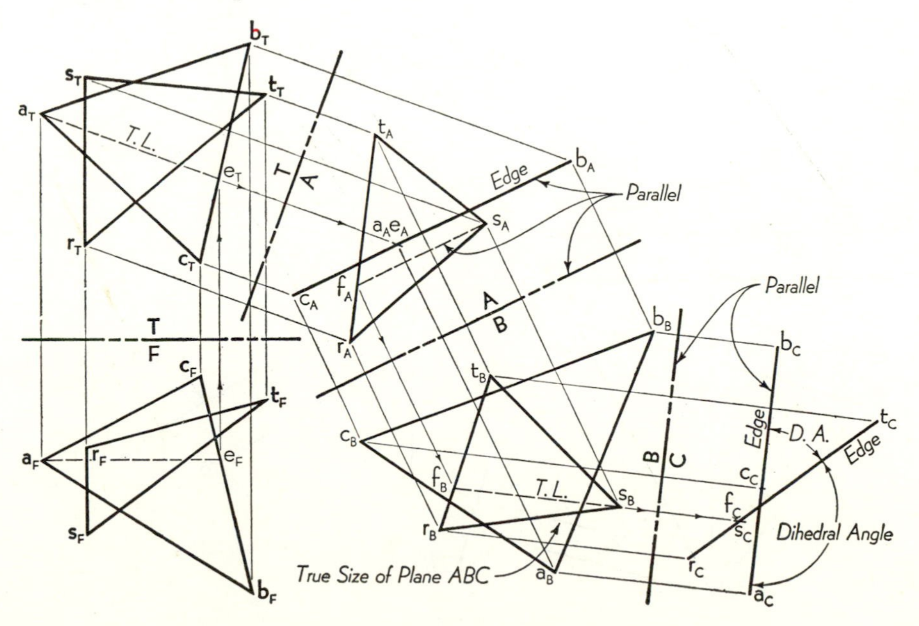

Fig. 4·45. Dihedral angle (line of intersection not given).

4·40. Angle between a line and a plane (edge-view method)

Analysis. In Art. 3·7 the slope angle of a line was defined as the angle between that line and a horizontal plane. It was observed then that the slope angle could be seen only in a view that showed the line as a *true length* and the horizontal plane as an *edge*. Analogously, the *line-plane angle* between any oblique line and any oblique plane requires the same condition: the line must appear *true length* in the *same view* that shows the *plane as an edge*. Such a view can be obtained in the following order:

> *Show the plane first as an edge, second in true size, and third as an edge again where the given line appears true length.*

Problem. In Fig. 4·46 the given plane is *ABC* and the given line *MN*. It is required to determine the angle between the line and the plane.

Construction. Following the procedure outlined above, view *A* is drawn first to show the plane *ABC* as an edge. This view *cannot* show the desired angle between line *MN* and plane *ABC* because line *MN* is not true length in view *A*. View *B* is selected next to show the plane in its true size; but (as suggested in Art. 4·39) it is not necessary actually to outline the plane here, and only line $m_B n_B$ is located. That view *B* need

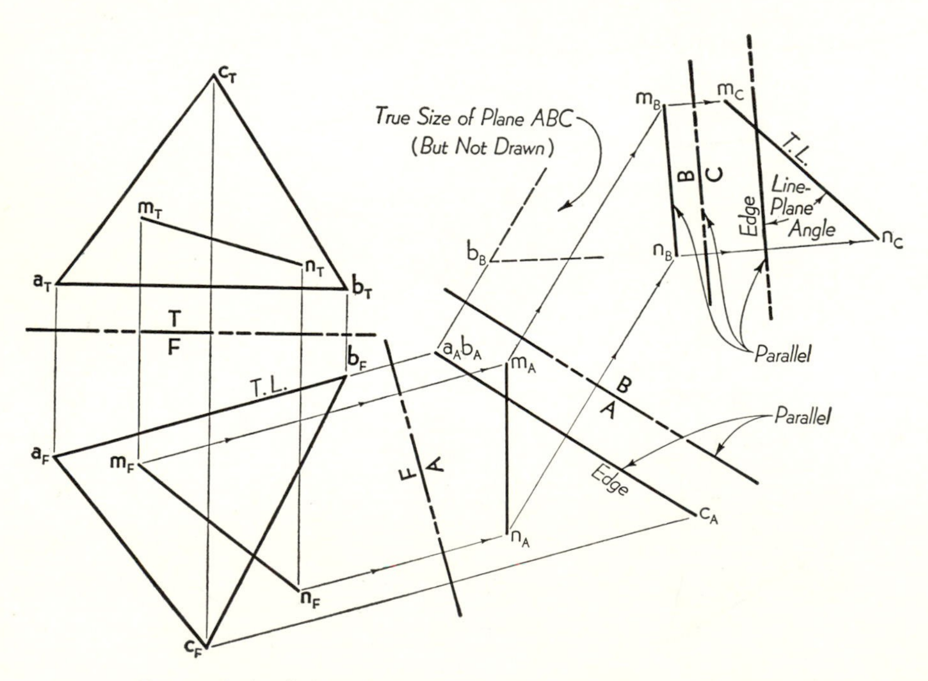

Fig. 4·46. Angle between a line and a plane (edge-view method).

not be completely drawn becomes apparent when view C is drawn. The direction of reference line B–C is determined solely by the line m_Bn_B, to which it has been taken parallel in order that line m_Cn_C shall be true length. The plane ABC must appear as an edge again in view C parallel to B–C, and it may therefore be located by measurement and drawn indefinite in length. The required line-plane angle appears as shown in view C, and it is obvious that the value of the angle does not depend upon the location of individual points in the plane ABC.

Common applications. In the general case illustrated in Fig. 4·46 the edge-view method required three new views. In many common applications, however, the given plane may appear as an edge in one of the given views, thereby reducing the number of new views needed. In Fig. 4·47, for example, an oblique pipe intersects two vertical walls. The angle that the pipe makes with each wall is required. Since the rear wall appears true size in the front view, only the front-adjacent true-length view A is needed to obtain the angle with the rear wall. The side wall appears as an edge in the front view (and in the top view); hence the left-side view is needed to show the wall in true size. View B then shows the pipe in true length, the side wall as an edge, and the required angle with this wall.

Summary. In all cases the method is the same: there must always be given or obtained first an *edge* view, then a *true-size* view, and finally an *edge* view again, which shows the line in *true length.*

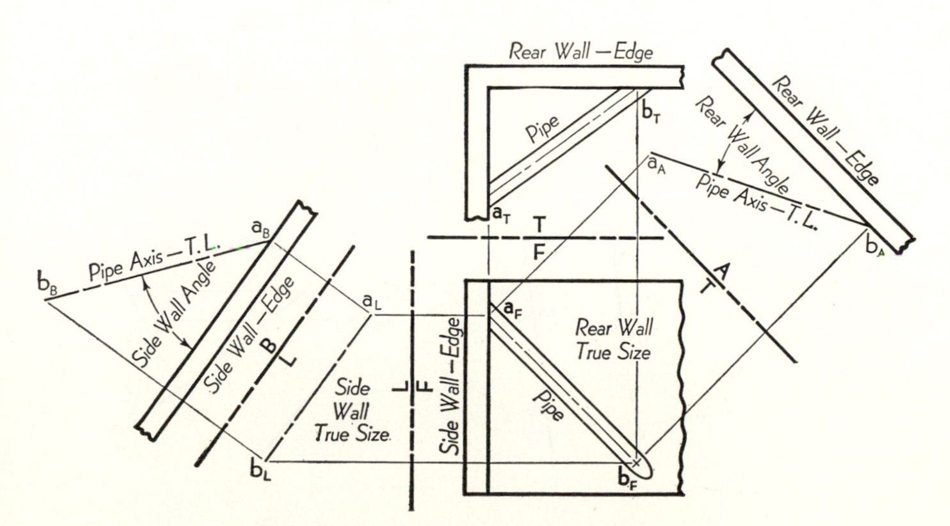

Fig. 4·47. Line-plane angles for vertical planes.

4·41. Angle between a line and a plane (complementary-angle method)

Analysis. If a line makes an angle L with a plane and an angle C with a perpendicular to that plane, then angles L and C are *complementary angles.* This principle is illustrated in the pictorial sketch of Fig. 4·48, where line MN is shown making an angle L with plane $ABCD$. A perpendicular to the plane from N intersects the plane at Q, and PQ is the projection of NP on the plane. Then triangle PQN is a right triangle, and angles L and C are complementary. The angle between a line and a plane may therefore be found by *first* determining the *complementary angle* between the line and a perpendicular to the plane. The solution thus involves two steps:

1. *From any point on the line draw a perpendicular to the plane.*
2. *Construct a true-size view of the plane containing the given line and the perpendicular.*

Problem. In Fig. 4·48 the given plane ABC and the line MN are the same as in Fig. 4·46. The line-plane angle is required.

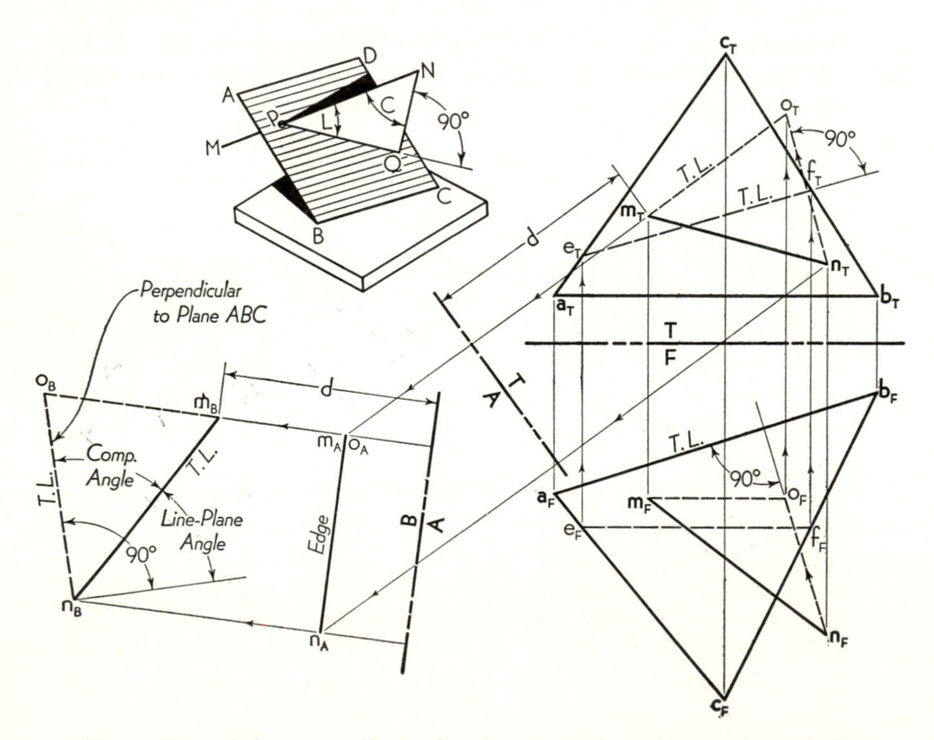

Fig. 4·48. Angle between a line and a plane (complementary-angle method).

Construction. From point N a perpendicular to the plane has been constructed by the two-view method of Art. 4·32. It is *not* necessary to determine the point where this perpendicular intersects the plane; point O may be selected anywhere along the perpendicular. In the plane MNO the angle at N between lines MN and NO is the complement of the desired angle between the given line and plane. It only remains now to show this angle in its true size.

Point O was conveniently selected to make line MO true length in the top view, and view A shows the plane MNO as an edge. In order to keep the true-size view B within reasonable limits on the paper the reference line A–B has been taken behind view A; this does not change the system of alignment or measurement (note distance d) and may be done when necessary but should be avoided if possible (see Art. 2·4). In view B the complementary angle is between the given line m_Bn_B and the perpendicular n_Bo_B as shown. By constructing a 90° angle the complementary angle may be graphically subtracted from 90° to give the required line-plane angle. This method thus yields a solution with only two new views in the general case.

In the complementary-angle method it is important to observe that as soon as the perpendicular to the plane is established the original given plane is ignored and that it is only the plane of the given line and the perpendicular which is shown in the auxiliary views.

PROBLEMS. Group 45.

4·42. Location of a solid on a plane surface

In engineering practice it is frequently necessary to attach brackets, clips, hangers, bearings, etc., to certain inclined or oblique surfaces. In general, this can usually be done most easily in auxiliary views that show the plane surface as an edge and as a true size. Having placed the solid object in the correct position on the plane in the auxiliary views, it may then be drawn in the given views by the usual method of alignment and measurement.

Problem. In Fig. 4·49 the problem is to design and attach to the canted airplane-fuselage bulkhead a bracket of the type shown in the pictorial sketch in stage 1. When riveted in place on the bulkhead surface this bracket serves as an anchorage for the yoke end of a tie rod, which is pinned to the bracket as shown by the dash outline in the picture. The given top and front views (stage 1) show only the given bulkhead surface, the center line CD, which lies on the surface, and the line AB, which represents the center line of the tie rod. The requirements of the problem are threefold: (1) to determine the values of the *bulkhead* and *center-line*

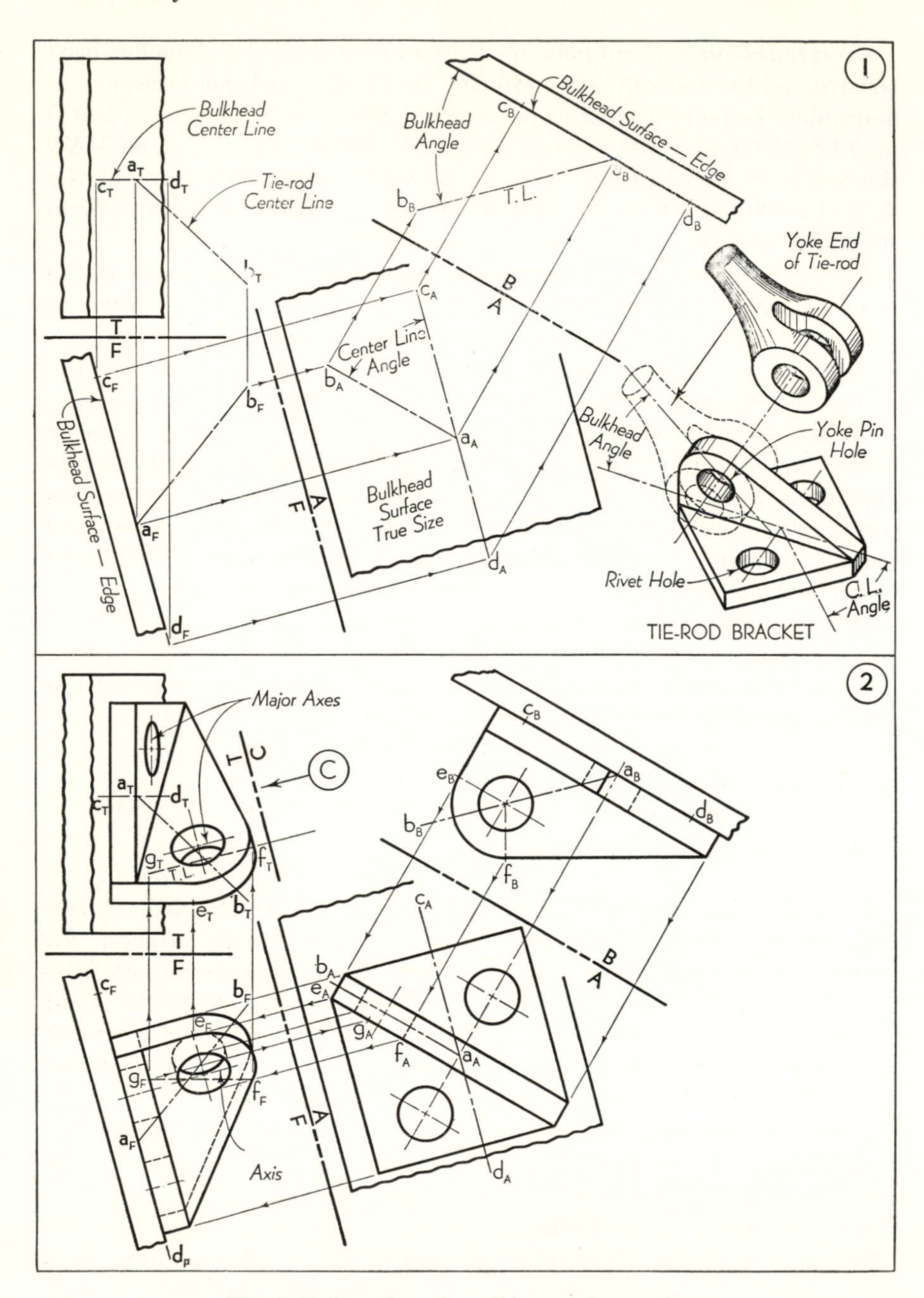

Fig. 4·49. Location of a solid on a plane surface.

angles, (2) to design a simple bracket having these angular measurements, and (3) to show the mounted bracket in the top and front views.

Stage 1. *Establish new views to determine the two angles.*

Since the bulkhead surface already appears as an edge in the front view, the view A is drawn to show this surface in its true size. In this view the base of the bracket will also appear true size, and the center-line angle will lie between lines c_Ad_A and a_Ab_A. View B is drawn next to show the tie-rod center line AB as a true length and the bulkhead surface as an edge again. In this view the angle between the tie-rod center line and the bulkhead surface is shown as the bulkhead angle.

Stage 2. *Design the bracket in the new views and then return all points and lines to the given views.*

In view B the bracket is designed in correct proportion by locating the yoke-pin hole on center line AB and aligning the two rivet holes with point A. The bracket is then drawn in view A to complete its representation. With the design of the bracket completed in views A and B, it is only necessary then to complete the front and top views by alignment and measurement of various points on the object (hidden lines have been omitted in the top view).

The points of tangency, e_B and f_B, of the circular arc in view B should be accurately determined and points E and F then located in the given views as an aid in drawing the elliptical curves. The ellipses that appear in the top and front views can be constructed most easily by determining the major and minor axes as explained in Art. 4·17. In the front view, for example, the major and minor axes of the yoke-pin ellipses are perpendicular and parallel, respectively, to reference line F–A. In the top view the major axis of each yoke-pin ellipse must be parallel to some true-length line, such as FG, in the plane of the circle. For this purpose the horizontal line f_Fg_F is assumed. Point g_A is then located on the edge view of the circles, and point g_T is located by measurement to establish line f_Tg_T. A partial auxiliary view (not shown) in the direction of arrow C will show the plane of the circle as an edge and is useful to determine the length of the minor axis of each yoke-pin ellipse. The rivet-hole circles appear edgewise in the front view; hence the ellipse major axis will be perpendicular to reference line T–F.

PROBLEMS. Group 46.

5. REVOLUTION

5·1. The basic principle of revolution

In the first four chapters of this text it was a cardinal principle that the given object was never moved from its given position (see Art. 2·8). Each view showed the object as it would appear to the observer when viewed from various directions. In other words, the observer *changed his position* or viewpoint to obtain each view, but the object *remained at rest.* In this chapter it will be shown that it is entirely feasible, and frequently convenient, to move the object while the observer remains at rest.

The movement of the object must, of course, be precise and rigidly defined: the motion shall be one of *rotation about a definitely located axis.* The *axis* may be any straight line, specified or assumed, but the position of this axis must always be definitely established before the object can be revolved about it. To attempt to revolve an object before establishing the axis is as absurd as trying to draw an auxiliary view before the direction of sight has been determined.

Let us assume that the point X given in Fig. 5·1 is to be revolved. This statement is meaningless until we specify an axis, and we shall therefore add "about the line AB as an axis." For simplicity we shall assume the line AB vertical and located as shown in Fig. 5·1. The perpendicular distance from point X to the axis AB is the distance R; and if point X revolves about axis AB, it must follow the circular path of radius R. The sketch at the right in Fig. 5·1 shows this circular motion of point X pictorially. Thus the circular path of point X lies in a plane that is *perpendicular* to axis AB, and point X can *never* move in a direction *parallel* to the axis.

In the top view of the multiview drawing the axis AB appears as the point a_Tb_T, and here the circular path appears as a true circle of radius R. In the front view the axis AB appears as the true-length line a_Fb_F, and here the circular path of point X must appear as a line or edge perpendicular to a_Fb_F.[1] We may now revolve the point X through any desired angular displacement and correctly locate its new position in both views. For example, if point X is to be revolved 150° in a clockwise direction as shown in Fig. 5·1, then x_T is first displaced along the circular path to its new position x_T^R. At the same time the corresponding movement of point x_F in the front view is in a direction perpendicular to the axis a_Fb_F (along the edge view of the circular path). The new position of x_F is therefore x_F^R directly below x_T^R. The points x_T^R and x_F^R are the top and front views, respectively, of the new position of point X.

Summary. The procedure outlined above for revolving any given point about a given axis has been greatly facilitated by the fact that the axis appears as a point in one view and as a true length in the adjacent view. If the axis AB had been an inclined or oblique line, then the circular path of point X would have appeared as an ellipse in one or both of the views. To avoid such complication, we shall always perform the revolution process *only* in those views which show the *axis* as a *point* and

[1] In the illustrations in this text the axis will be distinguished by a small circle when appearing as a point and by a heavy long-and-short dash line when appearing as a line.

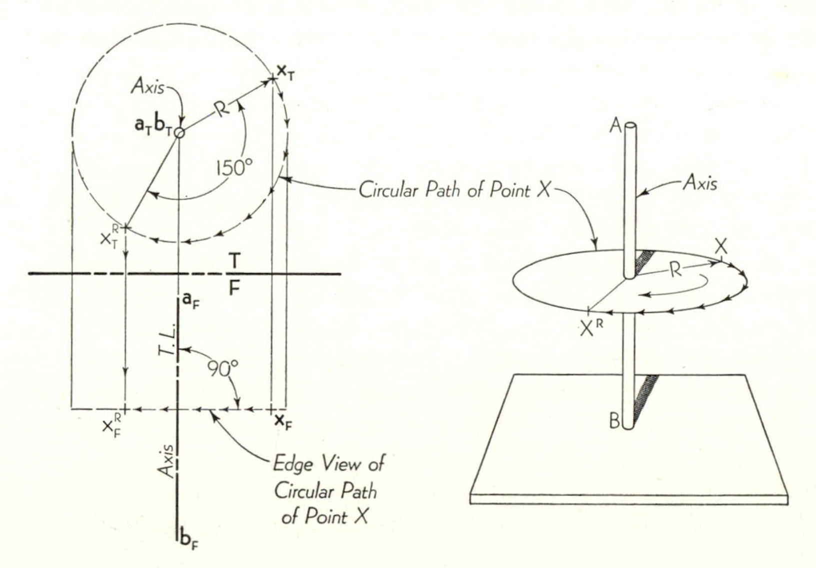

Fig. 5·1. A point revolving about a vertical axis.

as a *true length.* We may therefore state the following rule as the basic principle of revolution:

▪ ***Rule 16. Rule of revolution.*** *The circular path of any point revolving about any axis always appears as a circle in the point view of the axis and as a line perpendicular to the axis in the true-length view of the axis.*

5·2. Revolution of a point about an oblique axis

Analysis. The path of any point revolving about any axis is always a circle with its center on the axis. But the plane of this circle is always perpendicular to the axis, and therefore if the axis appears obliquely, the circle will appear elliptical. The construction of such ellipses is tedious and, fortunately, unnecessary. The revolution process can be performed easily in auxiliary views that:

Show the axis as a true length and as a point.

Problem. In Fig. 5·2 line AB is the given axis. About this axis point X is to be revolved 90° counterclockwise when viewed from the B end of the axis (this would be *clockwise* if viewed from the A end).

Construction. View A has been constructed to show the axis AB as a true length and view B to show the axis as a point. In view B the distance from the axis a_Bb_B to the point x_B is the radius R of the circular

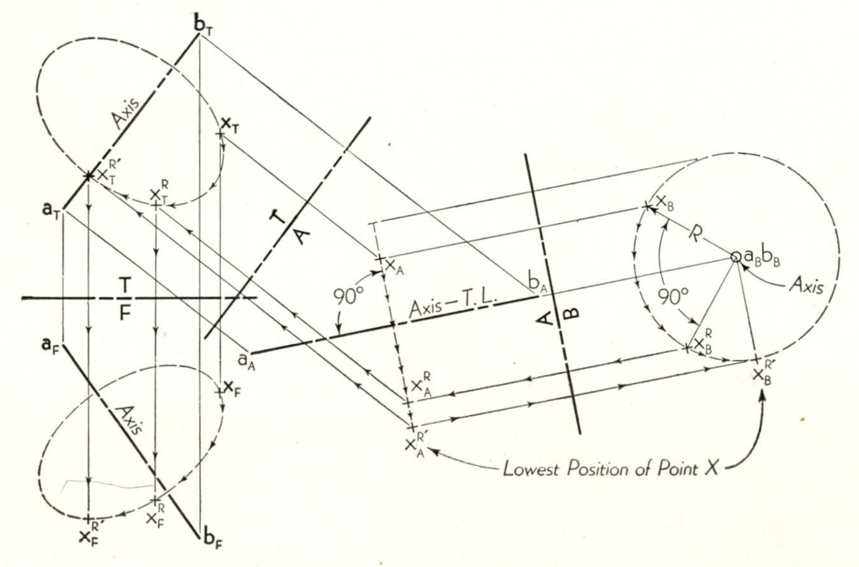

Fig. 5·2. A point revolving about an oblique axis.

path. The circle shown in view B appears as a line in view A perpendicular to the true-length axis $a_A b_A$, and therefore this line is so drawn through point x_A. It should always be kept in mind that this is not just a straight line but a circle appearing edgewise; as such, its length is obviously limited to $2R$, the diameter of the circle.

In the auxiliary views point x_B revolves 90° counterclockwise to the new position x_B^R, while simultaneously point x_A moves perpendicular to the axis to the position x_A^R. Revolved points, like any other points, must also conform to the rules of alignment and similarity, and point x_A^R must lie on the same parallel as x_B^R. With the revolved position of point X now definitely established in views A and B, points x_T^R and x_F^R may readily be located by alignment and measurement. The ellipses in Fig. 5·2 are shown only to emphasize the position of the circular path in these views; they are *not* used to locate x_T^R and x_F^R.

Other specifications. The required revolved position of the given point may be specified in various ways. Assume, for example, that point X in Fig. 5·2 is to be revolved to its lowest position. In this case attention should first be focused on view A rather than view B, for view A is an elevation view. In view A it is apparent that the lowest point on the circular path is at point $x_A^{R'}$. Point $x_B^{R'}$ must lie on the circle in view B directly opposite $x_A^{R'}$, and the required point is thus established in these two views. Alignment and measurement locate $x_T^{R'}$ and $x_F^{R'}$.

PROBLEMS. Group 47.

5·3. Revolution of a line about an oblique axis

Analysis. The revolution of a given line about an oblique axis may be performed by revolving any two points on the line. Each of the two points selected may be revolved as described in Art. 5·2, but each point must be revolved through exactly the *same angular displacement.* It must be remembered that the process of revolution alters the *position* of the object, but it must not change its size or shape. Thus the true length of the revolved line must be exactly the same as that of the given line regardless of the amount of rotation. Failure to observe this equality of angular displacement is a common error in the work of beginners.

Case 1. Degree of revolution specified. Figure 5·3 shows the revolution of line XY through an angle of 90° clockwise (when viewed from B toward A) about the profile axis AB. Since the given axis does not appear as a point or as a true length in the given views, the first step in the solution is to construct auxiliary views A and B to correct this deficiency. The complete process of revolving line XY into its new position may now be performed entirely in views A and B, after which the revolved position can be shown in the given views simply by alignment and measurement.

In view B the extremities of the line, x_B and y_B, move clockwise along the circular paths shown. Point x_B rotates through the required angle of 90° to its new position at x_B^R, and point y_B rotates through the *same* angle to position y_B^R. Line $x_B^R y_B^R$ is the required revolved position of line XY in view B. In view A the circular paths of points X and Y appear edgewise, and therefore points x_A and y_A move perpendicular to the axis $a_A b_A$ to the revolved positions x_A^R and y_A^R. Line $x_A^R y_A^R$ is the required revolved position of line XY in view A. The revolved position is now completely established in views A and B and may be located in the top and front views.

It is useful to note that revolution never alters the *relative position* of points *in view B*. Lines $x_B y_B$ and $x_B^R y_B^R$ differ only in position; they are equal in length. Obviously the same *cannot* be said of view A.

Case 2. Revolved position specified. Figure 5·4 shows the same line XY revolved about axis AB until it appears as a profile line. In this case the number of degrees of rotation required is unknown. But any profile line will appear true length in view A, and in view B it will therefore appear parallel to reference line A–B. Thus line $x_B y_B$ must be revolved until it is parallel to A–B. For this purpose a perpendicular is drawn from the axis to line $x_B y_B$, thus locating point p_B, the point on line XY closest to the axis. In all revolved positions, line $x_B^R y_B^R$ will always

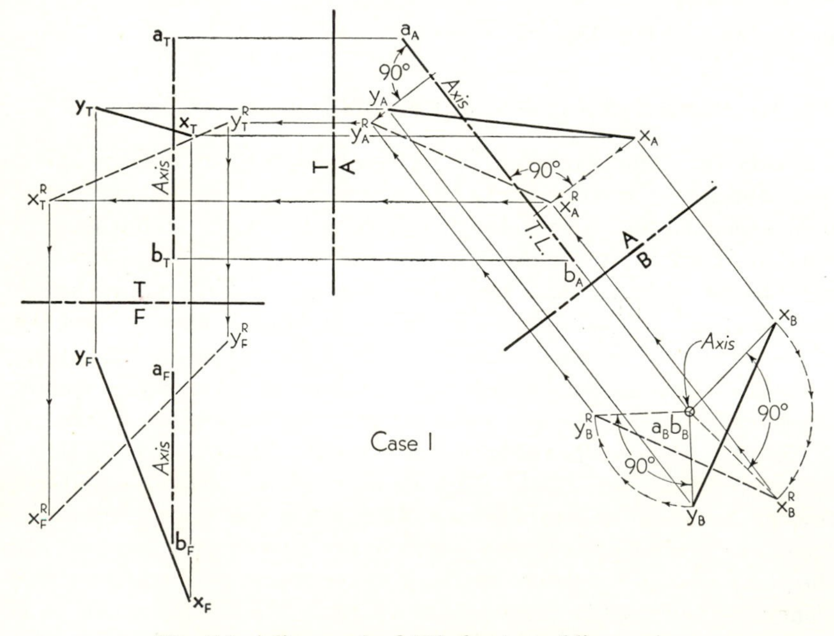

Fig. 5·3. A line revolved 90° about an oblique axis.

be tangent to the circular path of point p_B, and the distance $x_B^R p_B^R$ will always equal $x_B p_B$. Hence, if point p_B is first revolved to position p_B^R, then the revolved position $x_B^R y_B^R$ can be drawn through point p_B^R parallel to reference line A–B. In view A, points x_A^R and y_A^R are located on their respective paths perpendicular to axis $a_A b_A$. The top and front views of the line in this position are as shown.

PROBLEMS. Group 48.

5·4. Assumed axes

In the preceding articles the basic principle of revolution has been illustrated by revolving points and lines about certain definitely specified axes. These axes were given as oblique lines in order to demonstrate the most general case. Although revolution about a specified oblique axis is sometimes required, it happens more frequently in the solution of problems that the axis is not specified but must be assumed by the draftsman in accordance with the requirements of the problem.

In succeeding articles certain typical problems will be solved by the method of revolution; in other words, revolution will be employed to alter the position of the object, and in each case the direction and location of the axis of revolution will be assumed to accomplish the desired change

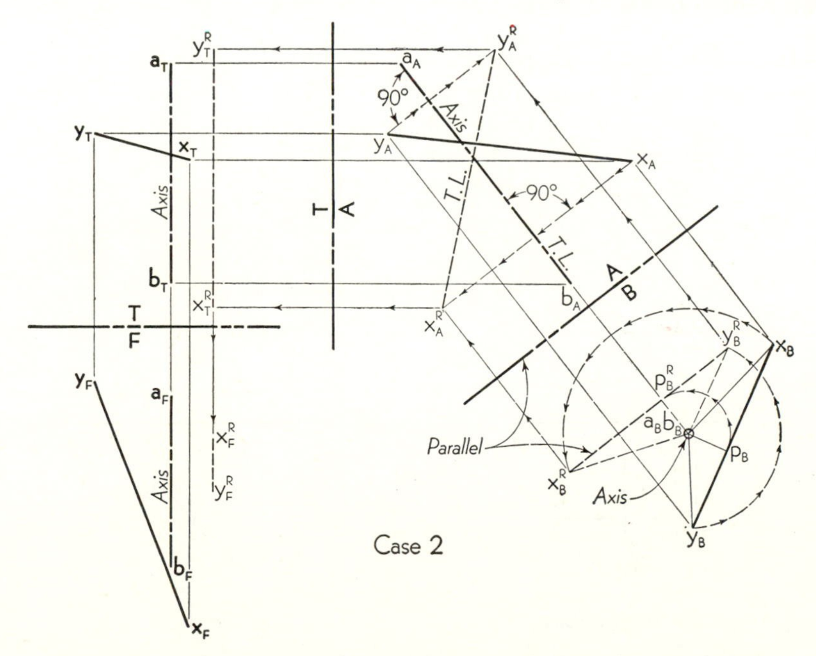

Fig. 5·4. A line revolved about an oblique axis into a profile position.

of position. But, however it may be necessary to assume the axis, the actual process of revolution will always be performed in views that show the axis as a point and as a true length (Rule 16).

Two particular positions of the axis are especially convenient: the *vertical* axis, which appears as a point in the top view and as a true length in the front view, as shown in Fig. 5·5(*a*); the *point-front* axis, which appears as a point in the front view and as a true length in the top view, as in Fig. 5·5(*b*). Revolution about either of these axes requires no additional views. It may be of interest to observe at this point that *any* problem may be solved by successive and alternate revolution about these two types of axes; this is multiple revolution—possible but seldom practical, and not recommended.

In studying the subsequent articles the student should observe that the choice of the axis is the most important element in the solution. Choosing the axis in revolution problems is analogous to choosing the reference lines in auxiliary-view problems; in either case an incorrect choice can yield only an incorrect answer. *How* and *why* the axis is to be selected must always be the primary consideration. To keep this thought foremost we shall always select, draw, and clearly label the assumed axis before beginning the actual revolution phase of the problem.

5·5. True length of a line

Analysis. If a line is parallel to the reference line in one view, it will appear true length in the adjacent view (Rule 6*b*). It has been observed that revolution of a line about *any* axis alters its position but does not change its length (Art. 5·3). In Fig. 5·4, for example, line XY was revolved about axis AB until it assumed a position parallel to the refer-

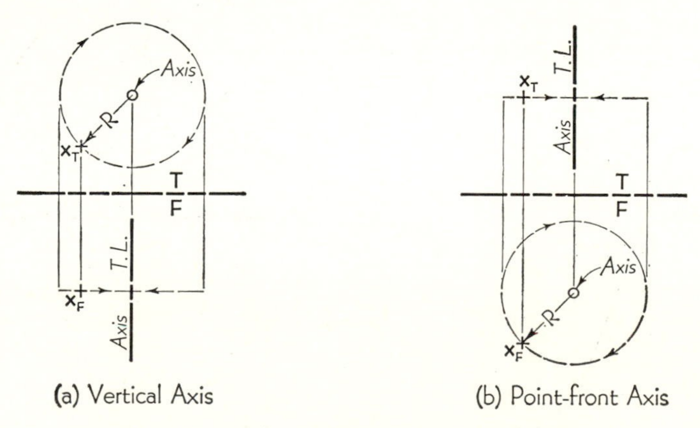

Fig. 5·5. Axes especially useful in revolution.

ence line in view *B*. In the adjacent view *A* line *XY* appeared true length.

By employing an assumed *vertical* or *point-front* axis a line can be revolved in the given views until it is parallel to reference line *T–F*. It will then appear true length in the adjacent view.

Problem. In Fig. 5·6 the oblique line *AB* is to be revolved until it appears true length in one of the given views.

Construction: vertical axis. A vertical axis will appear as a point in the top view and may be located anywhere, but the solution will be simplified if the axis intersects the given line at some point such as *A*. This choice of axis is indicated in Fig. 5·6 by a small circle about point a_T and by drawing a vertical line through point a_F in the front view. The pictorial illustration at the right shows the given line *AB* and the assumed vertical axis.

When the line *AB* revolves about the axis, point *B* must follow the indicated circular path; point *A*, being directly on the axis, conveniently remains fixed. Revolving point b_T to either b_T^R or $b_T^{R'}$ brings the line a_Tb_T into a position parallel to reference line *T–F*. In this position the line *AB* must now appear true length in the adjacent front view. Here the point b_F moves perpendicular to the axis to either of the revolved positions b_F^R or $b_F^{R'}$, and the dash lines $a_Fb_F^R$ or $a_Fb_F^{R'}$ each show line *AB* in its true length. Obviously, either position yields the desired solution.

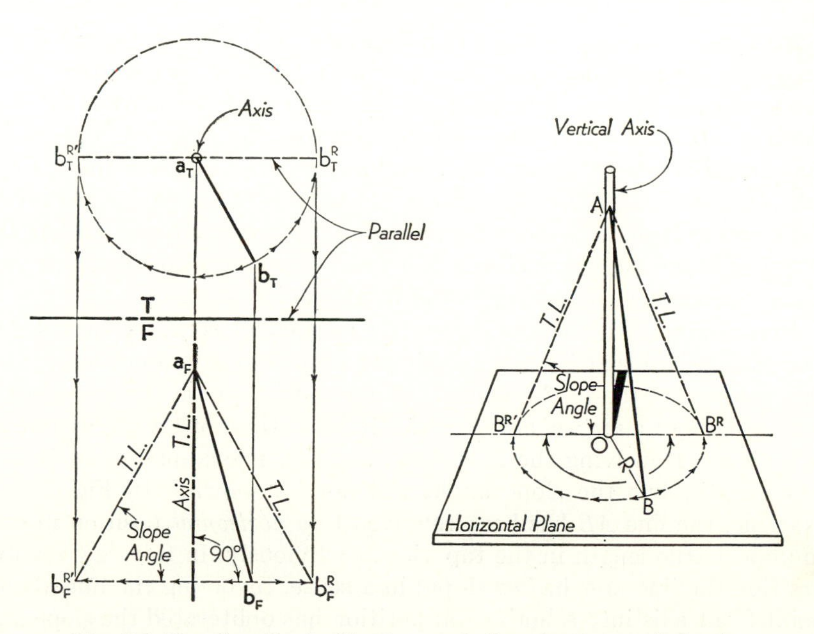

Fig. 5·6. True length of a line by revolution about a vertical axis.

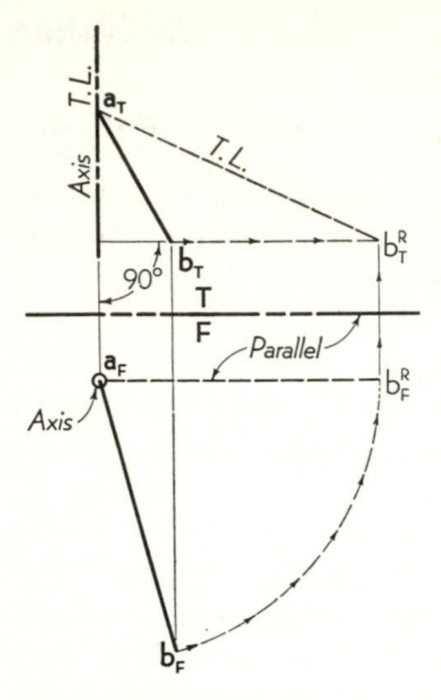

Fig. 5·7. True length of a line by revolution about a point-front axis.

Construction: point-front axis. Figure 5.7 shows the same line AB revolved about a point-front axis. The axis has again been assumed to pass through point A. In this case the path of point B appears as a circle in the front view, and b_F should be moved along this path to the position b_F^R in order to make line $a_F b_F^R$ parallel to the reference line. In the top view the movement of point b_T is, as always, perpendicular to the axis, and the point b_T^R is thus established directly above b_F^R. Line $a_T b_T^R$ must now be the true length of line AB and should agree exactly with the true length found in Fig. 5·6.

For the purpose of finding the true length of a series of lines the method of revolution described above is more efficient than the auxiliary-view method, and we shall find it particularly adapted to development work (Art. 10·11).

5·6. Slope of a line

Analysis. The slope angle of a line may be shown in its true size by revolving the line about a *vertical axis* until the line appears true length in the elevation view. The pictorial drawing in Fig. 5·6 shows that when the line AB is revolved about the vertical axis the right triangle AOB is unchanged in size or shape, and therefore the slope angle of the revolved line $AB^{R'}$ (or line AB^R) is the same as the slope angle of the given line AB.

Construction. In the multiview drawing of Fig. 5·6 line AB has been revolved about the vertical axis to appear true length in the front view. The slope angle also appears in the front view as shown.

Discussion. In the auxiliary-view method of solution (Art. 3·7) it was emphasized that the slope angle can appear *only* in an *elevation* view; this is equally true in the method of revolution where the line must be revolved about a *vertical* axis in order to show the slope angle in an *elevation* view. Revolving the line about any other axis may show it in its true length, but the slope angle will not be shown. In Fig. 5·7, for example, the line AB has been revolved to a *horizontal* position in order to appear true length in the top view. Obviously, in this new revolved position the line now has *no* slope; in a sense, revolving the line about a point-front axis into a horizontal position has obliterated the slope angle we are seeking.

5·7. Line of given bearing, slope, and true length by revolution

This problem was solved by the auxiliary-view method in Art. 3·8. It can now be solved, using only the given views, by revolving the line about a *vertical axis.* The construction is simply the reverse process of that described in the two previous articles.

Problem. The required line AB is the center line of a tunnel which starts at a given point A, has a bearing of N 40° E, is 112 ft long, and slopes downward from A to B with a 35 per cent grade. (This problem is the same as that in Art. 3·8.)

Stage 1. *Lay off the bearing from point A.*

Stage 2. *Lay off per cent grade and true length in the front view.*

From point a_F a line of 35 per cent grade downward is constructed. On this line the 112-ft true length is laid off to locate point b_F^R. Since this line is true length in the front view, it must be parallel to T–F in the top view, and is so drawn to locate b_T^R. Line AB^R is now a revolved position of the required line.

Stage 3. *About a vertical axis revolve the line to the required bearing position.*

Through point A a vertical axis is assumed, and line AB^R is revolved about this axis until it coincides with the bearing line. Thus point b_T^R moves to b_T, and b_F^R moves (perpendicular to the axis) to b_F. Line AB is the required line. (Compare Fig. 3·12.)

PROBLEMS. Group 49.

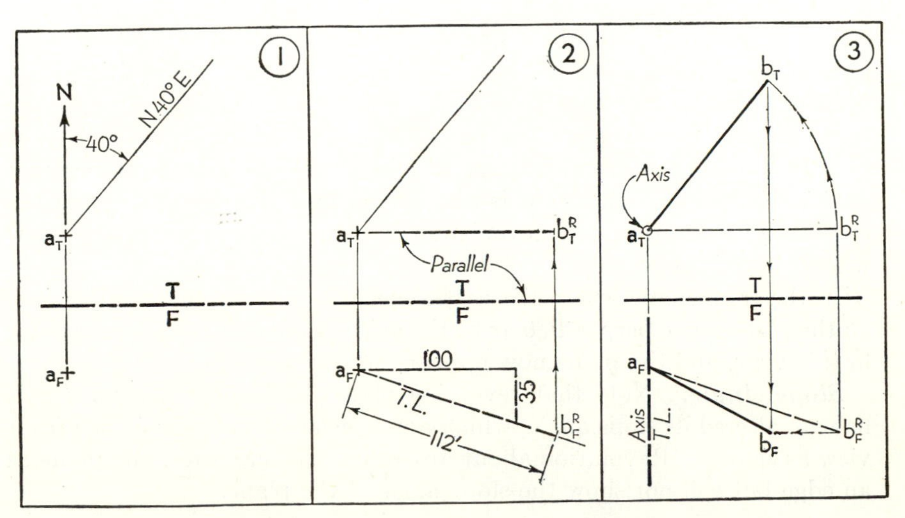

Fig. 5·8. Line of given bearing, slope, and true length by revolution.

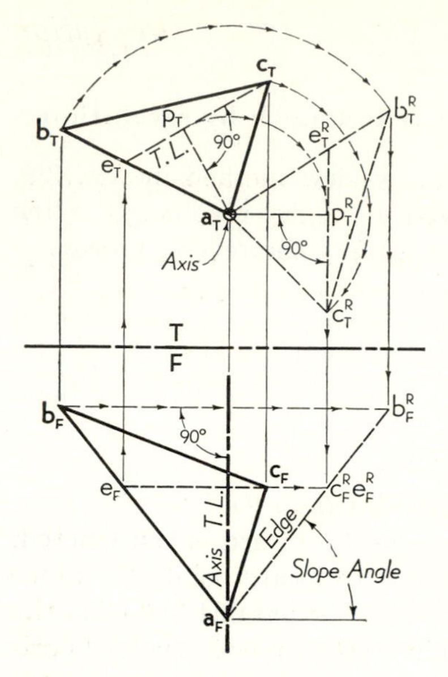

Fig. 5·9. Plane as an edge by revolution about a vertical axis.

5·8. Plane as an edge

Analysis. A plane surface may be revolved about any axis by revolving each of the lines of the plane through the same angular displacement. If any line on the plane appears as a point in one view of the revolved position, the plane will then appear as an edge. But the line on the revolved plane would then be true length and perpendicular to the reference line in the adjacent view. If this line was also true length before the plane was revolved, then the axis must appear as a point in the true-length view. Therefore, to make a plane appear as an edge in a specified view:

1. *Assume a line on the plane that is true length in the adjacent view.*
2. *Assume an axis that appears as a point in the same view.*
3. *Revolve the plane until the true-length line appears as a point.*

Problem. In Fig. 5·9 revolve plane *ABC* until it appears as an edge in the front view.

Construction. Assume the *horizontal* line *EC* in the plane, and select a *vertical* axis through point *A*. Line a_Tp_T is drawn perpendicular to line e_Tc_T to locate point p_T, and the line a_Tp_T is then revolved to position $a_Tp_T^R$, parallel to reference line *T–F*. Line e_Tc_T must now take the position $e_T^Rc_T^R$, perpendicular to reference line *T–F*. Point b_T, revolving through the same angular displacement, moves to b_T^R. The triangle $a_Tb_T^Rc_T^R$ is now exactly the same in size and shape as triangle $a_Tb_Tc_T$ because the *relative position* of points in this view is not altered by any amount of rotation. Obviously, revolution can never show a plane as an edge in the view where the axis appears as a point.

In the front view, where the axis appears as a true length, all points on the plane move perpendicular to the axis. Line *EC* appears as a point in this view, and the plane now appears here as an edge.

Slope Angle. Note that revolving plane *ABC* about a *vertical* axis has not altered its slope angle, which now appears in true size in the front view as shown. Revolution about any other axis may show the plane as an edge but will not show the slope angle of the plane.

PROBLEMS. Group 50.

5·9. True size of a plane

Analysis. A plane may be shown in its true size by revolving it until it is perpendicular to the direction of sight in one of the views. But when the plane appears true size in one view, it must appear as an edge, parallel to the reference line, in all adjacent views. To achieve this result:

Revolve the plane about an axis that lies in the plane and appears true length in the view where the true size is required.

Problem. In Fig. 5·10 revolve plane ABC until it appears true size in the top view.

Construction. The axis of revolution must be true length in the top view; hence the horizontal line EC has been selected as the axis. But the axis must also appear as a point in order to apply Rule 16, and hence view A has been drawn to show axis EC as a point. In this point view of the axis the plane appears as an edge that must be revolved until it is parallel to reference line T–A. Points a_A and b_A follow the circular paths shown to the new positions a_A^R and b_A^R. In the adjacent top view, points a_T and b_T must move perpendicular to the axis to the positions a_T^R and b_T^R. Points E and C both lie on the axis and do not move. The dash-line triangle $a_T^R b_T^R c_T$ shows plane ABC in its true size as required, and it is not necessary to show the revolved plane in the front view.

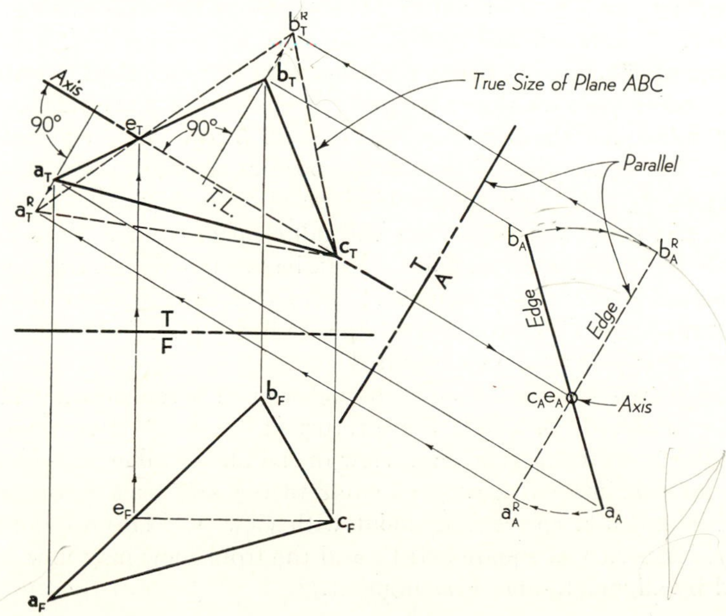

Fig. 5·10. True size of a plane by revolution.

Summary. A plane may be revolved to appear true size in any view if the axis is assumed in the plane to appear true length in that same view. To revolve plane *ABC* until it appears true size in the front view, for example, the axis must be true length in the front view and must appear as a point in a front-adjacent view.

PROBLEMS. Group 51.

5·10. Counterrevolution

Counterrevolution is the process of revolving newly established points and lines back into the original position from a previously revolved position. If, for example, a plane has been revolved to appear in its true size, then, in this new position, angles on the plane may be bisected, shortest distances located, etc. But if these bisectors or shortest distances must be shown in the original position of the plane, then they must be *counterrevolved.* Counterrevolution is therefore simply the reverse of revolution.

Problem. In Fig. 5·11 locate a 1¼-in. square in the plane *ABCD*. The center of the square must be 1⅜ in. from point *A*, 1 in. lower than point *B*, and two sides of the square must be horizontal. (This problem is the same as that in Art. 4·15 and Fig. 4·17.)

Construction. The horizontal line *EF* has been drawn in the plane 1 in. below point b_F, and point *O*, the center of the square, must lie on this line. To show plane *ABCD* true size in the top view, it must be revolved about axis *EF*, hence the edge view *A* is required. But if line *EF* is used as the axis, then the revolved position of plane *ABCD* will be superimposed on the top view (as in Fig. 5·10). Less confusion will result if the axis is taken in a *parallel* position through point *D*. Using this axis the plane can be revolved parallel to reference line *T–A* as shown, and the revolved position will fall clear of the given top view.

In the revolved position, point o_T^R can be located 1⅜ in. from point a_T^R as specified. The 1¼-in. square $r_T^R s_T^R t_T^R v_T^R$ can now be constructed with center at o_T^R and with two sides parallel to line $e_T^R f_T^R$. The revolved position of the square is also located in view *A* as shown. The square must now be counterrevolved back into the original given position of plane *ABCD*. Points *R*, *S*, *T*, and *V* are returned in view *A* along the circular paths shown to the original edge view of the plane, while in the top view these same points move perpendicular to the axis back into the given top view and into correct alignment with view *A*. This establishes the required top view of square *RSTV*, and the front view may now be completed by alignment and measurement.

PROBLEMS. Group 52.

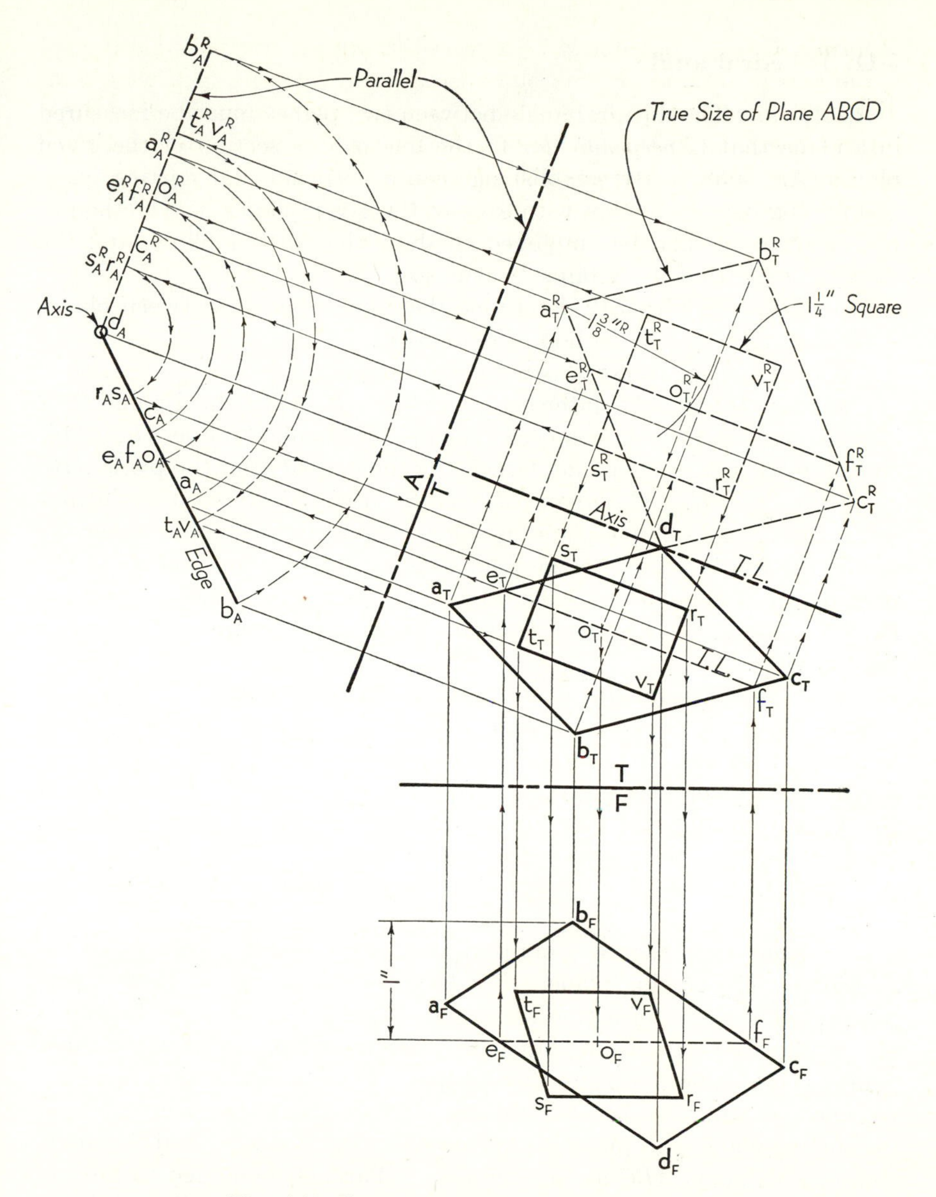

Fig. 5·11. Counterrevolution of a plane figure.

5·11. Dihedral angle

Analysis. The dihedral angle between two planes must be measured in a plane that is *perpendicular* to the line of intersection of the given planes (Art. 4·38). By establishing such a perpendicular cutting plane and finding its intersection with each of the given planes, the method of revolution may then be employed to show the cutting plane, and the dihedral angle that it contains, in true size.

Problem. In Fig. 5·12 determine the dihedral angle between planes *ABD* and *ABC* by the method of revolution.

Stage 1. *The process illustrated pictorially.*

Planes *ABD* and *ABC* intersect along line *AB*. The third plane *DPR* is a cutting plane, which has been taken perpendicular to line *AB*. This cutting plane intersects plane *ABD* along the line *DP* and the plane *ABC* along the line *PR*. The angle *DPR* is the required dihedral angle; but since it lies in an oblique plane, it will not appear true size in either the top or the front view. As the final step, therefore, the plane *DPR* has been revolved about the horizontal axis *DS* into the horizontal position DP^RR^R as shown. The true size of the dihedral angle can now be seen in the top view.

Stage 2. *Show the line of intersection as a true length.*

Since the cutting plane must appear as an edge before it can be revolved to appear true size (Art. 5·9), view *A* has been constructed to show the line of intersection of the given planes as a true length. (View *A* could also be a front-adjacent view, in which case the cutting plane would be revolved to appear true size in the front view.)

Stage 3. *Establish a cutting plane perpendicular to the line of intersection.*

(The front view has been omitted as it is no longer necessary to the solution.) The cutting plane *C-P* is assumed as an *edge* in view *A* perpendicular to the intersection line a_Ab_A. This plane *must* be *perpendicular* to the intersection line, but it may intersect line *AB* at any desired point. To assume the plane through point *D* as shown is simply a convenience, as will soon become apparent. In view *A* the intersection of the cutting plane with plane *ABD* appears as the line d_Ap_A and the intersection with plane *ABC* as the line p_Ar_A. Parallels extended to the top view locate these same lines there. These two intersection lines *DP* and *PR* are the lines that form the required dihedral angle, and they should always be clearly established in the two adjacent views before attempting any revolution.

Stage 4. *Revolve the cutting plane until it appears true size.*

The axis must appear as a point in view *A* and as a true length in the top view. By choosing the axis through point *D* only points *R* and *P*

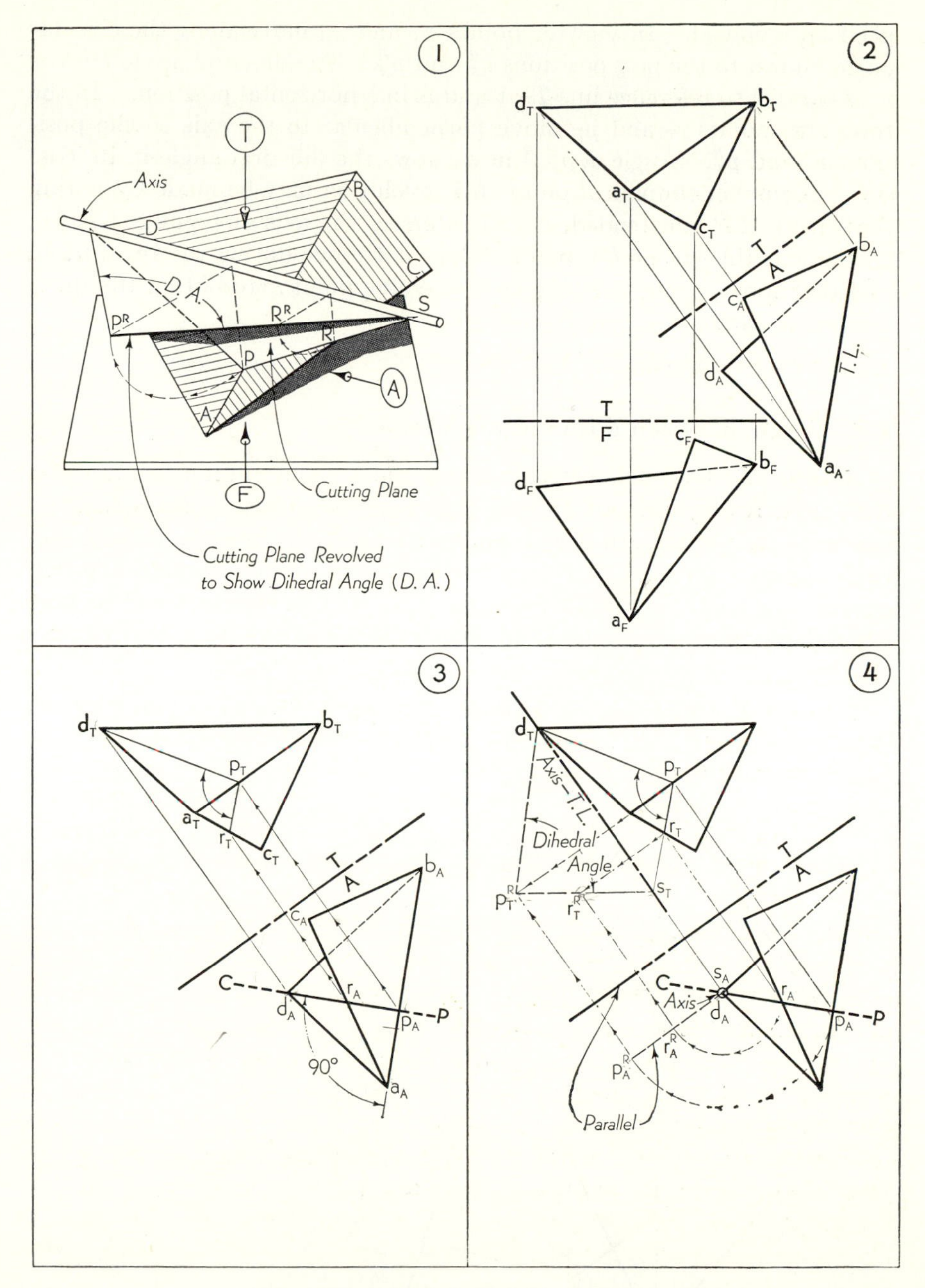

Fig. 5·12. Dihedral angle by revolution.

need be revolved. In view A points r_A and p_A move along the circular paths shown to the new positions r_A^R and p_A^R. The dihedral angle DPR is now parallel to reference line T–A and is in a horizontal position. In the top view, points r_T and p_T move perpendicular to the axis to the positions r_T^R and p_T^R. Angle $d_Tp_T^Rr_T^R$ now shows the dihedral angle in its true size. A further simplification of this revolution may be made by noting that, if line PR is extended, it will intersect the axis at point S. Being on the axis (like point D), point S does not move during the revolution, and therefore it is only necessary to revolve point P to obtain the dihedral angle DPS.

PROBLEMS. Group 53.

5·12. Angle between a line and a plane

Analysis. The angle between a line and a plane will not be altered if the line is revolved about an axis that is *perpendicular to the plane.* If the line is revolved until it appears true length, the line-plane angle will then appear true size. An axis perpendicular to an oblique plane will appear as a point only when the plane appears true size, and will be true length only when the plane is an edge. This method therefore requires that we:

Show the plane as an edge and as a true size. Then revolve the line into true length about an axis perpendicular to the plane.

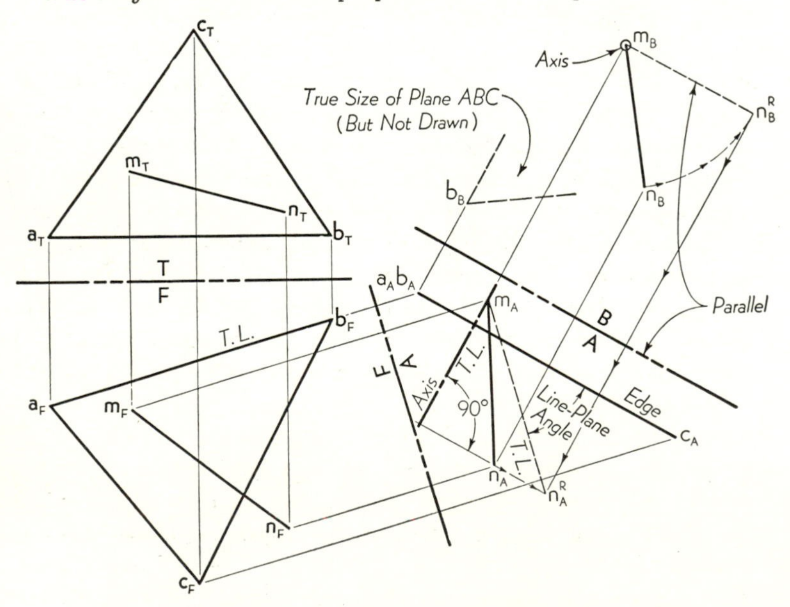

Fig. 5·13. Angle between a line and a plane by revolution.

Problem. In Fig. 5·13 the line *MN* and plane *ABC* are the same as those shown in Fig. 4·46. The line-plane angle is required.

Construction. View *A* has been drawn to show the plane as an edge and view *B* to show it in true size, but note that it is unnecessary to outline the plane in view *B*. The axis of revolution may now be assumed as a point in view *B* and therefore as a true length in view *A*. When line *MN* is revolved about this axis until it appears true length in view *A*, the line-plane angle then appears in view *A* as shown. It should be noted that the *relative* position of the line and plane has been altered; this can be done, without affecting the size of the line-plane angle, only by revolving the line about an axis *perpendicular* to the plane.

Summary. The problem of this article illustrates a fact that has undoubtedly already been noted by the reader: *revolution saves one view.* Comparison of the revolution methods illustrated in Arts. 5·5 to 5·12 with the corresponding auxiliary-view methods will show that in each case the process of revolution is simply a substitute for the final auxiliary view. The closing articles of this chapter will be devoted to the solution of a group of problems that cannot be solved by the exclusive use of auxiliary views but that require methods analogous to revolution.

PROBLEMS. Group 54.

5·13. Cone-locus of a line

The path of a point moving so that it always satisfies a stipulated condition is called the *locus* of the point. For example, the locus of a point moving so that it is always equidistant from two points *A* and *B* is a plane, bisecting and perpendicular to the straight-line segment *AB*. The locus of a point that is equidistant from each of three points *A*, *B*, and *C* (not in a straight line) is a straight line; this line is the common intersection line of three planes that bisect and are perpendicular to the line segments *AB*, *BC*, and *AC*. It may similarly be demonstrated that the locus of a point equidistant from each of four points *A*, *B*, *C*, and *D* (not in the same plane) is a single point. Many more examples could be given to illustrate the concept of locus, but it is the primary purpose of this article to consider the *locus of a straight line that intersects and makes a constant angle with a given line.*

To illustrate this locus assume that the vertical line *AO* in Fig. 5·14 is the given line. Then line *AB* is a line that intersects the given line at point *A* and that makes an angle *OAB* with the given line. When line *AB* is revolved about the vertical line, the angle *OAB* does not change, and therefore the conical surface swept over by line *AB* in one complete revolution is the locus of line *AB* when the angle *OAB* is held constant. We shall refer to such a conical surface as the *cone-locus* of the line *AB*.

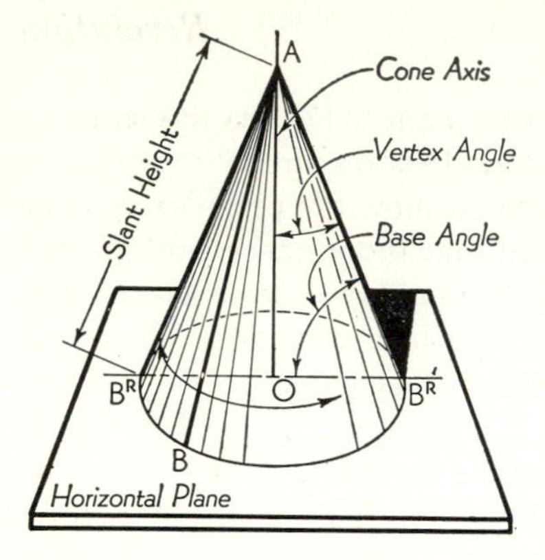

Fig. 5·14. Cone-locus of line AB.

In a multiview drawing the *circular base* of the cone would appear as a circle in the top view and as an edge in the front view.

We shall define the *vertex angle* of the cone as the true angle between line AB and the *cone axis.* This angle appears in true size only when line AB appears true length, as it does in either of the revolved positions shown. The true length of line AB is the *slant height* of the cone. A line drawn from the vertex to *any* point on the circular base will represent *one* possible position of line AB, and therefore the cone represents *all* possible positions of line AB when the angle between AB and the axis is constant. This cone is also the locus of a line passing through point A and making a constant *base angle* (the slope angle) with the horizontal plane.

When, in subsequent problems, it is required to locate a line that makes specified angles with certain other lines or planes, we shall find the solution by constructing *two* cone-loci of the required line.

5·14. Location of a line making given angles with each of two intersecting lines (special case)

Problem. In the multiview drawing of Fig. 5·15 two intersecting lines AO and BO are shown. Line AO appears as a point in the top view, and both the given lines AO and BO appear true length in the front view. The true angle between these lines is 35° as shown in the front view. Thus the two given lines have been assumed in a special position to simplify the solution; the general case will be discussed in the next article. It is required to locate a line (or lines) that shall pass through point O and make an angle of 20° with line AO and an angle of 25° with line BO.

Analysis. The locus of a line passing through point O and making an angle of 20° with line AO will be a right circular cone A with vertex at O and axis along the line AO. Similarly, the locus of a line making an angle of 25° with line BO will be a second cone B with axis along BO and vertex also at O. These two cones are shown pictorially in Fig. 5·15. Both cones have a common vertex O, and therefore the intersection of the two conical surfaces (if they do intersect) will be two straight lines such as XO and YO. But points X and Y must both lie on the circular bases of each cone. In other words, the circle of radius r, which forms the base of cone A, must actually intersect the circle of radius r', which forms the base of cone B. This is possible only if the slant height of cone A is *exactly the same* as the slant height of cone B. But if the slant heights R of the two cones are identical, then every point on the two circular bases is the same distance R from point O, and a sphere of radius R with center at O will exactly contain the two cones so that the two circular bases will lie on the surface of the sphere. In summary:

The two cone-loci must always have: (1) *a common vertex and* (2) *the same slant height.*

Construction. In the multiview drawing of Fig. 5·15 the given lines AO and BO will be the axes of the two cones. Since both axes appear true length in the front view, the bases of both cones will appear here as edges. To show the two cones in this view we shall first draw a circle with center at O and of any convenient radius R. The 20° vertex angle of cone A is constructed on both sides of line $a_F o_F$, and the sides of the cone are extended to the circle to establish the base of cone A. Cone B is drawn in the same way with line $b_F o_F$ as axis and with a vertex angle of 25°. The two intersections of the two cone bases now appear in the front view as the points x_F and y_F. In the top view the base of cone A will appear as a circle and is so drawn. The base of cone B would appear elliptical in this view, but fortunately it need not be shown since points x_T and y_T can be located on the circular base of cone A. The required line is therefore either line XO or line YO; either line may be extended, of course, as necessary or desired.

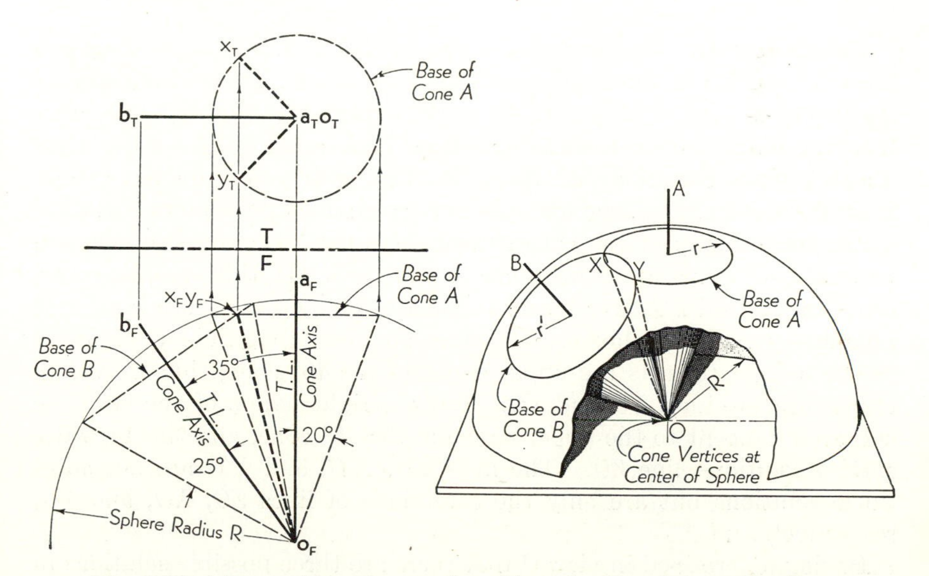

Fig. 5·15. A line making given angles with each of two intersecting lines.

Possible solutions. In the particular case shown in Fig. 5·15 there are *two* possible solutions. If, however, the sum of the vertex angles is less than the angle between the given lines, then the cones will not intersect and there will be *no* solution. If the sum of the vertex angles exactly equals the angle between the given lines, then the cones will be tangent and there will be only *one* solution. There may also be *three* or *four* solutions—as will be shown in Fig. 5·16—if the vertex angles are sufficiently large and each cone is extended to form a second nappe.

5·15. Location of a line making given angles with each of two intersecting lines (general case)

Problem. In Fig. 5·16 the two intersecting lines *AO* and *BO* are given in the top and front views. It is required to locate a line (or lines) that shall pass through point *O* and make an angle of 45° with line *AO* and an angle of 60° with line *BO*.

Analysis. This is basically the same problem as that discussed in the previous article except that the given lines *AO* and *BO* do not appear in the simplified position shown in Fig. 5·15. But *all problems* involving cone-loci can be reduced to the special case by constructing auxiliary views that will:

Show one cone axis as a point and in an adjacent view show both cone axes as true lengths.

Construction. The first step in the construction is to show the given lines *AO* and *BO* in the auxiliary views *A*, *B*, and *C*. Selecting either of the two lines—*AO* in this case—view *A* is drawn to show this line in its true length and view *B* to show the same line as a point. View *C* is then drawn to show the other line, *BO*, as a true length. We now have views *B* and *C*, which show the given lines in the simplified position of Fig. 5·15.

The second step of the construction is to establish the two cones in views *B* and *C* exactly as was done in Fig. 5·15. In this case, however, the lines *AO* and *BO* have been extended beyond *O*, and both *nappes* of each cone have been shown. It is now apparent that the left nappe of the 60° cone intersects the upper nappe of the 45° cone, thus providing two solutions—lines *XO* and *YO*. But the right nappe of the 60° cone is exactly tangent to the upper nappe of the 45° cone, and this provides a third solution—line *ZO*. The lines *Z'O*, *X'O*, and *Y'O* are not additional solutions but are only the extensions of lines *ZO*, *XO*, and *YO*, respectively.

Having determined in view *C* that there are three possible solutions in this case, each of the points *X*, *Y*, and *Z* may now be located in view *B* on the circular base of the 45° cone at the points x_B, y_B, and z_B. The three possible solutions, lines *XO*, *YO*, and *ZO*, are now completely established in views *B* and *C* and may now be located in the given views, where they are shown by the dash lines.

Also shown in Fig. 5·16, at the left of the original view *C*, is the same view redrawn to illustrate the case of four solutions. Here the vertex

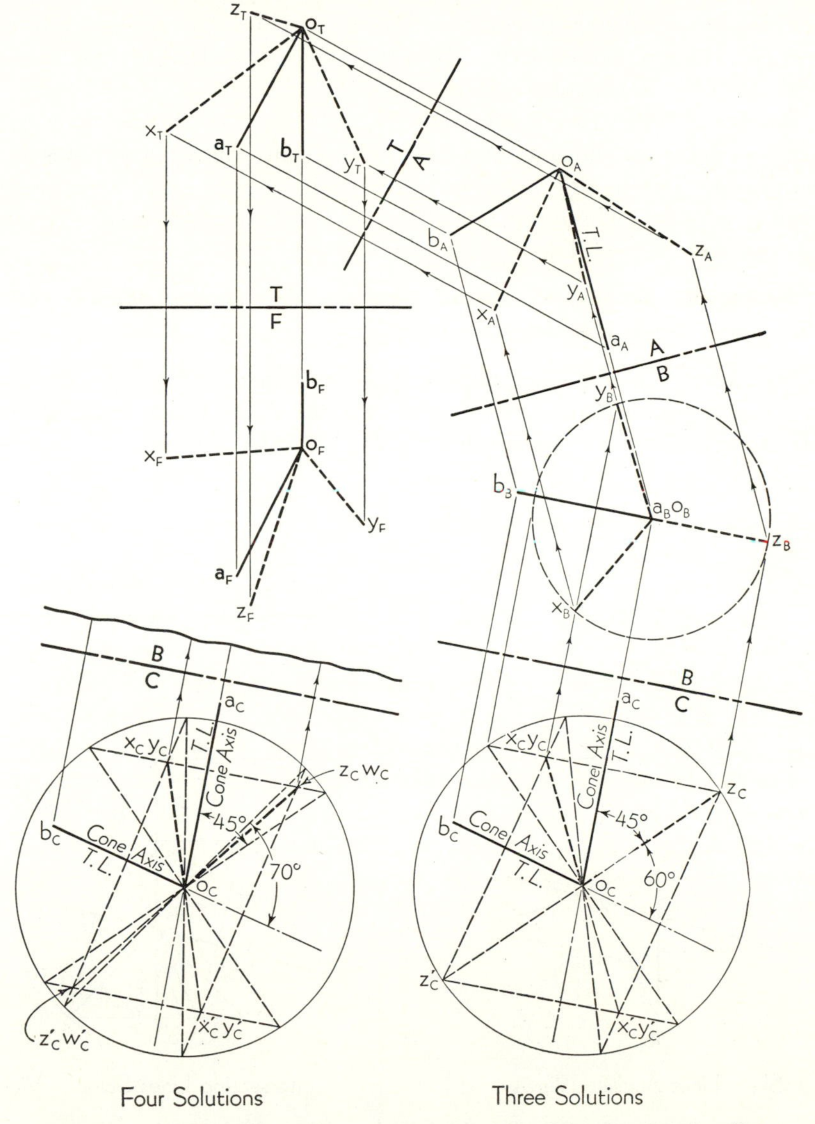

Fig. 5·16. A line making given angles with each of two intersecting lines.

angle of the 60° cone has been increased to 70°, and any one of the four lines *XO*, *YO*, *ZO*, or *WO* becomes a solution for this case.

PROBLEMS. Group 55.

5·16. Location of a line making given angles with each of two skew lines

This problem occurs in construction work when it is necessary to connect two skew pipelines with a third pipe using standard fittings that are available only in certain stock angles.

Analysis. The two given lines do not intersect; hence there is no common point for the cone vertices. But a solution for skew lines and a solution for intersecting lines that are parallel to the skew lines produce *parallel* results. Therefore, for skew lines the *direction* of the required line can be obtained if we:

First solve the problem of two intersecting lines each of which is parallel to the given skew lines.

Problem. In Fig. 5·17 the given skew lines are *AB* and *CD*. It is required to connect these two lines with a third line which shall intersect *AB* at an angle of 45° and intersect *CD* at an angle of 60°.

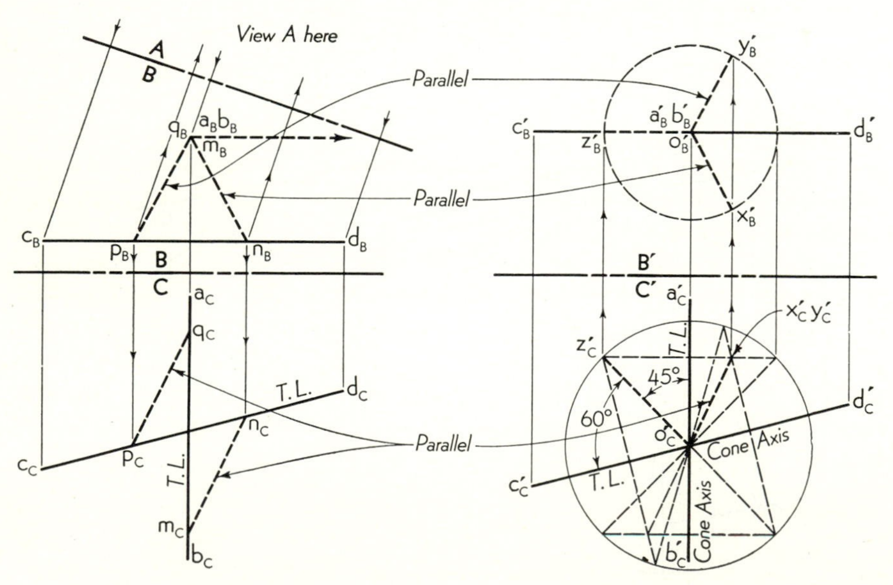

Fig. 5·17. A line making given angles with each of two skew lines.

Construction. Regardless of how lines AB and CD may be presented in the given views, the first step in the solution should always be the construction of auxiliary views that show the lines in *simplified position.* In general, three auxiliary views will be required, and the last two will show the given lines as in views B and C of Fig. 5·17. View A (not shown) was drawn to show line AB as a true length; view B shows line AB as a point; reference line B–C has been taken parallel to c_Bd_B to show both lines as true lengths in view C. Since the actual solution is performed entirely in views B and C, only these views have been shown in Fig. 5·17.

The second step of the construction may now be performed in any convenient area on the paper in the general vicinity of views B and C. In this area (at the right in Fig. 5·17) a new reference line B'–C' is drawn parallel to the original B–C line, and a point O' is assumed anywhere in views B' and C'. Through point O', lines $A'B'$ and $C'D'$ are drawn parallel to the given lines as they appear in views B and C. We now have a pair of intersecting lines, each of which is parallel to one of the given lines. In views B' and C' the two cone-loci may now be established with point O' as the common vertex point. This construction is identical with that shown in views B and C of Fig. 5·16.

It is now evident that three lines, $X'O'$, $Y'O'$, and $Z'O'$, make the required angles with the intersecting lines $A'B'$ and $C'D'$. But the required connecting line (or lines) in views B and C must have the *same direction* as the lines $X'O'$, $Y'O'$, and $Z'O'$ in views B' and C', and therefore it remains only to locate these lines in views B and C *parallel* to the corresponding lines in views B' and C'. The line MN, for example, has been drawn parallel to line $X'O'$. It should be noted here that line m_Bn_B must be drawn *first* through point a_Bb_B in order to locate point n_B on line c_Bd_B; point n_C can then be located on line c_Cd_C, and line m_Cn_C drawn parallel to $x'_Co'_C$ to locate point m_C.

Line MN is therefore one possible solution in this problem. Line PQ is a second solution obtained by drawing PQ parallel to $Y'O'$. But a third solution is impossible because a line through AB, parallel to $Z'O'$, will never meet line CD. There are, therefore, in this case, three solutions for the intersecting lines, but only two that are practical for the skew lines. In general, for any pair of skew lines there may be *no* solution, *two* solutions, or *four* solutions.

PROBLEMS. Group 56.

5·17. Location of line making given angles with each of two planes

Analysis. If a line passing through a given point always makes a constant angle with a plane, then the locus of the line is a conical surface whose axis is perpendicular to the plane. The constant angle between line and plane is called the *base angle* (see Fig. 5·14) and is the complement of the *vertex angle.* If the required line must pass through a given point O, then two cone-loci must be established with a common vertex at O, and the axes of the cones must be perpendicular to the given planes. To see these cone-loci in simplified position—one cone axis as a *point* and both cone axes in *true length*—we must:

> *Show one plane in true size, and in an adjacent view show both planes as edges.*

Problem. In Fig. 5·18 it is required to locate a line (or lines) that shall pass through the given point O, make an angle of 60° with plane $ABCD$, and have a slope angle of 45°.

Construction. The horizontal plane appears true size in the top view, and view A has been drawn to show both planes as edges. In view A the axis of one cone is drawn perpendicular to the horizontal plane through point O and the axis of the other cone perpendicular to plane $ABCD$. The sphere of convenient radius is then drawn, and the two cones are established within the sphere. It should be noted that in this problem the given angles of 45° and 60° become the *base angles* of

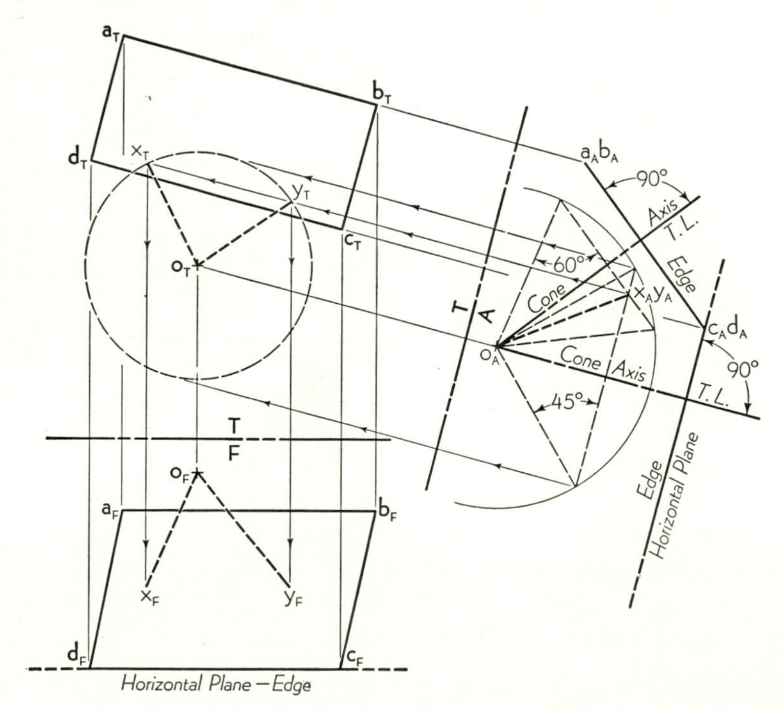

Fig. 5·18. A line making given angles with each of two planes.

the cones. In the top view the base of the 45° cone appears as a circle. Two solutions are now evident—the lines OX and OY. The second nappe of each cone has not been drawn here, since it should be apparent that this would yield no additional solutions. In general, however, problems of this type may have as many as four solutions.

General case. If both the given planes had been oblique it would then have been necessary to draw three additional views to obtain the same type of views employed in Fig. 5·18. Figure 4·45 shows such a sequence of views.

5·18. Location of a line making given angles with each of two planes (special case)

In this type of problem it frequently happens that the given planes are *perpendicular*, and one (or both) of the planes appears *true size* in the given views. In this special case the solution can be obtained without additional views.

Analysis. In the given true-size view one cone axis will appear as a point. Because the planes are perpendicular, the second cone axis will

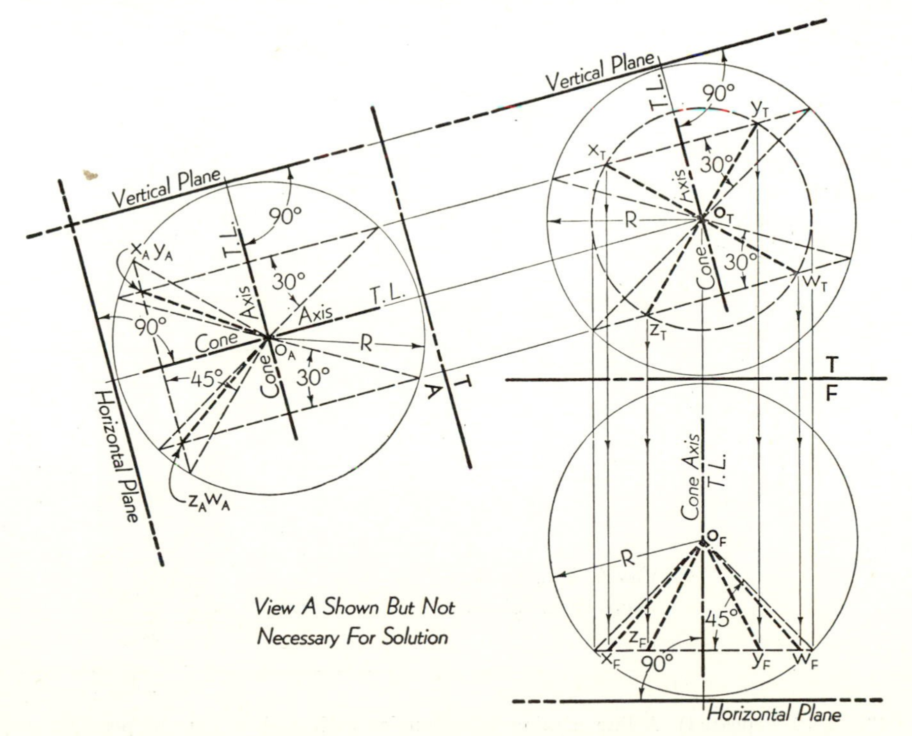

Fig. 5·19. A line making given angles with each of two perpendicular planes.

appear true length in this same view, and its circular base will appear as an edge. The intersection of the two cone bases will thus be apparent in this given view.

Problem. In Fig. 5·19 it is required to locate a line (or lines) that shall pass through point O and make angles of 45° and 30°, respectively, with the given horizontal and vertical planes.

Construction. By constructing view A this problem can be solved by the general method of Art. 5·17. But consider now only the given top and front views. The axis of the 45° cone will appear true length and vertical in the front view, and we may therefore assume a sphere of any radius R in that view and establish within it a cone having a base angle of 45°. The base of this cone will appear in the top view as a circle as shown. Also in the top view the axis of the 30° cone will appear true length and perpendicular to the given vertical plane. By drawing the *same* sphere of radius R in this view the cone with 30° base angles may be constructed; both nappes of this cone have been drawn. Both the 45° and 30° cones are now established within the same sphere, and points x_T, y_T, z_T, and w_T are the points of intersection of the cone bases.

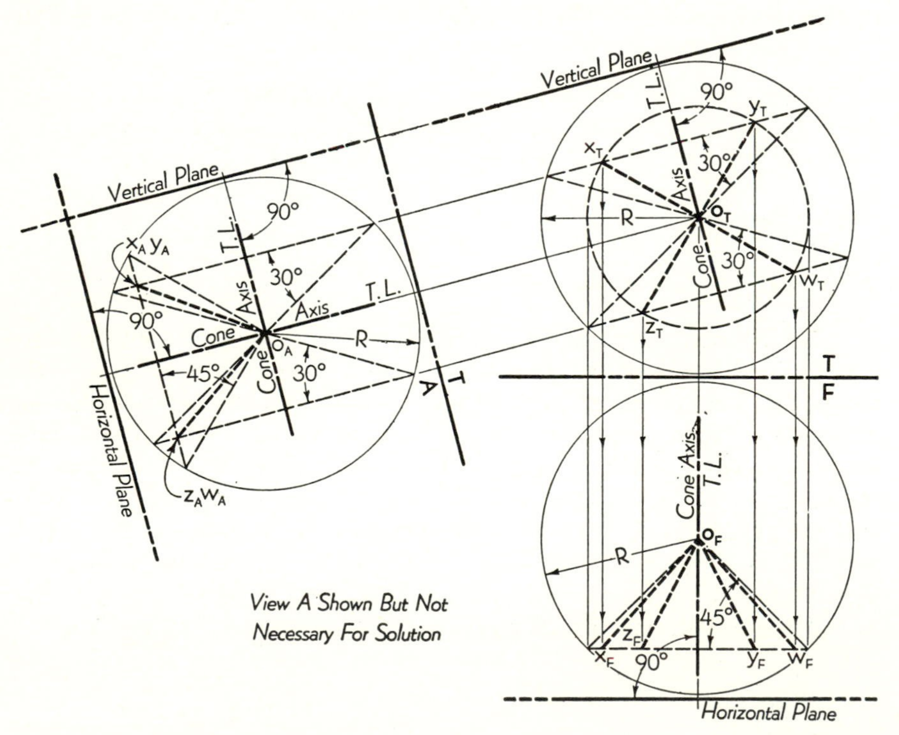

Fig. 5·19 (repeated). A line making given angles with each of two perpendicular planes.

These same points may then be located in the front view on the base circle of the 45° cone. Thus there are four solutions as shown.

Summary. If the *axes* of the two cone-loci are *perpendicular*, then the solution may be made in any pair of adjacent views if either view shows *one* of the axes as a *point* and the other as a *true length.* In other words, when the axes of the cone-loci are perpendicular, it is *not necessary* to obtain the view that shows *both axes* as *true lengths.*

PROBLEMS. Group 57.

5·19. Location of a line making given angles with a line and a plane

Analysis. The desired point of intersection of the given line and required line will be the common vertex point O. One of the cones must have as its axis the given line; the axis of the second cone must be perpendicular to the given plane. Then to see these axes in the simplified position:

> *Show the plane in true size, and in an adjacent view show the line in true length.*

Construction. This problem is so similar to those previously described that it has not been illustrated, but Fig. 4·46 is an exact illustration of the kind of auxiliary views discussed above.

PROBLEMS. Group 58.

5·20. Other locus problems

In the preceding articles the discussion has been limited to the location of a line making specified angles with other lines or planes. It has been shown that such problems may be solved by means of cone-loci. But, in general, the locus, or path, of a line, moving so that it always satisfies some stipulated condition, may be any one of a great variety of surfaces —plane, cylindrical, conical, warped, etc. The location of a plane may depend upon its tangency to a curved surface, and the locus of a point may frequently be a curved surface. Therefore, any further discussion of problems involving the consideration of locus must be reserved for subsequent chapters; we must first learn more about curved surfaces.

6.

SINGLE-CURVED SURFACES

6·1. The classification of surfaces

All regular surfaces may be divided into two major classes and further subdivided as follows:

1. ***Ruled surfaces.***
   a. Polyhedrons.
   b. Single-curved surfaces.
   c. Warped surfaces.
2. ***Double-curved surfaces.***
   a. Surfaces of revolution.
   b. Surfaces of evolution.

The above classification is based upon the fact that all these surfaces may be generated by the motion of a straight or curved line. In other words, a surface is the path of a moving line just as a line is the path of a moving point. Although surfaces form the boundary of solids, we may also consider the surface itself as a tangible thing; chutes, hoppers, and similar hollow objects whose walls are of negligible thickness are, for all practical purposes, purely surfaces.

Ruled surfaces are those which may be generated by moving a straight line.

Polyhedrons (many sides) are those objects whose surface is composed entirely of plane surfaces (see Arts. 4·28 to 4·31).

Single-curved surfaces are those which may be generated by moving a straight line in contact with a curved line so that any two consecutive positions of the generating line are either parallel or intersecting. The cone, cylinder, and convolute are the only examples of this class of surface.

Warped surfaces are those which may be generated by moving a straight line so that any two consecutive positions of the generating line are skew lines.

Double-curved surfaces are those which can be generated only by moving a curved line.

Surfaces of revolution are those which may be generated by revolving a line, straight or curved, about an axis. This special classification may include single-curved, warped, or double-curved surfaces.

Surfaces of evolution are those which can be generated only by moving a curved line, of constant or variable shape, along a noncircular curved path.

In engineering construction the single-curved surfaces occurring most frequently are the cylinder and the cone. We shall therefore devote considerable space to these particular surfaces. In the articles which follow it should be observed that any method of solution which is valid for the cylinder may also be applied to the prism and that the pyramid may be considered to be only a variant of the cone. Indeed, with a little imagination the cylinder can be considered as a cone whose vertex is at infinity; therefore all cone solutions are, with slight modifications, applicable to cylinders also.

6·2. Curved lines

A curved line is the path of a point that moves in a constantly changing direction. If the point moves so that it always lies in the same plane, then the resulting path is a *plane curve,* or *single-curved line.* If the various positions of the moving point do not lie entirely in one plane, the resulting path is called a *space curve,* or *double-curved line.*

Plane curves may be infinite in variety, but only a few are commonly used in engineering. Of primary importance are the *conic sections,* which include the *circle, ellipse, parabola,* and *hyperbola.*[1] These curves are discussed more fully in Art. 6·8. Other plane curves of practical importance are the *roulettes,* which include the *cycloid, epicycloid,* and *hypocycloid* (used for the tooth profile on cycloidal gears); the *spirals,* of which the *involute* is the most common (for tooth profiles on involute gears); and the *trigonometric curves,* especially the *sinusoid,* which occurs in periodic-motion studies.

Double-curved lines may also be infinite in variety, but the helix in its several forms is by far the most common (see Art. 6·27). The line of intersection of two curved surfaces is usually of double curvature.

[1] See Appendix for properties and methods of construction of conic sections.

6·3. Single-curved surfaces

Single-curved surfaces may be generated by a straight line moving so that it is always in contact with some curved line. The moving line that generates a surface is called the *generatrix*, or generating line. The curve that directs the motion of the generating line is called the *directrix*, or directing curve. But to obtain a single-curved surface the straight-line generatrix must also move so that any two consecutive positions of this line will lie in the same plane; otherwise, the surface will be warped. Single-curved surfaces are thus restricted to only three types, the *cone*, the *cylinder*, and the *convolute*.

Cones. A cone is generated by a straight line moving so that it is always in contact with a curved line and always *passes through a fixed point* (not in the plane of the curve) called the *vertex*. In Fig. 6·1(*a*) the directrix is the curved line *AB*, and the fixed point *V* is the cone vertex. The conical surface has been generated by moving the straight line *XY* to each of the consecutive positions shown. Each position that the generatrix assumes is called an *element* of the surface. The generatrix may extend indefinitely on both sides of the vertex, thus forming a cone of *two nappes;* line segment *VX* generates the *lower* nappe, and line segment *VY* generates the *upper* nappe. When the elements of the cone are terminated by a plane, horizontal (as shown) or oblique, the resulting curve of intersection is called the *base* of the cone.

Cylinders. A cylinder is generated by a straight line moving so that it is always in contact with a curved line and is always *parallel to its initial position*. In Fig. 6·1(*b*) the directrix is the curved line *AB*, and the cylindrical surface has been generated by moving the straight line *XY* from the initial position to each of the consecutive positions shown. Each position of line *XY* is an *element* of the cylindrical surface. The generatrix may be indefinite in length, but the elements are usually terminated by two planes, horizontal (as shown) or oblique, and the resulting curves are called the *bases* of the cylinder.

The directrix for the cone or cylinder may be either single-curved or double-curved. It may also assume any irregular shape as shown in Fig. 6·1; but ordinarily the base, which usually serves as the directrix, is a closed curve of regular shape, that is, circular or elliptical. In subsequent articles we shall deal primarily with cones and cylinders of the ordinary variety that commonly occur in practice, but we should also be able to recognize cones and cylinders when their aspect is less familiar.

Convolutes: tangent-line generation. A convolute may be generated by a straight line moving so that it is always *tangent to a double-curved line*. In Fig. 6·1(*c*) the directrix is the double-curved line *AB*. The surface shown has been generated by moving the straight line *XY* along the curve AB to each of the consecutive positions shown, keeping line XY always tangent to the curve at point X. Each position of the line XY is thus an element of the surface. The length of the generatrix XY may be fixed or variable; it may also extend on both sides of the point of tangency X, in which case the convolute formed is one of *two nappes.* The convolute shown in Fig. 6·1(*c*) has only one nappe.

The convolute may appear to be a warped surface, but it is actually single-curved, for any two consecutive elements taken very close together

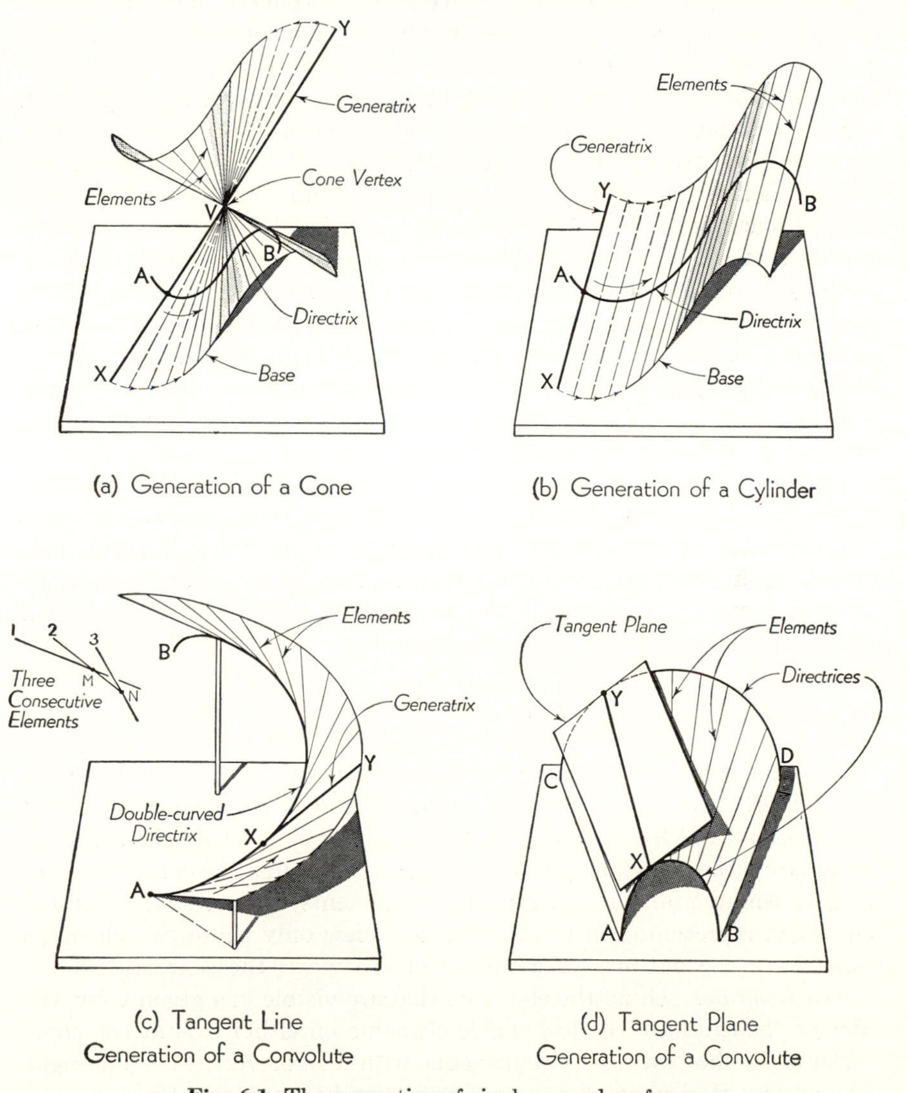

Fig. 6·1. The generation of single-curved surfaces.

will intersect, but a third adjacent element will intersect only one of the first two. The small detail drawing of *three consecutive elements* in Fig. 6·1(*c*) illustrates this. If elements 1 and 2 are very close together, they will intersect at some point *M* but the next element 3 will intersect element 2 at some other point *N* and will not intersect element 1 at any point.

Convolutes: tangent-plane generation. A convolute may also be generated in a second way, which is of considerable practical value. If a plane is placed so that it is tangent to each of two curved lines that do not lie in the same plane, then the straight line connecting the points of tangency will be one element of a convolute surface joining the two curves. By rolling the plane along the curves to successive positions of tangency, additional elements may be located to define the surface. In Fig. 6·1(*d*) the *directrices* are the two dissimilar curves *AB* and *CD*. When the *tangent plane* is placed so that it is tangent to curve *AB* at point *X* and tangent to curve *CD* at point *Y*, the straight line *XY* becomes an element of the surface. The tangent plane thus touches the convolute surface along the entire length of element *XY*, and the surface is therefore a ruled surface of single curvature. A surface generated in this manner is called an *envelope* of the successive positions of the tangent plane.

The directing curves *AB* and *CD* may be either plane or space curves of any shape. If, however, the shape of the curves is such that a series of consecutive elements are parallel, then that portion of the surface will be cylindrical. If a series of consecutive elements all pass through a common point, then that portion of the surface will be conical. Thus a surface generated by a tangent plane may be cylindrical, conical, convolute, or a combination of all three. Convolute surfaces may be infinite in variety, but a few of the more important types will be discussed in detail in later articles of this chapter.

6·4. Representation of cones

Cones are represented in multiview drawings by showing in each view the *base curve*, the *vertex*, and the *extreme elements.* If the base curve is of regular shape and has a definite center, the axis connecting the center of the base and the vertex may also be shown. Although the conical surface actually contains an infinite number of elements, it is sufficient for the purpose of representation to show in each view only those two elements which form the outlines or contours of the view; these are called the *extreme elements.* Of all the elements that are visible in a given view, the extreme elements are the *last visible* elements on either side of the cone.

Figure 6·2 shows a right circular cone with its axis vertical. The right circular cone is also called a *cone of revolution* because its surface may be generated by revolving the generatrix about the axis of the cone. In the top view the vertex v_T falls inside the base curve, and therefore every element drawn from point v_T to a point on the base curve will be visible in the top view. Thus there can be no extreme (or last) elements in this view. In the front view, lines v_Fa_F and v_Fb_F outline the cone and are the extreme elements in this view. These same two elements appear in the top view at v_Ta_T and v_Tb_T, indicating that exactly half of the cone surface is visible in the front view.

Figure 6·3 shows an oblique circular cone—circular because the base curve is circular.[1] In the top view, lines v_Tc_T and v_Td_T are extreme elements. By locating points c_F and d_F on the base curve in the front view, it is found that element *VC* is hidden here and element *VD* is visible. Any element drawn from the vertex to a point on the upper half of the base circle (such as element *VB*) will be visible in the top view. Conversely, element *VA* will be hidden in the top view. In the front view the extreme elements are lines v_Fa_F and v_Fb_F drawn tangent to the base circle. Radii of the circle drawn perpendicular to each of these elements determine the exact points of tangency, a_F and b_F. Thus in the front view somewhat less than half of the elements of the cone will be visible. It should now be noted that those elements which are extreme in one view are *not* the extreme elements in an adjacent view.

[1] Because a *right section* of this cone would be an ellipse, some authorities prefer to call it an elliptical cone; the author prefers the more common designation of all cones and cylinders according to the shape of their bases.

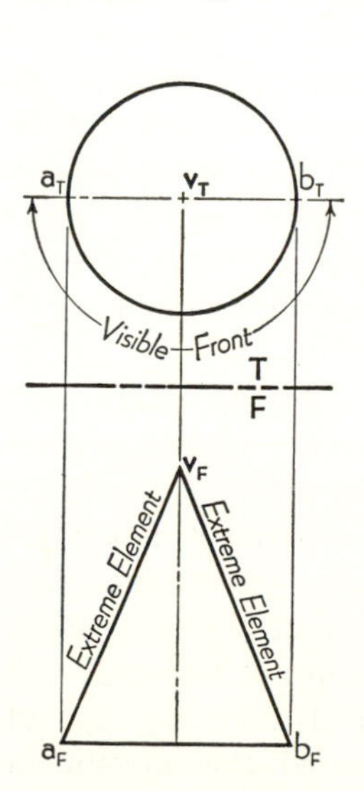

Fig. 6·2. Right circular cone.

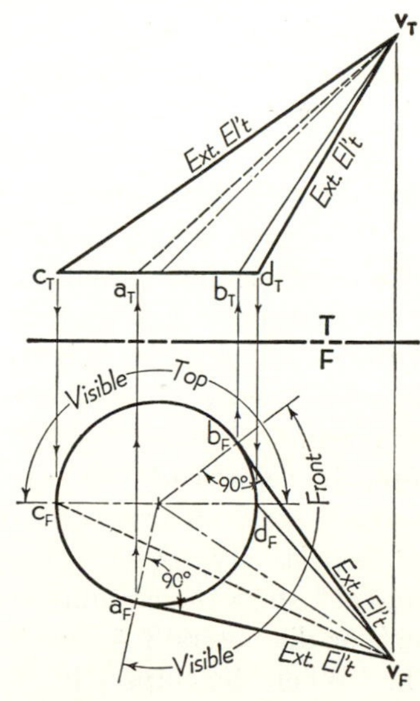

Fig. 6·3. Oblique circular cone.

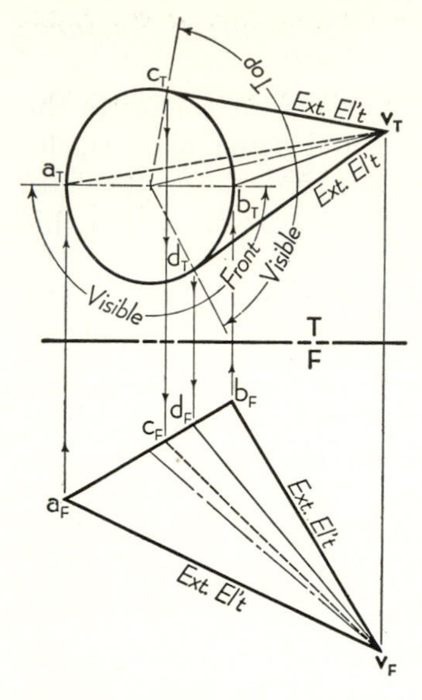

Fig. 6·4. Oblique circular cone.

Figure 6·4 shows an oblique circular cone whose base circle appears edgewise in one view but elliptical in the other. This case is very similar to that shown in Fig. 6·3. The extreme elements of the front view, VA and VB, have been located in the top view to show that exactly half of the elements will be visible in the front view. The extreme elements in the top view, v_Tc_T and v_Td_T, are drawn tangent to the ellipse (see Appendix, Art. A·13, for exact methods of determining such points of tangency). As indicated in the top view, any element drawn from a point on the shorter arc of the base between points C and D will be visible in the top view. To be visible in both the top and the front view an element must touch the small arc of the base between points B and D.

6·5. Principal views of a right circular cone with inclined axis

In Fig. 6·2 the right circular cone appears in its simplest position. It is sometimes necessary, however, to draw such a cone with its axis in an inclined position.

Analysis. The circular base of the cone will appear elliptical in the principal views. Since the cone axis is perpendicular to the plane of the base, the *major axis* of each ellipse (true-length lines in the plane) will appear *perpendicular* to the cone axis in each view (Rule 15). The direction of the ellipse axes is therefore known, and the length of each *minor axis* can be determined by either the edge-view method of Art. 4·17 or the two-view method of Art. 4·18.

Problem. In Fig. 6·5 it is required to draw the top and front views of a right circular cone whose axis is the given line VA and whose base diameter is given.

Edge-view method. In Fig. 6·5(a) the true-length major diameters 1–2 and 5–6 are drawn perpendicular to the cone axis in the top and front views, respectively. Views A and B, which show the cone axis in true length, each show the circular base as an edge at right angles to the axis. The minor diameters 3–4 and 7–8 are obtained from the edge views as shown. When the ellipses have been constructed, the extreme elements are drawn in each view from the vertex V *tangent* to the ellipses (*not* to the ends of the major axis). Note that the hidden portion of the top-view ellipse extends only between the points of tangency T and is somewhat less than half of the perimeter of the ellipse.

Two-view method. In Fig. 6·5(*b*) the true-length major diameters 1–2 and 5–6 are drawn perpendicular to the cone axis in each view and then represented in the adjacent views parallel to the reference line. The minor axes can now be established as in Art. 4·18 (also Appendix, A·8). Ellipses and extreme elements are drawn as described in the edge-view method above.

6·6. Location of a point on a cone

In Art. 4·5 it was shown that in order to locate a point on a given plane surface it was first necessary that the point lie on some line in that plane. This is a cardinal principle that is equally applicable to the location of

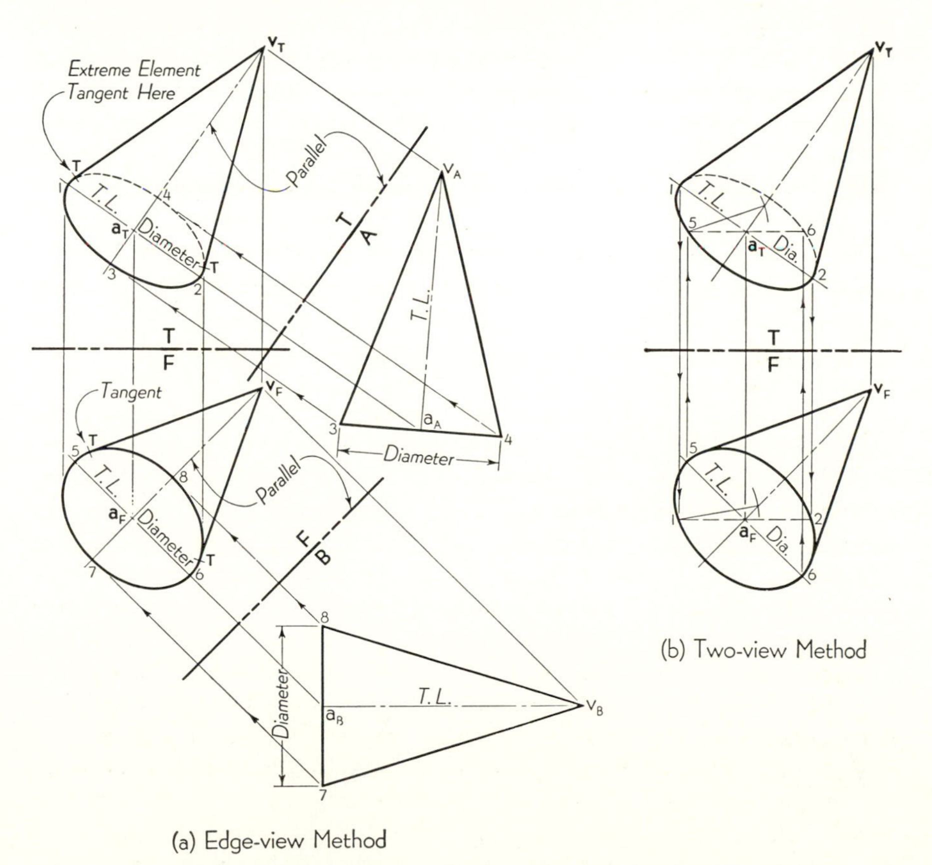

Fig. 6·5. Principal views of a right circular cone with inclined axis.

points on any single-curved, warped, or double-curved surface. Stated in general terms:

> *A point can be located on a surface only by first locating some line or element of that surface which contains the point.*

A straight-line element is the simplest line that can be drawn on a cone and in general is the preferred line to use in locating a point on the surface.

In Fig. 6·6(a) assume that the point x_F is given as a *visible* point on the right circular cone; it is required to locate x_T. From point v_F an element is drawn through point x_F, to intersect the base at point a_F. Because point x_F is given as visible, line v_Fa_F must be visible and point a_T is therefore located on the front half of the circular base in the top view. Line v_Ta_T is drawn as the top view of element VA. Point x_T is now located on line v_Ta_T directly above point x_F.

Assume again in Fig. 6·6(a) that point y_T is given and that it is required to locate y_F on the surface of the cone. Element VB, drawn through point Y, is a profile line; hence point y_F cannot be located by alignment only. But element VB can be revolved about the axis of the cone as shown to occupy the position of extreme element in the front view. Since point Y lies on element VB, point y_T will also revolve to y_T^R and y_F^R will lie on the extreme element. When element VB is counterrevolved to its original position, point y_F^R moves to y_F, thus locating the required point. The reader should observe that this particular construction is

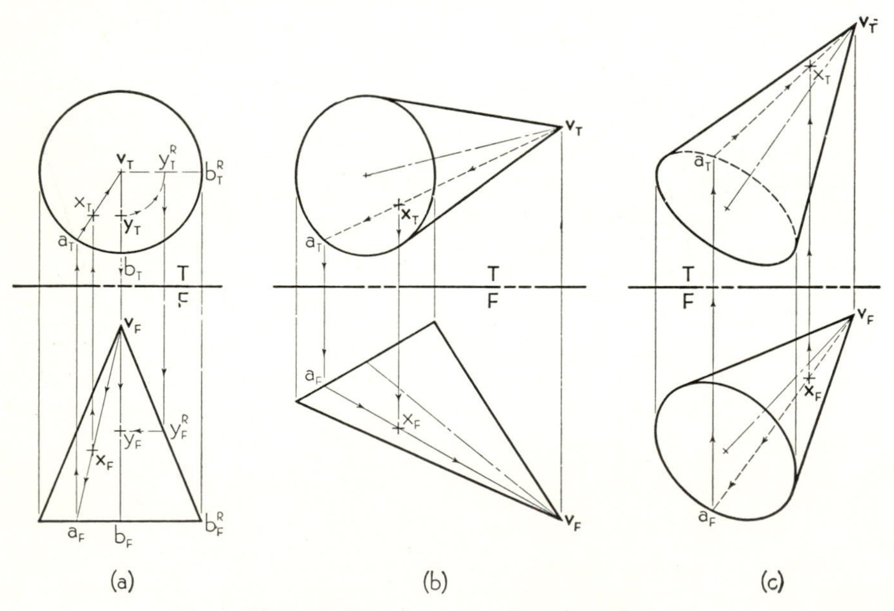

Fig. 6·6. Location of a point on a cone.

peculiar to a right circular cone; a profile element of an *oblique* cone may also be revolved about a *vertical* axis, but it will *not* then occupy the position of extreme element.

In Fig. 6·6(*b*) assume that point x_T is a point on the surface of the oblique circular cone and that it is required to locate x_F. Point X must be on the underside of the cone, and x_T should therefore be considered as hidden. An element drawn from v_T through x_T is extended to point a_T on the base curve. Since point a_T is on that part of the base which will be visible in the front view (compare Fig. 6·4), element VA will also be visible in the front view. Point x_F can now be located on v_Fa_F, where it is a visible point.

When the base of the cone does not appear as an edge in either view, as in Fig. 6·6(*c*), special care must be exercised to align properly points on the base curve in one view with those in an adjacent view. Assume, for example, that point x_F is given as a hidden point on the cone. The element drawn through point x_F must be hidden and is therefore drawn to point a_F on the base. A parallel extended from a_F to the top view intersects the curve in that view at two points, thus posing the question: Which point is a_T? To answer this question we must note first that point A is on the lower part of the base as evidenced by the position of a_F. In the top view the lower part of the base curve must be the hidden part; hence a_T lies on the hidden line. Point x_T can now be located on the hidden element v_Ta_T.

PROBLEMS. Group 59.

6·7. Intersection of a plane and a cone

Analysis. The intersection line of a plane and a cone will be a single-curved line of regular or irregular shape depending upon the shape of the base curve. By assuming a number of elements on the cone the intersection point of each element with the given plane may be located. Through this series of points a smooth continuous curve may then be drawn to establish the required plane section.

This analysis suggests a procedure basically the same as that used to determine the intersection of a plane and a polyhedron (Arts. 4·29 and 4·30). We can find the intersection point of *each assumed element* with the given plane by either of two methods:

1. *By showing the plane as an edge (as was done in Fig. 4·35).*
2. *By using cutting planes (as was done in Fig. 4·36).*

Because it will be necessary to locate a large number of points, the *first* method (edge-view method) will be more efficient despite the fact that it requires an additional view.

Problem. In Fig. 6·7 the top and front views of an oblique elliptical cone are given. The cone is intersected by plane *ABC*, and it is required to show the curve of intersection in the given views.

Construction. To employ the edge-view method plane *ABC* must be shown as an edge in either view *A* or view *C*, but view *A* is definitely preferable. Because the base plane of the cone appears true size in the top view, it will appear as an edge in any top-adjacent view such as *A*. View *C*, on the other hand, would show the base curve as an ellipse whose outline would have to be plotted point by point in view *C*. We therefore elect to draw view *A*. If the cone base were inclined as in Fig. 6·4, it would, in general, appear elliptical in either a top-adjacent or a front-adjacent view.

In view *A* the base, vertex, and extreme elements of the cone are drawn,

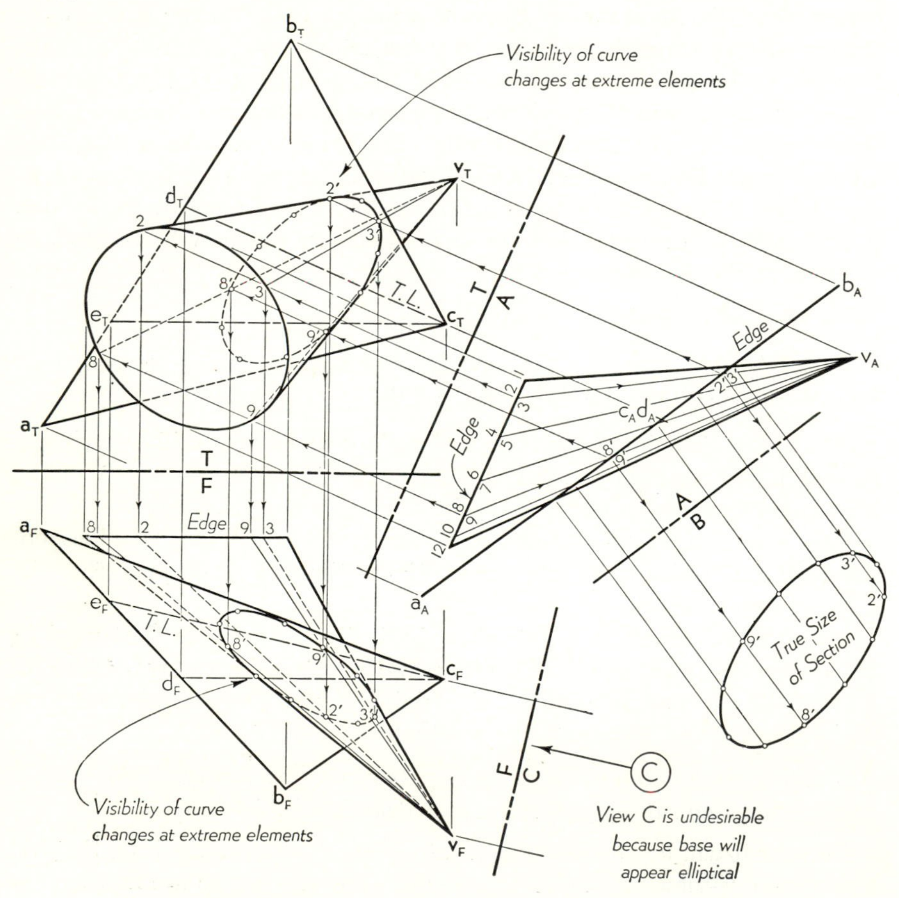

Fig. 6·7. Intersection of a plane and a cone (edge-view method).

and the plane *ABC* appears as an edge. By selecting a series of points (1, 2, 3, etc.) along the edge view of the base in view *A* and connecting these points to the vertex v_A we shall then have a number of elements on the cone. Note that, except for the extreme elements, each assumed point actually represents two points on the base curve and that for each visible element in view *A* there is a second hidden element directly behind it. At least 12 elements should be used, and frequently 24 or more are desirable for greater accuracy. Observe also that the elements should be spaced more closely in the vicinity of the extreme elements in order to obtain approximately equal spacing of points along the curve of intersection.

Points on the base, such as 2, 3, 8, and 9, may now be located in the top and front views and each element drawn in each view in correct visibility. With all elements now shown and numbered (for convenience) in each view we may locate the point where each element intersects plane *ABC*. These points are located first, of course, in view *A*, where elements 8 and 9, for example, are seen to intersect the plane at points 8′ and 9′. By alignment, points 8′ and 9′ may also be located on the corresponding elements in the top view and then in the front view. When all such intersection points have been located in each view, the irregular curve should be used to connect them in a smoothly continuous curve. These curves should be exactly *tangent* to the extreme elements in each view, and the points of tangency (such as point 2′ in the top view) should be carefully noted since they are the points where the visibility of the curve changes.

If it is desired to show the true size of the plane section, this may be done by drawing a new auxiliary view, such as *B*. After the intersection points have been located in view *A* and in the adjacent top view, they may be located in view *B* by alignment and reference-line measurement.

6·8. The right circular cone and conic sections

The curve formed by the intersection of a plane and a *right circular cone* is called a *conic section.* According to the angle that the plane makes with the base of the cone, the conic section may be an *ellipse,* a *parabola,* or a *hyperbola.* Each of these three cases is illustrated in Fig. 6·8.

When, as in Fig. 6·8(*a*), the cutting plane makes an angle *A* with the base of the cone that is *less* than the angle *B* between the elements and the base, the curve of intersection is an ellipse. When, as in Fig. 6·8(*b*), the cutting-plane angle *A* is exactly *equal* to the element angle *B,* the curve is a parabola. When, as in Fig. 6·8(*c*), the cutting-plane angle *A* is *greater* than the element angle *B,* the curve is a hyperbola.

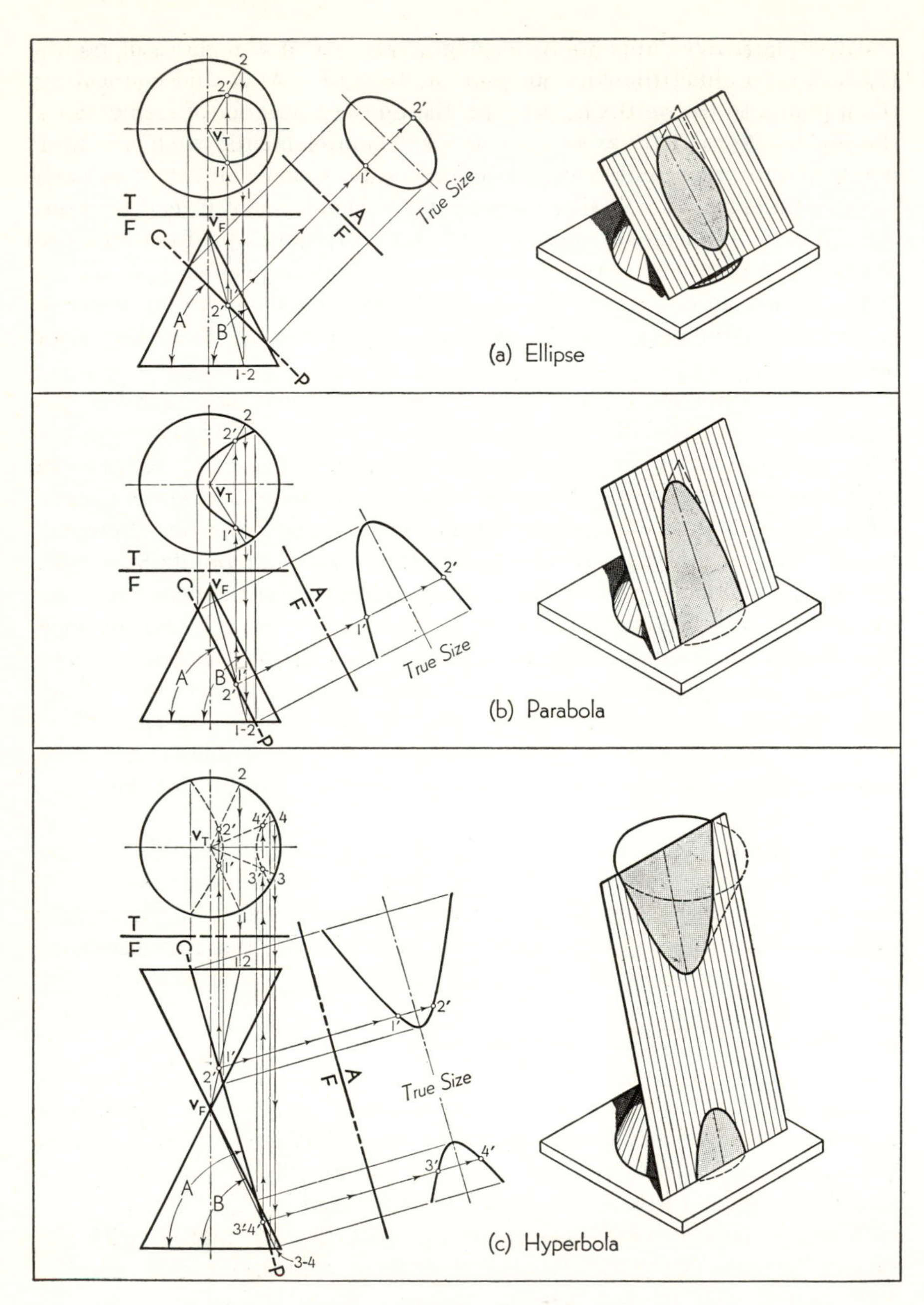

Fig. 6·8. The conic sections.

Construction. The method of constructing the conic sections in Fig. 6·8 is exactly the same as that of Art. 6·7. A series of elements should be assumed on the cone and the intersection of each element with the cutting plane found. In each case in Fig. 6·8 two elements V-1 and V-2 have been assumed to locate the intersection points 1′ and 2′ in each view. When the cutting-plane angle A is greater than angle B, the plane will then cut both nappes of the cone and the resulting hyperbola thus has two separate curves.

The conic sections may be constructed from the actual intersection of a plane and a right circular cone, as shown in Fig. 6·8, but they may also be constructed by the methods of plane geometry. Since the latter procedure is frequently necessary, these methods have been given in the Appendix.

Special cases. Two other positions of the cutting plane are of particular importance to subsequent discussion. When the cutting plane is parallel to the base of the cone, the intersection will be a *circle.* When the cutting plane passes through the vertex of the cone (and intersects it), the intersection will be a pair of *intersecting straight lines.* These two lines will be two elements of the cone. This is obviously the simplest possible intersection of a cone and a plane, and we should now note that this type of intersection will always be obtained for *any* kind of cone when the cutting plane passes through the vertex.

PROBLEMS. Group 60.

6·9. Intersection of a line and any surface

The intersection point of a line and a plane has been found by first choosing a cutting plane that contained the line and intersected the plane (Rule 14). This same general principle may also be employed to determine the intersection of *any* line and *any* surface. For example, in Fig. 6·9 the irregular curved surface *ABC* is intersected by the line *MN*. Then *any* cutting plane that contains line *MN* will intersect surface *ABC* along some line *RS*. Line *RS* may be straight, circular, or irregularly curved, but the intersection point *P* of line *MN* and surface *ABC* must obviously be at the intersection of lines *MN* and *RS*. Let us therefore restate Rule 14 as a general rule:

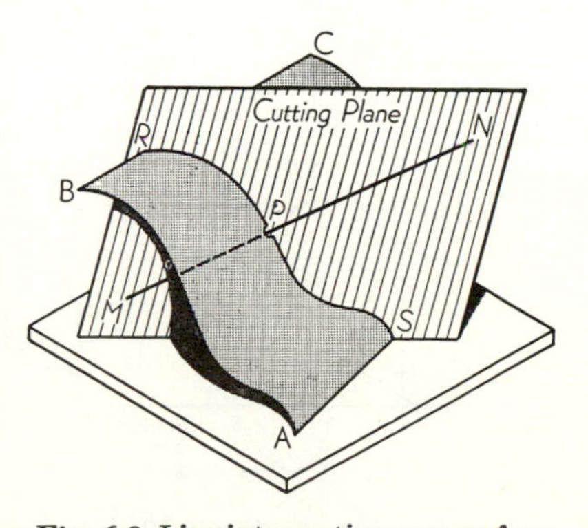

Fig. 6·9. Line intersecting any surface.

▪ ***Rule 17. Rule of a line intersecting any surface.*** *The intersection of a line and* ***any surface*** *must lie on the intersection line of the given surface and a cutting plane that contains the line.*

Since *any* cutting plane containing the given line will serve, we shall shorten the solution in subsequent problems by choosing a particular cutting plane that intersects the given surface in the simplest possible way.

6·10. Intersection of a line and a cone

Problem. In Fig. 6·10 the points where line *MN* intersects the surface of the oblique circular cone are required.

Analysis. An infinite variety of cutting planes could be assumed to contain line *MN*, and all but *one* would intersect the cone in a curved line. The single exception would be the plane passing through the *vertex* of the cone; this plane would intersect the cone along two straight line elements. Hence for simplicity:

Choose a cutting plane that contains the given line and the cone vertex.

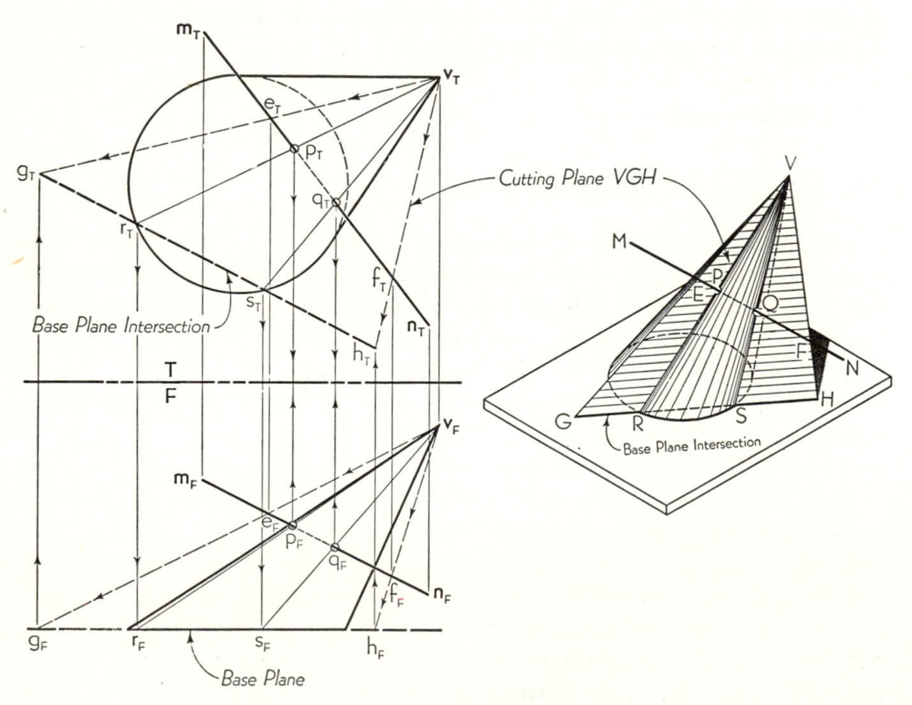

Fig. 6·10. Intersection of a line and a cone.

To form a cutting plane through vertex V, it is only necessary (as shown in the picture) to draw a straight line from point V to any assumed point on line MN, such as point E. The intersecting lines MN and VE will then define the cutting plane. Any additional lines, such as VF, may also be assumed in the cutting plane.

The next step is to determine which two elements form the intersection of the plane and cone. For this purpose the cutting plane must be extended until it intersects the plane of the cone base. In the pictorial illustration, line VE has been extended to intersect the base plane at point G; line VF extended intersects the base plane at point H. The line GH is therefore the intersection of the cutting plane and the base plane of the cone. Since line GH intersects the curve of the cone base at points R and S, the elements VR and VS are the elements of intersection. These two elements intersect line MN at points P and Q, and these are the required intersection points of line MN and the cone.

Construction. The same procedure may now be observed in the multiview drawing. The assumed lines v_Fe_F and v_Ff_F are extended to intersect the base plane at points g_F and h_F. Points g_T and h_T may then be located in the top view on v_Te_T and v_Tf_T extended to establish the *base-plane intersection line.* This line cuts across the base curve at points r_T and s_T to determine the elements v_Tr_T and v_Ts_T. These same two elements may also be located in the front view at v_Fr_F and v_Fs_F. The required intersection points P and Q are thus located on line MN in each view. As a final check on the accuracy of the work, points p_F and q_F should be directly below points p_T and q_T. The visibility of points P and Q must, of course, agree in each view with the visibility of the elements VR and VS on which they lie.

Although lines VE and VF were selected at random to establish points G and H, it should now be apparent that any other lines in the plane VMN, such as VM, VN, or MN, could also be extended to intersect the base plane and thus establish other points on the line GH. The position of line GH is not affected by this freedom of choice.

PROBLEMS. Group 61.

6·11. Representation of cylinders

Cylinders are represented in multiview drawings by showing in each view the *base curve* (or curves) and the *extreme elements.* For regular cylinders the axis should also be shown. We shall find it useful to imagine that cylinders are only a special kind of cone—a cone whose vertex is *so far away* from its base (at infinity) that all the elements may be considered parallel. This concept will enable us to see the very close similarity that exists between cone problems and cylinder problems.

Right circular cylinders. Figure 6·11 shows four views of a right circular cylinder with its axis vertical. The top and front views show the cylinder most advantageously; views *A* and *B* show the cylinder as it would appear when viewed obliquely, and here the bases appear as ellipses. We should especially note here that the *major axis* of each ellipse is always *perpendicular* to the *cylinder axis* and is always *equal* to the true *diameter D* of the cylinder. (Compare Fig. 3·20, and recall Rule 10.) From this observation it follows that:

> *The perpendicular distance between the extreme elements of a right circular cylinder always equals the cylinder diameter.*

Since right circular cylinders occur so commonly in practice, we shall find this a very useful fact to remember.

For problems involving the location of a point at a fixed distance from a given line we should also observe that the surface of a right circular cylinder is the locus of all such points when the cylinder axis is the given line. A good example is the locus of the center of a sphere that is always tangent to a given line.

Oblique circular cylinders. Figure 6·12 shows various views of an oblique circular cylinder. In this case the axis of the cylinder appears

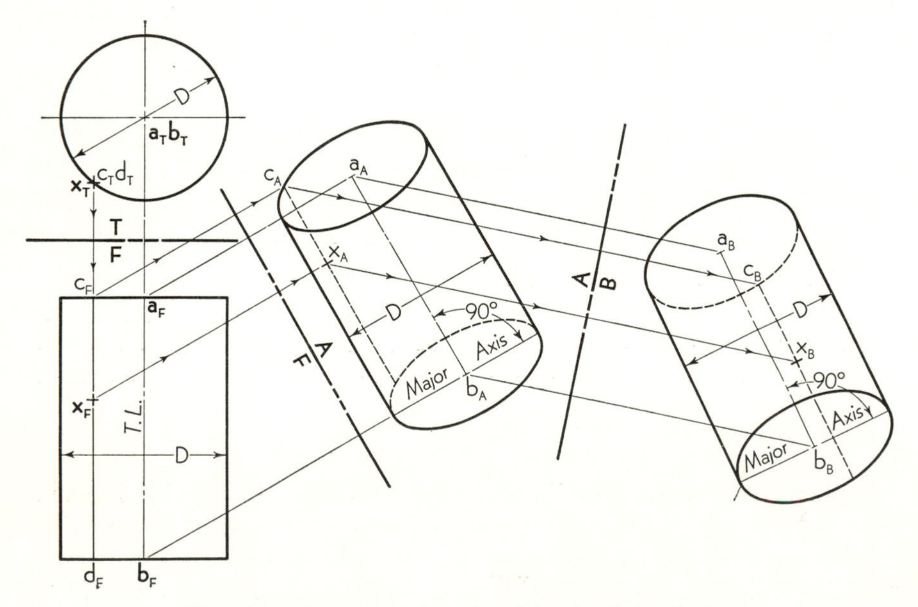

Fig. 6·11. Representation of a right circular cylinder.

neither as a true length nor as a point in the top and front views; yet it is frequently desirable to have such views (for development, for example). Views *A* and *C* both show the cylinder axis in true length; but because the base circles of the cylinder appear true size in the top view, they appear edgewise in top-adjacent view *A*. Thus view *A* is obviously the easier true-length view to construct. If it should be necessary to construct view *C*, then we should note that the major axis of each ellipse in this view is perpendicular to the reference line *C–F* and equal to the diameter *D* of the base. If this fact is forgotten, the ellipses can always be constructed by plotting a series of points, on the curve, like point *C*.

Views *B* and *D* both show the axis of the cylinder as a point; and since every element also appears as a point, the cylindrical surface appears edgewise in these views. Views *B* and *D* are thus identical and show the true cross-sectional area and perimeter of the cylinder; a right section of the cylinder has this same shape. The position of the major axis $r_D s_D$ in view *D* may be determined by noting that it is true length in view *D*; $r_C s_C$ must therefore be parallel to reference line *C–D* in view *C*.

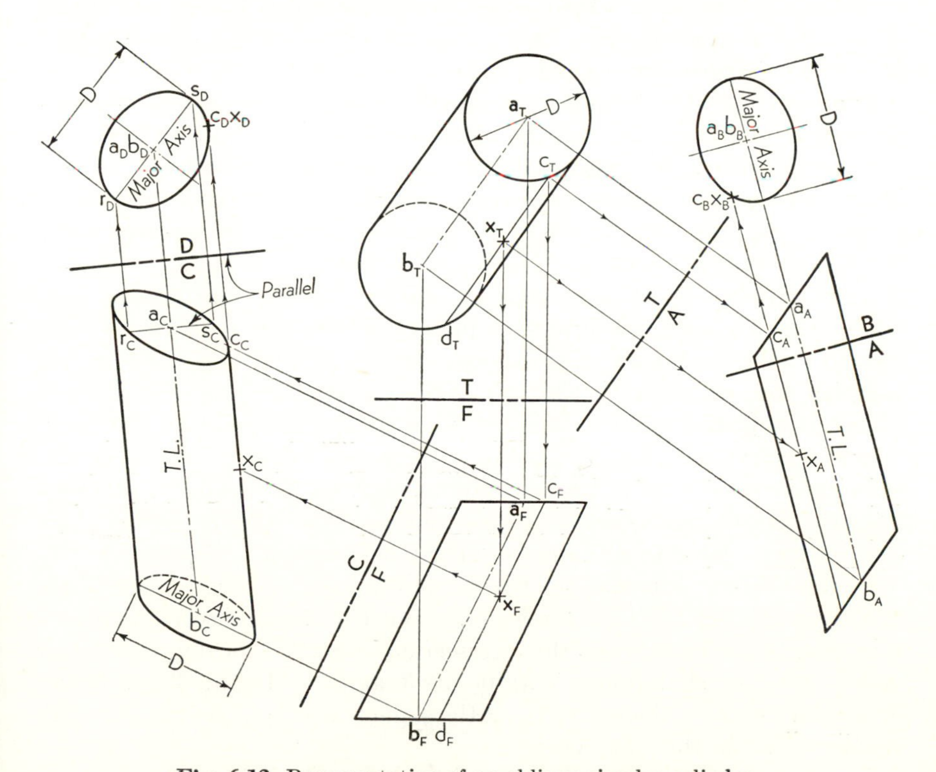

Fig. 6·12. Representation of an oblique circular cylinder.

6·12. Location of a point on a cylinder

As in the case of a cone (Art. 6·6), a point on the surface of a cylinder can be located in any view by first locating the straight-line element that passes through that point. In Fig. 6·11 point X lies on element CD as shown in the top and front views. To locate element CD in views A and B it is only necessary to locate points c_A and c_B on the base circle; the element must always be parallel to the cylinder axis. Points x_A and x_B may now be located on the element by alignment. Figure 6·12 similarly illustrates the location of an element CD and point X in the various views of an oblique cylinder.

PROBLEMS. Group 62.

6·13. Intersection of a plane and a cylinder

Analysis. By assuming a series of elements on the cylinder the intersection point of each element with the given plane may be determined. As in the case of a cone (Art. 6·7), the element intersection points can be found by either the edge-view method or the cutting-plane method. The fact that all elements of a cylinder are parallel suggests the use of a series of parallel cutting planes. Thus, although either method can be used here, the cutting-plane method will be illustrated.

Problem. In Fig. 6·13 it is required to determine the curve of intersection of the plane ABC and the oblique cylinder.

Construction. We may begin by selecting any element on the cylinder. Let us first choose element 1, which is one of the extreme elements in the front view. Element 1 is then located in the top view. We must now select a cutting plane—either vertical or edge-front—that will contain element 1; the edge-front plane C-P-1 has been used in Fig. 6·13. This cutting plane intersects plane ABC along line EF, and e_Tf_T is located in the top view. The intersection of e_Tf_T and element 1 in the top view locates point 1′, which is one point on the required curve of intersection in that view. Point 1′ is an important point in the front view since it establishes the point of tangency of the curve and the extreme element; this is also the exact point where the curve changes from visible to hidden. On the opposite side of the cylinder, point 2′ on element 2 is located in a similar manner (construction not shown).

Element 3 is an extreme element in the top view; to locate point 3′ on this element, C-P-2 has been selected. But C-P-2 will also contain element 4; and therefore when the intersection line GH is located in the top view at g_Th_T, we shall obtain both points 3′ and 4′. Points 3′ and 5′ are important points in the top view for the same reason that points 1′ and 2′ are important in the front view.

To complete the curve of intersection accurately it is only necessary now to assume additional parallel cutting planes. Because all cutting planes will be parallel, all intersection lines such as *EF* and *GH* will also be parallel. The cutting-plane method is therefore especially suitable for cylinders.

The visibility of the extreme elements should now be determined. Element 1, for example, will be visible in the front view from point 1 to point 1′, but at point 1′ it passes through plane *ABC* and will therefore be hidden from point 1′ to f_F. That element 1 passes in *front* of line a_Fb_F at point e_F may be verified by noting that point e_T is far *behind* element 1 in the top view (Rule *d*, Art. 2·9).

PROBLEMS. Group 63.

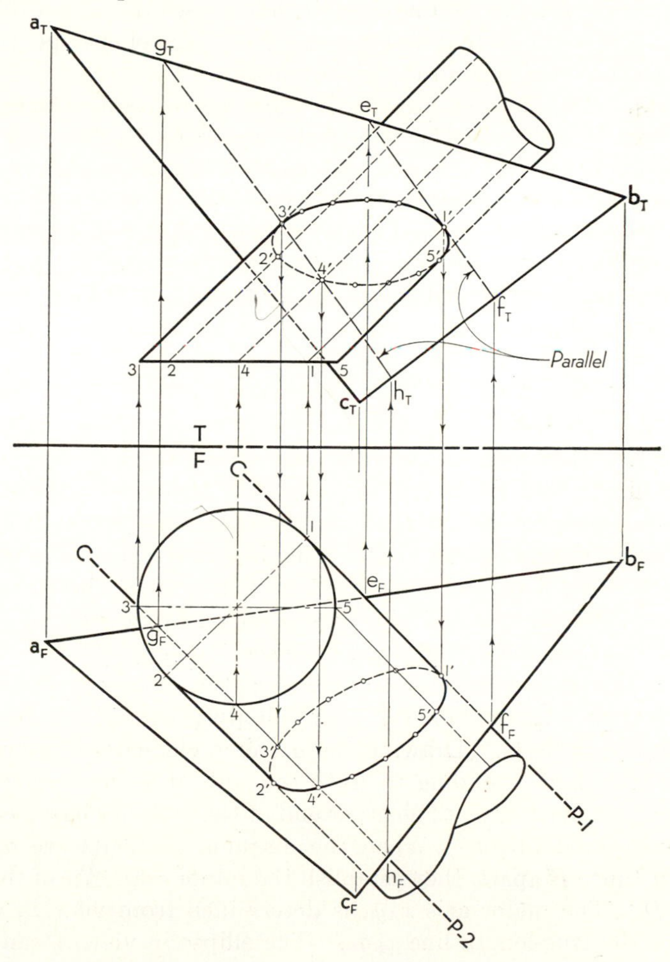

Fig. 6·13. Intersection of a plane and an oblique cylinder (cutting-plane method).

6·14. Intersection of a plane and a right circular cylinder

Pipes and ducts for the transmission of gases and fluids are usually right circular cylinders; and since these pipes frequently pass through walls and floors obliquely, we should give special attention to this problem. In such cases the diameter of the pipe and the direction of its axis are usually known. To solve this problem by the method of the previous article would require some base curve from which elements might be assumed, but such a curve may not be immediately available. The following problem is a typical case.

Problem. In the given top and front views of Fig. 6·14, line AB represents the axis of a pipe of diameter D that passes through a thin vertical wall. It is required to determine the true shape of the opening in the wall and to show the pipe in the given views.

Analysis. The pictorial drawing in stage 1 is introduced here to demonstrate that the intersection of a right circular cylinder and a plane oblique to its axis is an ellipse whose *minor diameter* is always *equal* to the *diameter* D of the cylinder and whose *major axis* M is *greater* than D. Since the magnitude of the major axis depends upon the angle between the plane and the cylinder axis, this angle becomes the key to the solution. Thus, the *position* and *size* of the major axis of the elliptical opening will be found if we:

> *Obtain a new view showing the true size of the line-plane angle between the given plane and the cylinder axis.*

Stage 1. *Construct views to obtain the line-plane angle.*
View A shows the wall in true size and will also show the true size of the elliptical opening. View B shows the pipe axis in true length, and the wall again appears as an edge. This is exactly the same procedure discussed in Art. 4·40 and illustrated in Fig. 4·46. The desired line-plane angle between wall and pipe axis is shown in view B.

Stage 2. *Draw the cylinder in each view.*
Because this is a right circular cylinder, the perpendicular distance between extreme elements in each view will always equal the diameter D of the pipe (Art. 6·11). Drawing the extreme elements in view B, distance D apart, and extending them to the wall at x_B and y_B completes this view of the cylinder. (The other end of the cylinder has been drawn as a conventional break.) When the extreme elements are drawn in view A, distance D apart, they establish the minor diameter of the ellipse equal to D. The major axis x_Ay_A is determined from view B, where it appears as the true-length line x_By_B. The ellipse in view A can now be drawn by the trammel method to complete the view and show the true size of the opening.

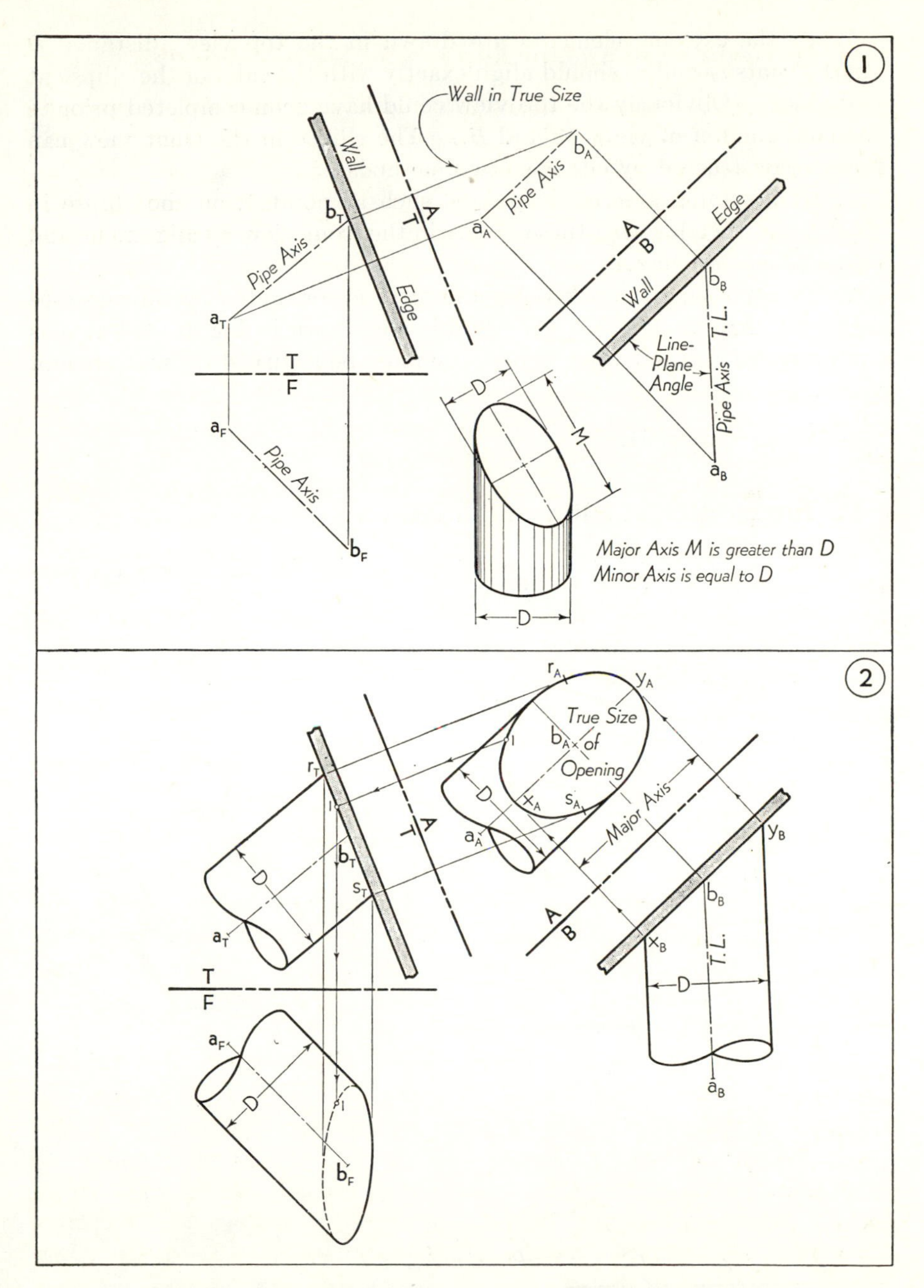

Fig. 6·14. Intersection of a plane and a right circular cylinder.

When the extreme elements are drawn in the top view, distance D apart, points r_T and s_T should align exactly with the sides of the ellipse at r_A and s_A. (Obviously the top view could have been completed prior to the construction of views A and B.) The ellipse in the front view can now be constructed by either of two methods:

1. By assuming a series of points, such as point 1, on the ellipse in view A and then locating these points in the front view by alignment and measurement as shown.

2. By employing the parallelogram method for multiview ellipses (see Appendix, Art. A·10). This method is not shown in Fig. 6·14, but it is the preferred method when the ellipse must be returned through several views.

PROBLEMS. Group 64.

6·15. Intersection of a line and a cylinder

Problem. In Fig. 6·15 the points where line MN intersects the surface of the oblique circular cylinder are required.

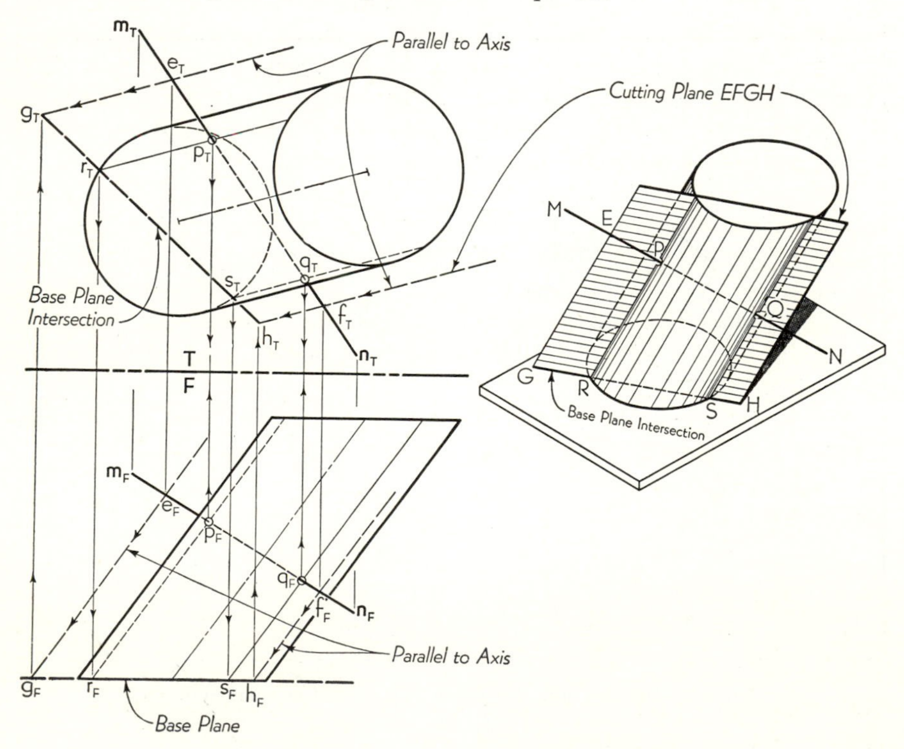

Fig. 6·15. Intersection of a line and a cylinder.

Analysis. A cutting plane that contains line *MN* must be selected. To avoid a curved line of intersection the cutting plane must be taken *parallel* to the cylinder *elements.* Such a plane will cut the cylindrical surface along two straight-line elements. We should therefore:

Choose a cutting plane that contains the given line and is parallel to the cylinder elements.

In the pictorial drawing of Fig. 6·15 point *E* is any random point on line *MN*, and through this point line *EG* has been drawn parallel to the cylinder. The two intersecting lines *MN* and *EG* now define the desired cutting plane—parallel to the cylinder and containing line *MN*. This plane must next be extended to intersect the base plane of the cylinder. For this purpose we note that line *EG* intersects the base plane at *G*; a second point *H* may be established by drawing another parallel line through another point on line *MN*, such as *F*. The two points *G* and *H* are sufficient to locate the *base-plane intersection line.*

If line *MN* were extended, it, too, would intersect the base plane on line *GH* extended, and it should be evident that any line in the plane of lines *MN* and *EG* could be similarly extended to locate a point on line *GH*. The position of line *GH* is not affected by this choice of lines. Line *GH* cuts the base circle at points *R* and *S*, thus determining the elements of intersection, and these two elements intersect the line *MN* at points *P* and *Q*, which are the required intersection points.

Construction. The sequence of construction in the multiview drawing of Fig. 6·15 is the same as that described for the pictorial drawing. The parallel lines drawn through the assumed points *E* and *F* could also have been extended to intersect the upper base plane. The correctness of points *P* and *Q* is beyond question if each point: (1) lies on line *MN*, (2) lies on an element of the cylinder, and (3) aligns perfectly between top and front views.

Figure 6·15 should be compared with Fig. 6·10 in order to observe the similarity of construction for cones and cylinders.

PROBLEMS. Group 65.

6·16. Tangent lines

Two lines, one straight and one curved, or both curved, are said to be *tangent* to each other when they *coincide* for a very short (infinitesimal) distance on either side of the point of contact. A straight line tangent to a single-curved line must therefore *lie in the same plane* as the curve. In Fig. 6·16(*a*), for example, the line *AB* is *not* tangent to the circle; the two lines *do intersect* at point T, but they *cannot* be considered tangent, for line AB does not lie in the horizontal plane of the circle. In Fig. 6·16(*b*), however, line AB *is* tangent to the circle; in the top view, a_Tb_T, which touches the circle at point t_T, is perpendicular to a radius of the circle, and the front view shows that line AB lies in the plane of the circle.

If the plane of the circle is inclined as in Fig. 6·16(*c*), line AB will be tangent to the circle only if a_Tb_T is tangent to the ellipse[1] in the top view and also lies in the plane of the circle as shown in the front view. When the circle does not appear as an edge in either view, as in Fig. 6·16(*d*), then the line and circle must appear tangent at point T in both views. In such cases an edge view of the circle should be drawn to more accurately establish line AB in the plane of the circle.

Two curved lines are tangent to each other if they are both tangent to the same straight line at a common point of tangency. If the tangent curves are both single-curved, then the common line of tangency must lie in the plane of each curve and must therefore be the line of intersection of the two planes. In Fig. 6·17 the two circles are tangent to each other because they are both tangent to line AB at the point T. A new view showing line AB as a point would show each circle as an edge and would also show the dihedral angle between the two planes (Art. 4·38).

Although the above discussion has been confined to circles, the principles outlined here are equally applicable to any plane curves. A double-curved line does not lie in a single plane, but a line tangent to such a curve may be constructed by drawing it tangent to the curve in each view (the helical convolute of Art. 6·28 illustrates this case).

[1] See Appendix, Art. A·13, for construction of tangents to ellipses.

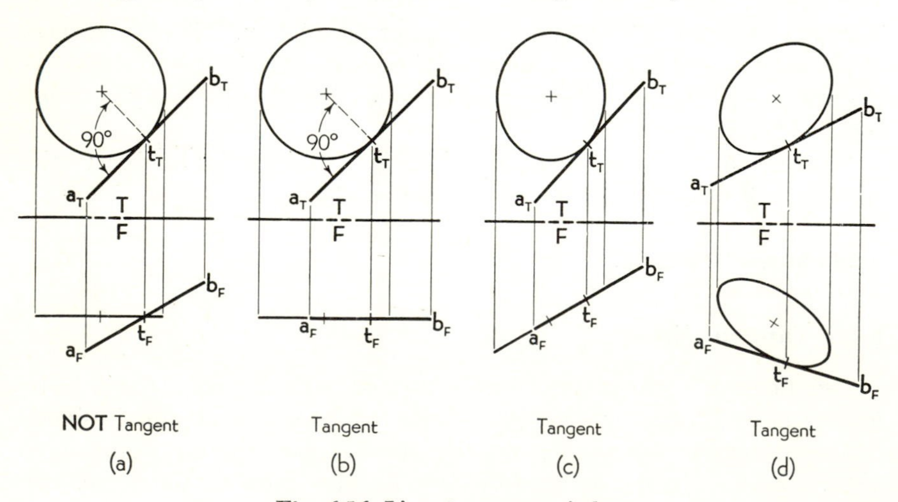

Fig. 6·16. Lines tangent to circles.

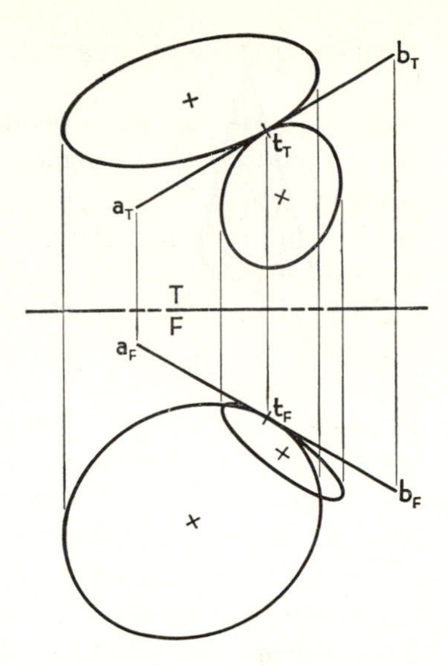

Fig. 6·17. Tangent circles.

6·17. Lines tangent to a surface

Any line is tangent to a surface if it is tangent to some line on that surface. In Fig. 6·18(*a*), for example, the line *AB* is tangent to the conical surface at point *X* because it is also tangent to the plane elliptical section at that same point. Line *AB* must, of course, lie in the plane of the ellipse. Line *CD* is tangent to the cone at point *Y* on the base curve and must therefore lie in the plane of the base.

In Fig. 6·18(*b*) lines *AB* and *CD* are similarly tangent to the cylinder at points *X* and *Y*, respectively, because each is tangent to a curve on the surface of the cylinder. To show that this principle of lines tangent to a surface is generally applicable, Fig. 6·18(*c*) shows lines *AB* and *CD* tangent to a sphere; each line is tangent to a curved line (a circle) that lies on the surface of the sphere.

6·18. Planes tangent to single-curved surfaces

A plane may be defined by a pair of intersecting lines. Therefore, any two straight lines tangent to a surface at a common point on that surface will define a plane which is tangent to the surface at that same point. For a single-curved surface, one of the two lines passing through the required

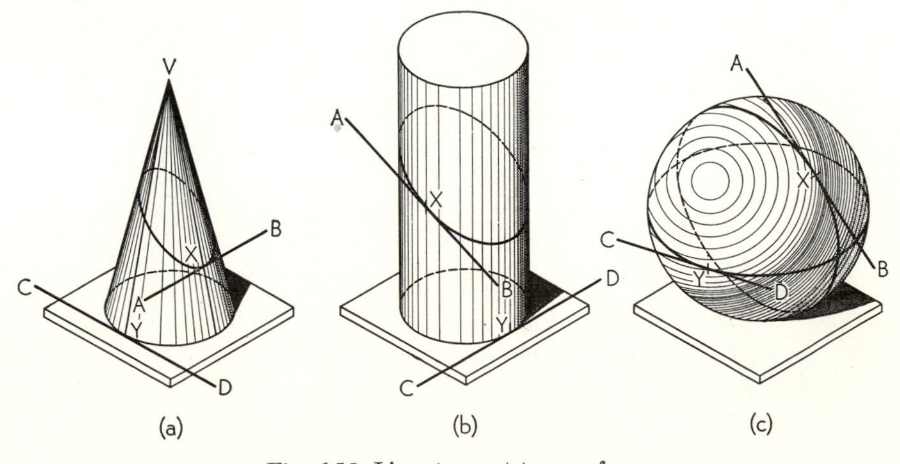

Fig. 6·18. Lines tangent to a surface.

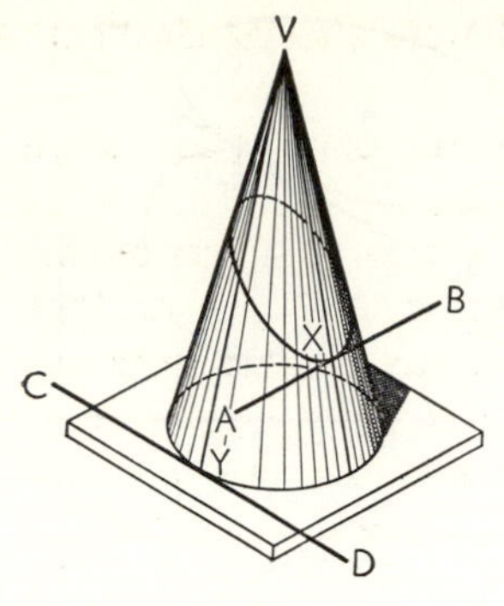

Fig. 6·19. Tangent planes.

point of tangency may be a straight-line element of that surface. In Fig. 6·19, for example, line AB is tangent to the cone at point X; element VX also passes through point X; therefore the plane of intersecting lines AB and VX is tangent to the cone. This tangent plane touches the cone all along the element VX, and element VX is therefore called the *element of tangency*. Similarly, line CD and element VY define a plane that is tangent to the cone along element VY. In the tangent-plane problems that follow this fact is basic:

The element of tangency and one intersecting tangent line will establish a plane tangent to a single-curved surface.

6·19. Location of a plane tangent to a cone through a given point on the cone

Analysis. An element through the given point will be one line in the tangent plane. The second line in the tangent plane must be tangent to the base curve. Thus, in the pictorial illustration of Fig. 6·20, point X is the given point on the cone, and VXT is the element of tangency.

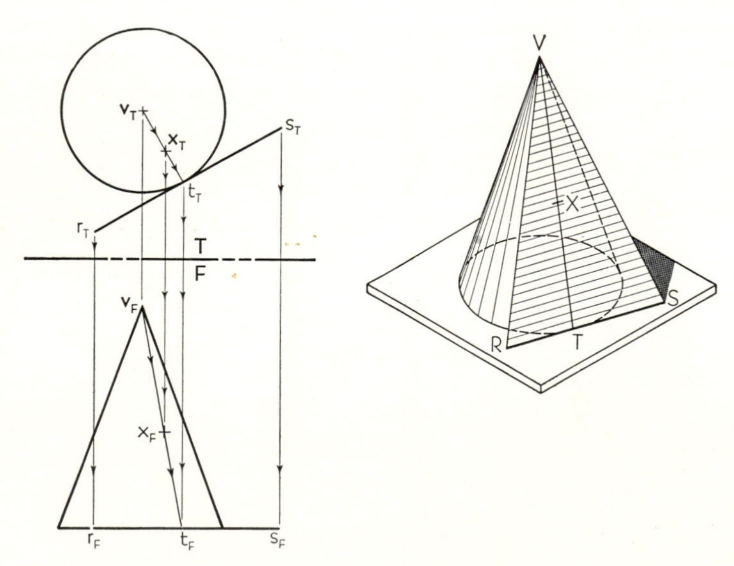

Fig. 6·20. Plane tangent to a cone through a given point on the cone.

Line *RST* is tangent to the base curve at *T*; hence plane *VRS* is the required tangent plane.

Construction. In the multiview drawing point x_T is given on the right circular cone. By drawing element *VXT* first in the top view, then in the front view, x_F is located. Line r_Ts_T, of any desired length, is drawn tangent to the circle at point t_T (perpendicular to radius v_Tt_T). Line r_Fs_F must lie in the horizontal plane of the base, and is so drawn. The required plane is now completely established by the intersecting lines *VT* and *RS*.

PROBLEMS. Group 66.

6·20. Location of a plane tangent to a cone through a given external point

Analysis. The required tangent plane must contain the *given external point,* and it must also contain the *vertex* of the cone. The line connecting these two points must therefore be one line in the tangent plane. Thus, in the pictorial illustration of Fig. 6·21, a line has been drawn from the given external point *X* to the vertex *V*. The second line in the tangent plane must be drawn tangent to the base curve and must also intersect line *VX*. Since this tangent line must lie in the same plane as the base curve, it is necessary to extend line *VX* to intersect the base plane

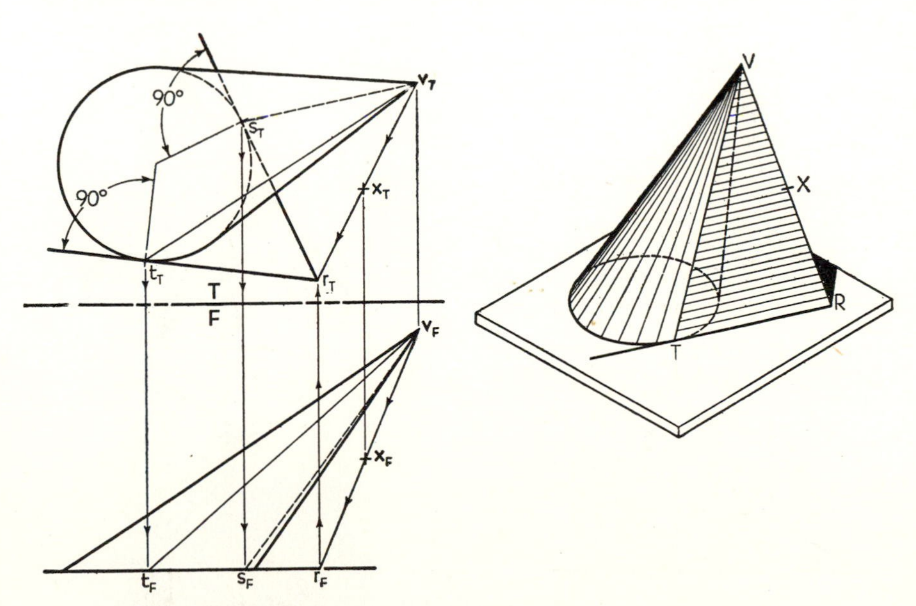

Fig. 6·21. Plane tangent to a cone through a given external point.

at R. The tangent line must now be drawn from point R tangent to the base at point T. Element VT, a third line in the required plane, is the element of tangency.

Construction. In the multiview drawing, line VX is drawn first and extended to intersect the base plane at point R. From r_T, tangent lines may be drawn to *either* side of the base circle as shown at r_Tt_T and r_Ts_T. There are therefore two possible solutions, plane VRT or plane VRS. (Only plane VRT is shown in the picture.) The points of tangency t_T and s_T should be accurately located by drawing radii of the circle perpendicular to the tangent lines. Locating points t_F and s_F on the base in the front view establishes the tangent lines here. The elements of tangency VT and VS are usually desired and should be shown.

Three cases. Depending upon the position of point X, the following cases may occur: (1) Point R may fall inside the base curve, in which case there can be no solution, for point X must then be inside one of the two nappes of the cone. (2) Point R may inconveniently fall beyond the limits of the paper, in which case the cone should be cut by a new plane closer to the vertex and the tangent line should then be drawn from the new position of point R to this new base curve. (3) Line VX may be parallel to the base plane, in which case the two possible tangent lines will be *parallel* to line VX.

PROBLEMS. Group 67.

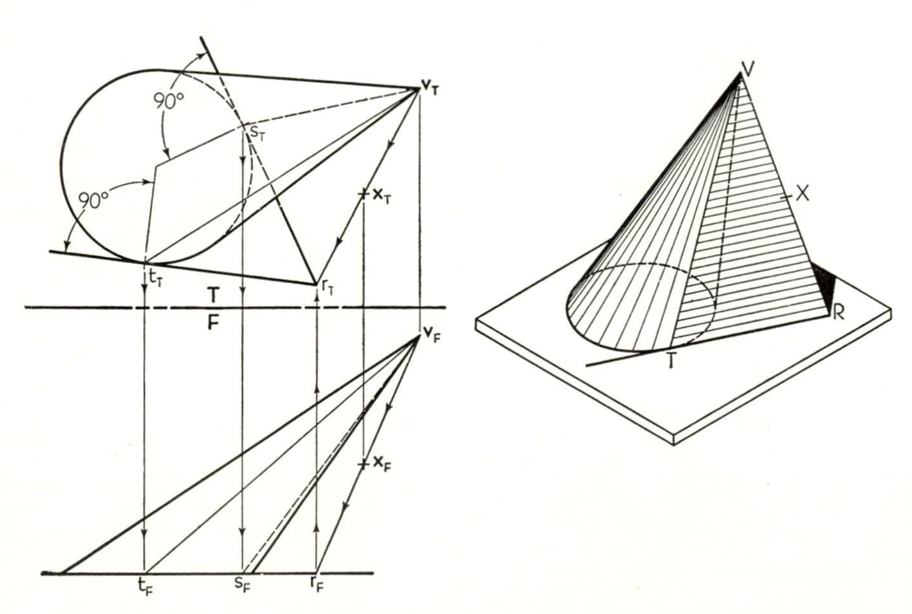

Fig. 6·21 (repeated). Plane tangent to a cone through a given external point.

6·21. Location of a plane tangent to a cone and parallel to a given line

Analysis. The required tangent plane must contain a *line parallel to the given line* and it must also contain the *vertex* of the cone. Therefore, a line must be drawn through the cone vertex parallel to the given line. This is one line in the required plane. Thereafter, the solution is identical with that of the previous article. The line through the vertex is extended to intersect the base plane of the cone, and the tangent line is drawn from this intersection point tangent to either side of the base curve. In general, there will be two solutions.

Problem. In Fig. 6·22 the oblique circular cone has its vertex at V and the center of its base at C, and it is required to locate a plane tangent to this cone and parallel to the line AB. This problem has been selected to illustrate not only this particular case but also the general method to be employed when the base of the cone does not appear as an edge in one of the given views.

Construction. Through vertex V a line is drawn parallel to the given line AB. This is one line in the required tangent plane. A second line

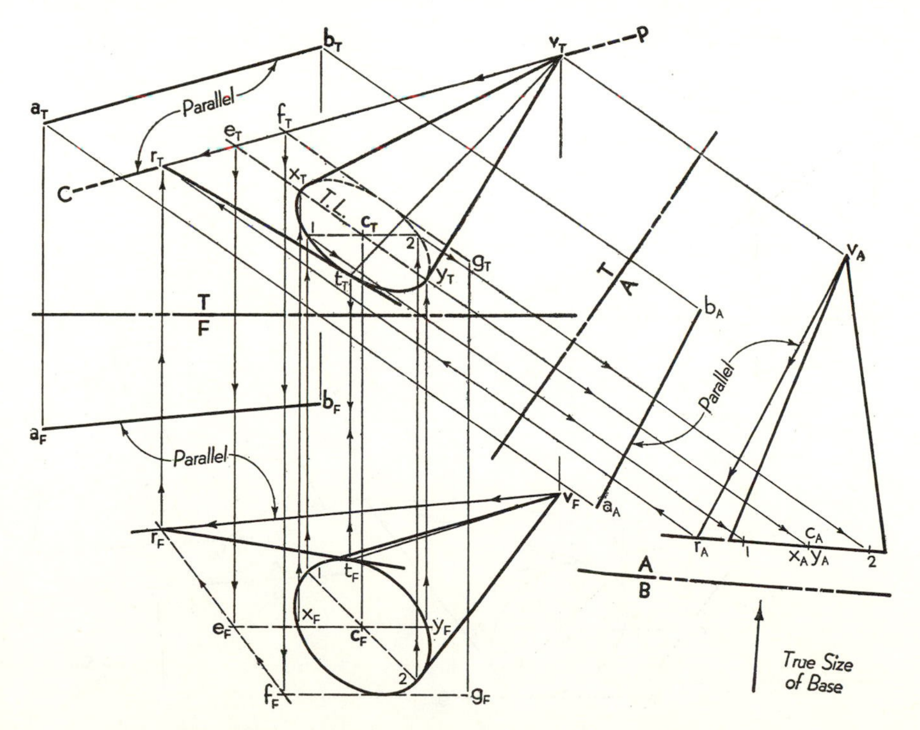

Fig. 6·22. Plane tangent to a cone and parallel to a given line.

must now be drawn tangent to the base curve, and for this purpose the vertex line must be extended until it intersects the base plane at R. Point R can now be located by either of two methods:

1. *A new view may be drawn to show the base plane as an edge.*

To show the base circle as an edge in view A the horizontal diameter $x_F y_F$ has been assumed. Then $x_T y_T$ will be true length; and because the base is circular, it will also be the major axis of the ellipse in the top view. By assuming any additional points, such as 1 and 2, on the base curve the plane of the base may then be established as an edge in view A. From point r_A the position of r_T and r_F can be found.

View A is also useful if it is necessary to confirm the fact that the base curve is single-curved. Although not essential, a true-size view B is useful to locate the point of tangency T more accurately.

2. *A cutting plane may be used to solve in the given views.*

To locate point R by the cutting-plane method it is first necessary to extend the plane of the base circle. Line XY, for example, can be extended indefinitely beyond the base curve, and a second line, FG, parallel to XY, can be drawn tangent to the curve. The vertical cutting plane shown intersects these extended lines at E and F to establish the

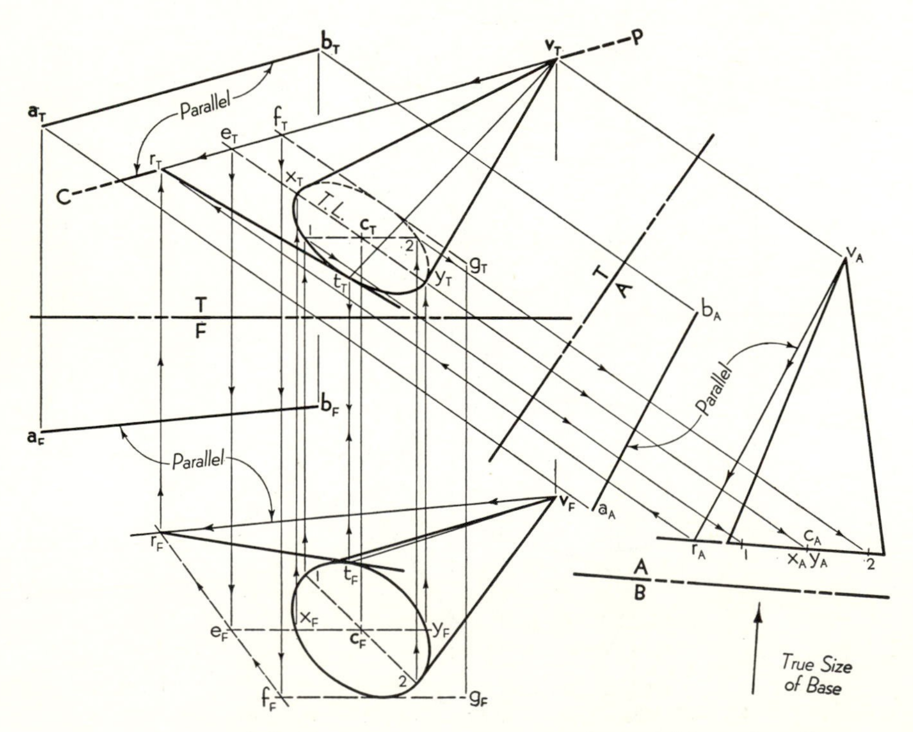

Fig. 6·22 (repeated). Plane tangent to a cone and parallel to a given line.

line of intersection between the cutting plane and the base plane, and points r_F and r_T are thus located.

When point R has been determined by either of the above methods, it only remains to draw the tangent line RT. As a check on the accuracy of the work, tangent point t_T must lie directly above tangent point t_F. Plane VRT is the required tangent plane. A second solution (not shown) may be obtained by drawing the tangent line RT on the opposite side of the base curve.

PROBLEMS. Group 68.

6·22. Location of a plane tangent to a cylinder through a given point on the cylinder

Analysis. The tangent plane can be defined by two intersecting lines: an *element* through the given point and a *line tangent* to the cylinder base curve.

Construction. In Fig. 6·23 the hidden point x_F is given on the oblique circular cylinder. Because the cylinder bases appear as edges in both the given views, some new view, such as view R, must be drawn to show the shape of the base curves. The element of tangency XT can then be located in each of the three views as shown.

The second line in the tangent plane is established in view R as line r_Rs_R tangent to the right base circle at point t_R. (It could also be drawn

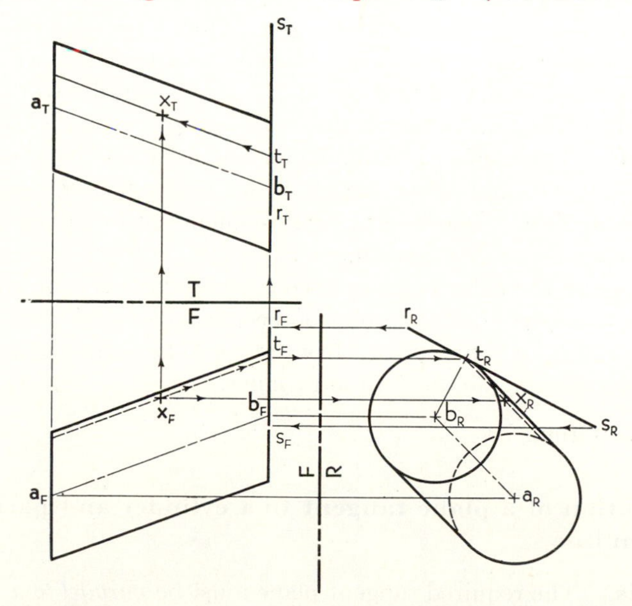

Fig. 6·23. Plane tangent to a cylinder through a given point on the cylinder.

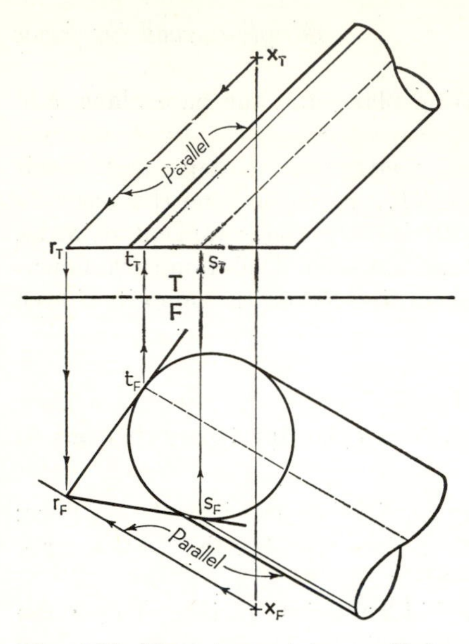

Fig. 6·24. Plane tangent to a cylinder through a given external point.

tangent to the left base circle.) When line RS has been located in the given views as shown, plane $XTRS$ is the required tangent plane. The similarity of solution for a cone or a cylinder should be noted (see Art. 6·19).

PROBLEMS. Group 69.

6·23. Location of a plane tangent to a cylinder through a given external point

Analysis. The required tangent plane must contain the *given external point,* and it must also contain a line that is parallel to the cylinder elements. Through the given point, therefore, a line must be drawn parallel to the cylinder. This is one line in the required tangent plane. The second line must intersect the parallel line and be tangent to the cylinder.

Construction. In Fig. 6·24 the given cylinder is described by one base and the direction of the elements, and the external point X is given as shown. A line is drawn first through point X parallel to the cylinder elements, and this line is then extended to intersect the base plane at point R. From point R either line RT or line RS may be drawn tangent to the base curve, and there are therefore two solutions, either plane XRT or plane XRS. The two possible elements of tangency are drawn through points T and S as shown.

If we imagine the cylinder to be a cone whose vertex is an infinite distance from the base, then this solution becomes identical with that for a cone (Fig. 6·21), for a line drawn from point X to such a distant vertex would necessarily be parallel to the cylinder.

PROBLEMS. Group 70.

6·24. Location of a plane tangent to a cylinder and parallel to a given line

Analysis. The required tangent plane must be *parallel to a trial plane* that contains the given line and is parallel to the cylinder. The pictorial drawing in Fig. 6·25 illustrates this analysis. Here the required tangent plane must be parallel to given line AB. Then through points A and B (or any other assumed points on AB), lines are drawn parallel to the cylinder. The plane so formed is the trial plane, and it is parallel to the required plane. When the trial plane is extended, it intersects the cylinder base plane along the line EF. But if the parallel required plane intersects this same base plane, its intersection RS will be parallel to the intersection line EF. Thus the tangent line RS and the element of tangency TU may be established.

Construction. In the multiview drawing lines are drawn through points A and B parallel to the cylinder elements. These lines are then extended to intersect the upper inclined base plane at points E and F, and the trial-plane intersection line e_Tf_T is thus established in the top view. Line RS of the required plane must be parallel to line EF of the trial plane, and therefore r_Ts_T is drawn parallel to e_Tf_T and tangent to the base curve at t_T. The element of tangency is line TU, and the required plane is plane $RSTU$. A second solution (not shown) is obtained when the tangent line is drawn on the opposite side of the base curve.

PROBLEMS. Group 71.

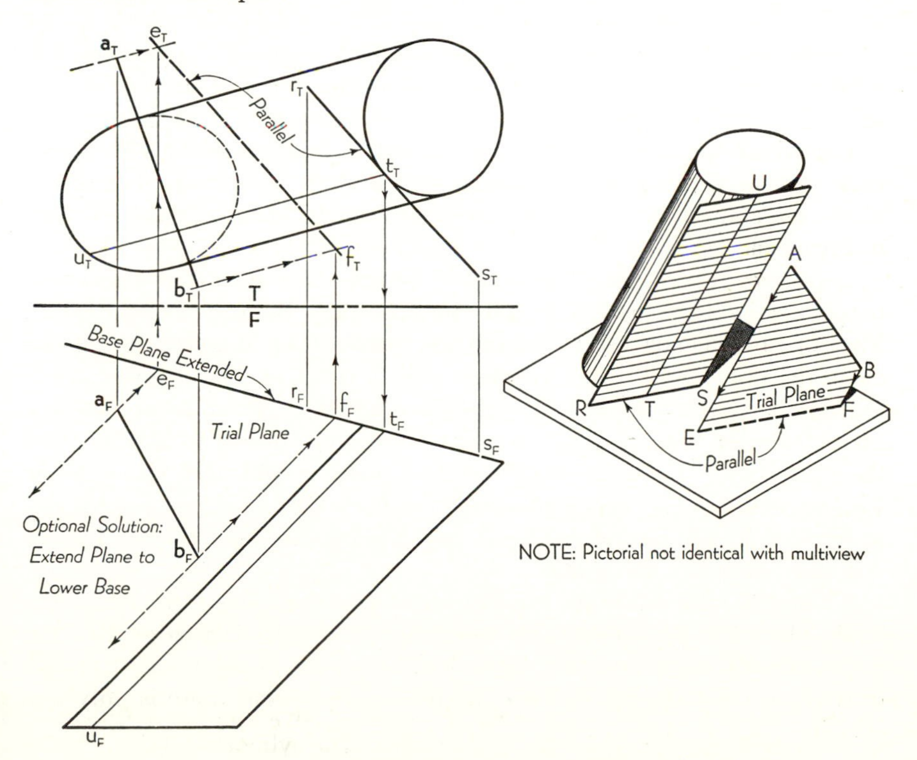

Fig. 6·25. Plane tangent to a cylinder and parallel to a given line.

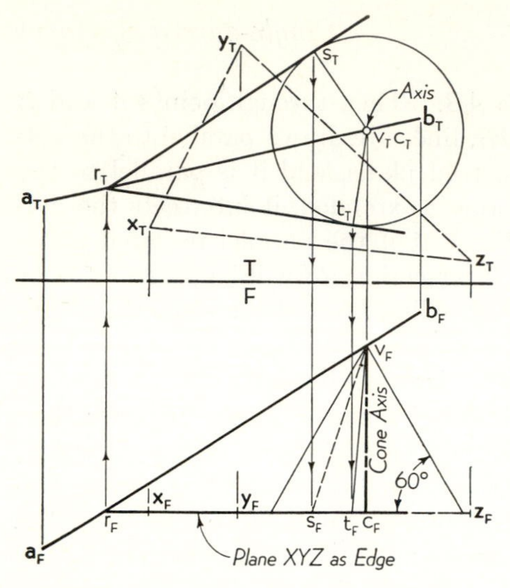

Fig. 6·26. Plane through a line and making a given angle with a given plane.

6·25. Location of a plane through a given line making a given angle with a given plane

Analysis. If a plane is tangent to a right circular cone (as in Fig. 6·20), then the dihedral angle that this tangent plane makes with the base plane of the cone is exactly the same as the *base angle* of the cone. A right circular cone therefore provides a convenient means of establishing any desired dihedral angle between two planes: the *base angle* of the cone must equal the desired dihedral angle, and the *axis* of the cone must be *perpendicular* to the given plane. Since the required plane must contain the given line, the cone *vertex* must be at some point *on the line.*

Problem. In Fig. 6·26 the given line is *AB*, and the given plane is *XYZ*. It is required to locate a plane that contains line *AB* and makes an angle of 60° with plane *XYZ*. Here the plane *XYZ* appears true size in the top view and as an edge in the front view; if plane *XYZ* had been given as an oblique plane, we should only need to construct two auxiliary views, edge and true size, to reduce the problem to the situation shown in Fig. 6·26.

Construction. From any point *V* on line *AB* the cone axis is drawn perpendicular to plane *XYZ*, and the base angle is made equal to 60°. The cone base is now established in the front view and is drawn as a circle in the top view. It only remains now to construct the required plane through line *AB* tangent to this right circular cone. From point *R*, where line *AB* intersects plane *XYZ*, tangent lines can be drawn to either side of the cone base. There are therefore two possible solutions, either plane *VRS* or plane *VRT*. The elements of tangency *VS* and *VT* may be drawn or omitted in this case, since the required planes are fully defined by line *AB* and either line *RS* or *RT*.

If point *R* falls inside the base circle of the cone, this means that the angle which line *AB* makes with the given plane *XYZ* is greater than the given dihedral angle and that there can be no solution. If point *R* falls exactly on the base curve, these angles are equal and there is only one solution.

PROBLEMS. Group 72.

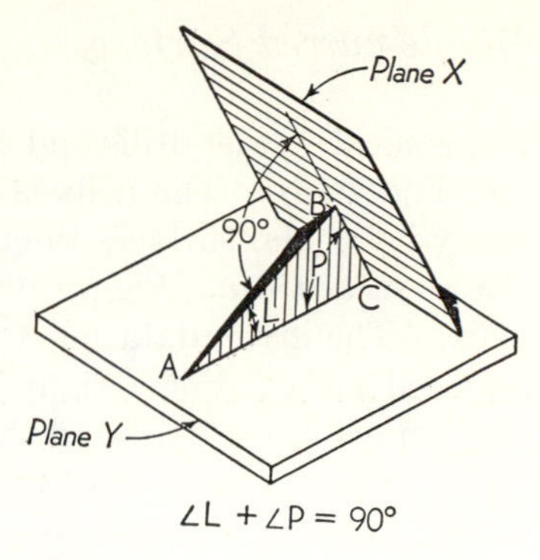

Fig. 6·27. A line perpendicular to a plane.

6·26. Location of a plane through a given point making given angles with each of two planes

Analysis. Figure 6·27 shows a line AB that is perpendicular to a plane X. The line AB makes an angle L with plane Y, and the plane X makes an angle P with plane Y. The triangle ABC is a right triangle, and therefore angles L and P are complementary. In general, then, if a line and a plane are perpendicular to each other, they make complementary angles with any other plane. Therefore, the problem of this article may be solved in two steps:

1. *Construct a line whose angles with the two given planes are complementary to the given angles.*
2. *Construct the required plane perpendicular to this line.*

Problem. Assume that it is required to locate a plane through a given point X and making angles of 23° and 42° with planes A and B, respectively.

Construction. *Step* 1. Assume any random point O and construct through this point a line that makes 67° (90 − 23) with plane A and 48° (90 − 42) with plane B. This construction would be similar to that shown in Figs. 5·18 or 5·19, and therefore no new illustration has been provided. As stated in Art. 5·17, there may be as many as four solutions, and any one of the four lines may be accepted as the result of step 1.

Step 2. Having established a line through point O that is perpendicular to the required plane, we may now draw the required plane through point X, as shown in Fig. 4·39 (Art. 4·33).

PROBLEMS. Group 73.

6·27. The helix

The helix is a double-curved line generated by a point that moves at a uniform rate around and also parallel to an axis. If the generating point is at a fixed distance from the axis, the helix is *cylindrical;* if the distance from the axis varies uniformly, the helix is *conical.* In either case the distance that the point moves parallel to the axis during one revolution is called the *lead,* or *pitch.* If *clockwise* rotation of the point causes it to move *away* from the observer, the helix is right-hand.

The most common application of the helix is the familiar screw thread, but it will also be recognized in screw conveyors, coil springs—cylindrical and conical—twist drills and spiral milling cutters, circular stairways, etc. (see Fig. 7·15). The helix is thus the *directing line* for the generation of many helicoidal surfaces in engineering.

Construction. Figure 6·28 shows the construction of a cylindrical helix. The given data must include the *diameter, lead,* and *hand* of the required helix. For a right-hand helix beginning at point 0 the curve must move upward around the cylinder in a *counterclockwise* direction as seen in the top view (the curve advances *toward* the observer in the top view). In one complete revolution the curve must move upward a distance equal to the lead. For any fractional amount of rotation the upward advance must be a corresponding fraction of the lead.

In Fig. 6·28 the circumference of the cylinder has been divided into 12 equal divisions (16 or 24 divisions is usually better). In the front view the lead has been divided into the *same* number of equal divisions. Twelve points on the helix may now be located in the front view. Point 2, for example, represents a rotation of 2 divisions, or one-sixth of a turn, from the starting point; the upward advance of point 2, must, therefore, be 2 divisions, or one-sixth of the lead. Point 2 is thus located in the front view as shown. When each of the 12 points has been similarly

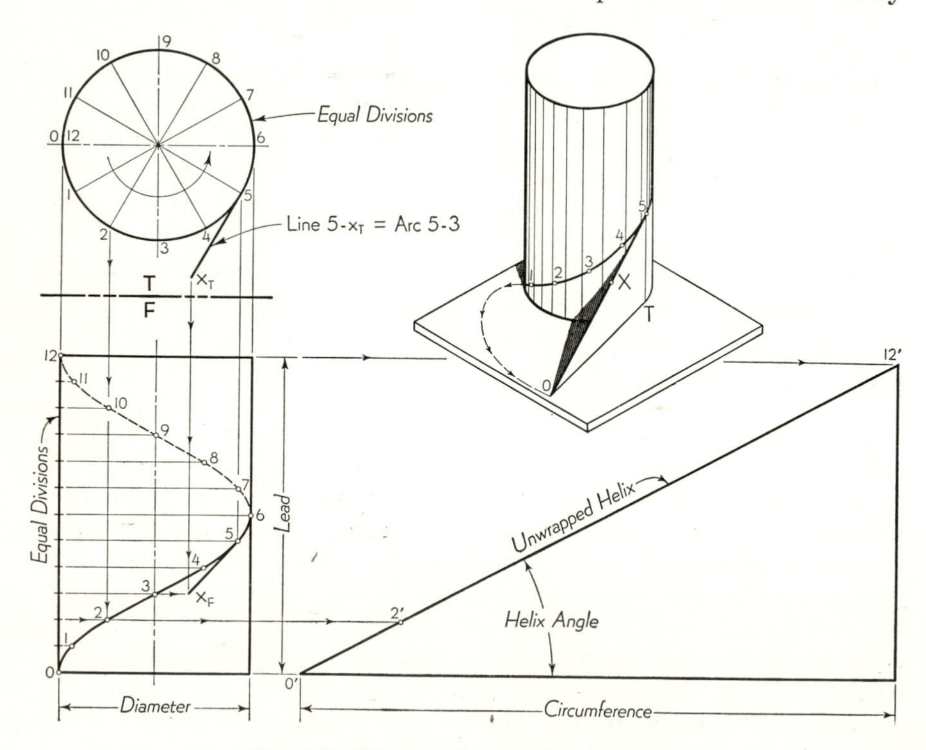

Fig. 6·28. The right-hand cylindrical helix.

located, a smooth curve can be drawn through them. That part of the curve from 0 to 6 is on the front half of the cylinder and is therefore visible; from 6 to 12 the curve must be drawn as a hidden line.

Helix angle. The right triangle shown at the right of the front view has been drawn to illustrate the important fact that the helix is a curve of *constant slope.* The base of the triangle is equal to the circumference of the cylinder, and the height is equal to the lead. The hypotenuse of this triangle is then equal in length to the true length of one turn of the helix, and the angle between the hypotenuse and the base represents the constant slope of the helix. The length of the helix and its *helix angle* may thus be accurately calculated from this triangle. We should particularly note and remember that

$$\textit{Tangent of the helix angle} = \frac{\textit{lead}}{\textit{circumference}} = \frac{\textit{lead}}{\pi\ \textit{diameter}}$$

If this triangular area is wrapped around the cylinder, as shown in the pictorial illustration, it will be seen that the hypotenuse of the triangle will coincide with the helix, thus demonstrating the truth of the above statements.

Tangent to a helix. To construct a tangent to the helix at some point, such as point 5, it is useful to observe that the tangent line must have the same slope as the helix. Thus, in the picture, tangent line $5–X$ coincides with the unwrapped helix, and if the length $5–X$ equals the helix length from 5 to 3, then point X must be at the same elevation as point 3. In the top view, line $5–x_T$ is drawn tangent to the circle at point 5, and the length $5–x_T$ is made equal to the arc length 5–3 (see Appendix, Arts. A·4 and A·5). In the front view, x_F must be at the same height as point 3 on the helix. Tangent line $5–x_F$ is thus located more accurately than if it were drawn tangent simply by eye.

6·28. Representation of a helical convolute

The helical convolute is generated by a straight line moving so that it is always tangent to a helix (see definition of convolute in Art. 6·3). This surface may be represented by showing the given helical directing line, and a series of straight-line elements to represent the consecutive positions of the generating line. In order to bound the surface it is also customary to show the curve connecting the outer ends of the elements [see Fig. 6·1(*c*)].

Problem. In Fig. 6·29 the left-hand helix AC is given. The generating line AB, of given length, shall extend downward from the point of tangency (thus generating only the lower nappe). It is required to show one turn of the helical convolute in the given views.

Construction. When the given helix has been constructed as described in the previous article, the helix angle should be determined. This may be done graphically as shown in Fig. 6·28, or the angle may be calculated by formula. We may now proceed to draw 12 elements of the convolute, corresponding to the 12 divisions used to draw the helix.

The first element drawn should always be one that appears *true length.* In Fig. 6·29 point A is on the front center of the cylinder, and therefore element AB will appear true length in the front view. The slope angle of line AB will also appear here and must be equal to the helix angle; hence $a_F b_F$ is drawn sloping downward at the helix angle and equal to the given length of the generating line. Line $a_T b_T$ is then drawn tangent to the

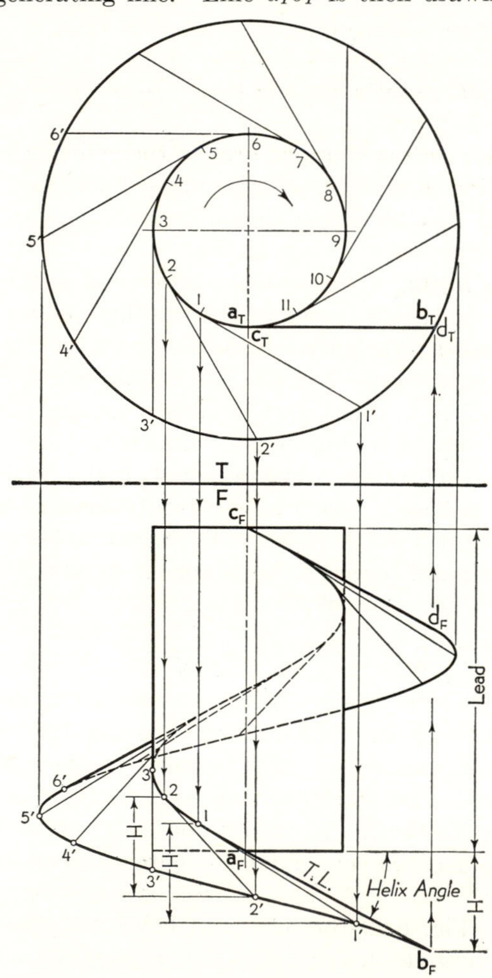

Fig. 6·29. Left-hand helical convolute (lower nappe only).

cylinder at a_T, directly above a_Fb_F. Since every element must have the same true length and slope as element AB, they will all have the same length in the top view. By drawing a second circle through point b_T the rest of the elements can be drawn tangent to the inner circle and terminated by the outer circle. The top view is now completed.

To show the 12 elements in the front view we need only observe that the difference of elevation between points A and B is distance H. This difference of elevation must be constant for all elements. Point 1′ must be distance H below point 1, point 2′ distance H below point 2, etc. Points 1′, 2′, 3′, etc., may thus be established in the front view directly below the corresponding points in the top view, but with each point located distance H below points 1, 2, 3, etc., on the helix. When points 1′, 2′, 3′, etc., are connected with a smooth curve, it will be found that this outer curve BD is also a helix of *greater diameter* than the given helix but having the *same lead.*

The helical-convolute surface drawn in Fig. 6·29 lies entirely in the annular space between two concentric cylinders. In this form it is sometimes used as the rotating blade in a screw conveyor. The helical convolute is not entirely satisfactory for this purpose, however, for its surface is not perpendicular to the axial direction of flow of the materials being conveyed. The *right helicoid* (Art. 7·17) provides such a surface, but it is more difficult to manufacture.

6·29. The helical convolute extended to a horizontal plane

The right-hand helical convolute shown in Fig. 6·30 is fundamentally the same as that shown in Fig. 6·29, except that each of the elements has been extended to intersect the horizontal base plane of the cylinder. Reference to Fig. 6·28 will show that the tangent line 5–0 in the pictorial illustration is actually the same as element 5–5′ in the pictorial of Fig. 6·30. The length of the tangent 5–0 in Fig. 6·28 must be equal to the length of the helix from 0 to 5, and the base length 0–T must be equal to that portion of the cylinder circumference between points 0 to 5.

Construction. In Fig. 6·30 the three-quarter-turn helix AB is given. In the top view all elements will be tangent to the cylinder, but their lengths will vary. The top-view length of element 5–5′, for example, must be equal to the arc length in the top view from 0 to 5 around the cylinder. If the divisions of the circle are taken small enough, then the chordal distance from 0 to 1 may be taken as equivalent to the distance along the arc and we may therefore lay off, along line 5–5′, five such spaces to locate point 5′ in the top view. Since point 5′ lies in the horizontal plane, it may now be located in the front view as shown and element 5–5′ is now established in both views. Each of the other elements is located in similar manner, remembering that each succeeding element in the top view will be one space longer than the previous one. The curve connecting points 1′, 2′, 3′, etc., in the top view is the involute of the circle.

Plane intersection. Assume that the convolute surface is intersected by a horizontal cutting plane as shown in Fig. 6·30. By noting the point, such as 7″, where each element intersects the cutting plane we can construct the curve of intersection in the top view. It will be found that this curve is also an involute of the circle.

Plane tangent. Let element 8–8′ be the desired element of tangency of a plane tangent to the surface. The line *RS*, tangent to the base involute at point 8′, is a second line in the required tangent plane. A third line *PQ* may also be drawn tangent to the upper involute at point 8″. Plane *PQRS*, shown also in the picture, is the required plane.

Tangent-plane generation. This helical-convolute surface can also be generated as the *envelope* of a series of tangent planes (see Art. 6·3). In Fig. 6·1(*d*) a convolute was generated by rolling a plane in contact

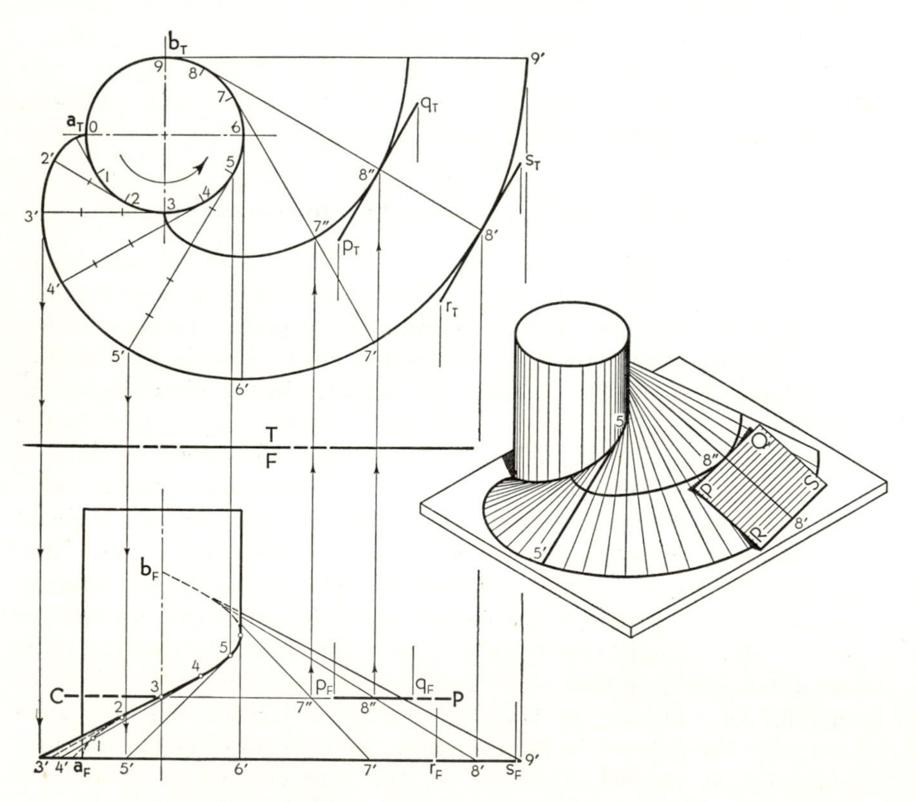

Fig. 6·30. Right-hand helical convolute extended to a horizontal plane.

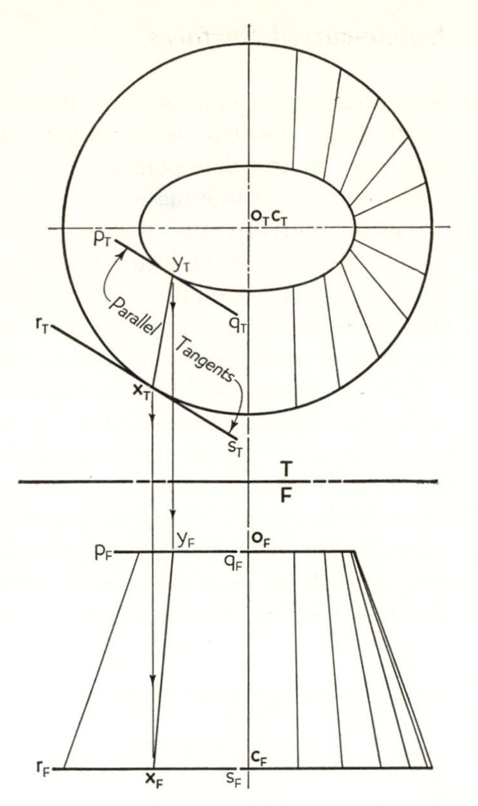

Fig. 6·31. Tangent-plane convolute: bases in parallel planes.

with two curved lines. To generate the helical convolute in this same manner requires that the two directing curves shall be involutes of the same circle lying in parallel planes. For many purposes the tangent-plane method of generating a convolute surface is the more useful. The practical application of this method will be discussed in the next two articles.

PROBLEMS. Group 74.

6·30. Tangent-plane convolutes: bases in parallel planes

It is frequently necessary to design a surface to connect two dissimilar curved lines. In Fig. 6·31, for example, the ellipse with center at O lies in one horizontal plane, and the circle with center at C lies in a lower parallel plane. If these two curves represent two adjacent openings, then the connecting surface is called a *transition piece*. To facilitate the flow of gases, fluids, or granular material from one opening to the other the transition piece should be as smooth as possible. For ease of manufacture it should also be a single-curved surface. A convolute surface is ideal under these conditions.

Construction. Proper representation of the desired convolute surface requires the location of a series of uniformly spaced straight-line elements, as shown on the right half of the surface in Fig. 6·31. On the left side of the surface is shown the construction for locating one such element, XY. Point X, on the circle, may be selected at random or may be one of a number of equally spaced points on the circumference. Now any plane tangent to the circle at point X must contain the tangent line RS, and this line necessarily lies in the horizontal plane of the circle as shown at r_Fs_F. But if the same tangent plane is also tangent to the ellipse, then the tangent line PQ, which lies in the parallel plane of the ellipse, must be parallel to line RS. Thus, if p_Tq_T is drawn parallel to r_Ts_T, it will be tangent to the ellipse at point y_T. The element XY is therefore the element of tangency, and it is also one element on the required convolute surface [compare Fig. 6·1(*d*)]. By applying the same construction to other points on the circle, additional elements may be obtained in the same manner.

The location of point y_T may be estimated simply by eye, but for accurate work it should be located by the construction given in Art. A·13*c* of the Appendix. If points, such as *Y*, are assumed initially on the ellipse, then the construction given in Art. A·13*a* should be used to establish the tangent *PQ* accurately. For convenience in developing the surface (Art. 10·20) the elements should be equally spaced on one of the two curves.

The method of establishing elements on the convolute surface in Fig. 6·31 is applicable to all cases where the given curves lie in parallel planes. The given curves need not be symmetrical, and each curve may lie anywhere in its plane. The openings may also be partly or entirely bounded by straight lines. Under such conditions a portion of the surface may sometimes be cylindrical, conical, or plane, but the method of construction remains the same.

6·31. Tangent-plane convolutes: bases in nonparallel planes

Figure 6·32 shows a convolute surface designed to connect two circular openings that lie in nonparallel planes. The tangent-plane method has again been employed, but the construction is complicated by the fact that the intersection of the tangent plane with each base plane is no longer parallel. The key to the solution is the *line of intersection of the two base planes*, for each tangent plane must either intersect or be parallel to this line.

Construction. In Fig. 6·32 a series of elements are shown on the rear half of the surface. On the front side only two elements, *AB* and *EF*, are shown, but the construction for each is given.

Step 1. Because the lower base curve is circular, with center at *O*, view *A* has been drawn to show this base in true size. This will simplify the location of tangent points on this base.

Step 2. The two base planes have been extended to intersect at p_Fq_F. This "key" line *PQ* is the line of intersection of the two base planes and should always be located in each view before attempting to establish any of the elements.

Step 3. To locate element *AB* select any point *A* on the upper base circle. A tangent drawn at point a_T is extended to point z_T, its intersection with line p_Tq_T. Line *AZ* represents the intersection of the tangent plane with the upper base plane. But the tangent plane must also be tangent to the lower circle, and it must also intersect line *PQ* at point *Z*. From z_T this second tangent line can be drawn tangent to the ellipse at point b_T, and line AB will be one of the required elements. But point B can be located more accurately in view A. Point z_A can be located on line p_Aq_A by measurement; and when the tangent z_Ab_A is drawn here, the perpendicular radius o_Ab_A will locate b_A with excellent accuracy. Point b_F and then b_T may now be located by alignment.

Additional elements may be located in similar manner, but each pair of tangent lines must always intersect on the line PQ. Care must be exer-

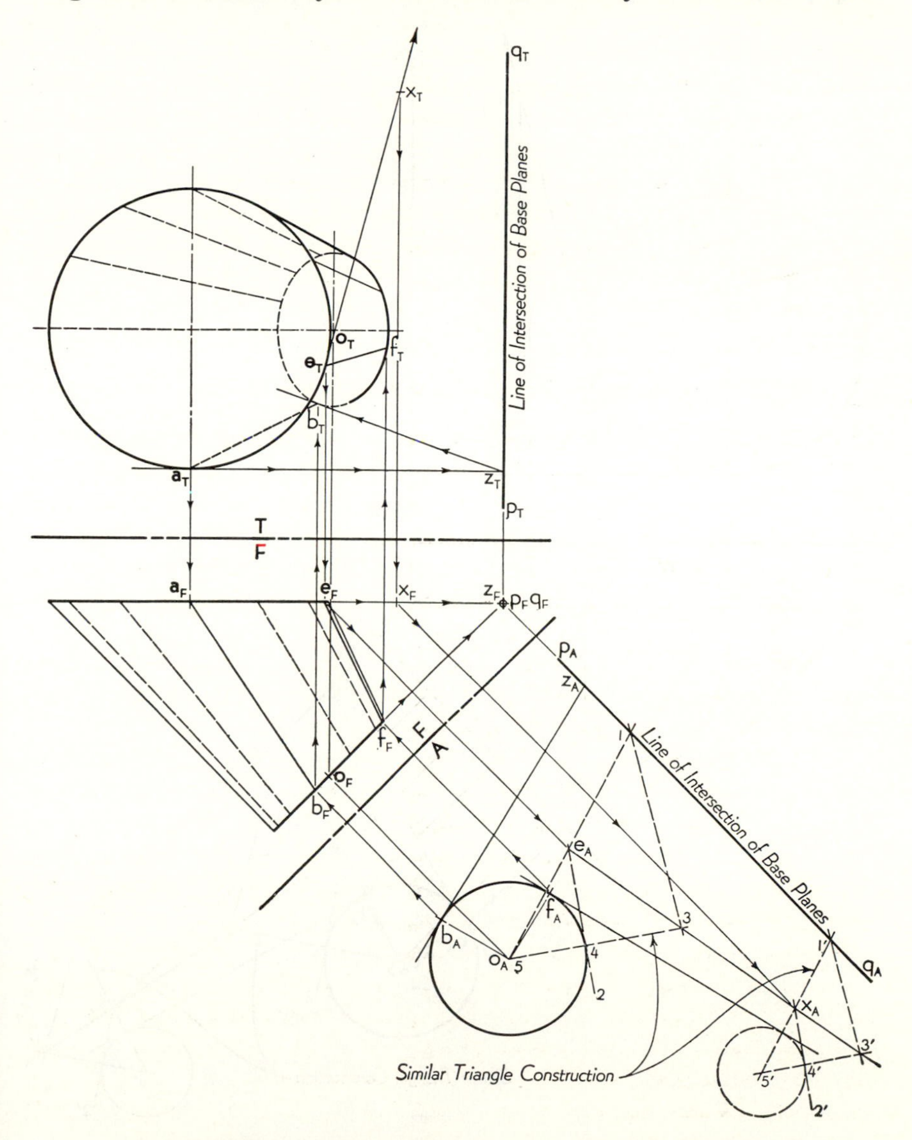

Fig. 6·32. Tangent-plane convolute: bases in nonparallel planes.

cised in drawing tangents in view A to ensure that they are drawn on the *proper side* of the curve. If point b_A had been located on the opposite side of the circle, then element AB would extend across the opening of the transition piece.

Element EF. This element has been selected to illustrate the procedure when the tangent plane intersects line PQ beyond the limits of the paper. At any point on the tangent line drawn from e_T a point x_T is

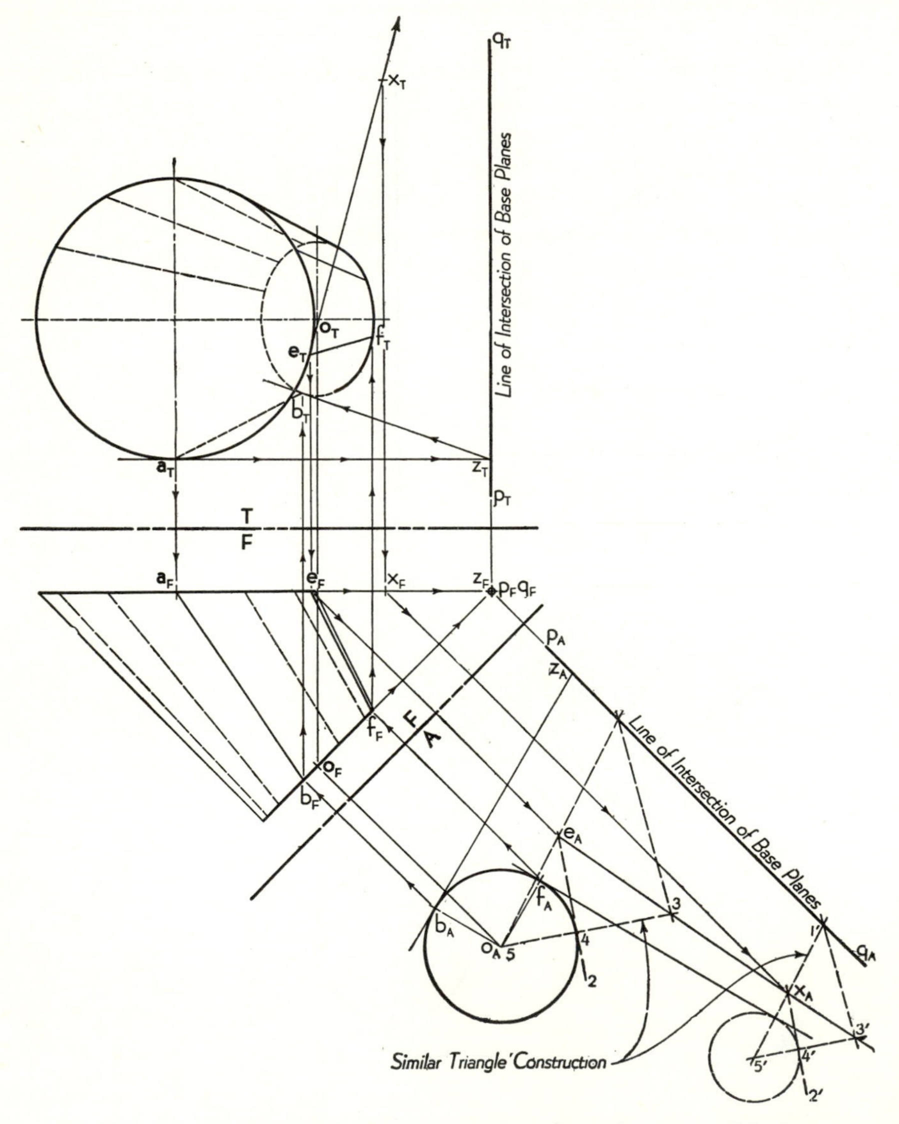

Fig. 6·32 (repeated). Tangent-plane convolute: bases in nonparallel planes.

assumed. Point X lies in the upper base plane, of course, and therefore points x_F and x_A are located as shown. Point e_A is also located in view A to establish the tangent line e_Ax_A. A line must now be drawn tangent to the circle through the inaccessible intersection of lines e_Ax_A and p_Aq_A.[1]

The solution is based upon the construction of two similar triangles as follows: Through points e_A and o_A a line is drawn to intersect p_Aq_A at point 1; line e_A–2 is then drawn tangent to the circle at point 4; line 5–4 is extended to intersect e_Ax_A at point 3, and line 1–3 is drawn to complete the triangle 1–3–5.

Starting at point x_A (or any other point on e_Ax_A) a similar triangle 1′–3′–5′ is constructed by drawing line 1′–5′ parallel to line 1–5, 1′–3′ parallel to 1–3, and 3′–5′ parallel to 3–5. Point 5′, the center of the smaller circle, is thus established. Drawing line x_A–2′ parallel to e_A–2 locates point 4′, and establishes the radius of the small circle. The required tangent line may now be drawn tangent to both circles to locate point f_A. Points f_F and f_T are located by alignment and measurement, and element EF is complete.

PROBLEMS. Group 75.

[1]A similar construction is also shown in the Appendix, Art. A·3.

7. WARPED SURFACES

7·1. What warped surfaces are

Warped surfaces are those surfaces which may be generated by moving a *straight line* so that any two consecutive positions of the generating line are *skew lines* (Art. 6·1). In contrast, the consecutive elements of single-curved surfaces are either *parallel* (cylinders) or *intersecting* (cones and convolutes). As we shall see in Chap. 10, single-curved surfaces are developable, that is, they may be formed by *rolling* or *curving* a plane sheet of metal; warped surfaces cannot be developed. This is an important distinction between these two classes of surfaces.

Warped surfaces find numerous applications in engineering, and some of these uses will be discussed in subsequent articles. Let us begin, however, with a simple problem that may be solved by means of a warped surface. Assume that it is desired to connect the lines AB and CD in Fig. 7·1 with a smooth continuous surface. Since lines AB and CD are neither parallel nor intersecting, we cannot use a plane or single-curved

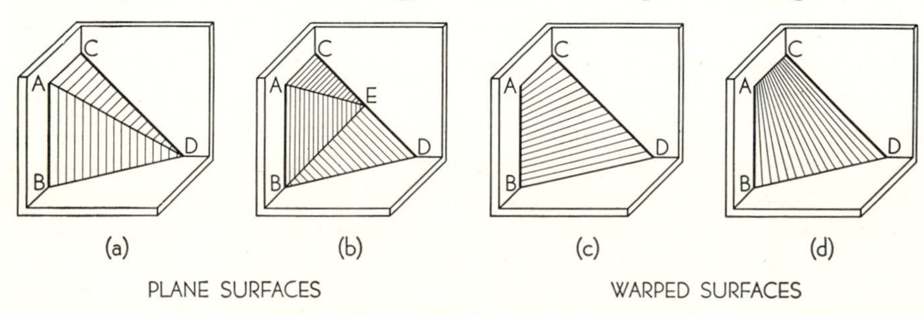

Fig. 7·1. Surfaces connecting two skew lines.

surface. By connecting points A and D with a straight line, as shown in Fig. 7·1(*a*), we can form two triangular plane surfaces ABD and ACD, but the connecting surface thus formed is not smoothly continuous. Division into three triangular areas, as shown in Fig. 7·1(*b*), is not much better, but by repeated subdivision into more and smaller triangles the connecting surface can be *approximated* by a series of very small triangular plane surfaces. The resulting surface would still be an approximation and not entirely satisfactory.

In Fig. 7·1(c) each of the lines AB and CD has been divided into 16 *equal* divisions, and these division points have been connected with straight lines. These lines are elements of a warped surface $ABDC$, and we now have a smoothly continuous surface connecting lines AB and CD. We may, of course, use smaller divisions on lines AB and CD and thus create more elements more closely spaced, but we should also observe that, no matter how close any two consecutive elements may be, that portion of the surface between them will not be plane. Exactly the same surface has been formed in Fig. 7·1(*d*) by dividing equally the lines AC and BD.

The warped surface shown in Fig. 7·1 is only one of an infinite variety. To understand what types of warped surfaces are possible we now need to consider the various ways in which a straight-line generatrix may be guided or constrained to move in order to generate a warped surface.

7·2. Classification of warped surfaces

The straight line that moves so as to generate a warped surface must always be guided or controlled by *three* directrices.[1] We may therefore divide warped surfaces into three broad classifications as follows:

1. The generatrix moves so that it is always in contact with *three line* directrices, straight or curved.

2. The generatrix moves so that it is always in contact with *two line* directrices and is always *parallel* to some *plane.*

3. The generatrix moves so that it is always in contact with *two line* directrices and always makes a *constant angle* with some *plane.*

The above classification is primarily one of convenience since *all* warped surfaces could be included in class 1 by the simple expedient of selecting any three lines on the surface as the three line directrices. Surfaces in class 2 could obviously be included also in class 3. For the practical purpose of constructing and representing the more common warped surfaces, however, we shall find this classification a useful one.

[1] In general, it may be said that the motion of *any* line which generates a surface must always be subject to three conditions of restraint.

Since the line directrices may be either straight or curved, we may further subdivide the above classification as shown in the following outline. The name of each surface or a representative of its class is given in boldface type. Some of the surfaces have no special name and have no important representative.

CLASSIFICATION OF WARPED SURFACES

1. Three line directrices.
   a. Three straight lines. ***Elliptical hyperboloid, hyperboloid of revolution.***
   b. Two straight lines and one curved line.
   c. One straight line and two curved lines. ***Warped cone, cow's horn.***
   d. Three curved lines.

2. Two line directrices and one plane director.
   a. Two straight lines. ***Hyperbolic paraboloid.***
   b. One straight line and one curved line. ***Conoid, right helicoid.***
   c. Two curved lines. ***Cylindroid.***

3. Two line directrices and a constant angle with a plane director.
   a. Two straight lines. ***Conchoidal hyperboloid*** (lines at right angles).
   b. One straight line and one curved line } ***Oblique helicoid.***
   c. Two curved lines }

Since no two consecutive elements of a warped surface may intersect or be parallel, certain obvious restrictions must be imposed upon the choice of the three directrices. For example, two straight-line directrices may not intersect, and a line directrix may not lie in the plane director. Such conditions may result either in an impossible situation or in the generation of a plane or single-curved surface.

All warped surfaces are *ruled surfaces* because all can be generated by a straight line, and therefore through any point on a ruled surface a straight line can always be drawn that will lie entirely in the surface. If only *one* straight-line element can be drawn through the point, then the surface is called *single-ruled.* If through any point on the surface *two* straight-line elements may be drawn, the surface is called *double-ruled.* The elliptical hyperboloid and the hyperbolic paraboloid are the only double-ruled surfaces possible (see Arts. 7·3 and 7·9).

Adequate representation of a warped surface in a multiview drawing usually necessitates showing not only the directrices and the bounding lines of the surface but also a number of evenly spaced elements.

7·3. The elliptical hyperboloid

Generation. The directrices for this warped surface must be three nonintersecting, nonparallel straight lines, no one of which is parallel to a plane parallel to the other two. If the three lines are all parallel to the same plane, then the surface becomes a hyperbolic paraboloid (see Art. 7·9).

Problem. Establish elements on the warped surface that has as its directrices the three skew lines *AB*, *CD*, and *EF* shown in Fig. 7·2.

Construction. The solution is simplified when one of the three lines appears as a point, as in Fig. 7·2. The point view also distinguishes this surface as an elliptical hyperboloid because any plane parallel to line *AB* would appear as an edge in the point view, and since c_Td_T is not parallel to e_Tf_T, such a plane obviously cannot be parallel to both lines *CD* and *EF*.

Every element of the surface must intersect each of the three lines *AB*, *CD*, and *EF*, and therefore every element must pass through point a_Tb_T. For uniformity of spacing, line e_Tf_T has been arbitrarily divided into eight equal divisions, and the top view of the warped surface is completed by drawing a straight-line element from each division point to point a_Tb_T. These elements may now be located in the front view by observing that element 2, for example, intersects line *EF* at point 2 and line *CD* at point 2′. Extending line 2–2′ in the front view locates point 2″, the point of intersection with line *AB*. For correct visibility of the surface

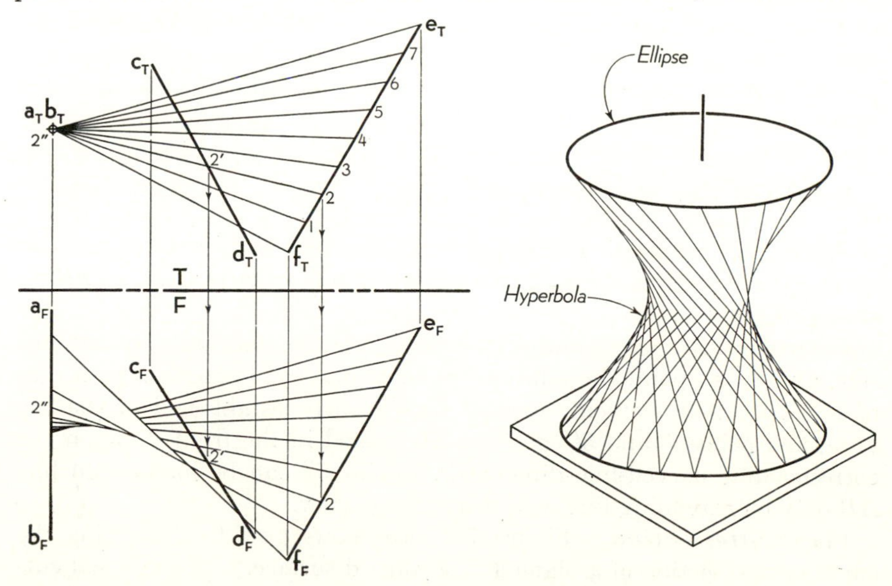

Fig. 7·2. Three straight-line directrices—the elliptical hyperboloid.

those elements nearest to the front should be drawn first; each succeeding element will then lie behind the previous one. In Fig. 7·2 the hidden part of each element has been omitted in the front view in order to emphasize the twisted shape of the surface.

The complete surface. The warped surface shown in the multiview drawing of Fig. 7·2 is actually only a small portion of a surface that in more complete form looks like the surface shown pictorially at the right. This hourglass-shaped surface is called an *elliptical hyperboloid* because planes perpendicular to its axis cut *ellipses* on its surface and planes parallel to its axis cut *hyperbolas*. Properly selected inclined planes will cut straight lines, circles, or parabolas from the surface, and therefore all plane sections of the elliptical hyperboloid are conic sections. This surface is also called the *hyperboloid of one nappe*. When the line directrices are so selected that the elliptical sections become circular, it is then called the *hyperboloid of revolution of one nappe*. The elliptical hyperboloid is of little practical value, but the more important hyperboloid of revolution will be discussed more fully in Art. 7·20.

Second generation. To show that the elliptical hyperboloid is a double-ruled surface, we may select at random any three of the elements shown in the multiview drawing of Fig. 7·2. Now let these three straight-line elements become the directrices, and let line *EF* be the generating line. As line *EF* moves in contact with these three new directrices, we shall then obtain a new set of straight-line elements. But every element of this *second generation* will intersect every element of the *first generation*, and therefore the two methods of generation form exactly the same surface. In the pictorial drawing the upper half of the surface shows only the elements of one generation, whereas the lower half shows both sets of elements.

7·4. Curved-line directrices

Problem. Establish elements on the warped surface that has one straight-line directrix *AB* and two curved-line directrices *CD* and *EF*, as shown in Fig. 7·3.

Construction. By viewing the three lines so that the straight line *AB* appears as a point the solution becomes identical in method with that of Fig. 7·2. Any element, such as 5–5″, can be established in the top view, and points 5 and 5′ can then be located in the front view on the corresponding curved-line directrices. Point 5″ can be located on line *AB* only by extending line 5–5′ in the front view.

Plane intersection. Figure 7·3 also shows how to determine the curve of intersection of a plane and a warped surface. The vertical cutting plane, for example, appears as an edge in the top view, and therefore in this same view the intersection of each element with the cutting plane can be seen. Each element intersection point can then be located in the front view on the corresponding element to establish the plane curve *XY*. Thus the intersection of a plane and a warped surface is found by the same basic method as that used for cones and cylinders.

Three curved-line directrices. The warped surface shown in Fig. 7·3 could also have been generated by employing the three curved-line directrices *CD*, *EF*, and *XY*. But if *any* three curved-line directrices are given, the process of locating each element becomes long and tedious and the establishment of a complete set of elements requires considerable time. Because such a surface very rarely occurs in practice, the construction is not shown here.

PROBLEMS. Group 76.

7·5. The warped cone

Generation. The warped cone is a practical example of a warped surface generated with one straight-line and two curved-line directrices. In this case the curved-line directrices are circular or elliptical and lie in nonparallel planes; the straight-line directrix is usually, but not necessarily, the line joining the centers of the two curves.

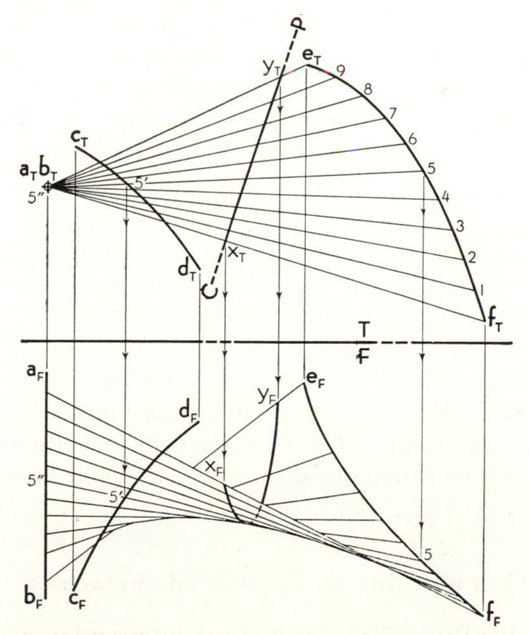

Fig. 7·3. One straight-line and two curved-line directrices.

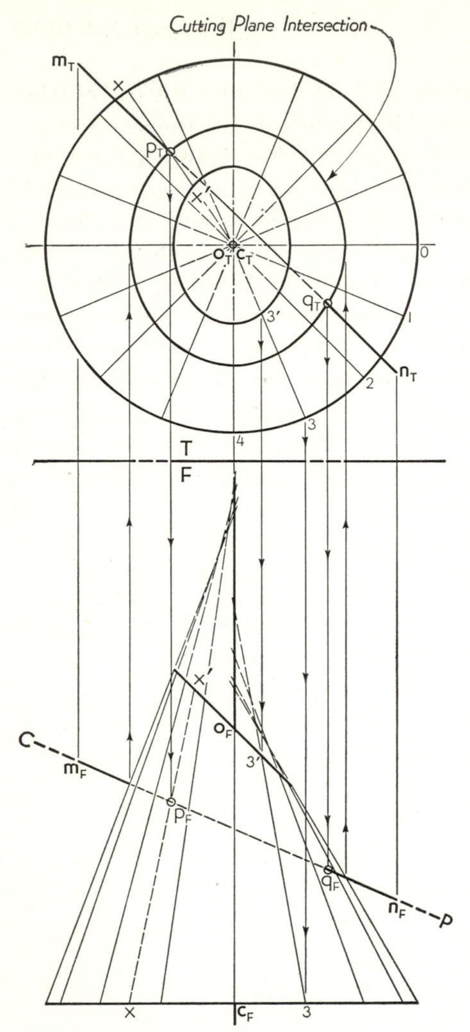

Fig. 7·4. The warped cone.

Problem. In Fig. 7·4 the curved-line directrices are circles, but one circle is horizontal with center at C, and the upper circle is inclined with center at O. Line OC is the given straight-line directrix.

Construction. In Fig. 7·4 line OC appears as a point in one of the views. This is essential to the solution. If line OC had appeared otherwise, the first step would have to be the construction of a new view showing OC as a point.

The elements must be established first in the top view, where OC appears as a point. For convenience, the large circle has been divided into 16 equal parts. From each division point, elements are then drawn to o_Tc_T. Each element may now be located in the front view by noting its intersection with each circle. Element 3–3′, for example, intersects the large circle at point 3 and the small upper circle at 3′. That this is no ordinary cone is quite apparent in the front view, where each extended element can be seen to intersect the axis at a different point.

Applications. This surface is commonly used as a transition piece to connect two dissimilar or nonparallel openings. To make this connecting surface of sheet metal, its development must be approximated by the method of Art. 10·21. The two openings could also be connected with a tangent-plane convolute surface which would be exactly developable (compare Fig. 6·32).

7·6. Intersection of a line and a warped surface

Analysis. By Rule 17 of Art. 6·9 the intersection of a line and *any* surface must lie on the intersection line of the given surface and a cutting plane that contains the line. To apply this method to a warped surface the surface must first be represented with a series of elements.

Problem. In Fig. 7·4 the points where line *MN* intersects the surface of the warped cone are required.

Construction. An edge-front cutting plane that contains line *MN* has been assumed. The curve of intersection of this plane and the warped cone must now be determined by noting the point of intersection of each individual element with the cutting plane. (This is exactly the same process as that described in Art. 7·4.) When each point has been established on the corresponding element in the top view, the points are connected with a smooth line to produce the intersection curve shown in Fig. 7·4. Line m_Tn_T intersects this curve at points p_T and q_T; these are the required intersection points, and they may now be located in the front view.

The visibility of points *P* and *Q* must agree with the visibility of the surface. Point p_T lies on a visible curve of the surface and hence is visible; but an element x–x' drawn through point p_T and then located in the front view would be hidden in that view, and therefore p_F is hidden. The visibility of point *Q* can be determined in similar manner.

PROBLEMS. Group 77.

7·7. The cow's horn (corne de vache)

Generation. The cow's horn is a second example of a warped surface generated with one straight-line and two curved-line directrices. In this case the curved-line directrices are two semicircles or two semiellipses that lie in parallel planes; the straight-line directrix is perpendicular to the plane of the curves and bisects their line of centers.

Problem. In Fig. 7·5 the curved-line directrices are the two semicircles with centers at *O* and *C*. The straight-line directrix *PQ* bisects line *OC* and is perpendicular to the planes of the semicircles.

Construction. Line *PQ* appears as a point in the front view, hence the elements should be drawn here first. For symmetry and uniform spacing the arcs from 0 to 8 and from 8′ to 16′ have each been divided into eight parts. The elements are then drawn from each division point to the center p_Fq_F. Then element 3–3′, for example, intersects the front circle at point 3 and the rear circle at point 3′. By locating these same points on the corresponding circles in the adjacent top and side views, the element may then be drawn in these views. Hidden elements have been omitted in the side view.

Applications. The cow's horn is sometimes used to form the *soffit*, or undersurface, of skew arches. The conoid (Art. 7·14) and the cylindroid (Art. 7·15) can be used for the same purpose when the front and rear *intrados*, or end curves of the soffit, are of dissimilar size or shape.

7·8. Location of a point on a warped surface

Analysis. The position of the point must be given in one of the views. To locate the point in an adjacent view, the point must lie on some *line* or *element* of the surface (Art. 6·6). Two cases then arise:

1. *The point may be given in a view in which any element may be assumed directly.*
2. *The point may be given in a view in which the elements must be derived from an adjacent view.*

Case 1. In Fig. 7·5 the given point on the surface is x_F, as shown in the front view. In this view a straight-line element, such as 4–4′, may be drawn through x_F and p_Fq_F. Locating points 4 and 4′ in the top view establishes the element and point x_T in that view.

Case 2. In Fig. 7·5 the given point on the surface is y_T, as shown in the top view. In this view the elements are not parallel; nor do they intersect at a common point. Hence we cannot accurately or conven-

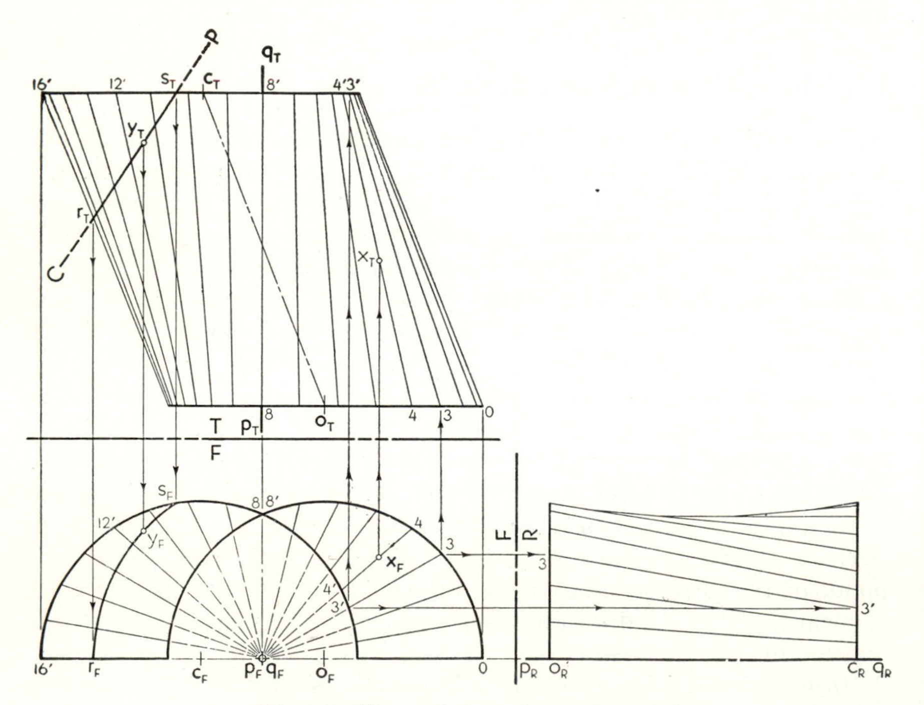

Fig. 7·5. The cow's horn (corne de vache).

iently establish a straight-line element of the surface through point y_T. Therefore, any cutting plane that contains point Y should be assumed. The vertical cutting plane intersects the warped surface along the curved line RS. Point y_F must now lie on the curve r_Fs_F in the front view directly below y_T. Actually, of course, only a segment of the curve r_Fs_F need be located in the vicinity of point y_F.

PROBLEMS. Group 78.

7·9. The hyperbolic paraboloid (plane director an edge)

Generation. The hyperbolic paraboloid is a warped surface whose straight-line generatrix moves so that it is always in contact with *two skew lines* and is always parallel to a *plane director*. The plane director may be assumed in any desired position, except that it must not be parallel to one of the line directrices.

Problem. Establish elements on the two hyperbolic paraboloids shown in Fig. 7·6. In each case the plane director appears as an edge in one of the given views, and lines AB and CD are the line directrices.

Construction. In Fig. 7·6(a) the plane director is horizontal, and therefore the desired elements are drawn first in the front view. Par-

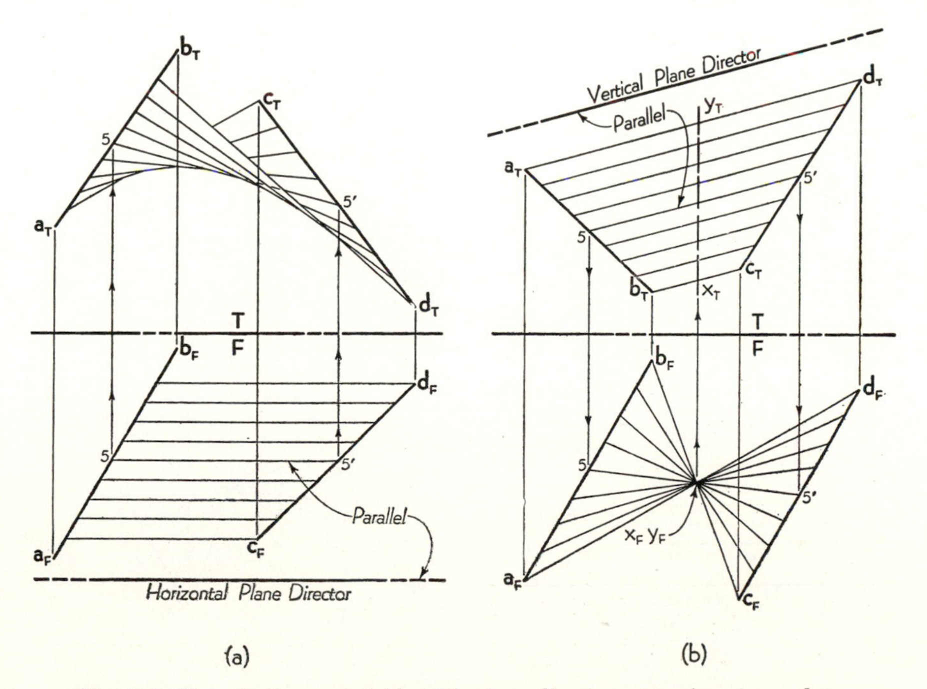

Fig. 7·6. Hyperbolic paraboloids with plane director appearing as an edge.

allels drawn from corresponding points, such as 5 and 5′, locate these same points on the line directrices in the top view to establish the element there. Because point b_F is farther from the horizontal plane director than d_F and a_F closer than c_F, it is impossible to utilize the whole length of line AB without extending line CD. As in previous examples, hidden elements have been omitted.

In Fig. 7·6(b) the plane director is vertical, and elements are therefore drawn first in the top view and are then located in the front view. (Note element 5–5′.)

Discussion. Because line a_Fb_F is parallel to line c_Fd_F, all elements in the front view pass through a common point x_Fy_F, and therefore the line XY is a third straight line that intersects all elements. This example thus demonstrates that the hyperbolic paraboloid can be generated with three straight-line directrices *if all three lines are parallel to the same plane* (see first paragraph of Art. 7·3).

7·10. The hyperbolic paraboloid (plane director oblique)

In Fig. 7·6(a) the elements in the front view were spaced equally to divide line c_Fd_F into eight equal parts. But that portion of line a_Fb_F

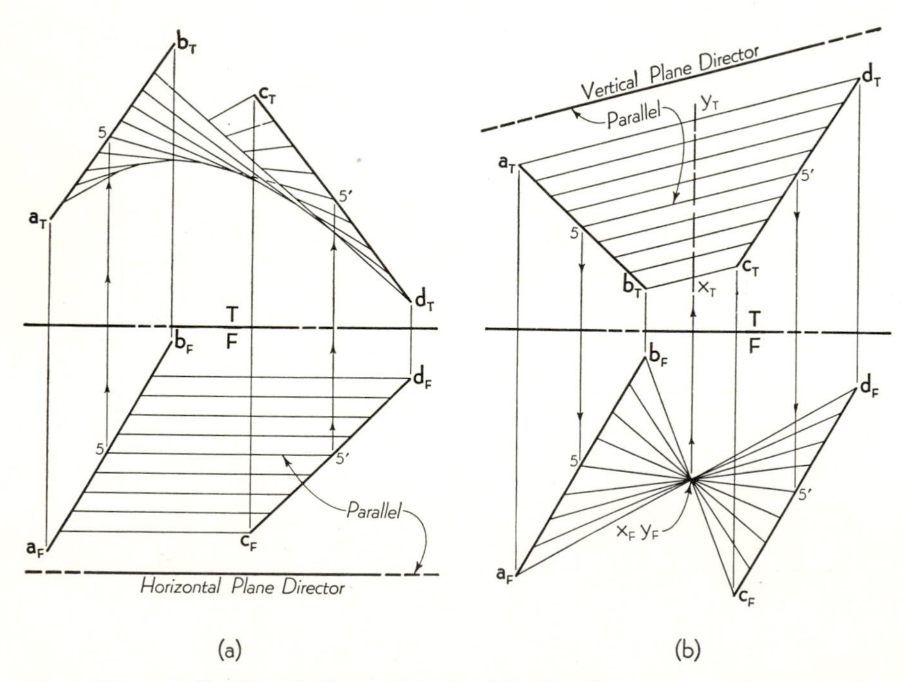

Fig. 7·6 (repeated). Hyperbolic paraboloids with plane director appearing as an edge.

which has been used is thus divided by the parallel lines into eight equal parts also. Similarly, in Fig. 7.6(*b*), the geometry of the figure shows that lines *AB* and *CD* each have the same number of equal divisions. By utilizing this equality of division the elements of a hyperbolic paraboloid may easily be drawn even though the plane director does not appear as an edge. The warped surface shown in Fig. 7·1 was constructed in this manner and is therefore a hyperbolic paraboloid.

Problem. In Fig. 7·7 the given skew-line directrices are *AB* and *CD*. The plane director is not shown, but it is required that the first element shall be line *AC* and the last line *BD*. In other words, the required surface shall be bounded by the warped quadrilateral *ABDC*.

Construction. Equally spaced elements are located as follows: divide line *AB* into any desired number of equal parts, and number the division points starting at point *A*; divide line *CD* into the *same number* of equal parts, and number the division points starting at point *C*; draw elements, such as 1–1′, 2–2′, etc., in each view by connecting similarly numbered points. Note that if either line is numbered in reverse sequence a different set of elements will be obtained; line *AD* would then be the first element and *BC* the last, and the resulting surface would not be the same as that shown in Fig. 7·7.

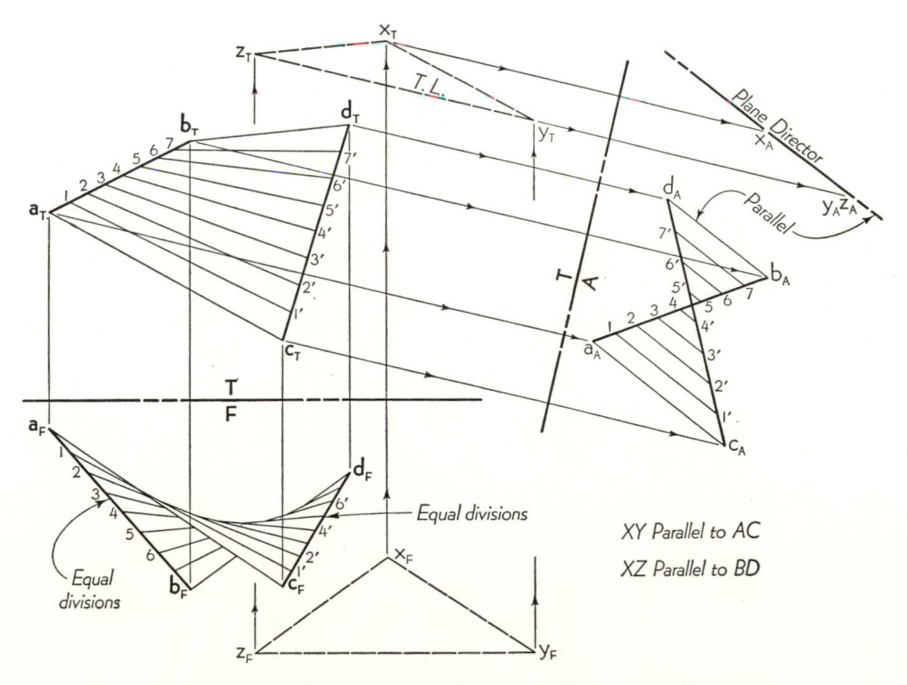

Fig. 7·7. Hyperbolic paraboloid with oblique plane director.

The plane director was not shown in Fig. 7·7; but since it must be parallel to the elements, its position can be determined. If we select any two of the elements shown, a plane director can then be constructed parallel to these two skew lines and this plane will then be parallel to all the elements. In Fig. 7·7 point *X* has been taken at random, and line *XY* has been drawn parallel to the first element *AC* and line *XZ* parallel to the last element *BD* (Art. 4·20). The plane *XYZ* is therefore a plane director for this hyperbolic paraboloid. To confirm the correctness of this solution, view *A* has been drawn to show plane *XYZ* as an edge. When the warped surface is also shown in this view, it is immediately evident that all the elements are parallel to the plane director.

7·11. The standard position of a hyperbolic paraboloid

In order to demonstrate some additional characteristics of the hyperbolic paraboloid, the surface shown in Fig. 7·8 has been placed in a standard position. In the top view the line directrices *AB* and *CD* appear parallel and of equal length.[1] The equally spaced elements connecting lines *AB* and *CD*, such as 2–2, 4–4, etc., must therefore appear parallel in the top view. But when the elements appear parallel, the plane director must appear as an edge (as demonstrated in Fig. 7·7) and *plane director* 1 must be the plane director for this surface.

As stated above, lines *AB* and *CD* have been considered the line directrices; and lines *AC* and *BD* are therefore the first and last elements. If, however, we let lines *AC* and *BD* be the line directrices, then the equally spaced lines 2′–2′, 4′–4′, etc., become a new set of elements of the warped surface. This is a second generation of the same surface and demonstrates that the hyperbolic paraboloid is a *double-ruled surface.* It also shows that this surface has a *second plane director* parallel to the elements of the second generation.

The pictorial illustration in Fig. 7·8 shows the saddlelike shape of this surface. Point *O*, where two horizontal elements—one of each generation—intersect, is called the *vertex.* The vertical line through point *O*—parallel to both plane directors—is called the *axis* of the surface.

7·12. The intersection of a plane and a hyperbolic paraboloid

The intersection of a plane and a hyperbolic paraboloid must be a single straight line, two intersecting straight lines, a hyperbola, or a parabola. In Fig. 7·8 any vertical plane parallel to either plane director will intersect the surface in a single straight line that will be one of the ele-

[1] By the construction of additional views any hyperbolic paraboloid may be so shown.

ments of the surface. The intersection of any other vertical plane, such as *C-P*-1, will be a parabola as shown in the front view. Points on the parabola are established (as described in Art. 7·4) by noting where each element of the surface intersects the cutting plane.

The horizontal plane *C-P*-2 is below the vertex *O* and intersects the surface in the two curves labeled *X*, which are the two equal branches of a hyperbola. An inclined plane *C-P*-3 above the vertex *O* will intersect the surface in a hyperbola also, but the two branches, labeled *Y*, will be unequal owing to the inclination of the plane. The horizontal plane that passes exactly through vertex *O* will intersect the surface in two intersecting straight lines, namely, the elements 6–6 and 6′–6′.

It is an interesting fact that the *contour curve* formed by the envelope of straight lines is always a *parabola* for any hyperbolic paraboloid in any position. Note, for example, the top view of Fig. 7·6(*a*) and the front views of Figs. 7·7 and 7·8 (see also Appendix, Art. A·15). The contour curve in the front view of Fig. 7·2 is not parabolic but elliptical.

7·13. Applications of the hyperbolic paraboloid

In Art. 7·1 a warped surface was designed to form a smoothly continuous surface connecting two skew lines (Fig. 7·1). The lateral surface of a concrete retaining wall, having sides that gradually change in slope,

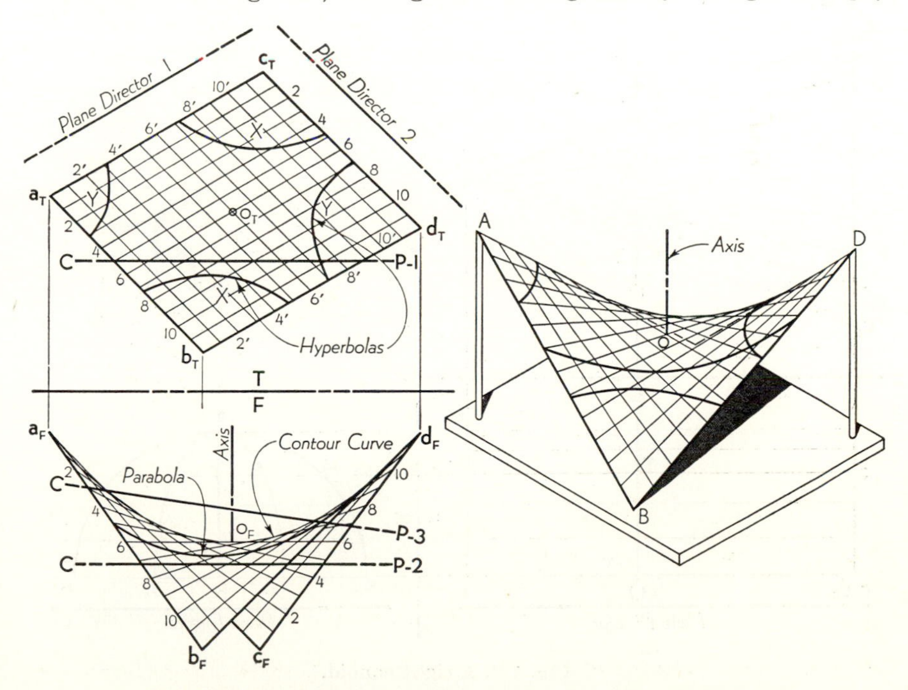

Fig. 7·8. Intersection of planes and a hyperbolic paraboloid.

would be a hyperbolic paraboloid. The surface is also used frequently to form a smooth transition from a vertical wall to a sloping wall in canals, irrigation ditches, flumes, and tunnels.

Since warped surfaces cannot be developed (Art. 7·1), they are most adaptable to use on machines or structures where the twisted surface can be carved, molded, or pressed into shape. To form the hyperbolic paraboloid in concrete the wooden form must be made of narrow strips of wood running in the direction of the elements of the surface. Similar strips parallel to the elements of the second generation are used to stiffen the form. When each strip is nailed or fastened to each cross strip, the wooden form becomes a strong rigid unit.

PROBLEMS. Group 79.

7·14. The conoid

Generation. The conoid is a warped surface whose straight-line generatrix moves so that it is always in contact with *one straight line* and *one curved line* and is always parallel to a *plane director*. If the straight-line directrix is *perpendicular* to the plane director, the surface is a *right conoid;* otherwise, it is an *oblique conoid.*

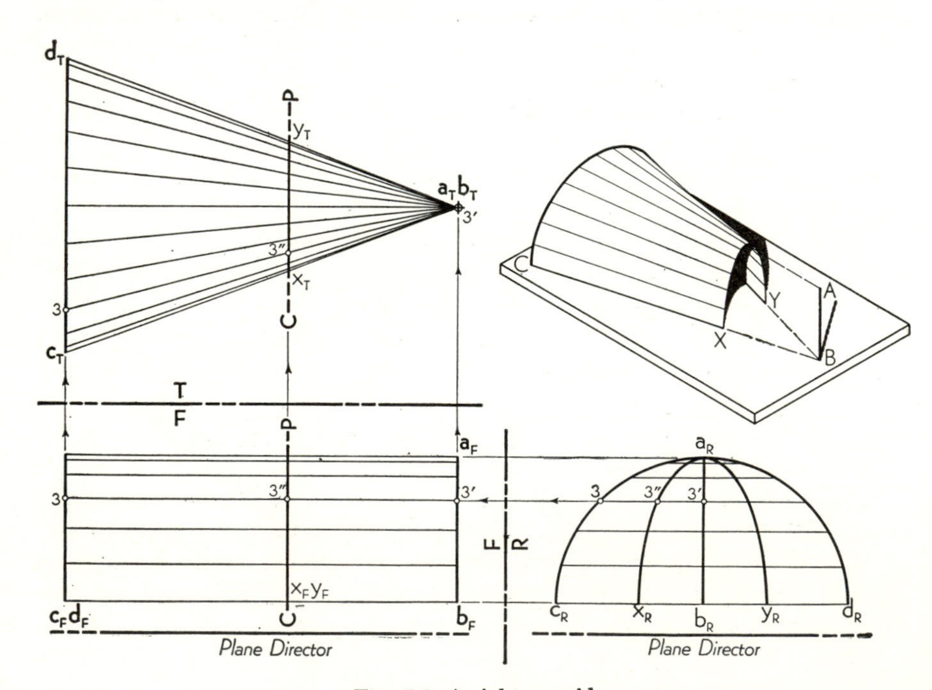

Fig. 7·9. A right conoid.

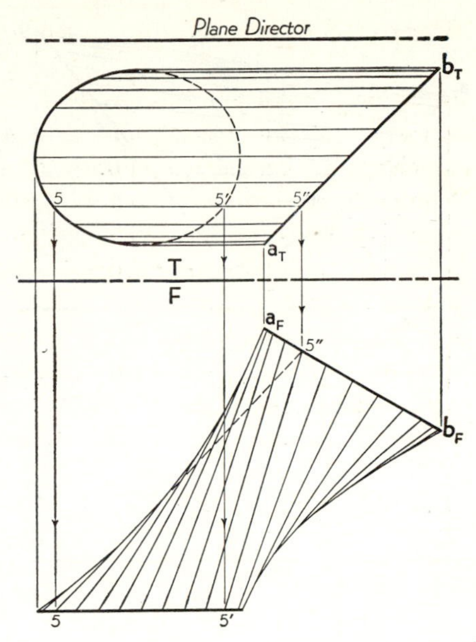

Fig. 7·10. An oblique conoid.

Problem. In Fig. 7·9 the *right conoid* has for its straight-line directrix the line AB and for its curved-line directrix the semicircular arc CD. The plane director is horizontal.

Construction. Because the plane director appears as an edge in both the front and right-side views, the elements will appear parallel in these views. For best appearance the elements have been spaced by dividing the arc $c_R d_R$ into equal parts. Any element such as 3–3′ may then be drawn in the side view parallel to the plane director. By locating point 3 on arc CD and point 3′ on line AB, in the front and top views, this element is established. Other elements are similarly drawn to complete the representation of the surface.

A vertical cutting plane parallel to the plane of the semicircle will intersect the right conoid in a *semielliptical* curve XY as shown in the figure. As in previous examples this curve is determined by finding the point of intersection of each individual element with the cutting plane. Element 3–3′, for example, intersects the cutting plane at point 3″. The conoidal surface between the two curves could be used as the soffit of an arched passageway connecting circular and elliptical openings.

Problem. In Fig. 7·10 the *oblique conoid* has a vertical plane director, and the line directrices are the ellipse and the line AB.

Construction. Elements are drawn first, of course, in the top view, where they appear parallel to the plane director. Element 5–5″ intersects the ellipse at point 5 and line AB at point 5″. Locating these points in the front view establishes the element 5–5″. A second element, 5′–5″, lies directly under element 5–5″; it, too, should be shown in the front view. Note that many of the elements, such as 5–5″, are partly visible and partly hidden in the front view.

A conoidal surface like that shown in Fig. 7·9 should not be confused with the cow's horn (Fig. 7·5); they are generated differently, and any resemblance is entirely superficial. The same is true of the oblique conoid (Fig. 7·10) and the warped cone (Fig. 7·4); the elements of the warped cone are not parallel to any plane.

PROBLEMS. Group 80.

7·15. The cylindroid

Generation. The cylindroid is a warped surface whose straight-line generatrix moves so that it is always in contact with *two curved-line directrices* and is always parallel to a *plane director.* The curved lines may be either plane or double-curved, but in practice they are most commonly circular or elliptical and lie in nonparallel planes.

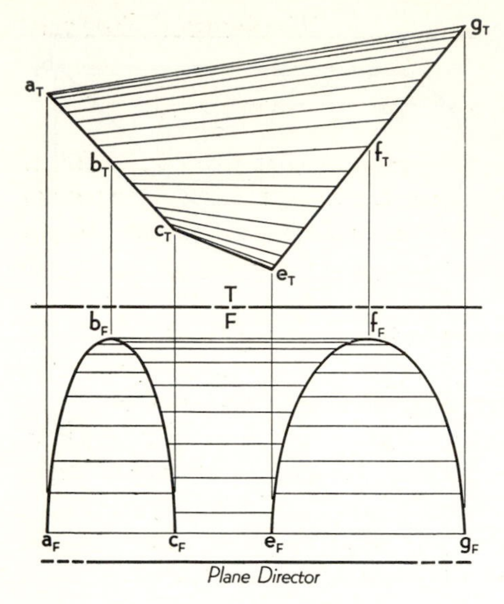

Fig. 7·11. A cylindroid.

Problem. In Fig. 7·11 the cylindroid shown has a horizontal plane director, and the curved-line directrices are the two elliptical curves *ABC* and *EFG.*

Construction. In the front view all elements will appear parallel to the plane director; hence they are drawn first in this view. Spacing of the elements along the curves is intentionally unequal in this case to obtain a more uniform appearance in the top view. When each element has been located in the top view, the representation of the surface is complete.

In order that a cylindroidal surface connecting two curved lines shall be continuous, it is essential that the extremities of the two curves shall be equidistant from the plane director. In Fig. 7·11, for example, points b_F and f_F must be equidistant from the plane director. If b_F is higher than f_F, it will be impossible to draw horizontal elements to that part of curve *ABC* which is above the level of point *F.*

Discussion. A portion of the surface of a hyperbolic paraboloid or of a conoid may sometimes appear to be a cylindroid. In Fig. 7·9, for example, that portion of the conoidal surface shown in the picture lies between two curved lines and can properly be called a cylindroid. Whether the cylindroid shown in Fig. 7·11 is also conoidal depends upon whether the elements extended would all intersect some straight line; to prove or disprove the existence of such a straight line would, in general, be very difficult or impossible. By intersecting a hyperbolic paraboloid with two random planes, that portion of the surface intercepted between the planes would be bounded by two curved lines and could properly be called a cylindroid. From a practical standpoint, however, it is of no importance whether a particular cylindroid may accidentally also be a conoid or a hyperbolic paraboloid.

7·16. Applications of the conoid and the cylindroid

Both the conoid and the cylindroid may be used to form the soffit of arched passageways connecting openings of dissimilar shape. The conoid is particularly suitable when it is necessary to connect a curved or vaulted ceiling to a flat one; the cylindroid is useful when the openings to be connected are in nonparallel planes.

Transition pieces (Art. 6·30) are frequently designed to employ these surfaces, but in the process of development and construction from sheet metal the warped surfaces are usually approximated with plane or single-curved surfaces (Art. 10·21). If the metal is formed under pressure between dies, it can be stretched or distorted to assume the exact shape of the warped surface.

Some portions of the hull of a ship, of an airplane fuselage, or of an automobile body can be designed as conoid or cylindroid surfaces. The advent of large presses in recent years has, however, caused a greatly increased use of double-curved surfaces. The body panels and fenders of the modern automobile are excellent examples of large double-curved surfaces formed in huge presses.

PROBLEMS. Group 81.

7·17. The right helicoid

Generation. In its general form the helicoid is a warped surface whose straight-line generatrix moves so that it is always in contact with each of *two coaxial helices* and makes a *constant angle* with their *axis.*[1] In this case the plane director is perpendicular to the axis of the helices, and the generatrix thus makes a constant angle with the plane director also. The generating line usually, but not necessarily, intersects the axis of the helices, in which case the axis may serve as a straight-line directrix and only one helical directrix is necessary. When the generating line is perpendicular to the axis, the surface is called a *right helicoid;* otherwise, it is an *oblique helicoid.* The right helicoid whose generatrix intersects the axis may also be classed as a *conoid* because its elements are all parallel to the plane director (see outline in Art. 7·2).

Problem. In Fig. 7·12 a *right helicoid* is shown in its simplest form. Here the generating line intersects and is perpendicular to the axis AB, and the helicoidal surface is assumed to extend only from the small inner cylinder to the outer helix CD.

[1] This is a limited definition since helicoids of varying pitch and varying radius, although uncommon, are also feasible, and in such cases the angle between the generatrix and the axis is also variable. Screw propellers are frequently of varying pitch.

Construction. The first step is to draw the helix CD, whose diameter, lead, and hand (right-hand, in this case) are given (Art. 6·27). The equal divisions of lead and circumference can then be employed to locate equally spaced elements. In the top view the elements, such as 6–6″, appear as radial lines connecting the axis a_Tb_T and the circle c_Td_T, but each element is terminated at the point where it intersects the inner cylinder (point 6′). In the front view the elements will all be horizontal and are so drawn from each corresponding point on helix c_Fd_F. Point 6′ may then be located on element 6–6″ in this view; and when each such point has been similarly located, it will be found that the curve e_Ff_F connecting these points is also a helix of the same lead as helix c_Fd_F. This inner curve will be a helix, regardless of the diameter of the inner cylin-

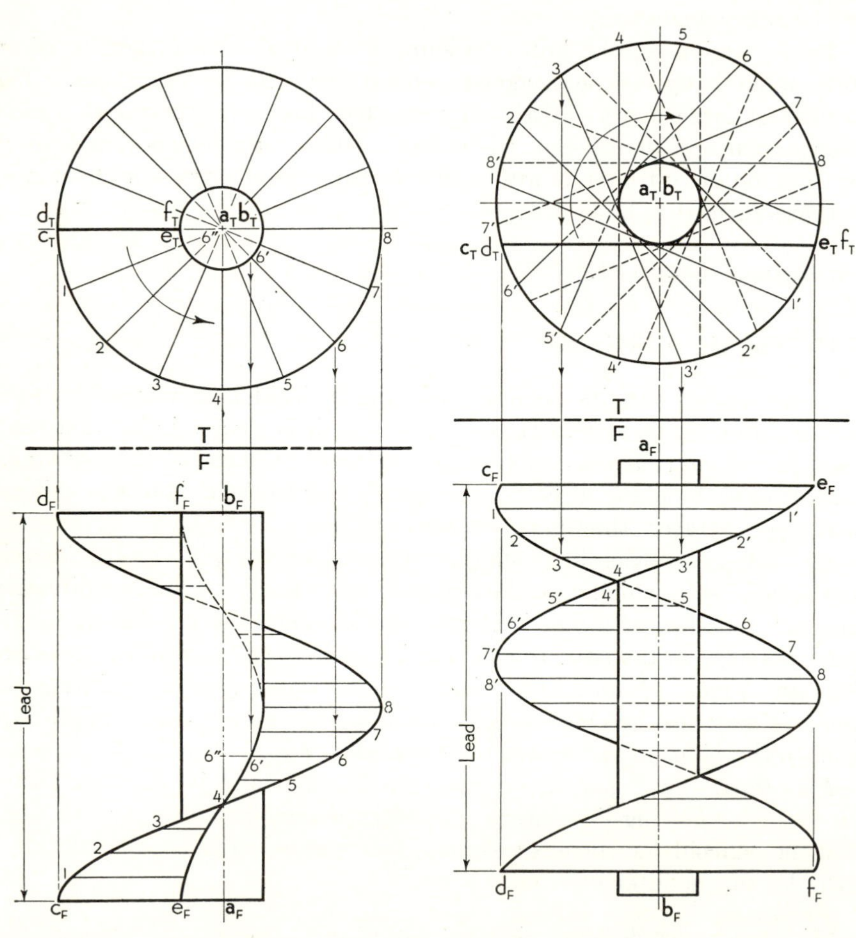

Fig. 7·12. A right helicoid—axis and generatrix intersecting.

Fig. 7·13. A right helicoid—axis and generatrix not intersecting.

der, and we therefore conclude that every point (except at the axis) on the generating line of a helicoid must follow a helical path.

Problem. In Fig. 7·13 the *right helicoid* is one whose generatrix does not intersect the helix axis but remains a constant distance from it. Each element is therefore tangent to the inner right circular cylinder. Assume that the axis *AB* is given, that line *CE* represents the initial position of the generatrix, and that the lead of the right-hand helicoid is given as shown.

Construction. As the generating line *CE* rotates in a clockwise direction, as shown in the top view, and moves downward, point c_F will follow the helical path that ends at d_F, and point e_F will follow the helical path that ends at f_F. By constructing these two helixes first, the horizontal connecting elements may then be drawn.

In the top view the inner directing cylinder is drawn tangent to c_Te_T, and the circular path of c_T and e_T is also drawn. (In the general case, points c_T and e_T may be at different distances from the axis and will therefore have separate circular paths.) Beginning at point c_T the outer circle is divided into equal parts, and the division points are labeled 1, 2, 3, etc. From each of these division points, elements are then drawn tangent to the inner circle and extending to the opposite side of the large circle. The opposite end of each element is labeled correspondingly 1′, 2′, 3′, etc., and these equally spaced points then represent the consecutive positions of e_T. Care must be exercised to draw all the tangent elements on the same side of the inner cylinder; going from c_T to e_T, the element passes to the right of the inner circle, and so must every other element. The visibility of the elements in the top view should be studied carefully, remembering that element *CE* is on top, 1–1′ is below *CE*, 2–2′ is below both *CE* and 1–1′, and 8–8′, for example, is below all other elements having lower numbers.

In the front view the lead distance c_Fd_F is divided into the same number of equal divisions as were used in the top view. Points 1, 2, 3, etc., are then located in the front view on the corresponding divisions of the lead to establish the helix c_Fd_F; points 1′, 2′, 3′, etc., similarly determine the helix e_Ff_F. Horizontal elements may now be drawn connecting the two helices to complete the representation of the surface. This right helicoid is not a conoid, but it could be classed as a cylindroid.

7·18. The oblique helicoid

Problem. In Fig. 7·14 the *oblique helicoid* is one whose generatrix intersects the axis at a constant angle. Assume that the helix *CD* with axis at *AB* is given, that the generating line slopes outward and down-

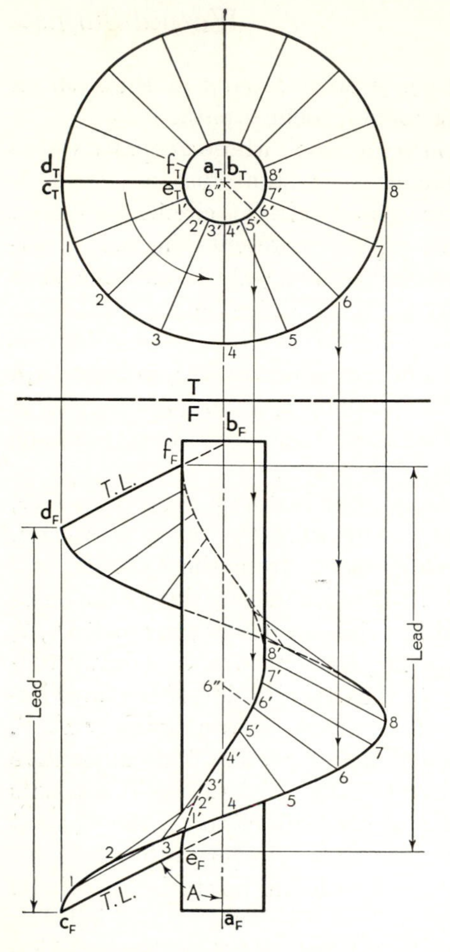

Fig. 7·14. An oblique helicoid.

ward at an angle A with the axis, and that the required surface shall be limited to the annular space between the inner and outer cylinders.

Construction. Equally spaced elements are located in the top view exactly as was done in Fig. 7·12. The first element drawn in the front view should be one which appears true length in that view. From c_F, for example, element c_Fe_F may be drawn at angle A with the axis to intersect the inner cylinder at e_F. Point e_F is therefore the starting point of the inner helix that ends at f_F and has exactly the same lead as the outer helix c_Fd_F. Points 1′, 2′, 3′, etc., may be extended down from the top view to construct this second inner helix. With both helices established the elements can then be drawn by connecting corresponding points on each curve. Visibility of the curves and elements in Fig. 7·14 should be studied carefully.

The construction for an oblique helicoid whose generatrix does not intersect the axis would be similar to that shown in Fig. 7·13 for a right helicoid, and therefore this case has not been illustrated. The top view, however, would be identical with that of Fig. 7·13, but in the front view the generating line c_Fe_F would be inclined at the required angle. The helical curve e_Ff_F would again have the same lead as helix c_Fd_F, but it would be higher or lower depending upon the inclination of line c_Fe_F.

7·19. Applications of the helicoid

Figure 7·15 shows a few of the more common practical applications of the helicoid surface. The square thread (a) is formed by cutting a helical groove of square cross section in a metal cylinder; each side of the groove is a right helicoid like that shown in Fig. 7·12. The 60° V thread (b) is made in similar manner except that the cutting tool is pointed; each side of the V-shaped groove is therefore an oblique helicoid like that shown in Fig. 7·14. The helical coil spring (*c*) is formed by wrapping a bar or wire around a cylinder; its inner and outer surfaces are cylindrical, but the other two surfaces are right helicoids.[1]

One type of twist drill is made by heating and then twisting a rectangular bar of steel into the shape shown in Fig. 7·15(*d*). The end is then pointed, and the cutting edges are ground to the proper angle for efficient drilling. The twisted surfaces of this drill are helicoids of the type shown in Fig. 7·13.

Figure 7·15(*e*) shows a spiral chute used to convey packages from one floor of a factory to a lower floor. The slide surface is a right helicoid, and the guardrail is cylindrical. The helicoid, like all warped surfaces,

[1] On engineering drawings the helical curves of screw threads and coil springs are usually drawn as straight lines to save time. This conventional representation is illustrated in texts on engineering drawing.

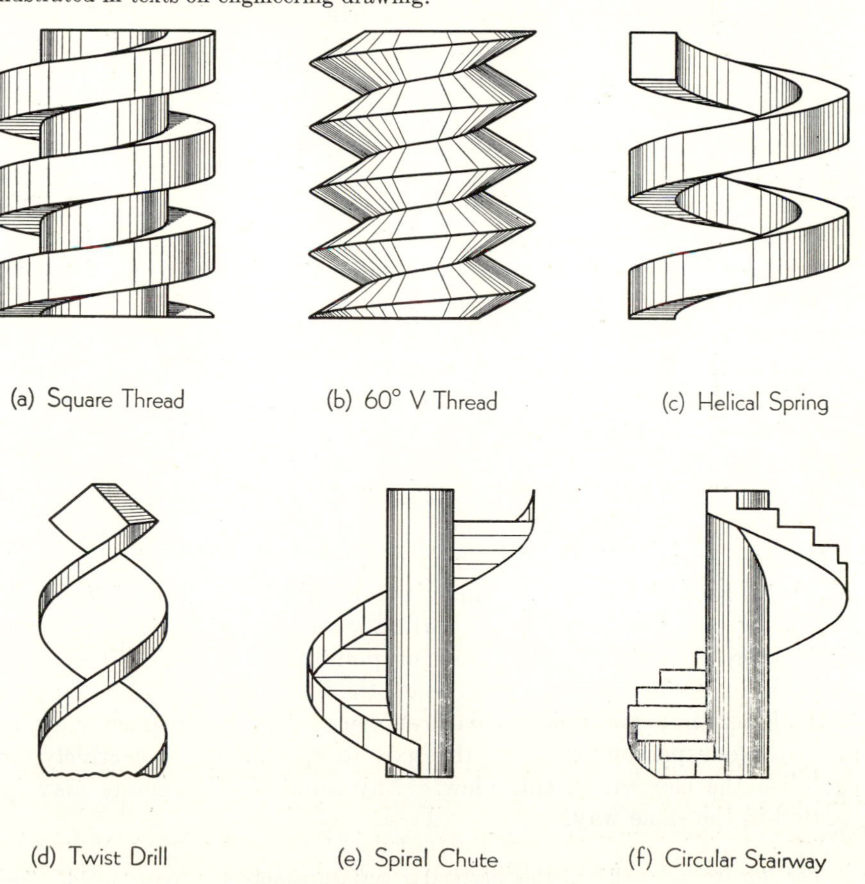

Fig. 7·15. Helicoid applications.

is undevelopable, but this chute surface can be closely approximated in sheet metal by constructing it in short sections, which are then riveted or welded together. A circular stairway like that shown in Fig. 7·15(*f*) can be made of concrete reinforced with steel bars; its undersurface is a right helicoid. Such a stairway can also be made of wood or metal by fabricating each tread and riser as a separate unit and then attaching these units to the central column.

PROBLEMS. Group 82.

7·20. Hyperboloid of revolution of one nappe

Generation. The hyperboloid of revolution of one nappe[1] is actually only a special case of the elliptical hyperboloid (Art. 7·3). Sections of the hyperboloid perpendicular to the axis will be circles instead of ellipses, and therefore a picture of this surface would be almost identical with that shown in Fig. 7·2. Although the hyperboloid may be generated (1) by a straight line moving so that it is always in contact with *three skew lines*, it is more useful to observe that it can also be generated (2) by a straight line *revolving* about a nonparallel, nonintersecting axis. If the generating line intersects the axis, a cone of revolution will result; if it is parallel to the axis, a cylinder of revolution is formed; the hyperboloid is therefore unique in that it is the only warped surface which is also a surface of revolution. It can also be generated in two other ways, (3) by a hyperbola revolving about its conjugate axis and (4) by a straight line moving so that it is always in contact with three circles that are perpendicular to, and have their centers on, a common axis.

Problem. In Fig. 7·16 the hyperboloid is to be generated by revolving line *CD* about line *AB* as an axis (second method described above). Represent the surface as shown in Fig. 7·16(*b*) by drawing a series of elements showing the generating line *CD* in its consecutive positions.

Construction. In the top view of Fig. 7·16(*a*) the inner circle is drawn tangent to line c_Td_T. This circle is called the *circle of the gorge*, and in the top view every element will appear tangent to it. The outer circle represents the circular path of points c_T and d_T. When c_T is revolved to any new position such as c_T^R, the line $c_T^Rd_T^R$ will still be tangent to the inner circle and point d_T will revolve to d_T^R. In the front view, c_F and d_F move perpendicular to the axis to c_F^R and d_F^R, respectively, to establish the element in this view. Any number of elements may be located in the same way.

[1] For brevity we shall call this particular surface simply the hyperboloid. The hyperboloid of revolution of two nappes is discussed in Art. 8·2.

In Fig. 7·16(*b*) the circular path of point c_T has been divided into 24 equal divisions, starting at point c_T. These points are numbered 1, 2, 3, etc. The corresponding positions of d_T are numbered 1′, 2′, 3′, etc. The construction of the top view is thus identical with that of the helicoid discussed in Art. 7·17 and shown in Fig. 7·14. Note that only one-half of each element in the top view is visible. In the front view only the visible part of each element has been shown.

Second generation. That the hyperboloid is a double-ruled surface is evident if we observe in Fig. 7·16(*a*) that this same surface could also be generated by revolving the line *EF* about axis *AB*. Line *EF* is therefore an element of the second generation; and if this second generation is superimposed on the first generation shown in Fig. 7·16(*b*), the sur-

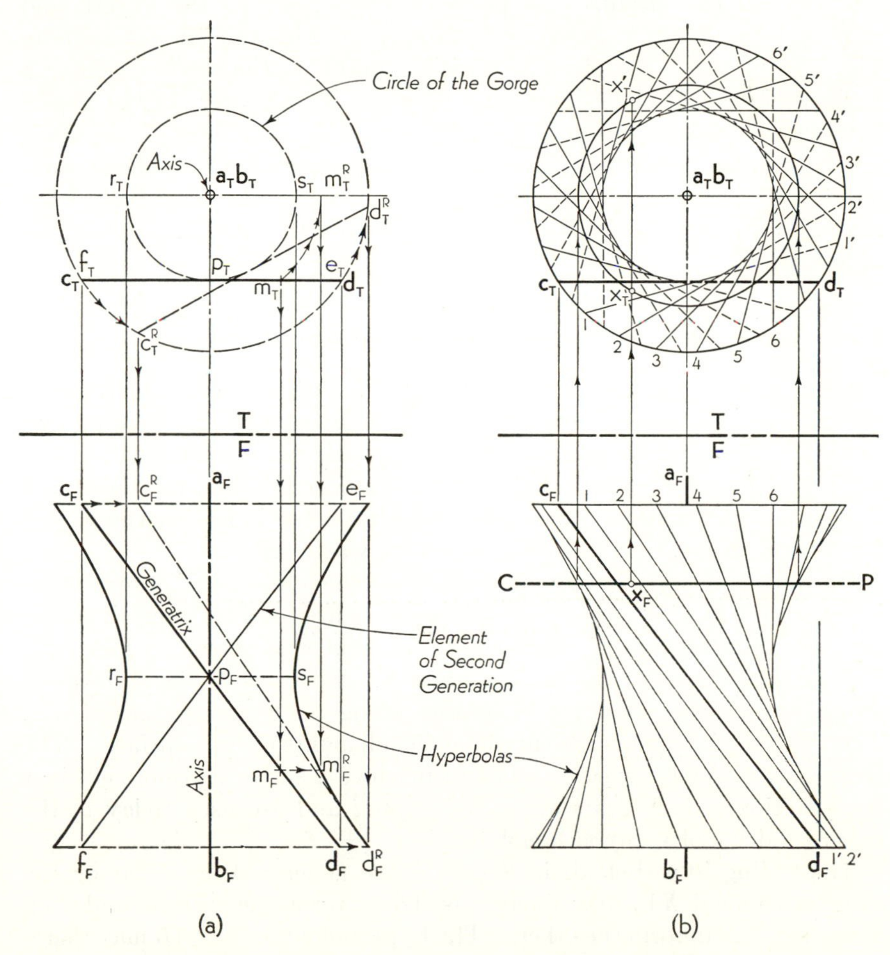

Fig. 7·16. Construction of the hyperboloid of revolution.

face would then look like the lower half of the elliptical hyperboloid shown in Fig. 7·2.

Contour curves. Figure 7·16(*a*) also shows how the contour curves of the hyperboloid may be obtained without drawing the elements. As line *CD* revolves, each point on it will successively become a point on the contour curve. Point *M*, for example, will appear on the contour curve in the front view when it appears on the horizontal center line in the top view. Thus, when m_T revolves to m_T^R, point m_F will move to m_F^R, a point on the hyperbolic contour. By selecting a series of points along line *CD* and revolving each in the same manner as point *M* the right branch of the hyperbola may be plotted. Revolving the same points to the left gives the left branch of the curve.

Point on the surface. A point on the surface of the hyperboloid may be located by the same methods suggested in Art. 7·8. Assume in Fig. 7·16(*b*) that x_F is given as a visible point on the surface. Then a horizontal cutting plane that contains point *X* will intersect the hyperboloid in a circle as shown in the top view. The diameter of this circle is shown in the front view where the cutting plane intersects the contour curves. Since x_F is a visible point, x_T must lie on the front half of the circle in the top view. Had x_F been given as hidden, then it would appear at x'_T.

7·21. Applications of the hyperboloid of revolution

Because the hyperboloid is a double-ruled surface it can easily be constructed as a latticework of straight rods or bars. The rods correspond to the elements of the surface; and when each rod of one generation is welded or fastened to each rod of the second generation, the result is a light but rigid structure. In this form the hyperboloid has been employed for towers, masts, reed furniture, and wire wastebaskets.

One of the most important applications of the hyperboloid of revolution is in the design of *skew gearing.* Skew gears are used to transmit power between two nonparallel, nonintersecting shafts as shown in Fig. 7·17. The pitch surfaces of the mating gears are frustums of the surfaces of two hyperboloids, which are in contact along the common element *XY*. The teeth of skew gears are cut along the elements of each surface. The teeth may also be curved or spiral to produce a smoother, more efficient tooth action. Such gears are called *hypoid* and are used today in the rear-axle drive of many automobiles.

The rolling hyperboloids in Fig. 7·17 are generated by revolving the tangent element *XY*, first about axis *AB* to form one surface, and then about axis *CD* to form the other. The hyperboloid on axis *AB* may therefore be constructed by the same methods as those shown in Fig. 7·16.

For the hyperboloid on axis CD it is necessary to construct a new view (not shown in Fig. 7·17) showing line CD as a point.

The position of the common element XY with respect to the two shafts determines the relative size and speed ratio of the mating gears. When line $x_F y_F$ is equidistant from both shafts and line $x_T y_T$ bisects the angle between the shafts, the mating gears will be equal in diameter. For more information on skew gearing, the student is referred to texts on mechanism or kinematics.

PROBLEMS. Group 83.

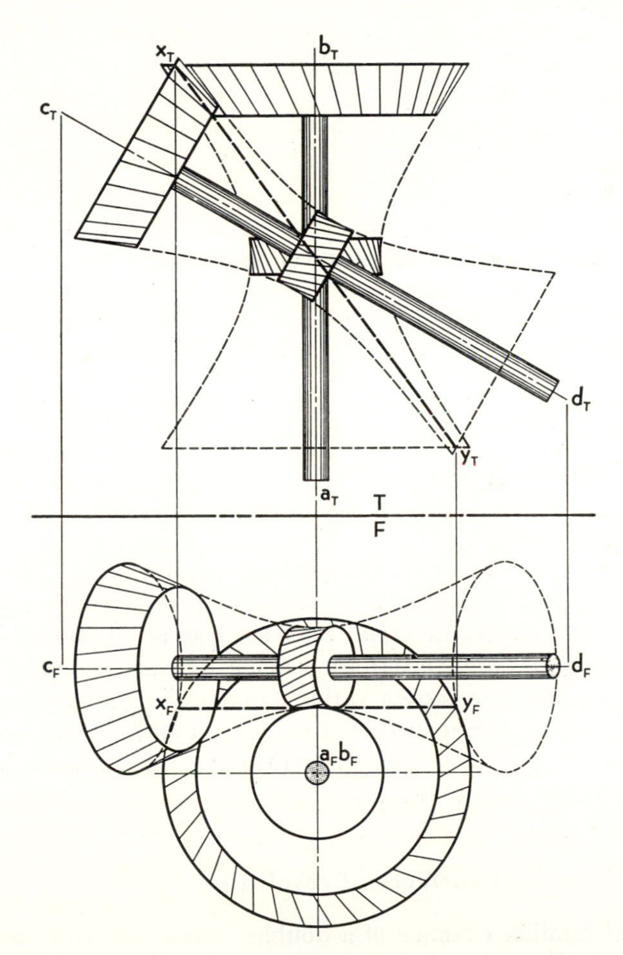

Fig. 7·17. Rolling hyperboloids form the pitch surfaces of skew gears.

8. DOUBLE-CURVED SURFACES

8·1. Classification of double-curved surfaces

Double-curved surfaces are those which can be generated *only* by moving a curved line (Art. 6·1). A double-curved surface has no straight-line elements, and therefore no two consecutive points on the generating curve may move in straight-line paths. The shape of the generating curve may be either constant or variable, and its motion may be guided by various line and plane directrices to produce an infinite variety of surfaces. For practical purposes double-curved surfaces may be divided into two classes as follows:

1. *Surfaces of revolution* are those which may be generated by *revolving* a curved line about an axis. The generating line is usually considered to lie in the same plane as the axis. Such surfaces are discussed in Art. 8·2.

2. *Surfaces of evolution*[1] are those which can be generated only by moving a curved line, of constant or variable shape, along a noncircular curved path. All double-curved surfaces that are not surfaces of revolution are therefore surfaces of evolution. Practical examples of such surfaces are discussed in Arts. 8·16 to 8·19.

8·2. Double-curved surfaces of revolution

The most familiar example of a double-curved surface of revolution is the sphere: its surface may be generated by revolving a circle about one of its diameters. The sphere is discussed more fully in Art. 8·3.

Figure 8·1 illustrates a few of the double-curved surfaces that may be

[1] A designation advocated by the author but not in general use.

obtained by revolving the conic curves about an axis. If an ellipse is revolved about one of its axes, the resulting surface is called a *spheroid.* When the ellipse is revolved about its major axis, the surface is a *prolate spheroid* (elongated sphere) and looks like a football. When revolved about its minor axis, it is an *oblate spheroid* (flattened sphere) and looks like a doorknob.

The *paraboloid of revolution* is the surface obtained by revolving a parabola about its axis. Polished reflectors of this form are used in searchlights; when the light source is placed at the focus, the reflected rays will be parallel to the axis.

The *hyperboloid of revolution of two nappes* is the surface obtained by revolving a hyperbola about its transverse axis. This surface is in two separate parts, corresponding to the two branches of the hyperbola. Revolving the hyperbola about its conjugate axis generates the *hyperboloid of revolution of one nappe;* this surface is not double-curved but warped and hence was discussed in Chap. 7. The hyperboloid of two nappes is of little practical value.

A *torus* is the surface obtained by revolving any curve about an axis that is not symmetrical with the curve. When the generating line is a

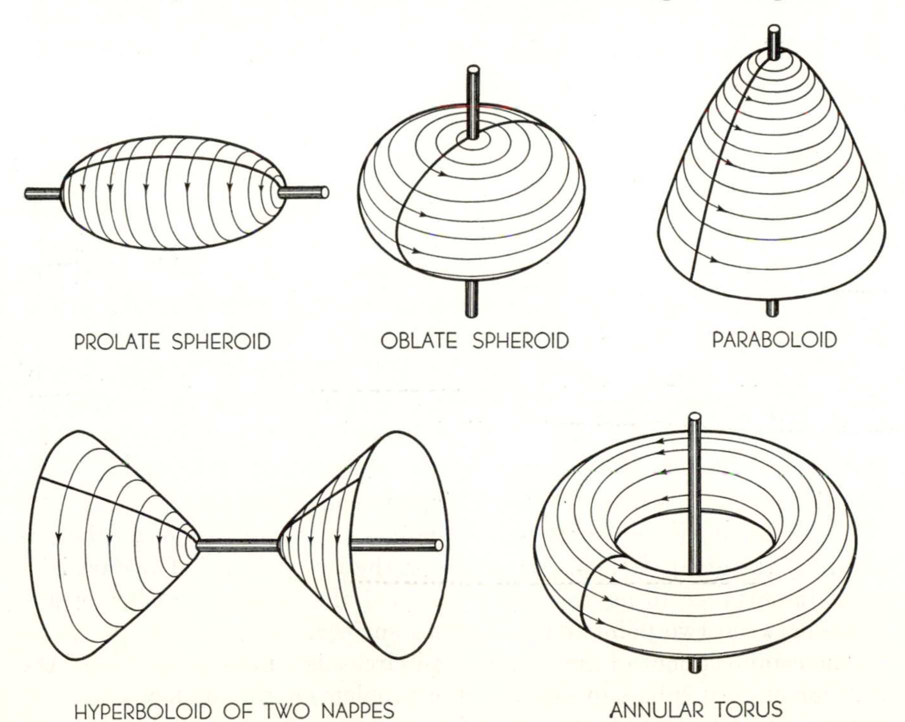

Fig. 8·1. Double-curved surfaces of revolution.

closed curve, especially a circle, the surface is called an *annular torus.* The usual form of the torus is that shown in Fig. 8·1, where the generating circle is outside the axis.

Double-curved surfaces of revolution may also be generated by curves of irregular form; vases, jars, spindles, etc., are examples of such surfaces.

Although double-curved surfaces are not developable, they may be approximated by the methods described in Art. 10·23. To duplicate a double-curved surface exactly, it must be cut, molded, or pressed into the desired shape. Surfaces of revolution are easily formed from the solid by turning in a lathe, or they may be made from sheet stock by spinning, a process by which the flat material is held against a form, while both revolve in a lathe, and is then gradually pressed into shape over the form by a smooth, hard tool.

The intersection of a surface of revolution by a plane perpendicular to its axis will be a circle. In Fig. 8·1 the motion of the generating curve has been emphasized in each case by showing a number of circles on the surface. These circles are *right sections* of the surface. Planes that contain the axis are *meridian planes*, and they intersect the surface in curves that are called meridian lines, or simply *meridians.* Every meridian line is identical in form with the generating line. In subsequent problems we shall find it very useful to employ these two types of plane intersections, especially the circular right section.

8·3. The sphere

To represent the sphere in any view it is necessary to show only a single circle whose diameter is equal to that of the sphere. This circle is the contour curve, or bounding line, of the sphere in each view. Thus, to construct a new view of a sphere, its center should first be established and the *contour circle* then drawn with this point as a center.

The intersection of a sphere and a plane is always a circle, regardless of the position of the plane. When the intersecting plane passes through the center of the sphere, the circle of intersection is called a *great circle* and its diameter is equal to that of the sphere. All other planes intersect the sphere in smaller circles, which are called *small circles.*

The shortest distance between two points on the surface of a sphere is an arc of the great circle passing through the two points. This fact is of great importance to navigators in planning the course of a ship or airplane between two points on the earth's surface.

The establishment of great and small circles is an essential part of the solution of most sphere problems. For problems involving loci, note that the surface of a sphere is the locus of all points in space at a fixed distance from a point at the center of the sphere.

Because the sphere is the simplest and most important double-curved surface of revolution, we shall consider first a number of problems involving this surface.

8·4. Location of a point on a sphere

Analysis. A point on a surface must always be located by *first* establishing some line on the surface that contains the given point. (Recall the method of locating a point on a plane, cone, cylinder, or warped surface.) The simplest line on a sphere is a circle, and a circle through the given point can be obtained if we:

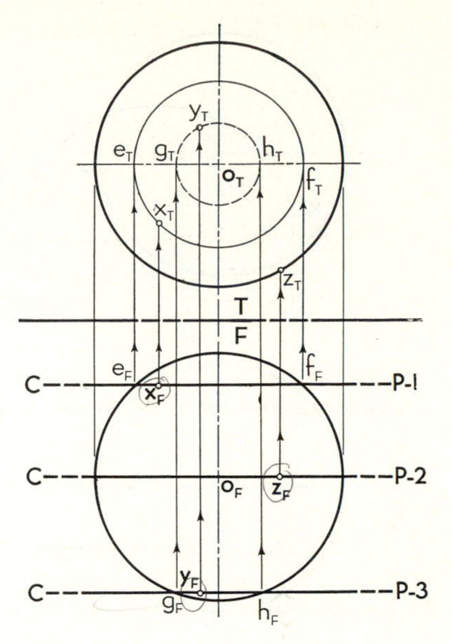

Fig. 8·2. Location of points on a sphere.

Choose a cutting plane that contains the given point and intersects the sphere in a circle that appears true size.

Problem. In Fig. 8·2 points x_F, y_F, and z_F are given points on the surface of the sphere. Points x_F and z_F are visible; y_F is hidden. The top view of the three points is required.

Construction. A horizontal cutting plane C-P-1 through point X intersects the contour circle in the front view at points e_F and f_F, and therefore EF is the diameter of the small circle. In the top view this small circle will appear true size with center at o_T. Because x_F was given as visible, point X must be on the front half of the sphere; hence x_T is located on that half of the small circle nearest to the front. The small circle is visible in the top view because it lies on the upper hemisphere, and thus point x_T is also visible.

Point y_T can be located by employing cutting plane C-P-3. In this case the small circle of diameter GH will be hidden in the top view, and y_T will be a hidden point. The fact that y_F is hidden means that point Y is on the rear half of the sphere and hence on the rear half of the small circle in the top view.

Cutting plane C-P-2, used to locate point z_T on the sphere, passes through point O and cuts a great circle from the sphere. Point z_T therefore lies on the contour circle in the top view. This case occurs frequently, hence we should remember that:

A point on the contour circle in one view will appear on a diameter parallel to the reference line in any adjacent view.

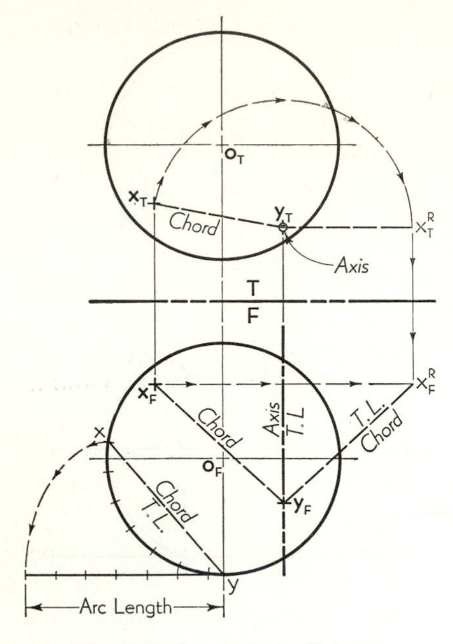

Fig. 8·3. Great-circle distance.

8·5. Great-circle distance between two points on a sphere

Analysis. The straight-line distance between the given points will be a *chord* of the great circle through the points.

Problem. In Fig. 8·3 determine the great-circle distance between points X and Y on the given sphere. The points have been located on the sphere by the method of Art. 8·4.

Construction. Draw the chord connecting points X and Y in each view. Obtain the true length of line XY by revolving it about an assumed axis as shown (Art. 5·5). At any convenient place on the contour circle of the sphere lay off the true-length chord as shown at xy. Then the arc length from x to y is the required great-circle distance. The arc may be rectified by the methods given in the Appendix, Arts. A·4 and A·5.

PROBLEMS. Group 84.

8·6. Intersection of a plane and a sphere

Analysis. The curve of intersection will be a circle. If the plane is generally oblique, the circle will appear as an ellipse in both the top and the front views. These ellipses can readily be determined by drawing new views that will show the plane as an edge. This is the same principle as that employed in Art. 4·17.

Problem. In Fig. 8·4 the given sphere with center at O is intersected by the plane $ABCD$. It is required to show the curve of intersection in the given top and front views.

Construction. To construct the ellipse in the top view the adjacent view A is drawn to show plane $ABCD$ as an edge. The sphere is also shown in view A by first locating point o_A and then drawing the contour circle with this point as center. It is now apparent that the plane intersects the sphere in a small circle whose diameter is D, the distance between points 3 and 4 in view A. Since points 3 and 4 lie on the contour circle in view A, they must lie on a diameter of the sphere parallel to reference line T–A in the top view (Art. 8·4). Distance 3–4 is therefore the minor axis of the required ellipse in the top view. The major axis 1–2 must equal D, the diameter of the small circle. In view A it will appear as a point (1,2) at the mid-point of line 3–4. (A line from o_A perpendicular to the plane should check this point.) The major and minor axes of the ellipse in the top view having been established, it can now be drawn by the trammel method.

The visibility of the ellipse can be determined from view A by noting that point 3 is on the upper half of the sphere and point 4 on the lower half; points 5 and 6, which lie on the diameter parallel to T–A in view A, must lie on the contour circle in the top view at 5 and 6. Points 5 and 6 are important because they establish accurately the two points on the ellipse where the visibility of the curve changes. It should be apparent that the construction described here is the same as that for locating a circle in a plane (Art. 4·17 and Fig. 4·19).

The best way to obtain the ellipse in the front view is by repeating the above process and using view B as shown in Fig. 8·4. Thus the ellipses in the top and front views may be constructed independently. As a check on the work the diameter D in view B should, of course, be the same as that in view A, and the extreme right and left sides of the two

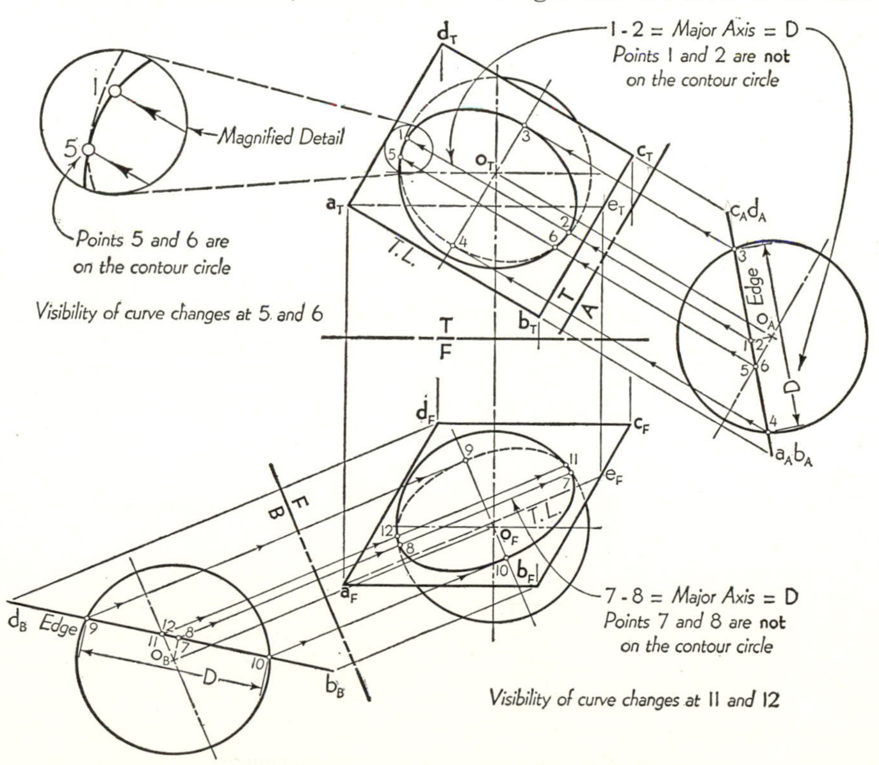

Fig. 8·4. Intersection of a plane and a sphere.

ellipses should align. Note again (as in Art. 4·17) that points 1, 2, 3, etc., in the top view do not necessarily correspond to any of the numbered points in the front view.

PROBLEMS. Group 85.

8·7. Intersection of a line and a sphere

Analysis. The sphere must be intersected by a plane that contains the given line (Rule 17). Any plane will serve, but for simplicity it should appear as an edge in one of the given views, and the intersection circle should be shown in true size. We should therefore:

Show a true size view of a cutting plane that contains the line.

Problem. In Fig. 8·5 the given sphere with center at O is intersected by the line MN. The points of intersection are required.

Construction. A vertical cutting plane through line MN intersects the sphere in a small circle whose diameter is $e_T f_T$. View A shows the cutting plane in true size, and here the small circle appears as a visible circle concentric with point o_A. At points p_A and q_A the line $m_A n_A$ intersects the small circle. These are the required points of intersection

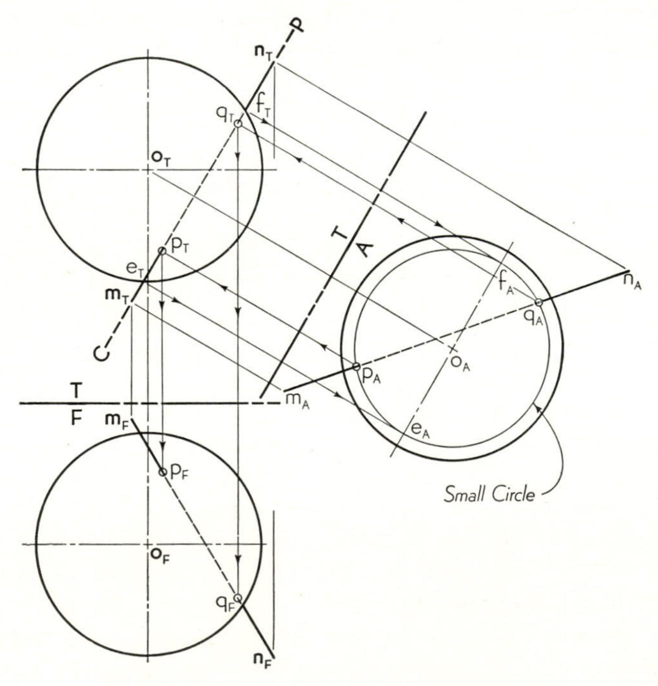

Fig. 8·5. Intersection of a line and a sphere.

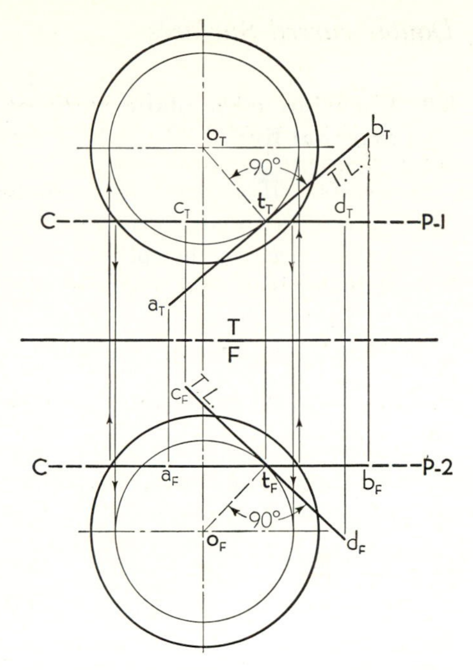

Fig. 8·6. Plane tangent to a sphere through a given point on the sphere.

in view A, and they may now be extended back to line m_Tn_T in the top view and then to line m_Fn_F. Point p_T is visible because both the front view and view A show that it lies on the upper half of the sphere. Conversely, and by similar reasoning, q_T is hidden. Point p_F is visible because the top view shows it to be on the front half of the sphere. Point q_F is hidden.

PROBLEMS. Group 86.

8·8. Location of a plane tangent to a sphere through a given point on the sphere

Analysis. A plane tangent to a sphere is perpendicular to the radius drawn to the given point on its surface. A plane perpendicular to a line can be constructed by the method of Art. 4·33. Therefore:

Construct the tangent plane perpendicular to the sphere radius at the given point on the surface.

Problem. In Fig. 8·6 construct a plane tangent to the sphere at the given point T.

Construction. If only t_T is given, then point t_F can be established by intersecting the sphere with cutting plane C-P-1 as shown. If only t_F is given, t_T can be found by using cutting plane C-P-2. The line OT is then a radius of the sphere, and the required tangent plane must pass through point T and be perpendicular to OT. The horizontal line AB (indefinite in length) is drawn perpendicular to radius OT (Rule 10) and is therefore one line in the required tangent plane. In similar manner, line CD, which appears true length in the front view, is drawn to form a second line in the required tangent plane. The two intersecting lines AB and CD therefore define the desired plane.

In Art. 6·17 it was stated that any line is tangent to a surface if it is tangent to some line on the surface. Figure 6·18(c) illustrated this concept for a sphere. Note now that lines AB and CD, in Fig. 8·6, are each tangent to a small circle of the sphere, thus proving that these lines are truly tangent to the sphere.

8·9. Location of a plane tangent to a sphere through a given external line

Analysis. If the given line is shown as a point, the tangent plane will then appear as an edge, passing through the given line, and tangent to the contour circle of the sphere. The point of tangency with the sphere is thus determined, and this point plus the given line defines the required plane. Two solutions are possible because the plane may be drawn on either side of the sphere. In short:

To see the tangent plane as an edge, show the given line as a point.

Problem. In Fig. 8·7 the line AB and sphere with center at O are given in the top and front views. It is required to show in these views the two planes through line AB tangent to the sphere.

Construction. Views A and B are drawn to show line AB as a point. Through point a_Bb_B two planes, each appearing as an edge, may now be drawn tangent to the sphere at points s_B and t_B. Radii of the sphere drawn perpendicular to the planes locate the points of tangency exactly.

To represent the tangent planes in the given views, any random lines in the planes could be selected (Art. 4·6). However, the points of tan-

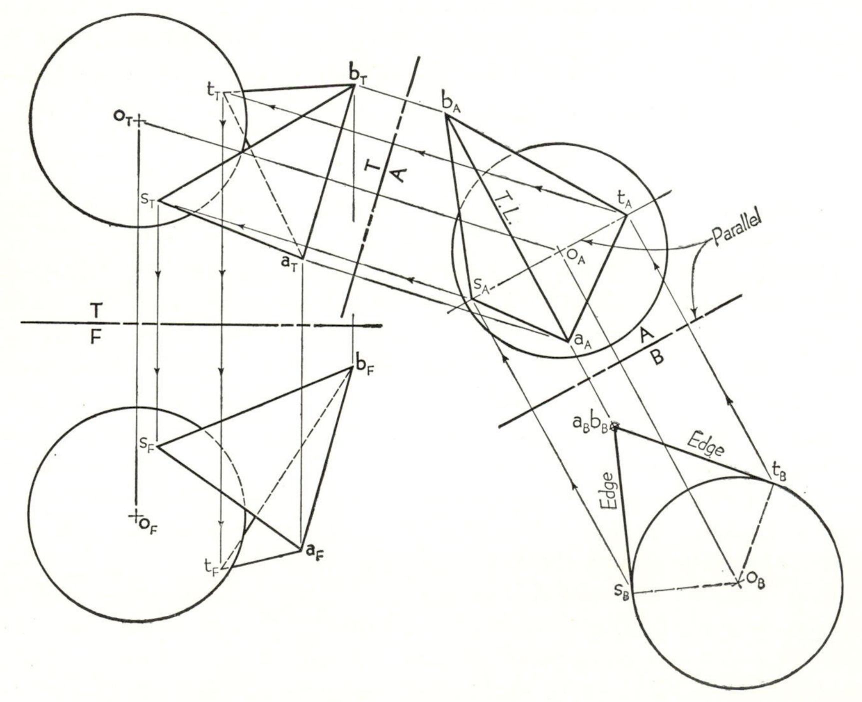

Fig. 8·7. Planes tangent to a sphere through a given external line.

gency are usually desired in the given views; hence the planes have been conveniently represented as the triangles *ABS* and *ABT*. Because points s_B and t_B lie on the contour circle in view *B*, points s_A and t_A must lie on the sphere diameter parallel to reference line *A–B* (Art. 8·4). Points *S* and *T* may now be located in the given views by alignment and measurement. Visibility of the tangent points in each view is determined, as in previous examples, by noting in an adjacent view whether the point is on the near or far half of the sphere.

PROBLEMS. Group 87.

8·10. Cone envelope of two spheres

If two spheres do not touch or intersect each other, then two cones may be constructed, each of which will envelop, or exactly contain, both spheres. Such a *cone envelope of two spheres* is a useful device, which may frequently be employed in the solution of problems that involve two or more spheres (see also Art. 10·16).

Figure 8·8 illustrates the general principle of the cone envelope. The given spheres, with centers at *O* and *P*, have been so assumed that the line connecting the sphere centers appears true length in the top view. Then in case (*a*) the right circular cone of two nappes has its vertex *V* on the line of centers between points *O* and *P* and is tangent to each of the spheres. To establish this cone it is necessary to draw in each view only the extreme elements tangent to the contour circles of the spheres.

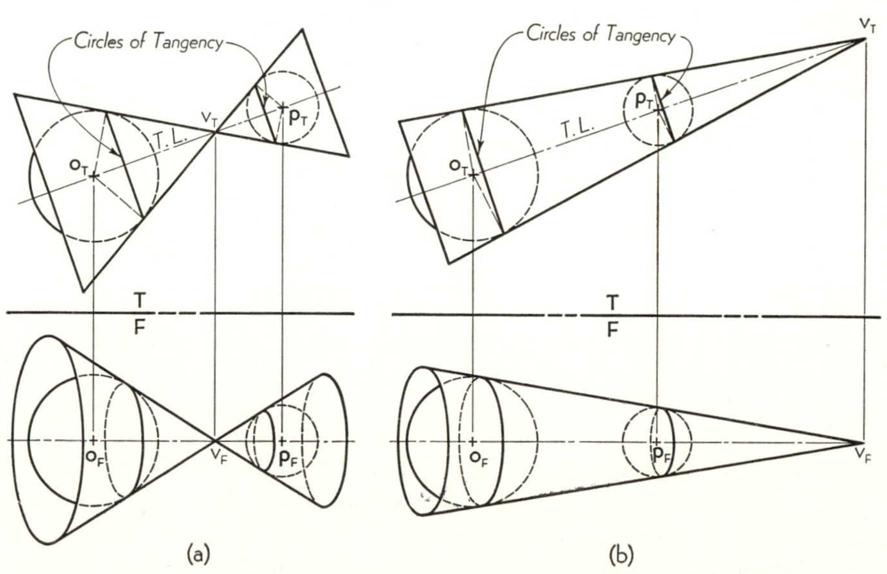

Fig. 8·8. Cone envelope of two spheres.

The intersection of the extreme elements on the line of centers locates the vertex V. In case (b) the procedure is the same; but here both spheres are contained in a single nappe of the cone, and the vertex V lies on line OP extended.

The *circles of tangency* between the cone and each sphere appear as an edge in the top view, where the cone axis is true length. Radii of the spheres drawn perpendicular to the extreme elements locate the circles. It is rarely necessary to show the base of the enveloping cone, as has been done in Fig. 8·8.

Problem 1. Assume that it is required to locate a plane tangent to each of two given spheres through a given point X.

Analysis. The required plane must be tangent to both spheres, hence it will also be tangent to the cone envelope of these spheres. The cone can be constructed with the vertex V in either of two possible positions [case (a) or (b)]. Then line XV will be one line in the required plane. In a view showing line XV as a point the plane can be established as an edge tangent to both spheres. Depending upon the position of point X, the number of solutions may vary from none to four.

Problem 2. Assume that it is required to locate a plane tangent to each of three given spheres.

Analysis. Selecting any two of the spheres a cone envelope can be constructed for this pair, thus establishing a vertex V. Two more vertex points may be located by drawing the cone envelopes of the third sphere and each of the first two. The three vertex points will lie in the same straight line, and this is one line in the required plane. A view showing this line as a point will show two possible tangent planes as edges. In this same view the points of tangency on each sphere can be located. If the three spheres are completely separated, then six different cone envelopes can be drawn to locate six cone vertices. In this general case there are eight possible tangent planes (see Prob. 88·5).

PROBLEMS. Group 88.

8·11. Representation of surfaces of revolution

The simplest views of a surface of revolution are those which show the axis as a point and as a true length. In Fig. 8·10, for example, the axis of each surface appears as a point in the top view, and here the contour curves are all circular. In the front views the axis appears true length, and the true shape of the generating curve is shown. Vertical planes through the axis of the spheroid or paraboloid produce *meridian lines* identical in size and shape with the contour outline of the front view. A meridian plane intersects the torus in two circles like those appearing in the front view.

All views that show the axis of a surface of revolution as a true length will be identical in contour and appearance. Therefore, when it is necessary to draw an auxiliary view of a surface of revolution, the view should, if possible, be taken adjacent to the view showing the axis as a point. Auxiliary views that show the axis foreshortened may be constructed by the method of *inscribed spheres,* illustrated in Fig. 8·9.

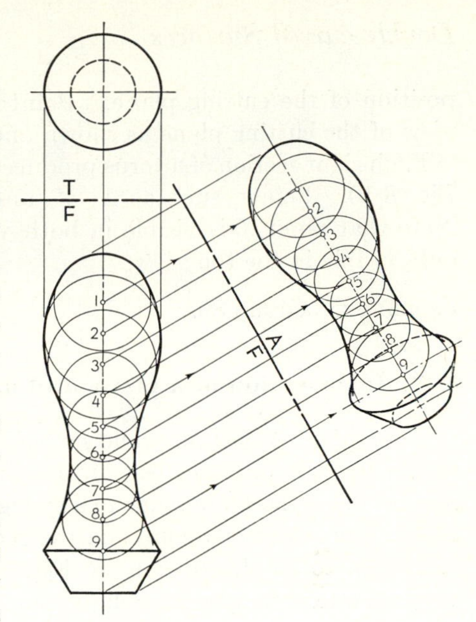

Fig. 8·9. Representation of double-curved surfaces of revolution.

Figure 8·9 shows the top and front views of a crank handle typical of those used on many machine tools. Assume that it is necessary to draw view A of this object. In the front view a series of circles are drawn, with centers on the axis, so that each circle is exactly *tangent* to the curved contour of the object. The centers of the circles need not be equally spaced along the axis. These circles now represent a series of spheres inscribed within the double-curved surface. The numbered centers 1, 2, 3, etc., may now be located along the axis in view A and the same spheres drawn on these centers as were used in the front view. The outline of view A is then drawn as a smooth curve that envelops the circles. The conical base of the handle is represented by the two ellipses. Note that the contour curve may not be tangent to the larger ellipse but may extend inside as happens in Fig. 8·9.

8·12. Location of a point on a surface of revolution

A point on a surface of revolution can be located most easily by choosing a *right section* that passes through the given point. In Fig. 8·10, for example, assume that point x_F is given as a visible point on the spheroid. Then a horizontal cutting plane through point X will intersect the surface in a circle that will appear hidden in the top view. Point x_T must lie on the front half (x_F is visible) of this circle as shown and will be a hidden point.

If point y_T on the paraboloid is given, then the circular right section must be drawn first in the top view. Parallels extending the diameter of this circle to the front view intersect the contour curve to establish the position of the cutting plane. Point y_F must then appear on the edge view of the cutting plane as shown and will be a hidden point.

Each right section of a torus produces two concentric circles as shown in Fig. 8·10. Points, such as *X*, *Y*, and *Z*, are located on these circles. Note that point *X* is visible in both views, whereas points *Y* and *Z* are only visible in the top view.

PROBLEMS. Group 89.

8·13. Intersection of a plane and a surface of revolution

Analysis. A cutting plane *perpendicular to the axis* of the surface of revolution will intersect the surface in a circle. The cutting plane will also intersect the given plane in a straight line. The intersection of circle and line provides two points on the required curve of intersection.

The pictorial illustration in Fig. 8·11 shows a spheroid with axis vertical intersected by an oblique plane *ABCD*. The horizontal cutting plane intersects the spheroid in a circle and intersects the plane along line *RS*. At points *X* and *Y* line *RS* intersects the circle on the spheroid, and therefore *X* and *Y* are two points on the required curve of intersection. By choosing at various levels a series of such cutting planes, all perpendicular to the axis, a sufficient number of points may be located to establish a smooth curve.

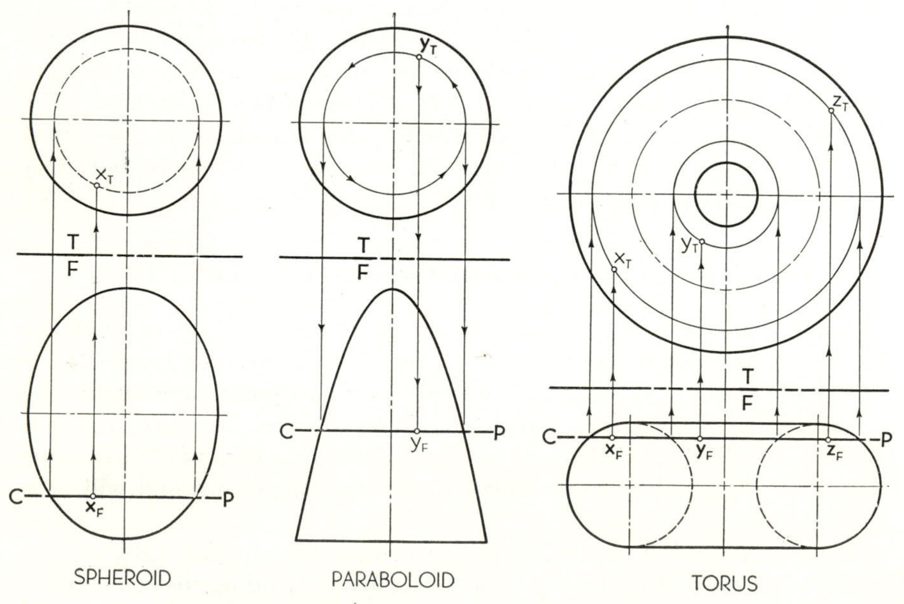

Fig. 8·10. Location of a point on a surface of revolution.

Construction. In the multiview drawing cutting plane *C-P*-1 has been assumed at any random level. The circle cut from the spheroid is shown in the top view. The line cut from the plane is *RS*. Line r_Ts_T intersects the circle at points x_T and y_T, thus establishing two points on the curve of intersection. Points x_F and y_F can then be located on the edge view of the cutting plane. When additional planes are selected, they should be assumed in pairs, symmetrically above and below center, to minimize the number of circles in the top view.

Special points. Although the solution can be completed by using only cutting planes perpendicular to the axis (and the majority of the points on the curve should be so obtained), certain important points should be located accurately by special methods. Points 1 and 2 are important because they are the highest and lowest points on the curve and also because they lie on the axis of symmetry of the curve in the top view. In top-adjacent view *A*, which shows the plane as an edge, the contour of the spheroid will be identical with that in the front view. In this view, points 1 and 2 are located as shown; and because they appear on the con-

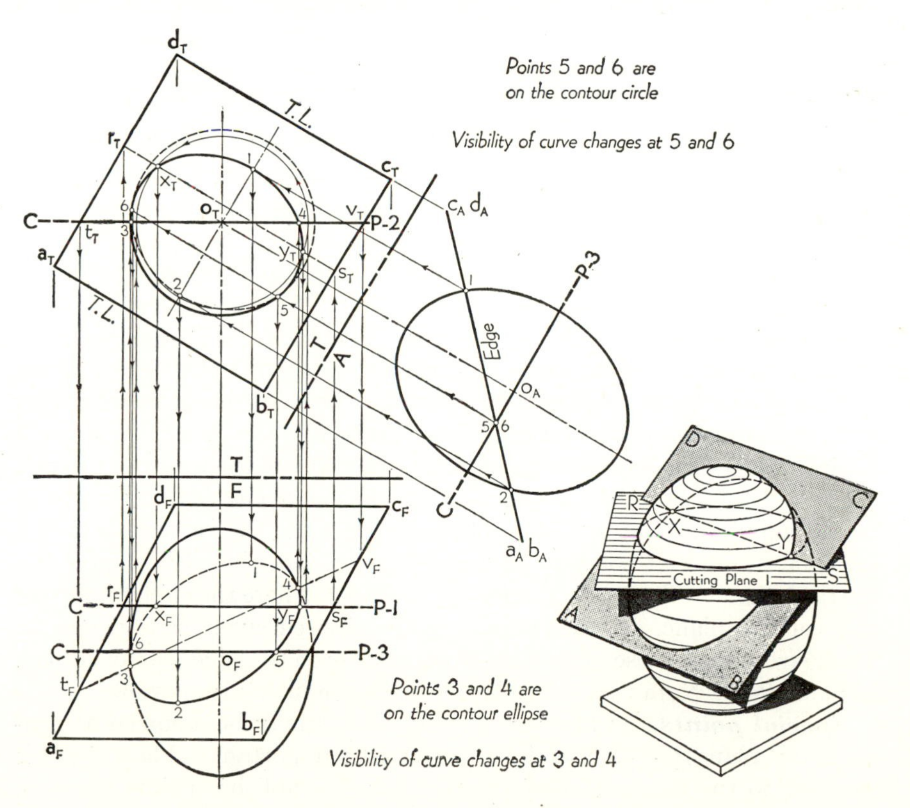

Fig. 8·11. Intersection of a plane and a surface of revolution.

tour curve in view A, in the top view they must lie on the spheroid diameter parallel to reference line T–A.

Visibility change points. Visibility of the intersection curve can change *only at points on the contour curves,* and these points should be located accurately with specially selected cutting planes. Cutting plane C-P-2, for example, is a meridian plane intersecting the spheroid in an ellipse. In the front view this ellipse appears as the contour curve. Cutting plane C-P-2 also intersects the given plane along the line TV, and therefore points 3 and 4, where $t_F v_F$ intersects the contour ellipse, are the last visible points in the front view.

Points 5 and 6 are the points in the top view where the curve changes visibility. They may be obtained by using cutting plane C-P-3, which passes through the center O of the spheroid, or we may refer to view A, where cutting plane C-P-3 again appears as an edge, and note that points 5 and 6 coincide in this view. In the top view these two points must, of course, lie on the outer contour circle.

PROBLEMS. Group 90.

8·14. Intersection of a line and a surface of revolution

Analysis. The surface of revolution must be intersected by a plane that contains the given line (Rule 17). As in previous line-intersection problems, the solution is shortened by employing a cutting plane that appears as an edge in one of the given views.

Problem. In Fig. 8·12 it is required to determine the points of intersection of the line MN and the given torus.

Construction. A vertical cutting plane C-P has been assumed through line MN, and the curve of intersection between this plane and the surface of the torus must now be plotted. This can be done, as in Fig. 8·11, by intersecting the vertical cutting plane and the torus by a series of horizontal cutting planes (perpendicular to the axis of the torus). Cutting plane C-P-1, for example, intersects the torus in two circles that are visible in the top view. But the intersection of C-P-1 and the vertical plane is a horizontal line that coincides in the top view with the edge view of the vertical plane. Therefore, points 1, 2, 3, and 4, where the circles intersect the vertical plane, are four points on the desired curve of intersection. In the front view these same four points appear on the edge view of cutting plane C-P-1. Note that only point 1 is visible in the front view.

Special points. Additional cutting planes may be assumed at random, but planes C-P-2, 3, and 4 are especially important. Plane C-P-2 is tangent to the top surface of the torus and establishes points 5 and 6, which are contour points in the front view. Plane C-P-3 cuts the largest and smallest possible circles from the torus, but the smaller circle does not intersect the vertical plane. Points 7 and 8 lie on the larger, or contour, circle in the top view, and they are also the extreme right and left points in the front view. Point 9 is located first in the top view by drawing a circle tangent to the vertical plane. Plane *C-P*-4 is then located at the proper height to cut a circle of this diameter from the torus, thus establishing point 9 in the front view. Note that cutting planes taken symmetrically above and below the horizontal center line produce circles of the same size, and therefore the curve of intersection is symmetrical with the horizontal center line. It is also symmetrical with respect to a vertical line through point 9.

The curve of intersection of the vertical plane and the torus intersects line *MN* at points *P* and *Q*, and these are the required intersection points. Point *P* is visible in both views, point *Q* hidden in both.

PROBLEMS. Group 91.

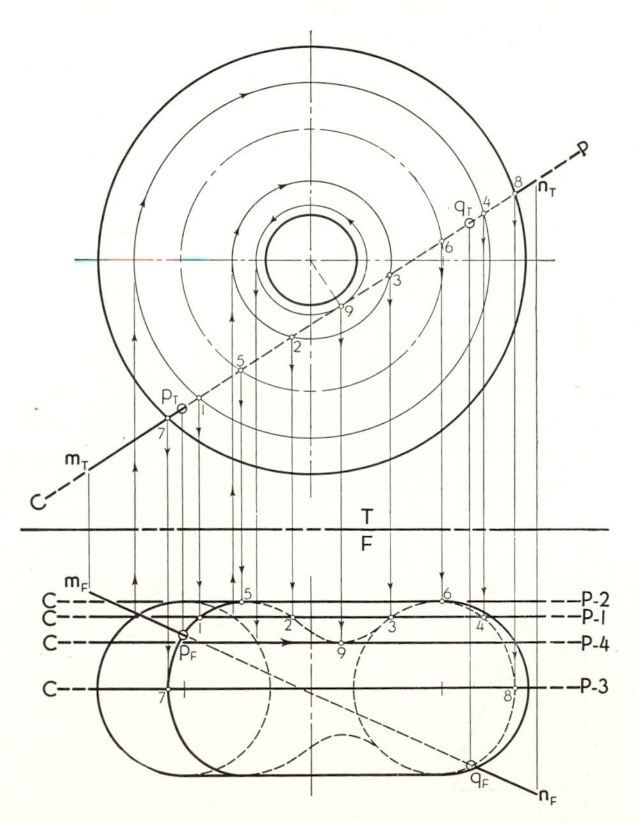

Fig. 8·12. Intersection of a line and a surface of revolution.

8·15. Location of a plane tangent at a given point on a surface of revolution

Analysis. The tangent plane may be defined by two straight lines that intersect at the given point, each line being tangent to the surface. But a line tangent to a surface must be tangent to some line on the surface (Art. 6·17). For a surface of revolution the two simplest lines on its surface are a right section and a meridian.

Problem. In Fig. 8·13 it is required to locate a plane tangent to the given paraboloid at point T on its surface.

Construction. In Fig. 8·13 the construction of the tangent plane can be observed in both the multiview drawing and the pictorial illustration. A right section of the surface at T appears as a circle in the top view. Line $a_T b_T$ is drawn tangent to the circle at t_T, and $a_F b_F$ must, of course, be horizontal (Art. 6·16). Line AB is one line in the required tangent plane.

In the picture a meridian line is shown passing through point T. This meridian line is in a vertical plane containing the axis, and therefore a line through T tangent to the meridian will intersect the axis at some point C. But the meridian line need not actually be drawn. If line CT is revolved

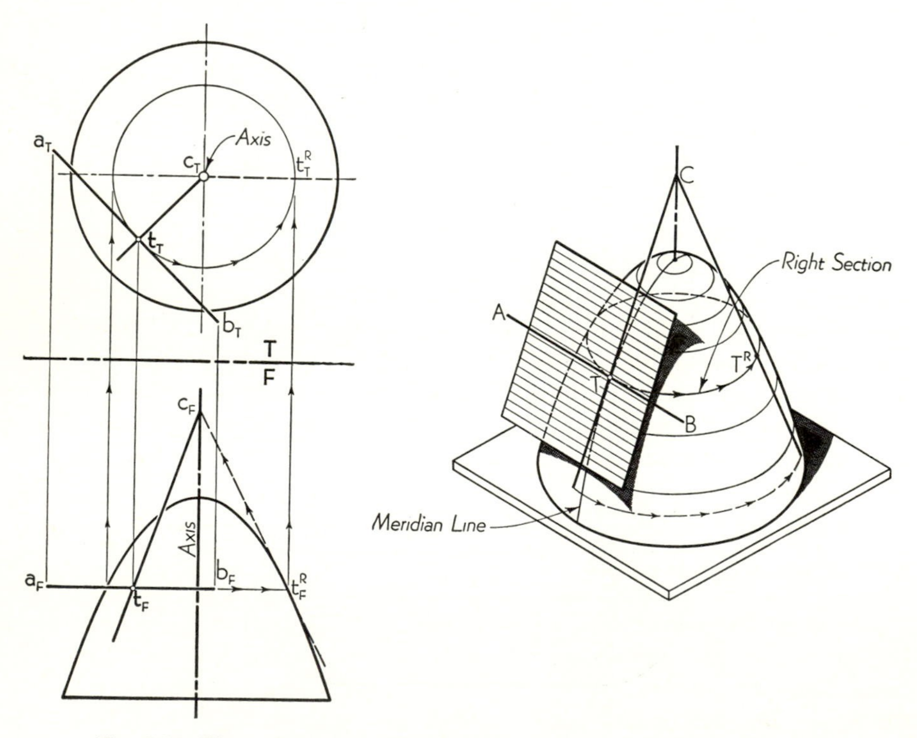

Fig. 8·13. Plane tangent to a paraboloid at a given point on the surface.

about the axis of the paraboloid until it appears true length, it will then appear tangent to the contour curve. Thus in the multiview drawing point t_T is revolved about the axis to position t_T^R, and point t_F^R then lies on the contour curve. A line drawn tangent to the contour at point t_F^R establishes the point c_F. When point T^R is counterrevolved to its given position T, point C on the axis does not move and hence line c_Ft_F can be drawn through points c_F and t_F to locate this second tangent line in the front view. We now have two intersecting lines, AB and CT, which define the required tangent plane.

PROBLEMS. Group 92.

8·16. Double-curved surfaces of evolution

In its general form the ellipsoid has three principal axes of unequal length and is therefore a typical surface of evolution. Its surface can be generated only by the motion of a *variable* ellipse, for example, by revolving a variable ellipse about one of its axes while the other axis constantly changes in length so that its extremities generate a second ellipse. It is because the double-curved surface in its general form necessitates a *changing* generatrix or a *changing* path that we have called it a surface of *evolution*.

Of more practical value, however, are the double-curved surfaces of the modern automobile body, of the airplane fuselage, of a ship's hull, and even of "streamlined" vacuum cleaners and bathroom scales. These smoothly rounded complex surfaces cannot be represented on paper simply by showing a contour outline in each view. The outlines must be supplemented by numerous additional contour lines representing the intersection of the surface by various cutting planes. Only in this way can the complex double-curved surface of general form be adequately described on paper. For many years the hulls of ships have been designed by this method of contour lines; and because many of the terms and methods of the shipbuilder have acquired widespread usage, we shall illustrate the representation of contoured surfaces with a drawing of a small ship.

8·17. Contoured surfaces

The three views shown in Fig. 8·14 constitute a *sheer drawing* of a ship. Except for arrangement and names these views correspond to the conventional top, front, and side views. The *sheer plan* is a view of the ship from the side. The *half-breadth plan* is a top view, traditionally placed below the sheer plan, and shows only one side of the ship because of symmetry.

The *body plan* is analogous to a front view except that on one side of the center line only the fore part of the ship is shown, and on the other side only the afterpart.

When the general outlines and over-all dimensions of the desired ship have been established, the designer then describes the curved surface of the hull by drawing in the *frame lines, water lines,* and *buttock lines.* Determining the proper shape for these curves is a matter of ship design involving displacement, center of buoyancy, etc., a subject that does not concern us here. We shall consider, then, only how these established lines describe the surface, how they are interrelated, and how a similar set of contour lines can be employed to describe any double-curved surface.

Each *frame line* shown in the body plan represents the curve of intersection of a transverse vertical plane with the surface of the ship. In Fig. 8·14 there are 11 such planes, equally spaced from bow to stern. In the sheer plan and in the half-breadth plan these planes appear edgewise at each numbered *station.* In the body plan only one symmetrical half of each curve is shown. Frame line 1 shows the transverse shape of the hull at station 1; frame line 5 shows the maximum width of the ship; frame line 11 shows a transverse section at the stern.

The *water lines* shown in the half-breadth plan represent the curve of intersection of a series of horizontal planes with the ship's hull. The height of each water-line plane, above and below the load water line (L.W.L.), is shown in both the sheer plan and body plan. The planes and lines are labeled 2 W.L., 3 W.L., etc.

The *buttock lines* appear in the sheer plan, and they represent the intersection of a series of longitudinal vertical planes with the surface of the ship. These planes appear edgewise in both the half-breadth plan and the body plan and are labeled A, B, C, corresponding to the letters on the buttock lines.

For the purpose of describing the shape of the ship any one of the three sets of lines would be sufficient. But the draftsman cannot be certain when he has established the desired frame lines that they accurately describe a smoothly continuous, or *fair,* surface. The water lines and buttock lines are therefore constructed from the assumed frame lines in order to see whether they, too, are fair curves. To draw water line 3 W.L., for example, we must note in the body plan where each frame line intersects the horizontal plane 3 W.L. Frame line 4 intersects plane 3 W.L. at point r in the body plan, and point r is observed to be distance x from the ship center line. In the half-breadth plan, point r is then located distance x from the center line at station 4 to establish one point on water line 3 W.L. When similar points have been located at each

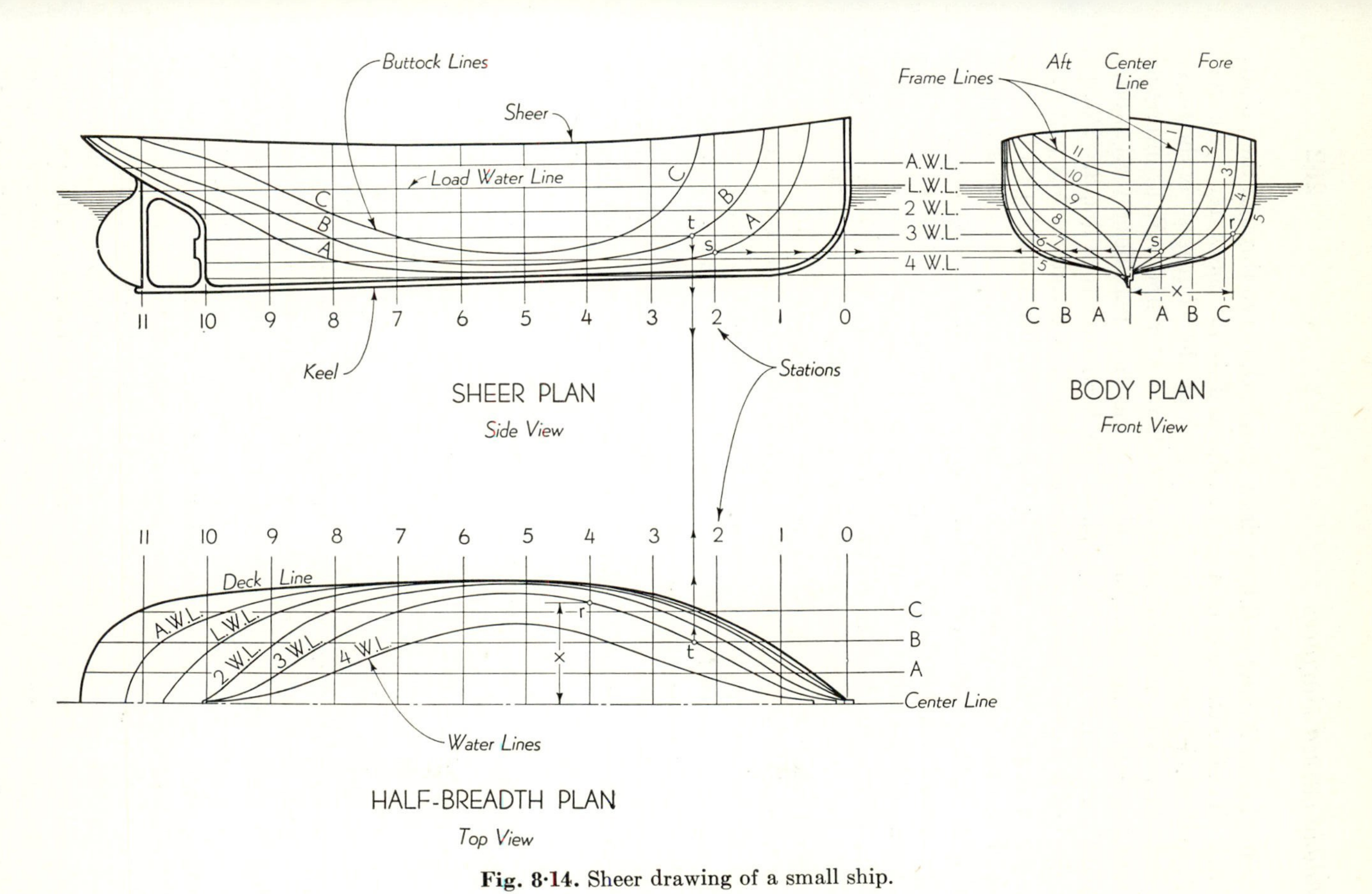

Fig. 8·14. Sheer drawing of a small ship.

station, it should be possible to draw a smooth curve through the points. If not, then the frame lines must be adjusted to fair the water-line curve.

The buttock lines provide a second check on the "fairness" of the ship's lines. At point t in the half-breadth plan, water line 3 W.L. intersects buttock plane B, and therefore point t in the sheer plan lies directly above at the intersection of water-line plane 3 W.L. and buttock line B. Point s in the body plan is the intersection of frame line 2 and buttock plane A. In the sheer plan directly opposite, point s establishes a point on buttock line A at station 2. The buttock lines should, of course, pass smoothly through all such points as s and t. If not, adjustment may be necessary in both water lines and frame lines. This process of changing, adjusting, and smoothing each set of curves is called *fairing* and must be continued until every point checks and the curves satisfy the eye of the draftsman.

The methods of the ship designer are equally applicable to an automobile fender or to any other double-curved surface. In general, the sequence of construction should be as follows: (1) Establish the necessary or desired contour outlines of the object in each of three views. (2) Intersect the surface with a series of parallel planes, and assume frame lines in these planes according to the desired shape at each station. (3) From the assumed frame lines construct water lines and buttock lines. (4) Correct and adjust all curves until the surface is fair and as desired.

8·18. Lofting

The actual construction of a ship or airplane begins in the mold loft. The floor of the mold loft is constructed of heavy planks which have been leveled, sanded, and painted white to produce a smooth surface on which accurate drafting may be done. On this floor the scale drawings are reproduced full size. Because the original drawings may shrink or expand slightly, points on the curves are established from dimensions given in a table of offsets. All lines are faired once more in this full-size layout, the curves being drawn with long slender strips of wood called *battens*, which can be bent to the shape of the curves.

The mold loft serves two purposes: (1) it provides a full-size check on the original engineering drawings, and (2) it produces the templates, patterns, and forms, in wood or metal, from which the ship or airplane can be constructed. The mold loft also produces scale models, and for any portion of the surface that presents difficulties it builds a full-size skeleton model called a *mock-up*. On an airplane, for example, a mock-up of the pilot's compartment enables the designers to study the placement of controls, instruments, equipment, etc.

8·19. Conic lofting

By the conventional method described in Art. 8·17 fairing the lines of a double-curved surface is a trial-and-error process, and the accuracy of the result greatly depends upon the eye of the draftsman. Long, sweeping curves of little curvature are particularly difficult to fair. By the method of *conic lofting* all curves become *conic sections,* susceptible of very precise graphical or mathematical determination. Because all conics are fair curves, and thus aerodynamically ideal, the aircraft industry has pioneered in the development and use of conic-lofting methods. The greater accuracy gained by using conics also makes it possible to dispense with the numerous water lines and buttock lines, and the desired surface can therefore be described with a minimum of lines.

As shown in Fig. 8·15, *five* point, or slope, conditions are necessary to determine a conic, and three cases are of interest. A conic may be drawn: (1) through any *five points,* no three of which lie in the same straight line, (2) through any *four points* and tangent to *a line* passing through one of the points, or (3) through any *three points* and tangent to *two lines* passing through two of the points. The construction of the conic for each case is explained in the Appendix (Arts. A·21 to A·23).

The three views of the forward body of an airplane fuselage show how the basic lines of this surface may all be developed as conics. In Fig. 8·15 all the curves have been determined and constructed by case 3—three points and two tangents. Assume that the cross-sectional shape of the fuselage at station A must be a circle of diameter $b_R d_R$ as shown in the body plan and that the half section at station D must be the outer frame line, where e_R is the point of maximum width and points g_R and m_R are any two accurately established known points on the curve. Then frame line D may be constructed as two conics, $a_R g_R e_R$ and $e_R m_R c_R$, each of which satisfies the conditions for case 3. The upper conic, for example, must be tangent to a horizontal line through point a_R, tangent to a vertical line through e_R, and must pass through the *shoulder,* or control, point g_R. The detailed construction of this curve is not shown in Fig. 8·15, but it would be exactly the same as that shown in Fig. A·23 in the Appendix.

The contour outline of the profile (side) view is given by the *upper center line* and the *lower center line.* The proper shape for these lines is a matter of aircraft design; but as soon as any five point or slope conditions have been established on each line, the lines may be faired by constructing them as conics. In this case the upper center line is constructed tangent to a horizontal line through point a_F, tangent to an inclined line through b_F, and passing through point p_F. Point p_F may be any point whose position on the curve has been accurately established by design. The lower

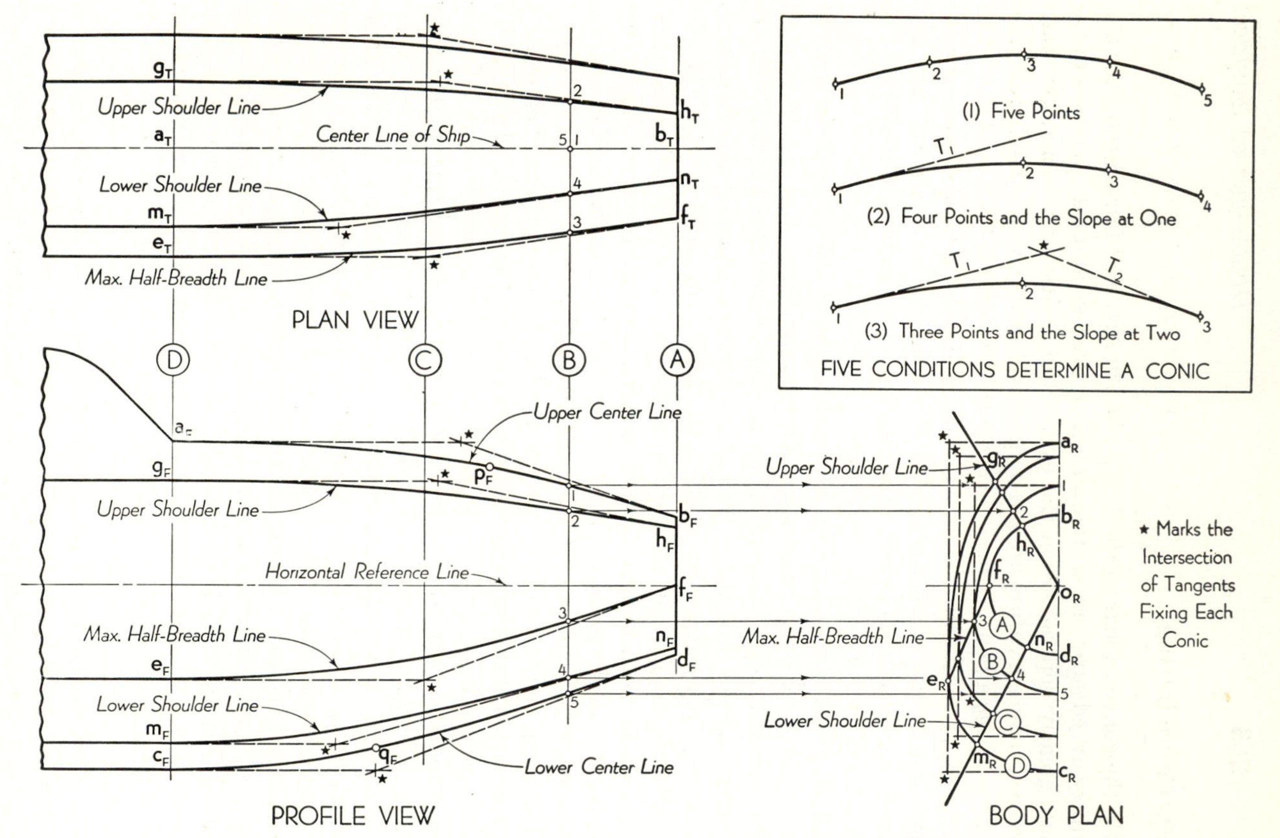

Fig. 8·15. Conic lofting of forward body of an airplane fuselage.

center line is similarly faired through points c_F, q_F, and d_F, and again q_F is any known point on the curve.

The maximum width of the body at station A is given by the location of point f_R on the circular section and at station D by point e_R. Points e_T and f_T may therefore be located in the plan view. The *maximum half-breadth line* may now be drawn in this view according to design and then faired as a conic, tangent to lines drawn through points e_T and f_T. Any known point on the curve may be selected as the control point. In the body plan the maximum half-breadth line has been assumed as a straight line, $e_R f_R$. The curve is thus established in two views, and it can now be located and faired in the profile view.

Only two more longitudinal lines are needed to define the surface of the body. These are the *upper* and *lower shoulder lines.* In the body plan these two curves are assumed to appear as straight lines passing through the shoulder points g_R and m_R and intersecting the axis at o_R. Points h_R and n_R are thus located on frame line A. In the profile view the upper shoulder line must pass through points g_F and h_F. It only remains, then, for the designer to fix the slope of the tangents at these two points and locate one more intermediate point on the curve. *Five* conditions are thus established, and the curve may be faired exactly. This curve could also be drawn through g_F, h_F, and any *three* intermediate points—again a total of five conditions (see Appendix, Art. A·20). The lower shoulder line is similarly established, and both curves may now be located and faired in the plan view. (Each curve is shown only once in the plan view because of symmetry.)

The five longitudinal curves shown in the profile view and the two frame lines at stations A and D completely determine the shape of the forward body, and a frame line at any other station, such as B or C, may now be determined precisely. Frame line B, for example, is determined by points 1, 2, 3, 4, and 5. These five points are found first in the profile view where the five longitudinal lines intersect station B. The points are then located on the corresponding curves in the body plan. Frame line B can now be drawn as two conics—1, 2, 3 and 3, 4, 5—with the same construction used for frame line D.

It should be observed that the method of conic lofting described here is not a means of *designing* a surface. It is only a method of *representing* the surface after the desired shape has been determined by the designer. In Fig. 8·15, 25 point, or slope, conditions define the surface, and their determination is a matter of design. Drawing accurate conic curves that satisfy these given conditions is conic lofting.

PROBLEMS. Group 93.

9. INTERSECTION OF SURFACES

9·1. Introduction

The intersection of a plane surface with a polyhedron, single-curved, warped, or double-curved surface has already been discussed in previous chapters. We shall now consider the more general case of establishing the line of intersection of any two surfaces. Such problems occur frequently in practice, and their solution may be of minor or major importance, but the competent draftsman should be able to draw the line of intersection for any case. The intersection of two holes in a casting or of certain fillets and rounds may be of minor importance, but the line of intersection is usually drawn for the sake of completeness or to clarify the drawing. Intersections of major importance, such as that of two pipes or ducts, of an airplane wing and fuselage, or of a hawsepipe (through which the anchor chain passes) and the ship's bow, must be accurately determined by the draftsman before the intersecting parts can be fabricated.

9·2. An outline of methods

Individual-line method. By assuming any line on one of the surfaces and then finding the intersection of this line with the second surface we may locate one point on the required line of intersection. The procedure may then be repeated with additional lines until enough points have been located to clearly establish the intersection. This method is applicable to any pair of intersecting surfaces, and fundamentally it is the basis for all other methods. The individual-line method will be discussed more thoroughly in Art. 9·3.

Auxiliary-cutting-surface method. In this method the given surfaces are intersected by a *third auxiliary surface.* If this third surface is properly selected, then its intersection with *each* of the given surfaces will be simple and easily located. Where these two simple intersection lines cross, we shall then have one or more points that lie on all three surfaces; hence these points lie on the line of intersection of the given surfaces. This principle has already been employed in several plane-intersection problems (see Figs. 4·32 and 8·11).

Cutting-plane method. The simplest auxiliary cutting surface is a *plane,* and most of the problems in this chapter will be solved by the cutting-plane method. The cutting planes may be vertical, edge-front, or generally oblique, but they should always be selected, if possible, to cut the given surfaces in either straight lines or circles. In many cases any one of several types of cutting planes may be feasible, and the draftsman should then choose that method, or combination of methods, which will yield the simplest and most accurate solution.

Cutting-sphere method. Under certain special circumstances a series of *concentric spheres* may be employed as the auxiliary cutting surfaces. Although of limited application, this method greatly simplifies the solution. The cutting-sphere method is illustrated in Art. 9·23.

Cutting-cylinder method. In this method a series of *parallel cylinders* are assumed as the cutting surfaces. The special type of intersection problem for which this method is suitable is discussed and illustrated in Art. 9·21.

When two surfaces intersect, either of two cases may occur: (1) One surface may pass through the second, in which case only the second surface is cut and the first surface exists uncut within the second. (2) Both surfaces may terminate at the line of intersection. Case 1 is analogous to that of two pipes one of which passes intact through a hole in the other; case 2 is analogous to that of two pipes joined so that fluids may pass uninterrupted from one pipe into the other. Since the second case is more common, we shall assume in all subsequent problems that each surface ends at the line of intersection.

9·3. Intersection of two prisms (individual-line method)

Analysis. The line of intersection of two prisms is a series of connected straight lines (an irregular space polygon). Each vertex of the intersection line represents the point where one edge of one prism intersects one face of the other prism. To locate these vertex points, then, we need determine only where *each* edge of *each* prism intersects the faces of the other prism. For this purpose it is obviously convenient to have one or both of the prisms appear endwise in the views so that the lateral faces will appear edgewise. On engineering drawings, prismatic surfaces are more apt to appear endwise then otherwise.

Problem. In Fig. 9·1 a vertical prism is intersected by an inclined prism. The intersection line is required.

Construction. The lateral faces of the vertical prism appear edgewise in the top view, and view *A* is added to show the lateral faces of the inclined prism edgewise. Each edge of each prism has been labeled at one end only. Referring now to the top view, it can be seen that edge v_T of the inclined prism intersects two faces of the vertical prism at points 1 and 2. Parallels extended to the front view locate points 1 and 2 on edge v_F in this view. Similarly, edge *S* intersects the vertical prism faces at points 3 and 4 and edge *T* at points 5 and 6. Edge *R* passes in front of the vertical prism.

In view *A* we now note where edges of the vertical prism intersect faces of the inclined prism. Edge *C* intersects at points 7 and 8, edge *A* at points 9 and 10. Edge *B* passes behind the inclined prism. Parallels from points 7, 8, 9, and 10 establish these same points on the corresponding edges of the vertical prism in the front view. There are therefore 10 points that determine the required line of intersection.

Visibility. To be visible in the front view a point must lie on a visible edge of one prism and on a visible face of the other; otherwise, it is a

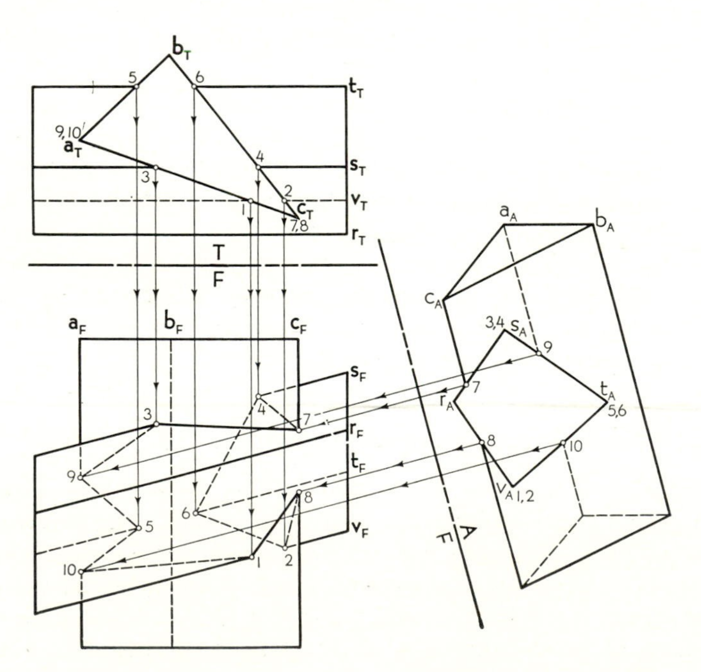

Fig. 9·1. Intersection of two prisms (individual-line method).

hidden point. Point 3, for example, lies on the front face of the vertical prism and on the visible edge S of the inclined prism; point 3 is therefore visible. Point 4 also lies on visible edge S (ignoring the presence of the vertical prism), but it is on a hidden face of the vertical prism and hence is hidden. Thus only points 1, 3, 7, and 8 are visible in the front view. Segments of the intersection line which connect visible points will be visible; hence only lines 1–8 and 3–7 are visible.

Point sequence. Connecting the points in proper sequence must now be done by careful inspection of the adjacent views. For this purpose all points have been labeled in all views. Starting from some visible point, such as 7, the line of intersection should be traced around each prism, observing the sequence of numbers in both the top view and view A. Note, for example, that, although both points 3 and 4 connect with point 9 in view A, reference to the top view shows that there is no line from 4 to 9 and hence only point 3 is connected to point 9. When all points have been connected in proper sequence, it is apparent that the intersection is one continuous closed line.

Edges. The front view is completed by extending each edge to the intersection line. Edges that join a visible part of the intersection line will be visible. Two edges that cross in the front view but do not actually intersect can be examined in the adjacent views to see which is in front of the other.

9·4. Intersection of two prisms (cutting-plane method)

Analysis. Both prisms must be intersected by a series of cutting planes. The solution is simplified if the planes are selected to cut the prisms *parallel to their edges.* The construction is further simplified if the lateral faces of one prism appear edgewise.

Problem. In Fig. 9·2 stage 1 shows the given prisms. One prism is vertical, and its lateral faces therefore appear edgewise in the top view. The edges of the inclined prism appear true length in the front view, but this is not essential to the solution. By the construction of one or more auxiliary views, any pair of intersecting prisms may be shown in this simplified position.

Stage 2. *A typical cutting plane illustrated pictorially.*

A vertical cutting plane passing through both surfaces has been selected to contain the upper edge RS of the inclined prism. This plane also cuts the lower surface of the inclined prism along a line TV, parallel to RS. The vertical prism is simultaneously cut on opposite sides along the vertical lines AB and CD. The four lines AB, CD, RS, and TV all lie in the cutting plane and intersect at points X, Y, Z, and W. These four points are therefore four points on the required line of intersection.

Stage 3. *The same cutting plane in the multiview drawing.*

The points and lines shown here duplicate those shown in stage 2. Note that the location of line t_Fv_F is facilitated by the fact that the right end of the inclined prism does not appear edgewise in the top view. If both ends appeared edgewise in the top view, then neither T nor V could be located by alignment, and it would be necessary either to draw an additional view or to cut off one end of the inclined prism at an angle as in the case presented here.

Visibility of the intersection points in stage 3 is easily determined from the visibility of the two lines that locate each point. Considering each prism independently, and ignoring the presence of the other, lines c_Fd_F and r_Fs_F are both visible and therefore point x_F is visible. Point y_F lies on a visible edge r_Fs_F of the inclined prism (if the vertical prism is ignored), but on a hidden line a_Fb_F of the vertical prism. The combined effect is to make y_F a hidden point. Similarly, z_F is hidden because it lies on a visible and a hidden line. Point w_F is at the intersection of two hidden lines and is therefore hidden. In summary, a point can be visible only if *both* the lines that locate it are independently visible. This is a general principle applicable to all intersection problems.

Stage 4. *Locate intersection points obtained from each cutting plane.*

Because the plane faces of the prisms intersect in straight lines, the cutting planes need only be taken through each intersecting edge of each prism. Thus, in Fig. 9·2, exactly six planes are required. The points in the front view have been numbered to correspond with the numbering of the cutting planes. Primed numbers label points on the right side of the vertical prism; unprimed numbers, those on the left. Note that all planes, except 1 and 6, yield four points. The visibility of each point is evident from the visibility of the intersecting lines that locate it. Points may also be distinguished, if desired, by different symbols—a small circle for visible and a small cross for hidden, for example. To avoid confusion, however, the cutting-plane intersection lines should be drawn lightly but in correct visibility.

Stage 5. *Draw the intersection line in proper sequence and visibility.*

Numbering the cutting planes consecutively from front to back and labeling the points correspondingly prevent error since the points can then be connected in numerical order. Only those line segments of the intersection which connect two visible points can be visible. There are two separate lines of intersection in this case, for the inclined prism passes completely through the vertical prism (compare Fig. 9·1). It will be noted that some of the points like 2′ and 5 simply complete the numerical sequence but do not form a vertex of the intersection line. The discerning student may omit such points.

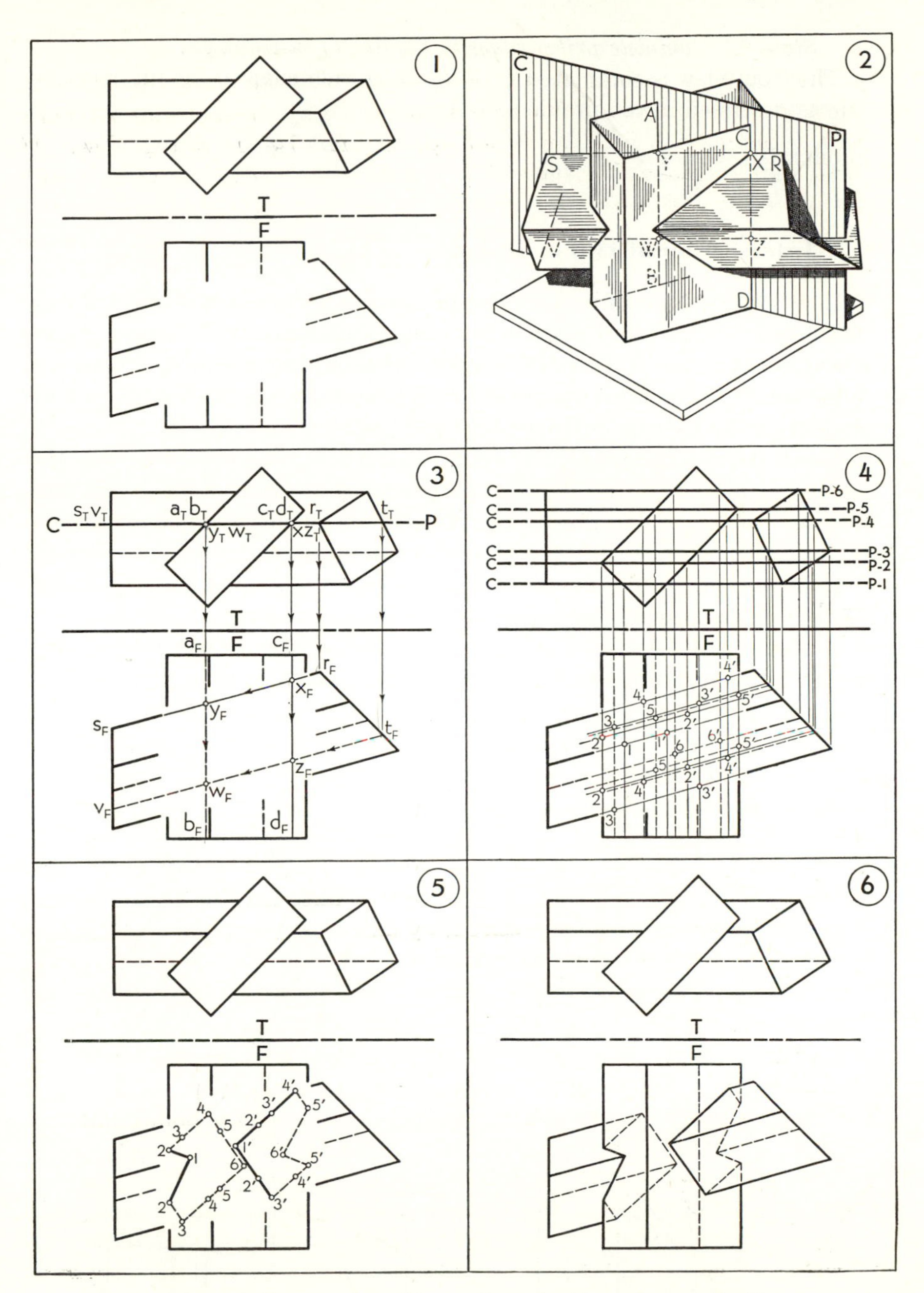

Fig. 9·2. Intersection of two prisms (cutting-plane method).

Stage 6. *Complete prism edges to the line of intersection.*

The front view is completed by extending each edge to the intersection line as described in the previous article.

PROBLEMS. Group 94.

9·5. Intersection of two cylinders

In general, the line of intersection of two single-curved surfaces will be a double-curved line, and only in certain special cases will it be a plane curve (see Art. 9·6). To establish the curve of intersection accurately, a large number of points located in orderly sequence is desirable, and we shall therefore use the cutting-plane method.

Analysis. To intersect two cylinders in the simplest possible way the cutting plane must be *parallel to both cylinder axes.* Such a plane will then intersect each cylinder along two straight-line elements, as shown in the pictorial drawing of Fig. 9·3. These four elements will intersect to locate four points on the required curve of intersection. Additional parallel cutting planes provide as many more points as desired. Note that cutting planes parallel to two cylinders will always appear edgewise in any view which shows either cylinder endwise.

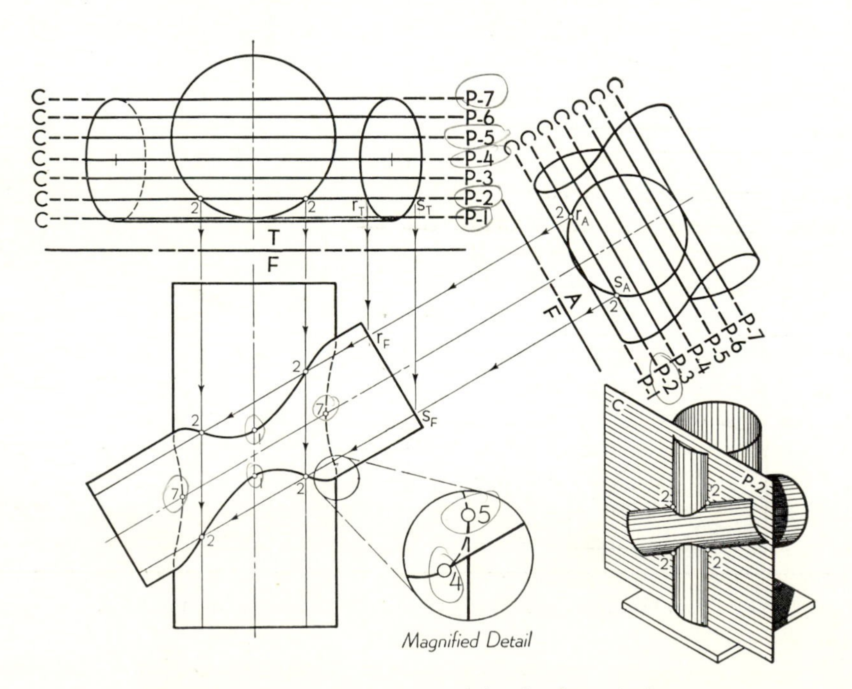

Fig. 9·3. Intersection of two right circular cylinders.

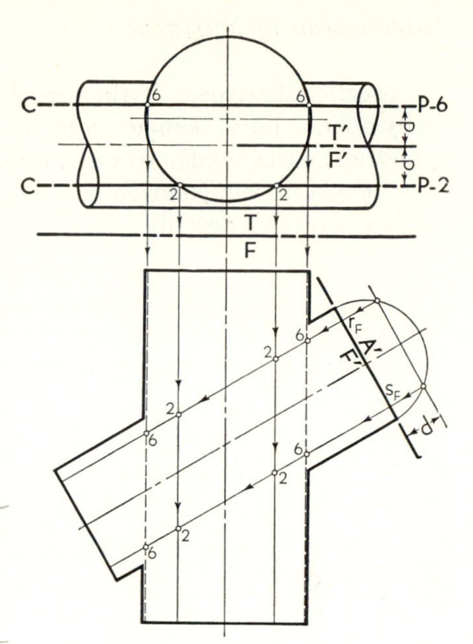

Fig. 9·4. Intersection points by semicircle method.

Problem. In Fig. 9·3 the curve of intersection of the two right circular cylinders is required.

Construction. A number of vertical cutting planes are selected as shown. These planes appear edgewise in both the top view and view A. Then cutting plane C-P-2, for example, cuts two elements from the vertical cylinder as shown in the top view and two elements from the inclined cylinder as shown in view A. All four elements are visible in the front view, and therefore the points labeled 2 are four visible points on the curve of intersection. Note that view A is not essential to the solution in this case, since the elements on the inclined cylinder could also be located by using points r_T and s_T. The circle in view A, however, is undoubtedly more accurate than the ellipse in the top view.

Special planes. Although the spacing of the cutting planes may be random, two important exceptions should be noted. C-P-1 and C-P-7 are each tangent to one of the cylinders and locate maximum points, 1 and 7, on each loop of the curve. C-P-4 passes through the center of the inclined cylinder and C-P-5 through the center of the vertical cylinder; hence two of the elements cut by each plane will be extreme elements in the front view. These planes thus locate the eight points where the curve is tangent to the extreme elements (two such points are shown in the enlarged detail).

Construction: semicircle method. When the inclined cylinder is *right circular*, as in Fig. 9·3, the construction can be simplified as shown in Fig. 9·4. Here a semicircle drawn on the end of the inclined cylinder serves as a partial auxiliary view, or substitute for view A. Then the displaced reference line T'–F' corresponds to reference line A'–F', and symmetrical distances such as d can be transferred from the top view to the semicircle as shown. The eight intersection points labeled 2 and 6 are thus quickly established by transferring only the single distance d. In this method it is obviously unnecessary to draw the elliptical cylinder ends in the top view; hence they have been represented here as conventional breaks.

The general case. Although the cylinders shown in Fig. 9·3 are right circular, the method illustrated there is applicable to cylinders of any cross-sectional shape. If the axes of both cylinders were given as generally oblique lines, then new auxiliary views could be drawn to show each cylindrical surface edgewise as in Fig. 9·3. For generally oblique cylinders, however, the solution can usually be shortened by using oblique cutting planes as described later in Arts. 9·13 and 9·14.

9·6. Typical cylinder intersections

The most common type of intersection encountered in practice is that of two cylinders, and the cylinders are most frequently circular in section. The intersecting cylinders shown in Fig. 9·5 have been drawn in the simplest position in order to demonstrate clearly the *four* basic forms that the curve of intersection may assume. Any view that shows one of the cylinders endwise (as in the top views of Fig. 9·5) will immediately indicate the form of the intersection curve. Obviously, the construction of the curve is greatly facilitated if the draftsman can know in advance what general form the curve will take.

The top view of Fig. 9·5(*a*) shows that each cylinder is only partly cut by the other, and the resulting intersection line is therefore *one continuous curve.* In Fig. 9·5(*b*), however, the smaller horizontal cylinder enters one side of the vertical cylinder, passes through it, and comes out on the other side, thus forming *two separate curves.* When, as in Fig. 9·5(*c*), the two

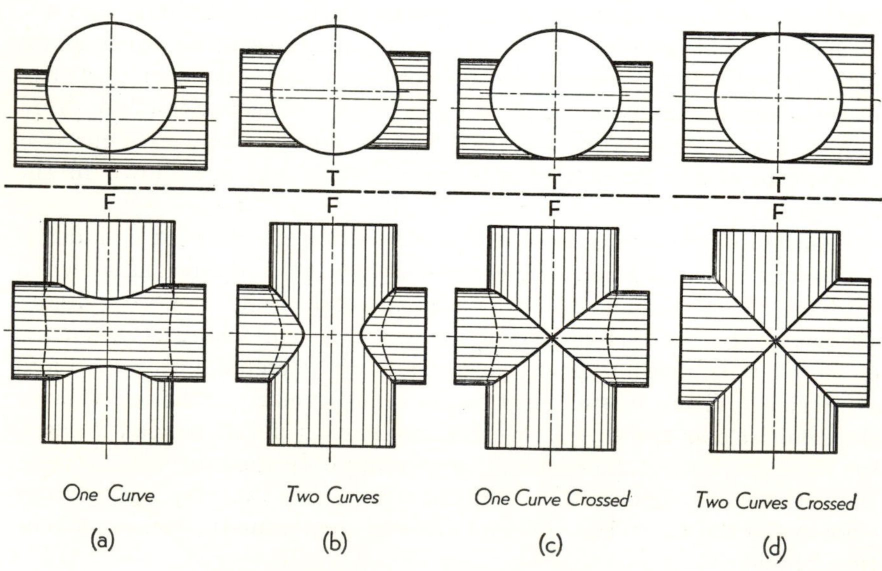

Fig. 9·5. Typical cylinder intersections.

cylinders are tangent to each other on one side (both cylinders tangent to a common cutting plane), then the curve is again *continuous and crossed* at the tangent point, forming a figure 8. If the two cylinders are of equal diameter and their axes intersect, then the cylindrical surfaces will be tangent on both sides as shown in Fig. 9·5(*d*) and the curve becomes *two intersecting plane curves* that are true *ellipses.* Because both cylinders appear true length in the front view, the ellipses appear edgewise, their planes bisecting the angle between the cylinder axes (see also Art. 10·16).

PROBLEMS. Group 95.

9·7. Intersection of a cylinder and a prism

Analysis. The intersection of a cylinder and prism may be determined by employing the same method used for two cylinders: cutting planes *parallel to the axes* of both surfaces will cut straight-line elements from each surface (Art. 9·5). In general, the intersection will be a curved line, continuous across each intersected face of the prism, but broken (or pointed) at each edge.

Problem. In Fig. 9·6 the curve of intersection of the oblique cylinder and prism is required. In this case the given surfaces are in a *special position:* the axes of cylinder and prism appear parallel in the top view, and therefore cutting planes parallel to both axes will appear edgewise in this view. Lacking this special feature, any cylinder and prism (or two cylinders or two prisms) can be reduced to this case by constructing a plane through one axis parallel to the other axis; when a new view is drawn to show this plane as an edge, the axes will then appear parallel (Art. 4·19).

Construction. Detailed construction has been shown for only *C-P*-6. This plane intersects the upper base of the cylinder at points m_T and t_T and establishes the corresponding pair of elements in the front view. At points r_T and s_T the plane also intersects the upper base of the prism, and the two elements cut from the prism are drawn in the front view. The four points, 6 and 6, 6′ and 6′, are thus located at the intersection of the cylinder and prism elements. All four points are hidden in the front view because both prism elements are hidden here. These same four points will appear in the top view on the edge view of *C-P*-6, and here one of the four points will be visible because elements from s_T and m_T are both visible in the top view.

Special planes. *C-P*-1 and *C-P*-8 are important because they are the first and last planes, respectively, that intersect both cylinder and prism. Also important are cutting planes like *C-P*-1 and *C-P*-5, which intersect the prism along its edges. *C-P*-1 intersects edge *C* and locates points 1 and 1′; *C-P*-5 intersects edge *A* and locates points 5 and 5′.

In order to determine the exact points in the front view where the curve of intersection is tangent to the extreme elements of the cylinder, *C-P*-2 and *C-P*-6 have been specially selected. *C-P*-2 is drawn through point n_T, and therefore one of the elements cut from the cylinder by this plane will be an extreme element in the front view. *C-P*-2 thus locates tangent points 2. In similar manner, *C-P*-6 is drawn through point m_T, thus intersecting the cylinder along the other extreme element in the front view, and locating tangent points 6′.

PROBLEMS. Group 96.

9·8. Intersection of two circular cones: bases parallel

Analysis. Two cones (or cylinders) having parallel circular bases may be intersected by cutting planes *parallel to the base plane.* Then, as shown

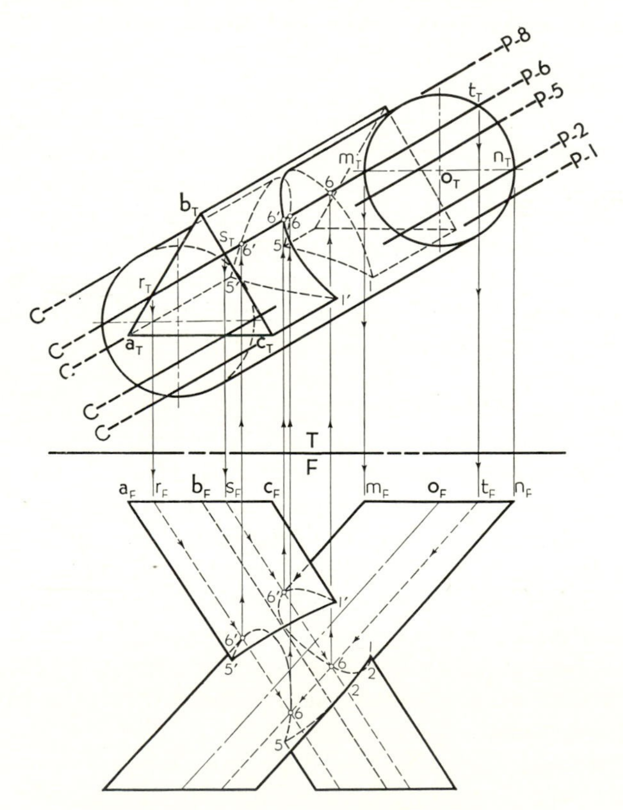

Fig. 9·6. Intersection of a cylinder and a prism.

in the pictorial illustration of Fig. 9·7, the cutting plane will intersect each cone in a circle. If these two circles intersect, points 4 and 4′ will be two points on the required line of intersection of the given cones.

Problem. In Fig. 9·7 the curve of intersection of the two right circular cones is required.

Construction. The cone bases are horizontal; hence a series of horizontal cutting planes must be employed. *C-P-*4, for example, cuts two circles from the cones, and when these two circles (labeled 4) are drawn in the top view, they intersect at points 4 and 4′ to establish two points on the curve of intersection in this view. Points 4 and 4′ can then be shown in the front view, appearing here on the edge view of *C-P-*4. Additional parallel cutting planes produce more points in the same manner.

Visibility. The visibility of each point depends upon the visibility of the two circles that locate it. In Fig. 9·7 all circles are completely visible on both cones in the top view (considering each cone independently), hence all points here are visible. In the front view only half of each circle will be visible, and therefore a point can be visible only if it lies on the front half of both circles. Referring to the top view, it is apparent that

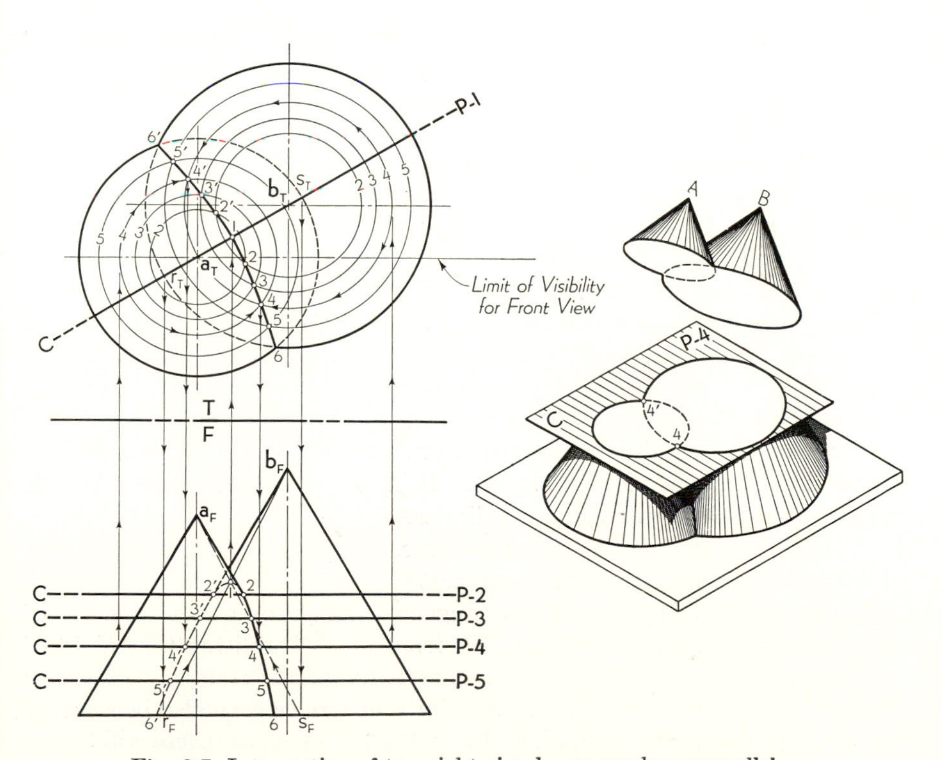

Fig. 9·7. Intersection of two right circular cones: bases parallel.

only those points on the front half of the foremost cone (2, 3, 4, 5, and 6) can be visible in the front view. The center line labeled "limit of visibility for front view" thus establishes the dividing line between visible and hidden points for the front view.

Special plane. Using horizontal cutting planes, the location of point 1 becomes a trial-and-error process of trying to find two circles in the top view that are exactly tangent to each other. This may be avoided by choosing the vertical cutting plane *C-P*-1, which passes through both cone vertices. Such a plane cuts two straight-line elements from each cone. The intersection of elements *AS* and *BR* locates point 1 as shown.

General application. This method of cutting circles is generally applicable to any two intersecting surfaces having parallel circular sections (Art. 9·22). It is highly desirable, of course, that the circles cut by the planes shall appear in some view as circles, and not as ellipses. Methods for determining the intersection of cones and cylinders of noncircular section are discussed in Arts. 9·13 to 9·18.

9·9. Intersection of a cone and a cylinder: axes parallel

Analysis. The cone and cylinder may be intersected by cutting planes in two different ways: (1) If the cone and cylinder are both circular, then planes *perpendicular to the axes* will cut circles from each surface. (2) Radial planes *containing the cone axis* will cut straight-line elements from each surface. Although each method is adequate for a complete solution, the best results are usually obtained by combining both methods.

Problem. In Fig. 9·8 the parallel axes of cone and cylinder are vertical, and both surfaces have circular bases. The required intersection line can therefore be obtained by either type of cutting plane.

Construction: horizontal planes. As shown in Fig. 9·8(*a*), *C-P*-5 intersects both cone and cylinder in a circle. Circles cut from the cylinder will be of constant diameter for all horizontal cutting planes, and therefore we need note only such points as 5 and 5′ where the cone circle intersects the cylinder in the top view. Points 5 and 5′ can then be located in the front view as shown. Point 5 is visible because it lies on the front half of both cone and cylinder.

Point 1, the highest point on the curve, may be obtained without trial by drawing in the top view the smallest cone circle that is tangent to the cylinder; point 1 is the point of tangency. Extending the diameter of this smallest circle to the cone in the front view establishes the height of point 1 in this view. In similar manner, the largest cone circle tangent to the cylinder locates point 9. Special cone circles may also be selected to pass through points 2, 3, 6′, and 7′ in the top view, for these will be the points in the front view where the curve of intersection is tangent to the extreme elements.

We should now note that circles slightly larger than the smallest cone circle or slightly smaller than the largest circle will intersect the cylinder at a very small angle. Therefore, in this region of the intersection, small drafting errors may result in relatively large errors in the curve, and points of questionable accuracy should be checked by employing the second type of cutting plane.

Construction: vertical radial planes. In Fig. 9·8(*b*) every vertical plane contains the cone axis, but each plane is assumed at a different angle so that the planes are radial with respect to the center of the cone. Then any plane, such as *C-P*-1, will cut two straight-line elements *VA* and *VB* from the cone and two vertical elements from the cylinder. By locating the four elements in the front view, points 1 and 1′ are found at their intersections. It can be seen that *C-P*-1 locates both the highest and the lowest points on the curve.

Additional planes may be selected at random, but *C-P*-3 is important because it locates points 3 and 3′ on the extreme elements of the cone in

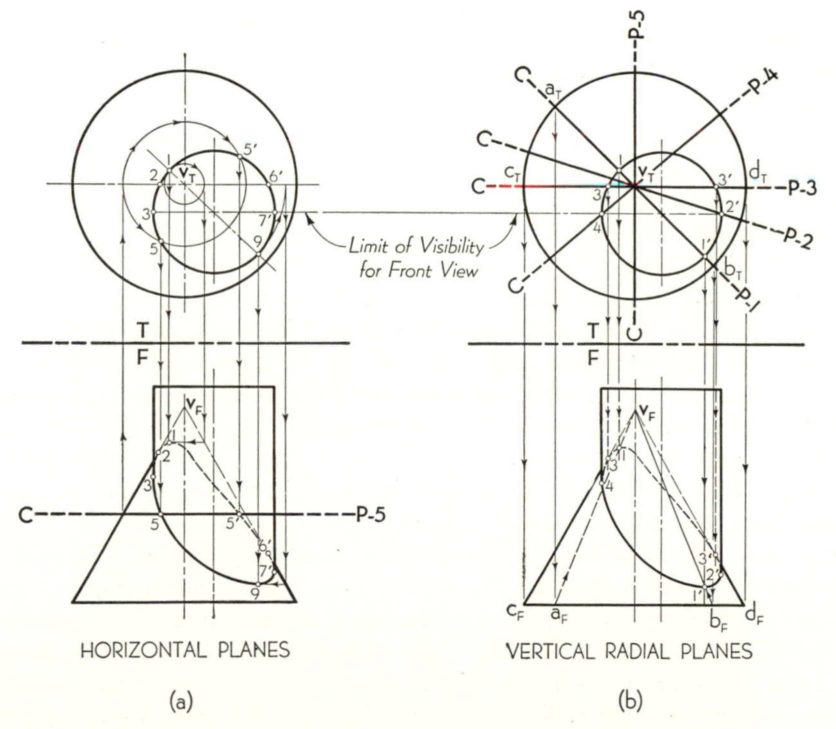

Fig. 9·8. Intersection of a cone and a cylinder: axes parallel.

the front view. Similarly, *C-P*-2 and *C-P*-4 are specially selected to give points 2′ and 4 on the extreme elements of the cylinder. Points obtained by *C-P*-5 cannot be located by alignment, and the revolution method shown in Fig. 6·6(*a*) should be used here. Points obtained by planes very close to *C-P*-5 should also be checked for accuracy by the revolution method.

Summary. Comparing the two methods shown in Fig. 9·8, it is apparent that each method tends to become inaccurate for certain portions of the intersection; hence a judicious combination of both methods usually produces the most accurate results. To use horizontal cutting planes as in Fig. 9·8(*a*) it is desirable that the cone be circular, but note that the vertical cylinder may be of any cross-sectional shape, even prismatic. Vertical radial planes may be employed as shown in Fig. 9·8(*b*) regardless of the shape of the cone or cylinder bases. It should be remembered, of course, that the methods of this article are limited to a cone and cylinder whose axes are parallel.

9·10. Intersection of a cone and a cylinder

Analysis. Any plane that passes *through the vertex* of the cone will intersect the conical surface in two straight-line elements. If the plane is also *parallel to the cylinder axis*, it will cut two straight-line elements from the cylindrical surface. Then, as shown in the pictorial illustration of Fig. 9·9, each of the two cone elements will intersect each of the cylinder elements to establish four points on the required curve of intersection. Each additional cutting plane must be assumed at a different inclination through the cone vertex, and therefore all such radial planes will appear edgewise only in a view that shows the cylinder endwise.

This analysis is also applicable if a pyramid is substituted for the cone or a prism for the cylinder.

Problem. In Fig. 9·9 a right circular cone is intersected by a horizontal cylinder. The intersection curve is required.

Construction. View *A* is drawn to show the cylinder axis as a point. Then all cutting planes must appear edgewise in this view, and all must pass through point v_A. *C-P*-4, for example, cuts the cylinder along its topmost element (labeled 4, 4′ in view *A*), and cuts it again along a lower element (similarly labeled 4, 4′). *C-P*-4 also cuts the cone along two elements appearing as line v_A–b_Ac_A in view *A*. In the top view the cone elements appear at v_Tb_T and v_Tc_T. The cylinder elements are also shown here, one visible, the other hidden. The intersections of the four elements then locate in this view four points marked 4 and 4′. By the customary rules of visibility, two of these points are visible, and two are hidden. The same four points may now be located in the front view on parallels extended to the corresponding cone elements $v_F b_F$ and $v_F c_F$. Each point in the front view may also be checked, of course, by reference-line measurements from view A.

Special planes. Choice of cutting planes in view A should begin with planes of special significance. C-P-1 and C-P-9 are the first and last planes in the series, C-P-1 being tangent to the cone on one side, and C-P-9 tangent to the cylinder on the other side. Inspection of the completed curve of intersection will show the critical position of points 1, 1′, 9, and 9′. Points that lie on the various extreme elements are also very desirable; planes through points 5 and 8, for example, locate such points on the cylinder.

Number of curves. Although the method discussed here requires an additional view, it has the advantage of indicating quickly the form of the intersection curve. View A shows that there will be one continuous curve of intersection because the cylinder cuts through only one side of the cone [compare the analogous case of two cylinders in Fig. 9·5(a)]. Had view A shown the cylinder entirely within the outlines of the cone (as in the

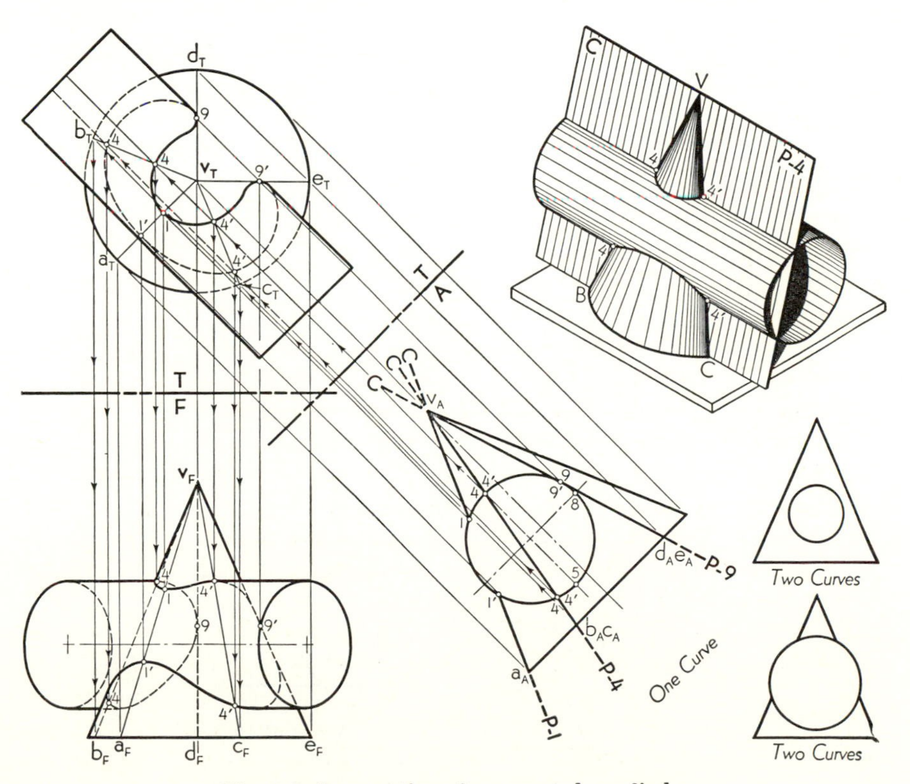

Fig. 9·9. Intersection of a cone and a cylinder.

upper small figure), this would have indicated two separate curves. Similarly, if the cylinder had been large enough to overlap both extreme elements of the cone in view *A* (lower small figure), this would again indicate two separate curves, caused by the cone passing in one side of the cylinder and out on the other side.

PROBLEMS. Group 97.

9·11. Intersection of a cone and a prism

Problem. In Fig. 9·10 a horizontal prism intersects a right circular cone. Select suitable cutting planes, and determine the curve of intersection.

Analysis. The position of the cone and prism in Fig. 9·10 is similar to that of the cone and cylinder in Fig. 9·9, and therefore inclined cutting planes could also be used here to cut straight-line elements from both cone and prism. But horizontal cutting planes will cut circles from the cone

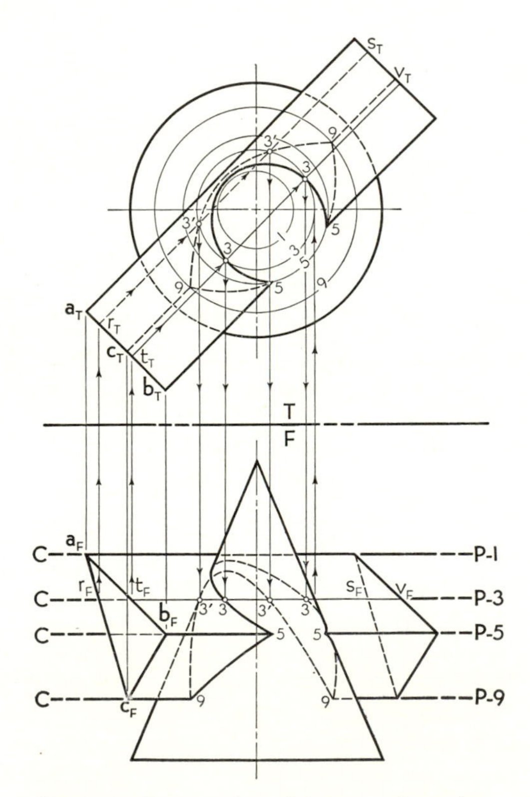

Fig. 9·10. Intersection of a cone and a prism.

and straight lines from the prism, and the intersection curve can thus be determined without an additional view.

Construction. A horizontal cutting plane, such as *C-P*-3, will cut a circle from the cone (circle 3 in the top view) and will cut the prism along the two parallel lines *RS* and *TV*. In the top view, line $r_T s_T$ cuts circle 3 in two points labeled 3′, and line $t_T v_T$ cuts the same circle in two more points labeled 3. These four points on the line of intersection may now be located in the front view as shown. Points 3 are both visible in the top view because they lie on a visible circle of the cone and also on a visible line of the prism. To be visible in the front view a point must be on the front half of the cone and also on a visible face of the prism.

Special planes. Cutting planes through each edge of the prism should be established first to locate the important cusps 5 and 9 on the curve. The fact that edge *A* of the prism does not intersect the cone is shown in the top view where circle 1 fails to intersect edge line a_T. Location of the topmost points on the curve need not be left to trial and error; they may be accurately established by using a vertical plane through the cone axis perpendicular to the prism, or a new view may be drawn to show the prism endwise. To avoid confusing the figure the detailed construction for these points has not been shown in Fig. 9·10.

PROBLEMS. Group 98.

9·12. Oblique cutting planes

In each of the previous intersection problems the cutting planes were selected to appear edgewise in one of the views. Planes of this type usually simplify the construction and should be employed whenever possible. In fact, most of the intersection problems that occur in engineering practice can be solved with such planes. There are, however, many intersection problems that can best be solved by using *oblique cutting planes.* In the articles that follow we shall show how oblique planes may be established, represented, and used to determine points on the line of intersection of oblique cylinders and cones.

9·13. Intersection of two oblique cylinders: common base plane

Analysis. Oblique cutting planes *parallel to both cylinders* will intersect the common base plane in a series of *parallel lines,* and each of these parallel intersection lines will locate one set of straight-line elements cut from the two cylinders by one cutting plane. The intersection of the elements establishes points on the intersection curve.

Problem. In Fig. 9·11 the axes of both cylinders are oblique, but the cylinder bases—one circular, one elliptical—both lie in a common horizontal plane.

Construction. The procedure can be observed in both the pictorial illustration and the multiview drawing. A point X is assumed anywhere in the vicinity of the given cylinders, and through this point two lines are then drawn parallel to the axes of the cylinders. (XM is parallel to AB, XN parallel to CD.) These two intersecting lines define a plane that is parallel to both cylinders. This *trial plane* intersects the base plane along the line MN. Then, as shown in the picture, all cutting planes parallel to the established trial plane will intersect the base plane in parallel intersection lines. Since it is understood that all cutting planes shall be parallel to the trial plane, we may represent them in the multiview drawing by showing only their line of intersection with the

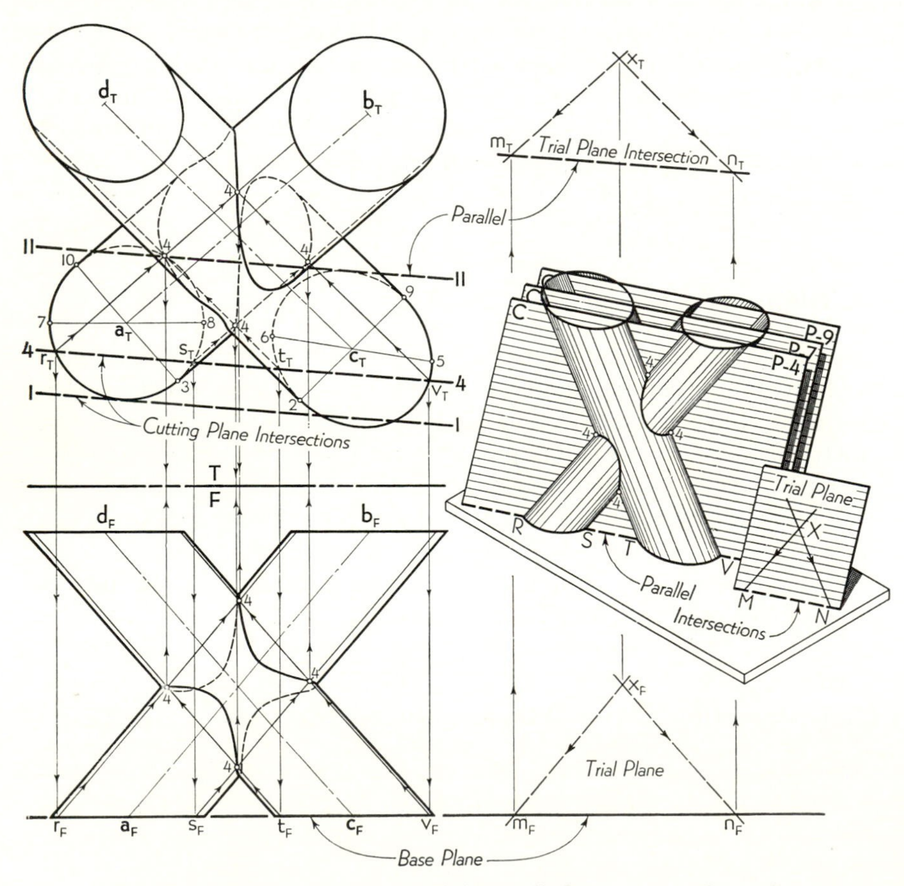

Fig. 9·11. Intersection of two oblique cylinders: common base plane.

cylinder base plane. Thus, the line 11-11, drawn parallel to m_Tn_T (the trial-plane intersection), represents the intersection of an oblique cutting plane with the common base plane of the cylinders. All additional cutting planes, such as 1–1 and 4–4, are similarly represented by their parallel intersection lines. These lines have been distinguished by drawing them as a series of long dashes.

Cutting plane *C-P*-4 intersects the base plane along the line 4–4, which intersects the base circle of one cylinder at points r_T and s_T and intersects the base ellipse of the other cylinder at points t_T and v_T. But since the cutting plane is parallel to the trial plane, it is also parallel to both cylinders and will therefore cut two straight-line elements from each. These four elements are drawn in the top view, and their intersections establish the four points labeled 4 on the curve of intersection. By locating points r_F, s_F, t_F, and v_F on the base plane in the front view these same elements can be drawn here and the same four points located in the front view. The accuracy of the work may now be checked by noting the alignment of the four points in the top view with those in the front view.

Visibility. Elements *R* and *V* are both independently visible in the top view and therefore locate a visible point on the curve. The other three points labeled 4 are hidden in the top view because each lies on a hidden element. In the front view all four intersecting elements are visible, thus yielding four visible points.

Form of the intersection curve. Intersection lines 1–1 and 11–11—each tangent to one cylinder, but cutting the other—represent the *first* and *last* cutting planes in the series. These two cutting planes should be the first established, for they immediately indicate the form of the intersection curve—one continuous curve in this case. As shown in Fig. 9·12, the intersection curve may assume any one of *four* basic forms depending upon the position of the first and last planes (compare Fig. 9·5). The shaded areas indicate the unintersected parts of each cylinder.

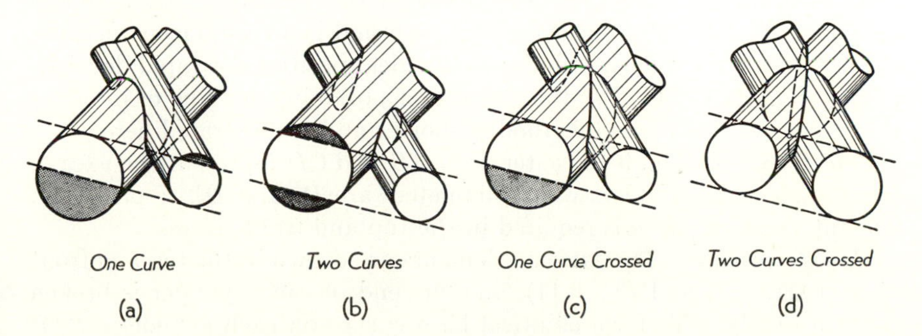

Fig. 9·12. Typical oblique cylinder intersections.

Special planes. Cutting-plane intersection lines drawn through points 3 and 10 on the circular base will accurately locate those points where the curve is tangent to the extreme elements of this cylinder in the top view. Intersection lines through points 2 and 9 on the elliptical base locate similar points on this cylinder in the top view. To establish points on the extreme elements in the front view will require intersection lines through points 5, 6, 7, and 8. Thus, in this particular problem, as many as 10 special planes may be desirable.

Summary. The method described here may be applied to any pair of cylinders (or prisms) whose bases lie in a common plane. It is desirable, although not essential, that the base plane appear edgewise in one of the views. It need not be horizontal as shown in Fig. 9·11 but may be inclined at an angle without affecting the method. If the bases lie in parallel planes instead of in the same plane, then one of the cylinders may be either extended or cut off to form a new base that lies in the same plane as the other base.

9·14. Intersection of two oblique circular cylinders

Many of the pipes and ducts encountered in engineering practice are circular cylinders. In such cases the diameter and axis (center line) of each cylinder are usually known, but base curves that lie in a common plane may not be immediately available.

Analysis. Two methods of solution are practical: (1) New auxiliary views may be drawn to show both cylinders in true length and each cylinder endwise, thus reducing the problem to the simple case shown in Fig. 9·3. (2) The cylinders may be intersected by a cutting plane to form elliptical base curves that lie in a common plane, and the solution may then be completed by the method of Art. 9·13.

The first method will, in general, require four additional views, with the added disadvantage that each point on the curve of intersection must be carried back to the given views if the curve is required there. If the cylinders must be developed, then this method is advantageous because the new views show each cylinder in the ideal position for rapid development (see Art. 10·9). For the purpose of obtaining the line of intersection, however, the second method is shorter and is illustrated here.

Problem. In Fig. 9·13 center lines AB and CD represent the axes of two intersecting cylinders whose diameters are D_1 and D_2, respectively. The intersection curve is required in the top and front views.

Construction. The extreme elements are drawn in the top and front views (Art. 6·11 and Fig. 6·11), and the end of each cylinder is broken conventionally. To form elliptical base curves on each cylinder a *horizontal* cutting plane is assumed to cut the cylinder axes at M and N.

This plane may be assumed at any desired level, but it should be horizontal in order that both ellipses may appear true size in the top view. The minor axis of each ellipse will equal the diameter of the cylinder, and the lengths of the major axes are obtained from views A and B, which show the *true angle* between each cylinder axis and the plane. (This is the method of Art. 6·14.) The two ellipses must be constructed accurately to avoid error in the intersection curve.

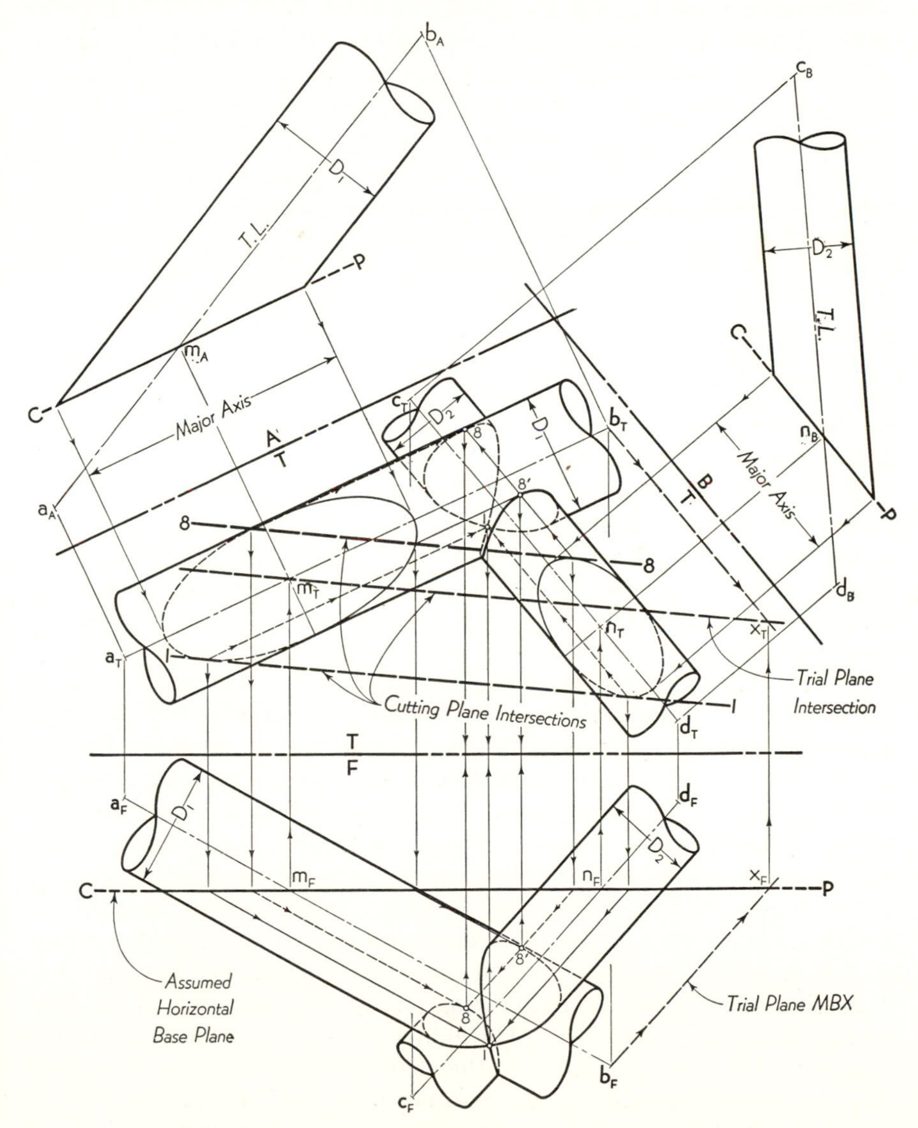

Fig. 9·13. Intersection of two oblique cylinders of revolution.

The solution may now be completed by employing oblique cutting planes in the manner described in Art. 9·13. The trial plane has been established in Fig. 9·13 by drawing line *BX* parallel to axis *CD*. Plane *MBX* is therefore parallel to both cylinders and intersects the assumed common base plane along the line *MX*. Eight parallel cutting planes were used to determine the curve of intersection, but only the first and last planes, *C-P*-1 and *C-P*-8 (represented by the intersection lines 1–1 and 8–8), have been shown. It should be noted that *C-P*-1 is tangent to both cylinders in this case, and therefore the line of intersection is one curve crossed at point 1 [compare Fig. 9·12(*c*)].

PROBLEMS. Group 99.

9·15. Intersection of two cones: common base plane

Analysis. To cut straight-line elements from each of two cones the cutting plane must pass *through both cone vertices.* Then, as shown in the pictorial illustration of Fig. 9·14, every cutting plane must contain the *vertex line AB.* Line *AB* extended intersects the common base plane at point *P*. Each plane will intersect the base plane in a straight line, and all such intersection lines must converge at the common point *P*. *Point P thus becomes a key point in the solution.*

Problem. In Fig. 9·14 the given cones are oblique, but the cone bases both lie in a common horizontal plane.

Construction. In the multiview drawing the vertex line *AB* has been extended to intersect the base plane at point *P*, the *point of convergence.* Any desired cutting plane may now be established simply by showing its line of intersection with the base plane—remembering, of course, that all such intersection lines must pass through point *P*. The dash line p_T6, for example, is the top view of the intersection of *C-P*-6 with the base plane of the cones. Intersection lines p_T1 and p_T8 are those for the first and last cutting planes, since any planes beyond these two will no longer intersect the elliptical cone base. It is thus evident at this stage that there will be two separate curves of intersection, for every element on the elliptical cone intersects the circular cone twice, thus passing in one side and out on the other [compare Fig. 9·12(*b*)]. In addition to those shown, we should also choose cutting planes that will cut the base curves at points 2, 3, 4, 5, and 7 to locate points on the various extreme elements.

Having selected the desired cutting planes, the elements cut by each plane can now be determined. *C-P*-6, for example, cuts the circular base at points *R* and *S* and establishes the elements *RB* and *SB* on that cone. The corresponding elements on the other cone are *TA* and *VA*. These four elements intersect in each view to locate the four points labeled 6 and 6′. Points in the front view should, of course, align with the corresponding points in the top view. Note that each element has been drawn in correct visibility in order to establish the visibility of the points by the usual rule for intersecting elements.

Special cases. If the vertex line is parallel to the base plane, then point *P* will be at infinity and the intersection lines will all be parallel

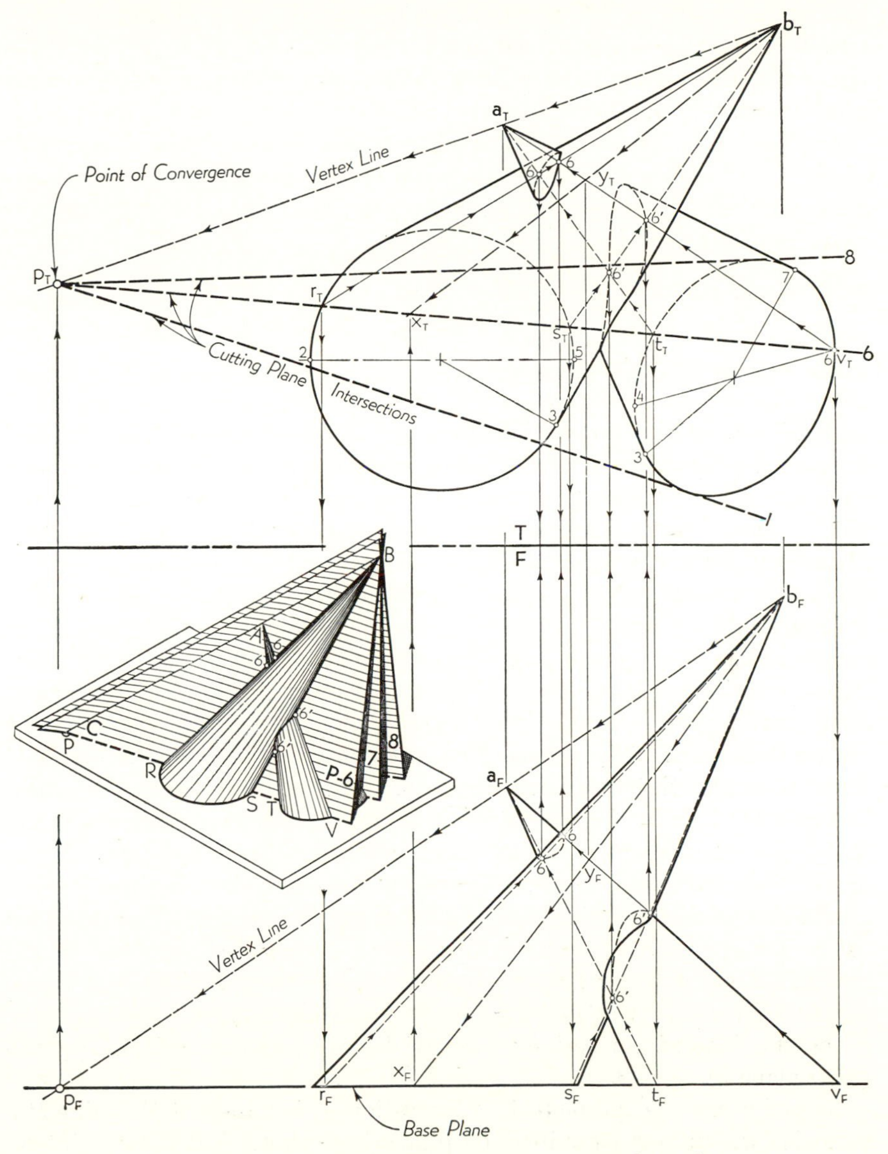

Fig. 9·14. Intersection of two cones: common base plane.

to the vertex line. In the event that point P falls inconveniently beyond the limit of the paper, any cutting-plane intersection line, such as p_T6, may be located as follows: (1) Arbitrarily choose any element on either cone, such as VA, thus forming a cutting plane VAB. (2) Choose any convenient point such as Y, on element VA. (3) Draw the line BY, and extend it to intersect the base plane at point X. (4) Then point X lies on the cutting-plane intersection line of plane VAB, and line XV may now be drawn to establish the other three elements, TA, SB, and RB, that lie in this cutting plane. Additional cutting planes may be similarly established.

Summary. The method described here may be applied to any pair of cones (or pyramids) whose bases lie in a common plane. The base plane may be in any position, but the solution is simplified when it appears as an edge in some view.

PROBLEMS. Group 100.

9·16. Intersection of two oblique cylinders: different base planes

Analysis. As in the previous example, oblique cutting planes *parallel to both cylinders* will cut straight-line elements from each cylinder. In this case, however, the cylinder bases are in different planes; hence the trial plane and each of the parallel cutting planes will intersect both base planes and form *two different* (but connected) *sets of parallel intersection lines.* Otherwise, the solution will be identical with that of Art. 9·13.

Problem. In Fig. 9·15 the elliptical base of one cylinder lies in an inclined plane (base plane 1), and the circular base of the other cylinder lies in a horizontal plane (base plane 2). The two base planes intersect along the line KG.

Construction. The procedure can be observed in both the pictorial illustration and the multiview drawing. The solution begins (as it did in Fig. 9·11) by selecting the point X at random. Through point X line XM is drawn parallel to axis AB and line XN parallel to axis CD. Plane MXN is thus the trial plane. We must now determine the intersection line of this trial plane with *each* of the given base planes. This requires *two points* in *each* base plane; hence point Y is selected anywhere along line XM, and line YP is drawn parallel to XN. (Any random line in the plane MXN will serve equally well.) This establishes two points, P and N, on the line of intersection of the trial plane and the horizontal plane (base plane 2).

Referring now to the pictorial illustration, it is apparent that line PN extended will intersect the inclined plane (base plane 1) at some point K, and this point K must lie on the line of intersection of the two base planes. But all parallel cutting planes, such as C-P-3 and C-P-9, will similarly intersect the *base-plane intersection line* at other points such as F and G, and therefore this line becomes a *key* line in the problem solution.

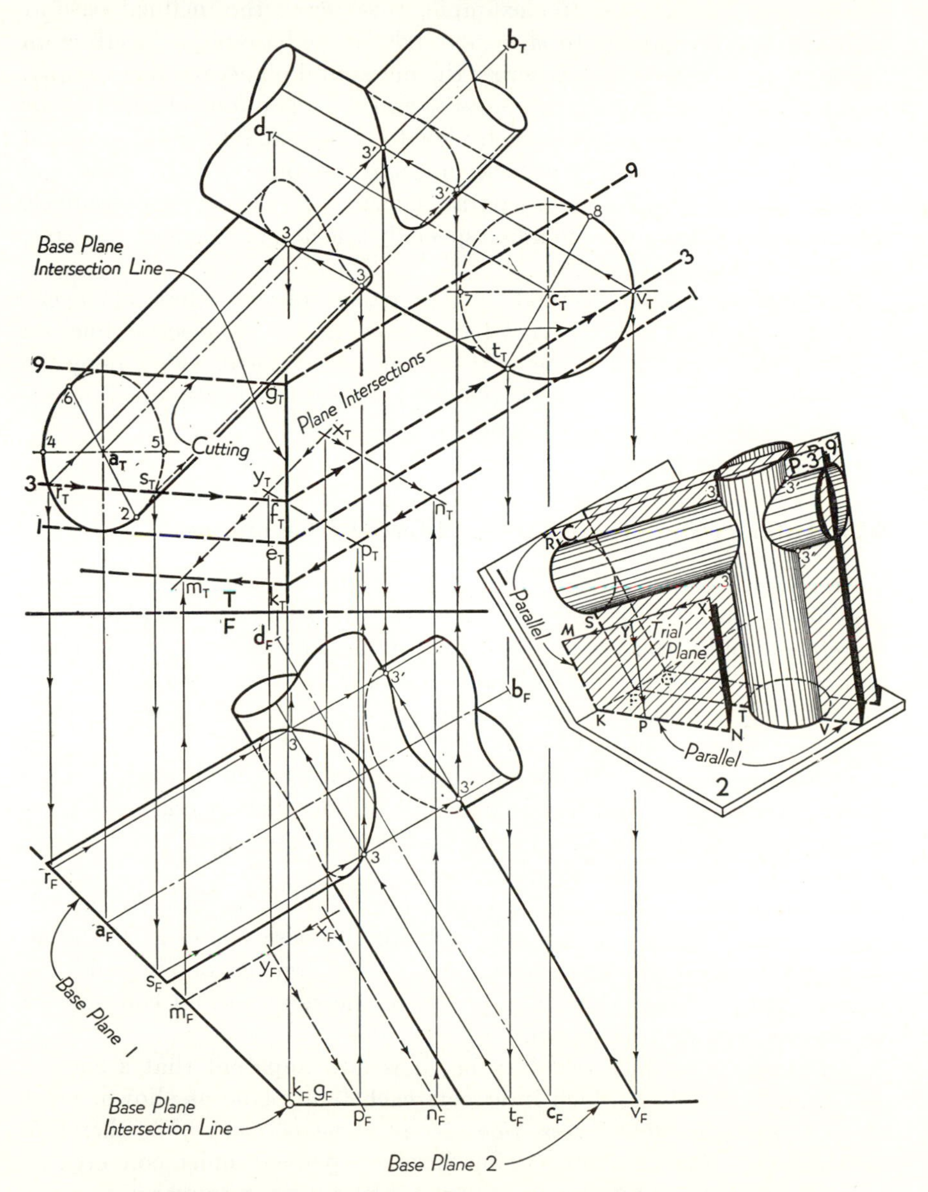

Fig. 9·15. Intersection of two oblique cylinders: different base planes.

In the top view the trial-plane intersection line runs from n_T to p_T to k_T along the horizontal base plane and then from k_T to m_T along the inclined base plane. Each cutting-plane intersection line must now be selected in the top view parallel to the trial-plane intersection line. The intersection line of *C-P*-3, for example, cuts across the inclined base at points r_T and s_T, parallel to m_Tk_T, extends to the base-plane intersection line at f_T, and then continues across the horizontal plane, parallel to k_Tn_T, to intersect the other base at points t_T and v_T. The four elements thus determined are then drawn in each view to locate the four points marked 3 and 3′. *C-P*-1 and *C-P*-9 are the first and last planes in the series and indicate that there will be two separate curves. As in previous examples, additional cutting planes should be specially selected to intersect the bases at points 2, 4, 5, 6, 7, and 8.

Summary. In the method of this article, observe that: (1) every cutting plane must intersect *each* base plane and (2) intersection lines on one base plane must join those on the other base plane at points along the intersection line of the two base planes; hence (3) *the base-plane intersection line is the key to the whole solution.*

9·17. Intersection of two cones: different base planes

Analysis. Cutting planes that cut straight-line elements from each cone must pass *through both cone vertices.* All such cutting planes will therefore contain the vertex line. But if the cone bases lie in different planes, then each cutting plane must intersect both base planes. There will thus be *two different* (but connected) *sets of intersection lines.* Otherwise, the solution will be identical with that of Art. 9·15.

Problem. In Fig. 9·16 the elliptical base of one cone lies in a vertical plane (base plane 1), and the circular base of the other cone lies in a horizontal plane (base plane 2). The right-side view is necessary, in this case, to show the shape of the elliptical base.

Construction. The solution begins (as it did in Fig. 9·14) by extending the vertex line to its intersection with the base *planes* at P and Q, since in this case there will be *two* points of convergence, one in each base plane. Points P and Q are key points in the solution and can now be located in each view as shown.

Referring to the pictorial drawing, it is now apparent that a cutting plane, such as *C-P*-2, which intersects each base plane as shown, must also cut the *base-plane intersection line* at some point F. We therefore conclude: (1) that all intersection lines on plane 1 must converge at point P, (2) that all intersection lines on plane 2 must converge at point Q, and finally (3) that each pair of intersection lines must be joined at a common point on the base-plane intersection line. With these three conditions in mind we can now proceed to establish the necessary cutting planes in the multiview drawing.

Consider the cutting plane *C-P*-2. Its intersection with one of the base planes may be assumed at random, but its intersection with the other plane will then be fixed. For example, we may draw at random the intersection line with plane 1 from p_R to f_R, thus intersecting the elliptical base at points r_R and s_R. But point f_R appears on the edge view of base plane 2; hence point *F* is a point on the line of intersection of the two base planes.

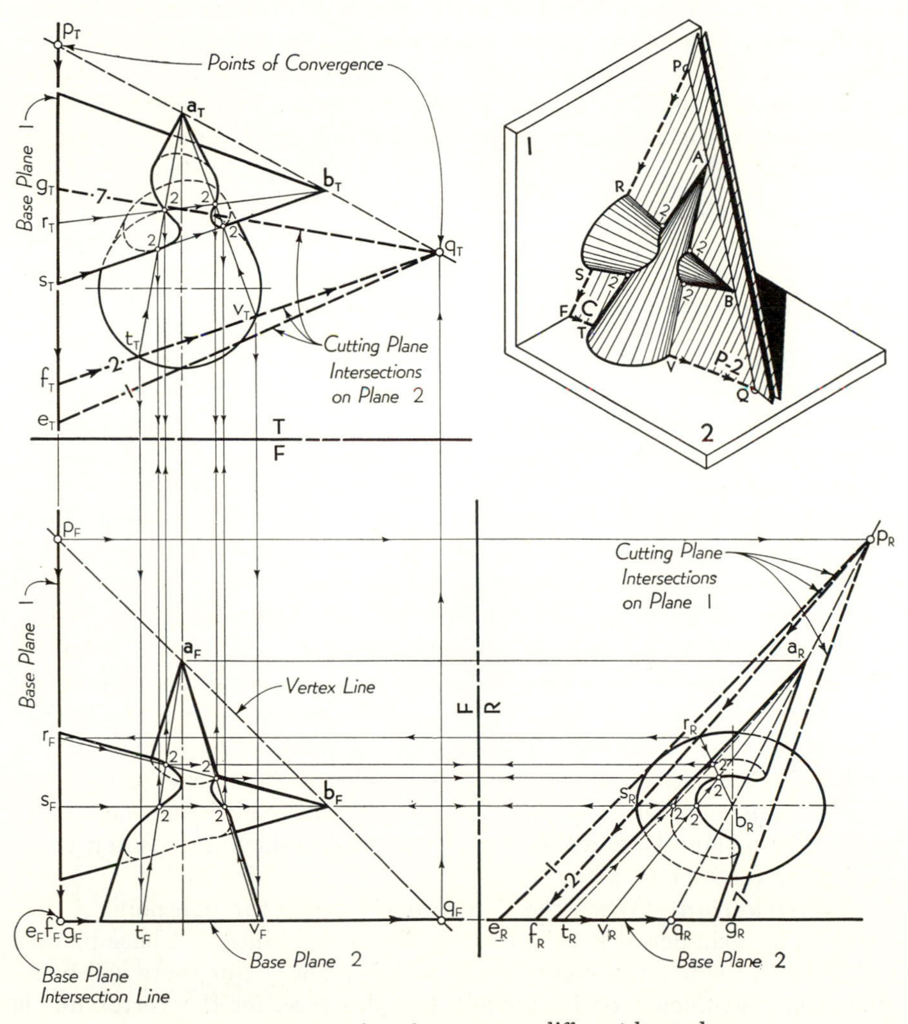

Fig. 9·16. Intersection of two cones: different base planes.

Therefore from point f_R the cutting-plane intersection line must now continue across the base plane 2 to point q_R. In the top view, f_T is located by reference-line measurement, and the cutting-plane intersection with plane 2 is drawn from f_T to q_T, intersecting the circular base at t_T and v_T. The cutting plane is now established, and two elements have been cut from each cone; r_Rb_R and s_Rb_R appear in the side view, and t_Ta_T and v_Ta_T appear in the top view. These four elements can now be located in each of the other views as shown, and their points of intersection establish the four points labeled 2. Note that these four points should check by alignment and measurement in all views.

C-P-1 and *C-P*-7 are the first and last planes that cut elements from both cones. These two planes cut the base-plane line of intersection at points E and G, respectively; point E was first established at e_R, whereas point G was found first at g_T. Since each plane is tangent to one cone but cuts the other, the curve of intersection will be one continuous curve as shown.

PROBLEMS. Group 101.

9·18. Intersection of an oblique cylinder and cone

Analysis. To cut straight-line elements from both cylinder and cone the cutting planes must be *parallel to the cylinder* and pass *through the cone vertex.* Then a line drawn through the cone vertex parallel to the cylinder becomes the *vertex line,* which lies in all cutting planes. As in the case of two cones (Arts. 9·15 and 9·17) this vertex line must be extended to intersect the base plane (or planes). The point (or points) so established will be points of convergence for the cutting-plane intersection lines, and the solution then proceeds as in previous examples.

To emphasize the analogy between this case and that for two cones, note in Fig. 9·14 that, if cone B were lengthened until its vertex was an infinite distance from its base, this cone would then become cylindrical and the line connecting vertex A to the very distant vertex B would then be parallel to the cylinder. This would alter the position of point P, but otherwise the solution would be identical with that for two cones.

Problem. In Fig. 9·17 the base of the oblique cone is in a horizontal plane (base plane 1), and the base of the horizontal cylinder is in a vertical plane (base plane 2).

Construction. When the vertex line is drawn through point A parallel to the cylinder, it is evident that this line will intersect base plane 2 at point P, the point of convergence on this plane. But there will be no point of convergence on base plane 1 in this case, for the vertex line is parallel to the plane. This means that cutting-plane intersection lines on base plane 2 will converge at point P but the intersection lines on base plane 1 will all be parallel to the vertex line.

We must now note the location of the line of intersection of the two base planes. In this case it coincides with the base-plane edge in each

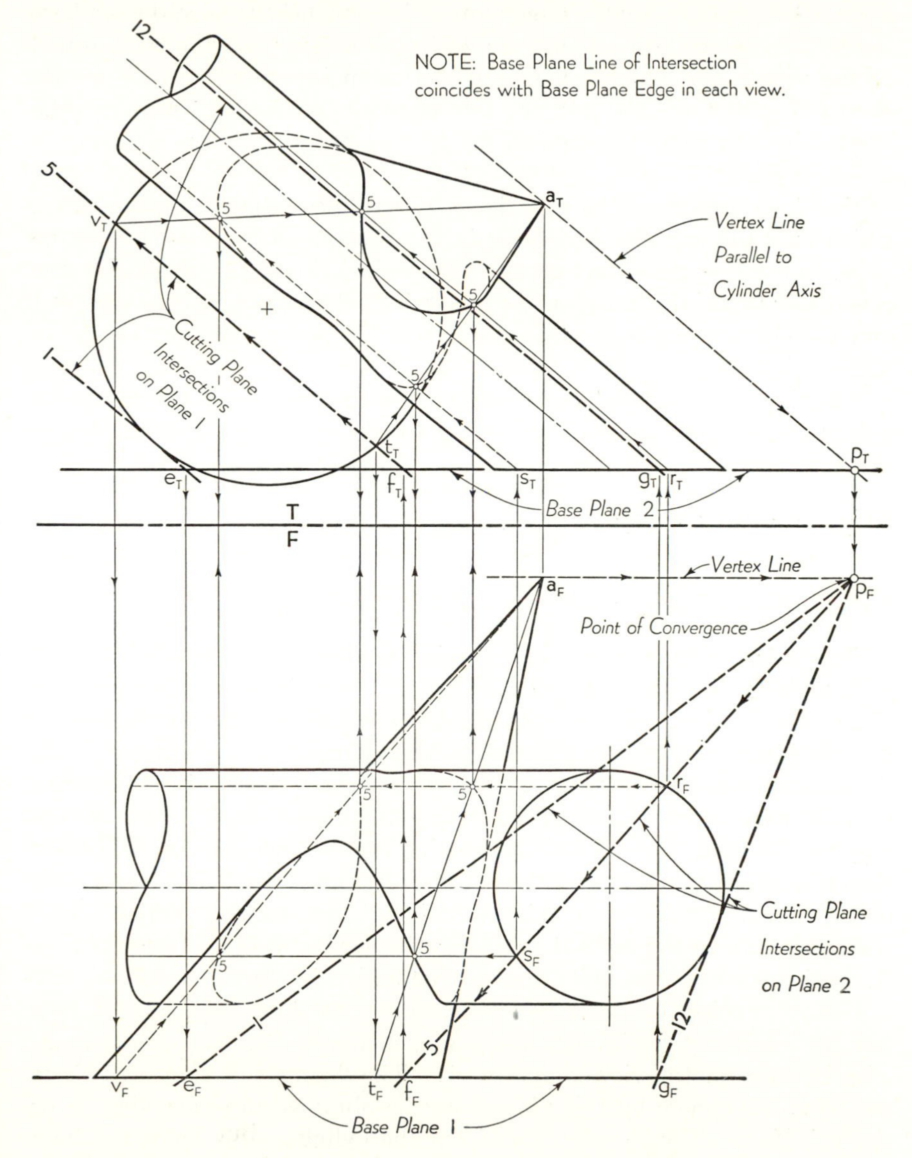

Fig. 9·17. Intersection of an oblique cylinder and cone: different base planes.

view. The intersection lines of any desired cutting plane may now be established. *C-P*-5, for example, has been assumed to cut base plane 2 along the line $p_F f_F$, thus intersecting the cylinder base circle at points r_F and s_F. But point F is on the base-plane intersection line and hence is the junction point of intersection lines in each plane. The top view of point F must appear on the edge view of base plane 2 at f_T, and from here the cutting-plane intersection line continues across base plane 1 parallel to the vertex line, cutting the cone base at points t_T and v_T. The two elements on the cylinder and the two elements on the cone may now be located in both views to establish the four points (labeled 5) on the curve of intersection.

Line $1e_T$, tangent to the cone base, determines intersection line $e_F p_F$, which cuts the cylinder base; this is the first plane in the series. Line $p_F g_F$, tangent to the cylinder base, determines line $g_T 12$, which cuts the cone base; this is the last plane in the series. From the position of *C-P*-1 and *C-P*-12 we conclude that there will be one continuous curve of intersection.

Selection of additional cutting planes and location of the resulting points are now accomplished according to the general routine outlined in previous examples.

PROBLEMS. Group 102.

9·19. Intersection of a sphere and a cylinder

Analysis. Any cutting plane passing through a sphere will intersect it in a circle, but it is highly desirable that the circle shall *appear as a circle* in some view. To cut straight-line elements from the cylinder the cutting plane need only be parallel to the cylinder axis. Hence, for the simplest solution we should have (or construct) a view which shows the cylinder axis in true length in order that the circles cut from the sphere may be true size. Another view, showing the cylinder endwise, will show the cutting planes edgewise. As noted in Art. 9·5 (Fig. 9·3), an end view of a cylinder is usually convenient but is not necessarily essential.

Problem. In Fig. 9·18 the sphere with center at O is intersected by the cylinder whose axis AB appears true length in the front view.

Construction. View A has been drawn to show the axis AB as a point. Then any vertical cutting plane, such as *C-P*-2, will appear as an edge in both the top view and view A. The diameter of the small circle cut from the sphere by this plane may be obtained in either view A or the top view. The elements cut from the cylinder may be determined either at r_A and s_A or at r_T and s_T. In the front view it is then apparent that element s_F on the cylinder intersects circle 2 to establish two points marked 2, but element r_F does not cut the circle and hence does not intersect the sphere. Additional parallel cutting planes may be selected at random, but note that *C-P*-4 and *C-P*-6 are necessary to establish extreme element points in the front view. *C-P*-7, tangent to the rear side of the cylinder, locates the two points labeled 7.

In Fig. 9·18 the special cutting plane *C-P*-1 has been taken perpendicular to the cylinder axis to establish the points labeled 1. This plane should suggest to the reader a second method of solution [compare Fig. 9·8(*a*)].

Summary. The methods described here are applicable to cylinders (or prisms) of any cross-sectional shape intersecting a sphere. If the axis of the given cylinder is generally oblique, then new views should be constructed to show the axis as a true length and as a point. If the given cylinder is a cylinder of revolution and its axis passes through the center

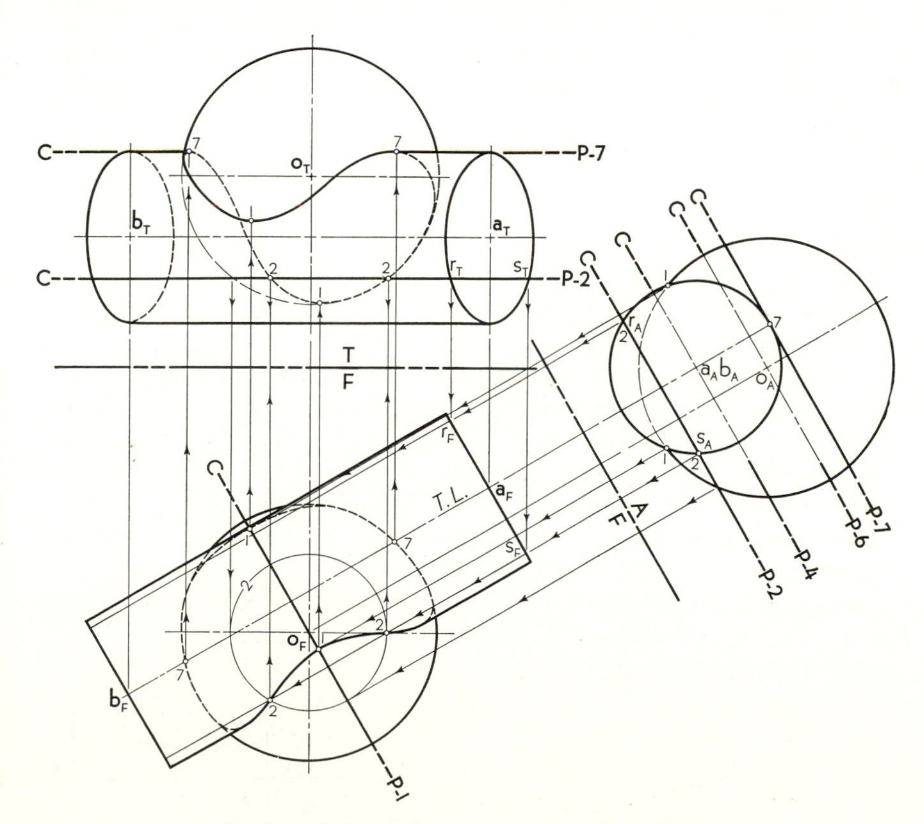

Fig. 9·18. Intersection of a sphere and a cylinder.

of the sphere, the curve of intersection will be two parallel circles (see sphere and cylinder in Fig. 9·21).

PROBLEMS. Group 103.

9·20. Intersection of a sphere and a cone: revolved cutting planes

Analysis. Cutting planes taken through the vertex of the cone will cut straight-line elements from the cone surface and circles from the sphere. But because the cutting planes are not parallel, it is impossible

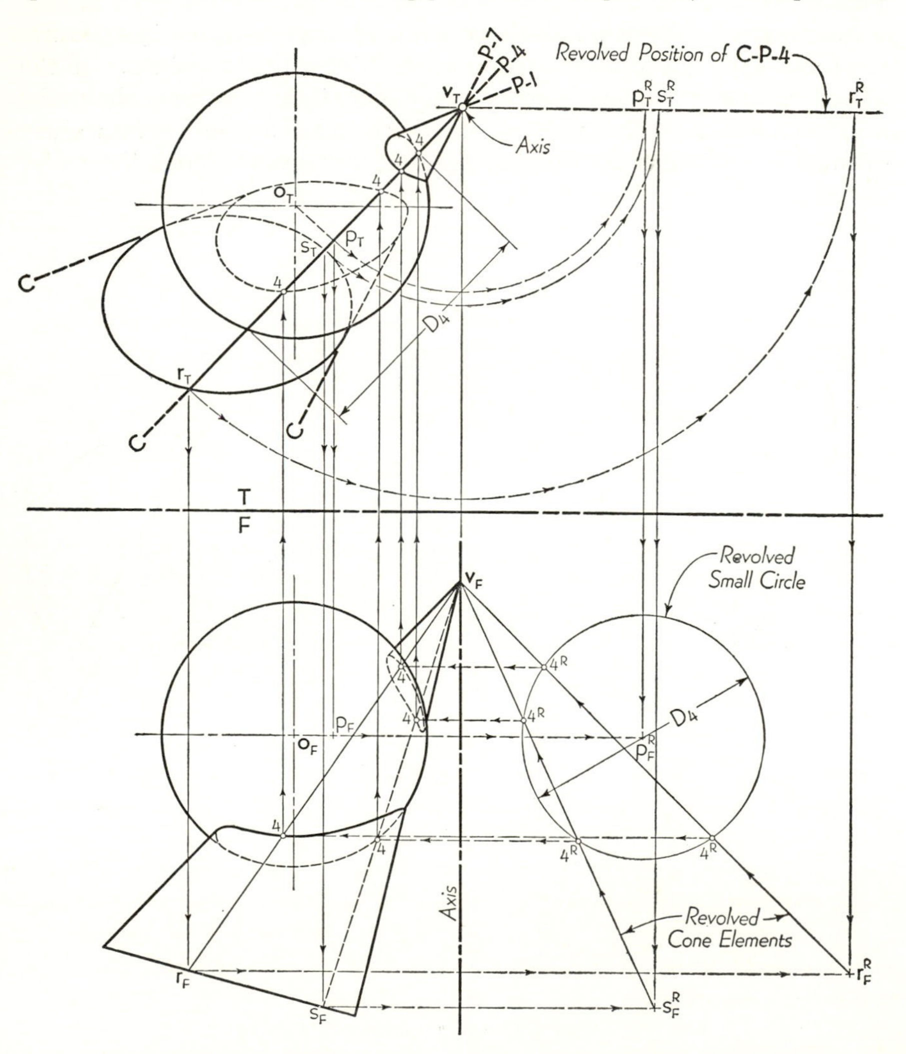

Fig. 9·19. Intersection of a sphere and a cone: revolved cutting planes.

to draw an auxiliary view showing all the circles true size in the same view. To draw a separate auxiliary view for each cutting plane would be tedious; hence we shall use the *method of revolution* to show each cutting plane in *true size* (Art. 5·9).

Problem. In Fig. 9·19 the sphere with center at O is intersected by an oblique cone with vertex at V.

Construction. A series of vertical cutting planes through vertex V will range from *C-P*-1 to *C-P*-7, as shown in Fig. 9·19. (Edge-front planes could also be used.) Then any plane such as *C-P*-4 will cut a small circle of diameter D_4 from the sphere. The center of this small circle is at p_T, located most readily by drawing o_Tp_T perpendicular to the cutting plane. *C-P*-4 also intersects the cone base at points R and S to establish the cone elements RV and SV. Thus the intersection of the cutting plane with each surface is now determined; it only remains to revolve the small circle and the cone elements until they appear true size.

A vertical line through vertex V has been selected as the axis of revolution. Then *C-P*-4 can be revolved about this axis until it appears parallel to reference line *T–F* in the top view. It will then appear true size in the front view. Points p_T, r_T, and s_T revolve to p_T^R, r_T^R, and s_T^R. In the front view, points p_F, r_F, and s_F move perpendicular to the axis (Rule 16) to the revolved positions shown. The revolved small circle, of diameter D_4, can now be drawn in true size with p_F^R as center. At the four points marked 4^R the revolved small circle intersects the revolved cone elements, and we thus have four points on the curve of intersection in this revolved position.

When the cone elements are revolved back to their original position, points 4^R counterrevolve to the positions labeled 4 on the respective elements in the front view. The same points can now be located in the top view by alignment as shown. The procedure described in this paragraph must now be repeated for each cutting plane until the desired number of points have been established on the intersection curve. The first and last planes, *C-P*-1 and *C-P*-7, should be revolved first in order to determine whether there will be one or two curves of intersection.

Summary. The method of this article is applicable to any cone (or pyramid) intersecting a sphere. If the cone base is circular, however, the solution may also be obtained by taking cutting planes parallel to the cone base; these planes will then cut circles from both cone and sphere, and the solution will be similar to that of Fig. 9·7. The idea of revolving a cutting plane is a general concept that can frequently be applied to specially selected cutting planes in order to obtain critical points on the curve of intersection.

PROBLEMS. Group 104.

9·21. Intersection of a surface of revolution and a cylinder: cutting cylinders

The intersection problems discussed thus far have all been solved by choosing a plane as the auxiliary cutting surface. But, as indicated in Art. 9·2, any type of surface can be employed to intersect the given surfaces; it is only necessary that the cutting surface shall intersect each of the given surfaces in a line or curve that can readily be established. In this article we shall show how a series of oblique cylinders may be employed as auxiliary cutting surfaces.

Problem. In Fig. 9·20 a paraboloid whose axis is vertical is intersected by an oblique cylinder whose base is elliptical.

Analysis. Since the paraboloid is a surface of revolution, every right section of this surface will be a circle. We may therefore choose any circular section of the paraboloid, as shown in the pictorial illustration at *E*, and construct a cutting cylinder having this circle as its base. But if the axis *EG* of the cutting cylinder is taken parallel to the axis *AB* of the given cylinder, then these two cylinders will intersect each other along

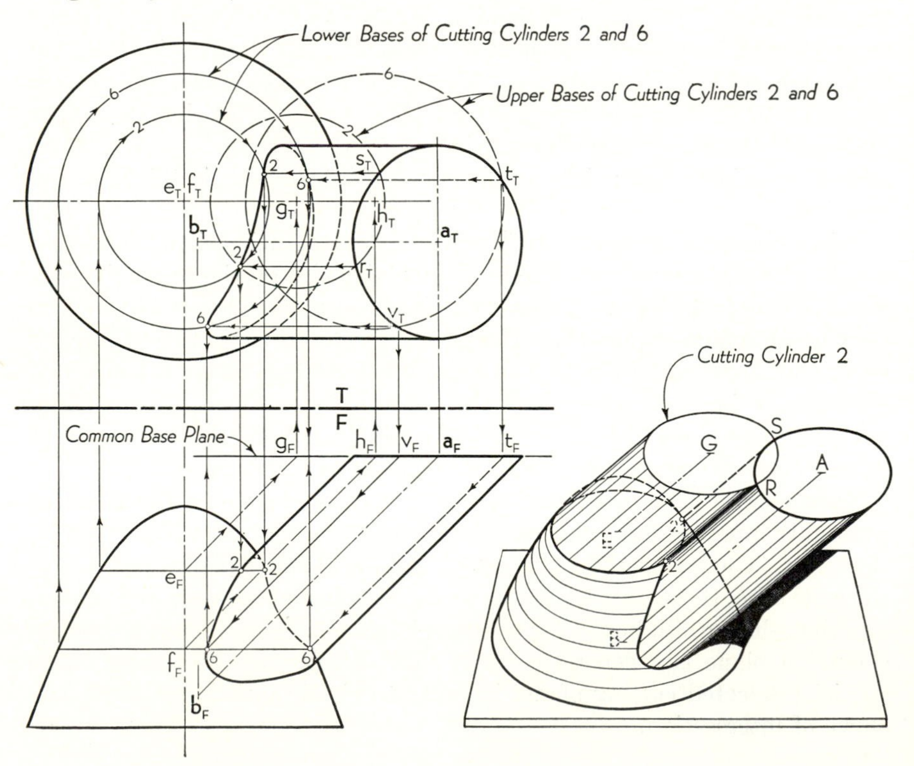

Fig. 9·20. Intersection of a surface of revolution and a cylinder: cutting cylinders.

the two straight-line elements labeled R and S. The intersection of these two elements with the base circle of the cutting cylinder will be two points on the required curve of intersection.

Construction. In the multiview drawing a right section of the paraboloid has been assumed at e_F. In the top view this circular section appears as the full circle labeled 2 and represents the lower base circle of the assumed cutting cylinder. The axis EG of the cutting cylinder must be parallel to the axis AB of the given cylinder and is so drawn in each view. But the upper base of the cutting cylinder must lie in the horizontal plane of the given cylinder base; hence point G must be the center of the upper base of the cutting cylinder. With g_T as center the dash circle labeled 2 is drawn of the same diameter as the lower base circle. Then the upper base circle of the cutting cylinder intersects the elliptical base of the given cylinder at points r_T and s_T, establishing two elements. When these elements are extended to the lower base circle in the top view, they locate two points (marked 2) on the curve of intersection. Parallels can now be extended to the front view to locate the same points there.

Additional cutting cylinders are assumed by choosing other right sections of the paraboloid as the cylinder bases. Cutting cylinder 6, for example, has been constructed on a lower, hence larger, base, and the two equal base circles (labeled 6) are shown in the top view. The axis of this larger cylinder is the line FH. Note that the intersection elements for any cutting cylinder can also be located in the front view in order to check the accuracy of the points in this view. Points 6 in the front view have been verified in this way.

Summary. To employ the cutting-cylinder method three conditions are essential: (1) The given surfaces must be a surface of revolution and a cylinder (or prism). (2) The axis of the surface of revolution must appear as a point in some view. (3) The base plane of the cylinder must be perpendicular to the axis of the surface of revolution. Condition 2 can always be satisfied, when necessary, by drawing additional views; condition 3 can be satisfied by extending or cutting off the given cylinder to form a new base.

9·22. Intersection of two surfaces of revolution: axes parallel

Analysis. The parallel axes of the two surfaces should appear as points in one of the views, given or constructed. Then cutting planes perpendicular to the axes will cut circles from both surfaces, and the intersection of these circles will establish points on the line of intersection. Thus the procedure is identical with that shown for two cones of revolution in Fig. 9·7.

9·23. Intersection of two surfaces of revolution: axes intersecting: cutting spheres

If the axes of two surfaces of revolution intersect, then the line of intersection may be obtained by employing a series of concentric spheres as the auxiliary cutting surfaces. This method is based upon the fact that a sphere whose center is on the axis of a surface of revolution will cut that surface in one or more circles. One of the pictorial drawings of Fig. 9·21 shows a cylinder of revolution intersected by a sphere whose center is on the cylinder axis. Because the cylinder axis is shown in true length, the

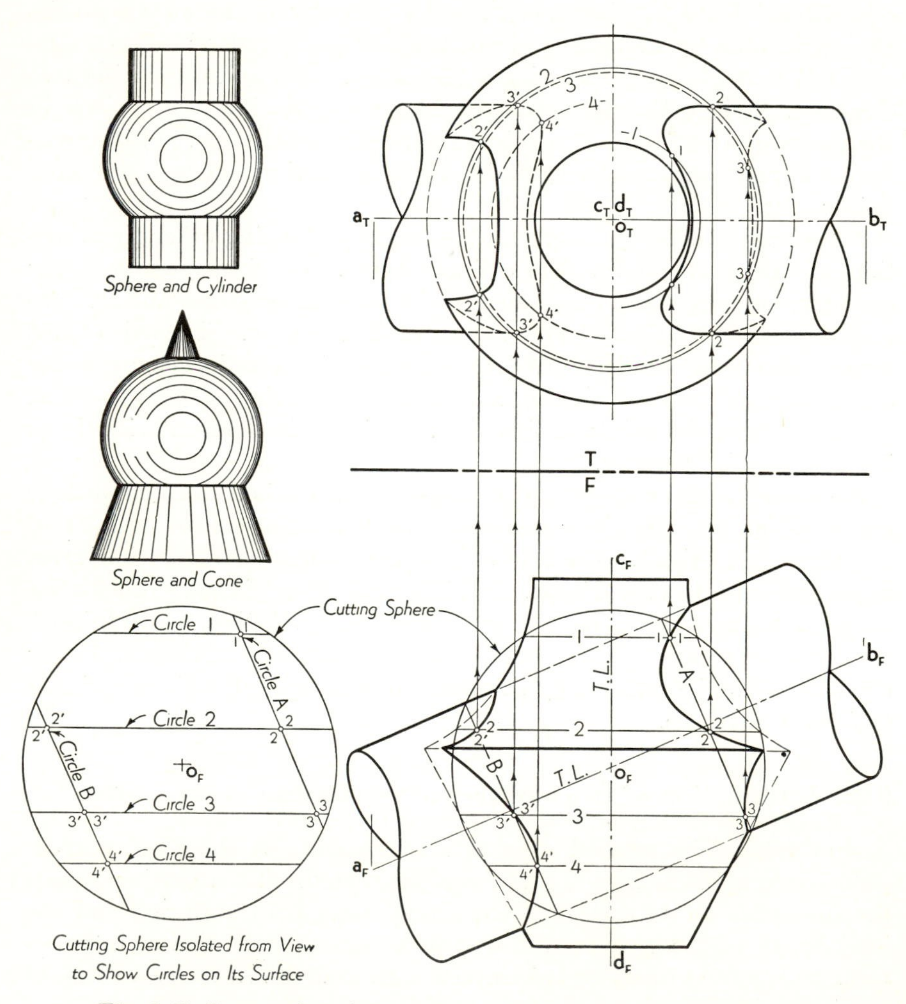

Fig. 9·21. Intersection of two surfaces of revolution: cutting spheres.

two intersection circles appear edgewise. The intersection of a sphere and a cone of revolution is also illustrated, and it should be evident that circular intersections would similarly be obtained for any surface of revolution.

Analysis. In order to cut circles from each of two surfaces of revolution the axes of these surfaces *must intersect*, and the center of the cutting sphere must be at the axis intersection point. Thus the method of cutting spheres is applicable only to *surfaces of revolution whose axes intersect.* Second, the solution is practical only when the axes of the given surfaces both appear true length in the same view, for only then will the circles of intersection appear edgewise and be easy to draw.

Problem. The top and front views in Fig. 9·21 show a cylinder of revolution (axis AB) intersecting a surface of revolution whose axis CD is vertical. The axes intersect at point O, and both appear true length in the front view; hence cutting spheres may be used.

Construction. With point o_F as center, a sphere of any diameter is drawn. (The sphere need not be drawn in the top view.) This sphere intersects the cylinder in two equal circles A and B, which appear edgewise in the front view as shown. The sphere also intersects the vertical surface of revolution in four horizontal circles labeled 1, 2, 3, and 4. To clarify the illustration the cutting sphere and the six circles that lie on its surface have been redrawn to the left of the front view. It is now evident that circle A on the cylinder intersects circles 1, 2, and 3 on the vertical surface of revolution, thus establishing 6 points (2 for each pair of circles) on the curve of intersection. Circle B intersects circles 2, 3, and 4 to provide 6 more points. These 12 points can be located in the top view on circles 1, 2, 3, and 4 of the vertical surface of revolution. Circles A and B on the cylinder would appear as ellipses in the top view, but it is unnecessary to draw them. Additional concentric spheres, of larger and smaller diameter, should be drawn to obtain more points on the curve of intersection.

PROBLEMS. Group 105.

9·24. Intersection of a contoured surface and a cylinder

In its general form a double-curved surface is represented by a series of contour lines (Art. 8·17). Simple lines, straight or circular, cannot be cut from such a surface, and it is therefore necessary to plot, point by point, the curved line of intersection of each cutting plane.

Problem. In Fig. 9·22 a cylinder of revolution (the hawsepipe, through which the anchor chain passes) intersects the deck and hull of a ship. The intersection of the hawsepipe with the plane surface of the

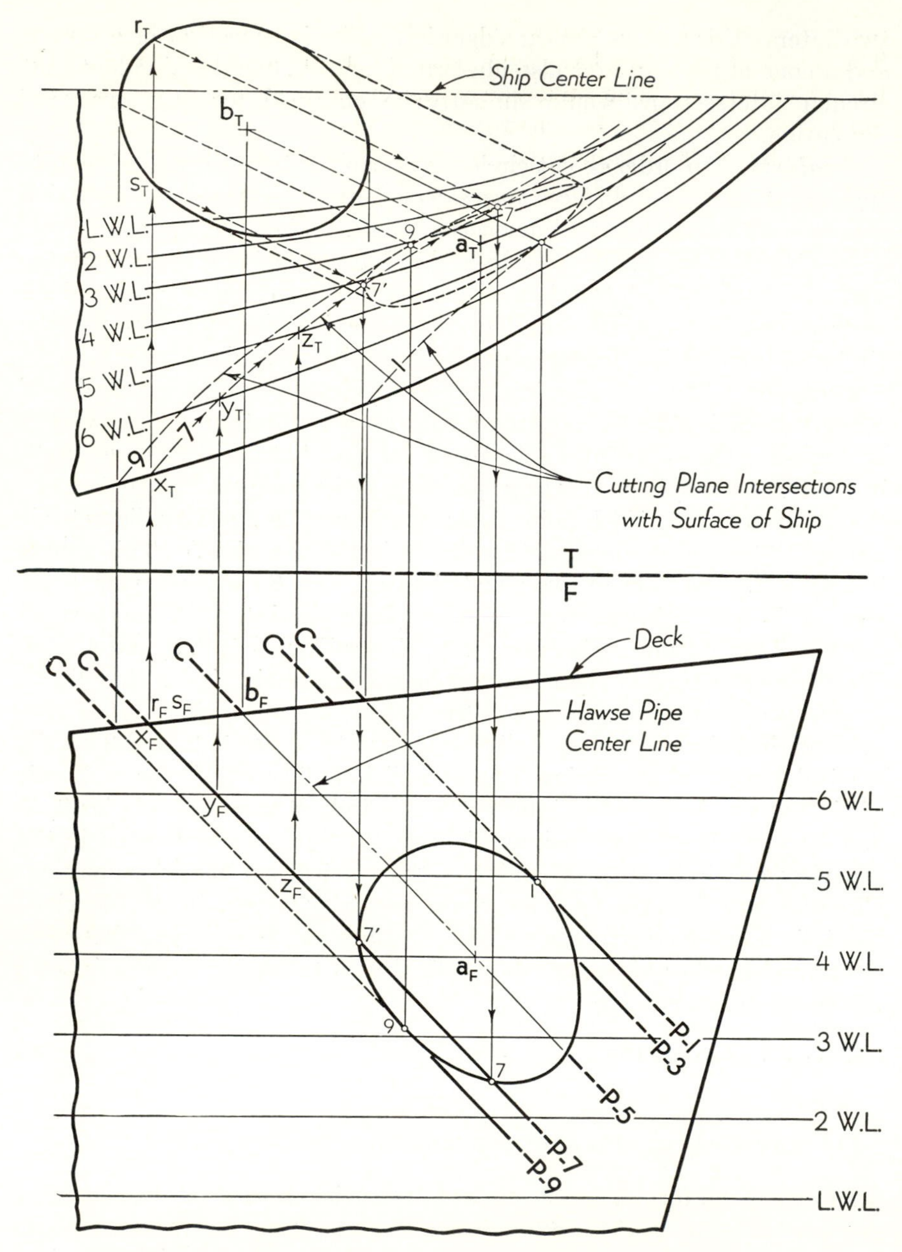

Fig. 9·22. Intersection of a contoured surface and a cylinder.

deck is an ellipse that can be constructed by the method of Art. 6·14. The intersection of the hawsepipe with the double-curved surface of the hull is required. (The diameter of the cylinder in Fig. 9·22 has been exaggerated to clarify the illustration.)

Analysis. Cutting planes parallel to the cylinder axis will cut straight-line elements from the cylinder and will cut the surface of the ship in a series of curved lines.

Construction. *C-P-*7 is a typical cutting plane that intersects the deck ellipse at points R and S and thus establishes two elements on the cylinder as shown in Fig. 9·22. The intersection of *C-P-*7 with the surface of the ship is determined by noting where this plane cuts each of the given water lines. At x_F the cutting plane intersects the edge of the deck, at y_F it intersects 6 W.L., at z_F it intersects 5 W.L., etc. Each of these points may be located in the top view on the corresponding water line and a smooth curve (dash line 7) drawn through the points. If this curve is not "fair" (Art. 8·17), this indicates that the water lines are inaccurate and that the surface of the ship has been poorly faired. Dash line 7 intersects elements r_T and s_T at points 7 and 7′, respectively, and thus locates two points on the required curve of intersection in the top view. Points 7 and 7′ are located in the front view by alignment. Additional points are located in the same manner by choosing other parallel cutting planes as shown in Fig. 9·22.

Tracing method. The intersection may also be determined by using horizontal cutting planes at each water-line level. Intersections on the surface of the ship thus correspond with the given water lines, but each such plane will then cut an ellipse from the cylinder. Since every ellipse will be identical in size and shape, however, it is necessary to construct only one such ellipse. A tracing of this ellipse can then be made, and by carefully shifting the position of the tracing the intersection of the ellipse with each water-line curve may be pricked through the tracing with a needle point. This method can be employed in any intersection problem that requires *duplicate but displaced curves* cut from a cylinder.

Warped surfaces. The methods discussed in this article may also be employed to determine intersections with warped surfaces, since each straight-line element of the warped surface can be treated as a contour line.

PROBLEMS. Group 106.

10. DEVELOPMENT OF SURFACES

10·1. Introduction

The development of a surface is the plane figure obtained by unfolding the surface onto a plane. Thus, in Fig. 10·1(*a*), we may imagine that the rectangular prism is enclosed in a wrapper of thin material, paper or foil. If this covering is opened along the edges as shown and each plane surface folded down to lie in the horizontal plane, then the flattened wrapper is a *development* of the prism. Viewed in true size the development appears as shown in Fig. 10·1(*b*). Here each rectangular area is the same *size* and *shape* as the corresponding surface on the prism. There is no stretching or distorting of the surface in the process of development; hence we can state the following rule:

▪ ***Rule 18. Rule of development.*** *Every line on a development shows the true length of the corresponding line on the surface.*

This is a cardinal principle that must be observed in constructing all true developments.

Only polyhedrons and single-curved surfaces can be developed. Polyhedrons are developable because they are bounded entirely by plane surfaces that may be laid down in the development in true size and in connected sequence. Single-curved surfaces (cone, cylinder, and convolute) are developable because each pair of consecutive straight-line elements lie in the same plane (Art. 6·3). Warped surfaces *cannot* be developed because no two consecutive elements can form a plane (Art. 7·1). Double-curved surfaces *cannot* be developed because these surfaces contain no straight lines (Art. 8·1). Warped and double-curved surfaces can, however, be approximately developed with sufficient accuracy for many practical purposes.

10·2. Sheet-metal developments

Many manufactured articles are made from sheet metal simply by cutting and bending the material to the desired shape. In each case a development of the surface of the object must be made first, either on paper to full or reduced scale, or directly on the flat surface of the metal. If a large number of parts are required, the development may first be reproduced as a metal pattern, or template, whose outline can be transferred to the flat stock. The *bend lines* [see Fig. 10·1(*b*)] are located by means of a prick punch inserted in the small holes drilled in the template. After cutting to the outline, the metal is bent in a machine called a *brake*, or curved between rolls, or pressed into shape between dies.

Edges that must be joined are then connected by soldering, welding, riveting, or seaming. Extra metal must be allowed for joining, the amount of lap depending upon the type of seam (see Art. 10·25). For economy and simplicity the seam should be taken on the shortest line, preferably on a line of symmetry. A development usually shows the inside surface in order that scribed lines and prick-punch marks on the metal may be concealed. For metals lighter than No. 24 gauge (0.0250 in. thick) the thickness of the metal can usually be neglected. For heavier sheet metals or steel plate a *bend allowance* must be made for curving or bending the thicker material (see Art. 10·26).

All the developments shown in succeeding articles are theoretical developments that do not include any of the above allowances. For practical applications, slight additions and modifications must be made for joining and bending.

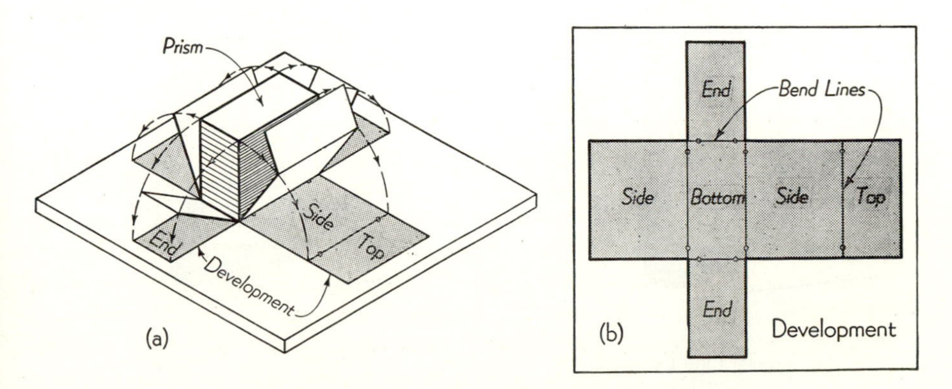

Fig. 10·1. Development of a rectangular prism.

10·3. Classification of developments

Developments may be divided into four classes, according to the type of surface or method employed, as follows:

1. Parallel-line developments are those obtained for prisms and cylinders.
2. Radial-line developments are those obtained for pyramids and cones.
3. Triangulation developments are those obtained by dividing the given surface into a series of triangular areas.
4. Approximate developments are those obtained for warped and double-curved surfaces.

Cylinders. Figure 10·2(a) shows pictorially the parallel-line development of a truncated right circular cylinder. The cylindrical surface, placed with its longest element against the plane, has been cut open along the short front element, and the surface has then been rolled back against the plane on either side to form the development. The right-section base circle becomes the straight line AB in the development, and the elements of the cylinder appear as a series of parallel lines perpendicular to the line AB. The upper elliptical end of the cylinder becomes the curved line in the development. These are the characteristics of *all* cylinder developments: the parallel elements remain parallel in the development, and right sections appear as straight lines perpendicular to the elements. The parallel lateral edges of a prism similarly appear parallel in the development.

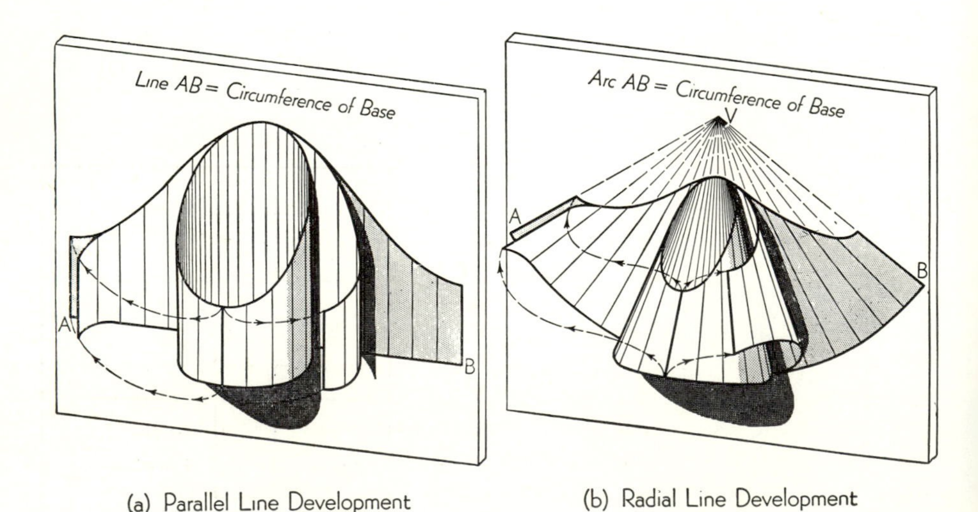

Fig. 10·2. Development of a cylinder and a cone.

Cones. Figure 10·2(*b*) shows pictorially the radial-line development of a truncated right circular cone. When the cone is placed with its longest element against the plane, the cone vertex *V* will also lie in the plane. By cutting the cone along its short front element its surface can be rolled back against the plane as shown. Thus, in the development, the cone elements become a series of radial lines having the vertex as their origin, and the right circular base of the cone becomes the arc *AB* with center at *V*. The elliptical section of the cone becomes an irregular curve in the development. The radial lines are characteristic of *all* cone developments, but the curve *AB* will be a circular arc only when the cone is right circular (all elements equal in length). The lateral edges of a pyramid similarly become radial lines in the development.

Triangulation developments. Any ruled surface may be developed by subdividing the surface between straight-line elements into two or more triangular areas. By determining the true length of each side of each triangle, all the triangles may be laid down in connected sequence to form the development. For polyhedrons the method of triangulation is exact; for single-curved surfaces the accuracy of the development increases as the triangles are taken smaller and more numerous; for warped surfaces the development can be only approximate since these surfaces are theoretically undevelopable.

Approximate developments. Double-curved surfaces can be approximated by dividing the surface into a number of narrow cylindrical or conical segments. Exact reproduction in sheet metal can be achieved only by "raising" or "bumping" (stretching) the metal with a hammer or by stamping it between form dies in a press (see Art. 10·23).

10·4. Development of a right prism

Problem. In Fig. 10·3 develop the lateral surface of the right prism.

Analysis. The lateral edges of the prism will appear as true-length parallel lines on the development, and these edges appear true length in the given front view. The length of the development will equal the perimeter, or true distance around a right section of the prism, and this distance is shown in the top view of the prism. Thus the given views provide directly all information needed for the development.

Construction. A right-section line is drawn directly opposite the base of the prism. This is called the *stretchout line*, and its length from 1 to 1 must exactly equal the perimeter of the prism base. Along this line the distances 1–2, 2–3, etc., taken from the top view, are laid off in sequence to establish the width of each face of the prism. The lateral edges (bend lines) may now be drawn perpendicular to the stretchout line at each division point, and the true length of each line can be obtained by extension from the front view as shown. Connecting points A, B, C, etc., with straight lines completes the development of the lateral surface of the prism.

Because the faces of the prism were taken in clockwise sequence from the top view, the development shows the inside surface of the prism. On *unsymmetrical* developments (hence unnecessary in Fig. 10·3) this fact should always be indicated by labeling the development "inside" or "bend up," "outside" or "bend down." Bend lines are distinguished by a small freehand circle (they represent the drilled holes in the template) at each end of the line.

If it is necessary to include the end surfaces of the prism in the development, they may be attached at any common edge. Surface $ABCDE$, for example, appears true size in view A but must be *turned over* before being transferred and attached at edge DE.

PROBLEMS. Group 107.

10·5. Development of an oblique prism

Problem. Figure 10·4 shows a diagonal offset connection used to connect two rectangular ventilating ducts that lie in different horizontal planes. A development of this oblique prism is required.

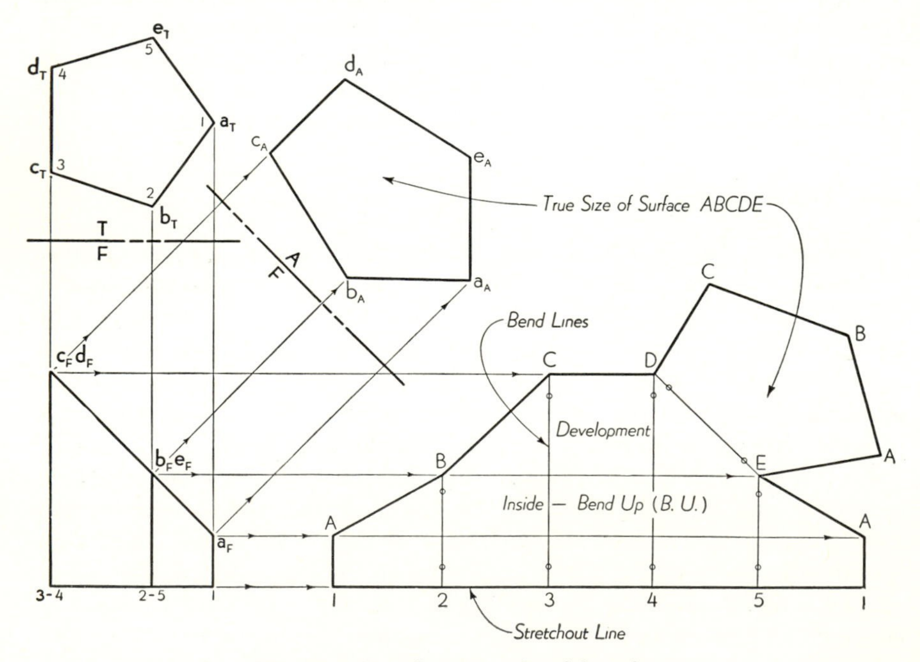

Fig. 10·3. Development of a right prism.

Analysis. Construction of the parallel-line development requires the true length of all lateral edges and the perpendicular spacing of these parallel edges along the stretchout, or right-section, line. Thus for the development of an oblique prism we should always have, or obtain:

1. *A view showing the true length of the lateral edges.*
2. *An end view (right section) showing the lateral edges as points.*

Construction. View A is drawn to show the prism edges in true length, and view B shows these edges as points. The development may now be constructed with measurements taken from views A and B, without further reference to the given views.

Since neither base is perpendicular to the edges, a right section has been arbitrarily created by cutting the prism with plane C-P. This cutting plane may be assumed anywhere in view A but must be perpendicular to

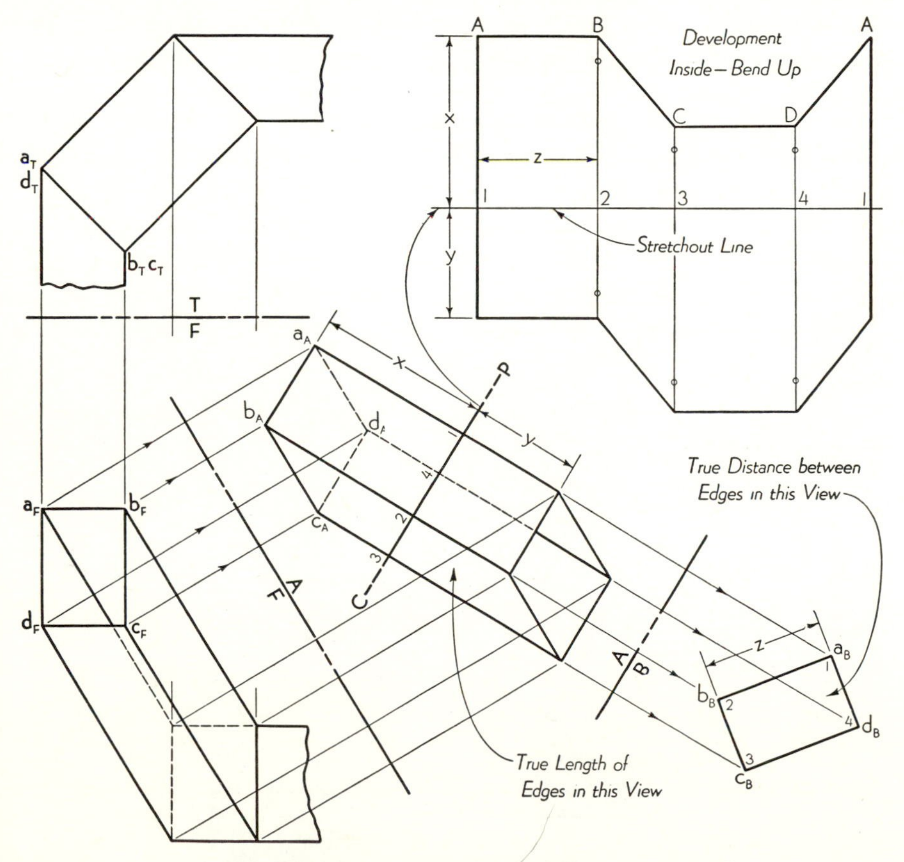

Fig. 10·4. Development of an oblique prism.

the lateral edges. The development may now be constructed in any convenient place on the paper. In Fig. 10·4 the stretchout line has been taken horizontal, and the spacing of the edges along this line is obtained from view B (note distance z). Then the length of each edge above and below the arbitrary cutting plane must be laid off correspondingly above and below the assumed stretchout line (note distances x and y). Connecting points A, B, C, etc., completes the required development. The development has been correctly labeled "inside—bend up," but careful inspection will reveal that the development is actually symmetrical and hence could be bent either way to form the offset connection.

10·6. Development of an intersected oblique prism

Problem. In Fig. 10·5 the oblique prism has been intersected by a rectangular prism. The line of intersection is shown in both views, but the rectangular prism has been indicated only in the front view. A development of the lateral surface of the oblique prism is required.

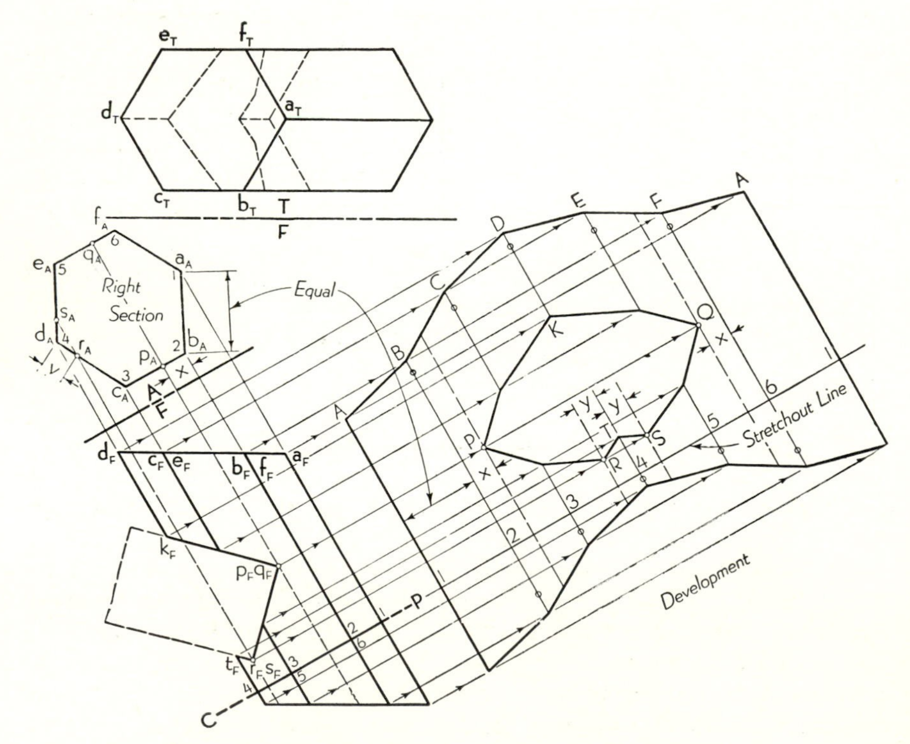

Fig. 10·5. Development of an intersected oblique prism.

Analysis. The whole surface of the oblique prism should be developed first as described in Art. 10·5. The outline of the intersection can then be located on the development.

Development. All lateral edges of the prism appear true length in the front view, and view A, which is an end view, or right section of the prism, is constructed to show the edges as points. Cutting plane C-P, perpendicular to the edges, establishes a right section of the prism, and the corresponding stretchout line has been placed adjacent to the front view for convenience in transferring the true length of the edges to the development.

Distances 1–2, 2–3, etc., along the stretchout line are obtained from the right section in view A; the length of each lateral edge is extended from the front view to the corresponding bend line in the development. The outline of the development is then completed by connecting points A, B, C, etc.

Intersection. The line of intersection can now be located on the development by noting in the front view where each face of the oblique prism has been cut. Edge d_F4, for example, is cut at k_F and t_F, and these two points, extended to the development, appear at K and T on bend line $D4$. Similarly, intersection points on edges c_F3 and e_F5 can be located on bend lines $C3$ and $E5$, respectively. Points on the intersection, such as p_F, q_F, r_F, and s_F, that do not lie on an edge of the oblique prism must be located in view A before their position in the development can be determined. In view A, p_A is distance x from edge b_A2, and therefore point P in the development must be distance x from bend line $B2$. Similarly, point r_A is distance y from edge d_A4; hence point R is this same distance from bend line $D4$.

Although this is an inside development, it has not been so labeled because it is obviously symmetrical about the line $D4$ and hence could be bent either up or down to form the oblique prism.

PROBLEMS. Group 108.

10·7. Development of the right circular cylinder

Problem. In Fig. 10·6 develop the lateral surface of the right circular cylinder.

Analysis. In Fig. 10·2(*a*) the development of a right circular cylinder is illustrated pictorially. In the development the cylinder elements appear as true-length parallel lines. The length of the development equals the circumference of the cylinder, and the divisions of the stretchout line show the distance between consecutive elements. The views given in Fig. 10·6 thus provide directly the necessary element true lengths and element spacing.

Construction. The circle in the top view has been divided into 12 equal parts, thus establishing the 12 equally spaced elements numbered from 0 to 11. The same elements are also drawn in the front view. The lower base of the cylinder is a right section, and in the development it will appear as a straight line; hence we may use it as the stretchout line. (If both ends of the cylinder are oblique, any right section may be selected for the stretchout line.)

The length and divisions of the stretchout line may now be established in either of two ways: (1) by calculating the exact length ($L = \pi D$),[1] and dividing this length into 12 equal parts; (2) by using the bow dividers (not a compass) to step off the chordal distance $a_T b_T$ along the stretchout line. The first method is obviously exact, but with 12 divisions the second method will give a development only about 1.1 per cent short. For 16 divisions the error will be about 0.6 per cent and for 24 divisions only 0.3 per cent; hence more than 24 divisions are rarely necessary in engineering work.

Having established the stretchout line and its divisions, the true length of each element can now be extended directly from the front view to the corresponding element in the development. Points A, B, C, etc., are connected with a smooth curve to complete the outline of the development. Note that lines $B1$, $C2$, etc., represent elements on a curved surface and

[1] See also Appendix, Art. A·6.

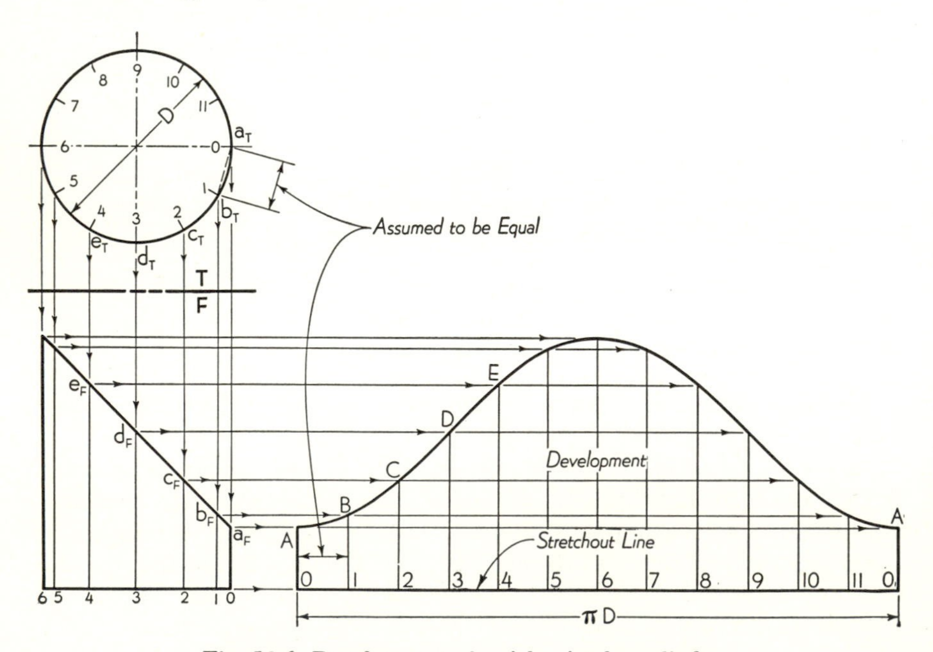

Fig. 10·6. Development of a right circular cylinder.

hence should *not* be marked with the bend-line symbol. The development was started at the shortest element for economy (Art. 10·2); and since it is symmetrical, it need not be labeled "inside."

PROBLEMS. Group 109.

10·8. Development of an oblique cylinder

Problem. In Fig. 10·7 the oblique cylinder has a horizontal circular base with center at A; the upper base with center at B is formed by the intersection of the cylinder with a vertical plane. A development of the cylindrical surface is required.

Analysis. The development of an oblique cylinder can be reduced to the simpler form shown in Fig. 10·6 if we first construct:

1. *A view showing the true length of the cylinder elements.*
2. *An end view (right section) showing the elements as points.*

Construction of the views. View A is drawn as a top-adjacent view to take advantage of the fact that the lower circular base will appear as an edge in this view. But the upper base is elliptical and will appear as

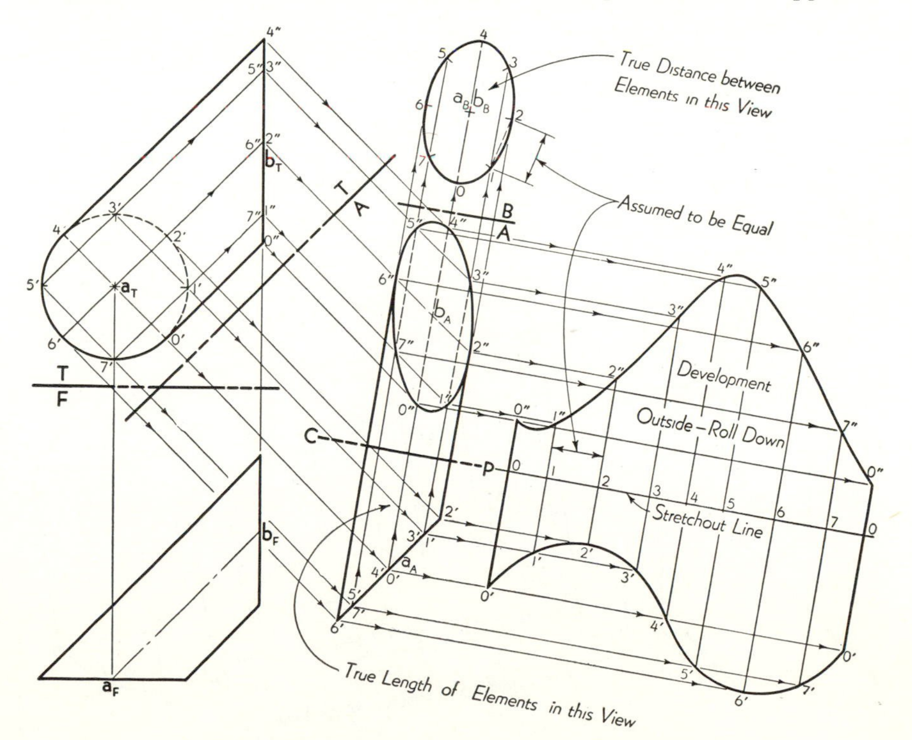

Fig. 10·7. Development of an oblique cylinder.

an ellipse in view *A*; hence it is necessary to assume a series of elements on the cylinder in order to construct the ellipse in view *A*. The circular base in the top view is divided into eight equal parts (a very inadequate number for practical use, but sufficient for illustration), and elements are drawn from each division point. Points $0'$, $1'$, $2'$, etc., on the base circle are then extended to view *A*, and the same elements drawn in this view. In the top view, the elements intersect the vertical plane at points $0''$, $1''$, $2''$, etc. When these intersection points are extended to the corresponding elements in view *A*, they establish the required ellipse. This phase of the construction is similar to that described in Art. 6·7. The elliptical end view *B* shows each of the selected elements as a point and establishes the required end view.

Construction of the development. Since neither base of the cylinder is a right section, the cutting plane *C-P* has been arbitrarily selected as shown, and the stretchout line has been conveniently assumed directly opposite it. (As shown in Fig. 10·4, the stretchout line can actually be placed anywhere, and element lengths, above and below the cutting plane, can easily be transferred to the development with dividers.) Because

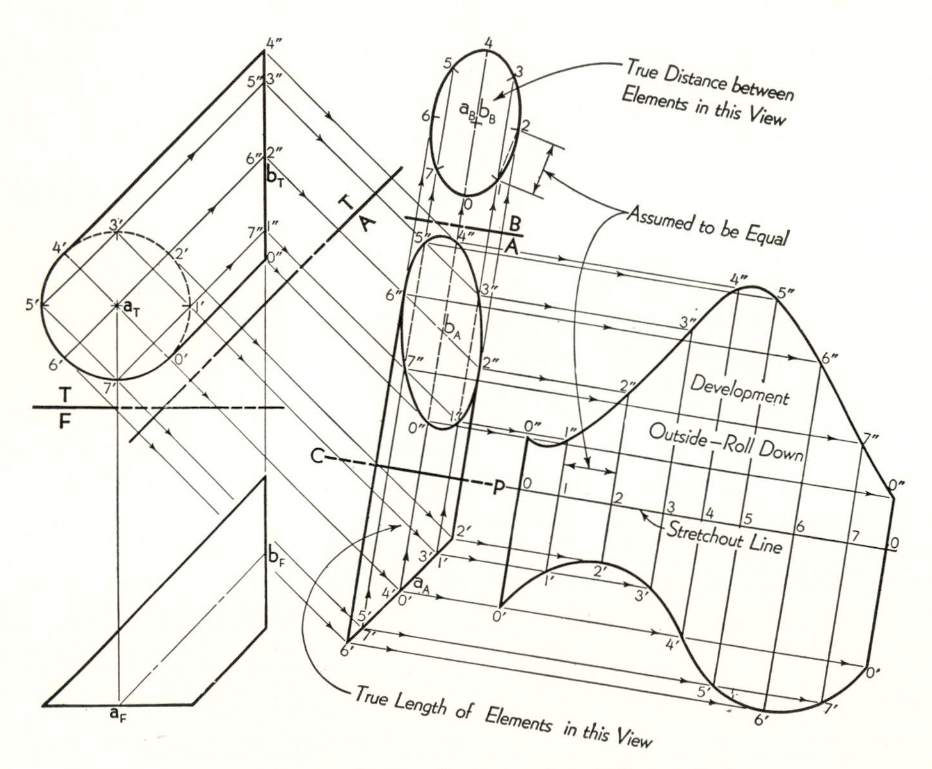

Fig. 10·7 (repeated). Development of an oblique cylinder.

the eight elements shown were initially assumed at equal intervals on the circular base, they are not equally spaced on the ellipse in view *B*; hence the distances from 0 to 1, 1 to 2, etc., on the ellipse must be individually transferred to the stretchout line (see also Appendix, Art. A·4). The true length of each element can then be extended to the development to locate points 0′, 1′, 2′, etc., on the lower curve and points 0″, 1″, 2″, etc., on the upper curve.

This development is not symmetrical; hence it *must* be labeled to indicate which side is toward the observer. That this is an *outside* development can be determined by comparing the sequence of the elements in the development and in view *A*. Element 0 in view *A* is on the near side of the cylinder, and elements 1 and 2 are progressively to the right of element 0. This is the same sequence shown in the development; hence the development from 0 to 2 shows the same side of the surface as is shown in view *A* from 0 to 2, namely, the *outside.*

If this development is rolled up instead of down, the resulting cylinder will incline to the left instead of to the right, as shown in the given views. In other words, a "right-hand" cylinder is obtained by rolling down, a "left-hand" cylinder by rolling up, and the two cylinders are as different as a right shoe and a left shoe. This can best be understood by actually cutting an unsymmetrical development out of heavy paper and then rolling it both up and down to observe the results.

10·9. Development of intersecting cylinders

Problem. The intersecting cylinders shown in Fig. 10·8 represent a common type of pipe connection called a *Y branch,* or *lateral.* For large pipe this connection can be made by cutting and welding the two pipes along the line of intersection. In order to establish the exact line of cut on each pipe the curve of intersection must first be determined, and then the two cylindrical surfaces must be developed.

Intersection. In Fig. 10·8 the curve of intersection has been determined by using a series of horizontal cutting planes, such as *C-P*-3, to cut straight-line elements from each cylinder. This part of the construction is very similar to that shown in Fig. 9·3. View *A*, which shows an end view of the branch pipe, is useful in this phase of the solution but not essential, since elements cut from this cylinder could, if desired, be established solely from the ellipse in the front view (see Art. 9·5). However, in order to develop the branch pipe, view *A is* essential.

Development. In the top view the elements of both cylinders appear true length; hence each cylinder may conveniently be developed adjacent to this view as shown. For the branch cylinder the length of the stretchout line is taken from view *A*, where this cylinder has been divided for convenience into 12 equal parts. Because the upper and lower halves are symmetrically alike, only half of the development (from 0 to 6) has been drawn. The length of each element is extended directly from the adjacent view to the corresponding element in the development. Connecting the ends of the elements with a smooth curve completes the development of the branch cylinder.

For the cylinder on axis AB the length of the stretchout line is taken from the circle in the front view; and since the intersection line lies only on the right side of the pipe, only this half has been developed. The elements used in this development are the same as those used to obtain the intersection curve; hence they are not equally spaced, and distances $0'$–$1'$, $1'$–$2'$, etc., in the front view must be transferred individually to the stretchout line. Extending points on the intersection line to the corresponding element in the development establishes the true developed shape of the required opening in the pipe.

Summary. The basic method of development for all cylinders, right or oblique, may be summarized as follows:

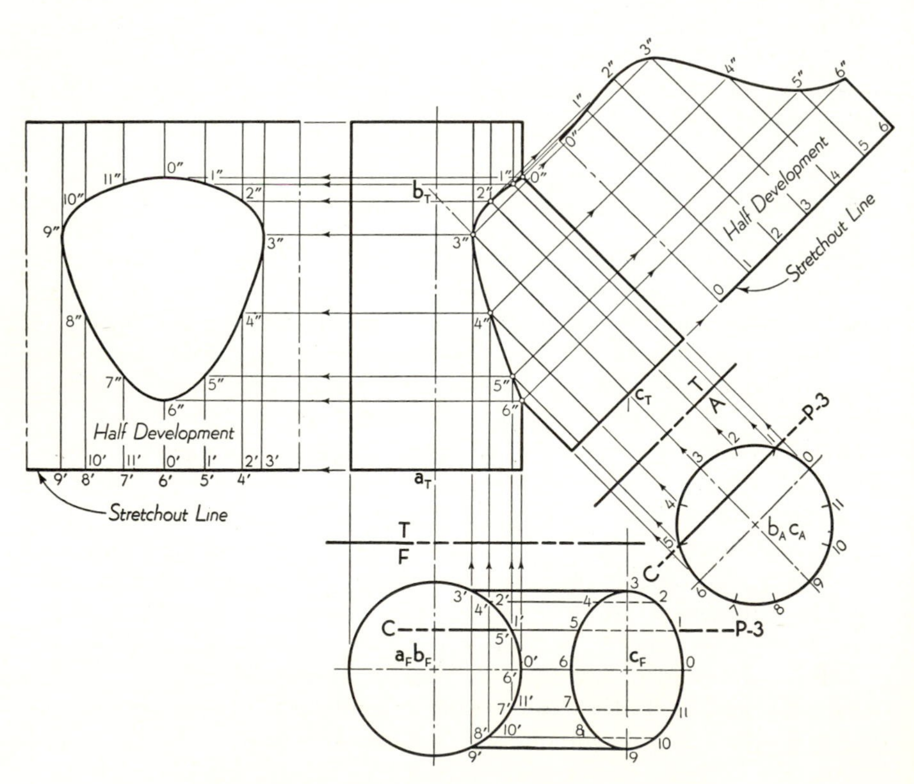

Fig. 10·8. Development of intersecting cylinders.

The dimensions of a cylinder development are always obtained from a true-length view and from the end view of the cylinder.

PROBLEMS. Group 110.

10·10. Development of a right pyramid

Analysis. All pyramids and cones have *radial-line* developments (Art. 10·3). The lateral edges of a pyramid or the elements of a cone appear on the development as straight lines that radiate from a focal point corresponding to the vertex of the pyramid or cone. In the development of a truncated pyramid or cone the simplest and most accurate solution is obtained by first developing the whole pyramid and then deducting from the whole the development of the vertex portion. This is a general procedure that should be followed whenever the vertex is available within the limits of the paper; when the vertex is inaccessible, the method of triangulation must be used (Art. 10·18).

Problem. In Fig. 10·9 develop the lateral surface of the truncated right pyramid.

Construction. Considering the whole pyramid, the six lateral edges are equal in length, but none appears in true length in the given views.

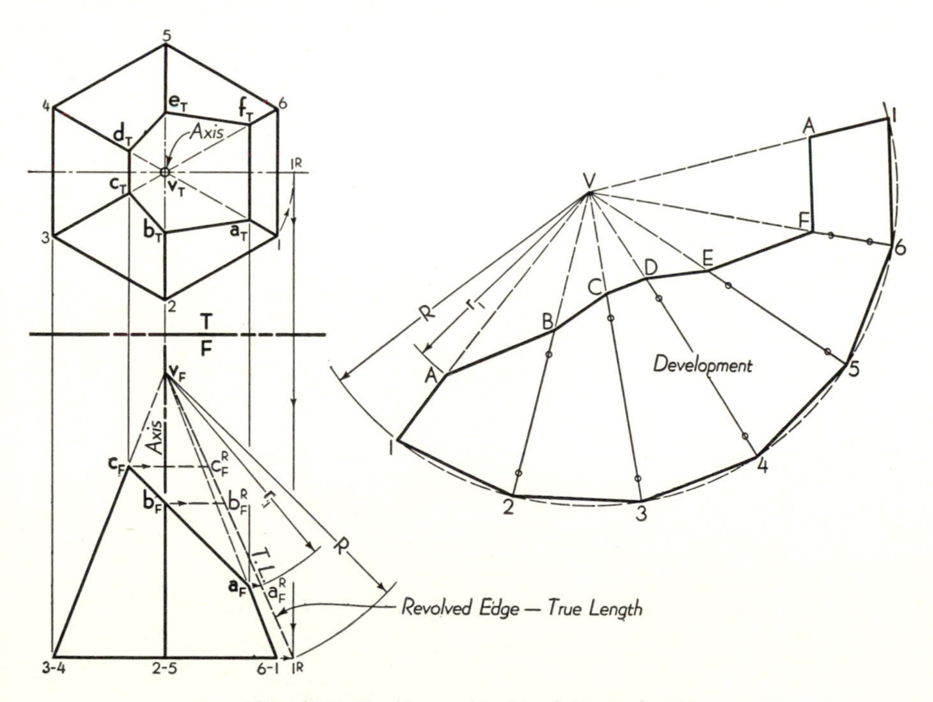

Fig. 10·9. Development of a right pyramid.

Using the axis of the pyramid as an axis of revolution, edge $V1$ can be revolved as shown to appear true length in the front view. Then line v_F1^R is the true length of all six lateral edges. In the development, point V, the focal point of all radial lines, is selected at any convenient spot, and an arc of radius R ($R = v_F1^R$) is drawn with V as center. The six sides of the hexagonal base appear true length in the top view; hence these distances may be transferred directly to the development, where they are marked off as the successive chords 1–2, 2–3, etc., along the arc. Drawing the radial lines $V1$, $V2$, etc., divides the development into six triangular areas corresponding to the six faces of the pyramid and completes the development of the whole pyramid.

The vertex portion of the truncated pyramid may now be deducted from the development simply by noting how much of each lateral edge has been cut off. Edge $V1$, for example, is cut off at point A, but the length VA must first appear in true length before it can be transferred to the development. If the line segment VA is revolved about the pyramid axis, its revolved position will coincide with the revolved position of edge $V1$ and point a_F will move horizontally to position a_F^R. The true-length distance $v_Fa_F^R$ (radius r_1) is marked off in the development along edge $V1$ to locate point A. If the other lateral edges are similarly

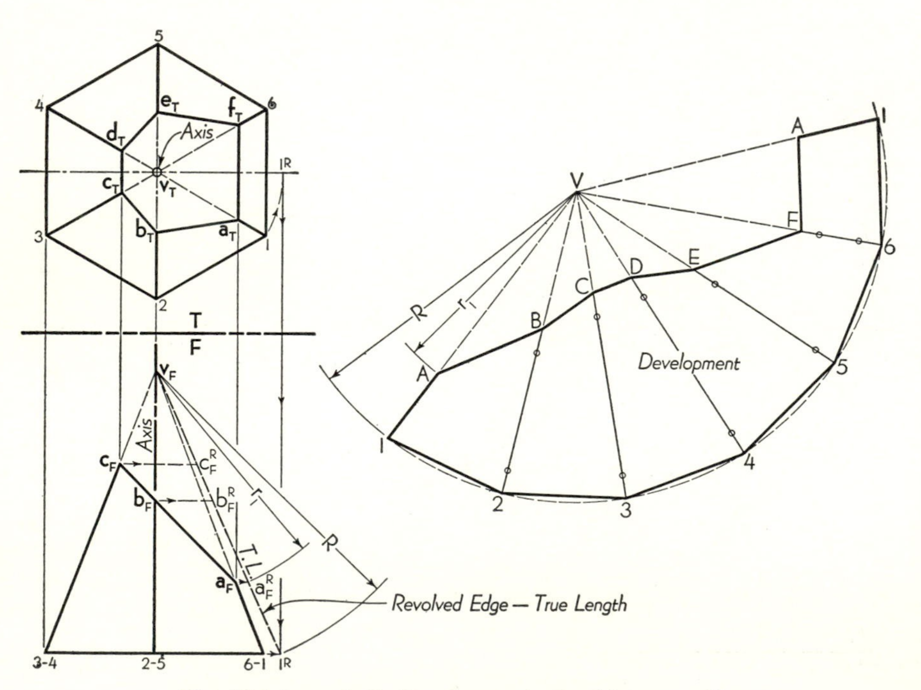

Fig. 10·9 (repeated). Development of a right pyramid.

revolved, they, too, will coincide with the revolved position of edge $V1$ and therefore points b_F, c_F, etc., will all move horizontally to positions b_F^R, c_F^R, etc., on this same line. Points B, C, D, etc., are located in the same manner as point A by marking off the true-length segments along the corresponding edges in the development. Connecting points A, B, C, D, etc., with straight lines completes the development of the truncated right pyramid.

PROBLEMS. Group 111.

10·11. Development of an oblique pyramid

Analysis. The method of developing an oblique pyramid is basically the same as that for a right pyramid, except that the lateral edges of an oblique pyramid are of various lengths, and therefore the true length of each edge must be individually determined. These true lengths can best be obtained by the method of revolution, and this method can be further simplified by reducing it to a *true-length diagram.*

Problem. In Fig. 10·10 develop the lateral surface of the truncated oblique pyramid.

True-length diagram. To obtain the true length of each lateral edge a vertical axis has been taken through vertex V. Then if edge $V1$

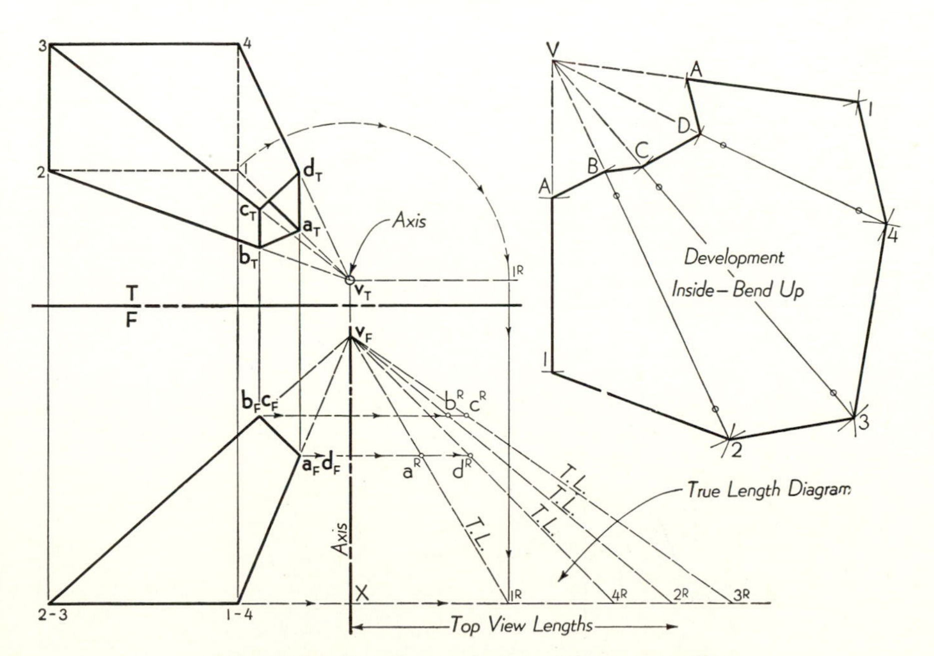

Fig. 10·10. Development of an oblique pyramid.

is revolved to the right about this axis as shown, it will appear true length at v_F1^R. But it should be apparent that it was not actually necessary to draw the arc from point 1 to 1^R and the parallel from 1^R in the top view to 1^R in the front view. The distance $X1^R$ in the true-length diagram is exactly the same as the distance v_T1 in the top view and could have been transferred directly from the top view with dividers, thus eliminating the arc and the parallel. The revolved true-length positions of edges $V2$, $V3$, and $V4$ have been established by this abbreviated procedure: $X2^R$ equals v_T2, $X3^R$ equals v_T3, etc. The true-length diagram thus consists of a series of triangles, each having a hypotenuse equal to the true length of one edge of the pyramid.

Development. The true length v_F1^R (the shortest edge) is taken from the true-length diagram and laid off in any convenient position $V1$. With true length v_F2^R as a radius and point V as a center a short arc is drawn to partly locate point 2 in the development. Then with point 1 as a center and the true length of edge 1–2 as a radius (all base edges appear true length in the top view) a second arc is drawn to intersect the first arc, thus establishing point 2. Connecting points V, 1, and 2 completes the development of one face of the pyramid. In similar manner, point 3 is located at the intersection of arcs drawn with V and 2 as centers, point 4 at the intersection of arcs drawn with V and 3 as centers, etc., all radii being taken as true-length distances.

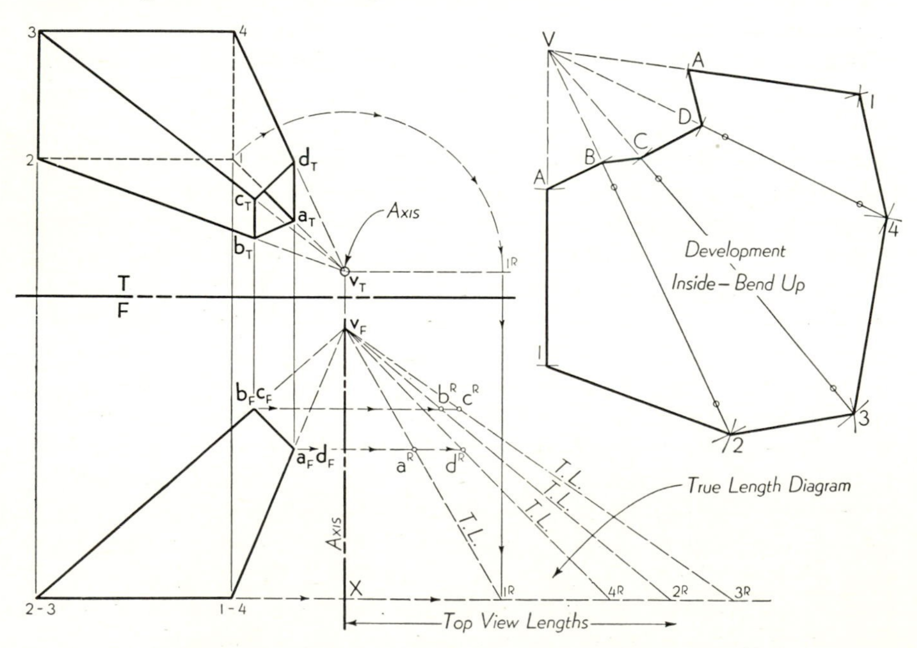

Fig. 10·10 (repeated). Development of an oblique pyramid.

The vertex portion may now be deducted by marking off on each radial line the true lengths *VA*, *VB*, *VC*, etc. But *VA* is a segment of edge *V*1, *VB* a segment of edge *V*2, etc.; hence the revolved position of these segments will coincide with the revolved positions of the edges, and a_F will move horizontally to position a^R on v_F1^R, b_F to b^R on v_F2^R, etc. Therefore v_Fa^R is the true length of *VA* and may be transferred from the true-length diagram to the development. Connecting points *A*, *B*, *C*, etc., completes the development of the truncated pyramid. Because this development is not symmetrical, it must be labeled as shown to indicate which side is toward the observer.

PROBLEMS. Group 112.

10·12. Development of a right circular cone

Analysis. In Fig. 10·2(*b*) the development of a right circular cone is illustrated pictorially. All elements of this cone are equal in length; hence the radial lines in the development are of constant length and equal to the slant height of the cone. Thus the base circle of the cone becomes a circular arc in the development, and the length of the arc should exactly equal the circumference of the cone base.

Problem. In Fig. 10·11 develop the lateral surface of the right circular cone.

Construction. The base circle is divided in the top view into 12 equal parts, thus establishing the 12 equally spaced elements numbered from 0 to 11. The same elements are also drawn in the front view. Extreme element v_F0 is true length and is the slant height *R* of the cone. Therefore, with *R* as a radius and any convenient point *V* as a center, an arc of indefinite length is drawn on the development. The length and divisions of this arc may now be obtained in either of two ways: (1) by calculating the angle α according to the formula given in Fig. 10·11, and then dividing the arc length into 12 equal parts; (2) by stepping off with bow dividers the chordal distances on the base circle as chords on the arc in the development as shown. The first method is exact, but the error in the second method will be less than the values given in Art. 10·7 for a cylinder. Radial lines drawn from *V* to each division point on the arc represent the 12 elements of the cone development.

The vertex portion of the cone may now be deducted from the whole cone development by marking off on each radial line the true length of the segment cut from that element. Element *V*2, for example, is cut at point *C*, and the true length of segment *VC* can be obtained by revolving it about the cone axis until it coincides with the extreme element v_F0. When so revolved, point c_F moves horizontally to c_F^R on v_F0, and the true length $v_F c_F^R$ (radius R_2) is then marked off on radial line $V2$ to locate point C in the development. Points A, B, D, etc., are similarly located and connected with a smooth curve to complete the development of the cone. The symmetry of the cone and its development make it unnecessary to label the development inside or outside.

Frustums of right circular cones occur frequently in engineering construction (see Fig. 10·15 for examples). When neither base of the cone is a right section, uniform spacing of the radial lines is best obtained by arbitrarily assuming a right section. The true length of each element above and below this section can then be marked off on the radial lines of the development as described above (see Art. 10·17).

PROBLEMS. Group 113.

10·13. Development of an oblique cone

Analysis. In the development of an oblique cone (or any cone other than a right circular cone) the radial lines will be *unequal* in length, and therefore the cone base (even though circular) will *not* develop as a circular arc. To develop such cones we must assume that the surface

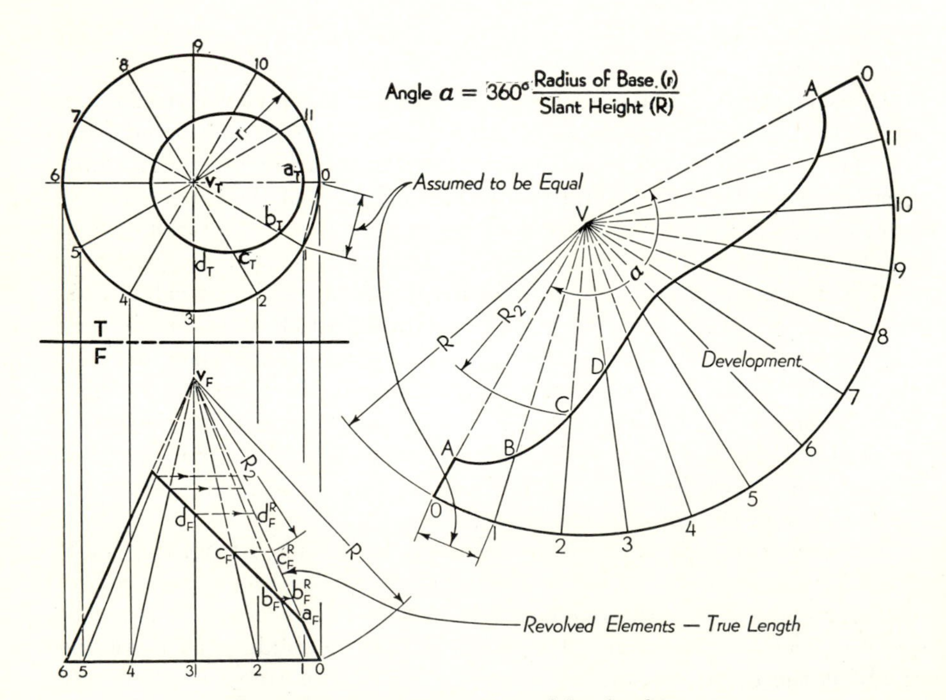

Fig. 10·11. Development of a right circular cone.

between any two consecutive elements is narrow enough to approximate closely a slender plane triangle; in other words, the curved portion of the base between two elements must be considered a straight line. By making this approximation (the error will seldom exceed 1 per cent) the cone may be developed as an oblique pyramid having a many-sided base.

Problem. In Fig. 10·12 the base curve of the oblique cone is elliptical, and the vertex portion has been cut off by a vertical plane. It is required to develop the cone surface between the vertical and horizontal bases.

Construction. The base curve of the cone is divided into 12 equal parts, and the 12 elements are drawn in each view. The curve in the front view is determined by the usual method of noting the points where each element intersects the vertical plane (Art. 6·7). Assuming the cone to be a 12-sided oblique pyramid, the development is now constructed by the same method described in Art. 10·11. The true-length diagram at the right of the front view is constructed to show the true length of each element from vertex to base. As in Fig. 10·10, the top-view length of each element is transferred to the diagram base line—$X3^R$ equals v_T3, $X4^R$ equals v_T4, etc.—to establish the true lengths v_F3^R, v_F4^R, etc. The base of the cone appears true size in the top view; hence chordal distances, such as 9–10, may be transferred directly to the development.

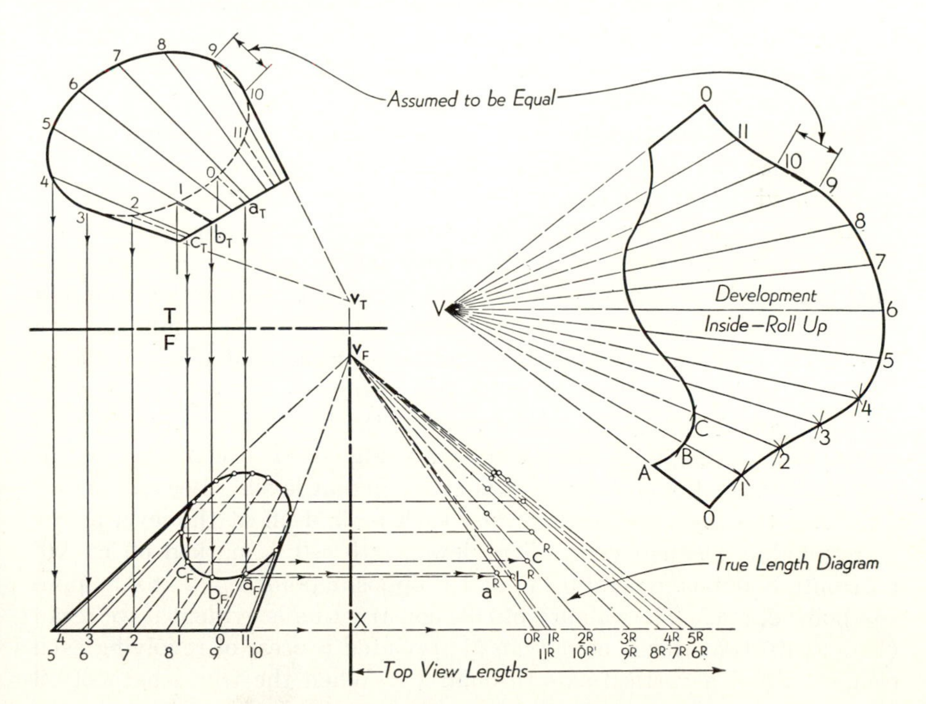

Fig. 10·12. Development of an oblique cone.

In the development the first radial line $V0$ may be placed as desired. Then with V as a center and radius equal to v_F1^R (the true length of the next element) a short arc is drawn to partly locate point 1. With 0 as a center and radius equal to the chordal distance 0–1 in the top view a second arc is drawn to intersect the first arc and locate point 1. In the same manner, points 2, 3, 4, etc., are successively located on arcs drawn from V and the previous base point. A smooth curve drawn through these points establishes the base curve in the development. Note that the *angular spacing* of the radial lines is not equal.

The curve of the upper cone base is obtained by marking off on each radial line the true length of the segment cut from each element. As in Fig. 10·10, these lengths may be taken from the true-length diagram: v_Fc^R, for example, is the true length of segment VC and is marked off on $V2$ in the development to locate point C. The development is obviously not symmetrical and is therefore labeled as shown.

10·14. Development of a conical offset connection

Problem. In Fig. 10·13 two parallel circular pipes of different diameters have been joined by a *conical connection.* The base of the cone may be designed to intersect the large pipe at any desired angle, but the connecting surface will be *conical* only if the planes of intersection with each pipe are parallel as shown.

Analysis. Since this surface is a frustum of an oblique cone, it may be developed by the method of Art. 10·13 with a few modifications necessitated by the fact that the cone base does not appear true size.

Construction. The vertex V is located by extending the extreme elements in each view. The circle, which represents the cone base in the top view, has been divided into 12 equal parts; but because the front and rear halves of the cone are symmetrical, complete construction has been shown for only the front half. From each numbered division point, elements are drawn in both views to the vertex.

The true-length diagram at the right of the front view is constructed on the same principle as the one shown in Fig. 10·12, except that in this case the base end of each element is at a different level. Because of this difference the top-view lengths must each be laid off at the level of the corresponding element end. Top-view length v_T0 is marked off at $X0'$ horizontally opposite point 0, v_T1 at $Y1'$ opposite point 1, v_T2 at $Z2'$ opposite point 2, etc. The validity of this construction is evident if we recall (Fig. 10·10) that this is merely an abbreviated process of revolving each element about a vertical axis through V. When the true length of all elements from vertex to base has been thus established, points a_F, b_F, c_F, etc., may then be extended to a′, b′, c′, etc., on the corresponding true-length line to obtain the true length of the segment cut from each element.

Since the large circle in the top view does not represent the true perimeter of the cone base, chordal distances such as 0–1, 1–2, etc., *cannot* be transferred directly to the development. The plane of the base is therefore revolved about a horizontal axis into the horizontal position shown, and the true elliptical shape of the base appears in the top view. Chords of the ellipse, such as 3^R–4^R, are true length and hence may be laid off in the development as the base of each triangle. The true size of the base could, of course, be shown in an auxiliary view if desired.

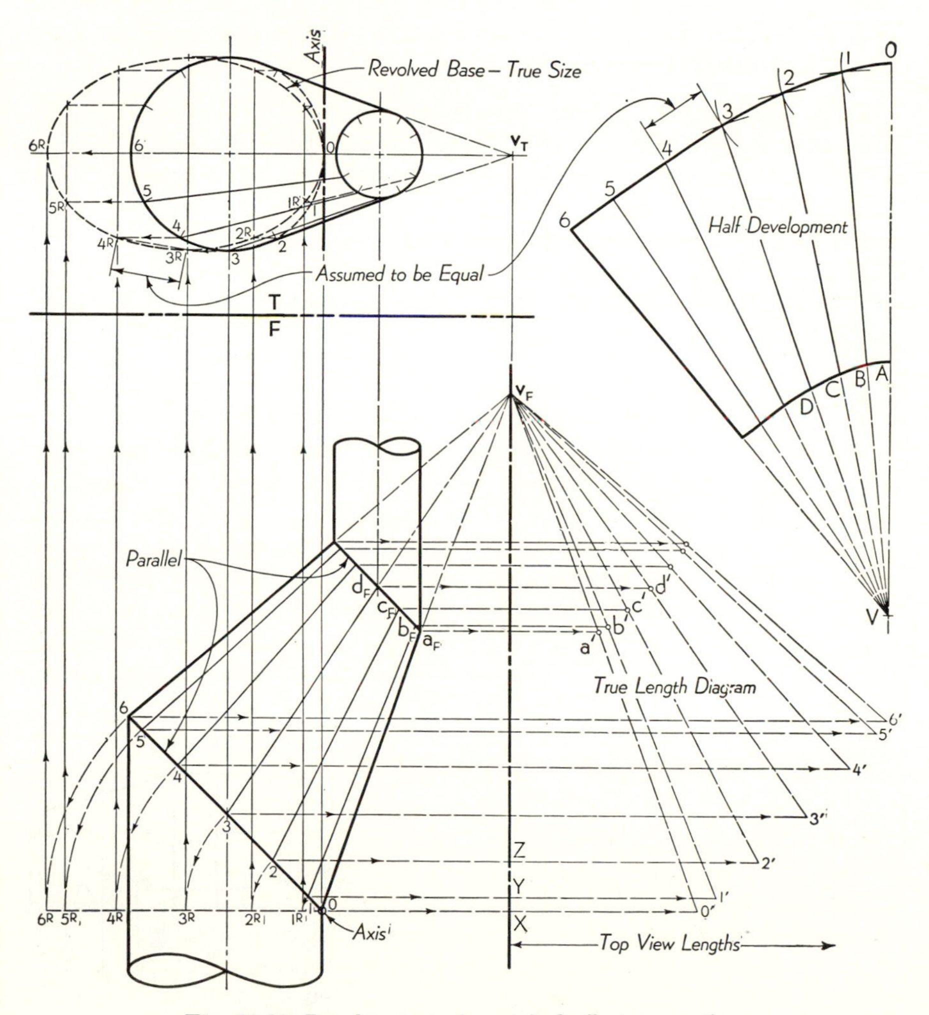

Fig. 10·13. Development of a conical offset connection.

All necessary true-length distances are now available, and the radial-line development can therefore be constructed as in Art. 10·13 by laying down each slender triangle in orderly sequence. As in Fig. 10·12, points 1, 2, 3, etc., are located consecutively at the intersection of arcs drawn with true-length radii. Points A, B, C, etc., are then located on the radial lines (as in Fig. 10·12) by transferring the true length v_Fa', v_Fb', etc., to the corresponding element in the development. The numbered and lettered points are connected with smooth curves to complete the development of one symmetrical half of the conical connection.

PROBLEMS. Group 114.

10·15. Cylindrical pipe elbows

Large sheet-metal pipe elbows, such as those used in heating and ventilating ducts, are made in cylindrical sections. By cutting each section at the same angle the development pattern for one section can be used for all sections.

Problem. Design and develop a 90° four-section elbow.

Design. Figure 10·14(a) shows the preliminary layout for an elbow of radius R, diameter D, and angle θ. For ordinary stovepipe the radius R is usually made as small as possible for compactness; for ventilating pipes,

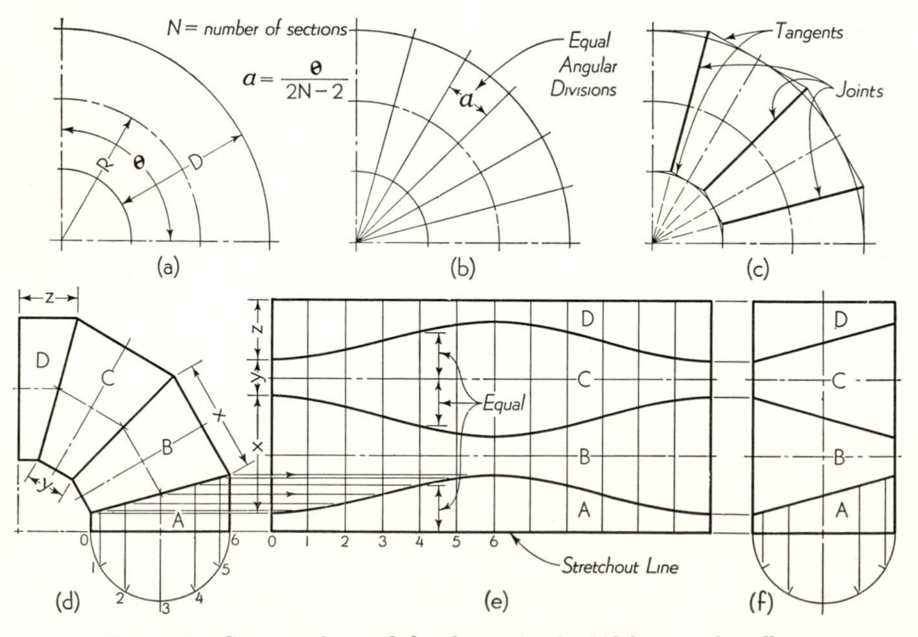

Fig. 10·14. Construction and development of a 90° four-section elbow.

R should be at least $1\frac{1}{2}D$ to reduce flow resistance. The angle θ (sometimes called *sweep angle*) need not be 90° but may have any desired value. Since this elbow must consist of four cylindrical sections as shown at (*d*), the angle θ must be divided into six equal angular divisions as shown at (*b*); in general, the number of divisions equals two less than twice the number of sections (note formula in Fig. 10·14). At (*c*) lines are drawn tangent to the inner and outer arcs perpendicular to *alternate* radial lines. The intervening radial lines thus become the *joints* (or *miter* lines) between each cylindrical section. The completed elbow, ready for development, is shown at (*d*).

Development. All sections of the elbow are right circular cylinders and may be developed by the method shown in Fig. 10·6. The development of section A has been conveniently placed at the right of the section, and the length and divisions of the stretchout line have been taken from the equally divided *construction semicircle*, which serves as a substitute for a top view. Sections B and C are each equal to two sections like A; hence these developments may be obtained by duplicating the dimensions of section A as shown. If the seams for sections A and C are taken on the short inner element and the seams for sections B and D on the long outer element, then the four developments can be compactly fitted together to form a rectangle as shown at (*e*). (In actual practice, the four developments must be separated enough to provide extra material for the girth seams.)

If section B of the elbow is revolved 180° and again joined to section A, its elliptical end will once more exactly coincide with the elliptical end of section A and the two sections will form one continuous cylinder. By similarly revolving sections C and D, the four sections become a single cylinder as shown at (*f*). Thick-walled seamless elbows can therefore be made by cutting a straight length of pipe into sections, reversing the alternate sections, and welding the sections back together again.

PROBLEMS. Group 115.

10·16. Plane intersections of cylinders and cones of revolution

Cylinders and cones that are tangent to and enclose (envelop) the same sphere will intersect each other in *plane curves*. These curves are *elliptical* but will appear as *straight lines* in that view which shows the axes of both intersecting surfaces in *true length*. For simplicity of manufacture, this fact is utilized—whenever possible—in the design of intersecting cones and cylinders. We should note that:

> *To envelop a sphere the cylinders and cones must be surfaces of revolution, and their axes must intersect.*

Figure 10·15 shows a number of examples of cylinders and cones of revolution intersecting in plane curves. In each case only one view is shown, and that is the view which shows both axes in true length. As shown at (*a*), (*b*), and (*c*), two intersecting cylinders may be designed to envelop a sphere whose center is at the intersection of their axes. At (*a*) and (*b*) the intersection is a single ellipse appearing edgewise (a miter line) as shown; at (*c*) the intersection consists of two half ellipses (1–2 and 2–3) established as shown. When a cone intersects a cylinder, as at (*d*) and (*e*), both cone and cylinder must be tangent to the sphere as shown and the miter line is located by connecting the intersections of the extreme elements.

Assume in Fig. 10·15(*f*) that two circular pipes of different diameters must be joined by a conical connection. Then if center lines a_Fb_F, b_Fc_F, and c_Fd_F all appear true length in this view, the connection can be designed as a cone of revolution. With b_F and c_F as centers a sphere is drawn tangent to each cylindrical pipe. The extreme elements of the cone are then drawn tangent to both spheres, the miter lines are drawn, and the design is completed. [Compare Fig. 8·8(*b*), which shows a cone enveloping two spheres.] Incidentally, the conical connector shown in Fig. 10·13 is *not* a cone of revolution.

The Y-branch air duct shown in Fig. 10·15(*g*) illustrates the case of three mutually intersecting cones. The three cones *A*, *C*, and *E*, with

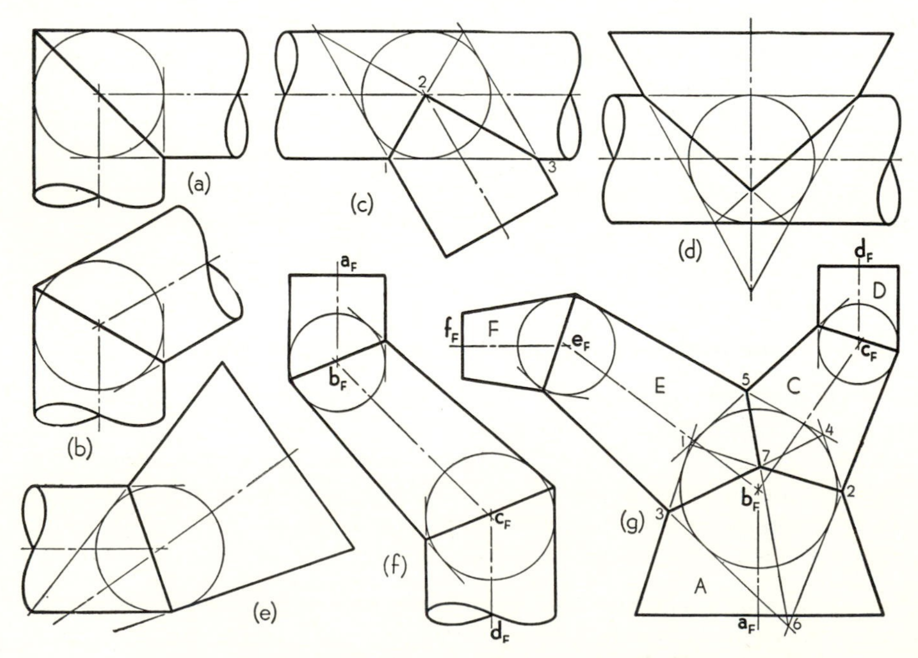

Fig. 10·15. Plane intersections of cylinders and cones of revolution.

center lines at a_Fb_F, b_Fc_F, and b_Fe_F, all envelop the sphere with center at b_F. All center lines are true length in this view, of course. Then the intersection of cones A and C (disregarding cone E) is the ellipse 1–2, of cones A and E the ellipse 3–4, and of cones C and E the ellipse 5–6. The combined effect of all three cones intersecting is therefore an intersection line that consists of three partial ellipses, 2–7, 3–7, and 5–7.

10·17. Development of cones and cylinders with plane intersections

Problem. Figure 10·16 shows a portion of the Y-branch air duct whose miter lines were established in Fig. 10·15(g). A development of the Y branch is required.

Analysis. The fact that all the connecting surfaces are cones and cylinders of revolution greatly simplifies the development. The truncated cones A and C can be developed by the method shown in Fig. 10·11 and the cylinder D by the method of Fig. 10·6.

Cone A. The extreme elements of cone A are extended to locate its vertex at V, and the construction semicircle drawn on the base of the cone is used (instead of a top view) to establish equally spaced elements. As a time-saving measure the vertex of the development has been taken

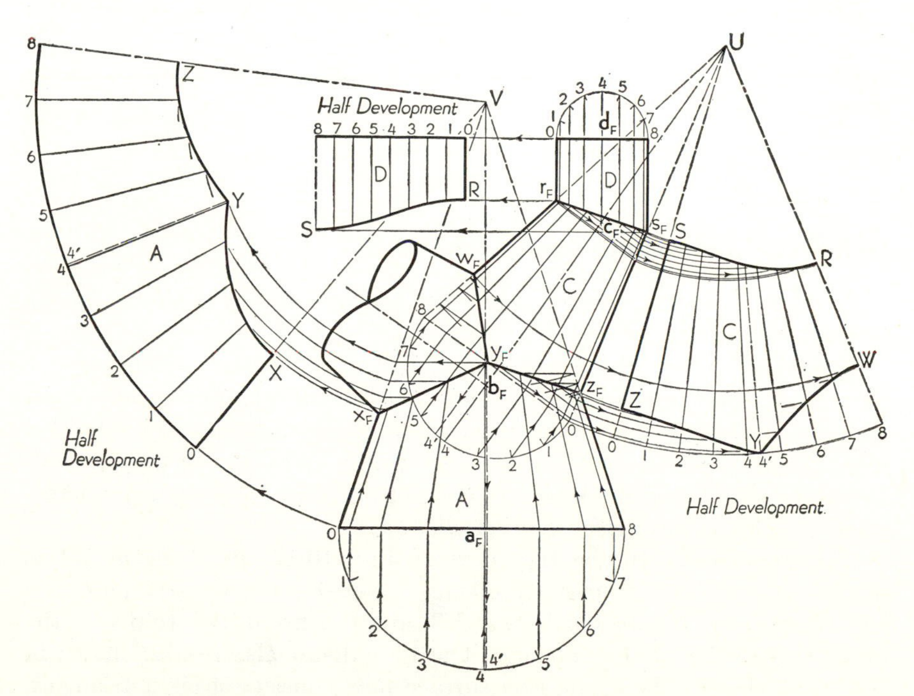

Fig. 10·16. Development of a portion of a Y-branch air duct.

to coincide with the vertex of the cone. Then the upper intersection end of each element on cone A need only be extended horizontally to the extreme element (on either side of the cone) and then revolved about V to its position on the corresponding radial line of the development. Note that the important point y_F does not fall on one of the equally spaced elements, and therefore the special element $4'$ is drawn to locate point Y exactly in the development.

Cone C. The development of cone C is similar to that for cone A, except that neither end of cone C is a right section and therefore it is necessary to assume a right section at some point on the cone axis. In this case the right section has been taken through point y_F, perpendicular to the axis b_Fc_F. Then the straight line 0–8 is the diameter of cone C at point Y, and the construction semicircle is drawn on this diameter as shown. In the development of cone C, distances along the development arc 0–8 correspond to distances on the construction semicircle. Note again on cone C that the upper and lower ends of each element must always be extended to an extreme element before being revolved about vertex U to the corresponding radial line. Point $4'$ is used to establish the exact position of point Y in the development.

Cylinder D. The method of developing cylinder D is identical with that shown for section A in Fig. 10·14.

Check. As a check on the accuracy of the three separate developments shown in Fig. 10·16, note that the length of curves SR on the cylinder D and on cone C should be exactly equal and that curves YZ on cones A and C should also be equal in length.

PROBLEMS. Group 116.

10·18. The method of triangulation

Problem. In Fig. 10·17 develop the sheet-metal hopper that connects two offset rectangular ducts.

Analysis. Since the connecting surface is neither prismatic nor pyramidal, previously described methods of development cannot be employed here. In such cases the various plane surfaces can be divided into triangular areas, and each triangle can be laid down in the development as soon as the true length of each of its three sides has been determined. This is the *method of triangulation.*

Construction. In the top view of Fig. 10·17 the *diagonal* b_T1 is drawn to divide the four-sided surface a_Tb_T2–1 into two triangles. In similar manner, the diagonals b_T3, d_T3, and d_T1 are drawn to divide the other three sides of the hopper. The four diagonals are also shown in the front view. The connecting surface now consists of eight triangles, and it remains only to determine the true length of each side of each triangle. Although each lateral edge could be revolved individually about a different vertical axis, to obtain its true length, it is more efficient to assume a common axis *RX* and mark off each top-view length as shown to form a true-length diagram. The true lengths of the diagonals have been obtained in similar manner, but to avoid confusion a separate diagram based on axis *SY* has been drawn. Lines *AB*, *BC*, etc., and 1–2, 2–3, etc., appear true length in the top view.

In the development, line *A*1, equal to true length *R*1′, is laid down first. Point *B* is then located by drawing an arc of true-length radius $a_T b_T$ from *A* as a center to intersect a second arc of radius *S*–*b*′1′ drawn from point 1 as a center. Point 2 is located at the intersection of arcs drawn from points *B* and 1 as centers, radius *B*2 being taken equal to *R*2′ and radius 1–2 being taken from 1–2 in the top view. The remaining triangles in the development are similarly drawn in proper sequence, radii being taken as needed from the top-view and the true-length diagrams. Care must

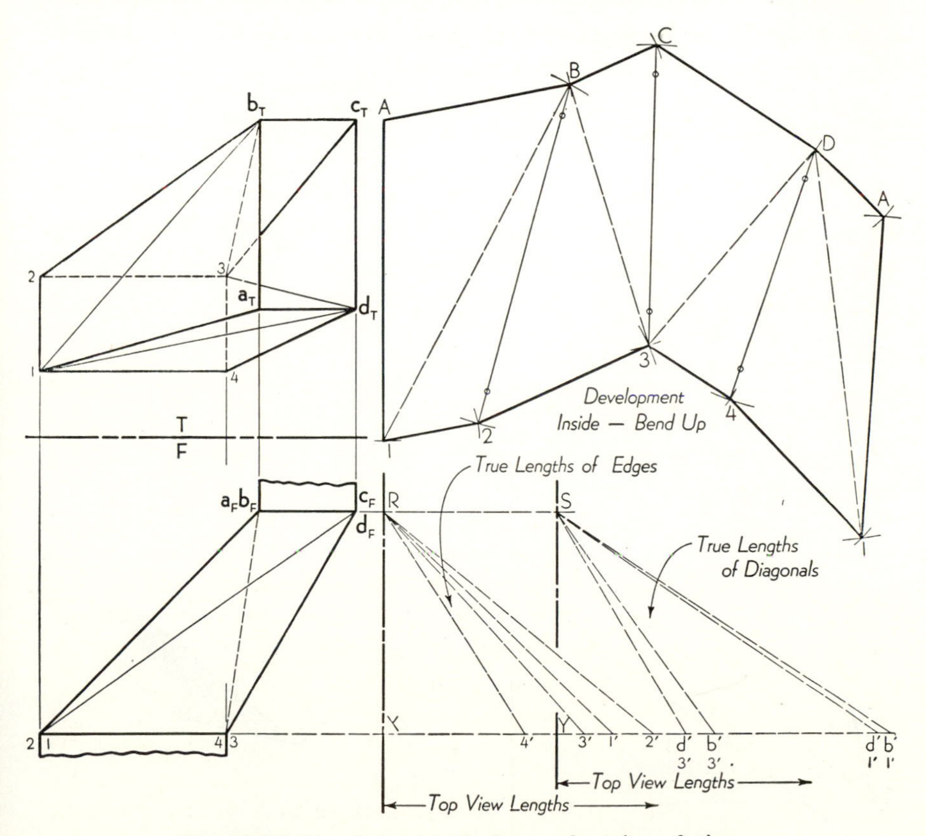

Fig. 10·17. Development of a hopper by triangulation.

be exercised to place the diagonals in the development in the same position assumed in the given views.

A pyramid whose vertex is not accessible can be developed by the above-described method of triangulation. Oblique prisms can also be developed by this method, but the auxiliary-view method of Art. 10·5 is preferable since it usually produces more accurate results.

10·19. Development of transition pieces

A sheet-metal connecting surface that joins two pipes of different size and shape is called a *transition piece.* In Arts. 6·30 and 6·31 it was shown that pipes having dissimilar curved openings can be connected with a convolute surface; the warped cone of Art. 7·5 is another form of transition piece.

Problem. In Fig. 10·18 design and develop a transition piece to connect the circular pipe to the rectangular duct.

Design. The connecting surface is composed of four triangular planes *P*, *Q*, *R*, and *S* and four partial oblique cones *E*, *F*, *G*, and *H*. Each tri-

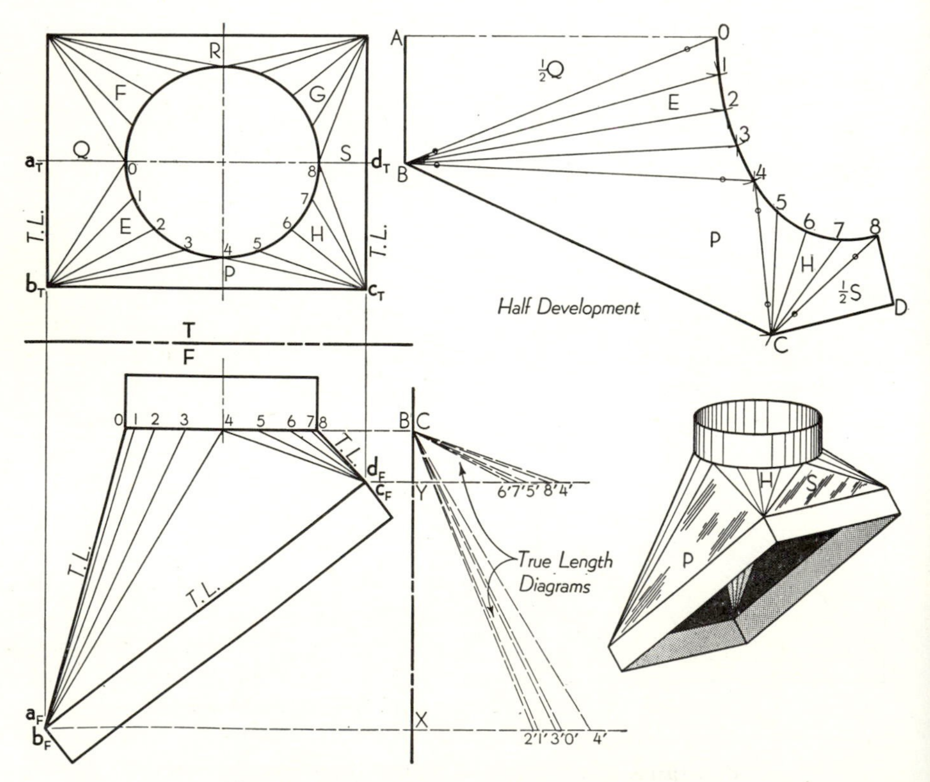

Fig. 10·18. Development of a transition piece.

angle has its base on one side of the rectangular opening and its vertex on the circle; each cone has its base on one-quarter of the circular opening and its vertex on one corner of the rectangle. By similarly employing plane surfaces and portions of cones and cylinders, simple developable transition pieces can be designed to connect a great variety of dissimilar pipes.

Development. The circle is divided into 16 equal parts, and cone elements are drawn to each of the four vertices. Because the surface is symmetrical about center line a_Td_T, only the front half has been developed. The true-length diagram constructed on axis BX gives the true length of elements 0, 1, 2, 3, 4 on cone E and that on axis CY the true length of elements 4, 5, 6, 7, 8 on cone H. The half triangle $AB0$ is developed by taking the true length of $A0$ from a_F0 in the front view, the true length of AB from a_Tb_T in the top view, and the true length of $B0$ from $B0'$ in the true-length diagram. The oblique cone E with vertex at B is then developed as a series of slender triangles as described in Art. 10·13. In triangle $BC4$, point C is located by drawing arcs with B and 4 as centers, noting that the true length of BC appears in the front view and that $C4$ is the first element of cone H. This procedure is continued to the seam at $D8$, thus completing one symmetrical half of the development.

PROBLEMS. Group 117.

10·20. Development of a convolute transition piece

Analysis. The tangent-plane convolute surfaces described in Arts. 6·30 and 6·31 are developable; but because the elements are not parallel and do not intersect at a common point, the surface must be developed by the method of triangulation. If the elements are sufficiently close together, then the curved lines that join the ends of adjacent elements may be assumed to be straight lines, thus dividing the surface into a series of very nearly plane quadrilaterals. Each quadrilateral may then be divided by a diagonal into two triangles. The development of the convolute surface is thus reduced to the orderly development of a series of triangles—a process identical with that described in Art. 10·18, except that the triangles are smaller and more numerous.

Problem. In Fig. 10·19 develop the convolute transition piece. This is the same surface whose elements were established in Fig. 6·31.

Construction. Although 24 elements, equally spaced on the circle, are shown, it is necessary to develop only one symmetrical quarter of the surface. Diagonals $A1$, $B2$, $C3$, etc. (dash lines), are drawn in each view to divide the surface between elements into two triangles. Two separate true-length diagrams are constructed to obtain the true lengths of elements and diagonals. Since both ellipse and circle appear true size in the top view, chordal distances in this view may be transferred as arc radii directly to the development.

Starting with element $A0$ in the development, point 1 is located at the intersection of arcs drawn from A and 0 as centers, point B on arcs drawn from A and 1, point 2 on arcs from B and 1, etc., the numbered and lettered points being located alternately and consecutively until the development is complete. If desired, a second quarter of the development can be simultaneously duplicated on the opposite side of element $A0$ to make a half development as shown.

10·21. Approximate development of a warped transition piece

Analysis. The convolute surface shown in Fig. 6·32 connects two circular openings that lie in nonparallel planes, but this surface can also be developed by the same triangulation method described in the previous article. It is common practice, however, to approximate the exactly

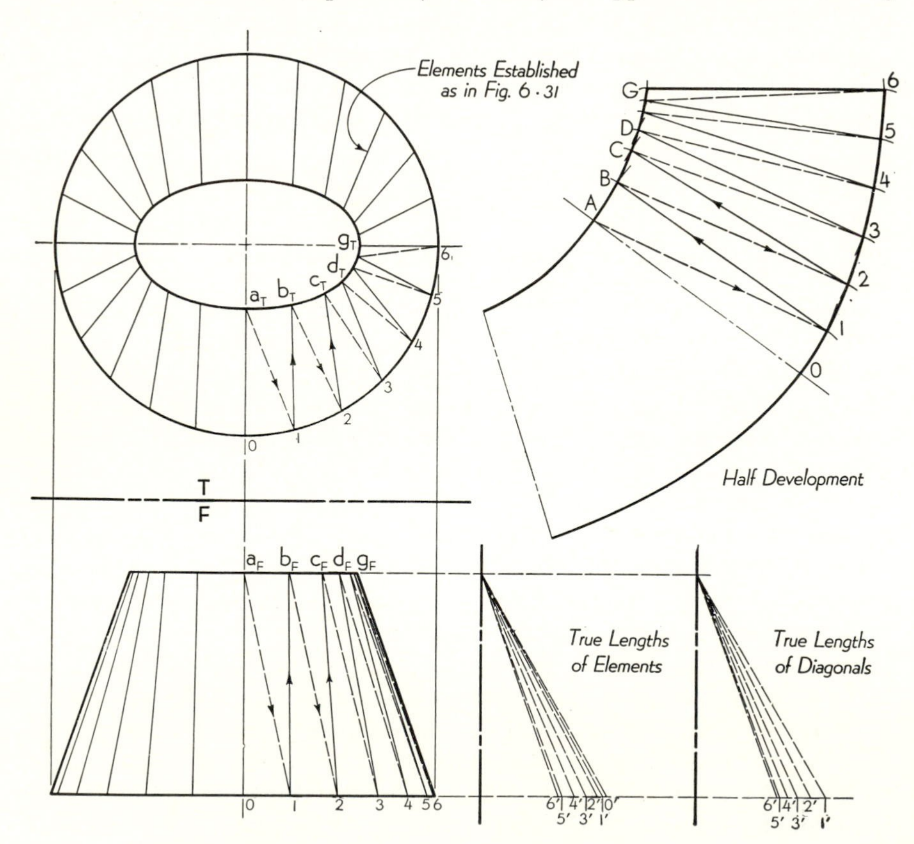

Fig. 10·19. Development of a convolute transition piece.

developable convolute by using elements that are equally spaced on both bases. This simplifies the work of establishing elements but produces a surface that is slightly warped. In most cases a development of this warped surface is a very close approximation to that for the convolute, and the slight deviations can usually be compensated on the job by a small amount of edge trimming and seam adjustment.

Problem. In Fig. 10·20 a transition piece connects two nonparallel circular openings (same as convolute of Fig. 6·32). Using elements equally spaced on each base, develop the warped surface.

Construction. Equal spaces on the inclined base circle are obtained (as in Fig. 10·16) from the construction semicircle, which serves as a substitute for an auxiliary view. Diagonals *B*0, *C*1, *D*2, etc., are drawn in each view between adjacent elements. The construction of the two true-length diagrams—one for elements, one for diagonals—is similar to that shown in Fig. 10·13: each top-view length must be laid off from axis *PQ* directly opposite the lower end of each element or diagonal. For element *C*2, for example, the top-view length c_T2 is laid off at *X*2′ to the *left* of axis *PQ*, directly opposite point 2 in the front view. For diagonal *D*2

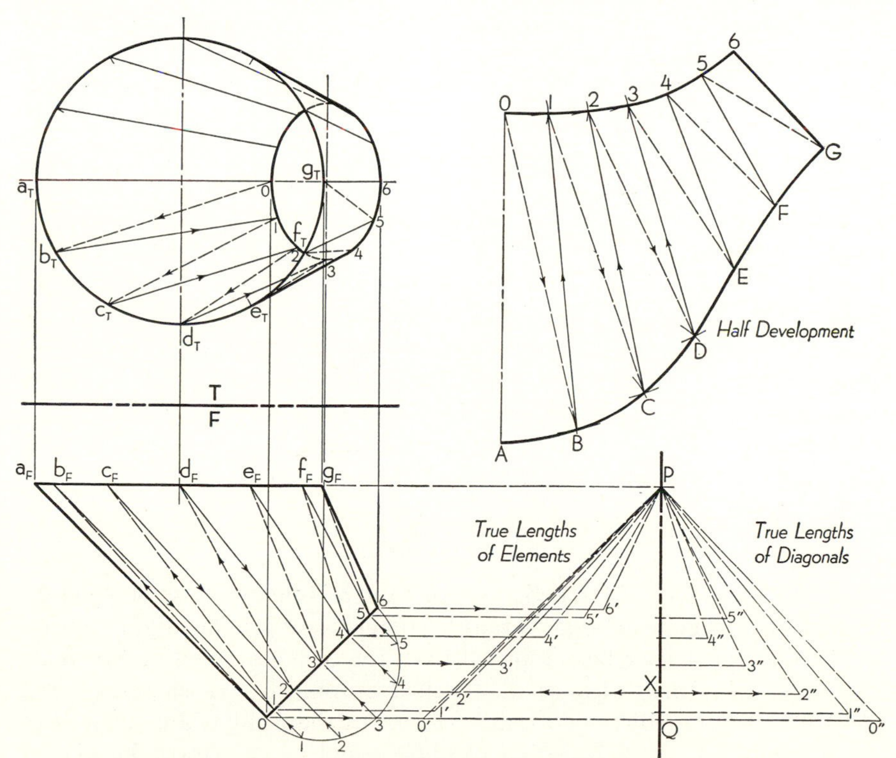

Fig. 10·20. Approximate development of a warped transition piece.

the top-view length d_T2 is laid off at $X2''$ to the *right* of axis PQ, again directly opposite point 2 in the front view. Construction of the development now proceeds, one triangle after another, in the same manner described in Art. 10·20. Note that distances 0–1, 1–2, etc., in the development must be taken from the construction semicircle where these distances appear true length. Distances AB, BC, etc., are taken directly from the top view of the large circle.

Summary. Many of the warped surfaces described in Chap. 7 may be approximately developed by the method of triangulation if the warped surface does not differ radically from a similar single-curved surface. Where exact development is required, it is preferable to design the surface as a convolute or as a combination of plane and single-curved surfaces. The method of triangulation is also useful in developing a frustum of a cone when the vertex of the cone is not accessible.

PROBLEMS. Group 118.

10·22. Development of the helical convolute

Analysis. The helical convolutes shown in Figs. 6·29 and 6·30 are developable surfaces. In Fig. 6·29 all elements are of equal length; and when this surface is flattened out into one plane, its development becomes a circular ring as shown in Fig. 10·21(*a*). The helical convolute of Fig. 6·30 has elements of constantly increasing length, and its development is that shown in Fig. 10·21(*c*). In both cases the helical directrix becomes a circular arc of radius R in the development, because the helix has a constant curvature R and a constant slope. The length of the circular arc—from A to C in Fig. 10·21(*a*), from A to B in Fig. 10·21(*c*)—is equal to the length of the helix. The straight-line elements that are tangent to the helical directrix of the convolute are also tangent to the circular arc in the development. Having observed these properties of helical convolute developments, we may now proceed to the details of construction.

Radius R. The helix radius of curvature R may be determined mathematically or graphically. Mathematically,

$$R = \frac{r}{\cos^2 \alpha}$$

where r is the radius of the helix cylinder and α is the constant slope angle of the helix. The helix angle α may be determined graphically, as shown in Fig. 6·28, or it may be computed from the formula given in Art. 6·27. The value of R may also be determined graphically, as shown in Fig. 10·21(*b*). In any convenient place construct the right triangle xyz with side xy equal to r and the adjacent angle equal to α. Draw the line zw perpendicular to the hypotenuse xz to intercept xy extended at w. Then distance xw equals R. Begin the development (in either case) by drawing a circle of radius R.

Helix length. The arc length AC in Fig. 10·21(a) [or AB in Fig. 10·21(c)] may be obtained graphically by making its length equal to the true length of the helix, where the helix length is the hypotenuse of a right triangle constructed as shown in Fig. 6·28. The same result can be obtained more accurately, however, by computing the angle θ, which subtends the arc, for

$$\theta = 360 \cos \alpha \qquad \text{(for one turn of the helix)}$$

where θ is in degrees and α is the helix angle. Thus the second step in the development (in either case) is to establish the arc length of the developed helix.

Development. In the helical convolute of Fig. 6·29, element AB appears true length at a_Fb_F; hence this true length may be laid off in the development at AB tangent to the inner circle. The last element CD, of the same true length, is drawn tangent to the circle at C. Then since all elements are of equal length, the outer circle of the development is drawn with radius OB from B to D. If desired, the inner circle may be divided equally and additional tangent elements drawn as shown in Fig. 10·21(a), but this is not essential.

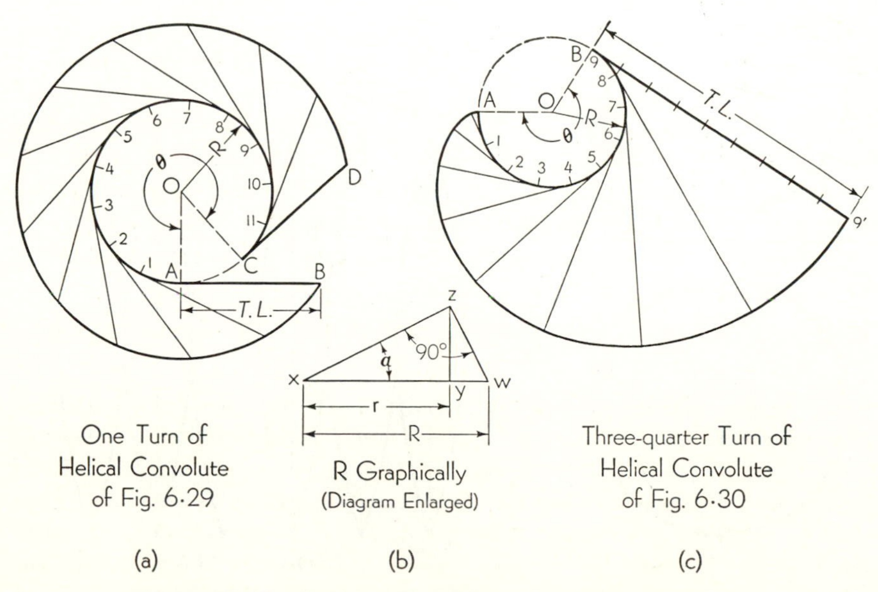

Fig. 10·21. Developments of helical convolutes (reduced).

In the helical convolute of Fig. 6·30 the longest element $B9'$ appears true length in the front view at b_F9'; hence this true length may be laid off in the development at $B9'$ tangent to the circle at B. But the true length of element $B9'$ equals the helix length from A to B (Art. 6·29), and therefore in the development arc AB equals line $B9'$. If arc AB and line $B9'$ are each divided into nine equal parts as shown in Fig. 10·21(c), then elements may be drawn tangent to the circle at each division point and the length of each element must be equal to the corresponding arc length. In other words, the length of the element at 1 equals arc $A1$, or one-ninth of $B9'$; the element at 3 equals arc $A3$, or one-third of $B9'$, etc. Curve $A9'$ is thus the involute of arc AB.

PROBLEMS. Group 119.

10·23. Approximate development of a sphere

Analysis. Double-curved surfaces are theoretically undevelopable. They may, however, be approximated by dividing the surface into a number of narrow cylindrical or conical segments. If a sphere is divided by meridian planes into a number of equal sections as shown in Fig. 10·22, then each section, or *gore*, may be developed as a portion of a cylinder.

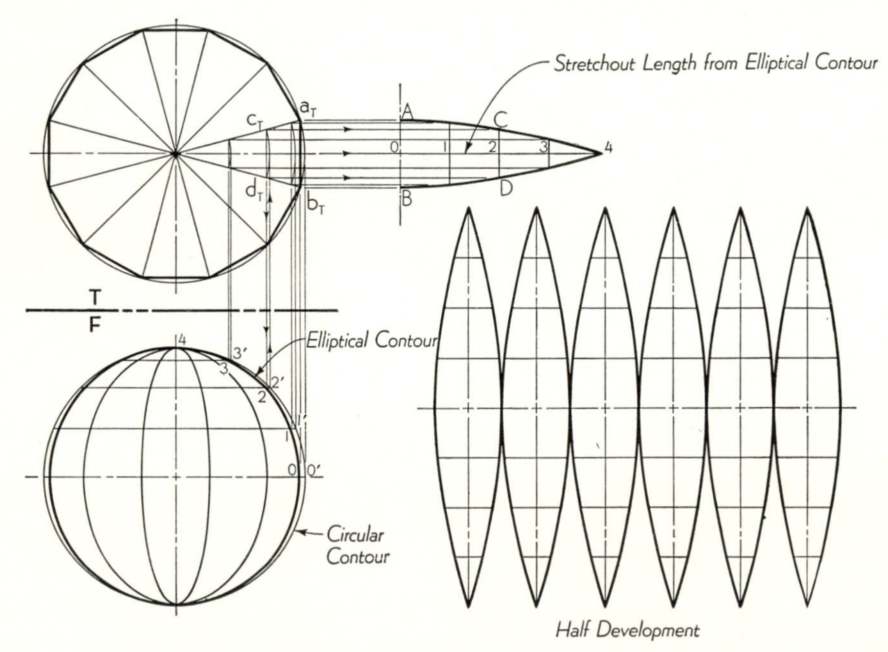

Fig. 10·22. Approximate development of a sphere: gore method.

If the spherical surface is divided by a series of parallel planes as shown in Fig. 10·23, then each *zone* may be developed as a frustum of a cone. In both cases the approximation improves as the number of gores or zones is increased.

Gore method. In Fig. 10·22 the sphere has been divided by vertical meridian planes into 12 equal gores, and each arc, such as a_Tb_T, is replaced by its chord. The circular contour of the sphere is thus replaced by an inscribed polygon in the top view and by an ellipse in the front view. By intersecting the spherical surface with a series of horizontal planes, points 1′, 2′, 3′ may be equally spaced (if desired) on the circular contour. Points 1, 2, 3, which represent elements on the cylindrical surface, are then located as shown to establish the elliptical contour of one gore. (The other elliptical curves in the front view are not necessary to the development.) Half of one gore may now be developed as shown at *A–B–*4. The center line of the gore becomes the stretchout line 0–4,

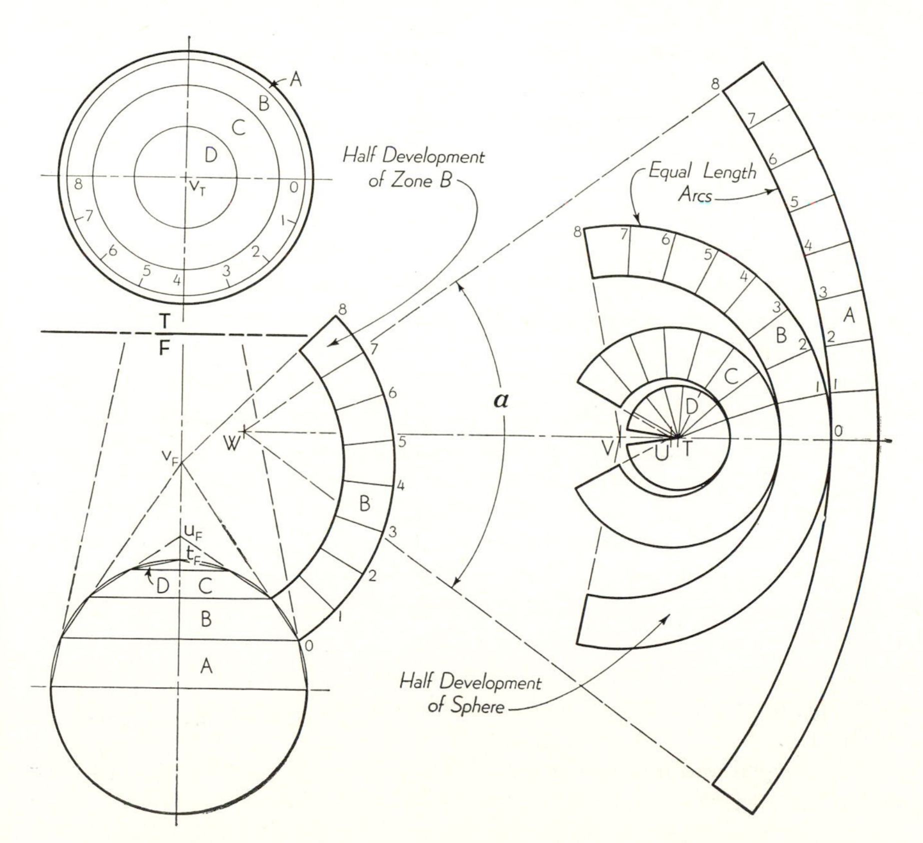

Fig. 10·23. Approximate development of a sphere: zone method.

whose length and divisions are taken from the elliptical contour. Cylinder elements, such as CD, are extended directly from the top view at c_Td_T. Having developed one symmetrical half of one gore (one twenty-fourth of the whole surface) duplicate gores may be arranged as shown in the half development.

Zone method. In Fig. 10·23 the upper half of the sphere has been divided by parallel planes into four zones A, B, C, and D of equal width. Replacing arcs by chords, each zone may be approximated as a cone frustum. Zone B, for example, becomes a frustum of the right circular cone whose vertex is at v_F and whose base is the equally divided circle in the top view. Half of zone B has been developed adjacent to the front view by using v_F0 as a radius for the outer arc and setting off on this arc distances 0–1, 1–2, etc., taken from the top view (method of Art. 10·12). All zones can be developed in the same manner, but they can be more compactly arranged as shown at the right in Fig. 10·23. Note that the outer arc 0–8 and its divisions on zone B are exactly equal in length to the adjacent inner arc on zone A. Note also that angle α can be calculated for each zone by means of the formula given in Fig. 10·11.

PROBLEMS. Group 120.

10·24. Geodetic lines

The shortest line on any surface connecting two points is called a *geodetic line* (or a geodesic). A cord stretched tautly between two points on a surface and contacting the surface along its whole length would follow the geodetic line. For plane or single-curved surfaces (the only exactly developable surfaces) geodetic lines become straight lines on the development. On such surfaces the shortest line between two given points on the surface may therefore be established as follows: (1) Develop the surface, and locate the two given points on the development. (2) Draw a straight line connecting the two points. (3) Note the various points where the geodesic intersects intermediate lines or elements on the development. (4) Locate these points on the corresponding lines or elements in the given views by following the reverse process of development. (5) Draw a smooth curve through the points. In some cases it may be necessary to rearrange the developed area, taking the seam on a different element, or edge, in order to decide which of two (or more) straight lines is shorter.

For warped and double-curved surfaces the above method is not applicable since these surfaces are not truly developable. For a sphere, however, the geodesic is an arc of the great circle passing through the two points (see Arts. 8·3 and 8·5).

10·25. Sheet-metal seams and joints

Figure 10·24 shows a few of the more common methods of joining sheet metal. Sheets thinner than 20 gauge (0.0375 in. thick) are usually seamed by folding together the edges of the joint, as shown in Fig. 10·24(*a*) to (*f*). Heavier sheets may be connected by solder or rivets, as shown in (*g*) to (*k*), and still heavier material by welding, as shown in (*m*) to (*p*).

Exposed edges of a duct are stiffened, and the dangerously sharp edges removed, by *single* or *double hemming* as shown at (*a*). The *wired edge* at (*b*), formed by wrapping the metal around a wire, serves the same purpose as hemming but requires an extra metal allowance on the pattern of about 2½ times the diameter of the wire. The *flat lock seam* shown at (*c*) is most commonly used for straight longitudinal seams. The *double lock seam* at (*d*) serves the same purpose but is stronger. The *double seam* shown at (*e*) is the standard method of joining the metal at corners. When the connecting edges have been bent and slipped together as shown

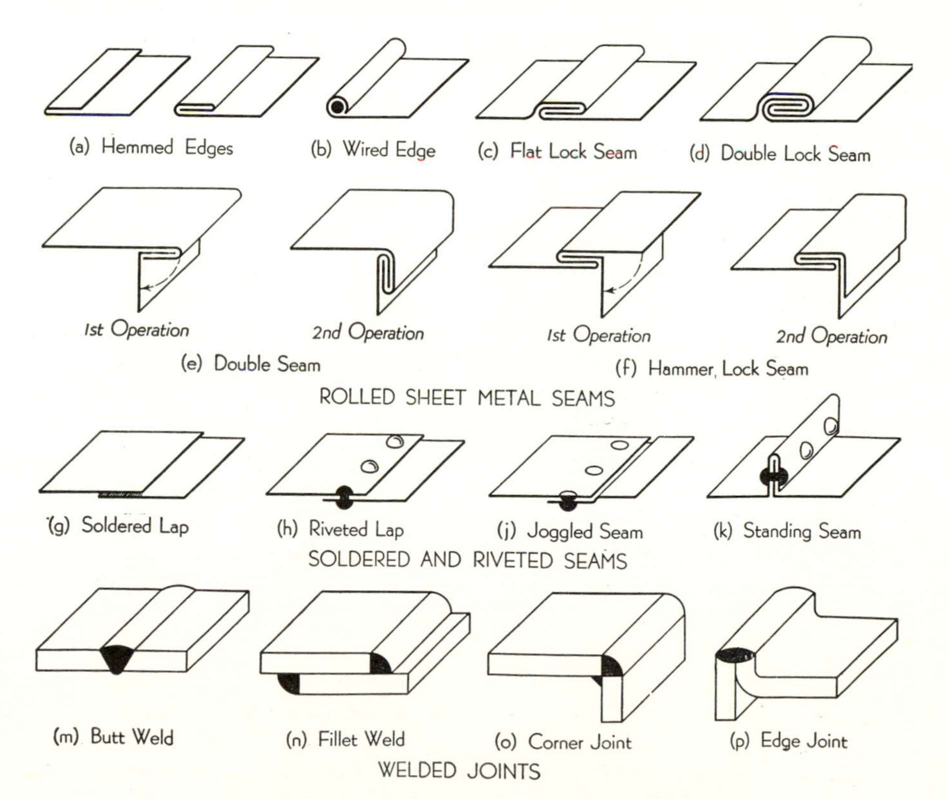

Fig. 10·24. Typical sheet-metal seams and joints.

in the first operation, the protruding edge is then rolled over to complete the seam as shown in the second operation. The *hammer lock seam* at (*f*) serves the same purpose as the double seam but is easier to form since it can be made up on the job by using a hammer.

Lap joints may be soldered, riveted, or spot-welded. In order to present a smoothly continuous outer surface, one of the edges may be *joggled* as shown at (*j*), and the rivetheads may be countersunk. *Standing seams* like that shown at (*k*) add considerable stiffness and are readily accessible in fabrication.

The four *welded joints* shown are typical of those in common use, but no attempt has been made here to show details of grooving and joining the edges. The darkened area on each weld indicates the metal added by the welder.

10·26. Bend allowance

Sheet metal of 24 gauge (0.0250 in. thick) or thinner can be bent to form a corner of such small radius that no allowance for the bend is usually necessary. Sheets of heavier gauge cannot be bent so sharply; hence provision must be made for a definite inside radius of bend ($1\frac{1}{2}$ to $2\frac{1}{2}$ times the metal thickness usually). The length of material allowed on the development for such bends is called *bend allowance.*

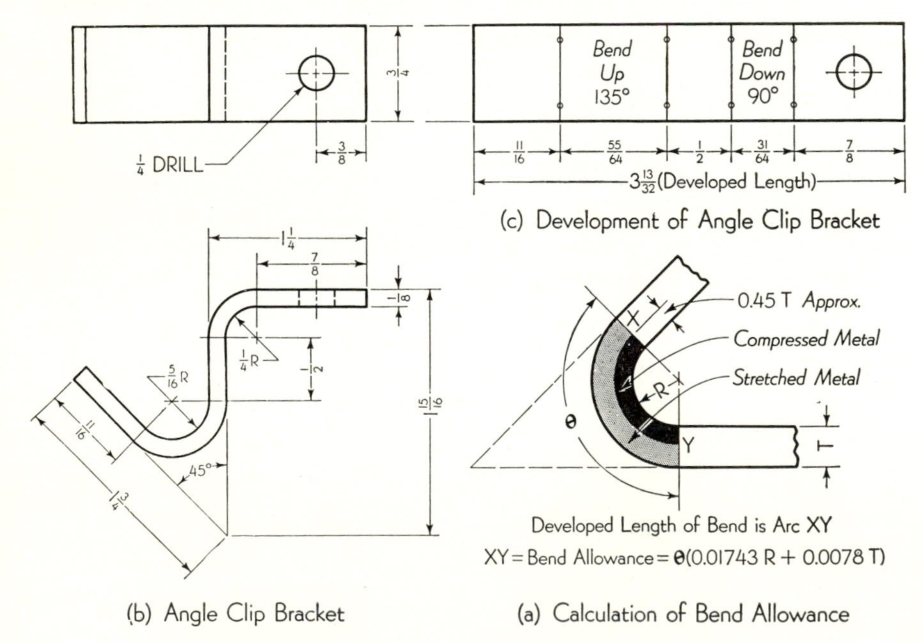

(c) Development of Angle Clip Bracket

(b) Angle Clip Bracket

(a) Calculation of Bend Allowance

Fig. 10·25. Application of bend allowance on a development.

Calculation of bend allowance. Figure 10·25(*a*) shows a metal plate of thickness T bent through an angle θ with an inside radius R. On the inside of the bend the metal is compressed, and on the outside it is stretched, but along the arc XY there is no change of length. Experiment has shown that arc XY occurs almost at the center of the material —at $0.45T$ from the inside approximately. Then the developed (or original) length of the bend is the length of arc XY, and

$$\text{Arc length } XY = \frac{\pi}{180}\theta(R + 0.45T) = \text{bend allowance}$$

where θ is in degrees and R and T in inches. Combining constants, this equation reduces to

$$\text{Bend allowance (B.A.)} = \theta(0.01743R + 0.0078T)$$

This formula may be applied to sheet steel, copper, aluminum, and most ductile metals. For bends of large radius it is usually accurate enough to take the length of the development equal to the center-line length of the metal. Thus for a circular cylinder the length of the development may be taken as the inside circumference plus 3.1416 times the thickness of the metal or as the outside circumference decreased by the same amount.

Problem. Develop the angle clip bracket shown in Fig. 10·25(*b*).

Construction. The lengths of the straight sections, $\frac{7}{8}$, $\frac{1}{2}$, and $\frac{11}{16}$ in., are dimensioned on the front view. (These distances may be measured or computed according to the desired degree of accuracy.) The bend allowance for the right-angle bend will be

$$\begin{aligned}\text{B.A.} &= 90[(0.01743)(0.25) + (0.0078)(0.125)]\\ &= 0.4806, \text{ or } \tfrac{31}{64} \text{ in.}\end{aligned}$$

The other bend allowance is similarly calculated and found to be $\frac{55}{64}$ in. Note that θ must be taken as 135° (180–45) in this second calculation.

The development may now be laid out as shown in Fig. 10·25(*c*), using the measured and computed lengths as shown.

PROBLEMS. Group 121.

11. VECTOR APPLICATIONS

11·1. Vector representation

Physical quantities may be classified as either scalar or vector. A *scalar* quantity is one having magnitude only, such as time, temperature, volume, mass, etc. A scalar quantity may therefore be described by a single real number, and can be represented graphically, as on charts and diagrams, by the magnitude of a line, area, or volume. The concept of direction is not involved in scalar representation.

A *vector* quantity is one having magnitude and direction, such as displacement, velocity, acceleration, momentum, force, torque, etc. The *speed* of an airplane, for example, is a scalar quantity that describes only the magnitude of its velocity; the *velocity* of the airplane is a vector quantity that describes not only its speed but also the direction of motion in space. A vector quantity can be represented graphically by a directed segment of a straight line—an arrow, or vector. The length of the vector must be proportional to the *magnitude* of the vector quantity, but the vector direction may be either parallel to, or coincident with, the *line of action* of the vector quantity. The *sense*, or direction, of the vector is indicated by an arrowhead.

When a vector is shown at the point of action, its direction coincident with the line of action of the vector quantity, it is called a *localized* vector. If the vector is moved to some convenient parallel position, it is called a *free* vector. Vector operations are independent of position; hence free vectors may be employed for the purpose of addition, subtraction, etc. In some applications, however, notably in force analysis, vectors must be localized on the line of action to completely determine the solution.

A system of vectors may be *concurrent*—all acting through a common point—*parallel*, or *nonconcurrent*—neither parallel nor concurrent. The system of vectors may also be either *coplanar*—all acting in the same or parallel planes—or *noncoplanar*—acting in nonparallel planes in space.

Since vectors are simply segments of straight lines, the principles of descriptive geometry are directly applicable to the graphical solution of all problems involving vector quantities.

11·2. Addition and subtraction of coplanar vectors

Two or more vectors can be added geometrically to obtain a single vector having an equivalent effect. This equivalent vector is called the *resultant*, and if the given vectors are coplanar, it lies in the same plane as the given vectors. For the simple purpose of addition or subtraction, vectors may be treated as *free* vectors and translated to *parallel* positions. The symbols ⇸ and → indicate vector addition and subtraction, respectively.

Two coplanar vectors can be added graphically by either of the following very similar constructions.

Parallelogram law. In Fig. 11·1 the magnitude and direction of vectors A and B are given as shown. Then at any point O parallel vectors are drawn, and the length of each vector is marked off to any convenient scale. A parallelogram, having A and B as sides, is completed, and the diagonal A ⇸ B is the resultant of vectors A and B. The magnitude of the resultant is measured to the same scale used for the given vectors, and its direction is as shown.

Note that the three vectors must be arranged either "tail to tail" or "head to head." Failure to observe this principle may result in choosing the wrong diagonal of the parallelogram.

Triangle law. In Fig. 11·2 the same vectors A and B are given. Beginning at any point O vector A is drawn parallel to the given vector

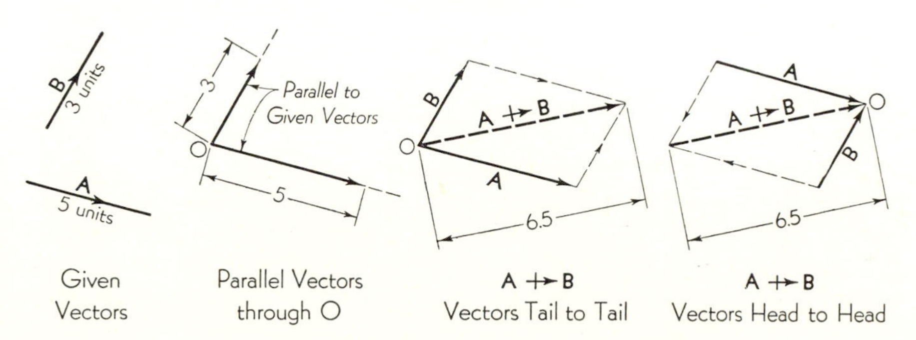

Fig. 11·1. Parallelogram law for vector addition.

A, and to scale length. From the head of vector A, vector B is drawn parallel to given vector B. Then the closing side of the triangle $A \nrightarrow B$ is the resultant of vectors A and B. If vector B is drawn first and vector A second, the resultant $B \nrightarrow A$ is the same as resultant $A \nrightarrow B$; thus the order in which the given vectors are drawn does not alter the resultant.

Note that the given vectors must be arranged "tail to head," but the direction of the resultant vector is from the beginning of the first vector to the end of the last.

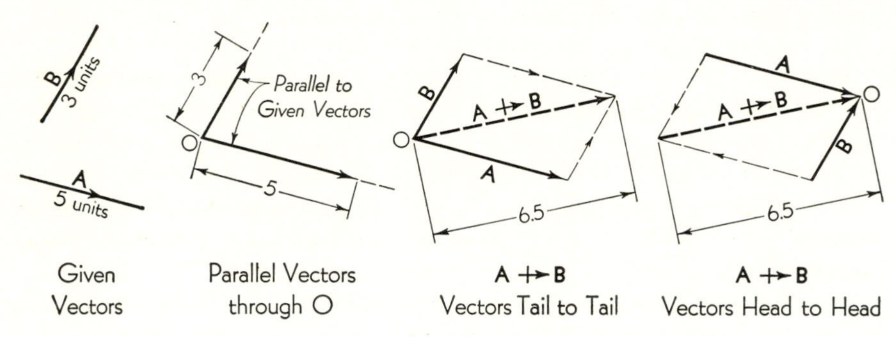

Fig. 11·1 (repeated). Parallelogram law for vector addition.

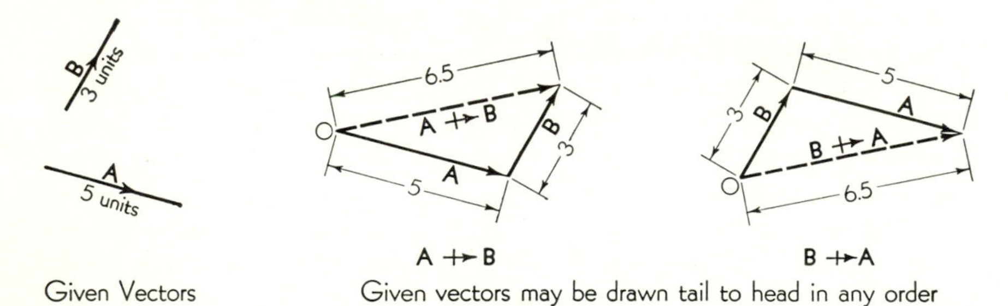

Fig. 11·2. Triangle law for vector addition.

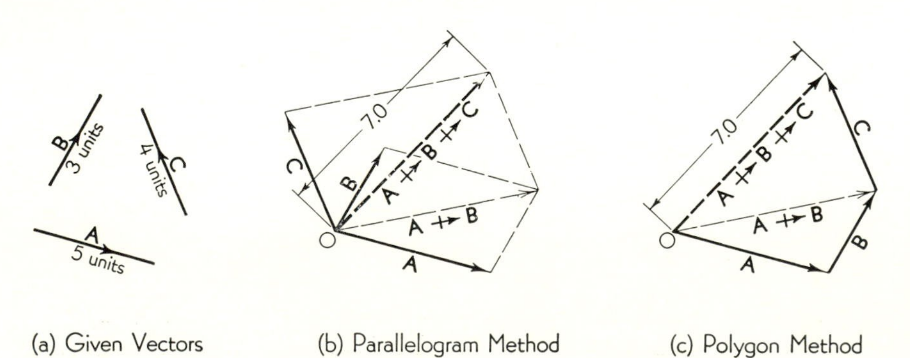

Fig. 11·3. Addition of three or more vectors.

Addition of three or more vectors. The parallelogram law or the triangle law can readily be extended to the addition of three or more vectors. In Fig. 11·3(a) the given coplanar vectors are A, B, and C. At any point O in Fig. 11·3(b) vectors A and B can be added by the parallelogram law to obtain the resultant $A \nrightarrow B$. This resultant can then be added to vector C in a second parallelogram to obtain the final resultant $A \nrightarrow B \nrightarrow C$. Note again that all vectors have their tails at the common point O. By successively applying the parallelogram law, any number of vectors can thus be reduced to a single resultant. It is immaterial in what order the vectors are combined.

Vectors A, B, and C can also be added by successively applying the triangle law to produce a polygon of vectors, as shown in Fig. 11·3(c). Here the given vectors are drawn consecutively (in any desired order), tail to head, until the chain of vectors is complete. The closing side of the polygon, drawn from the beginning of the first vector to the end of the last, is the resultant. Note that the polygon can be completed without drawing any of the intermediate resultants, such as $A \nrightarrow B$.

The polygon method obviously requires less construction than the parallelogram method, and is therefore preferred, especially when a large number of vectors must be added.

Subtraction of vectors. If a given vector B is reversed in direction, then the reversed vector is the negative of the given vector, or is equivalent to $-B$. One vector can therefore be subtracted from another simply by adding its negative. Vectorially, $A \rightarrow B$ is the same as $A \nrightarrow -B$.

In Fig. 11·4 vector B is to be subtracted from vector A. Vector B is first reversed in direction and then added to vector A. The addition can be made by either the parallelogram or triangle method, as shown. The resultant vector represents the difference $A \rightarrow B$. If vector A is subtracted from vector B, the resultant difference, $B \rightarrow A$, will be the negative of vector $A \rightarrow B$; in short, vectors $B \rightarrow A$ and $A \rightarrow B$ differ only in sense.

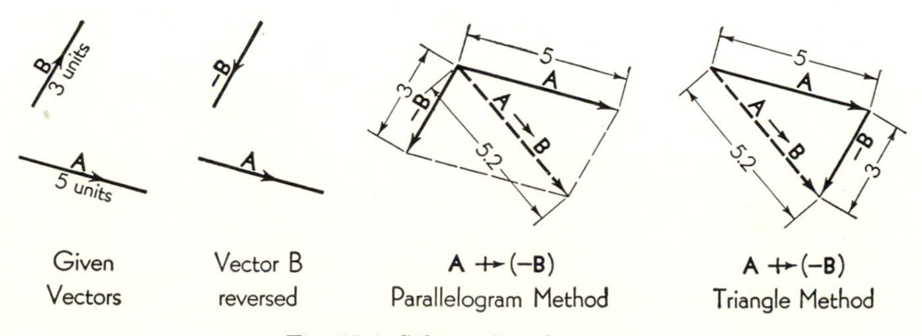

Fig. 11·4. Subtraction of vectors.

11·3. Resolution of a coplanar vector

It has been shown that two or more vectors can be added to form a single resultant vector. By reversing the process a single vector can be resolved into two or more *component* vectors. If the components are coplanar, then only *two* can be unknown in magnitude or direction. The most common case occurs when a vector must be resolved into two components having specified directions.

In Fig. 11·5(a) the given vector A must be resolved into two components having the directions OM and ON. The magnitude of each component is required. In Fig. 11.5(b) the directions OM and ON are extended to form the adjacent sides of a parallelogram whose diagonal is vector A. From the tip of vector A the parallel opposite sides are drawn to complete the parallelogram. Then A_M and A_N are the required components of vector A. The sense of components A_M and A_N is determined by noting that the tail of each vector is at the common point O.

Figure 11·5(c) shows the same problem solved by the triangle method. Through the tail of vector A a line is drawn parallel to direction OM, and through the head a second line is drawn parallel to direction ON. These intersecting lines complete the triangle and establish the components A_M and A_N. The sense of the components is determined by noting that they must be tail to head beginning at the tail of the given vector.

11·4. Velocity and relative motion

All motion is relative because the movement of a point or body can be described only with respect to some reference point or body, which may itself be fixed or in motion. Thus a seated passenger on a moving train

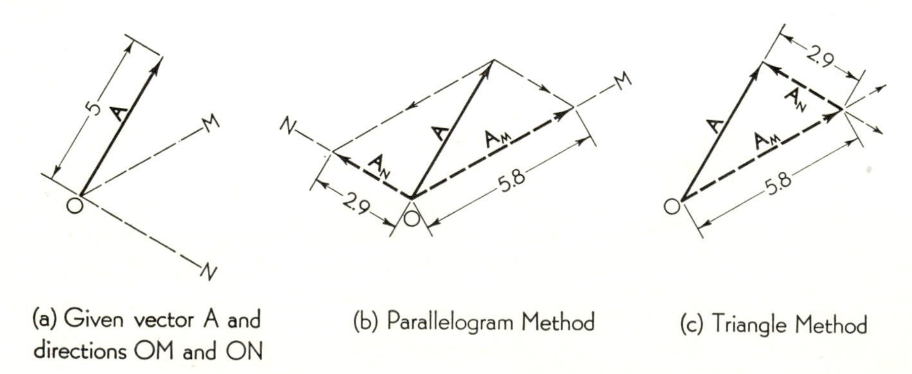

Fig. 11·5. Vector resolved into two coplanar components.

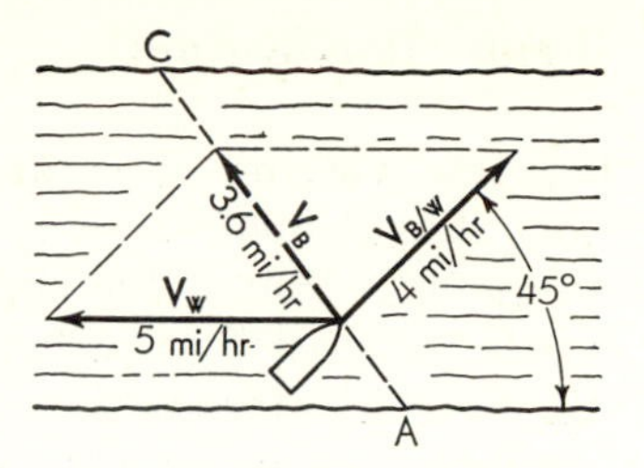

Fig. 11·6. Relative velocity.

has zero velocity with respect to the train but has the same velocity as the train with respect to the earth. For terrestrial problems we may assume the earth to be fixed, and the velocity of any body A with respect to a point on the earth would then be an *absolute* velocity V_A. The velocity of a point B on body A with respect to some point on body A would be the *relative* velocity of B with respect to A, or $V_{B/A}$. Then the absolute velocity of point B is the vector sum of V_A and $V_{B/A}$.

Expressed as an equation,

$$V_B = V_A \mathrel{+\!\!\!\!\rightarrow} V_{B/A} \qquad \text{or} \qquad V_{B/A} = V_B \mathrel{-\!\!\!\rightarrow} V_A$$

Example 1. Figure 11·6 shows a simple example of absolute and relative velocities. The current in a river is 5 mph, and a motor boat leaves the bank at point A headed upstream at 45° with the shore. The speed of the motor boat with respect to the water is 4 mph. Then the velocity of the water is an absolute velocity V_W, and the velocity of the boat is a relative velocity $V_{B/W}$. Adding these two velocity vectors by the parallelogram method gives V_B, the absolute velocity of the boat. The actual path of the boat is therefore from A to C at a speed of 3.6 mph.

Example 2. Figure 11·7 shows a second example of relative velocity between two moving bodies. At the moment of observation two ships, A and B, are located as shown in the figure. Ship A is moving at 11 knots in the direction AA'; ship B is moving at 14 knots in the direction BB'. The velocity of each ship is an absolute velocity and may be represented by the vectors V_A and V_B, as shown. Then the relative velocity of the two ships is the vector difference of the absolute velocities (see equation above). To subtract V_A from V_B vector V_A is reversed in direction and added to V_B (Art. 11·2). The resultant vector is $V_{B/A}$, the velocity of ship B with respect to ship A. Thus to an observer on ship A, ship B appears to be moving in the direction BC at a speed of 13 knots.

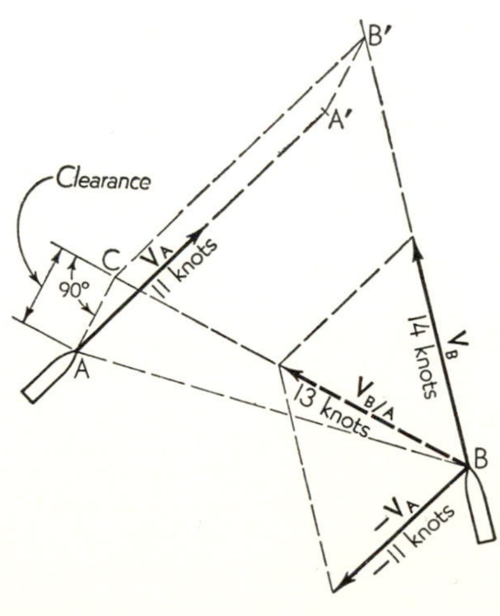

Fig. 11·7. Relative velocity and displacement.

The relative effect would be the same if ship A were motionless and ship B moved along line BC with an absolute velocity of 13 knots. Considered in this way it is apparent that ship B would pass in front of ship A with a clearance equal to distance AC. But if the ships continue at the given velocities, then the nearest approach will be at $A'B'$ ($A'B'$ is equal and parallel to AC). The time required for the ships to pass can be calculated by dividing the displacement (distance traveled) by the velocity in that direction: AA' divided by V_A, BB' divided by V_B, or BC divided by $V_{B/A}$. Note that distance and velocity scales are independent.

Example 3. Figure 11·8 shows a typical velocity determination in a simple mechanism. A rotating crank operates a slider through a connecting rod AB. Point A moves in a circular path while point B is constrained by the frame guides to move along the straight-line path. At the instant shown point A has a tangential velocity of 4 fps represented by the vector V_A. The velocity of point B is required.

The vector V_A can be resolved into two components: V_{AX} along the axis of the rod and V_{AY} perpendicular to the rod (Art. 11·3). Since points A and B at either end of the rod are always a fixed distance AB apart, then the velocity component along the X axis must be the same for each point: thus V_{BX} must equal V_{AX}. The velocity component V_{BY} must be perpendicular to the rod, and the parallelogram of vectors can now be completed at B to establish V_B as 3.2 fps.

The examples in this article are intended to serve only as illustrations of basic concepts of velocity and relative motion. More thorough and extensive treatment of this topic is usually reserved for courses in kinetics and kinematics.

PROBLEMS. Group 122.

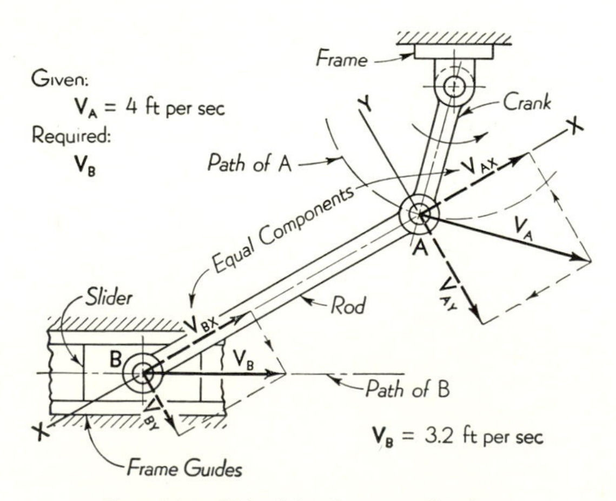

Fig. 11·8. Velocities in a mechanism.

11·5. Addition of space vectors

Vectors in space can be represented in a multiview drawing. The lengths, directions, and relationships of the vectors can then be determined by the standard methods of descriptive geometry.

Establishing vector length. To establish a given vector in space its direction must be shown in two views, and the true length of the vector must equal (to scale) the magnitude of the vector quantity. Figure 11·9 illustrates one quick method of establishing the length of a space vector.

Stage 1. *Establish vector direction in each view.*

Draw line OA, indefinite in length, parallel to the required vector direction in each view.

Stage 2. *Establish the true length of a segment of the given vector.*

Select any convenient point X on the line OA, thus establishing x_F and x_T. Construct the right triangle $o_T x_T x_A$ by transferring distance d from the front to the top view as shown. Note now that the right triangle is nothing more than a simplified version of a standard top-adjacent true-length view, and therefore $o_T x_A$ is the true length of vector segment OX.

Stage 3. *Establish the true length of the vector, and complete the views.*

Along the line $o_T x_A$ (extended if necessary) lay off to scale the true length, 5.7 units, of the given vector to establish a_A. Points a_T and a_F can now be located by alignment of the views as shown.

Addition of two space vectors. Figure 11·10 illustrates the method of adding two space vectors. The multiview *space diagram* on the left shows the location and direction of the given vectors A and B, but their

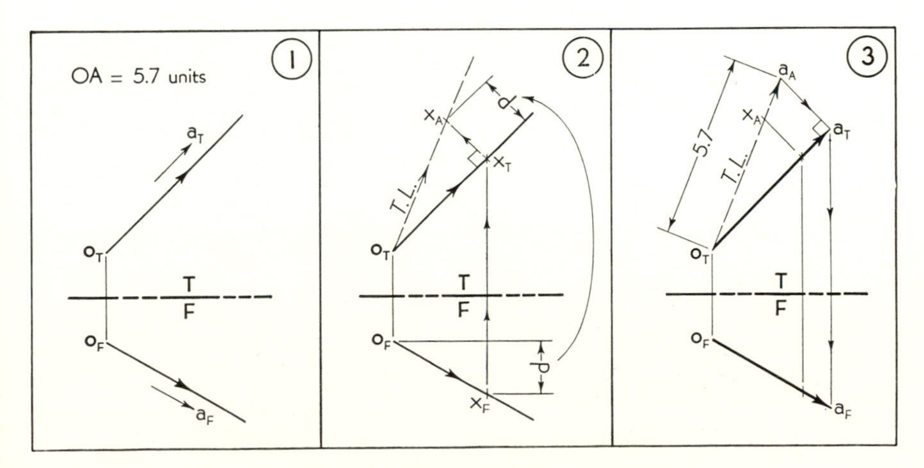

Fig. 11·9. Establishing the length of a space vector.

respective magnitudes of 6 and 5 units are not yet established in these views. Vectors A and B are not necessarily concurrent.

Construction of the *vector diagram* on the right begins at any convenient point O. Vector A, beginning at O, is drawn in the top and front views parallel to the directions shown in the adjacent space diagram. Since vector A is a frontal line, its true length of 6 units can be laid off in the front view, as shown, to locate the head of the vector at point A. Vector B begins at point A, and is drawn in each view parallel to the given space directions. Vector B is an oblique line; hence its true length of 5 units must be established by the method of Fig. 11·9, and point B is thus located as shown. Then line OB, the closing side of the space triangle, represents the top and front views of the resultant vector.

To obtain the true magnitude of vector $A \leftrightarrow B$, the true length of line OB must be found. This can be done most easily by constructing the right triangle $o_Tb_Tb_B$ (actually the equivalent of a top-adjacent view B). Note that distance e is the difference of elevation of points O and B, and the hypotenuse of the right triangle, 8.6 units, is the true length of the resultant.

11·6. Resolution of a space vector

If the components of a vector are noncoplanar, then *three* components can be unknown in magnitude or direction. A space vector can be

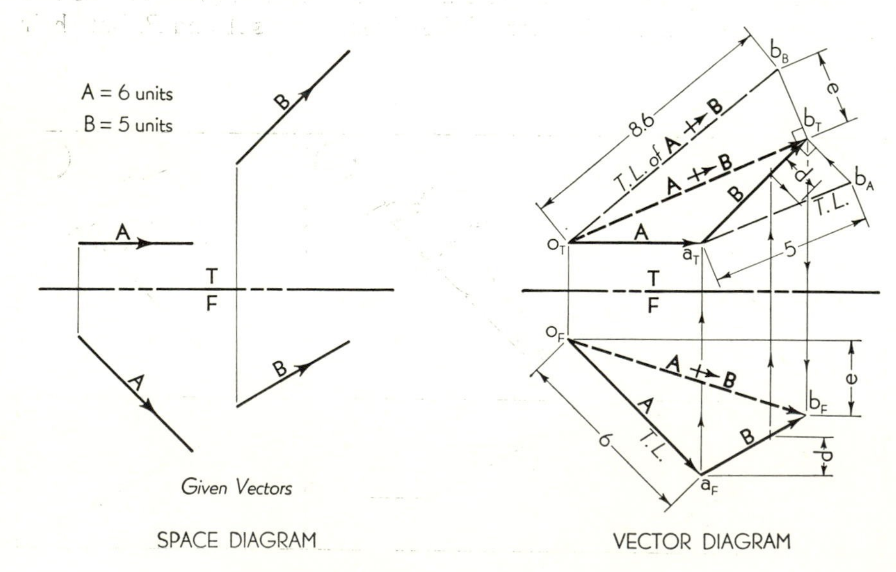

Fig. 11·10. Addition of two space vectors.

resolved into three noncoplanar components if the three component directions are specified. A common case is that of resolving a space vector into components along the axes of a rectangular coordinate system, as shown in Fig. 11·11.

In the multiview drawing of Fig. 11·11(*a*) the given vector A acts through the origin of the coordinate axes. Then a rectangular prism can be formed, as shown pictorially in Fig. 11·11(*b*), with vector A as its diagonal, and the three prism edges along the axes will represent the components A_X, A_Y, and A_Z. Figure 11·11(*c*) shows the same prism and components in the multiview drawing. Note again that all vectors have their tails at the common point O. Resolution of a space vector into rectangular components is a very useful operation because it permits replacement of a single oblique vector by three true-length vectors.

11·7. Force as a vector quantity

Force may be defined, briefly, as a *push* or a *pull*. A force acting on a body will: (1) change the velocity of the body if the force is unopposed, (2) induce opposing or reactive forces if the body is held at rest or constant velocity, (3) produce a deformation in the body, thus setting up internal stresses. We shall be concerned here only with the second effect: forces acting on a body at rest. This branch of mechanics is called *statics*.

A force has *three characteristics:* magnitude, direction, and point of application (or magnitude, line of action, and sense). Because it has magnitude and direction, force is a vector quantity, and because point of

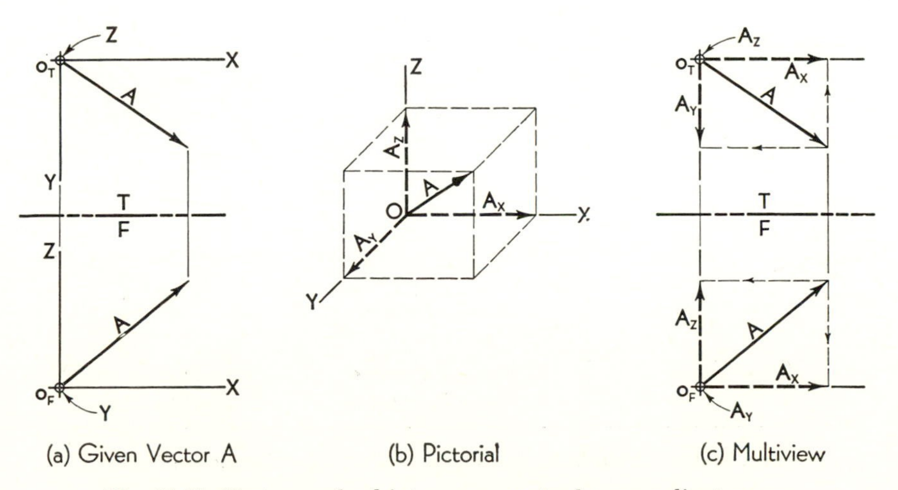

Fig. 11·11. Vector resolved into components along coordinate axes.

application is a characteristic, force vectors are *localized* vectors. For the operations of addition and subtraction, forces may be temporarily treated as free vectors, but any resultant force so obtained is not completely established until its point of application is also determined.

11·8. The principle of transmissibility of a force

The external effect of a force acting on a rigid body is not altered if the point of application is moved along the line of action of the force. This important fact is called the *principle of transmissibility* and is very useful in the solution of force problems. Figure 11·12 illustrates the application of this principle to a simple case of coplanar forces. An irregular body is acted upon by two forces F_1 and F_2 applied at points A and B,

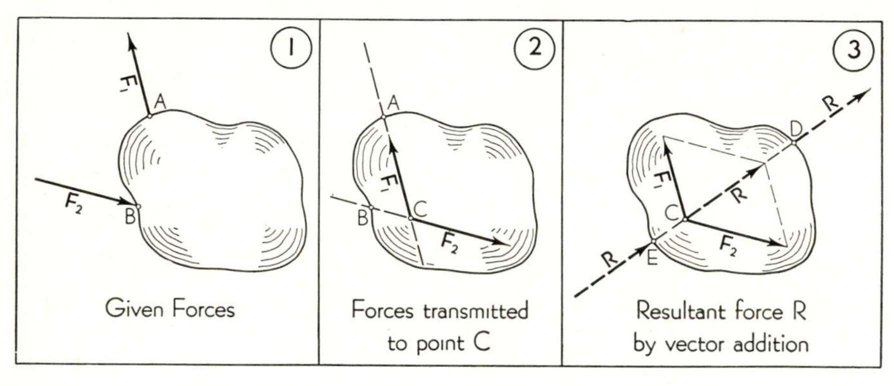

Fig. 11·12. The principle of transmissibility of forces.

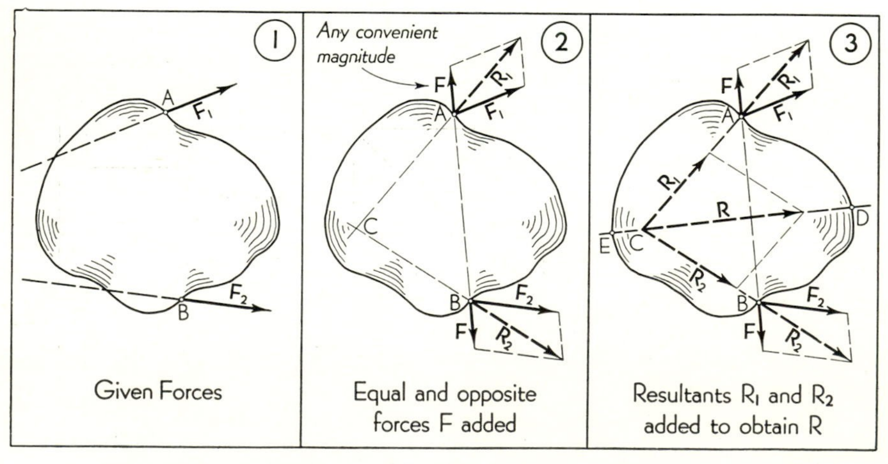

Fig. 11·13. Resultant by introduction of equal and opposite forces.

respectively, as shown in stage 1. By the principle of transmissibility these two forces can each be moved along its line of action until both act at the concurrency point C, as in stage 2. In stage 3 the force vectors F_1 and F_2 are added to obtain the resultant force R. This single force R can now replace forces F_1 and F_2, but it must act at some point, such as C, D, or E, along its line of action.

If the point of concurrency is inaccessible, the resultant force can be determined by the method shown in Fig. 11·13. The given coplanar forces F_1 and F_2, acting at points A and B as shown in stage 1, are concurrent at some inaccessible point to the left of the figure. If we apply two equal and opposite forces F at A and B acting along the line AB, as shown in stage 2, then the effect of the original forces is not altered because the two collinear forces F exactly neutralize, or cancel, each other. By vector addition forces F and F_1 can now be replaced by their resultant R_1, and F and F_2 by their resultant R_2. Forces R_1 and R_2 are concurrent at C, and can be transferred to that point, as shown in stage 3. Forces R_1 and R_2 are now added to obtain the final resultant R, which must act at some point along the line DE. Note that the added forces F can be of any desired magnitude; point C will always fall on the line DE.

11·9. Resultant of concurrent forces

Any system of concurrent forces can be replaced by a single resultant force acting through the point of concurrency. In the general case of noncoplanar forces, the force vectors must be represented and vectorially added in two adjacent views. Figure 11·14 illustrates the method of determining the resultant force for this general case.

The space diagram. The *space diagram* shows the direction of the three given forces acting through point O. For convenience here and in subsequent problems, Bow's notation will be employed to designate the forces. In this scheme of notation a capital letter is assigned to the space on each side of a vector, and each vector is then designated by the two letters on either side of it, reading in a clockwise direction. To avoid possible confusion in multiview drawings, only one view should be so labeled, as in the top view of Fig. 11·14; in adjacent views the sequence of vectors may be different, hence each vector should be labeled individually, as shown in the front view. The magnitudes of the given forces are listed on the drawing, but the vectors in the space diagram need not be drawn to scale.

The vector diagram. Following the order of notation, the *vector-diagram* construction begins at any convenient point A. Vector AB is drawn parallel to the directions shown in the adjacent space diagram, and its magnitude of 70 lb is laid off by the method of Fig. 11·9 to locate point B. Vector BC appears true length in the front view; hence its magnitude of 40 lb is laid off, as shown, to locate C. Vector CD, of magnitude 60 lb, is added to complete the chain of given vectors. Then AD, the closing side of the space polygon, is the resultant force, and its true magnitude of 120 lb is found by constructing the right triangle $a_T d_T d_A$ (compare the similar construction shown in Fig. 11·10).

To complete the solution the resultant R is drawn through point O in the space diagram parallel to the directions shown in the vector diagram. Note that the sense of R must be from A to D in order that it shall be equivalent to the vector chain $ABCD$. If the resultant R is reversed in sense, it then becomes the *equilibrant* of the given system, that is, a force that exactly opposes and holds in equilibrium the given forces.

PROBLEMS. Group 123.

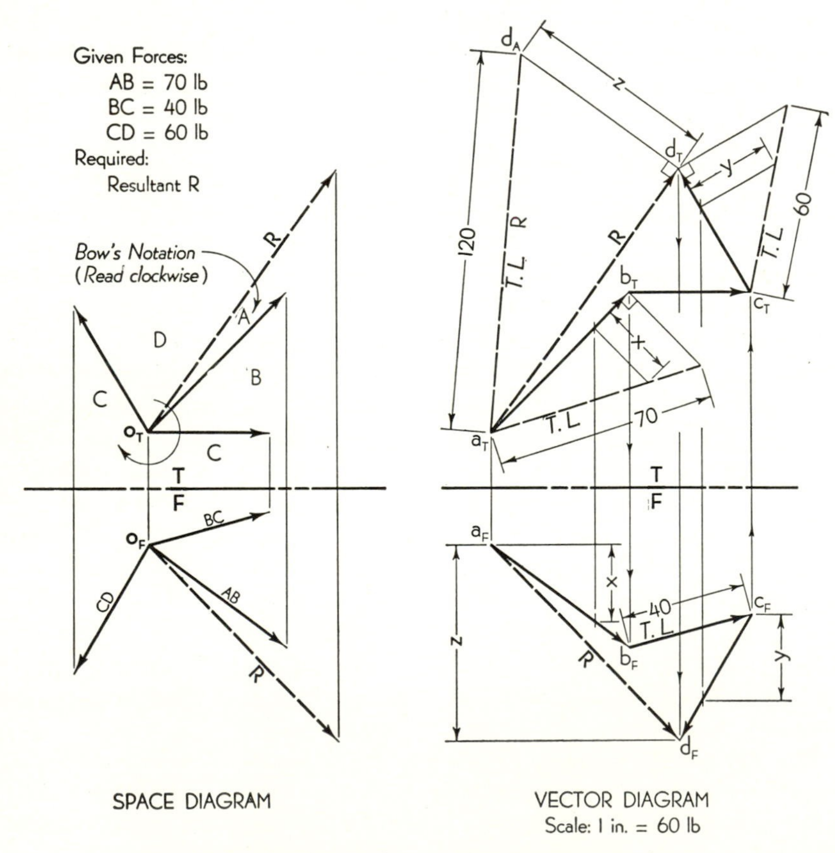

Fig. 11·14. Resultant of noncoplanar concurrent forces.

11·10. Resultant of two parallel forces

The resultant of two parallel forces may be either: (1) a single parallel force or (2) a couple. A *couple* consists of two equal and parallel forces, acting in opposite directions. The effect of a couple acting on a body is to produce rotation, and since a single force cannot cause rotation, a couple cannot be replaced by a single force. Couples and their vector representation are discussed in Art. 11·13.

The resultant of two parallel forces can be quickly obtained by the method shown in Fig. 11·15, which is simply a reapplication of the principle shown in Fig. 11·13. Three cases will be illustrated.

Case 1. In Fig. 11·15(*a*) the *unequal* parallel forces F_1 and F_2 act in the *same* direction. By introducing two equal and opposite forces F along the line AB the given forces can be replaced by resultants R_1 and R_2, which are concurrent at point C. The resultant R must also act through point C and must be parallel to F_1 and F_2, and its magnitude must equal $F_1 + F_2$. The magnitude of R can be obtained by arithmetical addition or by adding the scaled lengths of F_1 and F_2 graphically. Note that resultant R lies *between* the given forces in this case.

Case 2. In Fig. 11·15(*b*) the *unequal* parallel forces F_1 and F_2 act in *opposite* directions. Applying the same construction used in (*a*) above, the point of concurrency C is found to lie *beyond*, or to the left of, the *larger force* F_1. In this case the magnitude of R must equal $F_1 - F_2$ and must act through C in the direction of the larger force F_1, as shown.

Case 3. In Fig. 11·15(*c*) the *equal* parallel forces F_1 act in *opposite* directions and thus constitute a couple. When the previous construction is applied to this case as shown, it is found that the resultant forces R_1 are parallel and equal and also form a couple. The magnitude of a

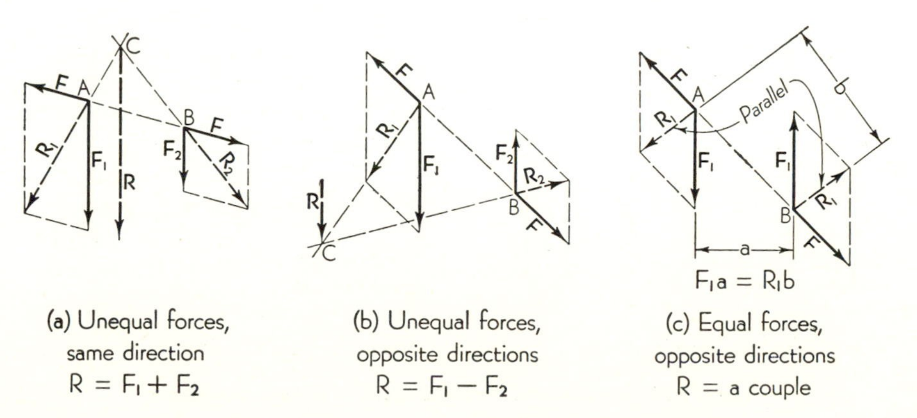

Fig. 11·15. Resultant of two parallel forces: three cases.

couple equals the force times the perpendicular distance between the forces, and it can now be observed that the resultant couple is equivalent to the given couple; that is, $F_1a = R_1b$.

11·11. Resultant of coplanar parallel forces

By the successive application of the method of Fig. 11·15 any number of parallel forces can be reduced to a single force or couple. However, when more than two forces are involved, it is more practical to use the method shown in Fig. 11·16. Since the resultant may be either a force or a couple, each case is illustrated.

Case 1. Resultant a force. In Fig. 11·16(*a*) the four given parallel forces are located to scale as shown in the space diagram. Using Bow's notation, the forces are labeled by lettering the spaces *A*, *B*, *C*, etc., on each side of each force, and the forces will be consistently read from left

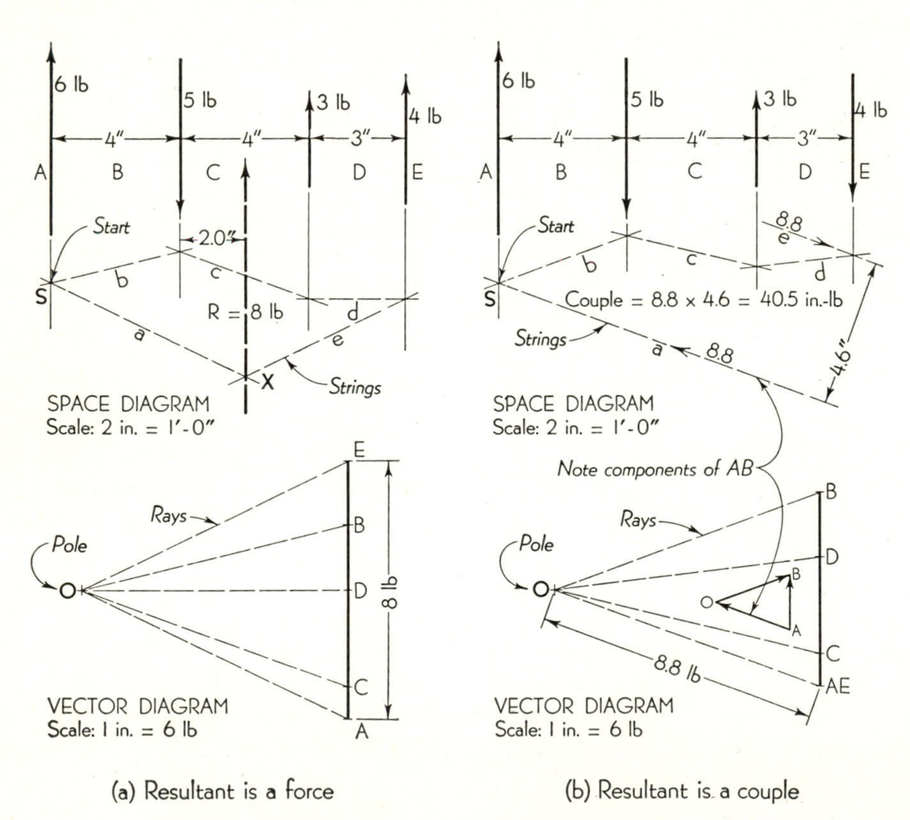

Fig. 11·16. Resultant of coplanar parallel forces: two cases.

to right. Then the vector diagram begins at any convenient point A, and the upward 6-lb force AB is laid off upward to scale to locate point B. From B the downward 5-lb force BC is laid off downward to locate C. Forces CD and DE are upward forces and hence are laid off consecutively upward to locate E. Then AE is the resultant force of 8 lb and acts upward. Note that arrowheads are omitted from these vectors because the order of the letters indicates the sense of the forces.

In order to locate the 8-lb resultant R in the space diagram select a *pole* O at any convenient point and draw the *rays* OA, OB, etc. The vector diagram now consists of a number of superimposed vector triangles, and each ray represents two equal and opposite components. Thus ray OD is one component of force CD, and DO is one component of force DE. Since vectors OD and DO are equal, opposite, and collinear, they cancel each other and cannot alter the original parallel-force system.

At any point S on the line of action of force AB start the *string polygon* by drawing *strings* a and b parallel to rays OA and OB. Extend string b to intersect force BC, and from this point draw string c parallel to ray OC. In similar manner draw string d and then e. Note that each string spans the correspondingly lettered space. Point X, where the first and last strings, a and e, intersect, is one point on the line of action of the resultant force R. Each string represents the line of action of two equal and opposite forces that cancel each other; hence the string polygon is simply a device for resolving each given force into two components, canceling all components except the first and last, and then combining these remaining components into the single resultant force.

Varying the position of pole O alters the shape of the string polygon, and moving point S changes its position, but point X will always fall on the line of action of the resultant force.

Case 2. Resultant a couple. In Fig. 11·16(b) the same force system is given except that the 4-lb force now acts downward. In the vector diagram point E now coincides with point A; hence the resultant force is zero. In the string polygon strings a and e are parallel and 4.6 in. apart. The resultant is therefore a couple whose forces act along the strings a and e. The magnitude of the couple forces is 8.8 lb, scaled from AO in the vector diagram as shown. The sense of force AO can be determined by considering one vector triangle of which AO is one component. Triangle AOB is sketched in miniature with arrowheads added to show that component AO must act from A to O as shown. The resultant couple therefore has a magnitude of 8.8 lb times 4.6 in., or 40.5 in.-lb, and acts in a clockwise direction.

If the strings a and e coincide, the resultant couple is zero, and the given forces are therefore in equilibrium.

11·12. Resultant of noncoplanar parallel forces

The string-and-ray-polygon method of the previous article can also be applied to noncoplanar parallel forces, but the construction must be performed twice—once in each of two adjacent views—in order to locate the resultant force in space. To illustrate the procedure, and also to show its application to a practical problem, the method will be used to determine the center of gravity of a solid. The gravitational forces acting on the various parts of a solid object actually constitute a parallel system of forces whose resultant is the sum of these forces and acts through the center of gravity of the whole object.

Problem. The homogeneous solid shown in Fig. 11·17 is composed of three simple geometric solids: part X, a 3-by-3-by-4 prism; part Y, a 9-by-8-by-2 prism; and part Z, a $4\frac{1}{2}$-by-6-by-2 triangular prism that has been removed (a negative volume). The gravitational force on each part is proportional to its volume (or weight) and acts through the center of gravity of the part. For existing solids the forces act downward; for a removed, or negative, solid the force acts upward.

Construction. In Fig. 11·17 the volume of each part has been calculated as shown and the total volume found to be 153 cu in. The cen-

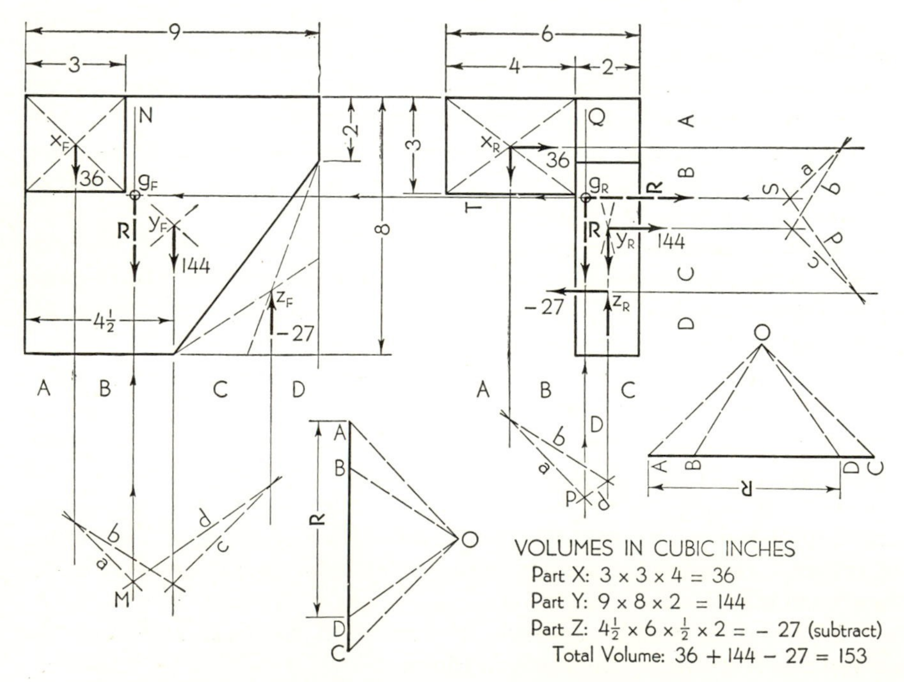

Fig. 11·17. Center of gravity (resultant of noncoplanar parallel forces).

ter of gravity of parts X and Y is at the geometric center of each prism; hence points X and Y can be located by drawing the diagonals of the prisms. The centroid of a triangle is located at the intersection of its medians (vertex to mid-point of side opposite), as shown at z_F.

In the front view the three gravitational forces—proportional to 36, 144, and -27, respectively—are shown acting vertically at points x_F, y_F, and z_F. Below this view a ray polygon is drawn to any convenient scale, and the corresponding string polygon is drawn on the extended action lines of the forces [as in Fig. 11·16(*a*)]. Strings *a* and *d* intersect at M; hence the resultant R must act along the line MN, and the center of gravity g_F is somewhere on this line.

In the side view the same three forces are again shown acting vertically through points x_R, y_R, and z_R. The ray polygon used for the front view can be used again here, but the string polygon is drawn on the extended action lines of the side view. Note that string *c* was not drawn because forces BC and CD have the same line of action in this view. Strings *a* and *d* intersect at P; hence the resultant R acts along line PQ, and g_R is somewhere on this line. Lines MN and PQ fix the line of action of R, but the height of point G is still unknown.

If the drawing is now rotated 90°, we can imagine that the side view is now the front view, and the front view now the top view. This is, of course, equivalent to actually rotating the object into a new position. In this position of the object the downward forces of gravity now appear to act to the right. The ray polygon is redrawn with its vectors parallel to the new direction, and the string polygon is drawn on the action lines that extend to the right. Strings *a* and *d* intersect at S; hence g_R lies on line ST. This fixes the position of g_R at the intersection of lines PQ and ST, and g_F can now be located by alignment. Point G is the center of gravity of the given solid.

The above method can also be used to locate the centroid of an area, in which case the ray and string polygons need only be drawn twice to fix the location of the centroid.

PROBLEMS. Group 124.

11·13. Vector representation of a couple

A *couple* consists of two equal and parallel forces acting in opposite directions. The perpendicular distance between the couple forces is called the *moment arm,* and the product of either force and the moment arm is the *moment* of the couple. The forces of a couple can be increased or decreased, rotated in the plane of the couple, or translated to a parallel plane without altering the moment, or turning effect, of the couple.

Figure 11·18(a) shows the effect of applying an 80-in.-lb couple to the pipe frame that is supported at the wall by two rivets 2 in. apart. If two 10-lb horizontal forces, 8 in. apart, are applied to the vertical pipe as shown, then this couple will induce a 40-lb force on each of the rivets. Two 20-lb forces, 4 in. apart, applied in the same plane will produce the same effect. The forces acting on the rivets would also be the same if two 8-lb forces, 10 in. apart, were applied to the horizontal pipe in a parallel plane. The four couples acting on the pipe frame are all equivalent because they have the same moment and sense of rotation and act in the same or parallel planes.

A couple thus has *three characteristics:* magnitude (or moment), sense of rotation, and direction of its plane of action. It may therefore be represented by a single vector perpendicular to the plane of the couple, as shown in Fig. 11·18(b). The length of the couple vector C is proportional to the moment Fa of the couple, and the vector points in the direction of advance of a right-hand screw that turns with the same sense of rotation as the couple. Note that a couple vector has no specific point of application, and may therefore be treated as a free vector.

11·14. Addition of couple vectors

A system of couples lying in the same or parallel planes (coplanar couples) can be combined into a single resultant couple whose moment is the algebraic sum of the individual moments. Noncoplanar couples can

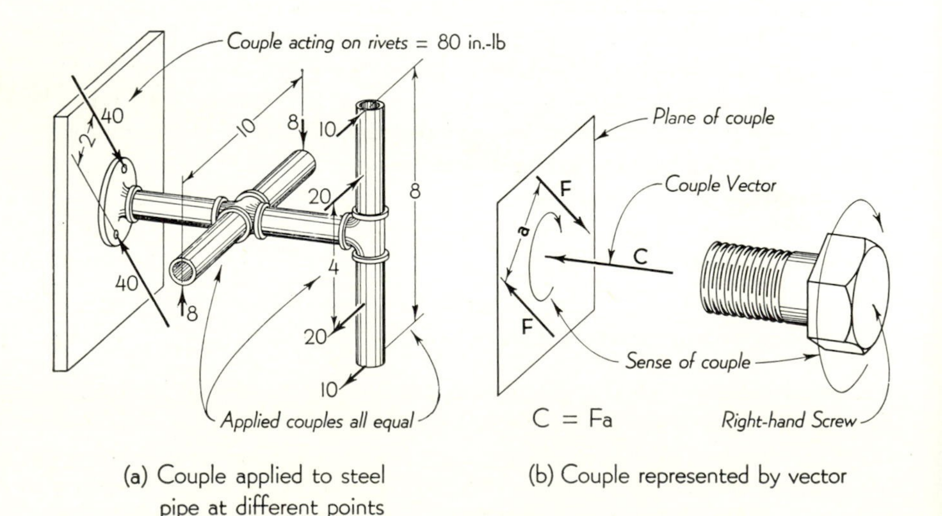

(a) Couple applied to steel pipe at different points

(b) Couple represented by vector

Fig. 11·18. Vector representation of a couple.

also be combined, but the resultant couple must be obtained by vector addition, as illustrated in Fig. 11·19. Here a wheel-operated gearbox is acted upon by three couples which must be combined into a single resultant couple. The views at (*a*) show a 500-in.-lb couple C_1 applied to the handwheel. Acting through a gear train, couple C_1 induces a 400-in.-lb couple C_2 in the horizontal shaft and a 600-in.-lb couple C_3 in the vertical shaft. The sense of rotation of each couple can be indicated by two parallel force vectors, as shown, or by a curled arrow. The combined effect of these three applied couples is to overturn the gearbox, and this is resisted by the four bolts in the base.

In Fig. 11·19(*b*) each of the three couples has been represented by a vector perpendicular to the plane of the couple. The true length of each vector must be proportional to the couple moment. The sense of the vectors is obtained by the right-hand screw rule: couple C_2, for example, is clockwise in the front view, and a right-hand screw turned in this direction would advance into the paper, or toward the rear of the gearbox; vector C_2 therefore appears as a point in the front view at (*b*) and points toward the rear in the top view.

In Fig. 11·19(*c*) the three vectors have been added vectorially to obtain the resultant couple vector C_R. The moment of C_R is 600 in.-lb as deter-

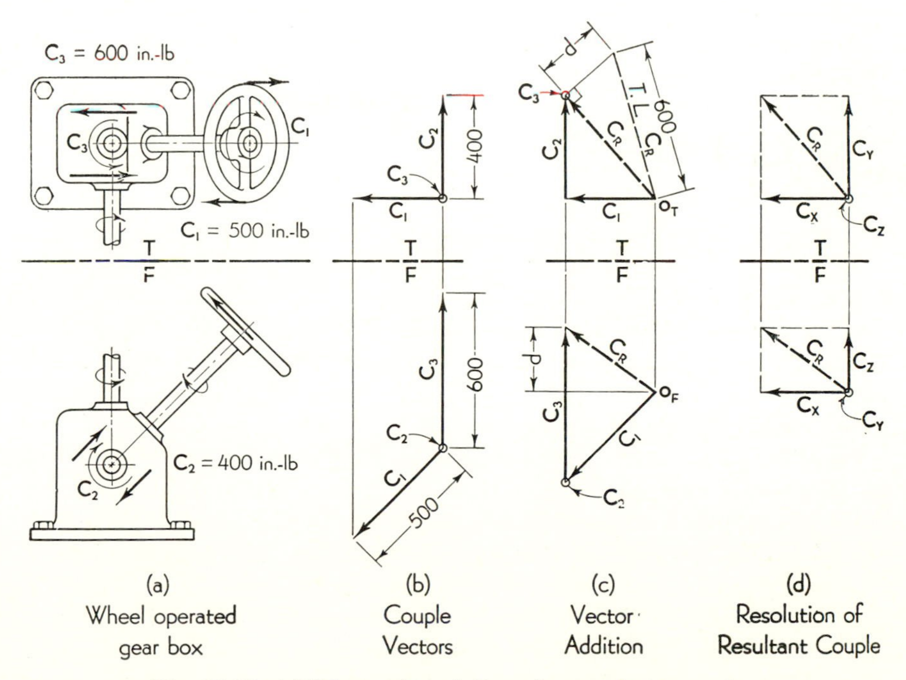

Fig. 11·19. Addition and resolution of noncoplanar couples.

mined from the true-length construction. The resultant of the three given couples is therefore a couple of 600-in.-lb moment acting in a plane perpendicular to vector C_R. Viewed in the direction of the vector, the sense of the couple is clockwise.

To better visualize the effect of couple C_R, it can be resolved into three component couples, as shown in Fig. 11·19(*d*) (compare Fig. 11·11). Then couple C_Z acts in a horizontal plane and tends to turn the gearbox counterclockwise about its vertical axis; couple C_Y acts in a frontal plane and tends to tip the gearbox to the right; couple C_X acts in a profile plane and tends to tip the box backward.

11·15. Resultant of noncoplanar nonconcurrent forces

This is the general case in which each given force may act at any point in space and in any direction. Such a system of forces can be reduced to *a single force,* acting through a given point, and *a couple.* (The resultant force and couple can be replaced by two concurrent forces if desired.)

The construction shown in Fig. 11·20 is based on the principle that *each view of a space system of forces can be treated as an independent coplanar system;* the coplanar resultants in each view can then be combined to obtain a single force and couple in space.

Problem. Three forces having magnitudes of 95, 65, and 80 lb, respectively, act in the positions and directions shown in stage 1 of Fig. 11·20. (The method can be extended to any number of forces.) These given forces are to be replaced by a single force, acting through point *O*, and a couple.

Construction

Stage 1. *Establish the given forces in the top, front, and side views.* Using any convenient space scale, locate the point of application (or any point on the line of action) of each force. Establish the direction of each vector and draw it to any convenient force scale using the method of Fig. 11·9. For clarity this preliminary phase of the construction is not shown in stage 1.

Stage 2. *Locate the coplanar resultant for the top view only.*

Because the force vector *CD* appears as a point in the top view, it has no effect in this view. The two remaining forces, *AB* and *BC*, can most easily be combined at the point of concurrency by the parallelogram construction shown. Then R'_T, the top-view resultant, can be scaled directly in this view as 73 lb, and it acts along the line *PQ*.

Stage 3. *Locate the coplanar resultant for the front view only.*

The resultant of the three forces shown in the front view can be found by consecutive application of the parallelogram construction, but for any

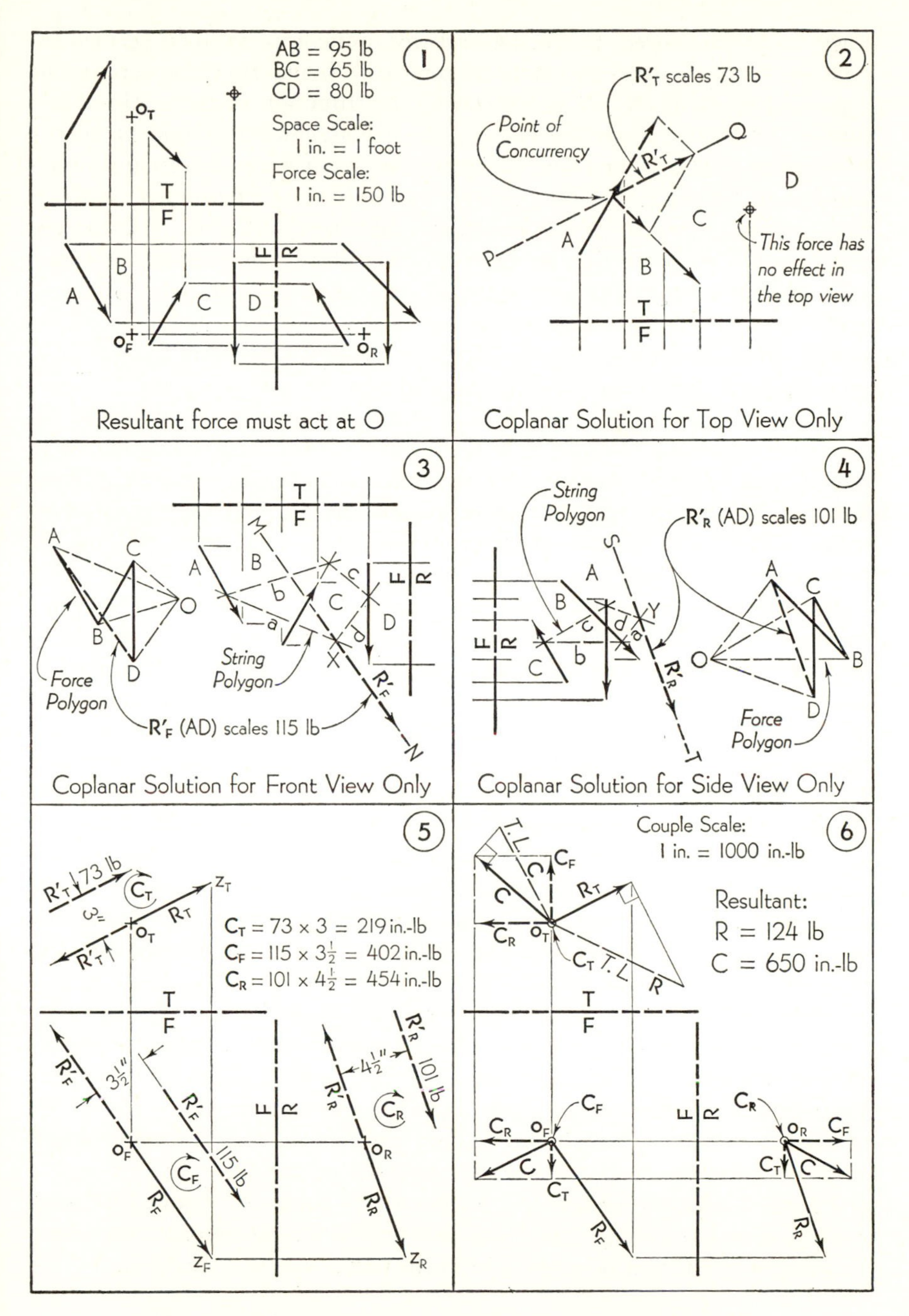

Fig. 11·20. Resultant of noncoplanar nonconcurrent forces.

large number of forces the string-and-ray-polygon method (described for parallel forces in Art. 11·11) is preferable. Using Bow's notation, the forces are labeled *AB*, *BC*, and *CD*, reading from left to right. The force polygon *ABCD* is then drawn anywhere conveniently adjacent to the front view. The closing side *AD* of the polygon scales 115 lb and represents the magnitude and direction of the front-view resultant R'_F. From any convenient pole *O* rays *OA*, *OB*, *OC*, and *OD* are drawn. The strings *a*, *b*, *c*, and *d* of the string polygon are then drawn parallel to the corresponding rays, beginning at any convenient point on force *AB*. Strings *a* and *d* intersect at *X*, a point on the line of action of R'_F. Then the resultant R'_F for the front view only is a force of 115 lb acting along the line *MN*.

Stage 4. *Locate the coplanar resultant for the side view only.*

In the side view the string-and-ray-polygon method is applied again, this time to locate the side-view resultant R'_R. Its magnitude of 101 lb is scaled at *AD* in the force polygon, and it acts along the line *ST*. Note that the determination of the coplanar resultant in each view is completely independent of that in the other two views.

Stage 5. *Calculate the coplanar couple in each view.*

In stage 5 the resultants obtained in each view have been retained, but for clarity the given forces have been omitted from the drawing. At o_T introduce two equal and opposite forces R_T and R'_T, equal and parallel to the top-view resultant R'_T. Then the equal, opposite, and parallel forces R'_T form a clockwise couple having a moment arm of 3 in. The moment of this couple is 73×3, or 219 in.-lb, clockwise. The top-view resultant R'_T is therefore replaced by a single force R_T acting through *O* and a couple C_T acting in a horizontal plane.

In similar manner the front-view resultant R'_F is replaced by a single force R_F and a couple C_F acting in a frontal plane; the side-view resultant R'_R is replaced by a single force R_R and a couple C_R acting in a profile plane. As a check on the accuracy of the construction, point *Z* should agree in alignment and measurement.

Stage 6. *Reduce to a single couple and a single force.*

The three couples C_T, C_F, and C_R are represented to scale by vectors acting through *O*. Their resultant is a single couple *C* acting in a plane perpendicular to the vector. The true length of the vector is 650 in.-lb, as shown. The single force *R* acts through *O* in the direction shown, and its magnitude is 124 lb. Thus the resultant of the three given forces is the force *R* and the couple *C*. If the given point *O* is moved to some other position, the magnitude and direction of force *R* remains the same, but the couple *C* will be altered in both magnitude and direction.

PROBLEMS. Group 125.

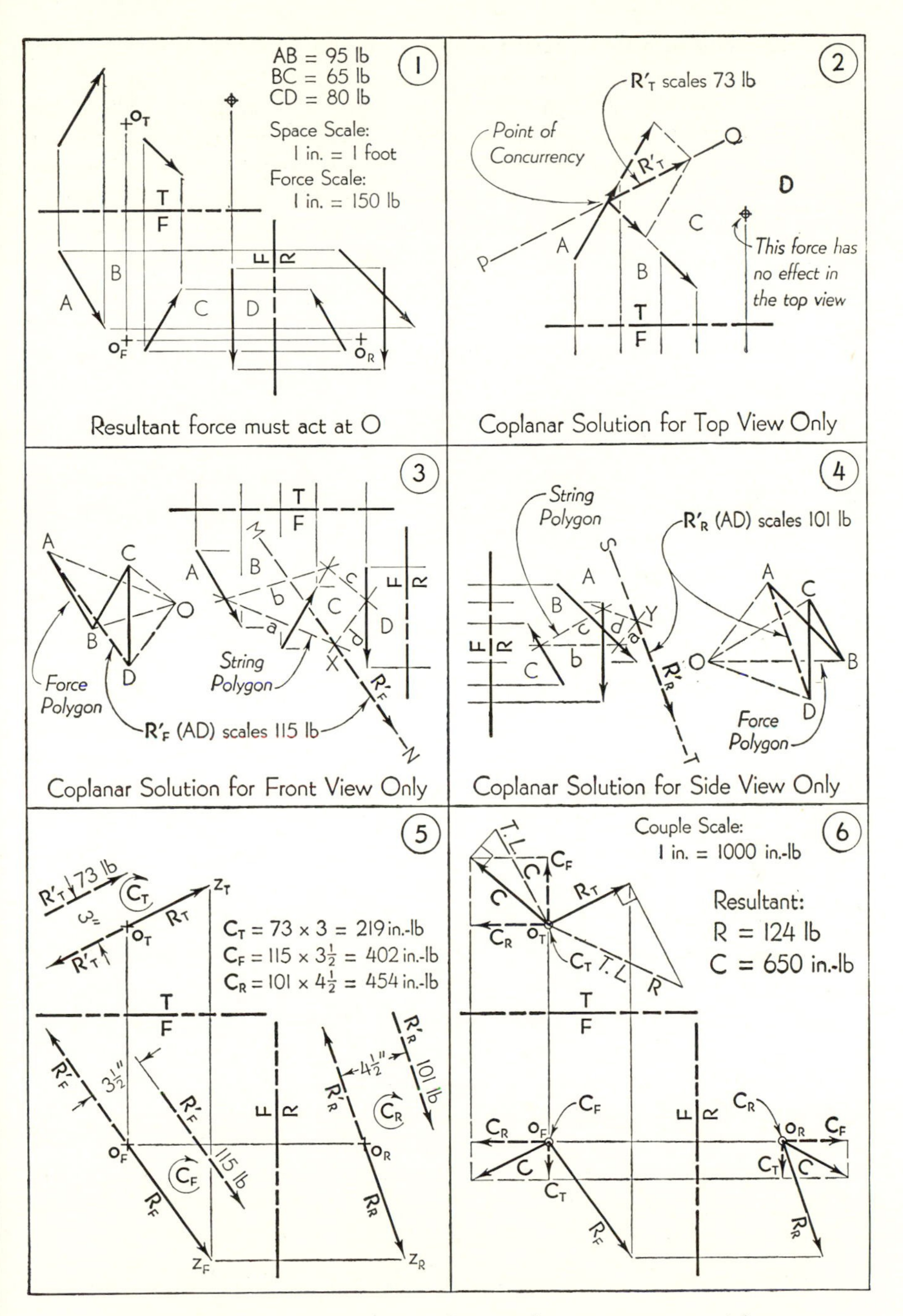

Fig. 11·20 (repeated). Resultant of noncoplanar nonconcurrent forces.

11·16. Forces in equilibrium

When a system of forces acts upon a rigid body and the resultant of this force system is zero, then the body is said to be in *equilibrium.* If the body is at rest, it will remain at rest under the action of these balanced forces. The graphical conditions for equilibrium are: (1) the *force polygon* must close to show that the resultant force is zero, and (2) the *string polygon* must close (the first and last strings must coincide) to show that the resultant couple is zero. Therefore when a body is known to be in equilibrium under the action of certain known and unknown forces, these graphical conditions can be employed to evaluate the unknown forces. In the examples that follow it will be shown, however, that the number of unknown quantities is strictly limited.

In the analysis of equilibrium problems it is essential that the body *on which* the forces act be clearly defined and that all of the forces acting on this body—known and unknown—be recognized and represented in the solution. The body may be the whole structure or any part of it, but this body must be a *free body*—isolated from all adjacent bodies. The effect of the adjacent bodies on the free body is then represented by force vectors. This free-body method will be illustrated by two simple examples.

Two magnitudes unknown. In Fig. 11·21(a) a weight is supported by two ropes. If we imagine the ropes are cut, then the isolated knot becomes a free body as shown at (b), and the *tension* forces in the cut ropes are represented with force vectors pointing *away* from the knot. Construction of such a free-body diagram should be the *first step* in the analysis of every equilibrium problem.

The force vectors T_1 and T_2 in Fig. 11·21(b) are known in direction but unknown in magnitude. But since the forces are in equilibrium, the

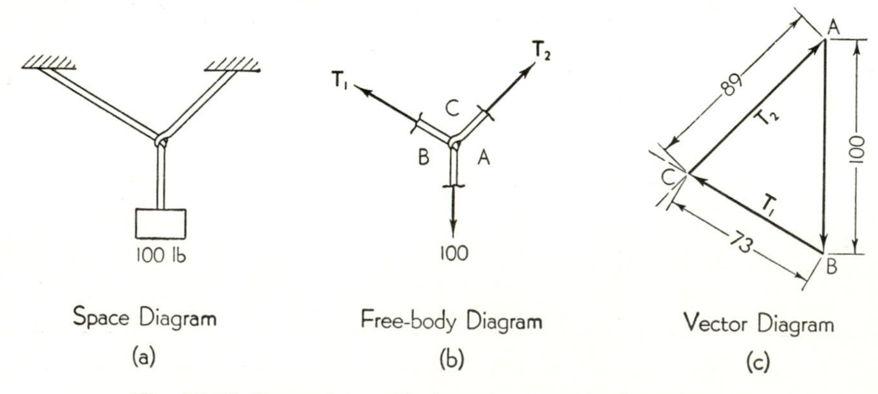

Fig. 11·21. Forces in equilibrium: two magnitudes unknown.

force triangle ABC must close. In the vector diagram at (c) the known vector AB is drawn first to scale. Vector BC is then drawn parallel to T_1 and CA parallel to T_2, thus establishing point C. The magnitudes of T_1 and T_2 can now be scaled as 73 and 89 lb, respectively. Note now that in a *concurrent coplanar* system of forces not more than *two* forces can be unknown in *magnitude.*

Two directions unknown. In Fig. 11·22(a) the ropes that support the 100-lb weight pass over frictionless pulleys and sustain weights of 60 and 80 lb, as shown. The tension in all three ropes is therefore known, but the values of angles X and Y for equilibrium are unknown. In the free-body diagram of the knot at (b) the 60- and 80-lb vectors are represented as wavy lines to indicate that their directions are unknown. In the vector diagram at (c) the known vector AB is again drawn first to scale. With radius equal to 60 lb an arc is drawn with B as center; with radius equal to 80 lb a second arc is drawn with A as center. These arcs intersect at C to complete the force triangle and give the angles X and Y as 37° and 46°, respectively. Note again that in a *concurrent coplanar* system of forces not more than *two* forces can be unknown in *direction.*

Summary. A coplanar force polygon may contain any number of known forces, but as these simple examples show, the polygon can be closed only if there are not more than *two unknown force quantities.* For a *general nonconcurrent coplanar* system the string polygon can also be employed, and the number of unknowns thereby increased to *three,* but this is the maximum number of unknowns for any body acted upon by coplanar forces.

When noncoplanar force systems are represented in a multiview drawing, *each view must be considered as a coplanar system* subject to the same restrictions noted above. The application to space problems of the above coplanar principles will be illustrated in the articles that follow.

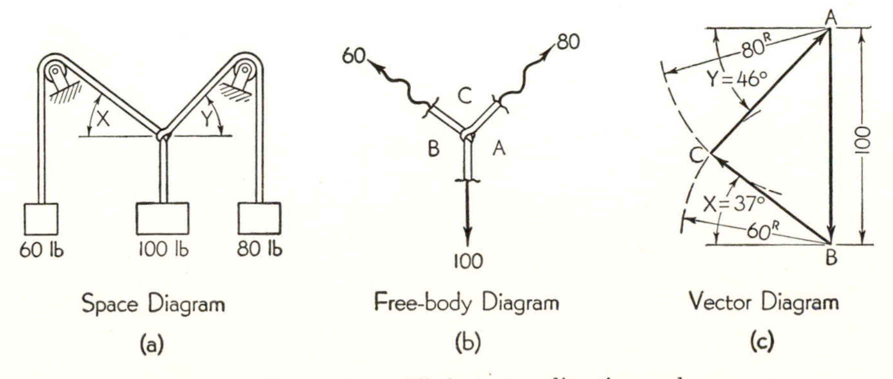

Fig. 11·22. Forces in equilibrium: two directions unknown.

11·17. Equilibrium of noncoplanar concurrent forces

If a noncoplanar concurrent system of forces acts on a body in equilibrium, then not more than *three* of the forces can be unknown in *magnitude*. In any single view of this system only *two* unknown force magnitudes can be evaluated by closure of the force polygon, but with two adjacent views a total of *three* unknown magnitudes can be determined. The solution should therefore begin in a view where the *three unknown forces appear as two* or where *only the three unknown forces appear*. This reasoning suggests the three following methods of solution.

One unknown force appears as a point. In Fig. 11·23(*a*) a mast and two guy wires are stressed by a 600-lb force acting as shown. At (*b*) the top section of the mast has been isolated as a free body, and the three unknown force vectors have been indicated without arrowheads. It is not necessary to assume or know the sense of these unknown forces; this will be revealed in the graphical solution. The mast force is vertical, but in the top view the mast is shown offset in order to apply Bow's notation in this view. The spaces are lettered clockwise beginning with the known force *AB* and *ending with the force DA* that appears as a point in this view. In the front view the vectors are labeled as shown.

In the vector diagram at (*c*) the known force vector *AB* (600 lb) is established to any convenient scale. In the top view vector *AD* appears

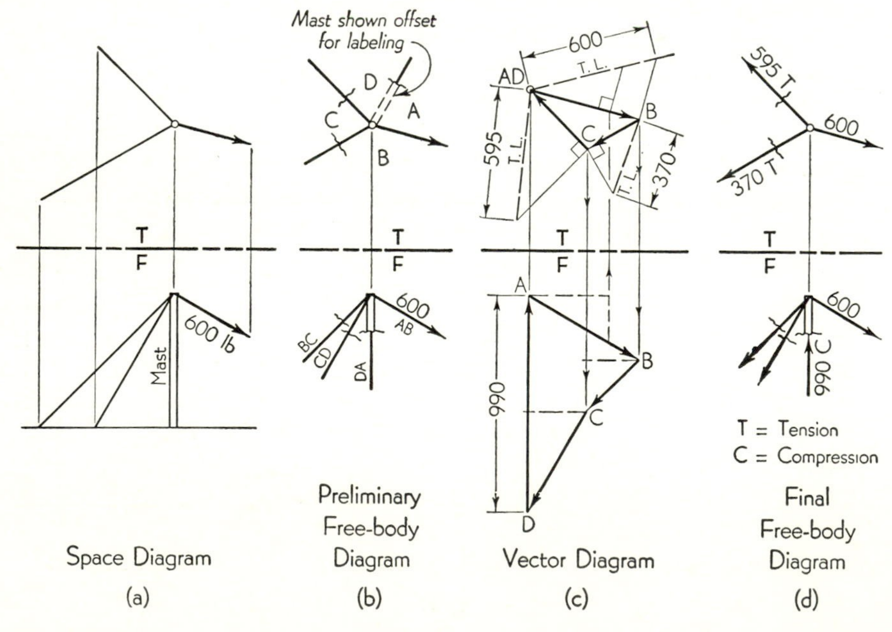

Fig. 11·23. Concurrent forces in equilibrium: one unknown force appears as a point.

as a point; hence the force polygon can be closed here by drawing the triangle *BCD* with sides parallel to the given force directions. In the front view vector *BC* can now be drawn in correct length, followed by vector *CD*, which must extend to the vertical vector *AD* and thus establishes point *D*. The magnitude of force *DA* is now found to be 990 lb and the true length of force vectors *BC* and *CD* to be 370 and 595 lb, respectively.

For equilibrium the vectors in the force polygon must all point in the same direction—tail to head—around the chain of vectors; hence the correct sense of each vector is determined and can now be shown in the free-body diagram as at (*d*). A vector pointing *away* from the free body indicates that the force is one of *tension* (T); a vector pointing *toward* the body indicates *compression* (C).

Although this method is obviously most suitable when one unknown force appears as a point in the given views, the method can also be applied to any case by drawing a new view in which one of the unknown forces appears as a point.

The known force appears as a point. In Fig. 11·24(*a*) two cables *OX* and *OZ* and a brace *OY*, support a 1,200-lb load. At (*b*) the joint *O* is isolated as a free body. At this stage arrowheads should be omitted

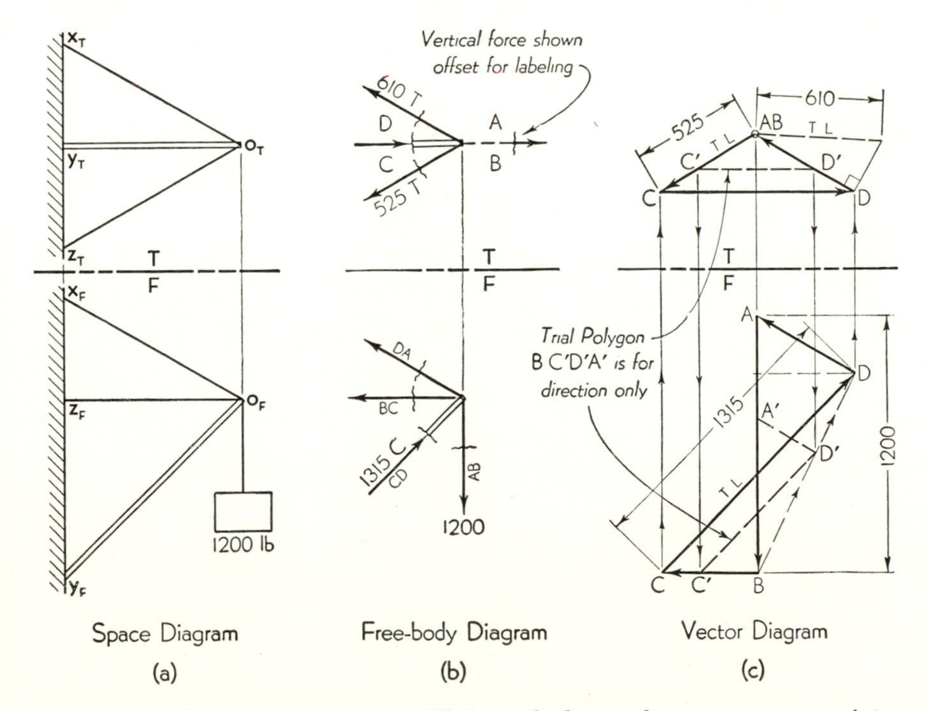

Fig. 11·24. Concurrent forces in equilibrium: the known force appears as a point.

from the unknown force vectors and Bow's notation applied beginning with the known force as AB (shown offset in the top view for convenience in labeling).

In the vector diagram at (c) the known force vector AB is established to any convenient scale. In the top view AB appears as a point; hence here the three unknown vectors form a triangle of *known shape* but *unknown size.* We must therefore draw first the *trial triangle* $BC'D'$ of any arbitrary size with sides parallel to the given force directions. In the front view this same trial polygon appears as $BC'D'A'$, and $A'B$ represents 1,200 lb to some arbitrary scale. To obtain correct scale the trial polygon can now be proportionately enlarged until A' coincides with A. In the front view extend line BD' and draw AD parallel to $A'D'$ to locate point D. Draw CD parallel to $C'D'$ to locate C. Enlarge the triangle in the top view to correspond with the enlarged front view. The magnitude and sense of each unknown force can now be determined and shown on the free-body diagram.

This method can be applied to any case by drawing a new view in which the known force (or resultant of all known forces) appears as a point.

Two unknown forces appear coincident. In Fig. 11·25(a) an inverted tripod structure supports a 1,000-lb force as shown. The force

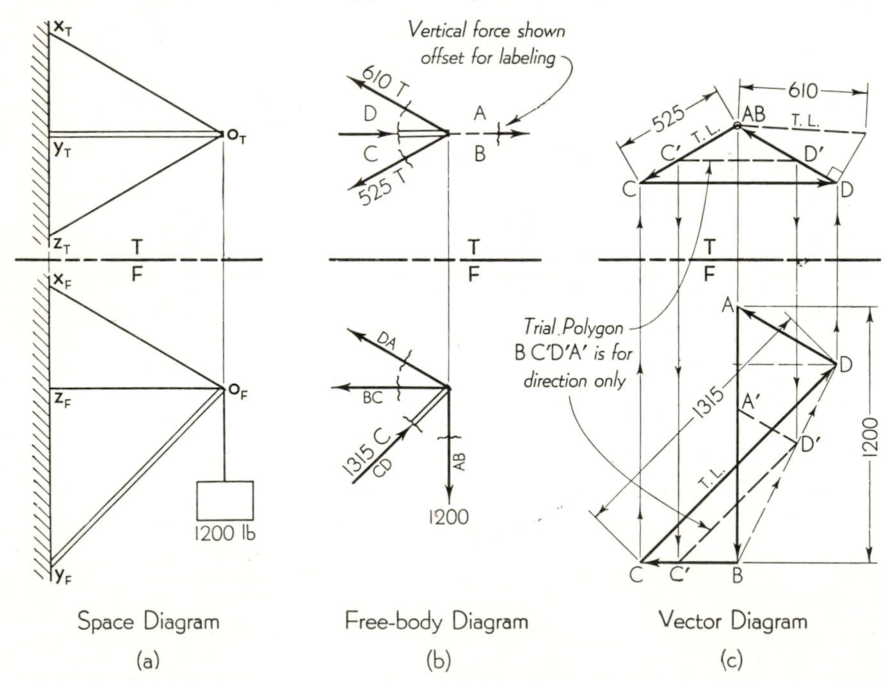

Fig. 11·24 (repeated). Concurrent forces in equilibrium: the known force appears as a point.

polygon for this structure cannot be closed in either the top or front view because three unknown vectors would appear in each view. But if a new view, such as A, is drawn to show the plane of two of the unknowns as an edge, then the force polygon will appear as a triangle in view A, and can be closed. The solution should therefore be performed in views A and T as shown in Fig. 11·25.

At (b) the joint O is isolated as a free body, and the coinciding vectors in view A are separated for ease in labeling. In applying Bow's notation, begin with the known vector and *reserve the last letter for the coinciding vectors*, which must be drawn last in the vector polygon.

In the vector diagram at (c) the known force vector AB is established to any convenient scale. Then in view A the vector triangle ABC can be drawn with sides parallel to the given force directions. The position of point D in this view is temporarily unknown, but vector CA represents

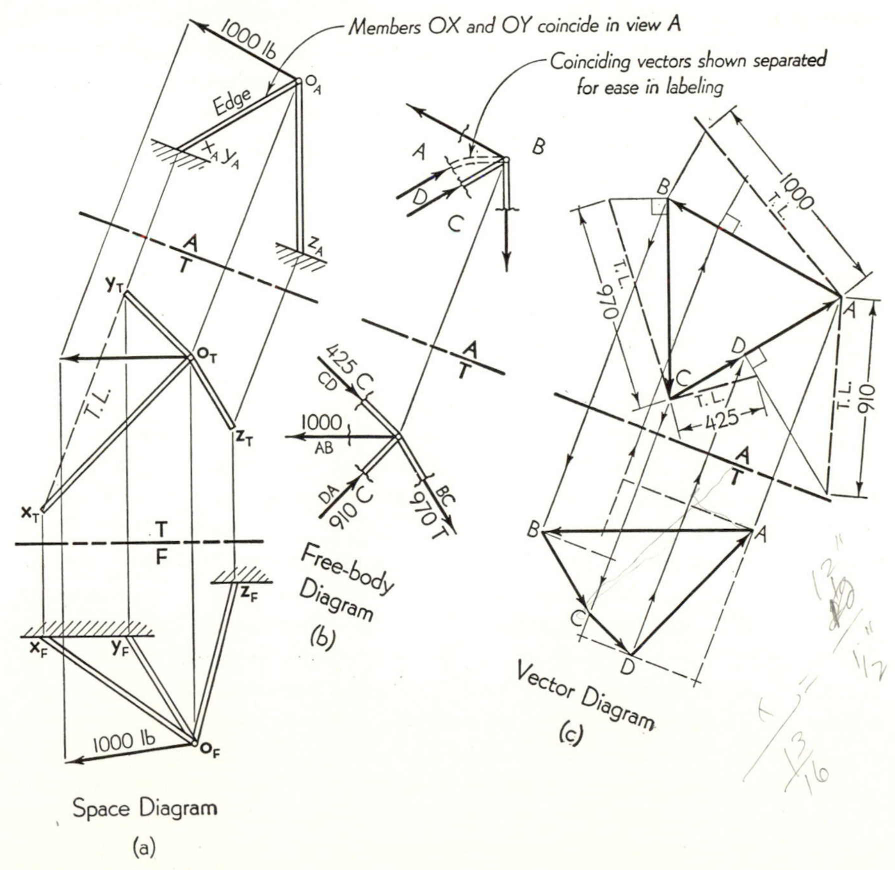

Fig. 11·25. Concurrent forces in equilibrium: two unknown forces appear coincident.

the combined effect of coinciding vectors CD and DA. In the top view of the force polygon vector BC can now be drawn in correct length to locate C. Vectors drawn from C and A parallel to the given top-view directions intersect at D and close the polygon in this view. By alignment D can now be located in view A. The magnitude and sense of each unknown force is determined as shown and then indicated on the free-body diagram.

Although this method is applicable to any case, it is obviously most suitable when two unknown forces appear coincident in one of the given views.

PROBLEMS. Group 126.

11·18. Equilibrium of noncoplanar parallel forces

If a body is held in equilibrium by a system of noncoplanar parallel forces, then *three* of the forces may be unknown in *magnitude.* Any such force system can be simplified by replacing all known forces by a single resultant force (or couple). Then the three unknown forces can be evaluated by solving repeatedly in views where the three forces appear as two, that is, in views where two of the unknown forces appear coincident. This procedure is illustrated in the following problem.

Problem. A triangular platform is supported at its three vertices A, B, and C as shown in Fig. 11·26. The platform sustains two concentrated loads of 850 and 500 lb located as shown. Determine the magnitude of each of the three vertical supporting forces.

Construction. Taking the entire platform as a free body, the vectors F_A, F_B, and F_C represent the removed supporting forces. The two given loads can be combined into a single load of 1,350 lb by introducing in the front view the equal and opposite forces F and thus locating point X on the line of action of the resultant force [compare Fig. 11·15(a)]. The 1,350-lb resultant must be coplanar with the given forces, and is so located in the top view.

In the right-side view line AC appears as a point, and forces F_A and F_C appear coincident. If we temporarily regard $F_A + F_C$ as a single unknown force, then F_B is the second of two unknown forces in this view. On the line of action of $F_A + F_C$ set off to scale 1,350 lb to locate M. From *any point* Y on the line of action of the 1,350-lb resultant extend lines through b_R and a_Rc_R as shown. Then draw MN parallel to Yb_R, and construct a perpendicular NO from N to $F_A + F_C$. Point O divides the 1,350-lb length into two segments of 890 and 460 lb that represent $F_A + F_C$ and F_B, respectively. Note that in this construction a single force (1,350 lb) has been resolved into two parallel components ($F_A + F_C$ and F_B) and the method is therefore simply the reverse of that shown in Fig. 11·15(a).

In view A forces F_B and F_C appear coincident, and by repeating the construction described above $F_B + F_C$ and F_A are found to be 810 and 540 lb, respectively. By subtraction F_C can now be evaluated as 350 lb.

PROBLEMS. Group 127.

11·19. The principle of concurrency

Let us consider the case of a rigid body held in equilibrium by a complex system of noncoplanar forces. This would be the general space condition in which the forces are neither concurrent nor parallel. Assume now that the applied forces are combined by finding the resultants of various groups of forces until the whole system has been reduced to only three forces.

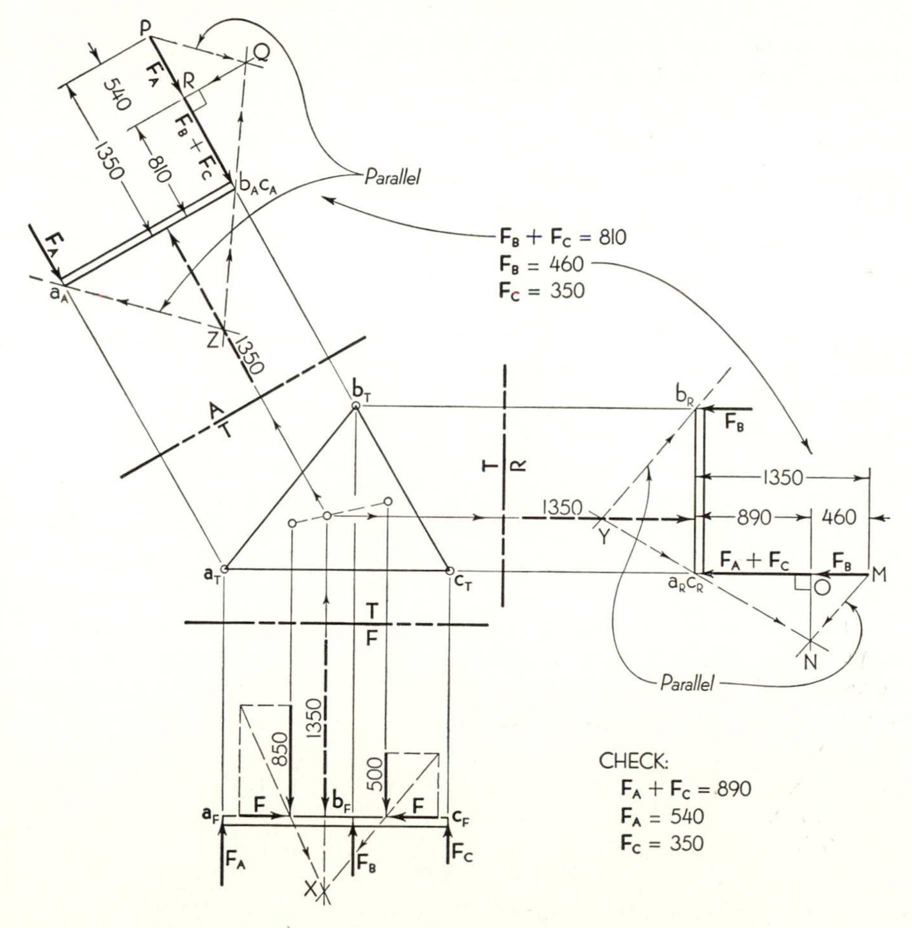

Fig. 11·26. Parallel forces in equilibrium: three unknowns.

Then to maintain equilibrium one of these three forces must be equal, opposite, and collinear with the resultant of the other two. This means that the three forces must lie in the same plane and pass through the same point (or be parallel). In the graphical solution of equilibrium problems this fact is of great value, and may be stated as a principle.

Three forces in equilibrium must be coplanar and either concurrent or parallel.

Figure 11·27 shows how this principle can be applied to greatly simplify the solution of a coplanar equilibrium problem. As shown in the space diagram at (*a*), a structure consisting of mast, crossbar, and brace supports a load of 1,000 lb. The weight of the members can be neglected. The lower end of the mast rests in a socket at 2, and the upper end fits freely in a hole at 1. The three members are pinned together at joints 3, 4, and 5. The direction and magnitude of the forces acting at each of the numbered points are required.

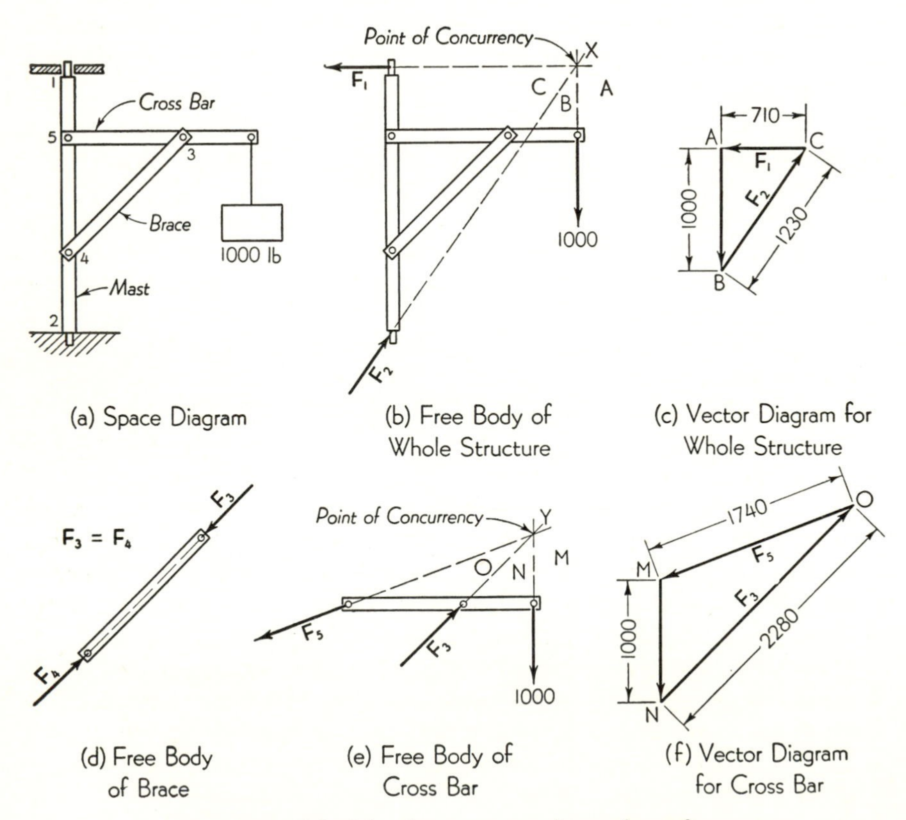

Fig. 11·27. Principle of concurrency for coplanar forces.

If the whole structure is taken as a free body as at (*b*), then it is apparent that there can be no vertical force at the upper end of the mast, and force F_1 must be horizontal. The direction of force F_2 is unknown. But only three forces act on this body; hence they must be concurrent. The lines of action of F_1 and the 1,000-lb force intersect at X, and therefore the line of action of F_2 must also pass through X. With the direction of F_2 now known, the vector diagram can be drawn as shown at (*c*) to determine the sense and magnitude of F_1 and F_2.

At (*d*) a free body of the brace shows that only two forces act on this member, and F_3 and F_4 are therefore equal and opposite and act in a direction parallel to the brace.

At (*e*) a free body of the crossbar shows that three forces act on this member and must therefore be concurrent. But the directions of F_3 and the 1,000-lb force are known, and their lines of action intersect at Y. The direction of force F_5 is thus fixed, and the vector diagram at (*f*) can now be completed to determine the sense and magnitude of forces F_3 and F_5.

Note now that the coplanar forces acting on any free body can be fully evaluated only if there are not more than *three unknown force quantities*: *two* magnitudes and *one* direction (as in Fig. 11·27) or *three* magnitudes, for example.

11·20. Equilibrium of the general noncoplanar force system

In a general noncoplanar force system that is in equilibrium there can be as many as six *unknown force quantities.* In a multiview drawing of such a force system each view can be considered a coplanar system to which the principle of concurrency can be applied to determine one force direction. A force polygon can also be drawn for each view to evaluate two more unknown quantities. Thus the forces in any view can be fully determined if not more than three unknowns appear in that view.

Problem. Figure 11·28 shows three views of a hoisting drum. The drum, supported on journal bearings at each end, is turned by a horizontal force P applied to the crank. The rope wrapped around the drum carries a load of 200 lb acting in the position and direction shown by the vector. The total weight of drum, axle, and crank is 50 lb acting at the center of gravity G. Determine force P, the axial thrust T on the drum, and the vertical and horizontal components of the bearing forces.

Free-body analysis. Assume that the end bearings are removed. Then the drum, axle, and crank constitute a free body held in equilibrium by the two known forces, force P, thrust T, and the bearing forces at A and B. The 200-lb force pulls the drum to the left, and is resisted at the left bearing by an axial force T. There is no axial force at the right bearing. The unknown bearing forces are represented by the horizontal components A_Y and B_Y and by the vertical components A_Z and B_Z. Although the correct sense of each unknown force is now shown on the free body, this is unnecessary in the initial stage of the solution; the force polygons will correctly indicate both sense and magnitude of the unknown forces.

Forces that appear as a point in any view have no effect on the coplanar equilibrium of that view. Forces that appear coincident, such as A_Y and B_Y in the side view, may be temporarily regarded as one combined unknown $A_Y + B_Y$. Then a survey of each view reveals the following unknowns:

Top view: T, A_Y, B_Y, and P—four unknowns.

Front view: T, A_Z, and B_Z—three unknowns.

Side view: $A_Y + B_Y$, $A_Z + B_Z$, and P—three unknowns (two combined).

The solution should therefore *begin in the front view*, where only T, A_Z, and B_Z are unknown. The side view can then be solved for P, $A_Y + B_Y$, and $A_Z + B_Z$. With P and T known, the top view is reduced to two unknowns and can now be solved for A_Y and B_Y. *Sequence* of solution is obviously very important in problems of this type, and a thorough analysis like that above is always desirable. Note, too, that *every force* must be shown on the free body; one force omitted or forgotten will invalidate the entire solution.

Front-view construction. The 200- and 50-lb force vectors appear true length in the front view, and should be laid off to scale here. The lines of action of these two forces intersect at C, where they can be combined into a single resultant force R_1 by the parallelogram method (compare Fig. 11·12). Then forces R_1 and B_Z extended meet at point X. If forces T and A_Z were vectorially combined into a single force $T \nrightarrow A_Z$, then this third force would also pass through the point of concurrency X, and the direction of $T \nrightarrow A_Z$ is fixed. If point X should fall beyond the edge of the paper, then the construction given in Art. A·2 of the Appendix can be employed.

Directly below the front view (or in any convenient space) the vector diagram for this view begins at any point o_F. (To avoid excessive notation, Bow's notation has been omitted in this illustration.) The two known forces are laid off consecutively from o_F, and the two unknown forces B_Z and $T \nrightarrow A_Z$ are then drawn parallel to their known directions to close the polygon. Vector $T \nrightarrow A_Z$ can now be resolved back into its two original components T and A_Z, as shown by the dash lines. The sense and scaled magnitude of each unknown force is shown on the diagram.

Side-view construction. The 200-lb force [labeled $(200)_Z$ in this view because it does not appear true length] and the parallel 50-lb force are combined into the single resultant force R_2 acting through point D [compare Fig. 11·15(a)]. Then forces R_2 and P extended meet at point Y.

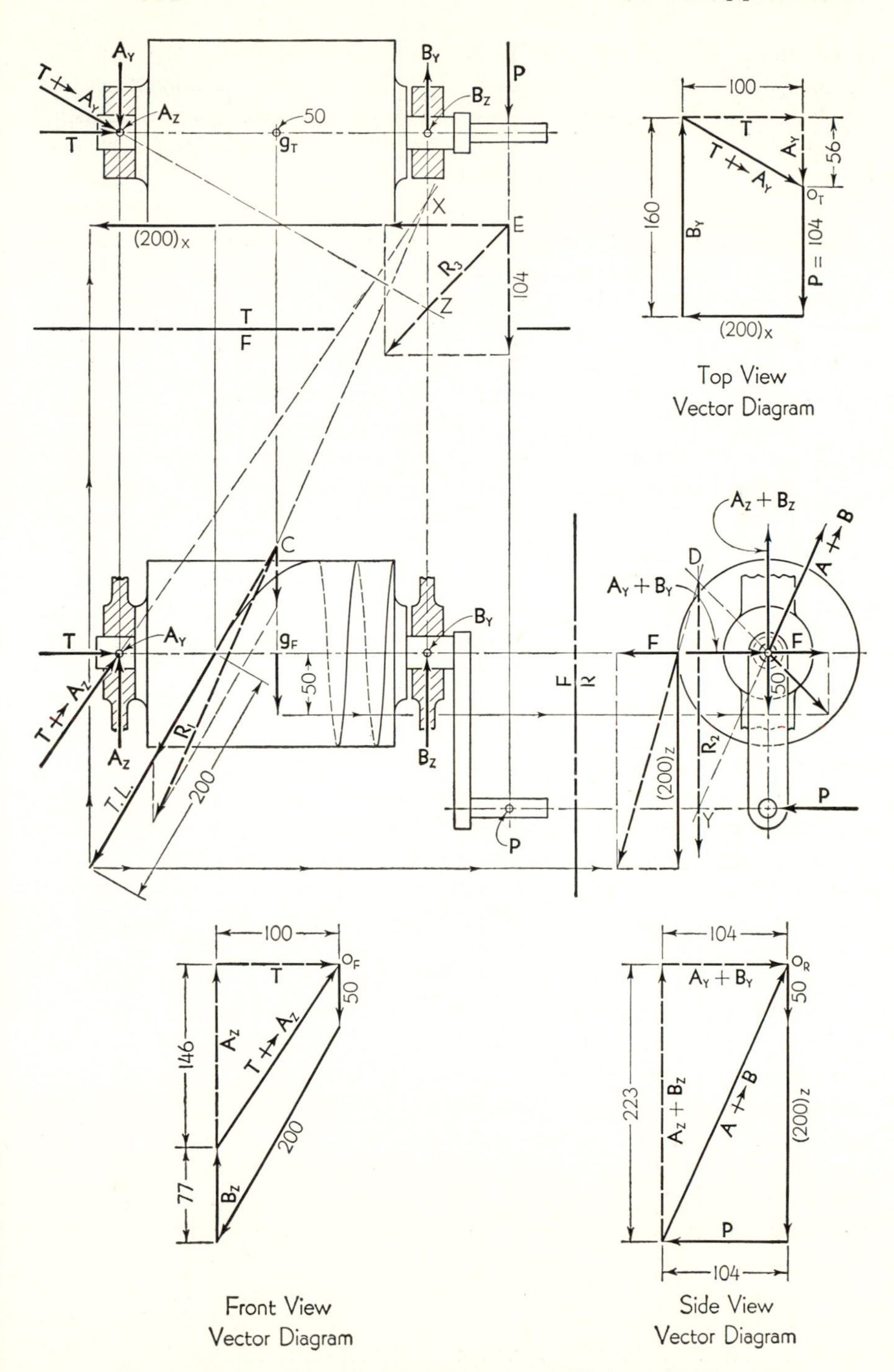

Fig. 11·28. General noncoplanar forces in equilibrium: six unknowns.

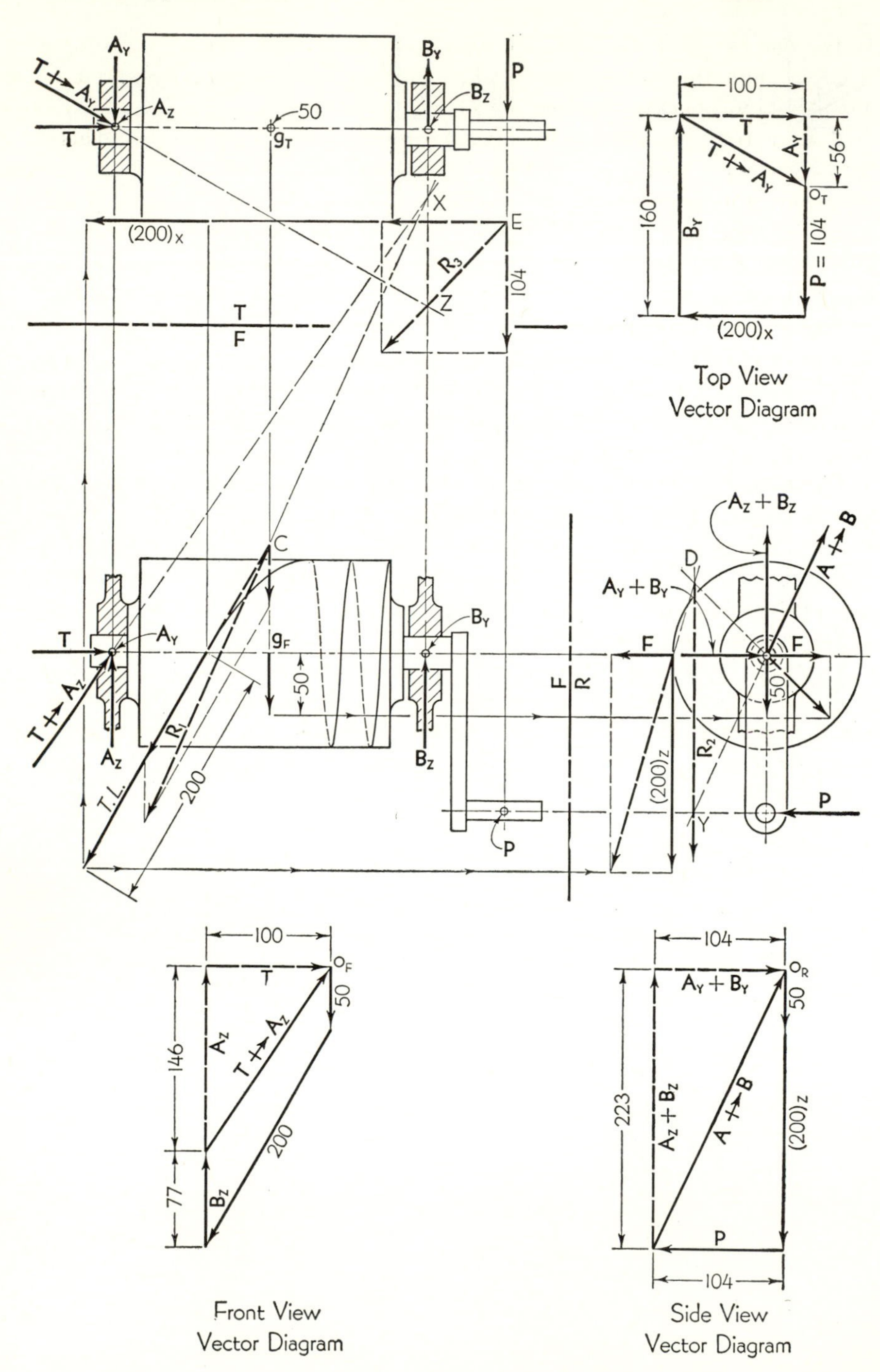

Fig. 11·28 (repeated). General noncoplanar forces in equilibrium: six unknowns.

If the forces $A_Y + B_Y$ and $A_Z + B_Z$ were combined into a single force $A \nrightarrow B$, then this third force would also pass through Y, the point of concurrency, and the direction of $A \nrightarrow B$ is fixed.

The vector diagram for the side view has been conveniently placed below the view (point o_R need not necessarily align with o_F). Here the force polygon is closed by the unknown forces P and $A \nrightarrow B$, and the vector $A \nrightarrow B$ can then be resolved into its two components $A_Y + B_Y$ and $A_Z + B_Z$. Vector $A_Z + B_Z$ scales 223 lb, which agrees with the separate values of 146 and 77 lb found in the front view for A_Z and B_Z.

Top-view construction. The 200-lb force [labeled $(200)_X$ in this view] and force P intersect at E, and since the value of P is now known to be 104 lb, these two forces can be combined into the single force R_3. Then forces R_3 and B_Y extended establish the point of concurrency Z, and the direction of $T \nrightarrow A_Y$ is fixed.

The vector diagram for the top view (arbitrarily placed to the right of the view) begins at point o_T with the known forces P and $(200)_X$, and is closed with the unknown forces B_Y and $T \nrightarrow A_Y$. Vector $T \nrightarrow A_Y$ resolves into the components T and A_Y. Forces A_Y and B_Y are now seen to be opposite in sense, but their algebraic sum agrees with the value of $A_Y + B_Y$ found in the side view.

The sense and magnitude of every unknown force is now shown in one or more of the three vector diagrams, and the diagrams should be compared to see that they are in agreement. The total force on each bearing can be determined if desired by vectorially combining A_Y and A_Z, and B_Y and B_Z.

PROBLEMS. Group 128.

12. GEOLOGY AND MINING APPLICATIONS

12·1. The surface of the earth

In the solution of geology, mining, and construction problems the engineer must work with the irregular contours of the earth's surface. Its hills and plains, ridges and valleys can all be represented accurately on a topographic map by *contour lines* (Art. 3·2, Fig. 3·4). In order to visualize the terrain represented by a topographic map it is necessary to study the shape and elevation of the various contour lines.

In Fig. 12·1, for example, the *contour interval* (difference of elevation between adjacent contour lines) is 100 ft, and each contour line is labeled to indicate its height above sea level. It is customary to draw every fifth contour line heavier than the others. A person walking along the 700-ft contour line would be following a *level* path. If he turned off this level path and walked toward the 1,000-ft contour, crossing each successive contour line, he would then be walking uphill.

The small circle inside the 1,200-ft contour is labeled 1,240 and represents the top of the hill. To the southeast of this hill is a second hill whose top is 1,080 ft above sea level. The lowest point between the two hills is somewhat more than 700 ft, while to the northeast and southwest the land drops away to less than 300 ft. Thus the land between the hills is in the form of a *saddle*.

Small dents in the contour lines represent minor ridges or ravines—ridges if the dents bulge toward lower contour lines, ravines, or water courses, if the dents bulge toward higher contours. The spacing of the contour lines indicates the slope of the ground. Contours close together indicate a steep slope; contours far apart indicate a gentle slope. Between adjacent contour lines the slope is assumed to be reasonably uniform, and it is therefore permissible when necessary to interpolate additional contour lines between the given lines, as shown by the 650-ft line in the lower right part of the map.

The *block diagram* in Fig. 12·2 is an isometric representation of the same terrain shown on the map in Fig. 12·1.

12·2. The crust of the earth

Below the topsoil and loose mantle rock that covers most of the earth's surface the rock formations lie in *layers*, or *strata*. These stratified formations are called *sedimentary* rocks because they were formed under water by the deposition, compacting, and cementing of sediment carried by rivers and ocean currents. Shale and slate were formed from mud and clay, sandstone from sand, limestone from the skeletons and shells of marine animals, and coal by the decomposition of vegetable matter.

During the period of sedimentation the stratified rock (with a few exceptions) naturally formed in continuous horizontal layers. But in

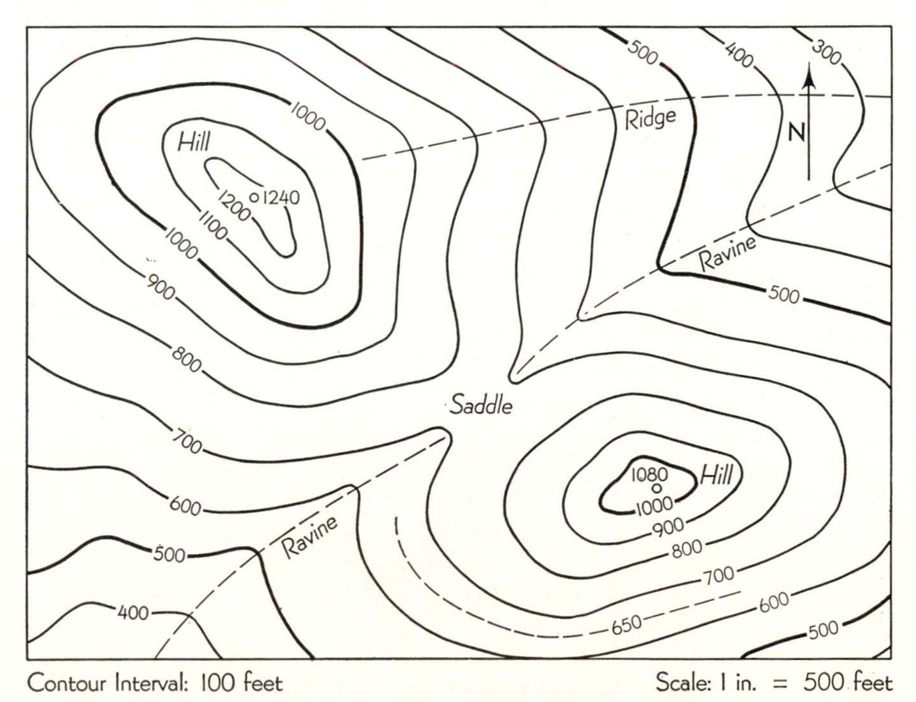

Fig. 12·1. A topographic map.

subsequent eras the upheaval and wrinkling of the earth's crust distorted these level layers, elevating some to mountainous heights, tilting many at steep angles, folding and fracturing them, so that today the originally horizontal strata are seldom level and are plane over limited areas only. Within such areas, however, it is practicable to assume that a bed of coal or ore is of uniform thickness and that it is bounded by two parallel planes called the upper and lower *bedding planes.* The block diagram in Fig. 12·2 shows how the strata of sand, coal, shale, and limestone may lie in inclined parallel planes below the eroded topsoil of the earth.

The continuity of the strata is frequently broken by cracks or fractures. When the fracture is followed by a displacement of one side of the bed relative to the other, it is called a *fault*; the plane of the fracture (if it is a plane) is called the *fault plane.* Note in Fig. 12·2 that the strata in front of the fault plane *ABCD* have been displaced upward, but subsequent erosion has obliterated any external evidence of the displacement. The trained eye of a geologist might discover the fault, however, by noting certain discontinuities along the line *BC*, where the fault plane intersects the earth's surface. This intersection line is called a *fault outcrop.*

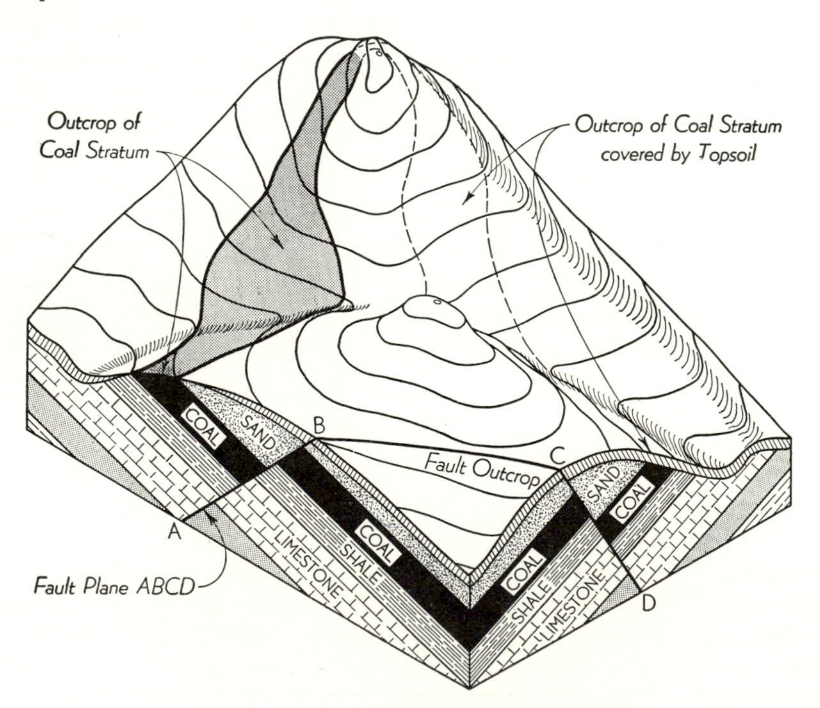

Fig. 12·2. Block diagram showing faulted strata.

Cracks or fissures in the rock that have been filled by mineral substances are called *lodes* or *veins*. Quartz, calcite, fluorite, the ores of gold, silver, lead, copper, etc., are a few of the minerals found in veins. Veins vary greatly in shape, depending upon the character of the fissure, but like sedimentary strata they frequently lie between parallel planes.

When a stratum of rock or vein of ore intersects the surface of the earth, the exposed edge is called an *outcrop*. Such outcrops are usually found on hillsides and cliffs or along the banks of a stream, where erosion of wind and water has uncovered the bedrock. In Fig. 12·2 the coal stratum is exposed (or only lightly covered by topsoil) on the left side of the hill; on the right side of the hill a considerable depth of topsoil conceals the outcrop.

By surface observations and a knowledge of the geologic formations of the area, a geologist or mining engineer can estimate the approximate position of the underlying strata. From drilled test holes, wells and nearby excavations, and geophysical methods (gravitational, magnetic, seismic, and electrical) additional data can be obtained concerning the underground structure. When all of this information has been compiled on a topographic map, the location and depth of three or more points on a particular bedding plane can be determined. Since three points fix the position of a plane, the location of the whole stratum is established, and therefore, within ascertainable limits, the mining engineer can plan the working of a vein of ore or a bed of coal by the methods of descriptive geometry.

When the subsurface strata are warped or curved, the surface of each stratum must then be described by underground contour lines, and these lines are called *structural contours*. Folded strata (strata that have been bent or wrinkled), salt domes, and oil closures must be represented by structural contours. Space limitations prevent further discussion of nonplane strata, but the interested student is referred to texts on *structural geology*.

12·3. Mining terminology

Figure 12·3 shows a cross section of the passages, or workings, commonly employed to mine the ore in an inclined vein. The terms shown in the figure are more accurately defined as follows:

Strike. The bearing of a horizontal line in the plane of the stratum.

Dip. The slope angle of the plane of the stratum.

Hanging wall. The rock surface above a vein of ore.

Footwall. The rock surface below a vein of ore.

Working. Any excavation, tunnel, or passage made in mining.

Shaft. A vertical, or approximately vertical, opening from the surface to the mine workings.

Collar. The horizontal timbering around the entrance to a shaft.

***Slope* (*or incline*).** An inclined shaft that follows the vein. It is usually "on the dip"; that is, it has the same slope angle as the vein.

Tunnel. A horizontal, or approximately horizontal, underground passage that is open to the surface at one or both ends.

Level. A horizontal passage, either in the vein or adjacent to it. Levels may be distinguished by their elevation or by the distance down the shaft.

Crosscut. A level that traverses country rock to connect one working with another.

Drift. A level lying entirely in the vein; hence a horizontal passage in the direction of the strike.

Raise. A passage driven upward from below.

Winze. A passage driven downward from one level to another.

12·4. Location of a stratum

If the strata in a given area are observed to be plane (or very nearly plane), then three points on the bedding plane of a stratum will fix the

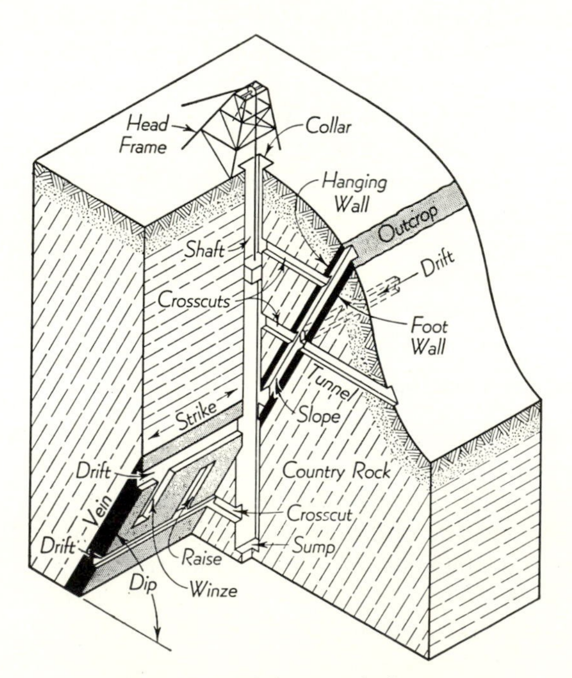

Fig. 12·3. Mining terminology.

location of that particular stratum. In Fig. 12·4(*a*), for example, points *A*, *B*, and *C* are known to lie on the upper bedding plane of a stratum of rock. The map shows the top-view location of each point, and the numbers in parentheses give the elevation of each point in feet above sea level. Point *A*, at an elevation of 1,000 ft, lies on the 1,000-ft contour line and hence is a point on the surface. Point *B* is similarly a surface point, and *A* and *B* are therefore points on the *outcrop line* of the upper bedding plane of the stratum. At an elevation of 610 ft, point *C* is obviously 90 ft underground at the bottom of a borehole drilled to locate a third point on the stratum.

In geology and mining problems it is more convenient to describe the position of a stratum in terms of its *strike* and *dip*. When three points have been established, as in Fig. 12·4(*a*), then the strike and dip of the stratum can be determined by either of two methods.

12·5. Strike and dip: conventional three-view method

Analysis. Assume any horizontal line on the plane; by definition (Art. 12·3) strike is the bearing of this line. Dip is the geologist's term for slope angle; hence an elevation edge view of the plane will show the dip.

Construction. In Fig. 12·4(*b*) the top view is a duplicate of the adjacent map data with contour lines omitted and line *BC* added. The

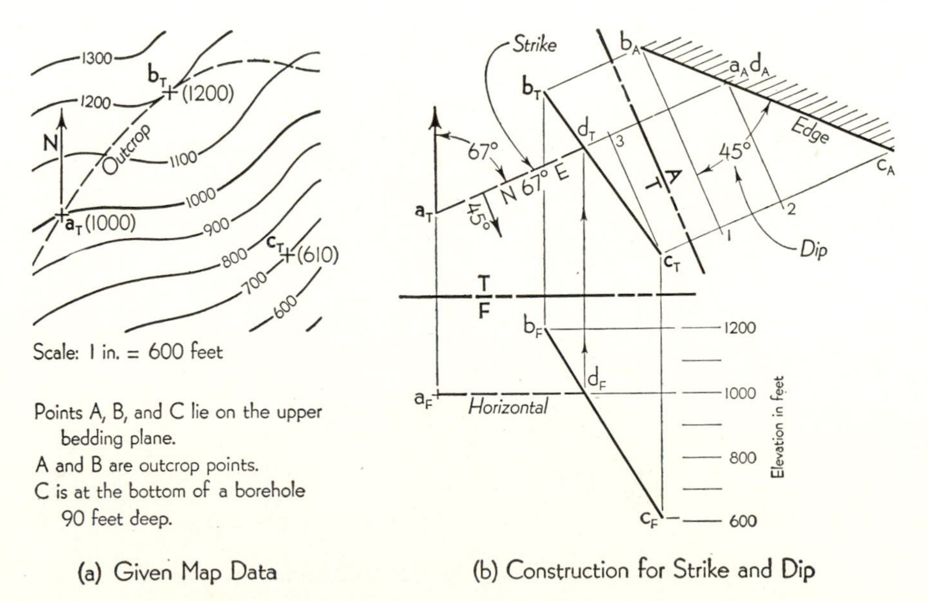

Fig. 12·4. Conventional three-view method for strike and dip.

front view has been introduced to show the respective elevations of points A, B, and C.

A horizontal line AD is assumed in the front view at $a_F d_F$ and shown in the top view at $a_T d_T$. The strike of the plane is the bearing of this line, N 67° E, and is lettered along the *strike line.* It is conventional to always measure strike from the north (not S 67° W).

View A shows the plane of the stratum as an edge, and here the slope angle of 45° is the dip (compare Fig. 4·13). To represent dip on a map a short arrow is drawn perpendicular to the strike line and pointing to the low side of the plane—in other words, downward on the dip. The numerical value of the dip is lettered beside the arrow, as shown in Fig. 12·4(*b*). Verbally, this dip would be described as 45° SE, that is, 45° downward in a *generally* southeasterly direction.

12·6. Strike and dip: geologist's three-point method

Analysis. Inspection of Fig. 12·4(*b*) shows that point D, which is intermediate in elevation between points B and C, divides the line BC into two segments that are proportional to the difference of elevation between A and B and between A and C. Therefore d_T can be located on line $b_T c_T$ by simple proportion and without reference to the front view.

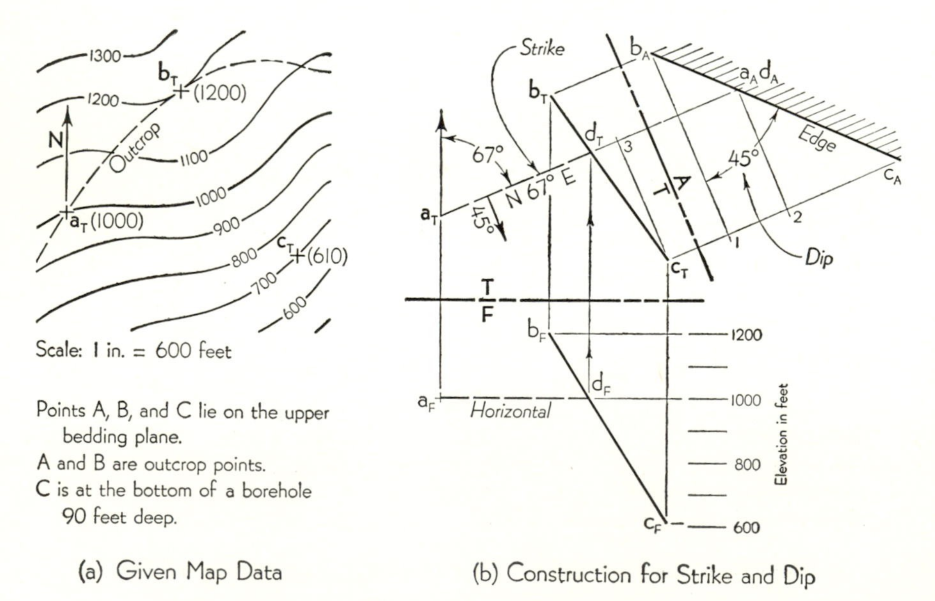

Fig. 12·4 (repeated). Conventional three-view method for strike and dip.

In view A of Fig. 12·4(b), right triangle b_Ac_A1 is similar to triangle a_Ac_A2. But distance a_A2 can be obtained from the top view at c_T3, and distance c_A2 is the difference of elevation between points A and C. Thus the right triangle that contains the dip angle can be constructed without either the front view or view A.

Construction. In Fig. 12·5(a) points A, B, and C, three points on the stratum, are given in the top, or map, view with their respective elevations. Line b_Tc_T is drawn to connect the *high* and *low* points. From b_T line b_Tx is drawn at any convenient angle with b_Tc_T. The difference of elevation between points B and C is 590 ft, and this distance is laid off on line b_Tx to *any convenient scale* to locate point 610. The difference of elevation between *high* point B and *middle* point A is 200 ft, and this distance is laid off on b_Tx to the *same scale* to locate point 1,000. A line is drawn from point 610 to c_T, and a parallel line from point 1,000 then locates d_T. Point D is now at the same 1,000-ft elevation as point A; hence a line joining the two points will be the required strike line.

Figure 12·5(b) shows separately how the dip can be determined after the strike line has been established. The constructions for strike and dip are usually superimposed but have been separated here for clarity. From the *low* point c_T a line c_T3 is drawn perpendicular to the strike line. (Line c_T3 is actually the top view of the dip line.) The difference of elevation between the strike line and the low point is 390 ft, and this distance is laid off along the strike line from point 3 (in either direction) to locate point 4. It is important to observe that this 390-ft distance must be

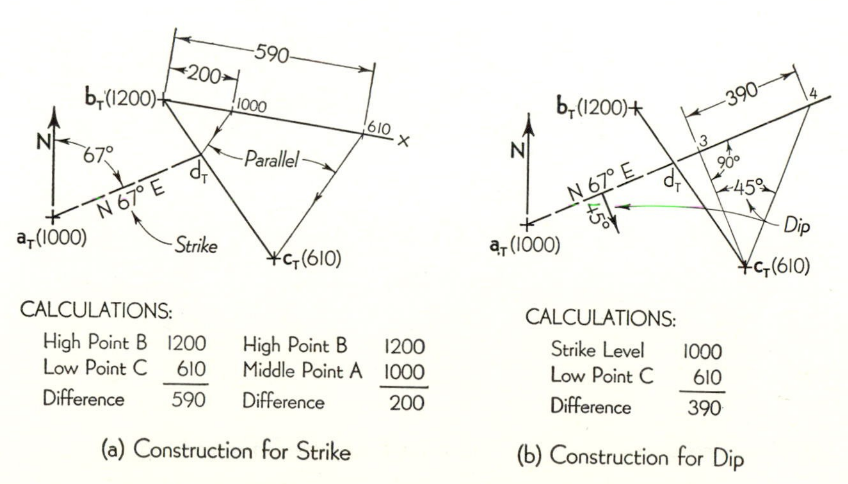

Fig. 12·5. Geologist's three-point method for strike and dip.

laid off to the *true scale of the map.* Connecting point 4 to c_T forms a right triangle that contains the dip angle as shown.

Since the distance 3–4 can be laid off on either side of point 3 without affecting the value of the dip, and since an equivalent construction could be made between the *high* point and the strike line, no significance should be attached to the direction of line c_T4. The direction of the dip—southeasterly in this case—should be determined by inspection.

PROBLEMS. Group 129.

12·7. Apparent dip

The *apparent dip* of a stratum is the slope angle of any line in the plane of the stratum other than the dip line. Since the dip line is the *steepest* line on the plane and is always *perpendicular* to the strike line, an apparent dip line is *not* perpendicular to the strike line and must have a slope angle *smaller* than the dip. An apparent dip line is described by stating the slope angle and bearing of the line as, for example, 19° N 77° E. The slope is taken as downward in the direction of the bearing.

When two apparent dip lines have been established, the strike and dip of the stratum can be determined.

Analysis. The two apparent dip lines can be drawn through any known point on the stratum. These two intersecting lines form a plane whose strike and dip can be determined.

Construction. In Fig. 12·6(*a*) a map shows two apparent dip lines and a point *A* that lie on a stratum. In Fig. 12·6(*b*) the construction for

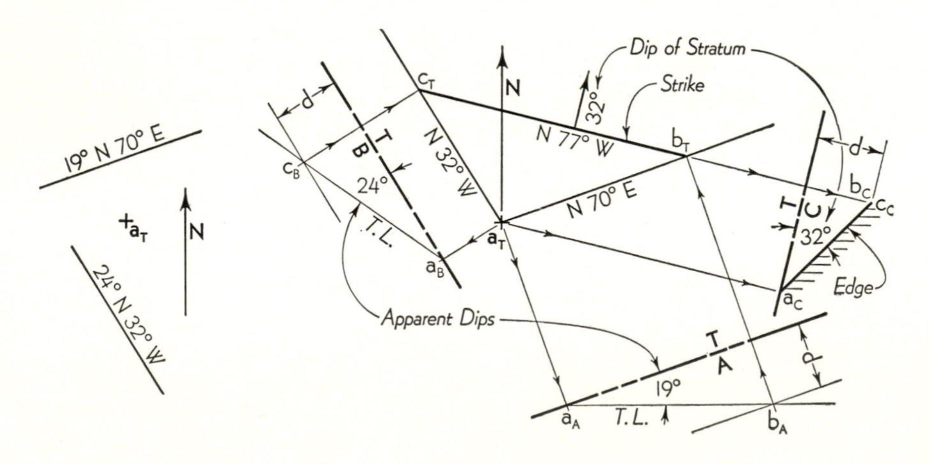

Fig. 12·6. Strike and dip from two apparent dip lines: three-view method.

obtaining strike and dip is shown. Through point a_T the apparent dip lines are drawn at the given bearings and extended indefinitely. View A is selected to show the N 70° E apparent dip line as a true length. Point a_A is conveniently taken on the reference line, and the dip line a_Ab_A is drawn here at the given slope angle of 19°. Point b_A is assumed at any convenient distance d from the reference line, and b_T is located in the top view.

View B is similarly drawn to show the true length and slope angle of the N 32° W apparent dip line, but c_B must be taken at the *same distance d* from the reference line as b_A. Points B and C are now at the same elevation below point A; hence BC is a horizontal line, and the bearing of b_Tc_T is N 77° W, the strike of the stratum.

View C shows plane ABC of the stratum as an edge, and the dip appears here as 32°. The dip arrow in the top view indicates that the stratum dips 32° in a northeasterly direction.

Geologist's method. Figure 12·7 shows the geologist's method of solving the same problem shown in Fig. 12·6. No detailed explanation is needed here, for a comparison of the two constructions reveals that they are geometrically the same. The geologist's more compact solution has been achieved simply by moving the three auxiliary views toward the top view until points a_A, a_B, and a_C all coincide with point a_T.

PROBLEMS. Group 130.

12·8. Strike, dip, and thickness from skew boreholes

When less than three points of outcrop are known on a stratum or vein of ore, additional points must be obtained by sinking boreholes from the surface to the vein. If the borehole is continued through the vein, then two points are established: one on the hanging wall and one on the footwall. A second borehole, skew to the first hole and some distance away,

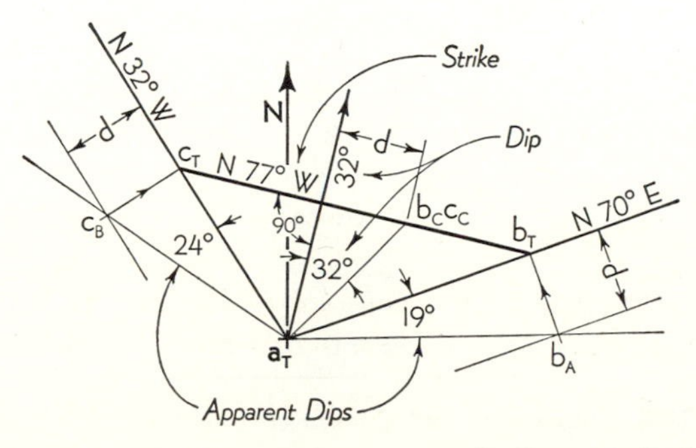

Fig. 12·7. Strike and dip from two apparent dip lines: geologist's method.

will establish two more points. Because the two boreholes are not parallel, the four points—two on the upper plane, two on the lower plane—are sufficient to determine the strike, dip, and thickness of the vein.

Analysis. Connecting the two points on the upper plane and the two points on the lower plane establishes two skew lines—one in each of the parallel surfaces of the vein. A plane passed through one line parallel to the other (Art. 4·19) will be one surface of the vein. Strike and dip can then be determined by the method of either Art. 12·5 or 12·6. Thickness of the vein will appear in an edge view of the vein.

Problem. A borehole was drilled S 47° W at a slope angle of 48°. Measured along the borehole, the vein was struck at a depth of 340 ft, the footwall at 750 ft. A second borehole was drilled S 74° E at a slope angle of 14°, and struck the vein at 260 ft, the footwall at 790 ft. The starting points X and Y of the boreholes are at elevations of 1,200 and 570 ft, respectively, and are located on a map at x_T and y_T, as shown in Fig. 12·8.

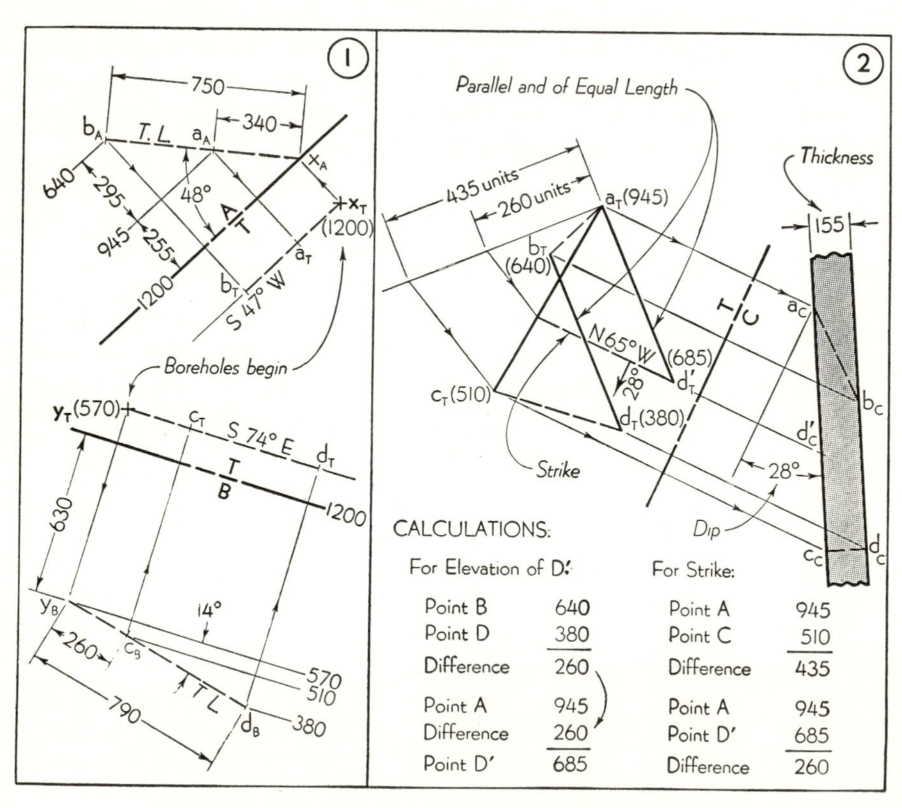

Fig. 12·8. Strike, dip, and thickness from skew boreholes.

Construction

Stage 1. *Establish the location and elevation of all points.*

From x_T lay off the bearing S 47° W of the first borehole. Construct view A to show the borehole in true length and at the given slope angle of 48°. Reference line T–A has been arbitrarily taken through x_A and at the 1,200-ft level. Along the true-length line mark off the given drill depths, 340 and 750 ft, from x_A to locate a_A and b_A. Then point A is 255 ft lower than point X, or at an elevation of 945 ft. Similarly, the elevation of point B is 640 ft above sea level. Locate a_T and b_T on the map view.

In similar manner construct view B and complete the map view of the second borehole. Note that point Y, at an elevation of 570 ft, is 630 ft lower than point X and is so located in view B.

Stage 2. *Determine strike, dip, and thickness from points on the vein.*

To clarify the illustration views A and B have been omitted in stage 2, but the elevations determined in stage 1 have been shown in parentheses. Draw line a_Tc_T and line b_Td_T to form two skew lines—one in each of the parallel surfaces of the vein. Draw line $a_Td'_T$ parallel to b_Td_T and of *equal length,* thus forming plane ACD'. Calculate the elevation of point D' by observing that since D is 260 ft lower than B, then D' must be 260 ft lower than A, or at an elevation of 685 ft.

The strike of plane ACD' can now be determined by the geologist's three-point method (Art. 12·6). The calculations and construction are shown in the figure, and the strike of the vein is found to be N 65° W.

View C is an edge view of plane ACD', the hanging wall of the vein. The parallel footwall also appears as an edge here, passing through points b_C and d_C. Dip and thickness of the vein can be obtained from this view as shown.

In stage 2 the strike, dip, and thickness could also have been obtained by the conventional three-view method (Fig. 12·5). In this method elevations are determined graphically from an elevation view (or views). The elevation view need not be a front view: in Fig. 12·8, for example, either view A or B will serve if all points are transferred to that view.

PROBLEMS. Group 131.

12·9. Location of workings in a mine

The various excavated passages used to mine a vein of ore are called *workings* (Art. 12·3). On small-scale mine drawings or maps these passages may be shown by a single line to represent the center line of the working. The vein of ore may be represented by a single strike line at a specified level and the dip of the vein. The location of workings, their true length and slope, and their intersection with the vein are mining problems that involve the same line and plane relationships discussed in Chap. 4, and may therefore be solved by the same methods. Mining data are customarily given on a map view; elevation, or top-adjacent, views are added as needed, but an elevation edge view of the vein is frequently most useful.

Slopes and drifts are passages lying within the plane of the vein and hence should be treated as lines in a plane (Art. 4·2). Shafts, tunnels, and crosscuts are passages that intersect the vein; hence an edge view of the vein will show the point of intersection (Art. 4·23). The shortest passages to a vein can be established by the methods of Arts. 4·9 and 4·10. To connect passages that are skew use the method of Art. 3·17, 4·21, or 4·22.

PROBLEMS. Group 132.

12·10. Intersection line between two veins of ore

The richest deposit of ore is usually found at the intersection of two veins. It is obviously desirable to locate and mine this portion of the deposit as soon as possible. When the strike and dip of each vein has been found, the intersection line can then be determined.

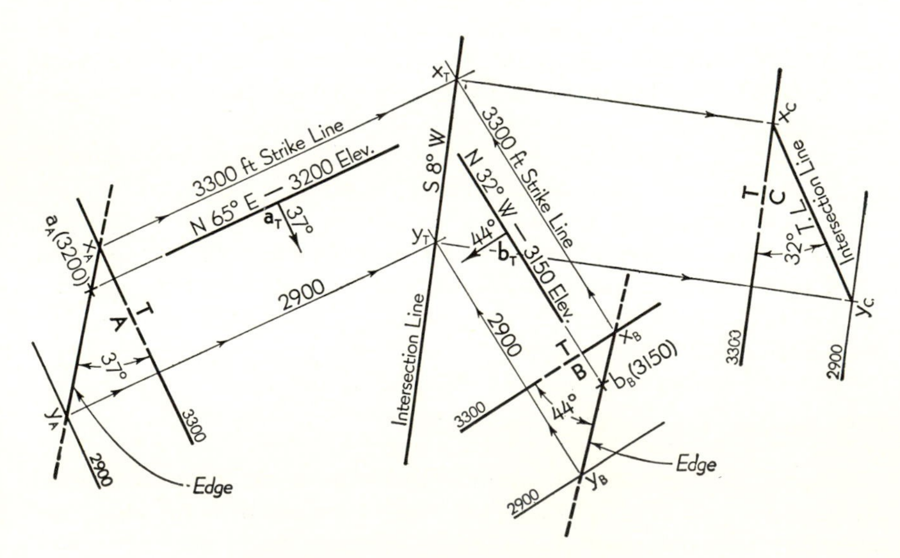

Fig. 12·9. Intersection line of two veins.

Analysis. An edge view of each vein will show the given strike line as a point. At the same elevation in each view, a new strike line can be assumed on each vein. Since the two new strike lines are at the same elevation, they must intersect at a point on the intersection line of the veins. At some other common elevation two more strike lines can similarly be assumed to locate a second point on the intersection line. These two points determine the required line of intersection of the veins.

Problem. In Fig. 12·9 point A, at an elevation of 3,200 ft, is on the outcrop of a vein that has a strike of N 65° E and dips 37° SE. Point B, at an elevation of 3,150 ft, is on the outcrop of a second vein that has a strike of N 32° W and dips 44° SW. The bearing and slope angle of the intersection line are required.

Construction. The location of a_T and b_T is shown, and the given strike lines have been drawn through these points. Reference line T–A, perpendicular to the strike line through a_T, has been arbitrarily assumed to represent a level of 3,300 ft. All other elevations must be measured from this level at the same scale as the given map. Through point a_A the edge view of the vein is drawn at the given dip. View B is similarly drawn to show the second vein as an edge. Note that reference line T–B also represents the 3,300-ft level.

In view A at the 3,300-ft level a strike line on the first vein will appear as a point at x_A. In view B at the same 3,300-ft level a strike line on the second vein appears at x_B. In the top view these two strike lines intersect at x_T, one point on the required intersection line. Similarly, strike lines at the 2,900-ft level appear at y_A and y_B and intersect at y_T. Line x_Ty_T, with a bearing of S 8° W, is the required intersection line between the two veins. In the true-length view C the slope angle of the intersection line is found to be 32° downward to the south (32° S).

12·11. Faults

A fracture in the earth's crust followed by a displacement of the strata on either side of the break is called a *fault*. Within a limited area the fracture is frequently a plane surface, as shown in Fig. 12·2, and the *fault plane* may therefore be described by its strike and dip just as strata are. The displaced strata may slide with *translatory* motion to a parallel position, or they may twist with a *rotational* motion to a skew position. Only the translatory case will be considered here.

In Fig. 12·10 a rectangular block of the earth's crust is shown to illustrate a translatory fault movement. The block has fractured along the fault plane AEF, and the right half of the block has moved downward and to the rear. The blocks are still in contact, but the point A has slipped along the fault plane to a new position A'. The following terms, illustrated in Fig. 12·10(a), are used to describe this fault movement:

Net slip. The distance (AA'), measured in the fault plane, between two formerly adjacent points.

Strike slip. The component (AB) of the net slip parallel to the strike of the fault plane.

Dip slip. The component (BA') of the net slip parallel to the dip of the fault plane.

Plunge. The slope angle (CAA') of the net slip line.

Slickensides. The smooth, striated, polished surfaces of the fault plane that are produced by the sliding of one surface on the other.

Striations. The scratches or minute grooves, parallel to the direction of fault movement, that appear on the slickensides.

Pitch. The angle that a line in a plane makes with a horizontal line in that plane. Thus the *pitch of the striations* is the angle (BAA') that they make with the strike of the fault plane.

After a fault movement has occurred, erosion may eventually level the area again, and the faulted region will then appear as in Fig. 12·10(b). On the ground surface the *offset* of the vein indicates that faulting has occurred here, but this external evidence does not show the *true direction* of slip. The fault shown here *could* have been caused solely by a displacement AB in the direction of the strike or solely by a displacement CD in the direction of dip. The fact that the net slip was actually in a diagonal direction can be determined only by the pitch of the striations.

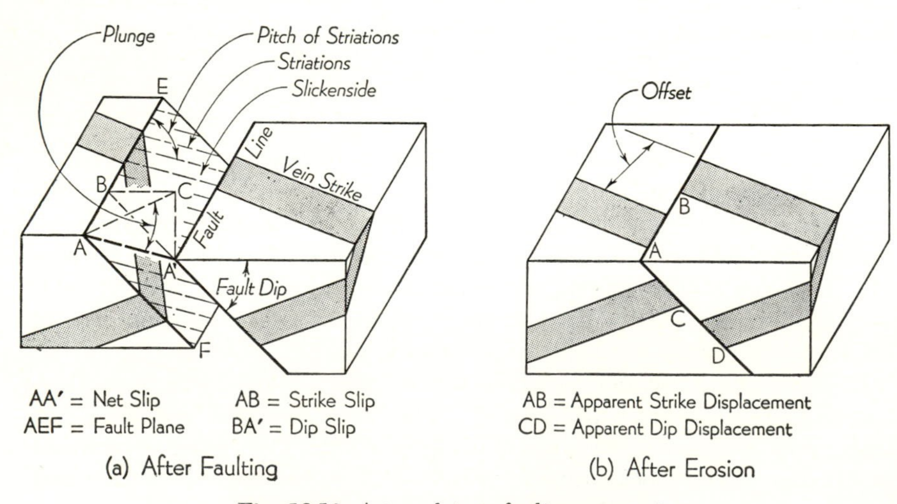

Fig. 12·10. A translatory fault movement.

A knowledge of fault movements is important in mining. A drift or gangway driven along a vein of ore may end abruptly at a wall of rock. The wall is a fault plane, and the miner must now determine the direction and distance of the movement in order that mining may be resumed in the faulted vein.

12·12. Displacement of a single faulted vein

A translatory fault movement displaces one portion of a vein to a *parallel* offset position. Then the strike and dip of the parallel faulted vein will be the same as for the original vein, and *one known point* on the faulted vein will fix its position. The direction of the fault plane can be described by its strike and dip. The *direction of displacement* will be indicated by the pitch of the striations on the fault plane. From these facts the *net slip* of the fault can be determined.

Analysis. An edge view of the original vein will also show the parallel faulted vein as an edge. A true-size view of the fault plane will show the pitch of the striations and hence the direction of displacement. This direction of displacement can then be shown in the vein edge view to determine the amount of slip.

Problem. From the bottom of a shaft at S (see Fig. 12·11) a drift is driven along a vein of ore at an elevation of 800 ft. The strike of the vein is N 30° W, and the dip is 38° NE. At A the drift ends at a fault having a strike of N 70° E and a dip of 55° NW. Striations on the slickenside are pitched 70° downward in a northeasterly direction. A borehole at C relocates the hanging wall of the faulted vein at an elevation of 770 ft. Find the net slip, strike slip, dip slip, bearing, and plunge of the fault displacement.

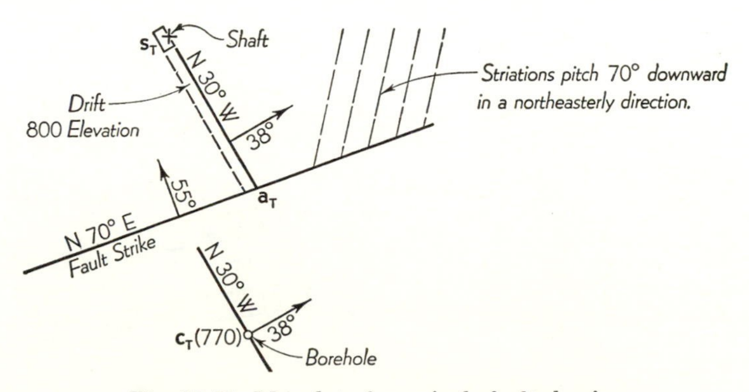

Fig. 12·11. Map data for a single faulted vein.

Construction. In Fig. 12·12 view A is an edge view of the offset veins given in Fig. 12·11. The hanging wall of the original vein passes through point a_A at the given dip of 38° northeasterly, and the hanging wall of the parallel faulted vein passes through c_A at the bottom of the borehole. Vein thickness can be shown here if desired.

View B is an edge view of the fault plane dipping 55° northwesterly. In view C the fault plane appears true size, and its strike line is perpendicular to reference line B–C. In this view the striations also appear true size at 70° to the strike line. Note that the 70° angle must be laid off in the direction shown by $a_C d_C$ in order that $a_T d_T$ will bear northeasterly as specified in this problem. Point d_C may be assumed at any convenient point along the striation.

Line AD is the *direction* (but not the distance) of displacement; hence the assumed point D must now be returned to each of the other views as shown. In the top view $a_T d_T$ shows the bearing of the displacement to be N 12° E. In view A, $a_A d_A$ is extended to meet the faulted vein at a'_A. Then A' is the displaced position of a point that was formerly adjacent to point A. Line AA' can now be returned to view C, where it will appear true length and equal to the *net slip*.

By projecting the net slip onto the fault strike line at $a_T b_T$ the *strike slip* is obtained in the top view. The *dip slip* appears on the edge view

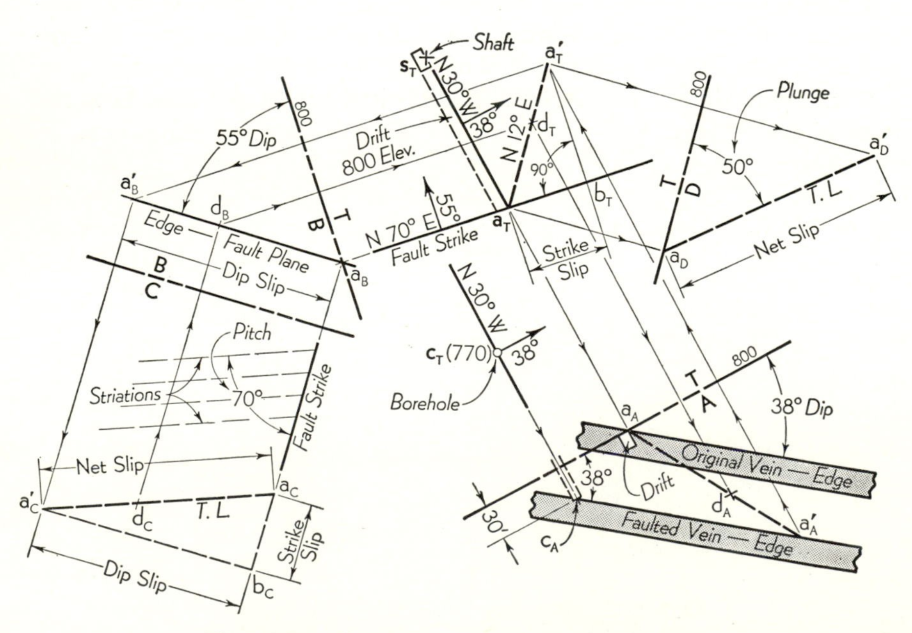

Fig. 12·12. Displacement of a single faulted vein.

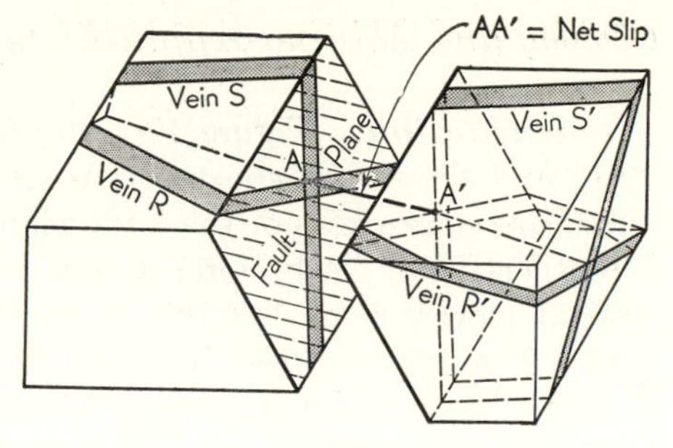

Fig. 12·13. Fault movement of two nonparallel veins.

of the fault plane. But note that these two components of net slip also appear in view C as two sides of a right triangle whose hypotenuse is the net slip. *Plunge* (slope angle of net slip) must be measured in the true-length elevation view D.

12·13. Displacement of two faulted nonparallel veins

If a fault plane intersects *two nonparallel veins* and the offset portions can be located, then the net slip of the fault can be determined without the pitch of the striations. In Fig. 12·13, for example, the original veins R and S intersect each other and the fault plane at a common point A. After a translatory fault movement the offset veins R' and S' similarly intersect at a point A' on the fault plane. Then A and A' are formerly adjacent points, and the line AA' represents the net slip.

Analysis. Five different planes—R, S, R', S', and F (the fault plane) —are involved in this problem. The intersection line of planes R and F will cross the intersection line of planes S and F to establish point A. Similarly, the intersection lines of R' and F and S' and F will cross at A'. The line joining A and A' is the required net slip. The solution can therefore be obtained by locating four plane intersection lines by the method of Art. 12·10.

Problem. In Fig. 12·14 r_T and s_T are outcrop points on the hanging walls of two veins R and S, and r'_T and s'_T are similar outcrop points on the offset portions of the same veins. Point f_T is a point on the fault plane. The strike, dip, and elevation at each point are given on the figure. Find the net slip, strike slip, dip slip, and plunge of the fault displacement.

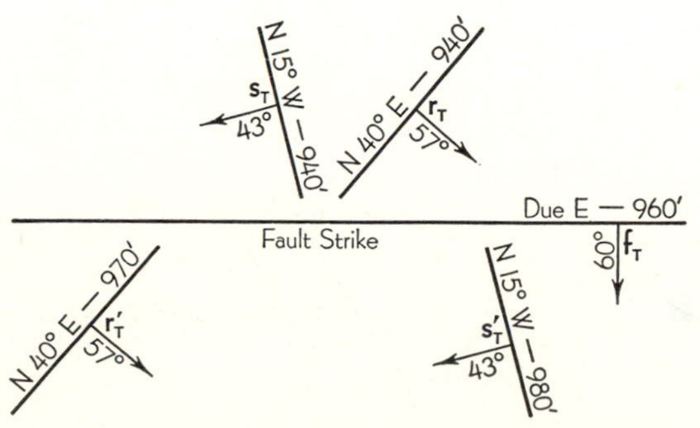

Fig. 12·14. Map data for two faulted nonparallel veins.

Construction. Figure 12·15 reproduces the map data given in Fig. 12·14 and shows the construction for fault displacement. Views A, B, and C are edge views of the fault plane, veins S and S', and veins R and R', respectively. All reference lines have been assumed at the same convenient elevation of 1,000 ft, and a 900-ft level is also shown in each view. Employing these two levels, BD is the intersection line of planes R' and F (compare Fig. 12·9). Similarly, CE is the intersection line of planes S' and F. Extending b_Td_T and c_Te_T establishes a'_T, the common point in planes R', S', and F.

The intersection lines of planes R and S with the fault plane will be parallel to those found for planes R' and S'; hence it is only necessary to locate one point, m_T and n_T, on each line. Then m_Ta_T is drawn parallel to b_Td_T and n_Ta_T parallel to c_Te_T to locate point a_T. Points A and A' both lie in the fault plane and hence appear at a_A and a'_A in edge view A.

The true-length top-adjacent view D of line AA' shows the net slip and plunge. The component of $a_Ta'_T$ parallel to the fault plane strike line is the strike slip. Distance $a_Aa'_A$, measured along the fault plane, is the dip slip.

PROBLEMS. Group 133.

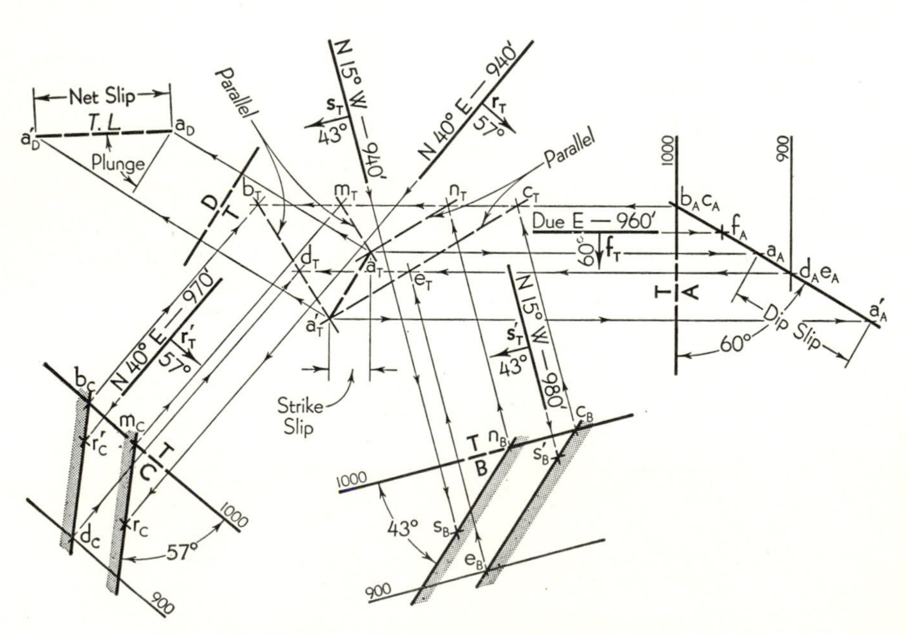

Fig. 12·15. Displacement of two faulted nonparallel veins.

12·14. Outcrop of a stratum or vein

When a stratum of rock or vein of ore intersects the surface of the earth, the exposed edge is called an *outcrop* (see Fig. 12·2). All or part of the outcrop may be lightly covered with topsoil, but if the position of the stratum is known, then the area of probable outcrop can be predicted. The most economical way to mine the seam or vein—by shaft, tunnel, incline, drift, or open pit—can then be planned.

Analysis. An elevation view that shows the stratum as an edge will also show the irregular surface contour lines as straight horizontal lines. In such a view it is immediately apparent where each contour line intersects the walls of the stratum. When these points of intersection are shown on the map view, they can be connected to outline the outcrop area.

Problem. On the topographic map in Fig. 12·16 an outcrop of coal has been discovered at point a_T on the 620-ft contour line. On the opposite side of the hills a second outcrop has been found at point b_T on the 580-ft contour. Closer examination reveals that points A and B both lie on the upper bedding plane (hanging wall) of the seam of coal. A vertical borehole drilled from point s_T on the surface (elevation 752 ft) struck the seam at point C at a depth of 86 ft (elevation 666 ft). Continued drilling struck the underlying rock (footwall) at point D at a depth of 110 ft (elevation 642 ft). Find the strike, dip, thickness, and outcrop lines for the seam.

Construction. In Fig. 12·17 the map data of Fig. 12·16 is reproduced. The location of three points, A, B, and C, on the upper bedding

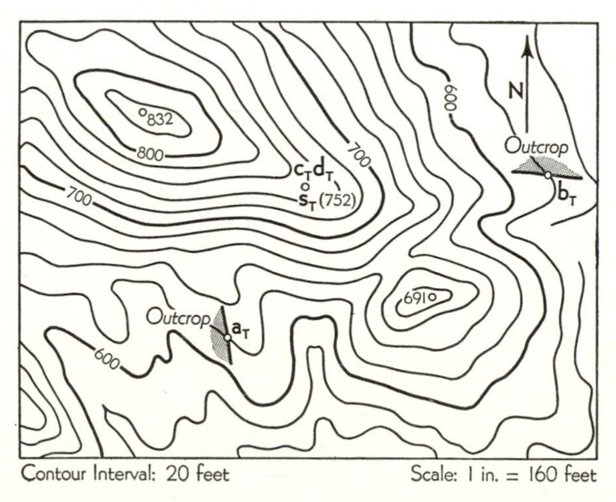

Fig. 12·16. Map showing points of outcrop for a bed of coal.

plane is known, hence the strike and dip can be determined by the method of either Art. 12·5 or 12·6. However, since an edge view will be needed to determine thickness and outcrop, the conventional three-view method of Art. 12·5 has been employed here. Line *AE*, a horizontal line in the plane *ABC*, shows the strike of the plane to be N 53° E. In view *A* plane *ABC* appears as an edge dipping 38° in a southeasterly direction. Point *D* lies in the parallel lower bedding plane; hence a line through d_A parallel to b_Ac_A establishes this lower plane. The thickness of the bed is the distance between these parallel planes.

In view *A* the contour lines again appear as horizontal lines (parallel to *T–A*), and therefore these lines are spaced at 20-ft intervals corresponding to the same levels shown in the front view. Since the coal bed appears as an edge in view *A*, we can readily note in this view where each contour line intersects both the upper and lower bedding planes. The 700-ft contour, for example, intersects the upper bedding plane in view *A* at point 1; a parallel extended to the top view intersects the 700-ft con-

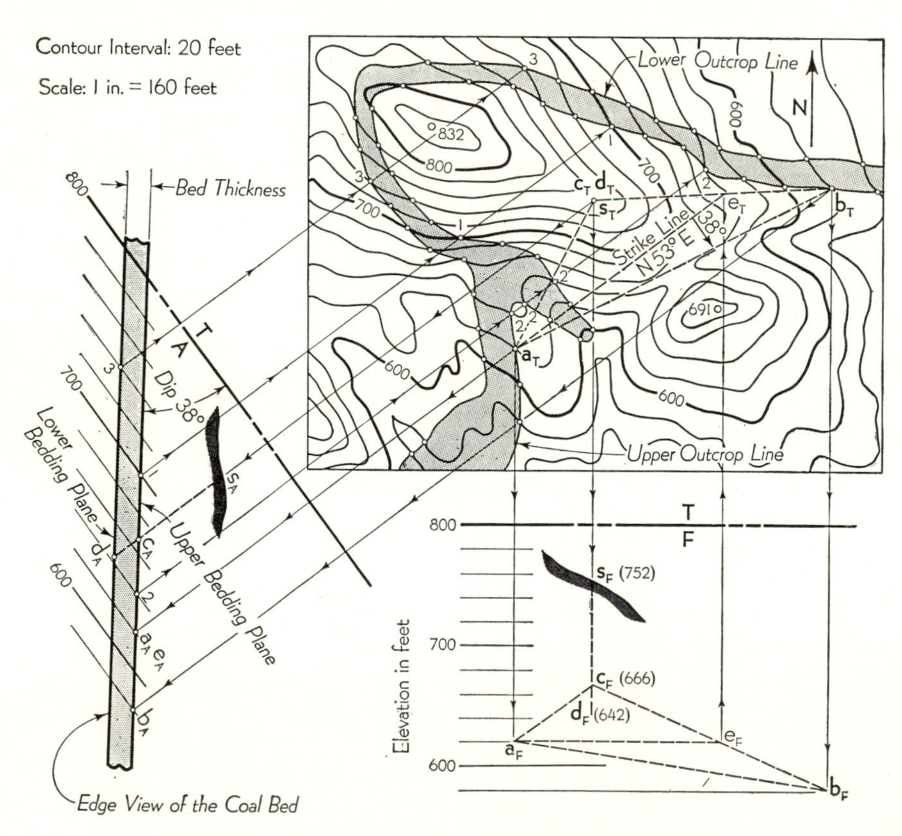

Fig. 12·17. Location of outcrop lines for a bed of coal.

tour on the map at two points labeled 1. These are two points on the probable line of outcrop of the upper bedding plane. The 640-ft contour intersects the upper bedding plane at point 2 in view A; a parallel extended to the top view locates four points of outcrop on the 640-ft contour. By continuing this procedure, points of outcrop can be located on each contour to establish the upper outcrop line. The lower outcrop line is determined in the same manner except that contour intersection points are taken on the lower bedding plane (note points 3 on the 740-ft contour line). The shaded area on the map represents the probable outcrop of the coal bed.

12·15. Profiles and sections

Profiles. A *profile* is a vertical section of the earth's surface taken along a given line on the surface. The given line may be either straight or curved, but the length of the profile must equal the true length of the line. On the topographic map of Fig. 12·18 the line $ABCD$ is the center line of a proposed highway. The station numbers 17, 18, 19, etc., along this center line represent horizontal distances in hundreds of feet. In

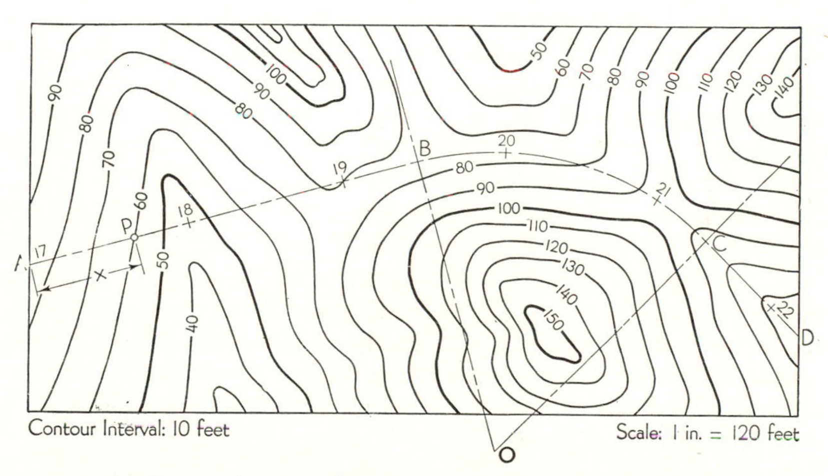

(a) Topographic Map of Proposed Highway along Line ABCD

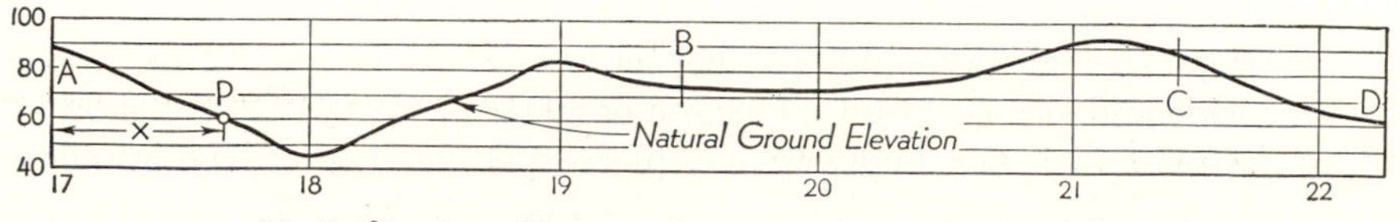

(b) Profile along Highway Center Line (Hundreds of feet)

Fig. 12·18. Construction of a profile.

the profile these distances are laid off along a horizontal base line to establish the true length of the section. The vertical scale, arbitrarily beginning at 40 ft elevation, has been divided into 10-ft intervals corresponding to the contour intervals.

Points on the profile curve may be established from the topographic map. On the map point P, for example, is on the center line AB, is also on the 60-ft contour line, and is distance x from station 17. On the profile point P is located on the 60-ft level, distance x from station 17. Thus by noting where each contour line crosses the line $ABCD$ on the map enough points may be determined to draw the profile. The elevation of points falling between two contour lines may be estimated by interpolation.

It is common practice on very long profiles to exaggerate the vertical scale, making it five, ten, twenty, or more times as large as the horizontal scale. Such exaggeration is imperative on profiles 50 miles long, for example, for the scale reduction on paper is so great that even large differences of elevation become almost imperceptible if the vertical and horizontal scales are equal, as in Fig. 12·18(b).

Sections. A *section* is a vertical section taken at right angles to the profile line. Figure 12.20(c) shows sections taken at three different station points along the completed highway. In each section the irregular dash line shows the slope of the ground before the highway was built. Line X–X on the map shows the location of the first section taken 95 ft from station 17 (labeled "station 17 + 95," or 1,795 ft from the zero station point).

The construction of a section is similar to that of a profile. A vertical scale of elevations corresponding to the contour intervals is established, and horizontal distances are measured on either side of a vertical center line. Then any point on the ground, such as Q in section X–X, may be located on the map at the point where the section cuts the 60-ft contour line. Transferring distance y from the map to the section locates Q on the same 60-ft level here and establishes one point on the ground line. Additional points at each contour level are located in the same manner to complete as much of the section as is desired.

12·16. Cuts and fills along a level road

The proposed highway, shown as center line $ABCD$ in Fig. 12·18, is to be 40 ft wide and at a constant elevation of 80 ft. Figure 12·20(a) shows the same map with the highway completed. Reference to the profile (repeated from Fig. 12·18) shows that between stations 17 and 19 the road crosses a valley; hence here the roadbed has been built up to the required 80-ft level by dumping earth to form a mound across the valley.

This earthwork construction is called a *fill,* and its cross-sectional shape is shown in section *X–X*. In the vicinity of station 21 the profile shows that the natural ground contour had to be cut down to the level of the highway. The section taken along line *O–Z* indicates the amount of earth that was excavated at this point. Such an excavation is called a *cut.* At section *O–Y* a cut was necessary on one side of the highway and a fill on the other side.

Angle of repose. If the sloping side, or embankment, of a cut or fill is too steep, the loose earth will slide downward until the slope of the embankment has been reduced to a definite maximum angle. This maximum angle of slope that the side of a pile of loose granular material can have without sliding is called the *angle of repose.* The value of the angle of repose depends upon the material. Fine sand, for example, when poured on a horizontal plane from a fixed point, piles up in the form of a right circular cone whose base angle is approximately 31°. The addition of more sand only enlarges the cone without altering the slope of its sides. Average values of angle of repose for other materials are as follows: dry earth, 39°; gravel, 39° to 48°; bituminous coal, 38°; iron ore, 35°. In engineering construction work, slopes are frequently given as a ratio. 1½ to 1, for example, indicates a slope angle whose tangent is 1/1½, or about 33°40′. Note that the first number in the ratio is the horizontal distance and the second number the vertical distance.

Lines of cut and fill. Figure 12·19 shows a typical cut and fill along a level road. The fill has been dumped at a slope of 1½ to 1, thus forming a sloping plane surface that intersects the natural ground surface along an irregular line called the *line of fill.* Contour lines taken at

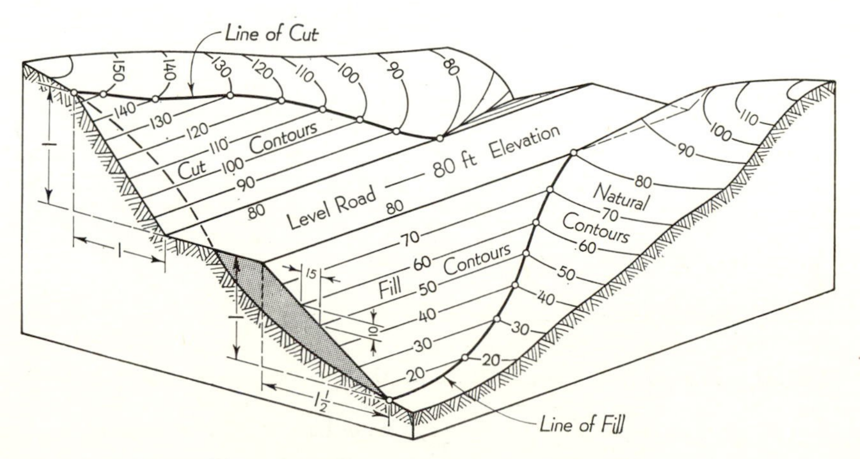

Fig. 12·19. Cut and fill along a level road.

10-ft vertical intervals on the sloping side of the fill will be parallel to the road and will be uniformly spaced 15 ft apart *horizontally* (15 is to 10 as $1\frac{1}{2}$ is to 1). Points along the line of fill can therefore be found by noting the intersection point of each fill contour with the natural contour of the same level. The *line of cut* can be determined in the same manner except that the parallel cut contours will be spaced 10 ft apart horizontally.

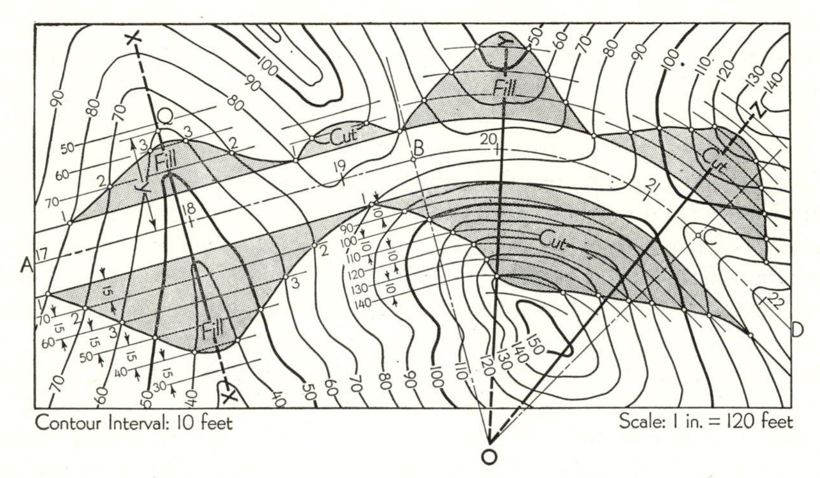

(a) Topographic Map of Highway Showing Cuts and Fills

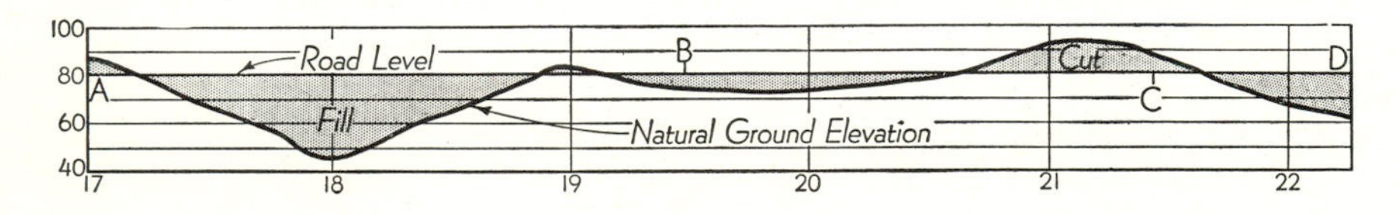

(b) Profile along Highway Center Line (Hundreds of feet)

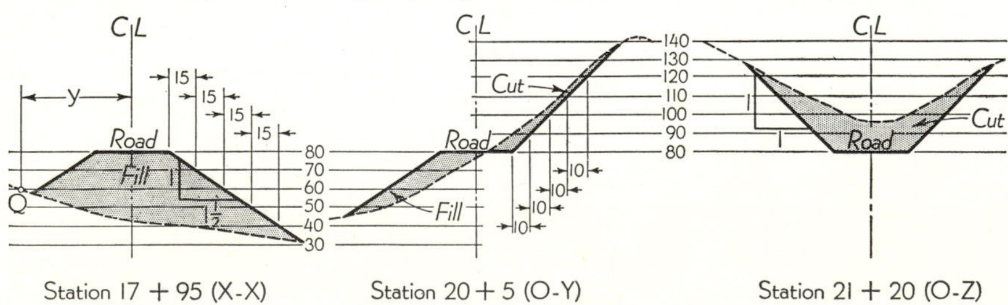

(c) Sections Perpendicular to Highway Center Line at Selected Stations

Fig. 12·20. Location of cuts and fills along a level highway.

Problem. In Fig. 12·20 the slope of all fills has been taken as 1½ to 1, a commonly used value for earthwork fills. The slope of all cuts is 1 to 1. The level road is at a constant elevation of 80 ft. To determine the width of right of way and the volume of earth to be moved the lines of cut and fill are required.

Construction. On the map between stations 17 and 19 (where the profile indicates a fill), fill-contour lines are drawn parallel to the highway, on each side, at 15-ft intervals. These fill contours are numbered consecutively to indicate their *decreasing* elevations. Then the two 70-ft fill-contour lines (one on each side of the highway) intersect the natural 70-ft contour at four points labeled 2. These are four points on the line of fill. Similarly, the intersection of the 60-ft contours locates four points labeled 3.

In those areas where the ground is higher than the highway a 1-to-1 cut must be made; hence here the cut contours must be spaced at uniform *horizontal* intervals of 10 ft. On the map these cut contours are therefore drawn at 10-ft intervals parallel to the highway and are numbered outward from the highway with *increasing* elevations. From *B* to *C* the highway center line is an arc with center at *O*; hence the sloping surface of a cut or fill in the sector *BOC* will be a portion of a right circular cone whose vertical axis is at *O*. Therefore between the two radii *OB* and *OC* the contour lines on the embankments must be drawn as equally spaced circular arcs with center at *O*. It should be noted that contour lines on all cuts are *above* the 80-ft road level and are spaced at 10-ft horizontal intervals because the slope of cut is 1 to 1. Contour lines on all fills are *below* the 80-ft road level and are spaced at 15-ft horizontal intervals because the slope of fill is 1½ to 1.

12·17. Cuts and fills along a grade road

In Fig. 12·21 the road has a uniform grade upward from an elevation of 50 to 100 ft in the distance shown. Fill dumped along the straight edge of this road will form a sloping plane surface, but the fill-contour lines will not be parallel to the road, as they are for a level road. Fill dumped from a single point *A* at the edge of the highway will form a cone having the same slope as the whole fill. Thus if the slope of fill is 1½ to 1, then a cone 10 ft high will have a base radius of 15 ft, a cone 30 ft high, a base radius of 45 ft, etc. The surface of the fill will be a plane *ABC*, containing line *AB* and tangent to the cone along element *AC*. If the slope-of-fill cone is 30 ft high, as shown, and its vertex is at 80 ft elevation, then the 45-ft radius base of the cone will be at 50 ft elevation. The 50-ft fill contour will be tangent to this cone base and meet the edge of the highway at the 50-ft level. All other fill contours will be parallel to this one and spaced at equal intervals along the edge of the highway as shown.

Cut contours can be established in a similar manner by using an inverted cone. The slope-of-cut cone shown in Fig. 12·21 has a slope of 1 to 1; hence a cone 50 ft high will have a base radius of 50 ft. The cone base is at an elevation of 100 ft, and therefore the 100-ft cut contour *DE* is tangent to this cone base and meets the edge of the highway at the 100-ft level. All other cut contours will be parallel to this one as shown.

Problem. The road shown in Fig. 12·22 has a 10 per cent grade; hence the contour lines on the road—50, 60, 70, etc.—are spaced at 100-ft intervals along the highway center line. The slope of fills is $1\frac{1}{2}$ to 1, of cuts, 1 to 1. The lines of cut and fill are required.

Construction. Contour lines south of the highway are generally lower than the road; hence a fill will be required here. Point *A* on the south edge of the highway is at 80 ft elevation, and this point is selected as the vertex of a slope-of-fill cone. The concentric circles with centers at *A* represent cone-base circles at 10-ft intervals of height; the cone 30 ft high has a base radius of 45 ft at an elevation of 50 ft. The 50-ft fill contour *BC* is drawn tangent to this cone base at *C*, and additional parallel, equally spaced fill contours are then added as needed.

Along the curved edge of the highway partial slope-of-fill cones must be drawn at regular intervals as at *D* and *E*, and the radii of the cone base arcs must increase in multiples of 15 ft. In this region the fill contours are drawn as parallel spiral curves tangent to successively larger arcs as shown.

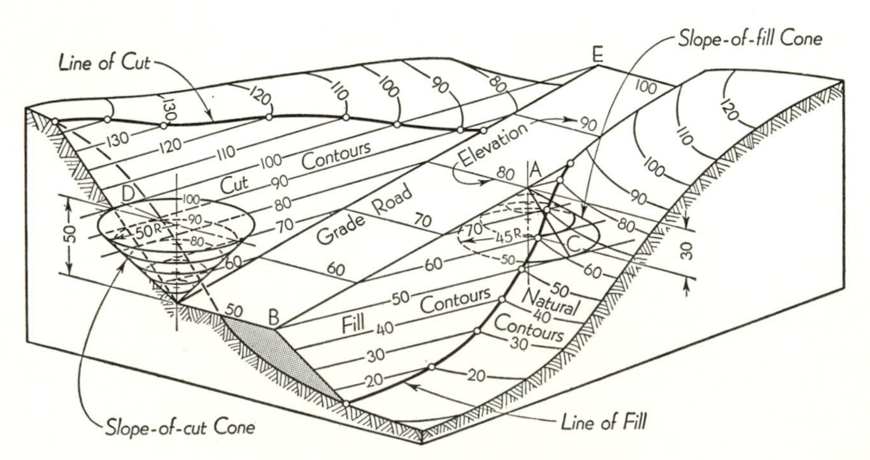

Fig. 12·21. Cut and fill along a grade road.

When a complete set of fill contours has been established, points on the line of fill, such as 2 and 3, are located at the intersection of each pair of fill and natural contours that have the same elevation.

In the area north of the road a cut is indicated; hence an inverted slope-of-cut cone is drawn at point *F*. Since the slope of cut is 1 to 1, all cone-base radii here will increase in multiples of 10 ft. The 80-ft cut contour *GH* is tangent to the 30-ft-radius cone base which is 30 ft above the highway at this point. Additional equally spaced, parallel cut contours are drawn as shown and then extended around the curve in parallel spirals tangent to the cone arcs at *J* and *K*. It is useful to observe that:

> *Along a grade road in the uphill direction, cut-contour lines converge toward the highway, fill-contour lines diverge from it.*

PROBLEMS. Group 134.

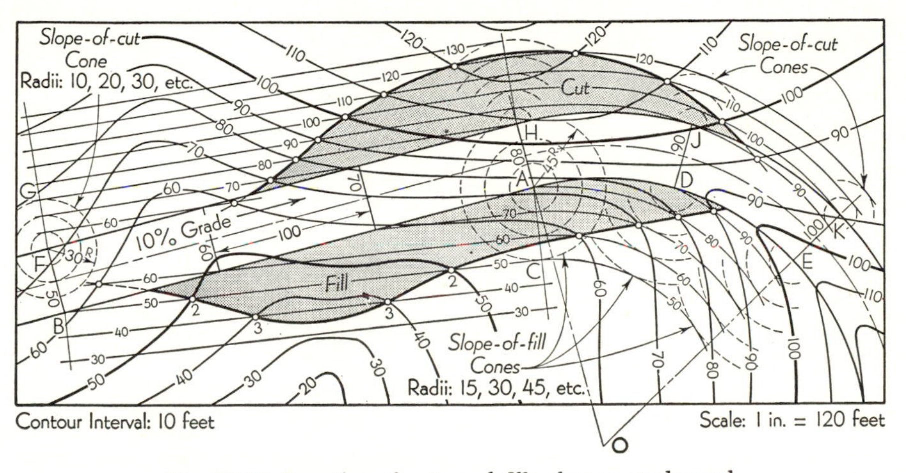

Fig. 12·22. Location of cuts and fills along a grade road.

PROBLEMS

Descriptive geometry is basically a "problem course," and a thorough mastery of the subject can be gained only by the actual solution of problems on the drawing board. The problems included in this chapter cover completely all the subjects discussed in the text; they have been divided into convenient topical groups for easy reference. Problem 75·6, for example, is the sixth problem in Group 75, Tangent-plane Convolutes, a subject discussed in Arts. 6·30 and 6·31.

Each group of problems is preceded by a general statement for all or part of the problems in that group. This general statement should always be read first. Each problem statement is preceded by one or more of the following four items, always given in the same order:

1. The problem number.
2. A figure number, which refers to a specific, fully dimensioned drawing of the given data.
3. The letter A, B, or C, which indicates the recommended sheet layout. All problems have been designed to fit one of the three sheets shown in Fig. 1. If the sheet layout is the same for all problems in the group, it is given in the group statement.
4. The scale of the drawing. If no scale is given, it is then understood that all dimensions are in inches and the drawing is to be made full size. When two scales are given, the first is recommended; the second, in parentheses, is a second choice.

When a problem is stated to be the same as a previous problem, this means that the same statement applies but that the figure reference is different. In a few cases the position of the drawing on the sheet has been indicated; in general, however, the draftsman is expected to plan the initial placement of the views.

Wherever possible, numerical answers have been given at the end of the problem statement. These answers are given simply to provide a check on the correctness of the solution and should not be accepted as mathematically exact. Most of these answers were obtained by graphical solution and hence are only approximately correct. Always record the dimension on your drawing—not the text answer.

In many of the problem figures a special scheme of dimensioning has been employed to simplify the drawing. Figure 8 is typical. Here the reference line (always *T–F* when not so labeled) serves as a base line for all layout dimensions. Distances between adjacent parallels, measured along the reference line, are given by numerals placed just above or below

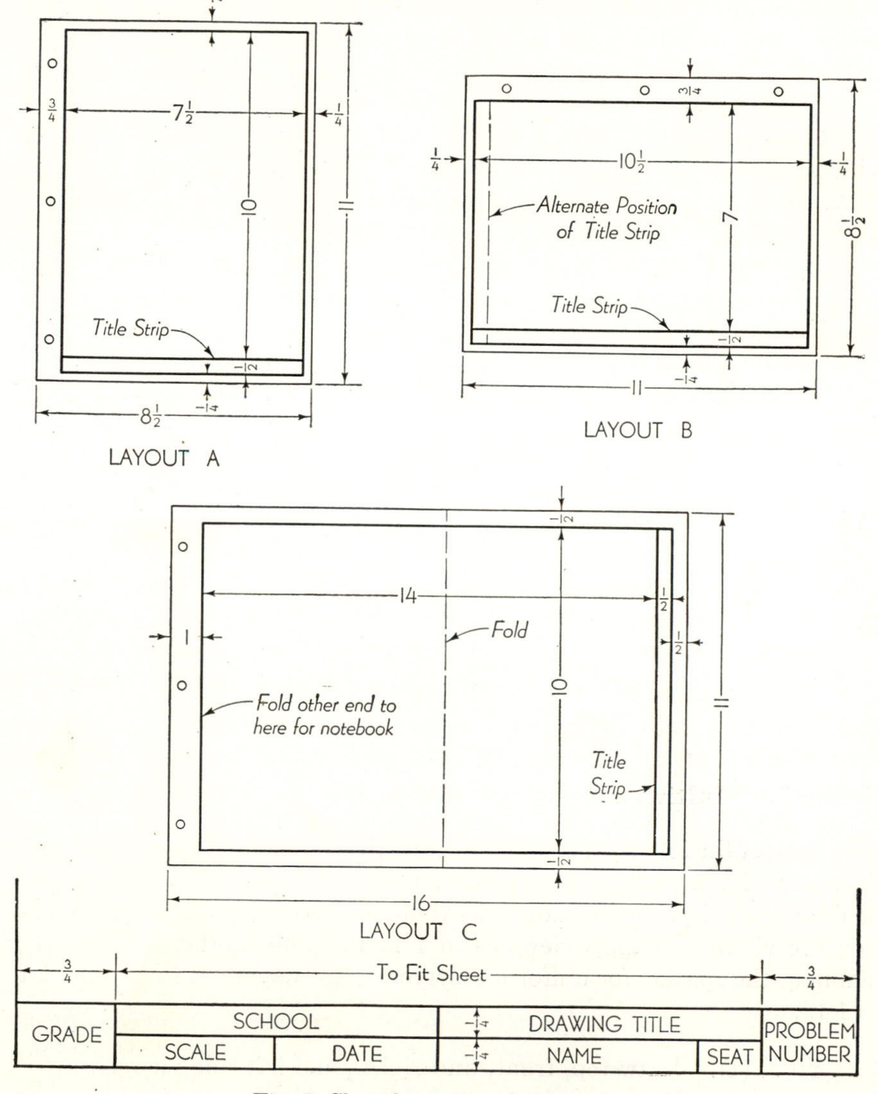

Fig. 1. Sheet layouts and title strip.

the line. These dimensions are lettered parallel to the reference line. Distances above and below the reference line, measured along the parallels, are given by numerals placed on the parallels. These dimensions are lettered at right angles to the reference line and always refer to the total distance from the reference line to the point on that parallel. When two or more points lie on the same parallel, they are dimensioned conventionally as in Fig. 8(3).

In the solution of problems the following procedure is recommended:

1. Read the *complete* problem statement carefully to be sure that you know exactly what is given and what is required.

2. Study the given figure before starting to lay it out on the sheet. It is frequently advisable to make a rough freehand sketch on scrap paper in order to decide what new views are needed, approximately where they will fall on the paper, and hence about where to place the given views on the sheet (see Fig. 3·25). Such advance planning *saves time.* Remember that the whole solution must lie on the paper.

3. Lay out the given views, and *check* all dimensions a second time. All problem figures are drawn to true scale proportion, and any serious error in layout should be evident by comparison with the figure in the text. Do not dimension the drawing.

4. Make the required solution, labeling neatly all points and reference lines.

5. Scale and record the required distances or angles, properly dimensioning each item to show where on the drawing the answer was obtained. In general, distances should be accurate to plus or minus 1/50 in., angles to plus or minus 15′.

6. When check answers are obtained, record the exact measurement in each case, not simply the duplicate of the first answer found. Graphical solutions are always subject to numerous small inaccuracies, and it is very unlikely that any two independently obtained answers will agree perfectly.

Group 1. Multiview drawings (Chap. 1)

In each of the following problems, draw the required views of the object shown in Fig. 2. Dimensions given are in inches, and all objects are to be drawn full size. Show hidden as well as visible lines. Do not dimension the views. Arrange views as in Fig. 1·8 (unless otherwise stated), planning the spacing for uniform margins. Use Layout B for Probs. 1·1 to 1·12.

1·1 FIG. 2(1). Draw top, front, and right-, and left-side views.
1·2 FIG. 2(2). Draw top, front, and right-, and left-side views.

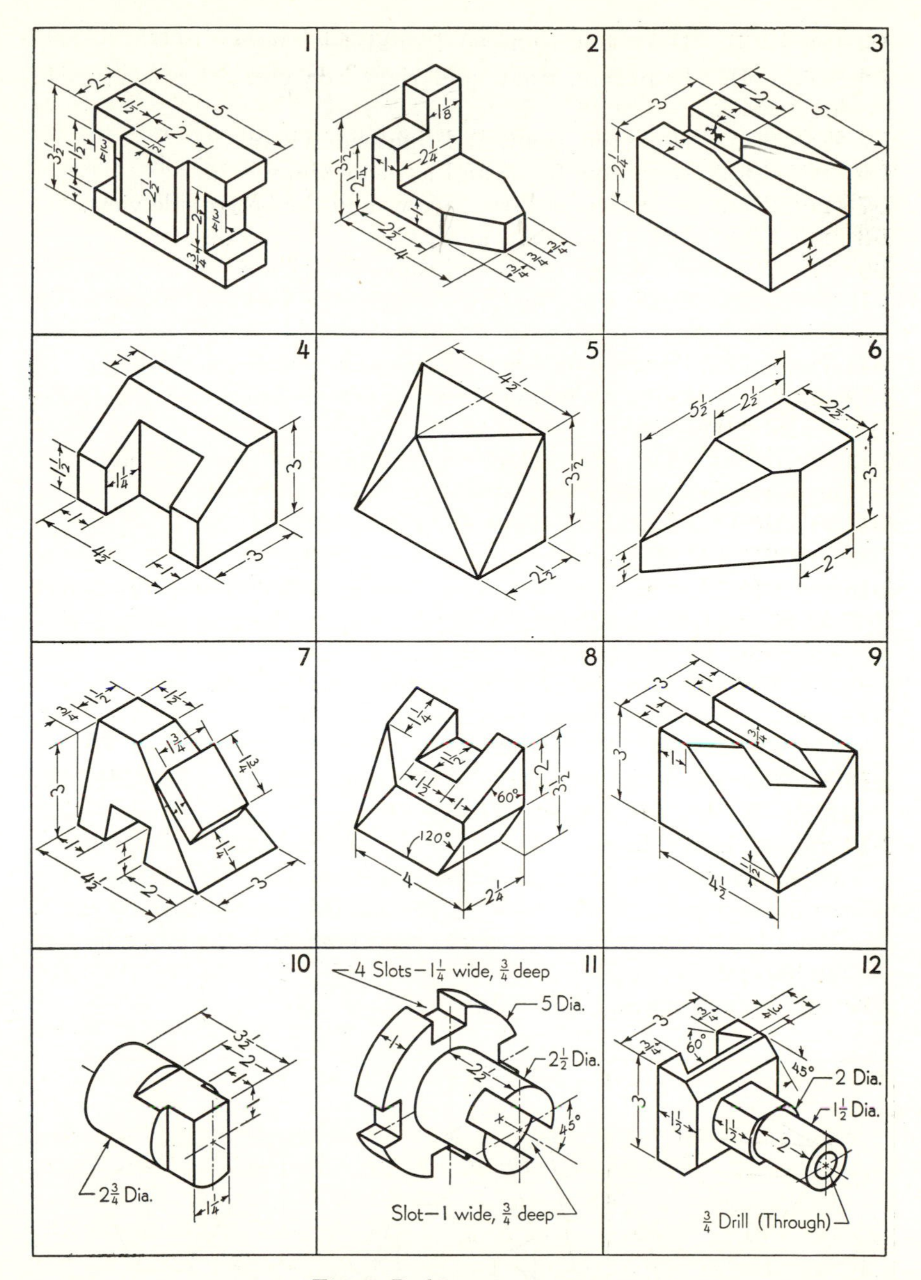

Fig. 2. Problems 1·1 to 1·15.

1·3 FIG. 2(3). Draw top, front, and right-side views.

1·4 FIG. 2(4). Draw top, front, and right-side views, arranged as in Fig. 1·9.

1·5 FIG. 2(5). Draw top, front, and right-side views.

1·6 FIG. 2(6). Draw top, front, and left-side views.

1·7 FIG. 2(7). Draw top and front views and the better side view.

1·8 FIG. 2(8). Draw top, front, and right-, and left-side views.

1·9 FIG. 2(9). Draw top, front, and right-side views.

1·10 FIG. 2(10). Draw top, front, and right-side views, arranged as in Fig. 1·9.

1·11 FIG. 2(11). Draw two views that completely describe the object.

1·12 FIG. 2(12). Draw as many views as are necessary to completely describe the object.

1·13 FIG. 2(2) C. Full size. Draw the six principal views of the object arranged as shown in Fig. 1·10.

1·14 FIG. 2(8) C. Full size. Same as Prob. 1·13.

1·15 FIG. 2(10) C. Full size. Same as Prob. 1·13.

1·16 FIG. 3(1). List in a column the following visible lines of the object shown, and in an adjacent column indicate the type of each line according to Fig. 1·14 (line *JH*, for example, is type 1): lines *AB*, *AK*, *KJ*, *AJ*, *BC*, *BM*, *MH*, *MN*, *HG*, *CN*, *NG*, *GF*.

1·17 FIG. 3(2). Same as Prob. 1·16, listing lines *AB*, *BC*, *CE*, *MN*, *AH*, *HJ*, *KJ*, *HG*, *JG*, *JF*, *KL*, *EF*.

1·18 FIG. 3(3). Same as Prob. 1·16, listing lines *AB*, *BC*, *BM*, *KJ*, *MN*, *KN*, *JH*, *HL*, *LE*, *DE*, *EF*, *JL*.

1·19 FIG. 3(1). List in a column the seven visible plane surfaces of the object shown, and in an adjacent column indicate the type of each surface according to Fig. 1·15. Surface *AKJ*, for example, is type 1. Read letters around each surface, beginning with the letter first in alphabetical sequence.

1·20 FIG. 3(2). Same as Prob. 1·19.

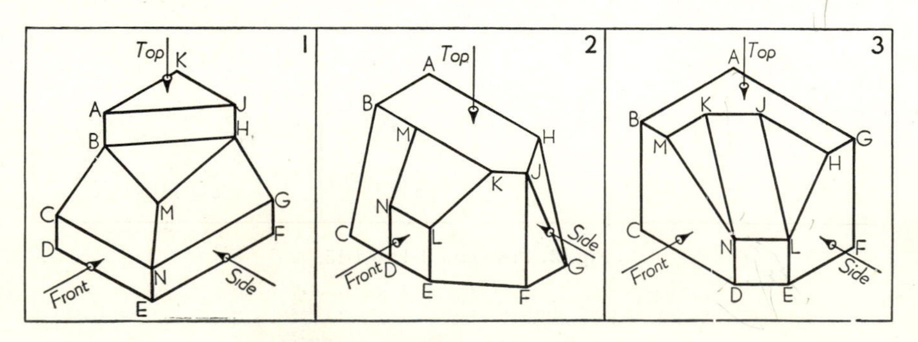

Fig. 3. Problems 1·16 to 1·21.

1·21 FIG. 3(3). Same as Prob. 1·19.

1·22 FIG. 4(1). Make a table like that shown, and list each of the lettered visible surfaces. Analyze the three views, noting alignment and similarity of configuration of the various surfaces, and then record by numbers the position of each lettered surface in the other two views. Surface *A*, for example, appears as line 7–9 in the front view and as line 16–17 in the side view.

1·23 FIG. 4(2). Same as Prob. 1·22.

1·24 FIG. 4(3). Same as Prob. 1·22.

1·25 FIG. 4(4). Same as Prob. 1·22.

Group 2. Construction of a third principal view (Arts. 2·2 to 2·4)

In each of the following problems, lay out the two given views full size, and construct the missing principal view. Space the views for uniform margins. Select and label reference lines as suggested in Art. 2·4. Label the more difficult surfaces and corners as an aid in the solution. Check surfaces by noting similarity of configuration, and check points by alignment and similarity. Use either Layout B or one-half of Layout A (except as noted).

2·1 FIG. 5(1). Construct the missing top view.

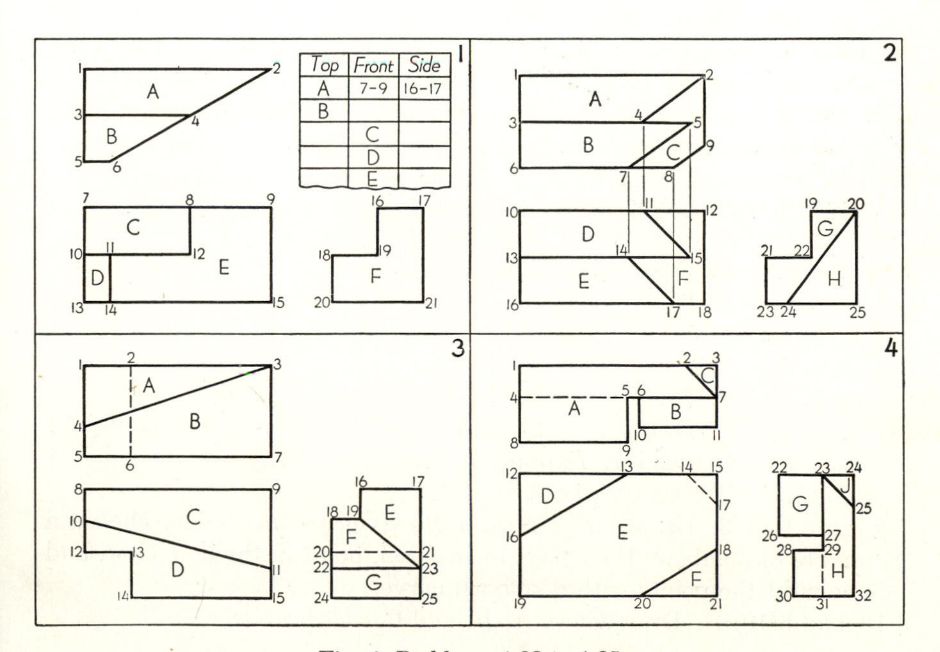

Fig. 4. Problems 1·22 to 1·25.

2·2 FIG. 5(2). Construct the missing right-side view.
2·3 FIG. 5(3). Construct the missing front view.
2·4 FIG. 5(4). Construct the missing top view.
2·5 FIG. 5(5). Construct the missing top view. (Note method of locating point x_F in the front view.)
2·6 FIG. 5(6). Construct the missing front view. (Missing dimensions in the top view may be obtained from the front view as construction progresses.)
2·7 FIG. 5(7). Construct the missing top view.
2·8 FIG. 5(8). Construct the missing top view.
2·9 FIG. 6(6) B. Draw the six principal views of the object.
2·10 FIG. 6(7). Construct the top and left-side views.
2·11 FIG. 6(8). Construct the right- and left-side views adjacent to the top view.
2·12 FIG. 6(11). Construct the right- and left-side views adjacent to the top view.
2·13 FIG. 6(12) B. Construct the right- and left-side views.

Group 3. Auxiliary views adjacent to principal views (Arts. 2·5, 2·6, 2·8, 2·9)

In each of the following problems, lay out the two given views full size, and construct the auxiliary views *A*, *B*, *C*, etc. The direction of sight for each auxiliary view is indicated by an arrow. Arrow directions not dimensioned are obviously parallel or perpendicular to a line on the drawing. Estimate the position of the new views, and then place the given views to avoid overlapping or running off the paper.

3·1 FIG. 6(1) A. Draw views *A*, *B*, and *C*.
3·2 FIG. 6(2) A. Draw views *A*, *B*, and *C*.
3·3 FIG. 6(3) A. Draw views *A*, *B*, and *C*.
3·4 FIG. 6(4) A. Draw views *A*, *B*, and *C*. (Object is a truncated pyramid.)
3·5 FIG. 6(5) A. Draw views *A*, *B*, and *C*.
3·6 FIG. 6(6) B. Draw views *A*, *B*, *C*, and *D*.
3·7 FIG. 6(7) B. Draw views *A* and *B*. (The given views should be separated to avoid overlapping of views *A* and *B*.)
3·8 FIG. 6(8) B. Draw views *A* and *B*.
3·9 FIG. 6(9) B. Draw views *A* and *B*. (On a curved line, choose a series of points on the curve, locate each point in the new view, and connect the points with a smooth curve.)
3·10 FIG. 6(10) B. Draw views *A*, *B*, and *C*.

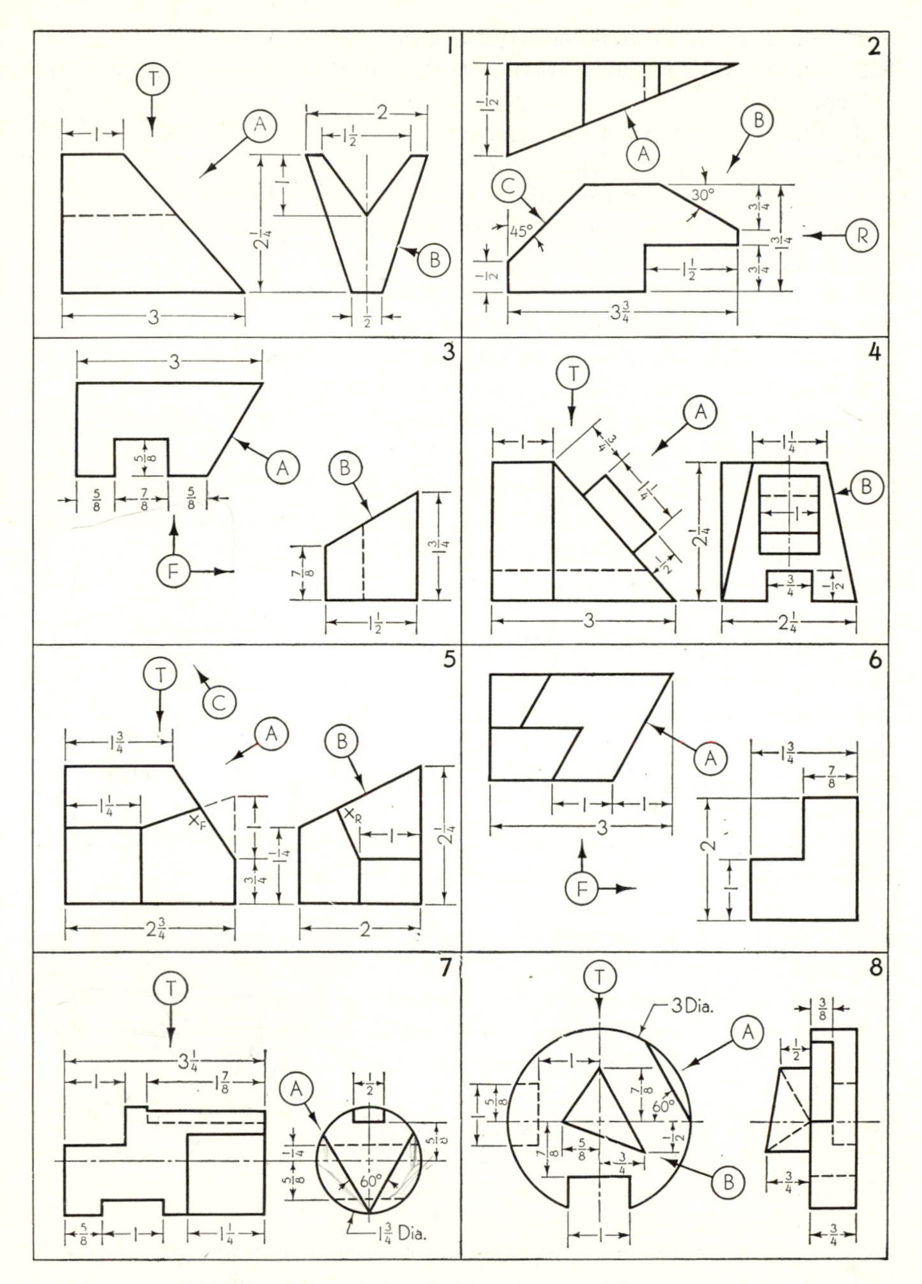

Fig. 5. Problems in Groups 2, 3, 15, and 16.

3·11 FIG. 6(11) A. Draw views *A*, *B*, *C*, and *L*.
3·12 FIG. 6(12) A. Draw views *A*, *B*, *C*, and *D*. (Object is a parallelepiped.)
3·13 FIG. 5(1) B. Draw front-adjacent view *A*.
3·14 FIG. 5(1) B. Draw side-adjacent view *B*.
3·15 FIG. 5(2) A. Draw top-adjacent view *A*.
3·16 FIG. 5(2) A. Draw front-adjacent view *B*.
3·17 FIG. 5(2) B. Draw front-adjacent view *C*.
3·18 FIG. 5(3) B. Move the given side view adjacent to the top view (as in Fig. 1·9), and then draw view *A*.
3·19 FIG. 5(3) B. Same as Prob. 3·18, but draw view *B*.
3·20 FIG. 5(4) B. Draw front-adjacent view *A*.
3·21 FIG. 5(4) B. Draw side-adjacent view *B*.
3·22 FIG. 5(5) B. Draw front-adjacent view *A*.
3·23 FIG. 5(5) B. Draw side-adjacent view *B*.
3·24 FIG. 5(5) B. Draw the top view, and then draw top-adjacent view *C* perpendicular to the vertical corner surface.
3·25 FIG. 5(6) A. Draw the front view, omit the side view, and then draw view *A*.
3·26 FIG. 5(7) B. Draw side-adjacent view *A*.
3·27 FIG. 5(8) B. Draw front-adjacent view *A*.
3·28 FIG. 5(8) B. Draw front-adjacent view *B*.

Group 4. Auxiliary-adjacent views (Arts. 2·7 to 2·9)

In each problem the direction of the consecutive auxiliary views *A*, *B*, and *C* has been indicated by giving the direction of the reference line between adjacent views. Draw the given views full size on Layout B, using the placement dimensions to locate the views from the *border* of the sheet and from the given reference line. Place each new reference line at the specified angle, and about ¼ in. from the adjacent view. Show hidden lines in all views. In Probs. 4·5 to 4·16 the arrangements of views and the objects have been interchanged as indicated.

4·1 FIG. 7(1). Draw views *A*, *B*, and *C*.
4·2 FIG. 7(2). Draw views *A*, *B*, and *C*.
4·3 FIG. 7(3). Draw views *A*, *B*, and *C*.
4·4 FIG. 7(4). Draw views *A*, *B*, and *C*.
4·5 FIG. 7. Arrangement of views as in (1), but use object of (2).
4·6 FIG. 7. Arrangement of views as in (1), but use object of (3).
4·7 FIG. 7. Arrangement of views as in (1), but use object of (4).
4·8 FIG. 7. Arrangement of views as in (2), but use object of (1).

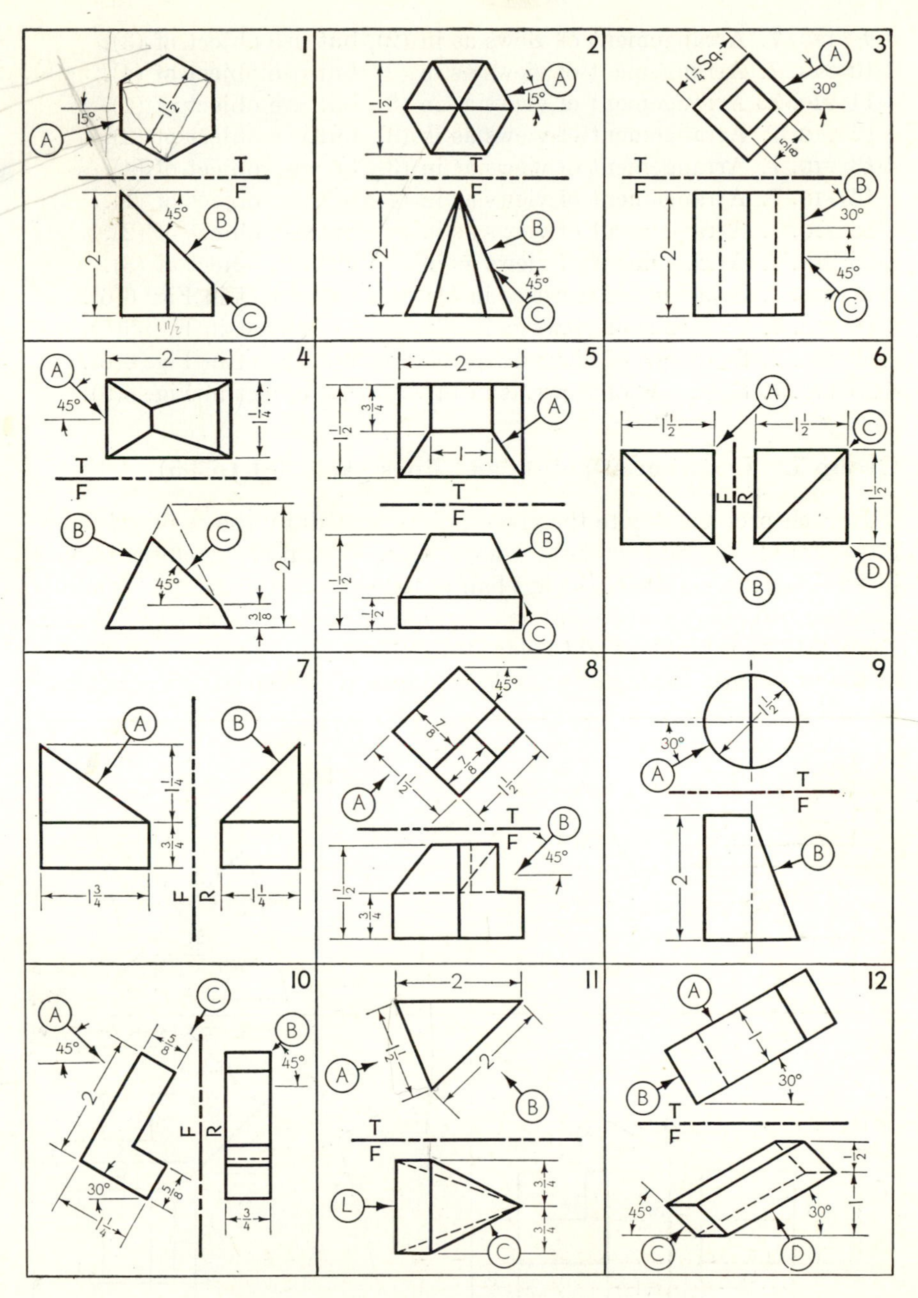

Fig. 6. Problems in Groups 2, 3, 15, and 16.

4·9 FIG. 7. Arrangement of views as in (2), but use object of (3).
4·10 FIG. 7. Arrangement of views as in (2), but use object of (4).
4·11 FIG. 7. Arrangement of views as in (3), but use object of (1).
4·12 FIG. 7. Arrangement of views as in (3), but use object of (2).
4·13 FIG. 7. Arrangement of views as in (3), but use object of (4).
4·14 FIG. 7. Arrangement of views as in (4), but use object of (1).
4·15 FIG. 7. Arrangement of views as in (4), but use object of (2).
4·16 FIG. 7. Arrangement of views as in (4), but use object of (3).
4·17 FIG. 7. Sequence of views as in (1), but use object in Fig. 6(6).
4·18 FIG. 7. Sequence of views as in (2), but use object in Fig. 6(6).
4·19 FIG. 7. Sequence of views as in (3), but use object in Fig. 6(6).
4·20 FIG. 7. Sequence of views as in (4), but use object in Fig. 6(6).

Group 5. Location of points and lines (Arts. 3·1 to 3·5)

In each problem locate the given points, and draw the required figure in the top, front, and right-side views. Use Layout B, placing reference line *T–F* in the middle of the left half of the sheet. In each problem it is necessary to assume one point as the origin of measurements (point *A* in Prob. 5·1, for example) and to place this point according to the conditions of the problem. Scale and record the required distances and bearings.

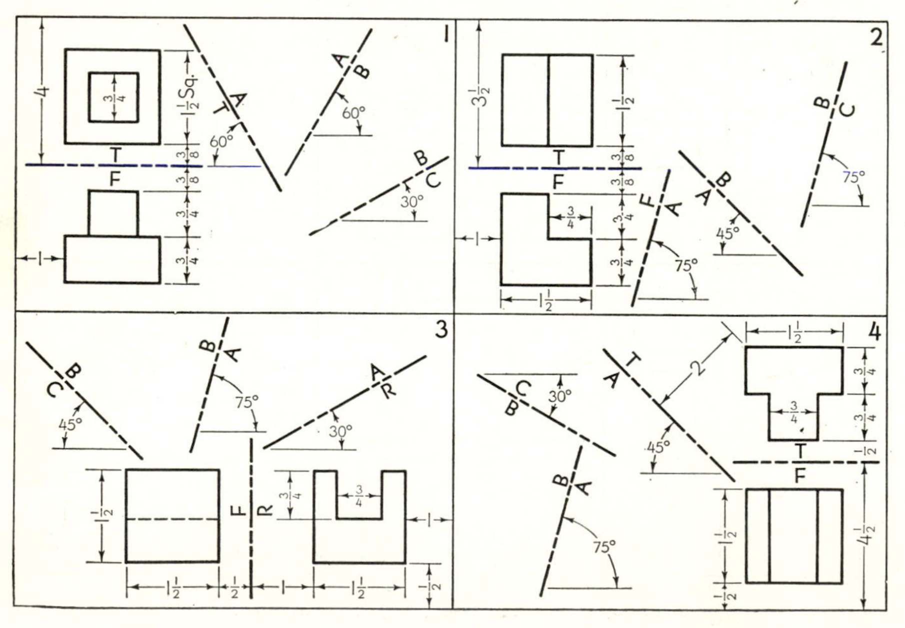

Fig. 7. Problems 4·1 to 4·20 (placement dimensions not to scale).

5·1 Given triangle *ABC*:

B is $1\frac{1}{4}$ in. to the left of, 1 in. behind, and $\frac{3}{4}$ in. below *A*.

C is $1\frac{1}{2}$ in. to the right of, $2\frac{1}{4}$ in. behind, and 2 in. below *A*.

D is on line *AB*, $\frac{5}{8}$ in. behind *A*.

E is on line *AC*, $\frac{3}{4}$ in. below *B*.

F is on line *BC*, $\frac{1}{2}$ in. to the right of *A*.

What is the bearing of line *DE*? State the position of *F* with respect to point *C*. ANS. N $60\frac{1}{2}°$ E; *F* is 0.45 in. in front of and 0.45 in. above *C*.

5·2 Given triangle *ABC*:

B is $2\frac{1}{2}$ in. behind and $2\frac{1}{4}$ in. above *A*.

C is $2\frac{1}{2}$ in. to the right of, $1\frac{1}{2}$ in. behind, and $\frac{1}{4}$ in. above *A*.

D is on line *AB*, 1 in. behind *A*.

E is on line *AC*, $\frac{3}{4}$ in. to the left of *C*.

F is on line *BC*, $\frac{3}{4}$ in. below *B*.

G is the mid-point of line *DF*.

What is the bearing of line *EG*? State the position of *G* with respect to point *A*. ANS. N 68° W; *G* is 0.47 in. to the right of, 1.57 in. behind, and 1.18 in. above *A*.

5·3 Scale: 1 in. = 50 ft. Given triangle *ABC*:

B is 135 ft N 40° W from *A* and 75 ft higher than *A*.

C is 160 ft S 70° E from *B* and 110 ft higher than *A*.

D is due north of *A* on line *BC*.

E is on *AC* at the same elevation as *B*.

F is on *AB* due west of *C*.

What is the bearing of line *BE*? What are the elevations of points *D* and *F* with respect to point *A*?

ANS. S $61\frac{1}{2}°$ E; *D* is 95 ft above *A*; *F* is 35 ft above *A*.

5·4 Scale: $\frac{1}{8}$ in. = 1 ft. Given triangle *ABC*:

B is 18 ft S 65° W from *A* and 19 ft higher than *A*.

C is N 40° E from *B*, due north of *A*, and 11 ft higher than *A*.

D is N 45° W from *A* and on line *BC*.

E is on line *AC*, 6 ft lower than *C*.

What is the bearing of line *DE*? What is the height of *D* with respect to *E*? ANS. N 77° E; *D* is 9 ft 3 in. higher than *E*.

5·5 Given the warped quadrilateral *ABCD*:

B is $1\frac{3}{4}$ in. to the left of, $\frac{3}{4}$ in. before, and 2 in. below *A*.

C is $2\frac{1}{2}$ in. to the left of, $2\frac{1}{2}$ in. before, and $\frac{1}{2}$ in. above *A*.

D is $\frac{1}{2}$ in. to the right of, 2 in. before, and 1 in. below *A*.

In each view connect the mid-points of each side of the figure to form a second quadrilateral. What is the shape of this second figure?

5·6 Using the same data as in Prob. 5·5, connect the four given points to form a tetrahedron, and show all lines in correct visibility.

5·7 Same as Prob. 5·5, but using the following points:
B is $\frac{3}{4}$ in. to the right of, $\frac{3}{4}$ in. before, and $1\frac{1}{4}$ in. above *A*.
C is $2\frac{3}{4}$ in. to the left of, $1\frac{1}{2}$ in. before, and $\frac{1}{2}$ in. above *A*.
D is $1\frac{1}{2}$ in. to the left of, $2\frac{1}{4}$ in. before, and 1 in. below *A*.

5·8 Using the same data as in Prob. 5·7, connect the four given points to form a tetrahedron, and show all lines in correct visibility.

5·9 Scale: $\frac{1}{8}$ in. = 1 ft. Given the pyramid *ABCD*:
B is 12 ft N 15° E from *A* and 8 ft higher than *A*.
C is 15 ft S 45° E from *A* and 20 ft higher than *A*.
D is 14 ft S 60° W from *A* and 10 ft higher than *A*.

Locate the mid-points of each edge of the pyramid, and label these points 1 to 6. Draw the three median lines in each triangular face, thus locating the centroids *w*, *x*, *y*, and *z* of each triangle. The medians of the solid, which are the four lines connecting these centroids to the opposite vertices, will intersect at a common point *P*. This point is the center of gravity of the solid. State the position of *P* with respect to *A*.

ANS. *P* is 5 in. to the right of, 18 in. before, and 9 ft 6 in. above *A*.

5·10 Scale: 1 in. = 50 ft. Same as Prob. 5·9, but using the following points:
B is 125 ft N 15° W from *A* and 60 ft lower than *A*.
C is 140 ft N 45° E from *A* and 30 ft lower than *A*.
D is 80 ft N 75° E from *A* and 130 ft lower than *A*.

ANS. *P* is 36 ft E, 60 ft N, and 55 ft lower than *A*.

5·11 Given the rectangular pyramid *VABCD*, whose vertex is *V*. Show all lines in correct visibility.
A is 2 in. to the right of, $\frac{1}{4}$ in. behind, and $1\frac{1}{2}$ in. below *V*.
B is $2\frac{1}{4}$ in. to the right of, $1\frac{1}{4}$ in. behind, and $\frac{1}{4}$ in. below *V*.
C is $1\frac{1}{4}$ in. to the right of, $2\frac{1}{4}$ in. behind, and $\frac{3}{4}$ in. below *V*.
D is 1 in. to the right of, $1\frac{1}{4}$ in. behind, and 2 in. below *V*.

5·12 Scale: $\frac{1}{4}$ in. = 1 ft. Same as Prob. 5·11, but using the following points:
A is 10 ft S 15° E from *V* and 5 ft lower than *V*.
B is S 45° W from *V*, N 75° W from *A*, and 5 ft lower than *V*.
C is S 75° W from *V*, N 15° E from *B*, and 5 ft lower than *B*.
D is S 45° E from *V*, N 15° E from *A*, and 5 ft lower than *A*.

Group 6. True length and slope of a line (Arts. 3·6, 3·7)

In Probs. 6·1 to 6·12 find the true length in inches and the slope angle in degrees for line *AB*. Check the true length by drawing another auxiliary view. Draw the given views full size on Layout A, placing *T–F* in the center of the sheet.

6·1 FIG. 8(1). ANS. 3.07 in., 22°.

6·2 FIG. 8(2). ANS. 2.67 in., 41°.
6·3 FIG. 8(3). ANS. 1.80 in., 34°.
6·4 FIG. 8(4). ANS. 1.95 in., 40°.
6·5 FIG. 8(5). ANS. 2.35 in., 25°.
6·6 FIG. 8(6). ANS. 2.26 in., 50°.
6·7 FIG. 8(7). ANS. 2.92 in., 31°.
6·8 FIG. 8(8). ANS. 2.36 in., 32°.

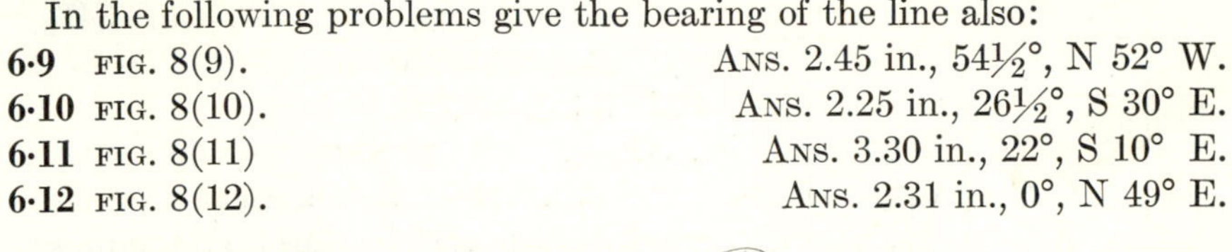

In the following problems give the bearing of the line also:

6·9 FIG. 8(9). ANS. 2.45 in., 54½°, N 52° W.
6·10 FIG. 8(10). ANS. 2.25 in., 26½°, S 30° E.
6·11 FIG. 8(11) ANS. 3.30 in., 22°, S 10° E.
6·12 FIG. 8(12). ANS. 2.31 in., 0°, N 49° E.

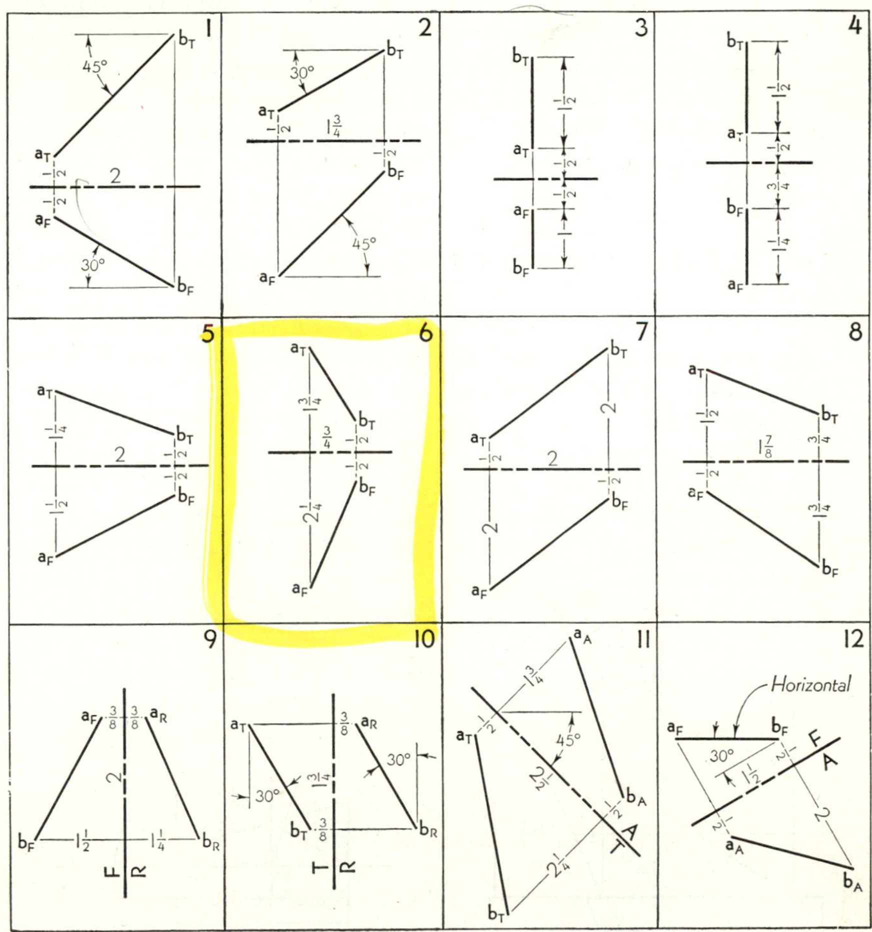

Fig. 8. Problems in Groups 6 and 8.

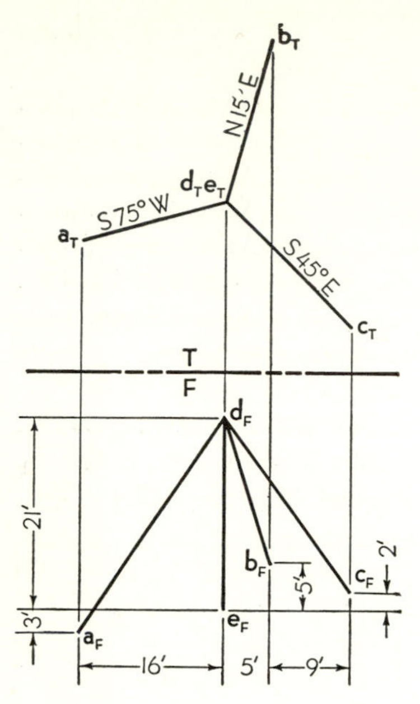

Fig. 9. Problem 6·13.

6·13 FIG. 9 A. Scale: $\frac{1}{8}$ in. = 1 ft. The mast *DE* is supported by three guy wires *AD*, *BD*, and *CD*, each anchored at a different level. Find the true length of each guy and the angle it makes with the mast. Show and label the mast in each view. ANS. *AD*, 29 ft 4 in., 34½°. *BD*, 25 ft 0 in., 50°. *CD*, 27 ft 5 in., 46°.

6·14 FIG. 10 A. Scale: $\frac{1}{8}$ in. = 1 ft. The shear-leg derrick shown is mounted on a barge and supports the derrick boom (not shown). Find the true length and slope angle of the leg *BC* and of the anchor cable *AC*. Show and label the deck in each view.
ANS. *BC*, 25 ft 6 in., 60°. *AC*, 41 ft 2 in., 32°.

6·15 FIG. 11 A. Scale: ½ in. = 1 ft. The framework shown is made of steel tubing welded together. To check the alignment of the finished unit the four diagonal distances *AB*, *CD*, *EF*, and *GH* are measured. Find the true length of each diagonal, and check its length in a new view. ANS. *AB*, 8 ft 0½ in. *CD*, 5 ft 11½ in. *EF*, 8 ft 1 in. *GH*, 4 ft 10 in.

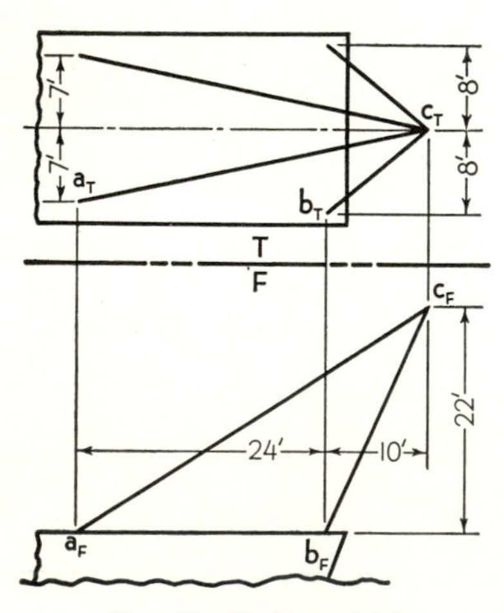

Fig. 10. Problem 6·14.

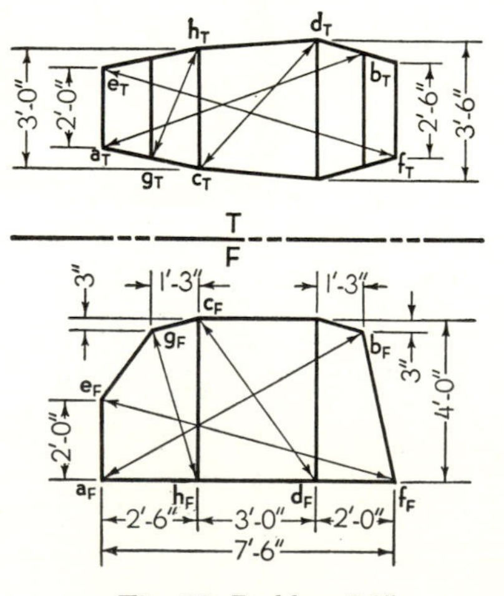

Fig. 11. Problem 6·15.

6·16 FIG. 12 A. Scale: 1 in. = 16 ft. A radio antenna mast *DE* is supported by three cables *AD*, *BD*, and *CD*. It is desired that all three cables shall make the same angle with the mast. Hence find (*a*) the true length and angle with the mast of cable *AD*; (*b*) what distance anchorage *B* must be from the base of the mast, and the true length of *BD*; (*c*) how far the given point *C* must be moved along the sloping ground toward or away from the mast, and the true length of *CD* after moving *C*. ANS. (*a*) 53 ft 0 in., 33½°; (*b*) 26 ft 6 in., 48 ft 0 in.; (*c*) 6 ft 0 in. away, 44 ft 6 in.

6·17 A. Scale: ¼ in. = 1 ft. A building 16 ft wide by 24 ft long is covered by a hip roof having a central ridge 8 ft long, commencing 6 ft from the right-hand end of the building. The right-end corner rafters have a slope angle of 45°. Draw the top and front views of the roof. (*a*) What is the height of the ridge above the plane of the eaves? (*b*) What is the length of corner rafters at each end? (*c*) What is the slope angle of the left-end corner rafter?

ANS. (*a*) 10 ft 0 in.; (*b*) 14 ft 1½ in., 16 ft 3 in.; (*c*) 38°.

6·18 A. Scale: ½ in. = 1 ft. A tripod rests on a level floor. One leg is 7 ft long and makes an angle of 50° with the floor. The other two legs are 6 ft 6 in. and 5 ft 9 in. long. In the top view the legs appear equally spaced (120° apart). Draw the top and front views of the tripod. What is the slope angle of each of the other two legs?

ANS. 55°30′, 68°45′.

6·19 FIG. 13 C. Scale: 3⁄16 in. = 1 ft (B, ⅛ in. = 1 ft). The bridge truss shown connects 12- by 16-ft openings in two adjacent factory build-

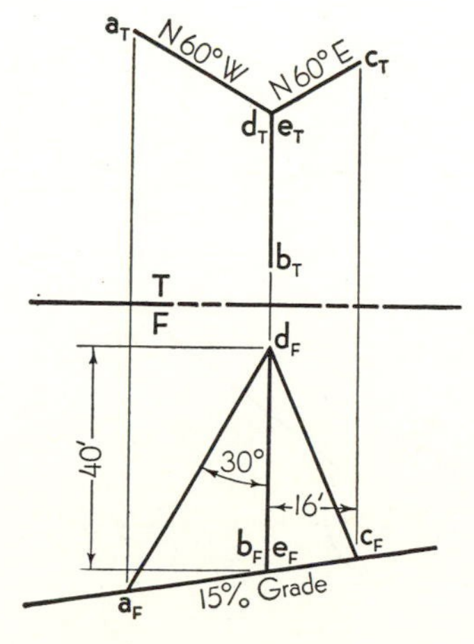

Fig. 12. Problem 6·16.

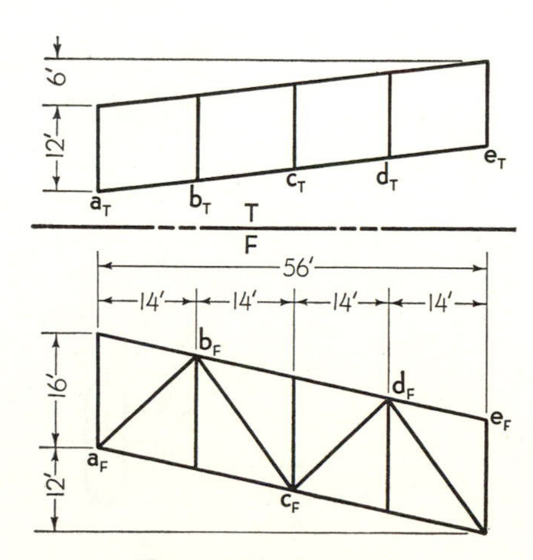

Fig. 13. Problem 6·19.

ings. Find the true length of the diagonal members *AB* and *BC*. What is the true length and per cent grade of member *DE*?

ANS. *AB*, 19 ft 2 in. *BC*, 23 ft 8 in. *DE*, 14 ft 5 in., 21 per cent.

6·20 FIG. 14 C. Scale: 1 in. = 1 ft (B, ¾ in. = 1 ft). A tripod-type airplane landing gear is shown. Find the true length and slope angle of each of the three struts. Note that front and right-side views are given. ANS. *AD*, 28¼ in., 58°. *BD*, 41 in., 36°. *CD*, 39 in., 38°.

6·21 FIG. 15 C. Scale: ¾ in. = 1 ft (B, ½ in. = 1 ft). A tripod-type airplane landing gear is shown. Find the true length and slope angle of each of the three struts. Front and right-side views are given.

ANS. *AD*, 42 in., 59½°. *BD*, 47½ in., 34½°. *CD*, 59½ in., 27°.

6·22 FIG. 16 C. Scale: 1 in. = 1 ft (B, ¾ in. = 1 ft). The framework shown is an engine mount for an aircraft radial engine. It consists of a tubular steel ring to which are welded eight straight tubes by means of fittings and brackets. The engine attaches to the ring, and the engine mount attaches to a bulkhead in the fuselage or nacelle.

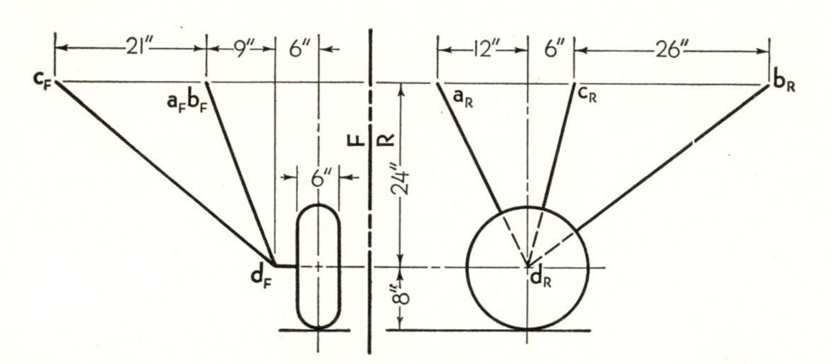

Fig. 14. Problem 6·20.

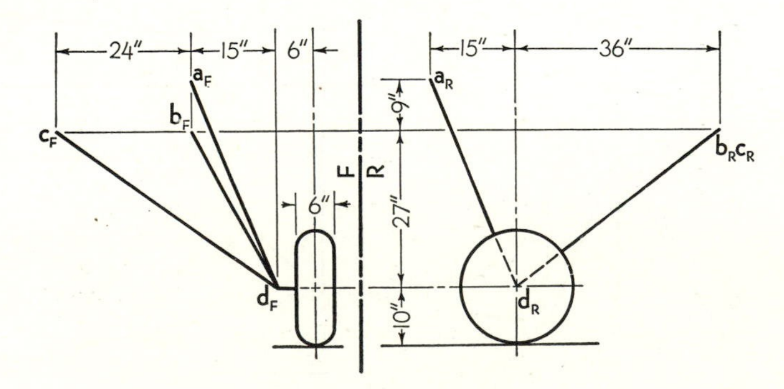

Fig. 15. Problem 6·21.

The lines shown in the figure represent the center lines of the ring and tubes. Find the true length of tubes *AB*, *AC*, *CD*, and *DE* and the angle that each makes with the vertical bulkhead. (HINT: Imagine the given front view to be a top view; then the vertical bulkhead becomes a horizontal plane.) ANS. *AB*, 31¼ in., 50°. *AC*, 30⅛ in., 53°. *CD*, 34⅝ in., 44°. *DE*, 33½ in., 46°.

6·23 FIG. 16 C. Scale: 1 in. = 1 ft (B, ¾ in. = 1 ft). Same as Prob. 6·22, but change the mount to fit on the canted bulkhead. Do not change the front view. Find the true length of each straight tube, but do *not* find the angles with the canted bulkhead.
ANS. *AB*, 33¾ in. *AC*, 32½ in. *CD*, 31⅞ in. *DE*, 30⅝ in.

Group 7. Lines of given bearing, true length, and slope (Art. 3·8)

In Probs. 7·1 to 7·6 draw the top and front views of the given line. In each case the bearing and slope is from *A* to *B*. Assume point *A* to the right or left on the sheet, and high or low, according to the conditions of the problem. Use Layout A.

7·1 Line *AB* is 2¾ in. long, bears N 60° E, and slopes downward at 35°.

7·2 Line *AB* is 2½ in. long, bears S 75° W, and slopes downward on a 25 per cent grade.

7·3 Line *AB* is 3 in. long, bears N 15° W, and slopes upward at 50°.

7·4 Line *AB* is 2⅝ in. long, bears S 60° E, and slopes upward on a 30 per cent grade.

7·5 Scale: 1 in. = 100 ft. A straight highway *AB* is 317 ft long, runs N 52°15′ W with a downward grade of 12 per cent.

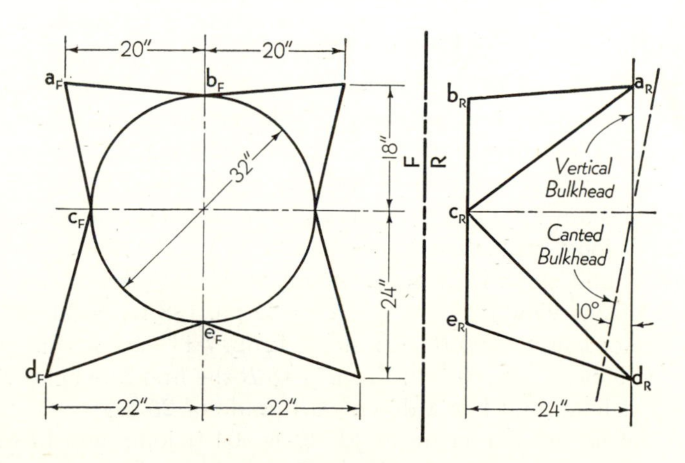

Fig. 16. Problems 6·22 and 6·23.

7·6 Scale: 1 in. = 50 ft. From point *A* at the bottom of a mine shaft a tunnel bears S 33°30′ W on a rising grade of 15 per cent for a distance of 140 ft.

7·7 A. Scale: ⅛ in. = 1 ft. From a point *A* an old sewer pipe runs N 41°30′ E on a downward grade of 10 per cent. From point *C*, which is 14 ft 10 in. east, 6 ft 4 in. north, and 20 in. lower than *A*, it is planned to run a new sewer N 54°15′ W on a downward grade of 18 per cent. Both pipes are 12 in. in diameter. Will the new pipe pass under or over the old one, and what will be the vertical clearance (if any) where they cross? ANS. Under; clearance 7 in.

7·8 A. Scale: 1 in. = 50 ft. From point *A* a highway runs S 39°30′ W on an upward grade of 12 per cent. From point *C*, which is 127 ft (map distance) S 75°50′ W from *A* and 36 ft higher in elevation than *A*, a railroad track runs S 23°30′ E on a rising grade of 5 per cent. What will be the vertical clearance between track and highway at the crossing? ANS. 23.5 ft.

7·9 A. Scale: 1 in. = 100 ft. Two mining tunnels start at a common point *A*. Tunnel *AB* is 306 ft long and bears S 54°30′ E on a downward slope of 19°. Tunnel *AC* is 224 ft long and bears N 60° E on a downward slope of 15°. What will be the length, bearing, and per cent grade of a new tunnel connecting points *B* and *C*?

ANS. 282 ft, N 10°0′ W, 15.5 per cent.

7·10 A. Scale: 1 in. = 500 ft. From an observation tower 400 ft above sea level a ship is sighted S 62°0′ E of the observer at an angle of depression of 18°15′. Three minutes later the same ship is S 49°30′ W at an angle of depression of 12°30′. Assuming that the ship held a steady course and speed, what was the bearing of its course, and what was its speed in knots? (1 knot = 6,080 ft per hr.)

ANS. S 76°30′ W, 8.4 knots.

7·11 B. Scale: 1 in. = 50 ft. An aerial tramway is planned to connect two points *A* and *D*, but because of intervening obstacles it is necessary to establish point *D* by running the survey in three lines. From *A* to *B* the first line bears N 55° E, is 135 ft long, and slopes upward at 15°. From *B* to *C* the second line bears N 33° W, is 115 ft long, and slopes upward at 22°. From *C* to *D* the third line bears S 75° W, is 149 ft long, and slopes upward at 9°. What are the true length, bearing, and slope in degrees of line *AD*? (It is not necessary to draw a front view.) ANS. 188 ft, N 36°30′ W, 33°.

7·12 B. Scale: 1 in. = 100 ft. Same as Prob. 7·11 except that the three survey lines run as follows: From *A* to *B* the first line bears S 70° W, is 265 ft long, and has a downward grade of 25 per cent. From *B* to *C* the second line bears N 30° E, is 304 ft long, and has a downward grade of 30 per cent. From *C* to *D* the third line bears S 75° E, is 226 ft long, and has an upward grade of 40 per cent. What are the true length, bearing, and per cent grade of line *AD*?

ANS. 170 ft, N 43°30′ E, 46 per cent.

Group 8. Line as a point (Art. 3·9)

In Probs. 8·1 to 8·12 draw a view in which the line *AB* appears as a point. Use Layout A.

8·1 FIG. 8(1).
8·2 FIG. 8(2).
8·3 FIG. 8(3).
8·4 FIG. 8(4).
8·5 FIG. 8(5).
8·6 FIG. 8(6).
8·7 FIG. 8(7).
8·8 FIG. 8(8).
8·9 FIG. 8(9).
8·10 FIG. 8(10).
8·11 FIG. 8(11).
8·12 FIG. 8(12).
8·13 FIG. 6(11) A. Draw a view of the pyramid in which the lower foremost lateral edge appears as a point.
8·14 FIG. 6(12) A. Draw a view of the prism in which the long lateral edges appear as points.

Group 9. Parallel and intersecting lines (Arts. 3·10 to 3·13)

9·1 FIG. 17(1) B. Line *AB* is one side of a parallelogram, and point *C* is the mid-point of the opposite parallel side. Draw the top, front, and right-side views of the parallelogram.
9·2 FIG. 17(2). Same as Prob. 9·1.
9·3 FIG. 17(3). Same as Prob. 9·1.
9·4 FIG. 17(4). Same as Prob. 9·1.

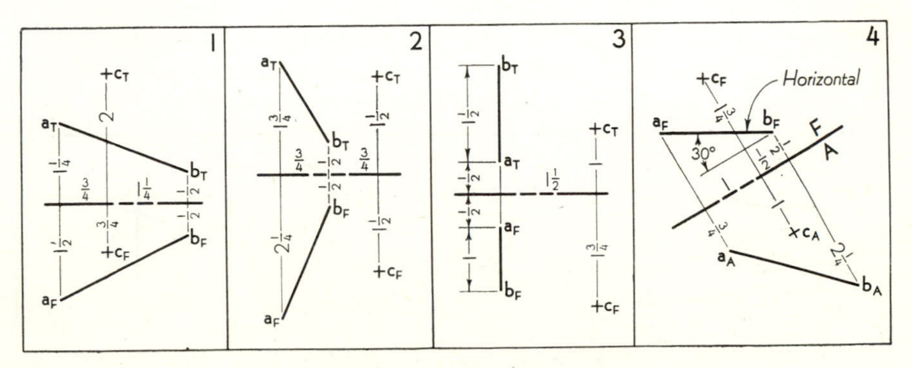

Fig. 17. Problems 9·1 to 9·4.

9·5 FIG. 18(1) B. Lines AB, AC, and AD are three adjacent edges of a parallelepiped. Draw the top, front, and right-side views of the solid. The given lines are not necessarily all visible.

9·6 FIG. 18(2). Same as Prob. 9·5.

9·7 FIG. 18(3). Same as Prob. 9·5.

9·8 FIG. 18(4) B. Draw the given views of lines AB, CD, and EF, and determine by construction which pairs of lines intersect, or would intersect if extended. Do not test CD and EF by extension.

9·9 FIG. 18(5). Same as Prob. 9·8.

9·10 FIG. 18(6). Same as Prob. 9·8.

9·11 FIG. 19(1) B. What is the true distance between parallel lines AB and CD? ANS. 1.43 in.

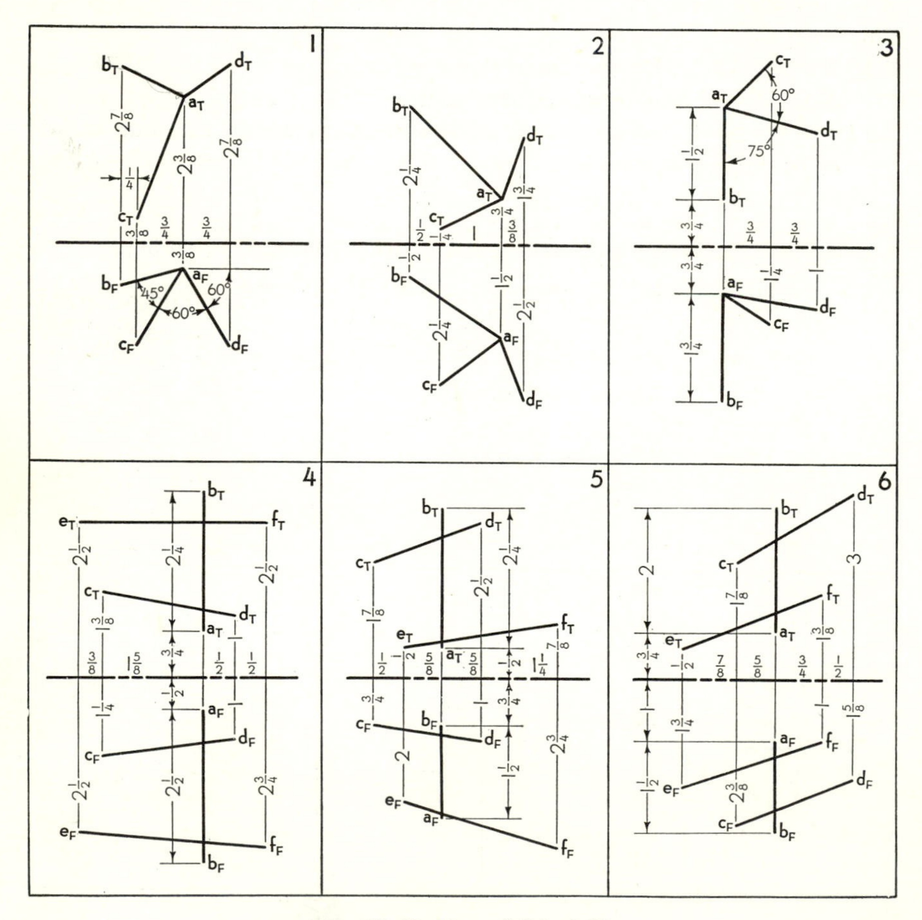

Fig. 18. Problems 9·5 to 9·10.

9·12 FIG. 19(2) A. Same as Prob. 9·11. ANS. 1.08 in.
9·13 FIG. 19(3) B. Same as Prob. 9·11. ANS. 1.78 in.
9·14 FIG. 19(4) A. Same as Prob. 9·11. ANS. 1.35 in.

Group 10. Perpendicular lines (Arts. 3·14, 3·15)

10·1 FIG. 20(1) A. Line *AB* is the axle of a 2-in.-diameter wheel whose center is at the mid-point of *AB*. The wheel has eight equally spaced spokes, one pair being horizontal. Draw the top and front views of axle and spokes. The rim of the wheel may be omitted.
10·2 FIG. 20(2). Same as Prob. 10·1.
10·3 FIG. 20(4) B. Same as Prob. 10·1.
10·4 FIG. 20(5) A. Same as Prob. 10·1, except that one pair of spokes must be true length in the front view (frontal), instead of horizontal.
10·5 FIG. 20(6). Same as Prob. 10·4.
10·6 FIG. 20(10). Same as Prob. 10·4.
10·7 FIG. 20(7) A. Given line *AB* and point c_T (disregard given c_F). Relocate c_F so that line *AC* will be perpendicular to *AB*. Also show in the given views a third line *AD* that is perpendicular to both *AB* and *AC*. Make line *AD* of the same true length as *AC*.
10·8 FIG. 20(5). Same as Prob. 10·7.
10·9 FIG. 20(6). Same as Prob. 10·7.
10·10 FIG. 20(8) A. Given line *AB* and point c_F (disregard given c_T). Relocate c_T so that line *AC* will be perpendicular to *AB*. Also show in the given views a third line *AD* that is perpendicular to both *AB* and *AC*. Make line *AD* of the same true length as *AC*.
10·11 FIG. 20(3). Same as Prob. 10·10.
10·12 FIG. 20(4). Same as Prob. 10·10.

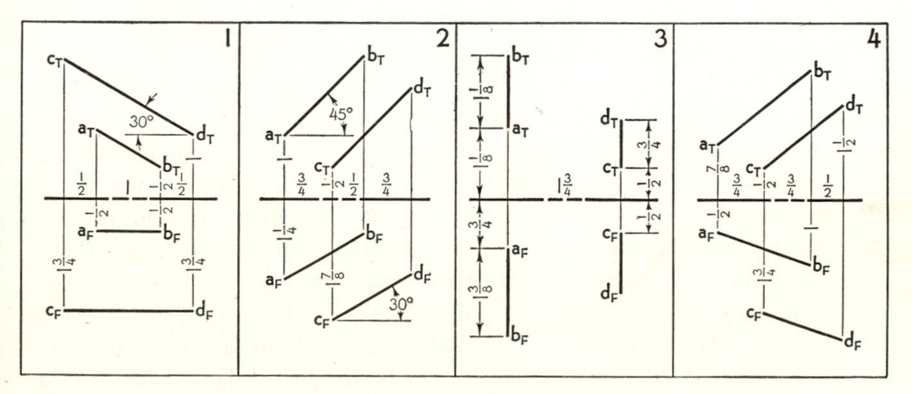

Fig. 19. Parallel lines. Problems 9·11 to 9·14.

Group 11. Shortest line from a point to a line (line method, Art. 3·16)

In Probs. 11·1 to 11·16 find the true length, bearing, and slope angle of the shortest line *CX* from point *C* to the line *AB*. Show line *CX* as a heavy line in all views.

11·1 FIG. 20(1) A. ANS. 0.89 in., N 61° W, 34°.
11·2 FIG. 20(2) A. ANS. 0.87 in., S 17° W, 40°.
11·3 FIG. 20(3) B. ANS. 1.69 in., N 74° W, 22½°.
11·4 FIG. 20(4) B. ANS. 2.17 in., N 75° E, 17½°.
11·5 FIG. 20(5) A. ANS. 1.10 in., S 7½° W, 24°.
11·6 FIG. 20(6) A. ANS. 1.19 in., S 69° W, 8°.
11·7 FIG. 20(7) A. ANS. 1.30 in., N 23° E, 34½°.
11·8 FIG. 20(8) A. ANS. 0.92 in., S 45° W, 54°.
11·9 FIG. 20(9) A. ANS. 1.46 in., N, 59°.
11·10 FIG. 20(10) A. ANS. 0.70 in., N, 50°.
11·11 FIG. 20(11) A. ANS. 1.75 in., S 12½° W, 42½°.
11·12 FIG. 20(12) A. ANS. 0.35 in., N 68° W, 58°.
11·13 FIG. 20(13) B. ANS. 0.40 in., S 79° E, 33°.
11·14 FIG. 20(14) B. ANS. 0.66 in., N 41° E, 33½°.
11·15 FIG. 20(15) A. ANS. 0.77 in., S 66½° W, 28½°.
11·16 FIG. 20(16) A. ANS. 0.88 in., N 41° W, 72°.

11·17 A. Scale: ¼ in. = 1 ft. A drainpipe runs through a room 10 ft wide, 12 ft long, and 8 ft high. The pipe enters the room at a point on the left wall 1 ft 9 in. below the ceiling and 2 ft 6 in. from the rear wall. It leaves the room at a point on the right wall 6 in. above the floor and 3 ft 3 in. from the front wall. From a point on the rear wall 1 ft below the ceiling and 7 ft 6 in. from the left wall a new drain runs to the given pipe and joins it in a 90° tee. How long is the new pipe from wall to drain? How far from the left wall, measured along the pipe, must the old drain be broken?

ANS. 5 ft 10 in., 5 ft 4½ in.

11·18 B. Scale: ¼ in. = 1 ft. A length of pipe, used as a brace, is anchored to the floor 6 ft 9 in. in front of a wall and attached to the wall (directly behind the floor anchorage) 8 ft 3 in. above the floor. From a point on the wall 4 ft 2 in. to the right and 10 in. above the floor a stiffening brace runs to the main brace, joining it in a 90° tee connection. How long must this second brace be? How far from the floor, measured along the main brace is the connection? If a side-outlet tee (side outlet perpendicular to plane of tee) is used at the joint, a third brace could be run to the left. How long would the third brace be, and where would it have to be anchored on the

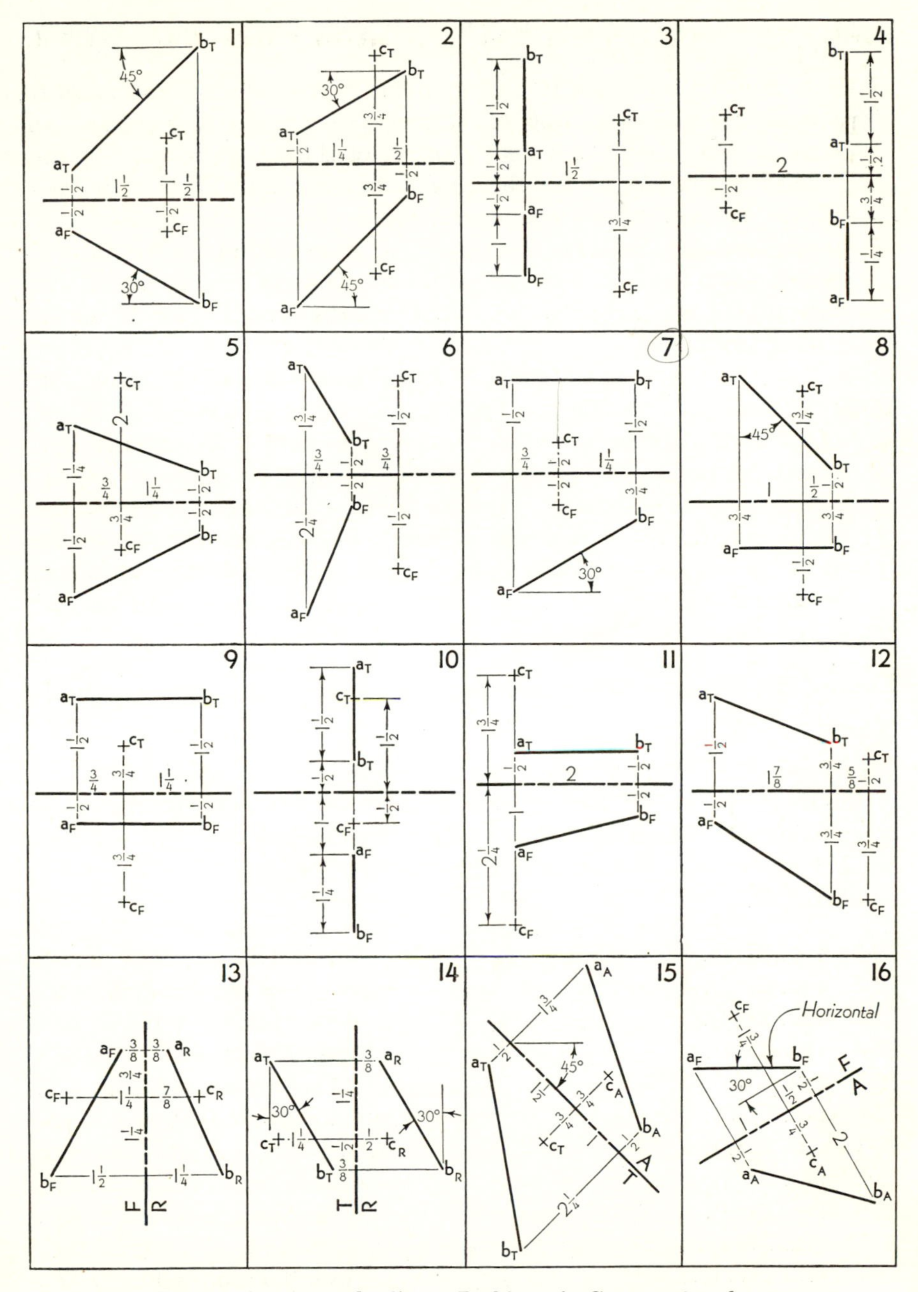

Fig. 20. A point and a line. Problems in Groups 10 and 11.

wall? ANS. 6 ft 4 in., 4 ft 11 in.; 7 ft 2 in., 5 ft 4½ in. to the left, 10 in. above the floor.

11·19 A. Scale: 1 in. = 200 ft. From point A at an elevation of 1,260 ft a mining tunnel runs N 36°45′ W on a 28 per cent rising grade. At a point 380 ft west and 280 ft north of A a shaft is sunk to the 1,200-ft level. What are the length, bearing, and slope in degrees of the shortest tunnel connecting the bottom of the shaft to the given tunnel? How far from point A, measured along the tunnel, will the two tunnels join? If either the slope or the bearing of the new tunnel is in error by ½°, what will the error in feet be at the end of the tunnel?

ANS. 226 ft, N 72°45′ E, 50°45′; 420 ft; error 2 ft.

11·20 A. Scale: 1 in. = 500 ft. From a main tower at A, at an elevation of 2,250 ft, a transmission line runs N 41° E uphill on a 45 per cent grade. A building is located 1,120 ft east and 255 ft north of point A, at an elevation of 2,585 ft. Locate the shortest connecting line from building to main line, and find its length, bearing, and per cent grade. How far from A, measured along the main line, must the connection be made? ANS. 675 ft, N 52° W, 10 per cent; 990 ft.

Group 12. Shortest line between two skew lines (line method, Art. 3·17)

In Probs. 12·1 to 12·12 find the true length, bearing, and slope angle of the shortest line YZ between the given lines AB and CD (disregard given point X). Show line YZ as a heavy line in all views. Make a freehand sketch of your proposed solution, and then choose either Layout A or Layout B.

12·1 FIG. 21(1). ANS. 1.57 in., S 45° E, 0°.
12·2 FIG. 21(2). ANS. 0.48 in., N 27° W, 49°.
12·3 FIG. 21(3). ANS. 1.21 in., S 16° E, 53°.
12·4 FIG. 21(4). ANS. 0.80 in., S 20° W, 35°.
12·5 FIG. 21(5). ANS. 0.62 in., S 16° E, 32°.
12·6 FIG. 21(6). ANS. 0.95 in., S 14½° E, 54½°.
12·7 FIG. 21(7). ANS. 1.03 in., N 51½° E, 47°.
12·8 FIG. 21(8). ANS. 0.97 in., N 12° E, 53°.
12·9 FIG. 21(9). ANS. 1.24 in., E, 45°.
12·10 FIG. 21(10). ANS. 0.58 in., S 25½° W, 36½°.
12·11 FIG. 21(11). ANS. 0.58 in., S 40° E, 47°.
12·12 FIG. 21(12). ANS. 2.00 in., E, 0°.

12·13 A. Scale: 1 in. = 100 ft. From point A, at an elevation of 1,740 ft, a tunnel bears N 45° E on a downward grade of 35 per cent. A

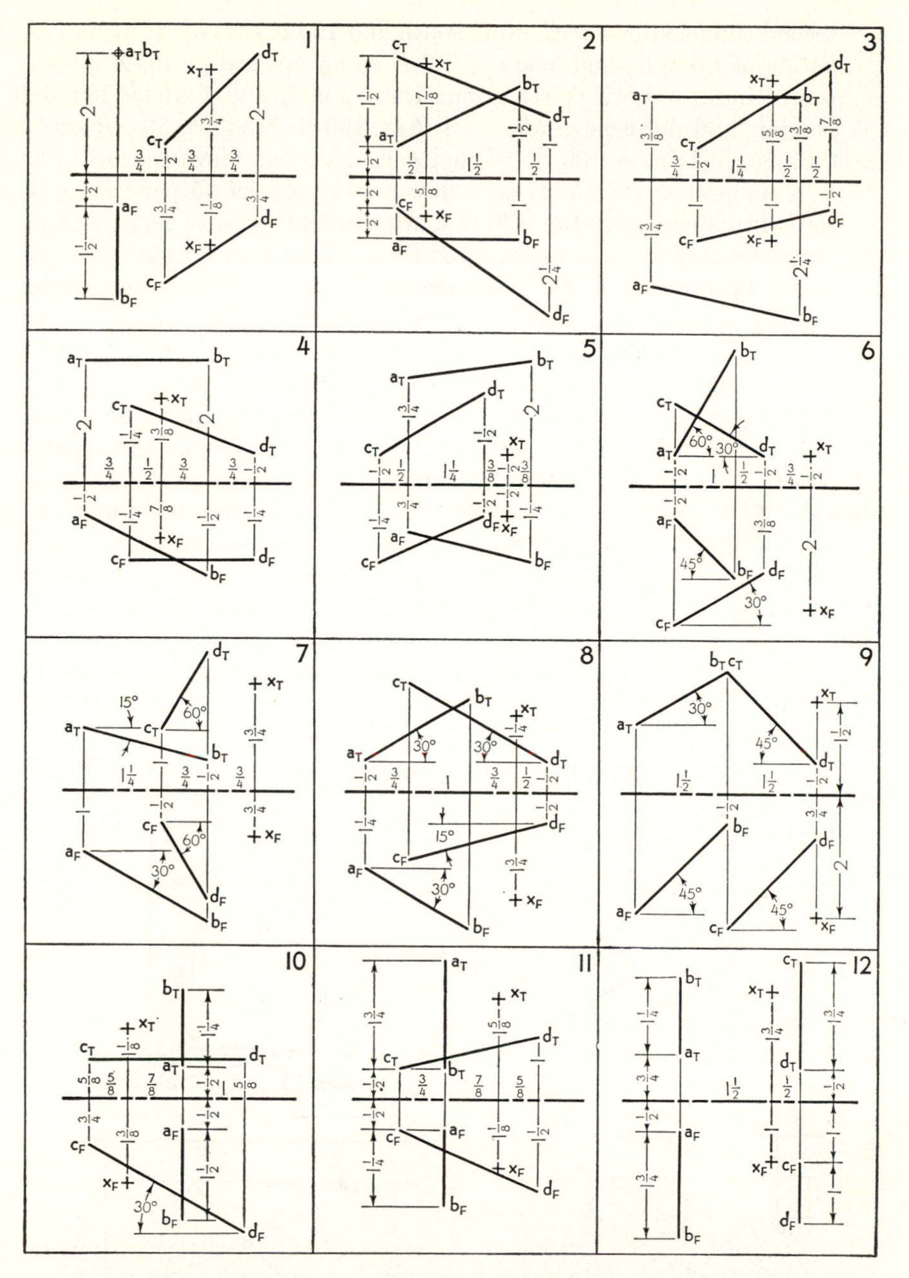

Fig. 21. Two skew lines and a point. Problems in Groups 12 and 13.

second tunnel starts at C, 40 ft south and 190 ft east of A, at an elevation of 1,650 ft, and bears N 30° E on an upward grade of 39 per cent. Locate the shortest connecting tunnel, and find its length, bearing, and per cent grade. ANS. 130 ft, S 53° E, 39 per cent.

12·14 B. Scale: 1 in. = 100 ft. From point A, at an elevation of 2,100 ft, a tunnel bears S 45° E on a downward grade of 45 per cent. A second tunnel starts at C, 150 ft south and 50 ft west of A, at an elevation of 2,200 ft, and bears S 75° E on a downward grade of 15 per cent. Locate the shortest connecting tunnel, and find its length, bearing, and per cent grade. ANS. 165 ft, N 1½° E, 154 per cent.

12·15 FIG. 22 A. Scale: 1 in. = 1 ft. The center lines of two 6-in.-diameter pipes are located as shown. How much clearance (if any) is there between the two pipes? How much must pipe A be raised or lowered vertically in order that the two pipes shall just touch each other? ANS. Clearance 2½ in.; raise pipe A 3½ in.

12·16 FIG. 23 C. Scale: ⅛ in. = 1 ft. The cable AB passes through one section of an open tower whose structural members are represented by the lines of the figure. What is the clearance distance between the cable and the nearest structural member? Identify this nearest member as XY in all views. To distinguish members on the near side from those on the far side, show the latter as dash lines in each auxiliary view. ANS. Clearance 14 in.

12·17 FIG. 24 A. Scale: 1 in. = 1 ft. A wire runs from point A as shown and passes through a rectangular hole in a steel plate. What is the

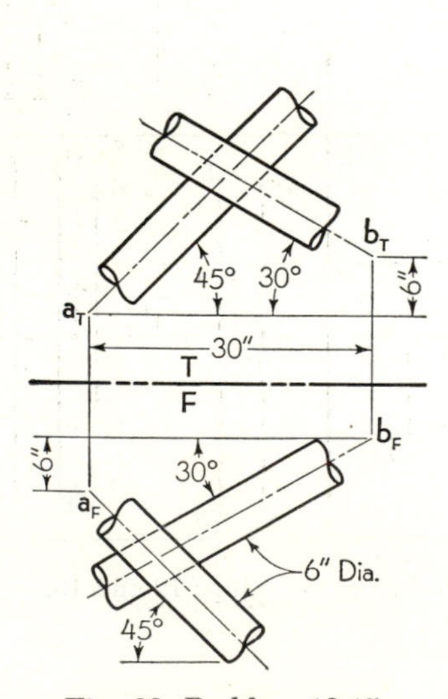

Fig. 22. Problem 12·15.

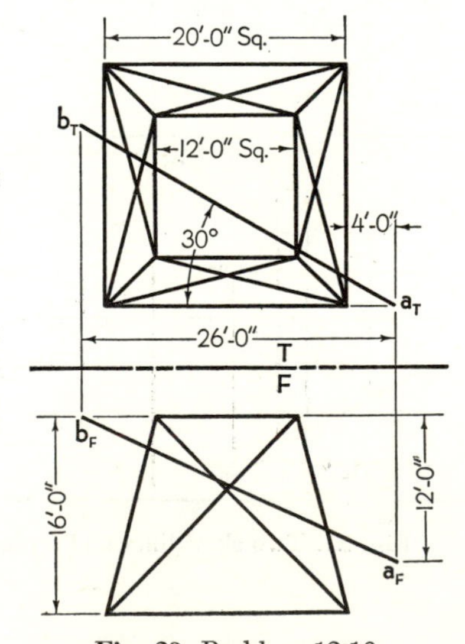

Fig. 23. Problem 12·16.

minimum distance from the wire to the edge of the hole? Show this distance as a dash line in all views. ANS. 3⅛ in.

12·18 FIG. 25 A. Scale: 1 in. = 1 ft. A wire runs from point *A* as shown and passes through a circular hole in a steel plate. What is the minimum distance from the wire to the edge of the hole? Show this distance as a dash line in all views. ANS. 2¼ in.

Group 13. Line through a point and two skew lines (Art. 3·18)

In each problem locate the straight line *YZ* that connects lines *AB* and *CD* and passes through the point *X*. Show line *YZ* as a heavy line in all views. Use either Layout A or Layout B according to the space requirements of your planned solution.

13·1 FIG. 21(1).
13·2 FIG. 21(2).
13·3 FIG. 21(3).
13·4 FIG. 21(4).
13·5 FIG. 21(5).
13·6 FIG. 21(6).
13·7 FIG. 21(7).
13·8 FIG. 21(8).
13·9 FIG. 21(9).
13·10 FIG. 21(10).
13·11 FIG. 21(11).
13·12 FIG. 21(12).

Group 14. Principal views of objects with inclined axes (Art. 3·19)

In each problem construct the top and front views of the given object as it would appear when its longitudinal axis is inclined to coincide with the given line *AB*.

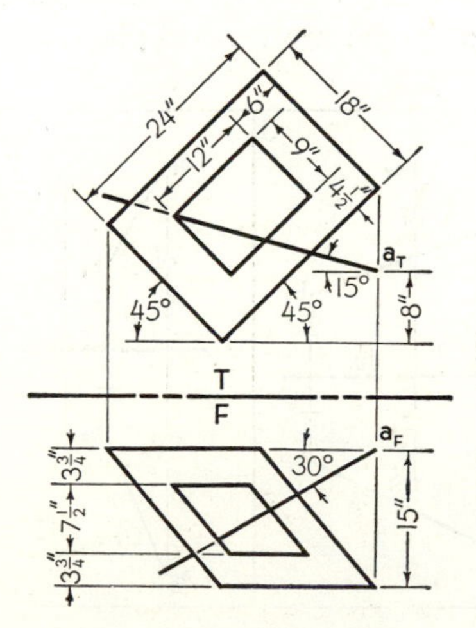

Fig. 24. Problem 12·17.

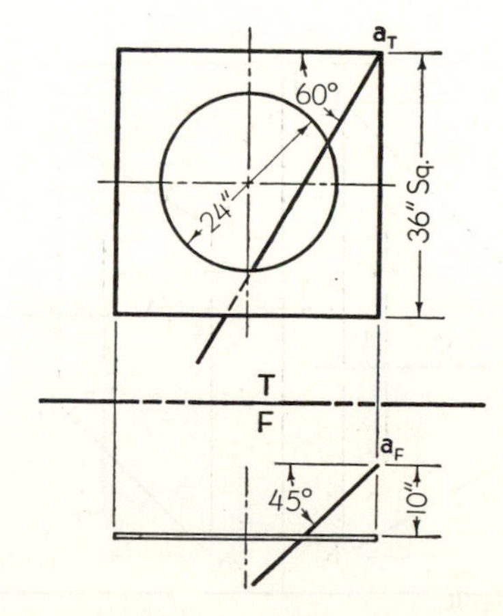

Fig. 25. Problem 12·18.

14·1 FIG. 26(1) A. A right hexagonal prism with altitude equal to line *AB*. The hexagonal bases are 1½ in. across flats and have two edges horizontal.

14·2 FIG. 26(1) A. A right hexagonal prism with altitude equal to line *AB*. One longitudinal edge of the prism passes through point *C*.

14·3 FIG. 26(1) A. A right octagonal pyramid with base at *A* and altitude equal to 2 in. The octagonal base is 1½ in. across flats and has two edges frontal.

14·4 FIG. 26(2) A. A right square prism with altitude equal to line *AB*. One longitudinal edge of the prism passes through point *C*.

14·5 FIG. 26(2) A. The hollow prism shown in Fig. 6(3) with the center of one base at *B*. Two edges of the base are horizontal.

14·6 FIG. 26(2) A. A right circular cylinder, 1½ in. in diameter, with altitude equal to line *AB*.

14·7 FIG. 26(3) A. A right hexagonal prism with altitude equal to line *AB*. One longitudinal edge of the prism passes through point *C*.

14·8 FIG. 26(3) A. The hexagonal pyramid shown in Fig. 6(2) with vertex at *B*. Two edges of the base are frontal.

14·9 FIG. 26(4) B. A right circular cylinder with altitude equal to line *AB*. Point *C* lies on the lateral surface of the cylinder.

14·10 FIG. 26(4) B. The truncated hexagonal prism shown in Fig. 6(1) with the lower base at *A*. The upper slanting base faces upward, and the highest and lowest edges of this base are horizontal.

14·11 A. A right circular cylinder, 2 in. in diameter, with an altitude of 1¾ in. Axis *AB* is a profile line sloping 35° downward to the rear. Place a_T and a_F each 1 in. from *T–F*.

14·12 A. Same as Prob. 14·11, but draw a right octagonal prism, 2 in. across flats. Two lateral faces of the prism are vertical planes.

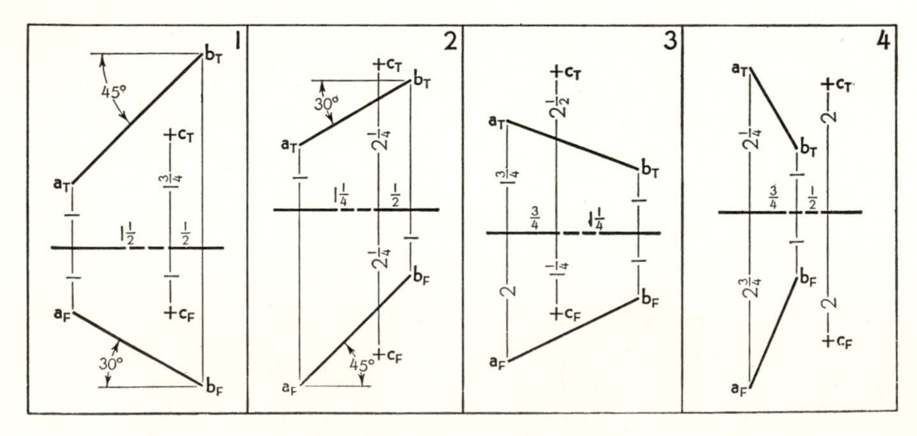

Fig. 26. Problems in Group 14.

14·13 FIG. 26(1) A. The truncated cylinder shown in Fig. 6(9) with the upper semicircular base at A. The slanting surface faces upward, and the straight edge is horizontal.

Group 15. Auxiliary views in a prescribed direction (Art. 3·20)

In each problem draw top and front views of the given object, and then construct an auxiliary view showing the object as it would appear when the direction of sight is inclined at angles R and S as shown in Fig. 3·27. Show an upward-pointing arrow in each view.

15·1 FIG. 7(1) A. $R = 30°$, $S = 30°$.
15·2 FIG. 7(2) A. $R = 45°$, $S = 45°$.
15·3 FIG. 7(3) A. $R = 60°$, $S = 30°$.
15·4 FIG. 7(4) A. $R = 30°$, $S = 15°$.
15·5 FIG. 6(1) A. $R = 20°$, $S = 35°$.
15·6 FIG. 6(2) B. $R = 15°$, $S = -45°$.
15·7 FIG. 6(3) B. $R = 30°$, $S = -30°$.
15·8 FIG. 6(4) A. $R = 15°$, $S = 15°$.
15·9 FIG. 6(5) A. $R = 60°$, $S = 15°$.
15·10 FIG. 6(6) A. $R = 45°$, $S = 30°$.
15·11 FIG. 6(7) A. $R = 30°$, $S = 30°$.
15·12 FIG. 6(8) A. $R = 15°$, $S = 30°$.
15·13 FIG. 6(9) A. $R = 45°$, $S = 30°$.
15·14 FIG. 5(2) C. $R = 60°$, $S = 15°$.
15·15 FIG. 5(4) C. $R = 40°$, $S = 25°$.
15·16 FIG. 5(6) C. $R = 35°$, $S = 20°$.
15·17 FIG. 5(8) C. $R = 75°$, $S = 45°$. Let the given front view be the top view, and construct a new front view. Draw the first auxiliary view far to the right in order to allow space for the second auxiliary view.
15·18 FIG. 2(10) C. (Turn sheet with long side vertical.) $R = 45°$, $S = 30°$. Draw the object in the top and front views with its axis vertical, standing on the circular end with the T-shaped surface facing frontward.

Group 16. Axonometric drawings (Art. 3·21)

In each problem draw the top and front views of the given object in the left half of the sheet, and then construct an axonometric drawing in the right half. Select the desired type of drawing from Fig. 3·29 (or that assigned by the instructor). Hidden lines may be omitted. Use Layout B.

16·1 FIG. 6(1).
16·2 FIG. 6(2).
16·3 FIG. 6(3).
16·4 FIG. 6(4).
16·5 FIG. 6(5).
16·6 FIG. 6(6).
16·7 FIG. 6(7).
16·8 FIG. 6(8).
16·9 FIG. 6(9). Construct the curved lines by choosing a series of points on each curve (preferably equally spaced), and then locate these points on the axonometric drawing by the usual method of offset measurements.
16·10 FIG. 5(1).
16·11 FIG. 5(2).
16·12 FIG. 5(3).
16·13 FIG. 5(4).
16·14 FIG. 5(5).
16·15 FIG. 5(6).
16·16 FIG. 5(7). See statement of Prob. 16·9.
16·17 FIG. 5(8). See statement of Prob. 16·9.

Group 17. Lines in a plane (Arts. 4·1 to 4·4)

In Probs. 17·1 to 17·9 the given plane (unlimited in extent) is fully defined by one set of lines. A second set of lines, also in the given plane, is shown in only one view. Draw the top, front, and right-side views of the plane, showing all lines in all views. Also show on the plane in each view a horizontal line *MN*, a frontal line *PQ*, and a profile line *XY*. Give the strike of the plane. Use Layout B.

17·1 FIG. 27(1). Lines *RST* in plane *ABC*. ANS. N 59° W.
17·2 FIG. 27(2). Lines *RST* in plane *ABC*. ANS. N 63° E.
17·3 FIG. 27(3). The square in the plane of intersecting lines *AB* and *CD*. ANS. N 43½° W.
17·4 FIG. 27(4). The parallelogram in plane *ABC*. ANS. N 75° E.
17·5 FIG. 27(5). Lines *RSTV* in the plane of parallel lines *AB* and *CD*. ANS. N 50½° E.
17·6 FIG. 27(6). Lines *RSTV* in the plane of parallel lines *AB* and *CD*. ANS. N 62½° E.
17·7 FIG. 27(7). The letter A in plane *ABC*. ANS. N 53½° W.
17·8 FIG. 27(8). The oblique surface of a milling fixture in plane *ABC*. ANS. N 62° E.
17·9 FIG. 27(9). The oblique surface of a milling fixture in plane *ABC*. ANS. N 80½° W.
17·10 FIG. 28 C. Scale: ¼ in. = 1 ft (B, 3/16 in. = 1 ft). In the corner of a courtyard a shed roof is to be constructed resting upon the beam *AB* and the post *C*. The height and incomplete location of a second post *D* are given. Post *C* is to be under the front edge of the roof, post *D* at the left front corner, beam end *A* under the left edge, and

the roof is to extend to the walls. Draw the top, front, and right-side views of the roof, showing its intersections with the walls and its left and front edges. How far is post D from the rear wall?

ANS. 9 ft 8 in.

Group 18. Points in a plane (Art. 4·5)

In Probs. 18·1 to 18·7 locate in the given plane (unlimited in extent) the specified points X, Y, and Z (disregard lines shown only in one view).

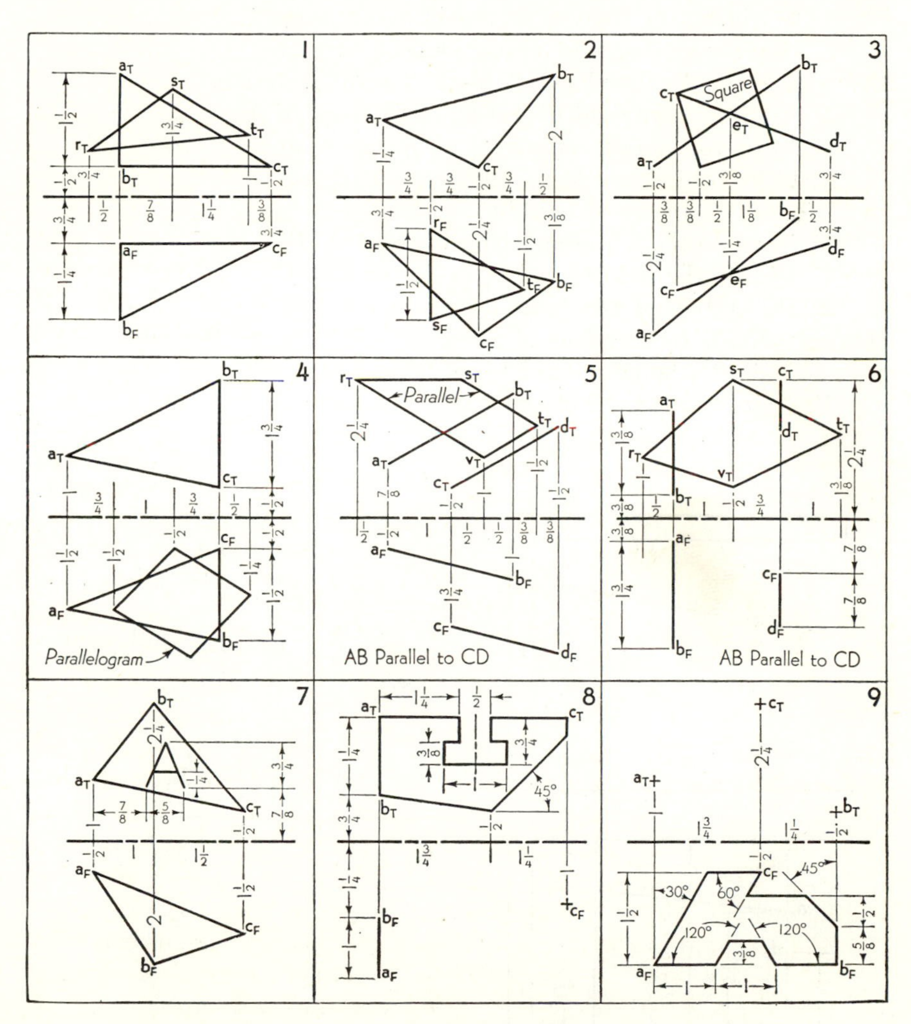

Fig. 27. Plane surfaces. Problems in Groups 17, 18, and 19.

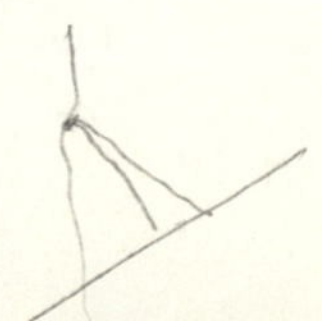

Draw the top and front views only, showing points *X*, *Y*, and *Z* in both views (all problems may be solved without additional views). Use Layout A or one-half of Layout B.

18·1 FIG. 27(1). In plane *ABC* locate:
X, 1 in. to the right of and 1 in. in front of *A*.
Y, 1½ in. to the right of and 1 in. lower than *A*.
Z, ¼ in. behind and ½ in. higher than *B*.

18·2 FIG. 27(2). In plane *ABC* locate:
X, ½ in. to the left of and ¾ in. in front of *A*.
Y, 1¾ in. to the right of and ¼ in. lower than *A*.
Z, ½ in. behind and ¾ in. higher than *C*.

18·3 FIG. 27(3). In the plane of lines *AB* and *CD* locate:
X, 1½ in. to the right of and ¼ in. in front of *A*.
Y, ½ in. to the right of and ¾ in. higher than *C*.
Z, ½ in. behind and ½ in. higher than *D*.

18·4 FIG. 27(4). In the plane *ABC* locate:
X, ¾ in. to the left of and 1 in. in front of *B*.
Y, 1 in. to the right of and ½ in. lower than *C*.
Z, 1⅛ in. behind and ¾ in. lower than *A*.

18·5 FIG. 27(5). In the plane of parallel lines *AB* and *CD* locate:
X, ¼ in. to the left of and ½ in. behind *D*.
Y, ½ in. to the right of and 1 in. lower than *A*.
Z, ¾ in. in front of and ½ in. lower than *B*.

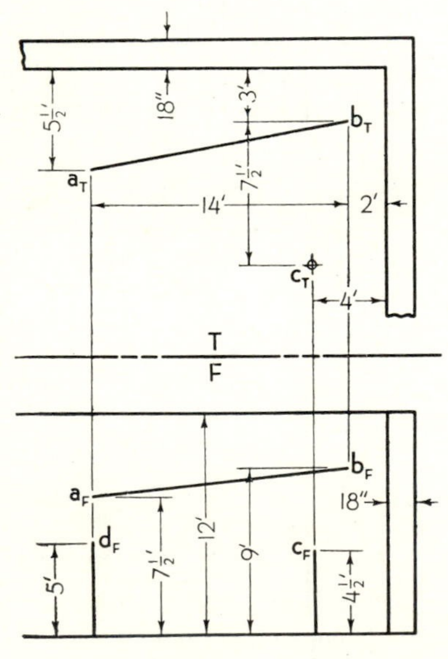

Fig. 28. Problem 17·10.

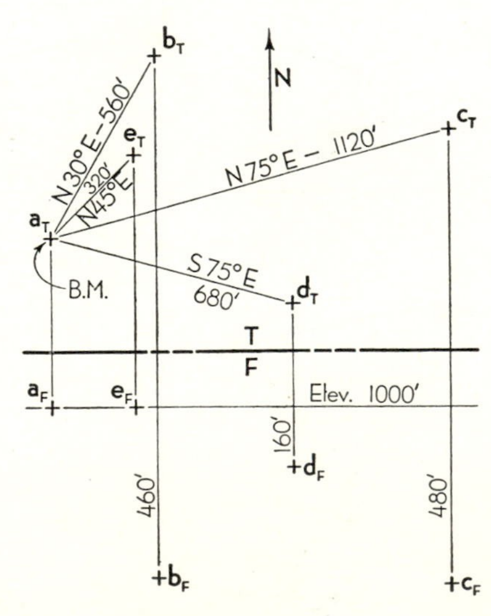

Fig. 29. Problems 18·8, 19·13, and 20·12.

18·6 FIG. 27(6). In the plane of parallel lines AB and CD locate:
D, without using additional views.
X, $1\frac{1}{4}$ in. to the right of and $\frac{1}{2}$ in. behind B.
Y, $\frac{3}{4}$ in. to the left of and 1 in. lower than A.
Z, $\frac{1}{4}$ in. in front of and $\frac{1}{2}$ in. lower than C.

18·7 FIG. 27(7). In the plane ABC locate:
X, $\frac{1}{4}$ in. to the right of and 1 in. behind A.
Y, $\frac{1}{2}$ in. to the right of and $1\frac{1}{4}$ in. higher than B.
Z, $1\frac{1}{2}$ in. behind and $1\frac{1}{2}$ in. lower than C.

18·8 FIG. 29 A. Scale: 1 in. = 200 ft. A vein of ore is known to lie below a level plateau whose elevation is 1,000 ft. Three vertical boreholes are made to strike the vein at B, C, and D. The figure shows the depth, and location from a bench mark A, of each borehole. What is the strike of the vein? From point E on the plateau a vertical shaft is to be cut to the ore vein. How deep will it be? What will be the per cent grade of an incline in the plane of the vein running due east from the bottom of the shaft?

ANS. N 75° W. Depth, 310 ft. Grade, 15 per cent downward.

18·9 FIG. 30 A. Scale: $\frac{1}{4}$ in. = 1 ft. A 6-ft-diameter vertical stack passes through a roof as shown in the top view. Complete the front view of the opening in the roof.

18·10 FIG. 31 A. Scale: $\frac{1}{4}$ in. = 1 ft. An oval sewer pipe passes through the inclined side wall of a ditch, and the opening in the wall is shown in the front view. Complete the top view of the wall opening.

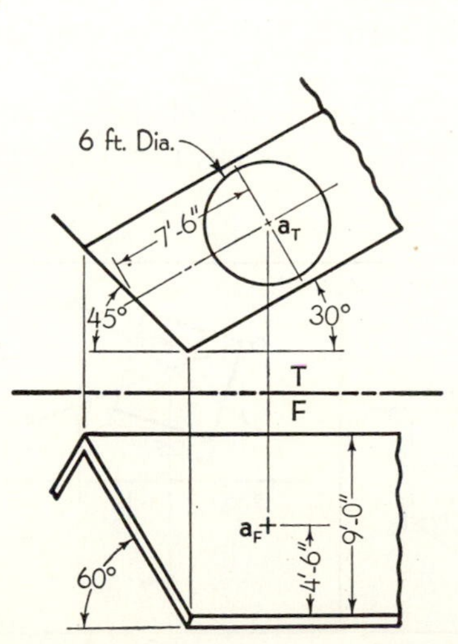

Fig. 30. Problems 18·9, 19·14, and 22·17.

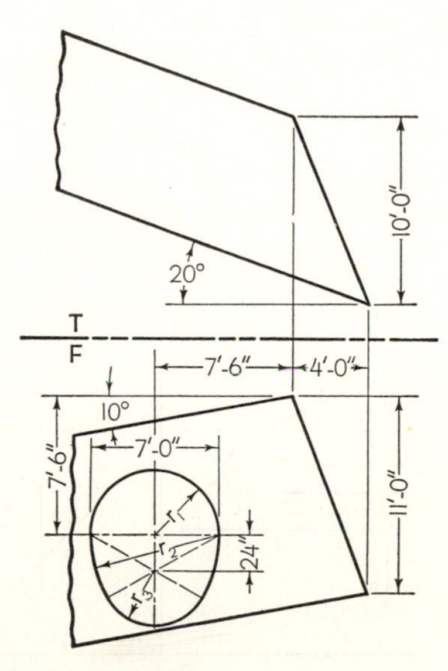

Fig. 31. Problems 18·10 and 22·18.

Group 19. Edge views, slope, and dip of a plane (Arts. 4·6 to 4·8)

In Probs. 19·1 to 19·9 draw a top-adjacent view showing the given plane as an edge, and measure the slope angle in degrees (disregard lines shown only in one view). Also, draw a front-adjacent view showing the plane as an edge. What is the significance of the angle in this view between edge and reference line? Use Layout B.

19·1 FIG. 27(1). Plane *ABC*. ANS. 44½°.
19·2 FIG. 27(2). Plane *ABC*. ANS. 48½°.
19·3 FIG. 27(3). Plane of intersecting lines *AB* and *CD*. ANS. 33½°.
19·4 FIG. 27(4). Plane *ABC*. ANS. 41°.
19·5 FIG. 27(5). Plane of parallel lines *AB* and *CD*. ANS. 53½°.
19·6 FIG. 27(6). Plane of parallel lines *AB* and *CD*. ANS. 55°.
19·7 FIG. 27(7). Plane *ABC*. ANS. 43°.
19·8 FIG. 27(8). Plane *ABC*. (It is not necessary to complete the front view.) ANS. 42°.
19·9 FIG. 27(9). Plane *ABC*. (It is not necessary to complete the top view.) ANS. 44½°.
19·10 FIG. 32(1) B. Plane *ABC* has a strike of N 60° E and a dip of 30° southeasterly. Let *RST* be a parallel plane ½ in. from and above *ABC*. Show both planes in correct visibility in top and front views.
19·11 FIG. 32(2) B. Plane *ABC* has a strike of N 75° W and a dip of 45° northeasterly. Let *RST* be a parallel plane ⅜ in. from and below *ABC*. Show both planes in correct visibility in top and front views.
19·12 FIG. 32(3) B. Plane *ABCD* has a strike of N 45° E and a dip of 30° southeasterly. Let the square be a parallel plane ½ in. from and above *ABCD*. Show both planes in correct visibility in top and front views.
19·13 FIG. 29 B. Scale: 1 in. = 400 ft. What is the dip of the ore vein located by points *B*, *C*, and *D*? ANS. 30° northeasterly.

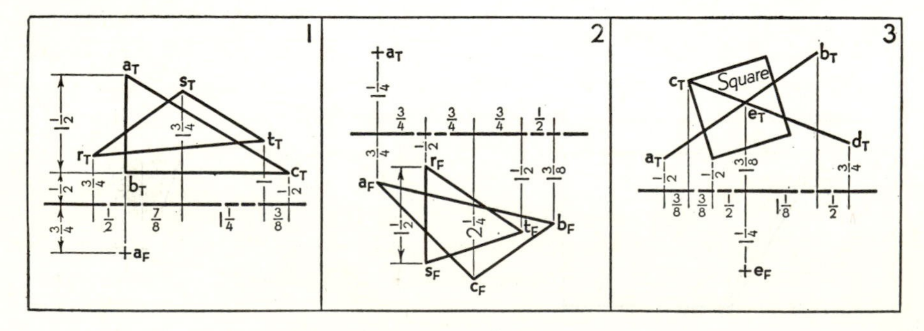

Fig. 32. Problems 19·10 to 19·12.

19·14 FIG. 30 B. Scale: $\frac{1}{4}$ in. = 1 ft. What is the pitch of the roof shown in the figure? Pitch is the slope of a roof expressed as the ratio of the height to the half span (or the *rise* to the *run*). Six inches rise in 12 in. run is written 6:12. ANS. $15\frac{3}{8}$:12.

19·15 B. Scale: 1 in. = 25 ft. The outcrop of a vein of ore appears on a level plateau. The strike of the vein is N 50° W, and its dip is 40° to the northeast. Fifty feet away another outcropping vein is found having the same strike and dip. What is the perpendicular distance between the two parallel veins? ANS. 32 ft.

19·16 On a square building having a hip roof, the hip rafter has a slope angle of 30°. What is the slope angle of the roof plane? ANS. 39°15′.

Group 20. Shortest line from a point to a plane (Art. 4·9)

In Probs. 20·1 to 20·9 find the shortest distance *XP* from point *X* to the given plane. Show line *XP* as a heavy line in all views. Use Layout B.

20·1 FIG. 33(1). Plane *ABC*. ANS. 0.89 in.
20·2 FIG. 33(2). Plane *ABC*. ANS. 1.25 in.
20·3 FIG. 33(3). Plane of parallel lines *AB* and *CD*. ANS. 0.88 in.
20·4 FIG. 33(4). Plane *ABC*. ANS. 1.72 in.
20·5 FIG. 33(5). Plane *ABC*. ANS. 1.15 in.
20·6 FIG. 33(6). Plane of parallel lines *AB* and *CD*. ANS. 0.77 in.
20·7 FIG. 33(7). Plane *ABC*. ANS. 1.14 in.
20·8 FIG. 33(8). Plane of parallel lines *AB* and *CD*. ANS. 0.96 in.
20·9 FIG. 33(9). Plane of intersecting lines *AB* and *CD*. ANS. 1.14 in.

20·10 FIG. 34 A. Scale: $\frac{1}{16}$ in. = 1 ft. Two masts *X* and *Y* stand adjacent to a building. One guy wire from the top of each mast must be attached to the nearest roof surface and should be as short as possible. Show the two guy wires from *X* and *Y* to the roof in all views, and record the length of each.
ANS. *X*, 22 ft 0 in. *Y*, 23 ft 4 in.

20·11 FIG. 35 A. Scale: 1 in. = 1 ft. The hopper shown is to be connected with a pipeline. The center line of the pipe will pass through point *X* and be run by the shortest possible distance to the nearest face. Show the pipe in all views, and record the length from *X* to the hopper. ANS. 18 in.

20·12 FIG. 29 B. Scale: 1 in. = 400 ft. The ore vein located by points *B*, *C*, and *D* is to be connected to point *E* on the surface by the shortest possible shaft. Show this shaft in all views, and record its length, bearing, and slope angle in degrees.
ANS. 276 ft, S 15° W, 60°.

20·13 What is the distance from a corner of a 2-in. cube to the plane of the three adjacent corners? ANS. 1.15 in.

20·14 What is the distance from a corner of a $1\frac{1}{2}$- by 2- by $2\frac{1}{2}$-in. block to the plane of the three adjacent corners? ANS. 1.08 in.

Group 21. Shortest grade line from a point to a plane (Art. 4·10)

In Probs. 21·1 to 21·9 find the shortest horizontal distance *XR* and the shortest line *XQ* of given grade, from point *X* to the given plane. Show lines *XR* and *XQ* as heavy lines in all views. Use Layout B.

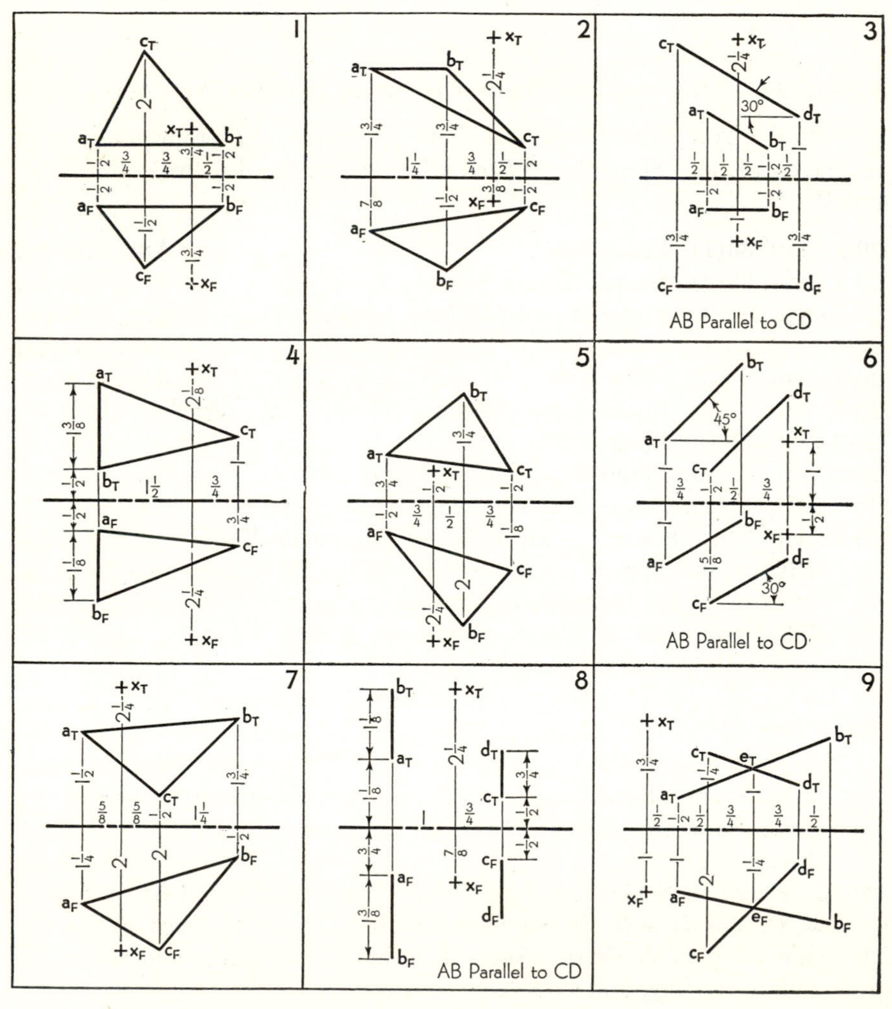

Fig. 33. A point and a plane. Problems in Groups 20 and 21.

21·1 FIG. 33(1). Plane *ABC*. Grade of *XQ* is 30°.
ANS. *XR*, 1.61 in. *XQ*, 1.00 in.

21·2 FIG. 33(2). Plane *ABC*. Grade of *XQ* is 60°.
ANS. *XR*, 1.54 in. *XQ*, 1.37 in.

21·3 FIG. 33(3). Plane of parallel lines *AB* and *CD*. Grade of *XQ* is 45°.
ANS. *XR*, 1.00 in. *XQ*, 0.91 in.

21·4 FIG. 33(4). Plane *ABC*. Grade of *XQ* is 45 per cent.
ANS. *XR*, 2.66 in. *XQ*, 1.90 in.

21·5 FIG. 33(5). Plane *ABC*. Grade of *XQ* is 35 per cent.
ANS. *XR*, 1.59 in. *XQ*, 1.26 in.

21·6 FIG. 33(6). Plane of parallel lines *AB* and *CD*. Grade of *XQ* is 25 per cent. ANS. *XR*, 1.16 in. *XQ*, 0.94 in.

21·7 FIG. 33(7). Plane *ABC*. Grade of *XQ* is 25°.
ANS. *XR*, 1.60 in. *XQ*, 1.20 in.

21·8 FIG. 33(8). Plane of parallel lines *AB* and *CD*. Grade of *XQ* is 60°.
ANS. *XR*, 1.22 in. *XQ*, 1.03 in.

21·9 FIG. 33(9). Plane of intersecting lines *AB* and *CD*. Grade of *XQ* is 30 per cent. ANS. *XR*, 1.33 in. *XQ*, 1.18 in.

21·10 B. Scale: 1 in. = 100 ft. An outcrop of ore has been found on a hill at point *A*, elevation 1,157 ft. At *B*, 135 ft N 81°0′ W of *A*, a vertical borehole strikes the vein at an elevation of 1,108 ft. A second vertical borehole at *C*, 181 ft S 19°30′ E of *A*, strikes the vein

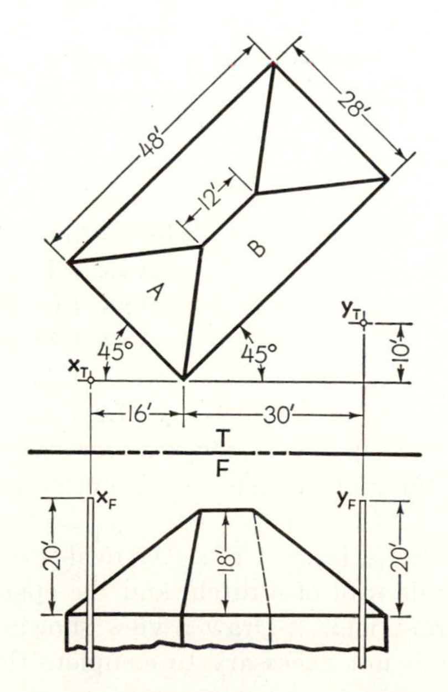

Fig. 34. Problems 20·10 and 22·19.

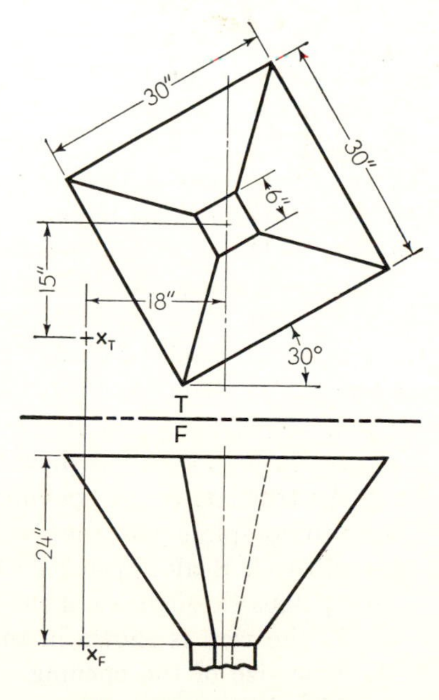

Fig. 35. Problem 20·11.

at an elevation of 1,059 ft. From point *E*, 325 ft S 51°30′ W of *A*, elevation 1,042 ft, it is planned to drive a tunnel to the vein on an upward grade of 15 per cent (to facilitate removal of ore). What are the length and bearing of the proposed tunnel? How much shorter would a horizontal tunnel be?

ANS. 214 ft, N 33° E. 38 ft shorter.

Group 22. True size of a plane (Art. 4·11)

In Probs. 22·1 to 22·16 draw a view showing the true size of the given plane surface. In some problems one of the given views is incomplete, but the true-size view can be constructed without completing the given view. Make a preliminary freehand sketch of your proposed solution (see Fig. 3·25) and then choose either Layout A or B.

Additional problems on true size of a plane may be selected from Figs. 27 and 33. Use Layout B.

22·1 FIG. 36(1). Plane *ABC*.
22·2 FIG. 36(2). Plane *ABCD*.
22·3 FIG. 36(3). Plane *ABC*.
22·4 FIG. 36(4). Plane *ABCD*.
22·5 FIG. 36(5). Plane *ABCD*.
22·6 FIG. 36(6). Plane *ABCD*.
22·7 FIG. 36(7). Face surface of wedge cam.
22·8 FIG. 36(8). Airport roof with directional marker.
22·9 FIG. 36(9). Milling fixture end surface.
22·10 FIG. 36(10). Dovetail stop surface.
22·11 FIG. 36(11). Roof flashing around a stack.
22·12 FIG. 36(12). Firing pin end surface.
22·13 FIG. 19(1). Plane of parallel lines *AB* and *CD*. What is the true distance between lines *AB* and *CD*? ANS. 1.43 in.
22·14 FIG. 19(2). Same as Prob. 22·13. ANS. 1.08 in.
22·15 FIG. 19(3). Same as Prob. 22·13. ANS. 1.78 in.
22·16 FIG. 19(4). Same as Prob. 22·13. ANS. 1.35 in.
22·17 FIG. 30 B. Scale: ¼ in. = 1 ft. A 6-ft-diameter vertical stack passes through a roof as shown in the top view. Draw a view showing the true size of the opening in the roof surface. (It is not necessary to complete the front view.)
22·18 FIG. 31 C. Scale: ¼ in. = 1 ft (B, 3⁄16 in. = 1 ft). An oval sewer pipe passes through the inclined side wall of a ditch, and the opening in the wall is shown in the front view. Draw a view showing the true size of the opening. (It is not necessary to complete the top view.)

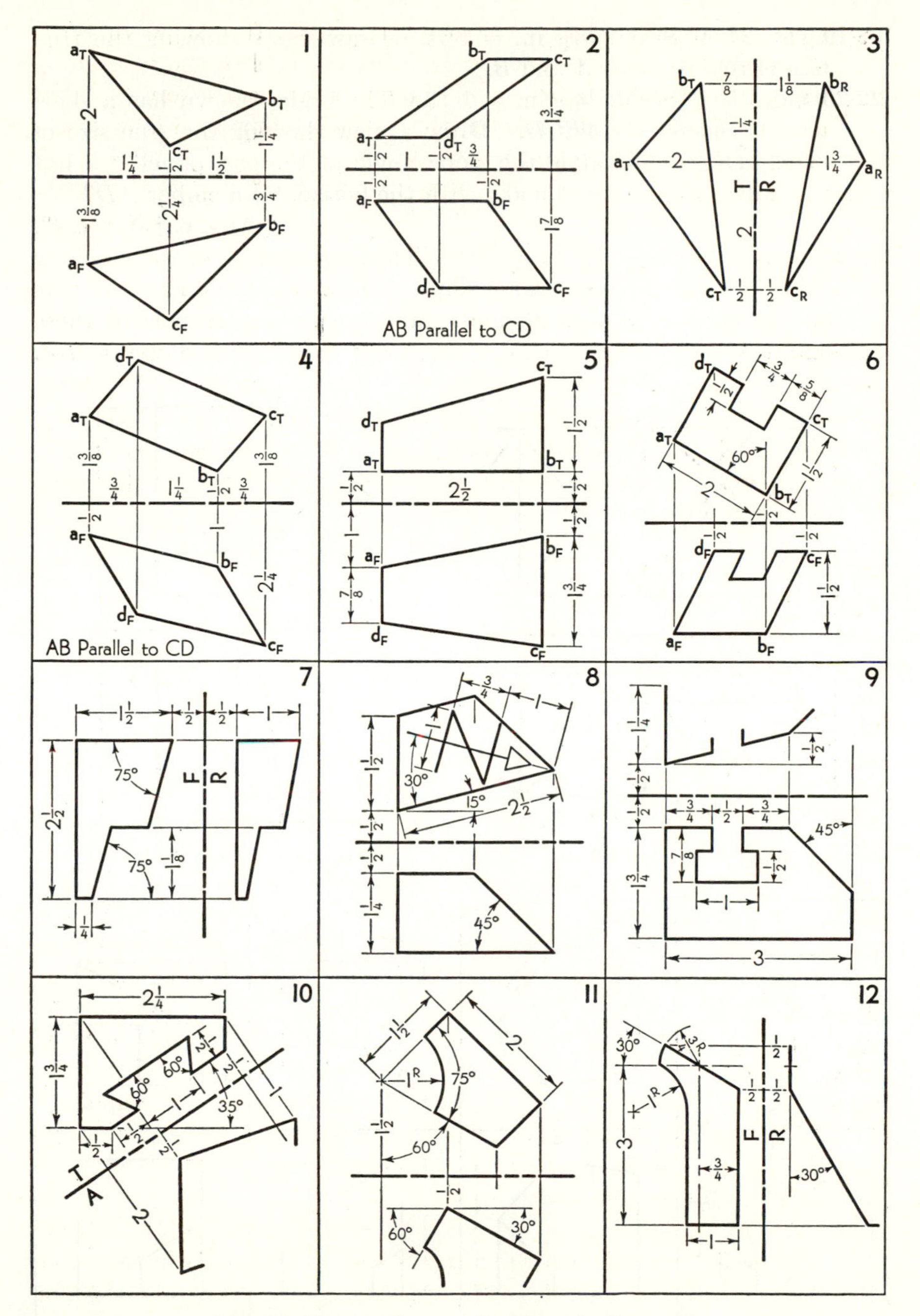

Fig. 36. Plane surfaces. Problems in Group 22.

22·19 FIG. 34 A. Scale: $\frac{1}{16}$ in. = 1 ft. Draw views showing the true size of roof surfaces *A* and *B*.

22·20 FIG. 37 A. Scale: $\frac{1}{16}$ in. = 1 ft. The bridge shown has a skew portal (end panel) *ABCD*. Draw a view showing the true size of panel *ABCD*. What is the slope angle of the end panel? What angle does the end post make with the horizontal member *AD*?

ANS. 63½°. 104°.

22·21 FIG. 38 B. Scale: ⅛ in. = 1 ft. The sloping sides of the concrete bridge pier all have the same slope. Draw views showing the true size of surfaces *A* and *B*, and find the slope in degrees of these surfaces.

ANS. Slope = 75°.

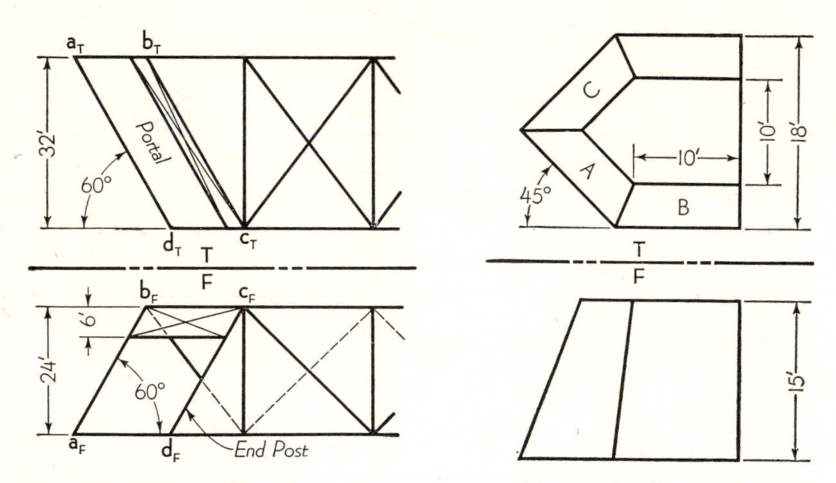

Fig. 37. Problem 22·20.

Fig. 38. Problem 22·21.

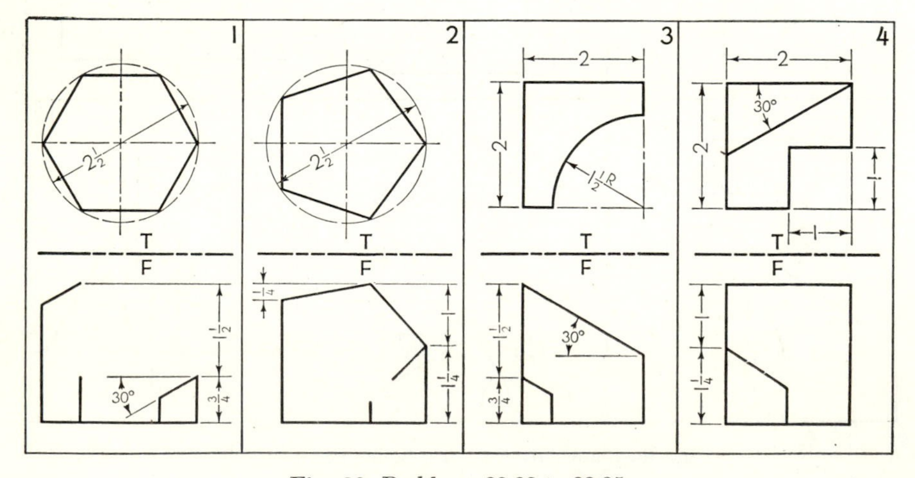

Fig. 39. Problems 22·22 to 22·25.

22·22 FIG. 39(1) B. Draw a view showing the true size of the upper sloping surface of the truncated prism. Show the entire prism in all views, and then use the edge view as an aid in completing the front view.

22·23 FIG. 39(2). Same as Prob. 22·22.

22·24 FIG. 39(3). Same as Prob. 22·22.

22·25 FIG. 39(4). Same as Prob. 22·22.

Group 23. Angle between two intersecting lines (Art. 4·12)

23·1 FIG. 27(3) B. Draw a view showing the true size of the angle between lines *AB* and *CD*, and measure the acute angle. Bisect angle *AEC*, and show the bisector in the given views.

ANS. Angle *AEC*, 53°.

23·2 FIG. 33(9). Same as Prob. 23·1. ANS. Angle *AEC*, 65½°.

23·3 FIG. 40 B. Scale: ⅛ in. = 1 ft. The structural members of one leg of a tower are shown. To detail the gusset plates the angles *BAC*, *BCA*, and *CBD* are required. Also find the corresponding angles for plane *ABE*. ANS. *BAC*, 26½°. *BCA*, 45½°. *CBD*, 71½°.

23·4 FIG. 41 C. Scale: 1 in. = 1 ft (B, ¾ in. = 1 ft). A tripod-type airplane landing gear is shown. Determine the angles *ADB*, *BDC*, and *ADC* between each pair of struts.

ANS. *ADB*, 42°. *BDC*, 41½°. *ADC*, 70°.

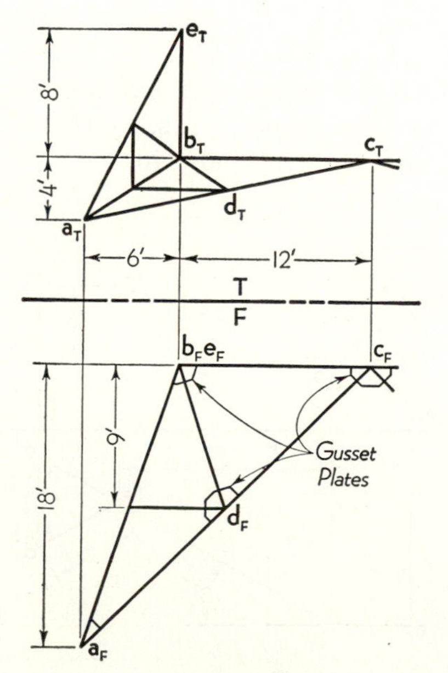

Fig. 40. Problem 23·3.

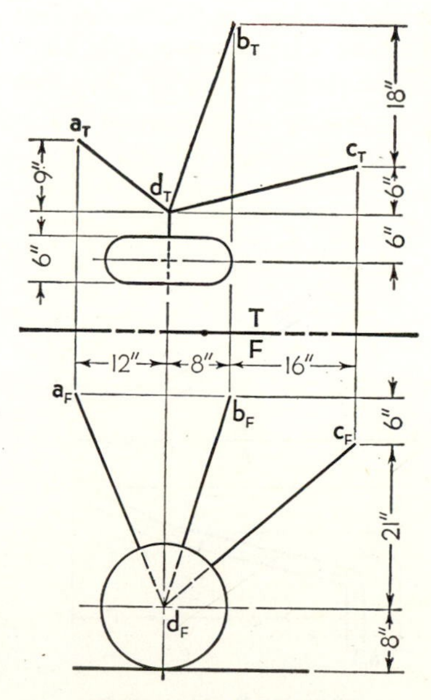

Fig. 41. Problem 23·4.

23·5 FIG. 14 C. Scale: 1 in. = 1 ft (B, ¾ in. = 1 ft). Same as Prob. 23·4.
ANS. *ADB*, 76°. *BDC*, 49°. *ADC*, 45½°.

23·6 FIG. 15 C. Scale: ¾ in. = 1 ft (B, ½ in. = 1 ft). Same as Prob. 23·4. ANS. *ADB*, 55°. *BDC*, 23°. *ADC*, 65½°.

23·7 FIG. 16 C. Scale: 1 in. = 1 ft (B, ¾ in. = 1 ft). The framework shown is an engine mount for an aircraft radial engine (see statement of Prob. 6·22). Find the angles between each pair of adjacent straight tubes at the intersection points *A*, *B*, *C*, *D*, and *E*. The mount is attached to the vertical bulkhead as shown.
ANS. *A*, 42½°. *B*, 79¼°. *C*, 80½°. *D*, 38¾°. *E*, 82°.

23·8 FIG. 16. Same as Prob. 23·7, but change the mount to fit on the canted bulkhead. Do not change the front view.
ANS. *A*, 39½°. *B*, 73°. *C*, 82½°. *D*, 42½°. *E*, 91°.

23·9 FIG. 42 B. Scale: 3 in. = 1 ft. A length of steel tubing is bent as shown in the figure. Determine the number of degrees the pipe is bent at *B* and *C*. ANS. *B*, 78°. *C*, 34°.

23·10 FIG. 43 B. Scale: 1½ in. = 1 ft. *AB*, *BC*, and *CD* are the center lines of three shafts connected at *B* and *C* by universal joints. Shaft *BC* is fixed, but shaft *AB* is free to swing 30° on either side of its center position as shown. Shaft *CD* is free to swing 30° above and below a horizontal position. What is the maximum deflection angle between shafts *AB* and *BC*, and in which extreme position does it occur? And similarly for shafts *CD* and *BC*?
ANS. *AB*, 55° in position 1. *CD*, 62° in position 2.

23·11 FIG. 44 C. Scale: 1½ in. = 1 ft (B, 1 in. = 1 ft). An X-frame stamping is shown. The angle *AEB* is required, but give hole center-to-center distances for ease of fabrication. ANS. *AEB*, 93°. *AB*, 28⅛ in. *BC*, 20⅜ in. *CD*, 25⅞ in. *AD*, 30⅜ in.

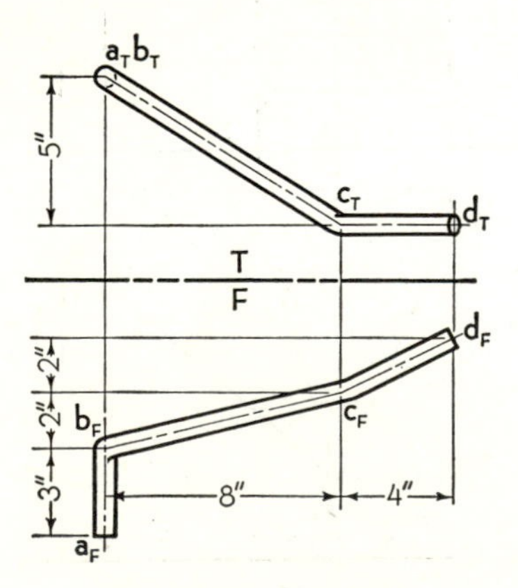

Fig. 42. Problem 23·9.

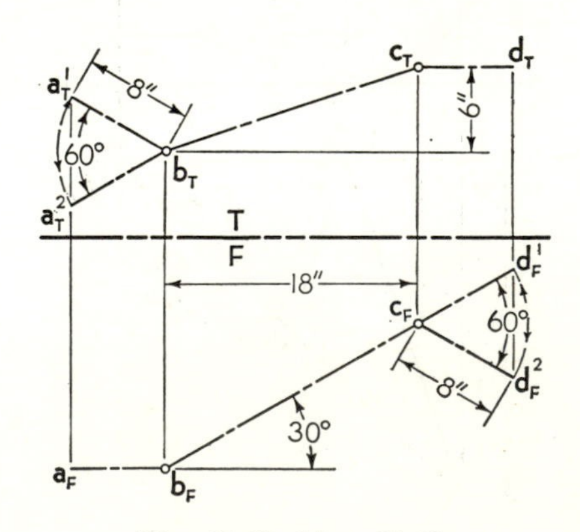

Fig. 43. Problem 23·10.

23·12 FIG. 45 C. Scale: 1 in. = 1 ft (B, ¾ in. = 1 ft). The ends of a wooden timber, 12 in. square, must be cut to fit against a wall and floor as shown. The completed top view shows that two faces, or sides, of the timber are in vertical planes. Draw a new view of the timber showing one of its vertical faces in true size, and dimension the *down-cut* and *heel-cut* angles. Draw another new view showing one of the sloping faces in true size, and dimension the *side-cut* angle. What is the over-all length of the timber? Complete the front view. ANS. Down cut, 55°, heel cut, 35°. Side cut, 39°. Length, 6 ft 5¼ in.

23·13 FIG. 46 C. Scale: 1½ in. = 1 ft (B, 1 in. = 1 ft). The ends of a 4-by-8 rafter must be cut to fit against the plate and ridge as shown. The completed front view shows that the two 4-in. faces of the timber appear edgewise in this view. (This is *not* a hip rafter like that shown in Fig. 47.) Draw a new view of the timber showing one of the 4-in. faces in true size, and dimension the *side-cut* angle. Draw another view showing one of the 8-in. faces in true size, and dimension the *down-cut* and *heel-cut* angles. What is the over-all length of the timber? Complete the top view.
ANS. Side cut, 55°. Down cut, 39°; heel cut, 39°. Length, 3 ft 11 in.

23·14 FIG. 47 C. Scale: 1 in. = 1 ft. The framing of one corner of a hip roof is shown. Lay out the two given views at the extreme left of the sheet (draw hip line $a_F b_F$, but leave hip and jack rafters incom-

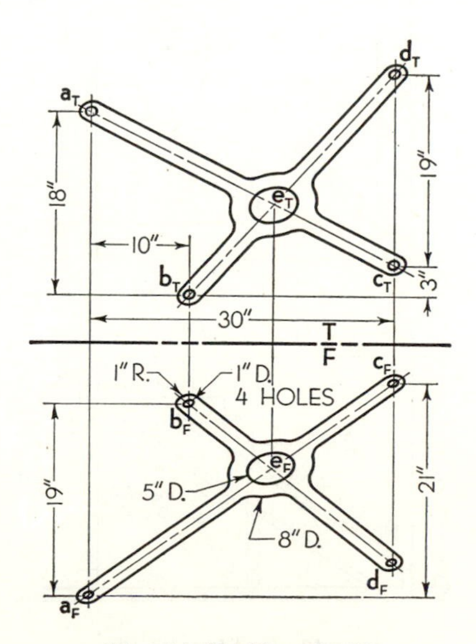

Fig. 44. Problem 23·11.

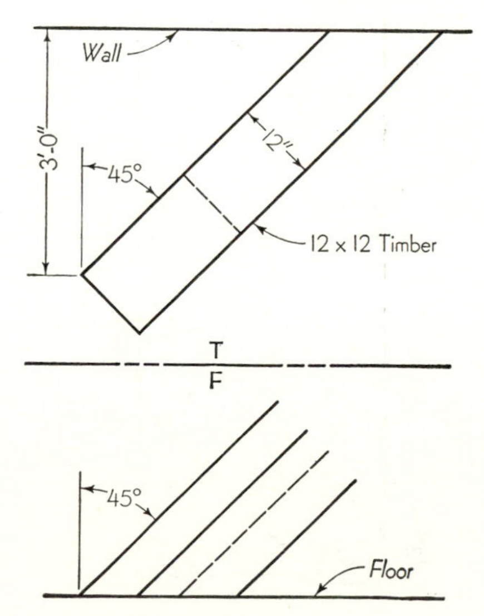

Fig. 45. Problem 23·12.

plete). Draw a new view of the hip rafter showing its 6-in. face in true size (note that line AB is slightly above upper face of rafter), and dimension the *down-cut* and *heel-cut* angles. Draw another view showing the 2-in. face of this rafter in true size, and dimension the *side-cut* angles. Complete the front view, and then draw true-size views of the jack rafter, dimensioning angles. What is the over-all length of each rafter? Ans. Hip rafter: down cut, 65°; heel cut, 25°; side cut, 42°; length, 7 ft 0 in. Jack rafter: down cut, 56°; heel cut, 34°; side cut, 40°; length, 2 ft 9¼ in.

Group 24. Shortest line from a point to a line (plane method, Art. 4·13)

In Probs. 24·1 to 24·16 find the true length, bearing, and slope angle of the shortest line CX from point C to the line AB by the plane method. Show line CX as a heavy line in all views.

24·1 FIG. 48(1) B. Ans. 0.89 in., N 61° W, 34°.
24·2 FIG. 48(2) A. Ans. 0.87 in., S 17° W, 40°.
24·3 FIG. 48(3) A. Ans. 1.69 in., N 74° W, 22½°.

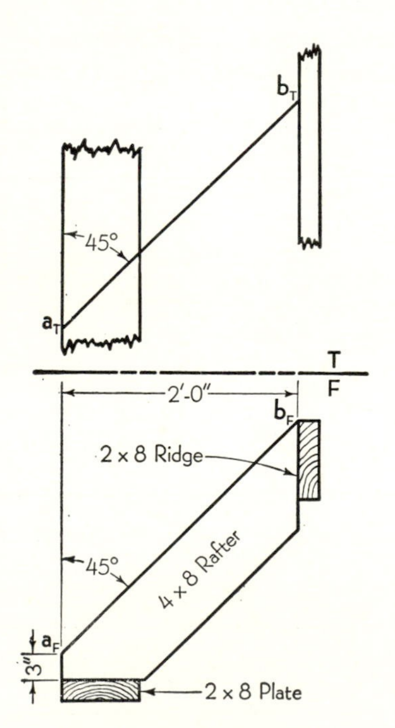

Fig. 46. Problem 23·13.

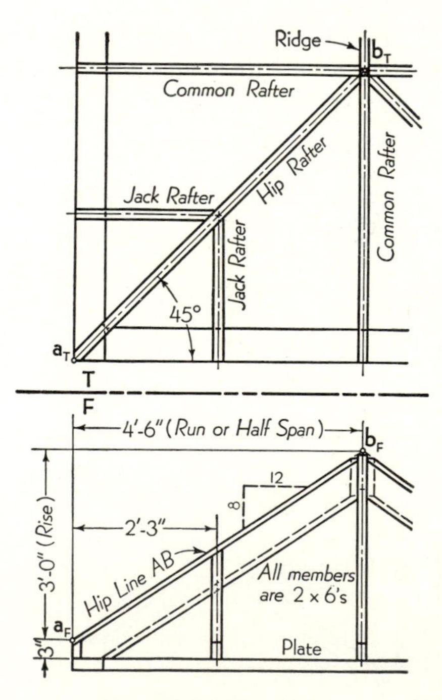

Fig. 47. Problem 23·14.

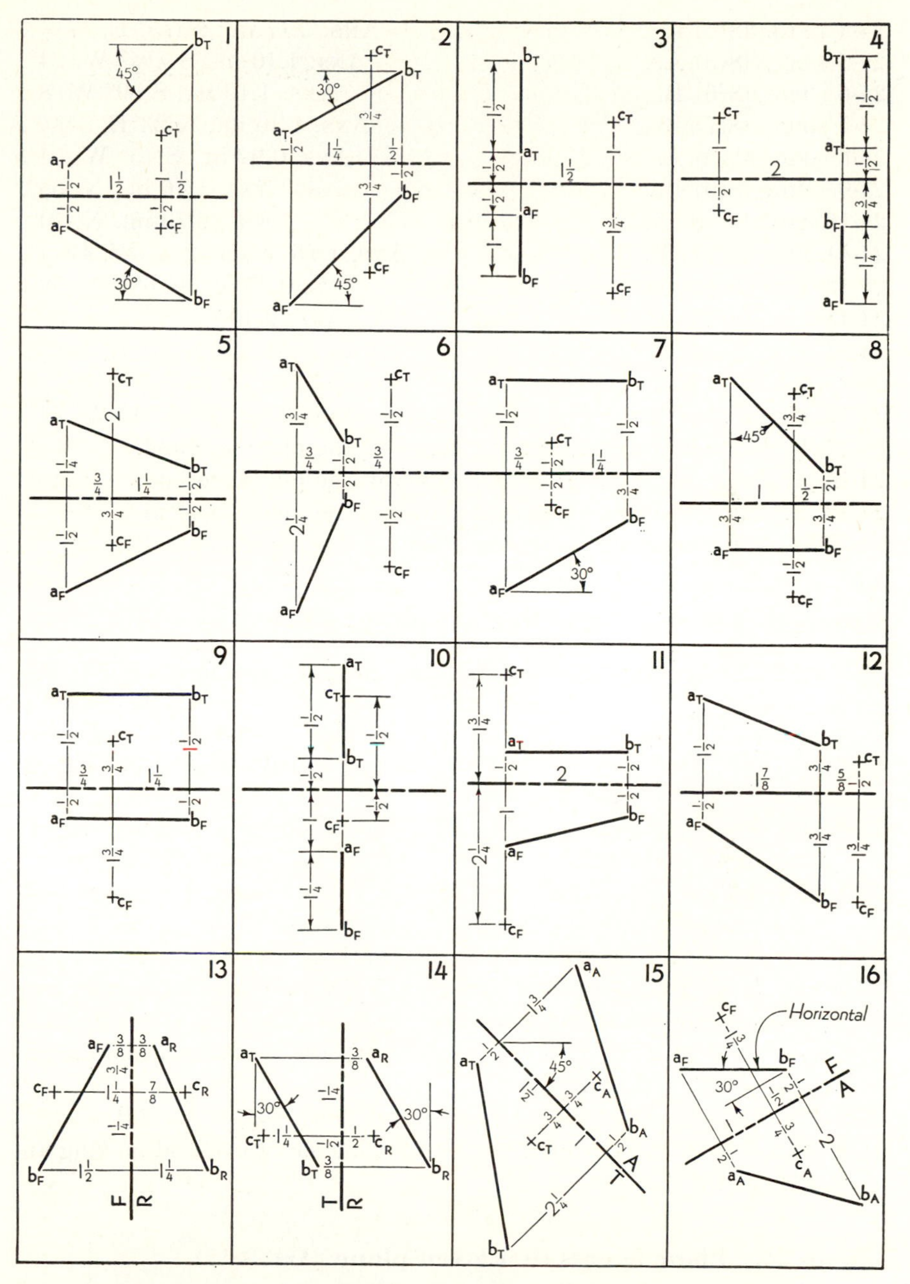

Fig. 48. A point and a line. **Problems in Groups 24, 25, and 39.**

24·4 FIG. 48(4) A. ANS. 2.17 in., N 75° E, 17½°.
24·5 FIG. 48(5) A. ANS. 1.10 in., S 7½° W, 24°.
24·6 FIG. 48(6) B. ANS. 1.19 in., S 69° W, 8°.
24·7 FIG. 48(7) B. ANS. 1.30 in., N 23°E, 34½°.
24·8 FIG. 48(8) A. ANS. 0.92 in., S 45° W, 54°.
24·9 FIG. 48(9) B. ANS. 1.46 in., N, 59°.
24·10 FIG. 48(10) A. ANS. 0.70 in., N, 50°.
24·11 FIG. 48(11) B. ANS. 1.75 in., S 12½° W, 42½°.
24·12 FIG. 48(12) A. ANS. 0.35 in., N 68° W, 58°.
24·13 FIG. 48(13) A. ANS. 0.40 in., S 79° E, 33°.
24·14 FIG. 48(14) A. ANS. 0.66 in., N 41° E, 33½°.
24·15 FIG. 48(15) A. ANS. 0.77 in., S 66½° W, 28½°.
24·16 FIG. 48(16) A. ANS. 0.88 in., N 41° W, 72°.
24·17 B. Same as Prob. 11·17, but solve by the plane method.
24·18 A. Same as Prob. 11·18, but solve by the plane method.
24·19 B. Same as Prob. 11·19, but solve by the plane method.
24·20 B. Same as Prob. 11·20, but solve by the plane method.

Group 25. Lines intersecting at a given angle (Art. 4·14)

In each problem show the required line (or lines) as a heavy line in the given views.

25·1 FIG. 48(1) B. Line from C to AB, sloping downward, and making an angle of 30° with AB.
25·2 FIG. 48(2) A. Line from C to AB, sloping upward, and making an angle of 45° with AB.
25·3 FIG. 48(3) A. An isosceles triangle CXY whose base XY lies on line AB and whose base angles are 75°.
25·4 FIG. 48(4) A. The steepest line, 2¼ in. long, from C to AB.
25·5 FIG. 48(5) A. An isosceles triangle CXY whose base XY lies on line AB and whose base angles are 60°.
25·6 FIG. 48(6) B. Line from C to AB, sloping upward, and making an angle of 60° with AB.
25·7 FIG. 48(7) B. The steepest line, 1½ in. long, from C to AB.
25·8 FIG. 48(8) A. Line from C to AB, bearing eastward, and making an angle of 45° with AB.

Group 26. Plane figures in a given plane (Art. 4·15)

In each problem the plane figure, whose true size, shape, and location are specified, lies in the given plane (assume plane to be unlimited in extent). Show this figure in the given views. Use Layout B.

26·1 FIG. 49(1). In plane *ABC* a 1-in. square whose center is ⅜ in. lower than *AB* and 1¼ in. from *B*. Two sides of the square are parallel to line *AC*.

26·2 FIG. 49(2). In plane *ABC* a hexagon, 1½ in. across flats, whose center is at *B*. Two sides of the hexagon are horizontal.

26·3 FIG. 49(3). In the plane of parallel lines *AB* and *CD* a 2-in. square whose center is ½ in. higher than *CD* and 1½ in. from *C*. Two sides of the square are frontal lines.

26·4 FIG. 49(4). In plane *ABC* the letter A, 1 in. high, ⅞ in. wide, crossbar ⅜ in. from the bottom. The vertex of the A is 1½ in. to the left

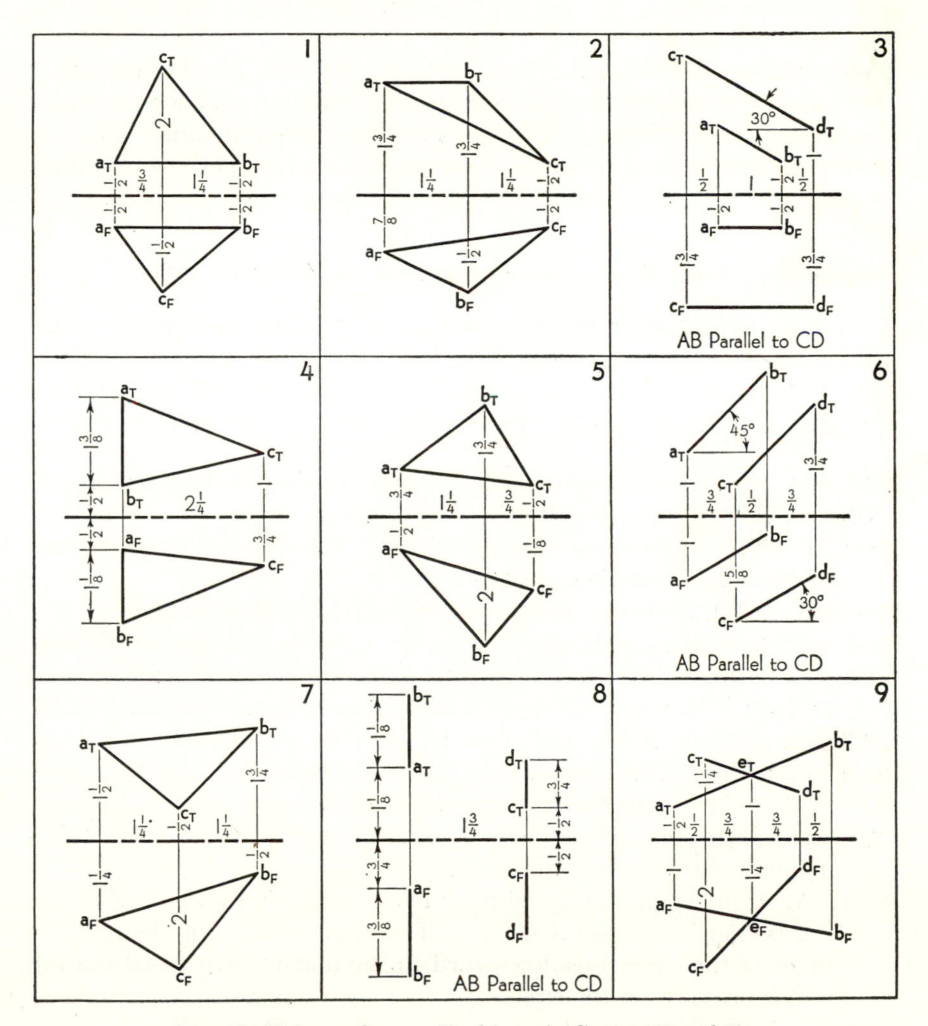

Fig. 49. Plane surfaces. Problems in Groups 26 and 27.

of point *C* and is 1 in. from point *A*. One leg of the A extended passes through *B*; the other leg intersects line *BC*.

26·5 FIG. 49(5). In plane *ABC* an equilateral triangle having 1¼-in. sides. Its center is equidistant from points *A*, *B*, and *C*, and one side is horizontal and higher than the opposite vertex.

26·6 FIG. 49(6). In the plane of parallel lines *AB* and *CD* a hexagon, 1½ in. across flats, whose center is equidistant from lines *AB* and *CD* and 1 in. from point *D*. Two sides of the hexagon are frontal lines.

26·7 FIG. 49(7). In plane *ABC* the letter M, 1 in. high and 1 in. wide, standing on line *BC* at its mid-point. The top of the letter M is toward point *A*.

26·8 FIG. 49(8). In the plane of parallel lines *AB* and *CD* a 1¼-in. square whose center is at the same elevation as point *D* and is equidistant from points *B* and *C*. Two sides of the square are horizontal.

26·9 FIG. 49(9). In the plane of intersecting lines *AB* and *CD* a rectangle, 1 by 2 in., with center at *E*. The 1-in. sides are profile lines.

Group 27. A circle in a given plane (Arts. 4·16 to 4·18)

In Probs. 27·1 to 27·9 the circle, whose size and location are specified, lies in the given plane (assume plane to be unlimited in extent). Show the circle in the given views. See Appendix, Arts. A·7 to A·11, for methods of drawing ellipses. Use Layout B.

27·1 FIG. 49(1). In plane *ABC* a 2-in.-diameter circle with center at *C*.

27·2 FIG. 49(2). In plane *ABC* a 1½-in.-diameter circle with center at *B*.

27·3 FIG. 49(3). In the plane of parallel lines *AB* and *CD* a circle tangent to these lines with center 1¾ in. from *C*.

27·4 FIG. 49(4). In plane *ABC* a circle inscribed in the triangle.

27·5 FIG. 49(5). In plane *ABC* a circle circumscribed about the triangle.

27·6 FIG. 49(6). In the plane of parallel lines *AB* and *CD* a circle tangent to these lines with center equidistant from points *A* and *B*.

27·7 FIG. 49(7). In plane *ABC* a circle inscribed in the triangle.

27·8 FIG. 49(8). In the plane of parallel lines *AB* and *CD* a circle tangent to these lines and passing through point *D*.

27·9 FIG. 49(9). In the plane of intersecting lines *AB* and *CD* a 2-in.-diameter circle with center at *E*.

27·10 A. A circle appears in the top view as an ellipse whose major axis is 2 in. long and bears N 60° E. The minor axis is 1 in. long. The plane of the circle dips downward to the northwest. Draw the top and front views of the circle.

27·11 A. A circle appears in the front view as an ellipse whose major axis is 2 in. long and slopes downward to the right at 45°. The minor axis is $1\frac{1}{4}$ in. long. The lower part of the circle is nearest to the front. Draw the top and front views of the circle.

27·12 FIG. 50 B. Scale: 3 in. = 1 ft. The steering gear shown is rotated by the tiller rope *ABC*. In order to have the rope stay on the steering-post quadrant it must approach the quadrant in the plane of its central section as shown at *AB*, and therefore a guide pulley must be installed at *B* to change the direction of the rope from *BC* to *AB*. Draw the top and front views of the central section of a 6-in.-diameter guide pulley.

27·13 FIG. 51 C. Scale: $1\frac{1}{2}$ in. = 1 ft (B, 1 in. = 1 ft). The two pulleys shown are connected by a belt 3 in. wide. In order to run a flat belt onto a pulley the center line of the approaching belt must lie in the mid-plane of the pulley. If the drive is to be reversible (run in either direction), the belt must approach and leave each pulley in its mid-plane. In Fig. 51 the belt must be run as shown, and an idler pulley must be installed at *B* to change the direction from *AB* to *BC*. (Point *B* may be higher or lower if desired.) Draw the front and side views of a 10-in.-diameter idler pulley, 4 in. wide, at *B*. Also, draw the belt in each view, neglecting its thickness. Is this drive reversible?

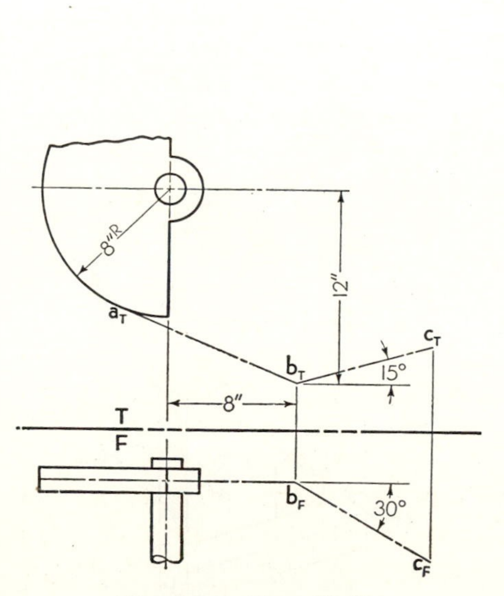

Fig. 50. Problem 27·12.

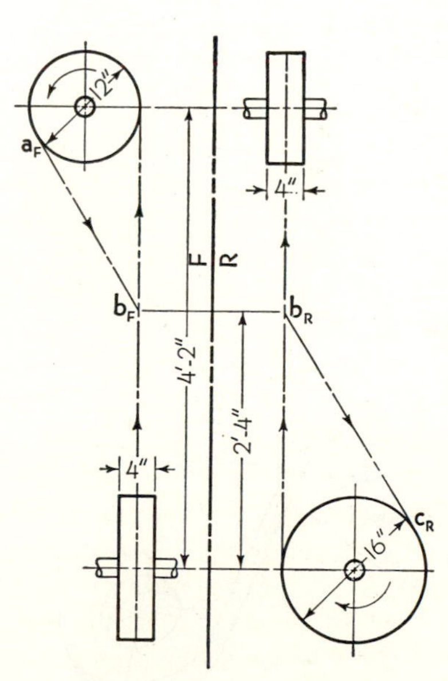

Fig. 51. Problem 27·13.

27·14 FIG. 52 A. Draw the top and front (note arrow *F*) views of the cable anchor bracket shown. In the necessary auxiliary views only the inclined part of the object need be shown.

27·15 FIG. 53 A. Same as Prob. 27·14, but using the angle bracket shown.

Group 28. Plane through one line and parallel to a second line (Art. 4·19)

28·1–28·12 FIG. 54(1) to (12). In each problem construct a plane that contains line *CD* and is parallel to line *AB*. As a check, draw an edge view of the plane. Use Layout B.

Group 29. Plane through a point and parallel to two given lines (Art. 4·20)

29·1–29·12 FIG. 54(1) to (12). In each problem construct a plane that contains the point *X* and is parallel to lines *AB* and *CD*. As a check, draw an edge view of the plane. Use Layout B.

Group 30. Shortest line between two skew lines (plane method, Art. 4·21)

30·1–30·14 Same as Probs. 12·1 to 12·14, except that the solution is to be made by the plane method.

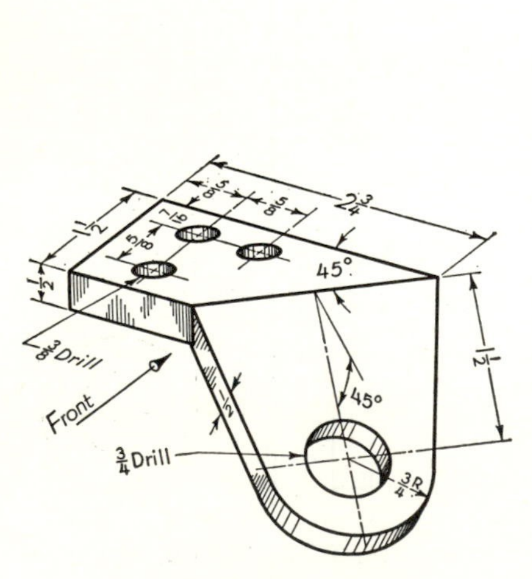

Fig. 52. Problem 27·14.

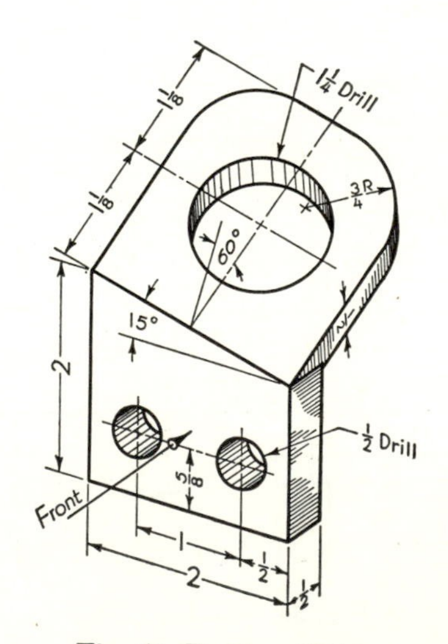

Fig. 53. Problem 27·15.

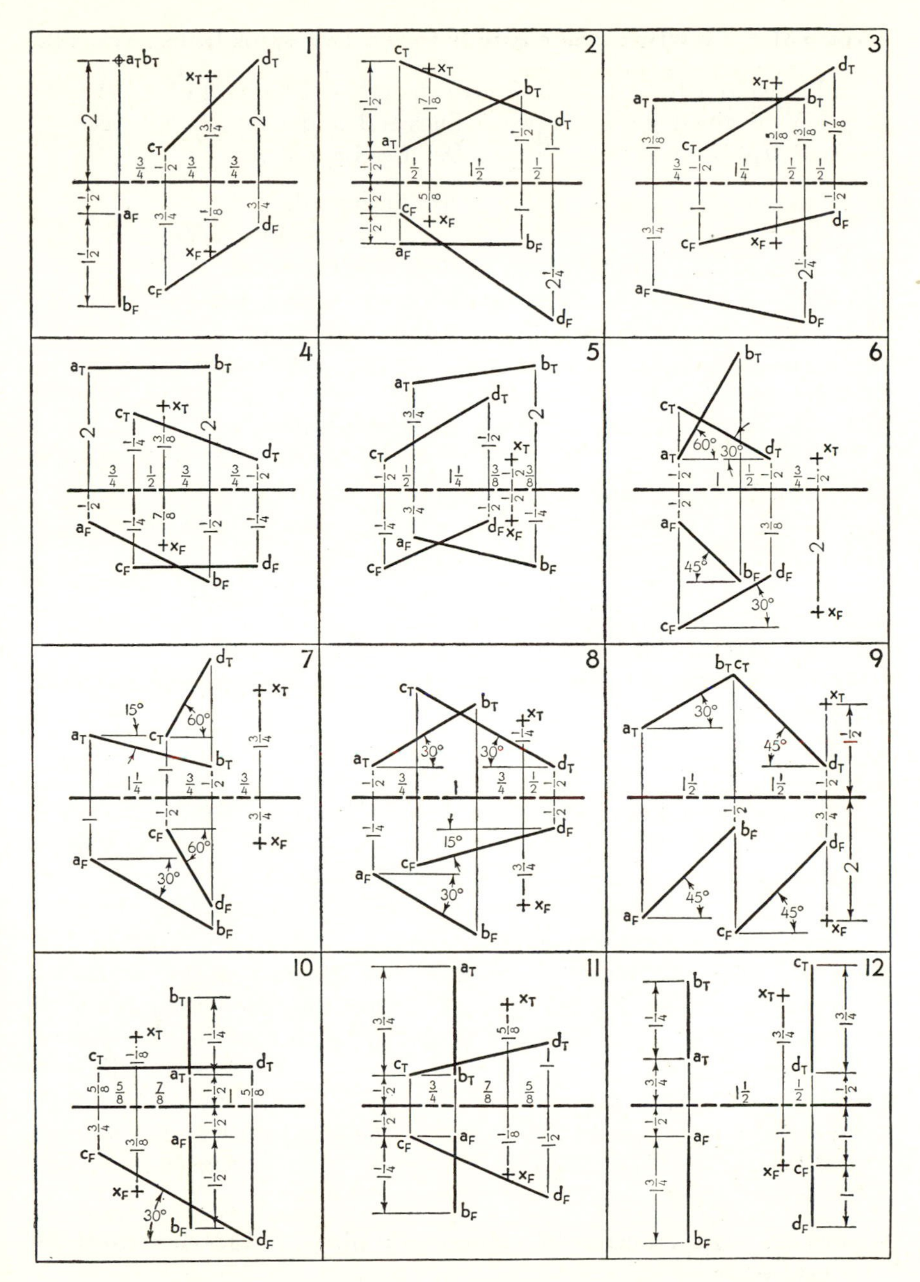

Fig. 54. Two skew lines and a point. Problems in Groups 28, 29, and 31.

Group 31. Shortest grade line between two skew lines (Art. 4·22)

In Probs. 31·1 to 31·12 find the true length and bearing of the shortest line *MN* of given grade, between lines *AB* and *CD*. In all cases the grade of *MN* is given from *AB* to *CD*. Use Layout A.

31·1 FIG. 54(1). Grade of *MN* is 30° downward.
ANS. 1.80 in., S 45° E.

31·2 FIG. 54(2). Grade of *MN* is 30 per cent upward.
ANS. 1.19 in., N 27° W.

31·3 FIG. 54(3). Grade of *MN* is 60° upward. ANS. 1.23 in., S 16° E.

31·4 FIG. 54(4). *MN* is horizontal. ANS. 0.98 in., S 20° W.

31·5 FIG. 54(5). *MN* is horizontal. ANS. 0.73 in., S 16° E.

31·6 FIG. 54(6). Grade of *MN* is 15° downward.
ANS. 1.20 in., S 14½° E.

31·7 FIG. 54(7). Grade of *MN* is 40 per cent upward.
ANS. 1.14 in., N 51½° E.

31·8 FIG. 54(8). Grade of *MN* is 15 per cent upward.
ANS. 1.38 in., N 12° E.

31·9 FIG. 54(9). *MN* is horizontal. ANS. 1.73 in., E.

31·10 FIG. 54(10). Grade of *MN* is 45 per cent upward.
ANS. 1.20 in., S 25½° W.

31·11 FIG. 54(11). *MN* is horizontal ANS. 0.85 in., S 40° E.

31·12 FIG. 54(12). Grade of *MN* is 30° downward. ANS. 2.30 in., E.

31·13 FIG. 54(5) A. Line *PQ* bears N 45° W, slopes 30° downward to the northwest, and connects lines *AB* and *CD*. Locate *PQ* in the given views, and find its true length. ANS. 0.68 in.

31·14 FIG. 54(6) A. Line *PQ* bears N 30° W, slopes 45° downward to the southeast, and connects lines *AB* and *CD*. Locate *PQ* in the given views, and find its true length. ANS. 0.97 in.

31·15 FIG. 54(7) A. Locate the shortest line *PQ* bearing due north and connecting lines *AB* and *CD*. What are the true length and slope angle of *PQ*? (The solution is analogous to that for the shortest horizontal line.) ANS. 1.21 in., 60°.

31·16 FIG. 54(8) A. Locate the shortest line *PQ* bearing due east and connecting lines *AB* and *CD*. What are the true length and slope angle of *PQ*? (See Prob. 31·15.) ANS. 1.18 in., 81°.

Group 32. Intersection of a line and a plane (edge-view method, Art. 4·23)

In Probs. 32·1 to 32·12 locate the point *P* where line *MN* intersects the given plane. Show line *MN* in correct visibility, assuming the given plane to be opaque within its defining lines. Use Layout B.

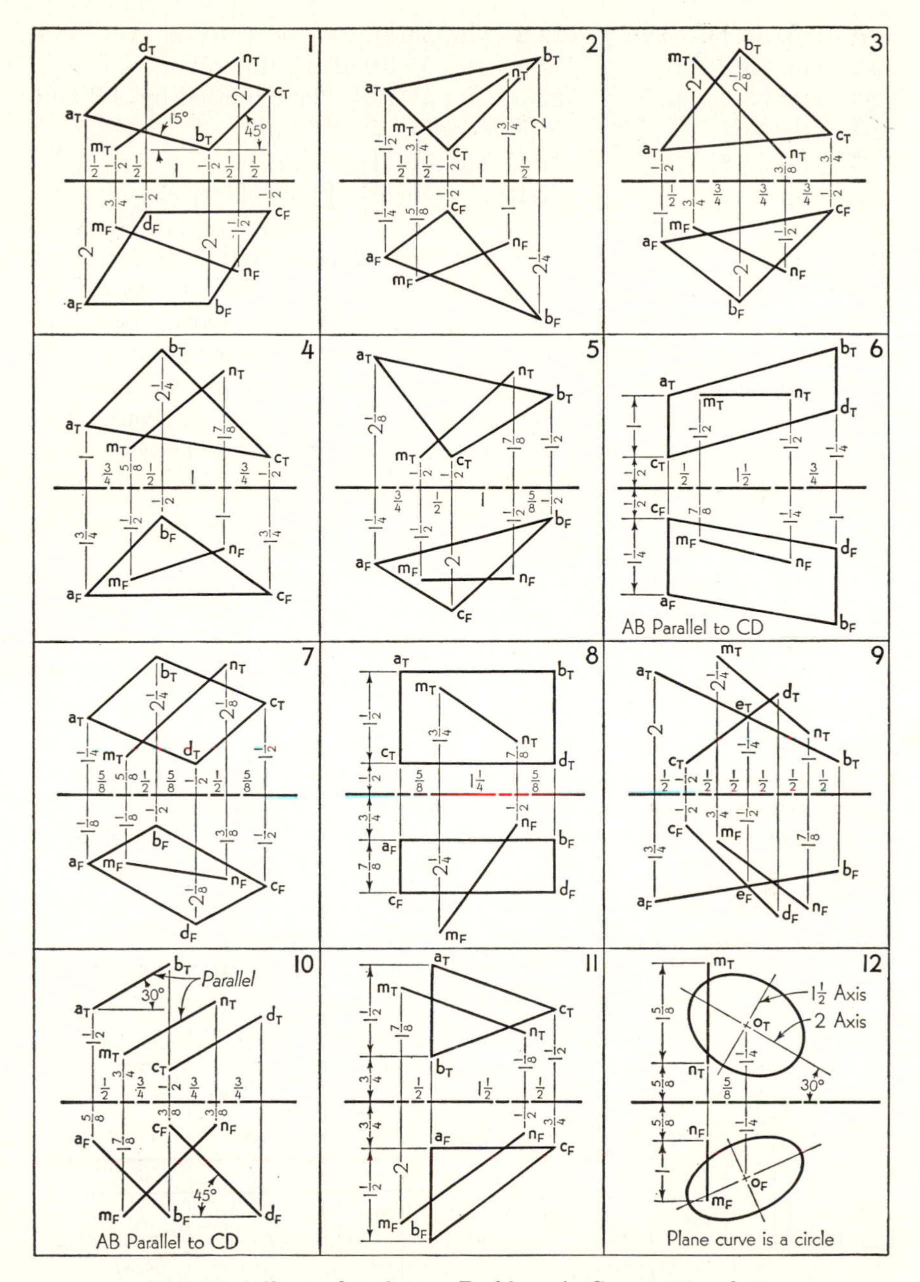

Fig. 55. A line and a plane. Problems in Groups 32 and 34.

32·1 FIG. 55(1).
32·2 FIG. 55(2).
32·3 FIG. 55(3).
32·4 FIG. 55(4).
32·5 FIG. 55(5).
32·6 FIG. 55(6).
32·7 FIG. 55(7).
32·8 FIG. 55(8).
32·9 FIG. 55(9). Plane unlimited.
32·10 FIG. 55(10). Plane limited by *AB* and *CD* only.
32·11 FIG. 55(11).
32·12 FIG. 55(12). Plane of circle.
32·13 FIG. 56 B. Scale: 3/16 in. = 1 ft. A shot was fired from a window in a building, breaking the bulb of a lamppost at *B* and striking the ground at *A*. From what window was the shot fired? Assuming that the floor level is 2 ft below the window sill, if the gun was held 1½ ft behind the front face of the wall, how far above the floor was it held? If a man whose eye level is 5½ ft above the ground was standing at *C*, could he have seen the shot fired? If he could not, locate a point *D* 12 ft in front of the building where he would have had to stand to just see the mouth of the gun.

ANS. Middle window. 9 ft 10 in. above floor.

32·14 FIG. 57 B. Scale: ¾ in. = 1 ft. A rectangular piece of cardboard is hung from the ceiling as shown in the figure. Find the compound shadow that will be cast on the ceiling by the light from the two electric light bulbs *P* and *Q*. Single-crosshatch the simple shadow, and double-crosshatch the compound shadow.
32·15 FIG. 58 A. Scale: ¾ in. = 1 ft. Find the shadow cast on the ground by the monument when the sun's rays are in the direction shown. Double-crosshatch the shadow, and single-crosshatch the visible shaded surfaces of the object. In conventional shading the

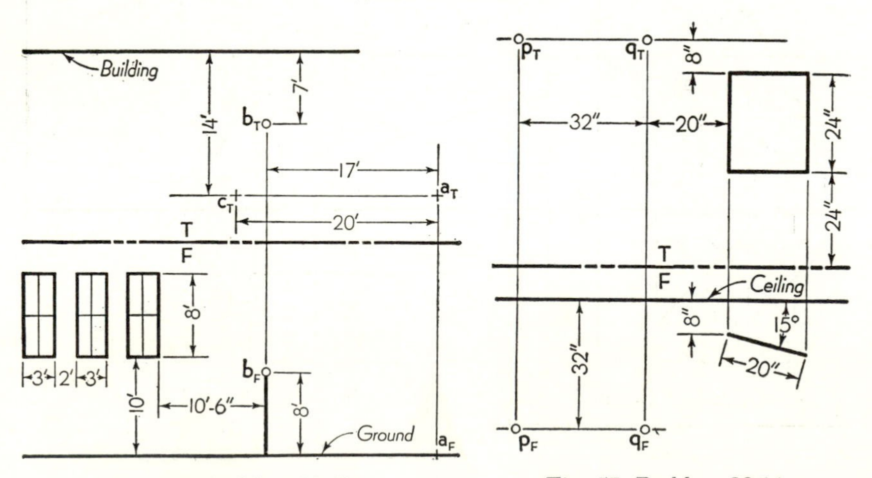

Fig. 56. Problem 32·13. **Fig. 57.** Problem 32·14.

sun's rays are assumed to be parallel and passing over the left shoulder of the observer in the direction indicated by a single light ray in Fig. 58. Thus the shadow of point A on the object may be located at S by assuming a single light ray through point A and then finding its intersection with the ground at S as shown. By repeating this process with other points on the object the entire shadow may be outlined. *Shadow* is that part of a surface from which light is excluded by another object. *Shade* is that part of the surface of an object not directly exposed to the light.

32·16 FIG. 59 A. Scale: 1½ in. = 1 ft. Find the shadow cast on the floor and wall by the pedestal, assuming that all light rays are parallel to the single ray shown. Double-crosshatch the visible shadow, and single-crosshatch the visible shade. Do not overlook the shadow cast on the pedestal column by the cap. (See statement of Prob. 32·15.)

32·17 FIGS. 60 and 61(1) C. Figure 60 shows one basic method of constructing a perspective drawing of an object. The object (a cube) is viewed by an observer (eye at E) through a transparent vertical plane (the *picture plane*). Light rays drawn from points on the object, such as A and B, to the observer's eye intersect the picture plane at P and Q. Points p_F and q_F are then two points on the perspective drawing, which lies on the picture plane and appears

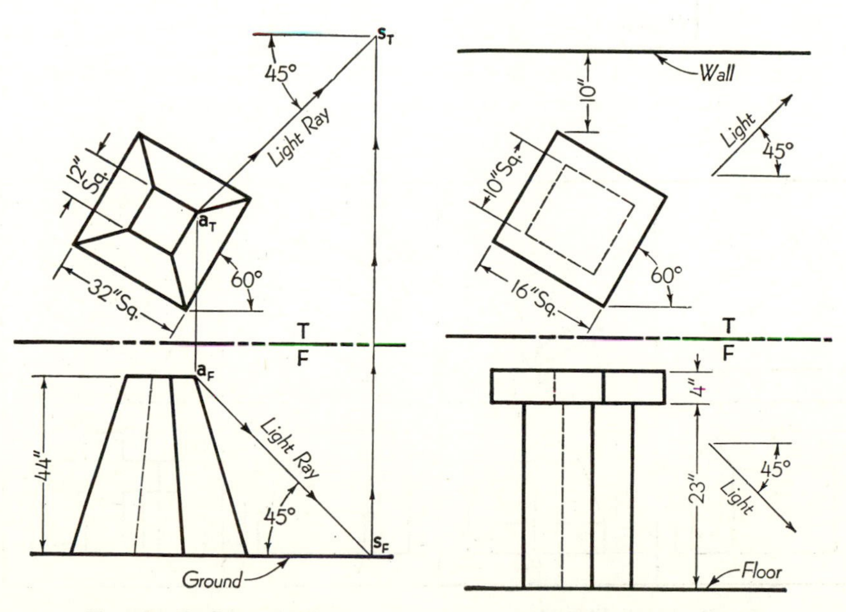

Fig. 58. Problem 32·15.

Fig. 59. Problem 32·16.

true size in the front view. By repeating this process with other points on the object the entire perspective drawing may be completed. Object, picture plane, and observer have been shown only in the top and side views in order to leave the front view clear for the perspective drawing. Reference lines *T–F* and *F–R* have been conveniently assumed to coincide with the edge views of the vertical picture plane.

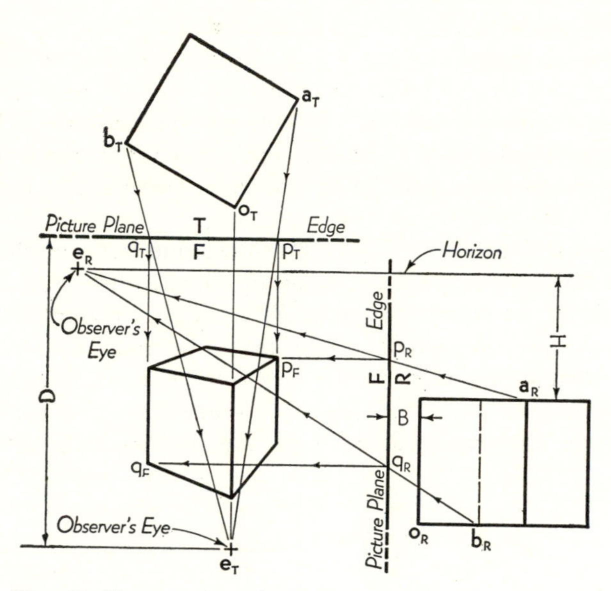

Fig. 60. Construction of a perspective drawing of a cube.

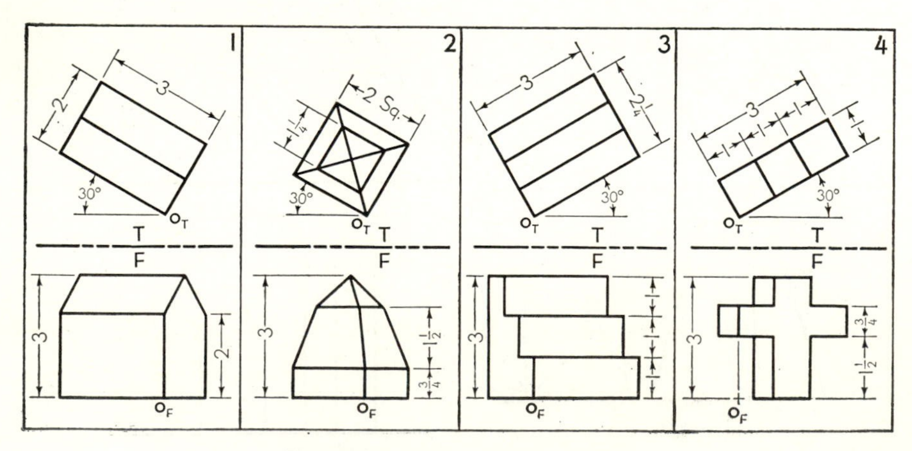

Fig. 61. Problems 32·17 to 32·20.

Using the layout shown in Fig. 60, substitute for the cube the object shown in Fig. 61(1), and construct a perspective drawing of this object. Let $D = 5$ in., $B = \frac{1}{2}$ in., and $H = 1$ in. The observer's eye is directly in front of point O on the object.

32·18 Same as Prob. 32·17, but substitute object shown in Fig. 61(2).

32·19 Same as Prob. 32·17, but substitute object shown in Fig. 61(3).

32·20 Same as Prob. 32·17, but substitute object shown in Fig. 61(4).

Group 33. Intersection line between two planes (edge-view method, Art. 4·24)

In Probs. 33·1 to 33·9 locate and label the intersection line PQ between the two given planes. When the planes overlap, show each plane in correct visibility, assuming the planes to be opaque within their defining lines. As a check, draw a second edge view. Use Layout B.

33·1 FIG. 62(1).

33·2 FIG. 62(2).

33·3 FIG. 62(3).

33·4 FIG. 62(4).

33·5 FIG. 62(5).

33·6 FIG. 62(6).

33·7 FIG. 62(7). Planes unlimited.

33·8 FIG. 62(8). Planes unlimited.

33·9 FIG. 62(9). Planes unlimited.

33·10 A. Scale: 1 in. = 50 ft. At a point A on a hillside an outcrop of a vein of ore has been located. From two other points B and C on the hillside vertical holes have been drilled to the vein, striking it at B' and C', the borings being 25 and 60 ft deep, respectively. A line RS marks the western termination of a level plateau whose elevation is 1,180 ft. Using point A, elevation 1,020 ft, as a bench mark, the points described above are located as follows:

B is 140 ft N 60° E of A, at an elevation of 1,160 ft.
C is 170 ft N 75° E of A, at an elevation of 1,170 ft.
R is 240 ft east of A, and line RS bears N 30° W.

Assuming that the vein of ore continues as a plane surface, determine whether it will outcrop on the plateau.

33·11 FIG. 33(1) B. Connect point X to points A, B, and C to form a tetrahedron $ABCX$. Any plane parallel to two opposite edges of a tetrahedron will intersect it in a parallelogram. Show the parallelogram cut from $ABCX$ by a plane $\frac{3}{8}$ in. from line AX. Show all lines in correct visibility.

33·12 FIG. 33(7) B. Intersect the tetrahedron $ABCX$ with a plane that will cut out a parallelogram. Two sides of the parallelogram must be 2 in. long and parallel to line BX. (See statement of Prob. 33·11.)

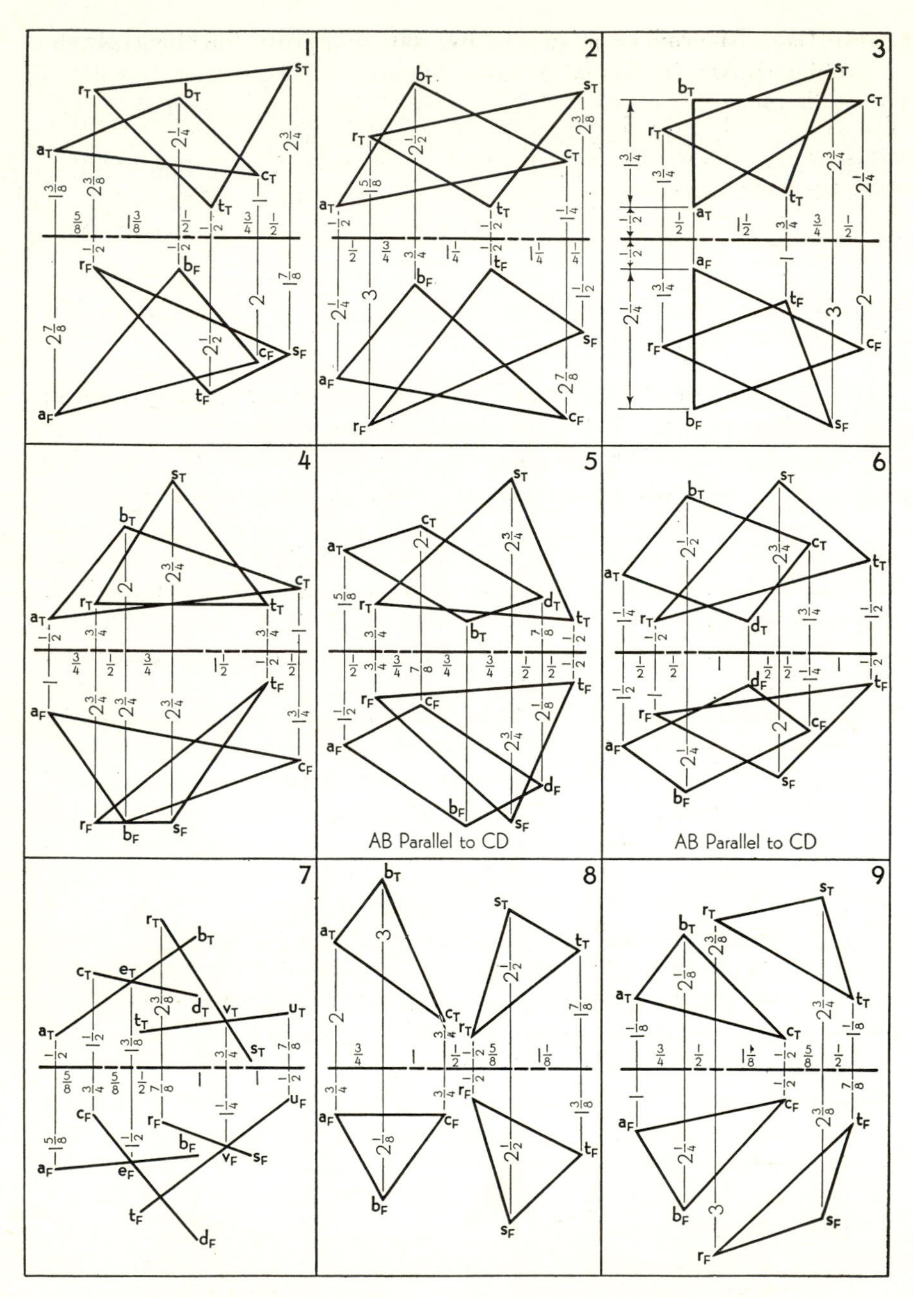

Fig. 62. Two intersecting planes. Problems in Groups 33, 35, 41, and 44.

Group 34. Intersection of a line and a plane (cutting-plane method, Art. 4·25)

In Probs. 34·1 to 34·11 locate the point P where line MN intersects the given plane. Show line MN in correct visibility, assuming the given plane to be opaque within its defining lines. As a check, use both a vertical and an edge-front cutting plane. Use either Layout A or one-half of Layout B.

34·1 FIG. 55(1).
34·2 FIG. 55(2).
34·3 FIG. 55(3).
34·4 FIG. 55(4).
34·5 FIG. 55(5).
34·6 FIG. 55(6).
34·7 FIG. 55(7).
34·8 FIG. 55(8).
34·9 FIG. 55(9). Plane unlimited.
34·10 FIG. 55(10). Plane limited by lines AB and CD only.
34·11 FIG. 55(11).

34·12 FIG. 63 A. Scale: 1½ in. = 1 ft. The bent floor plate $ABCD$ is intersected by a cable from point E, a column from point F, a brace from point G, and a curved pedal pivoted at H. Using only the given views, locate the centers of the four holes that must be cut in the floor plate.

34·13 FIG. 64 A. Scale: 1 in. = 10 ft. A wire is stretched between two masts from A to B. Assuming that the direction of the light is the same as that shown in Fig. 58 (see statement of Prob. 32·15 also),

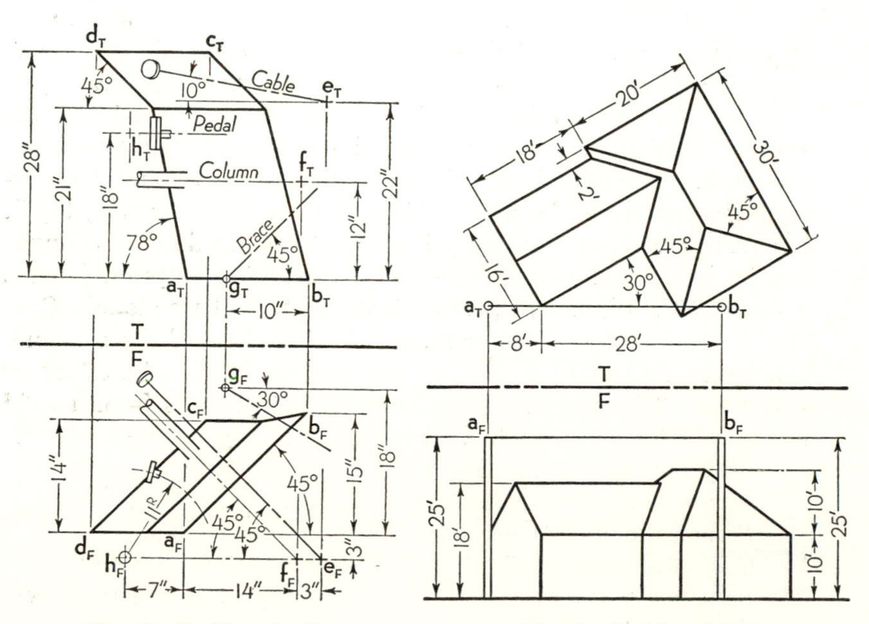

Fig. 63. Problem 34·12.

Fig. 64. Problem 34·13.

find the shadow of the wire and masts on the roof and sides of the building. Solve, using only the given views.

34·14 FIG. 54(5) A. Using only the given views, locate the straight line *YZ* that connects lines *AB* and *CD* and passes through point *X*. (HINT: Line *YZ* must lie in plane *XAB*.)

34·15 FIG. 54(7) A. Same as Prob. 34·14.

34·16 FIG 54(11) A. Same as Prob. 34·14.

Group 35. Intersection line between two planes (cutting-plane method, Arts. 4·26, 4·27)

In Probs. 35·1 to 35·9 locate four well-separated points *M*, *N*, *P*, and *Q* on the line of intersection of the given planes. Draw the line of intersection, and recheck any point that seems to be in error. When the planes overlap, show each plane in correct visibility, assuming the planes to be opaque within their defining lines. Use Layout A.

35·1 FIG. 62(1).
35·2 FIG. 62(2).
35·3 FIG. 62(3).
35·4 FIG. 62(4).
35·5 FIG. 62(5).
35·6 FIG. 62(6).
35·7 FIG. 62(7). Planes unlimited.
35·8 FIG. 62(8). Planes unlimited.
35·9 FIG. 62(9). Planes unlimited.

35·10 B. The strike of a vein of ore is N 45° E, and its dip is 60° to the southeast. A nearby stratum of rock has a strike of S 60° E and dips 45° to the southwest. Assuming the vein and rock stratum to be two plane surfaces, find the bearing and slope angle of the line of intersection. ANS. N 11½° E, 43½°.

35·11 A. The strike of a vein of ore is S 15° E, and its dip is 15° to the west. A similar vein nearby has a strike of N 75° W and dips 45° to the south. Assuming both veins to be plane surfaces, find the bearing and slope angle of the line of intersection. ANS. E, 14½°.

35·12 A. A pyramid has a horizontal base that is an equilateral triangle, 2½ in. on a side. The three lateral faces of the pyramid make angles of 50°, 60°, and 70°, respectively, with the base plane. Construct a top view and one or more elevation views of the pyramid. What is its altitude? ANS. 1.22 in.

35·13 B. Scale: 3 in. = 1 ft. A polyhedron is to be cut from a solid block 21 in. long, 9 in. wide, and 14 in. high. The left face is determined by the points *A*, *B*, and *C*, located as follows:

A is on the upper front edge, 4 in. from the left end.
B is on the upper back edge, 7 in. from the left end.
C is on the bottom front edge, 17½ in. from the left end.

The right face will be perpendicular to the top surface of the block, contain point *C*, and make an angle of 60° toward the left with the front surface of the block. The other faces of the polyhedron will be portions of the faces of the original block. Draw top, front, and right-side views of the polyhedron. How many plane faces has the polyhedron?

35·14 B. Scale: 1½ in. = 1 ft. Two plane surfaces are to be cut into a block 48 in. long, 26 in. wide, and 18 in. high. Points *A*, *B*, and *C* determine one plane; *R*, *S*, and *T*, the other. With respect to the front, top, and left-end faces of the block the points are located as follows:

A is 16 in. behind, 4 in. below, and 25 in. to the right.
B is 2 in. behind, 7 in. below, and 13 in. to the right.
C is 5 in. behind, 12 in. below, and 19 in. to the right.
R is 12 in. behind, 6 in. below, and 27 in. to the right.
S is 14 in. behind, 2 in. below, and 33 in. to the right.
T is 4 in. behind, 2 in. below, and 39 in. to the right.

Draw top, front, and right-side views of the cut block, showing the intersection of the two planes with its faces and with each other.

Group 36. Intersection of a plane and a polyhedron (**Arts. 4·28 to 4·30**)

36·1–36·9 FIG. 65(1) to (9). In each problem determine the plane section cut from the polyhedron by the plane *ABC* (indefinite in extent). Use the method of either Art. 4·29 or 4·30 or that specified by the instructor. Use Layout B for the edge-view method, Layout A for the cutting-plane method.

Group 37. Intersection of a line and a polyhedron (**Art. 4·31**)

37·1–37·9 FIG. 65(1) to (9). In each problem locate the points where the line that begins at *M* intersects the polyhedron. Extend the line to pass entirely through the solid, and show the line in correct visibility. Solve by using a vertical cutting plane, and check by using an edge-front plane. Use Layout A.

Group 38. Line perpendicular to a plane (**Art. 4·32**)

38·1–38·8 FIG. 66(1) to (8). In each problem draw a line *XY* (of any desired length) through point *X* perpendicular to the given plane. Solve, using only the given views. Use either Layout A or one-half of Layout B.

38·9 FIG. 67 A. Scale: 1 in. = 1 ft. Two braces *AB* and *CD* are connected to a wall and intersect at point *E*. Using only the given

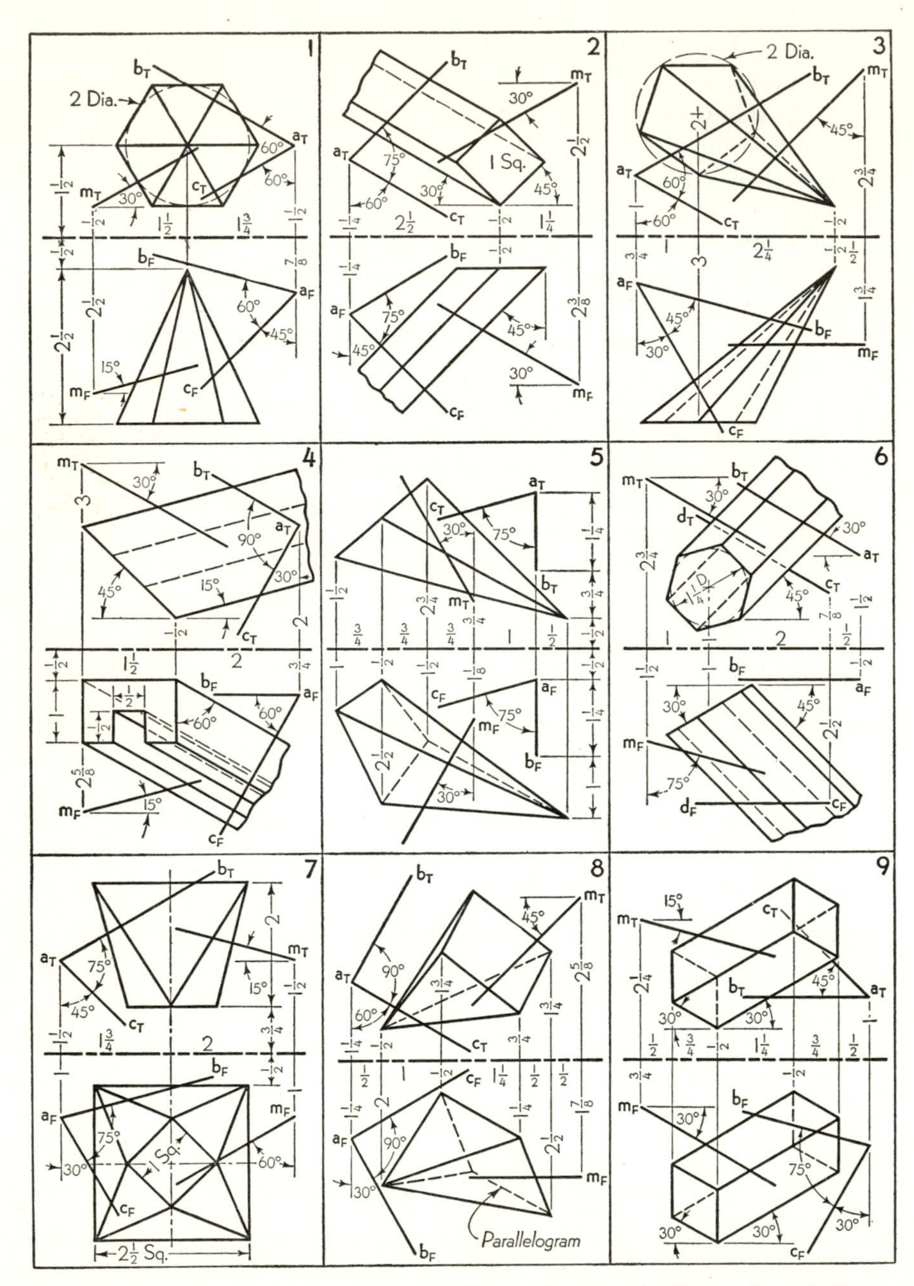

Fig. 65. A line, a plane, and a polyhedron. Problems in Groups 36 and 37.

views, locate a third brace at *E* perpendicular to both *AB* and *CD*. Extend this third brace to intersect the floor or wall, and locate the intersection point *P* with respect to point *B*.

ANS. *P* is 33 in. to the left of *B*.

38·10 FIG. 68 A. Through the block shown, a rectangular hole is cut perpendicular to the oblique surface. Using only the given views, draw the hole in the top and front views.

38·11 FIG. 69(1) A. Plane *ABCD* is the upper base of a tilted right prism. The lower base is parallel to the upper base, and the edge parallel to *AB* is ½ in. lower than *AB*. Using only the given views, draw the top and front views of the prism.

38·12 FIG. 69(2) A. The square lies in the plane of lines *AB* and *CD* and is the lower base of a right prism whose lateral edges are perpendicular to this base. The upper base of the prism is a horizontal surface ¼ in. higher than *B*. Using only the given views, draw the top and front views of the prism.

38·13 FIG. 69(3) A. The parallelogram lies in the plane *ABC* and is the lower base of a right prism whose lateral edges are perpendicular to this base. The upper base of the prism is a horizontal surface ¾ in. higher than *C*. Using only the given views, draw the top and front views of the prism.

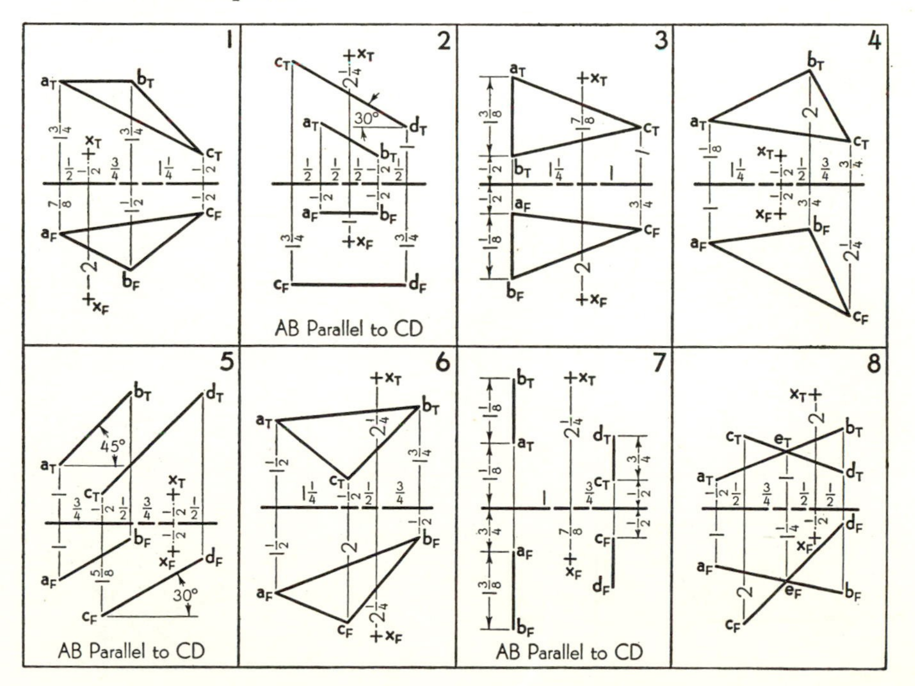

Fig. 66. A point and a plane. Problems in Groups 38 and 42.

Group 39. Plane perpendicular to a line (Art. 4·33)

In Probs. 39·1 to 39·13 construct a plane, represented as a triangle *RSC*, through point *C* perpendicular to line *AB*. Solve, using only the given views. Use either Layout A or one-half of Layout B.

39·1 FIG. 48(1).
39·2 FIG. 48(2).
39·3 FIG. 48(5).
39·4 FIG. 48(6).
39·5 FIG. 48(7).
39·6 FIG. 48(8).
39·7 FIG. 48(9).
39·8 FIG. 48(11).
39·9 FIG. 48(12).
39·10 FIG. 48(13).
39·11 FIG. 48(14).
39·12 FIG. 48(15).
39·13 FIG. 48(16).

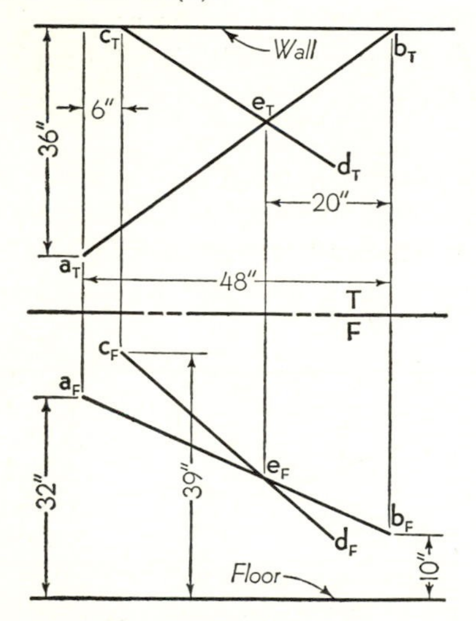

Fig. 67. Problem 38·9.

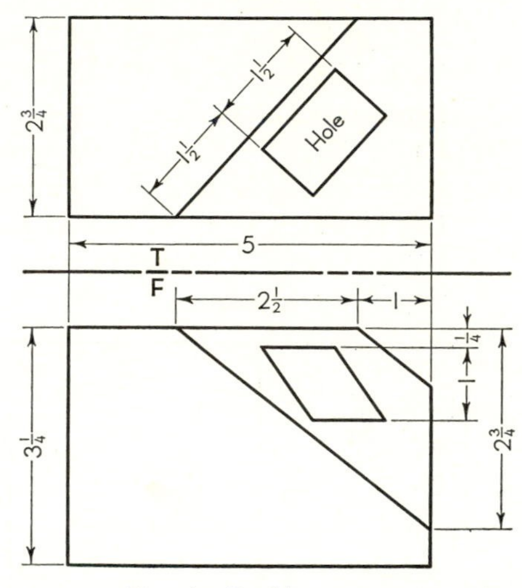

Fig. 68. Problem 38·10.

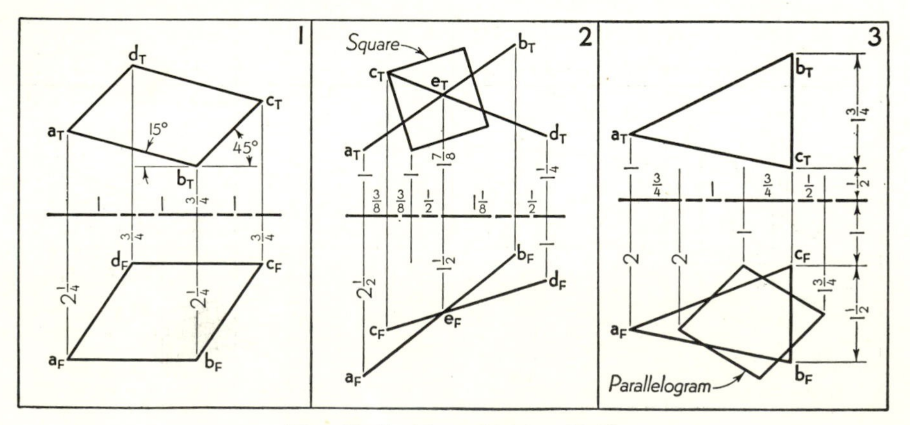

Fig. 69. Problems 38·11 to 38·13.

39·14 Why were Figs. 48(3), (4), and (10) omitted from the above group?

39·15 FIG. 70(1) A or one-half B. Point c_T is one view of a point C that is equidistant from points A and B. Locate c_F, using only the given views.

39·16 FIG. 70(2). Same as Prob. 39·15.

39·17 FIG. 70(3) A or one-half B. Point c_F is one view of a point C that is equidistant from points A and B. Locate c_T, using only the given views.

39·18 FIG. 70(4). Same as Prob. 39·17.

39·19 FIG. 70(5) A or one-half B. A point Z on line CD is equidistant from points A and B. Locate Z, using only the given views.

39·20 FIG. 70(6). Same as Prob. 39·19.

39·21 FIG. 70(7) A or one-half B. Construct a plane ORS perpendicular to line MN and passing through the mid-point O of line MN. Locate the intersection of planes ORS and ABC. Solve, using only the given views.

39·22 FIG. 70(8). Same as Prob. 39·21.

39·23 FIG. 65(2) A. Disregard the given plane ABC, and construct a new plane through point A perpendicular to the axis of the polyhedron.

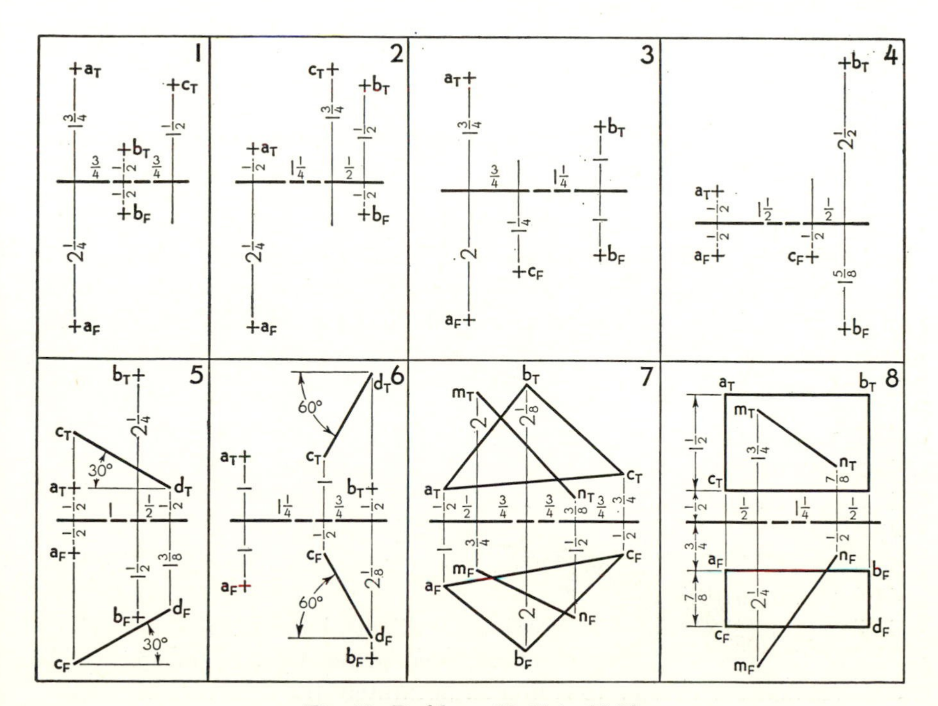

Fig. 70. Problems 39·15 to 39·22.

Determine the *right section* cut from the solid by the new plane. Solve, using only the given views.

39·24 FIG. 65(3). Same as Prob. 39·23.

39·25 FIG. 65(6). Same as Prob. 39·23.

Group 40. Plane through a line and perpendicular to a plane (Art. 4·34)

In each problem construct a plane, represented as a triangle *MNX*, through line *MN* perpendicular to the given plane. Find the strike of plane *MNX*. Solve, using only the given views. Use either Layout A or one-half of Layout B.

40·1 FIG. 71(1). ANS. N 71° E.
40·2 FIG. 71(2). ANS. N 73° E.
40·3 FIG. 71(3). ANS. N 52° W.
40·4 FIG. 71(4). ANS. N 42° E.
40·5 FIG. 71(5). ANS. N 47° E.
40·6 FIG. 71(6). ANS. N 71° E.
40·7 FIG. 71(7). ANS. N 52° E.
40·8 FIG. 71(9). ANS. N 84° W.
40·9 FIG. 71(10). ANS. N 57° W.
40·10 FIG. 71(11). ANS. N 74° E.

Group 41. Plane through a point and perpendicular to two planes (Art. 4·35)

In each problem draw the given views in the left half of the sheet. In the right half of the sheet assume any point *X*, and construct top and front views of a plane *XYZ*, through point *X*, and perpendicular to both of the given planes. Find the strike and dip of plane *XYZ*. Solve, using only the given views, except that an extra view may be used to obtain the dip of plane *XYZ*.

41·1 FIG. 62(1) B. ANS. N 41° E, 53° NW.
41·2 FIG. 62(2) B. ANS. N 26° W, 83° NE.
41·3 FIG. 62(3) B. ANS. N 48° W, 52° SW.
41·4 FIG. 62(7) B. ANS. N 63° W, 50° NE.
41·5 FIG. 62(8) B. ANS. N 77° W, 45° SW.
41·6 FIG. 62(9) B. ANS. N 74° E, 48° SE.
41·7 FIG. 62(4) C. Same as previous problems, but add right-side views of the given planes and of required plane *XYZ*. The extra side views are necessary to establish accurately the direction of perpendiculars, which are almost profile lines. ANS. N 55° E, 48° SE.

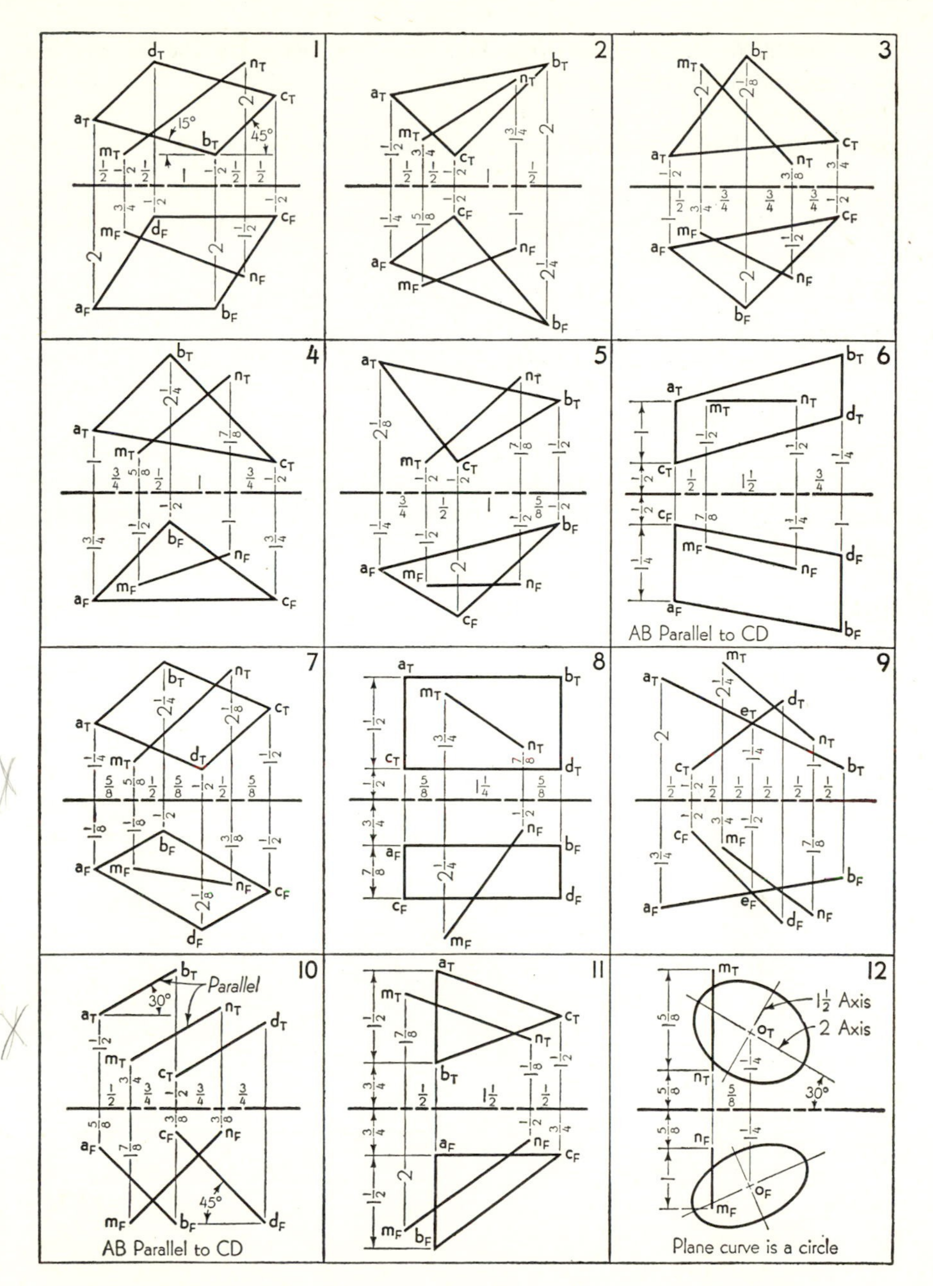

Fig. 71. A line and a plane. Problems in Groups 40, 43, and 45.

41·8 FIG. 62(5) C. Same as Prob. 41·7. ANS. N $2\frac{1}{2}°$ E, 87° E.

41·9 FIG. 62(6) C. Same as Prob. 41·7. ANS. N 83° E, 52° S.

Group 42. Projection of a point on a plane (Art. 4·36)

42·1–42·8 FIG. 66(1) to (8). In each problem locate the projection *P* of point *X* on the given plane. Solve, using only the given views. Use either Layout A or one-half of Layout B.

42·9 FIG. 72 A. Scale: $\frac{1}{4}$ in. = 1 ft. A ray of light from point *S* strikes an inclined mirror at its center *M* (the mirror is greatly enlarged for graphical accuracy). Locate, with respect to point *S* and the floor, the spot *R* where the reflected ray strikes the wall. Solve, using only the given views. (HINT: The light source *S* and its image *S'* are equidistant from the mirror on a perpendicular to the mirror. The reflected ray extended backward passes through points *M* and *S'*.) ANS. *R* is 9 ft 10 in. left of *S*, 12 ft 0 in. above floor.

42·10 Same as Prob. 42·9, except that *S* is 14 ft above the floor.
ANS. *R* is 2 ft 10 in. left of *S*, 6 ft 1 in. above floor.

42·11 FIG. 73 A. Scale: 1 in. = 50 ft. The figure shows an incline *BC* in the plane of a vein of ore. By means of borings a rich pocket has been located in the same vein at point *D*. Using only the given views, solve the following problems: (*a*) Locate the shortest tunnel *DX* from *BC* to *D*, and find the elevation of point *X*. (*b*) It is desired to ventilate tunnel *BC* at the 900-ft level by an air shaft perpendicular to the plane of the vein. Locate, with respect to the bench mark (B.M.), the point *Y* at elevation 1,000 ft where the

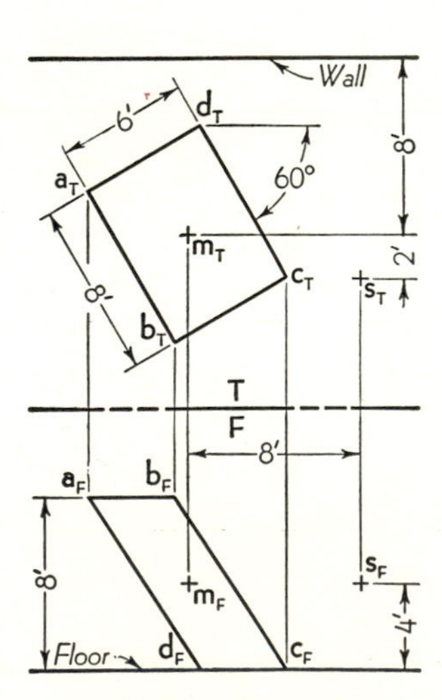

Fig. 72. Problem 42·9.

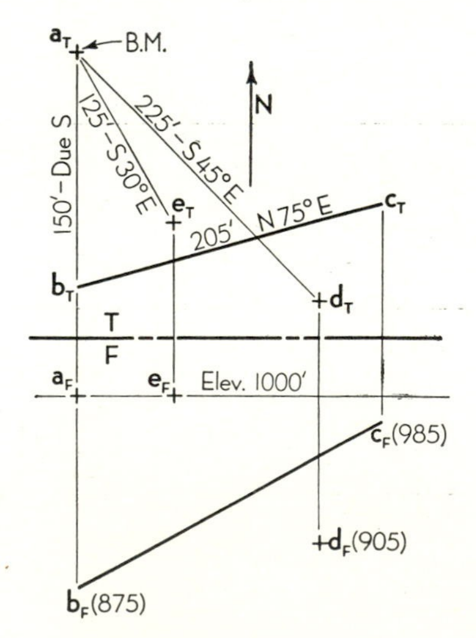

Fig. 73. Problem 42·11.

shaft should be started. (*c*) At point *E* a shaft is to be sunk perpendicular to the plane of the ore vein until it strikes the ore at a point *Z*. At what elevation will this point *Z* be?

ANS. *X*, 946 ft. *Y*, 253 ft S 4° E. *Z*, 973 ft.

Group 43. Projection of a line on a plane (Art. 4·37)

In Probs. 43·1 to 43·11 locate the projection *PQ* of line *MN* on the given plane. Solve, using only the given views. Check the accuracy of your work by noting the alignment of the intersection point of *MN* and the plane. Use either Layout A or one-half of Layout B.

43·1 FIG. 71(1).
43·2 FIG. 71(2).
43·3 FIG. 71(3).
43·4 FIG. 71(4).
43·5 FIG. 71(5).
43·6 FIG. 71(6).
43·7 FIG. 71(7).
43·8 FIG. 71(9).
43·9 FIG. 71(10).
43·10 FIG. 71(11).
43·11 FIG. 71(12).

43·12 FIG. 54(5) A. Locate the shortest line *YZ* between lines *AB* and *CD*. Solve, using only the given views. The true length and slope of *YZ* need not be determined.

43·13 FIG. 54(7) A. Same as Prob. 43·12.

43·14 FIG. 74 A. Scale: 1 in. = 1 ft. The bottom outlet of the hopper must be connected to the bin by a sheet-metal duct perpendicular to

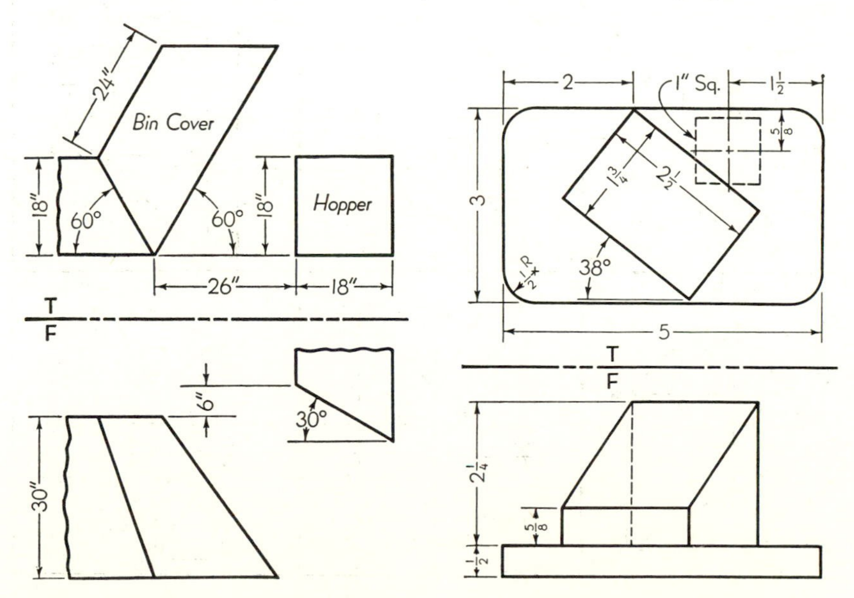

Fig. 74. Problems 43·14 and 44·16.

Fig. 75. Problem 43·15.

the sloping surface of the bin cover. Using only the given views, draw the connecting duct.

43·15 FIG. 75 A. Full size. Through the casting shown it is necessary to cut a straight hole, perpendicular to the sloping surface, that will intersect the bottom of the casting as the 1-in. square. Using only the given views, draw the hole in the top and front views. Will the hole be entirely within the solid casting?

Group 44. Dihedral angle (Arts. 4·38, 4·39)

In Probs. 44·1 to 44·9 determine the dihedral angle in degrees between the given intersecting planes, using the method of Art. 4·38. Use Layout B.

44·1 FIG. 76(1). Planes *ABC* and *BCD*. ANS. 143°.
44·2 FIG. 76(2). Planes *ABC* and *BCD*. ANS. 56°.
44·3 FIG. 76(3). Planes *ABC* and *BCD*. ANS. 24°.
44·4 FIG. 76(4). Planes *ABC* and *BCD*. ANS. 132½°.
44·5 FIG. 76(5). Planes *ABC* and *BCD*. ANS. 105°.
44·6 FIG. 76(6). Planes *ABC* and *X*. ANS. 49°.
44·7 FIG. 18(1). Planes *ABC* and *ACD*. ANS. 94½°.
44·8 FIG. 18(2). Planes *ABD* and *ACD*. ANS. 90°.

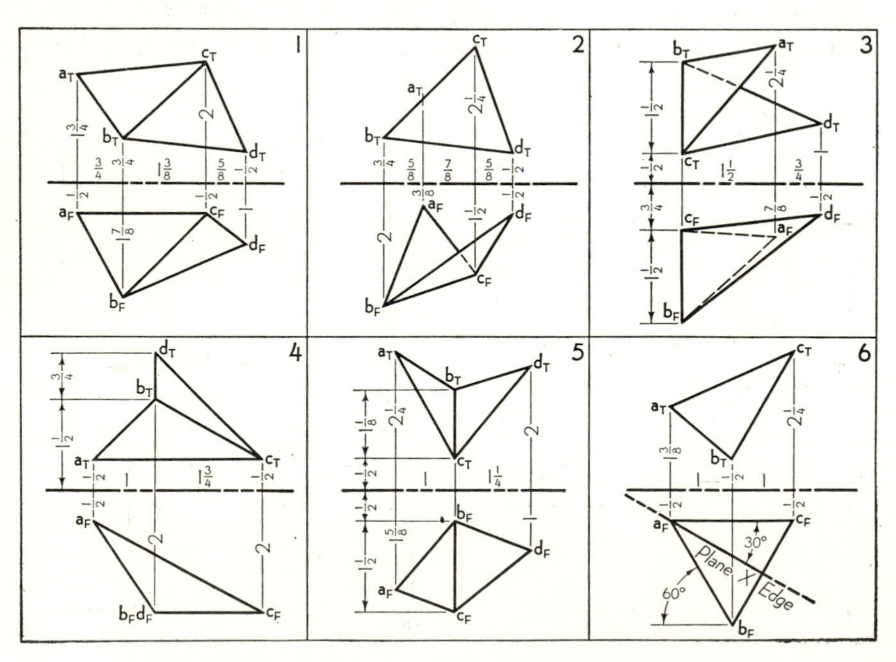

Fig. 76. Two intersecting planes. Problems in Group 44.

44·9 FIG. 18(3). Planes *ABC* and *ABD*. ANS. 56°.

In Probs. 44·10 to 44·15 determine the dihedral angle in degrees between planes *ABC* and *RST*, using the method of Art. 4·39. Use Layout C.

44·10 FIG. 62(1). ANS. 47°.
44·11 FIG. 62(2). ANS. 83°.
44·12 FIG. 62(3). ANS. 16°.
44·13 FIG. 62(4). ANS. 42½°.
44·14 FIG. 62(5). ANS. 83°.
44·15 FIG. 62(6). ANS. 6½°.

44·16 FIG. 74 B. Scale: ¾ in. = 1 ft. Find the dihedral angle between the bin cover and the adjacent surface. (Disregard the hopper.) ANS. 129°.

44·17 FIG. 77 A. Scale: 1/16 in. = 1 ft. Find the dihedral angle between the end panel *ABCD* and the vertical side panels of the skew bridge. ANS. 116½°.

44·18 FIG. 78 A. Scale: ⅛ in. = 1 ft. The sloping sides of the concrete bridge pier all have the same slope. Corners such as those between surfaces *A* and *B* and surfaces *A* and *C* are frequently covered by a steel angle iron embedded in the concrete. Determine the angle to which each of these special iron plates must be bent. ANS. *A–B*, 137°. *A–C*, 94°.

44·19 FIG. 79 C. Scale: 8× full size. One end of the firing pin of a certain rifle is shaped approximately as shown in the figure. For machining purposes it is necessary to have the dihedral angle between end surface *A* and surface *B* and between surface *B* and side surface *C*. Determine these two angles in degrees and then complete the lines in the side view. Dimensions are in inches. ANS. *A–B*, 145°30′. *B–C*, 72°0′.

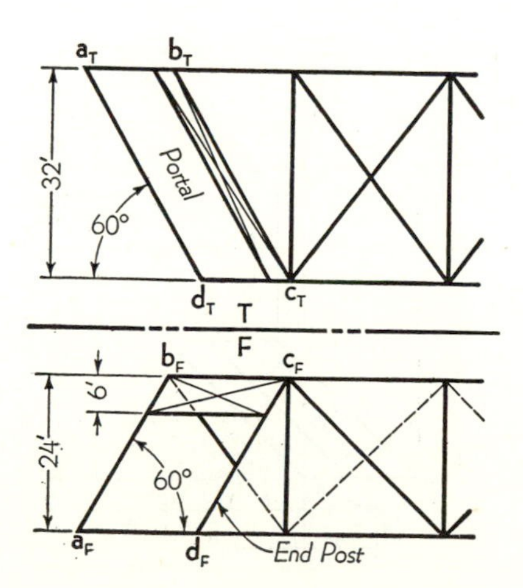

Fig. 77. Problem 44·17.

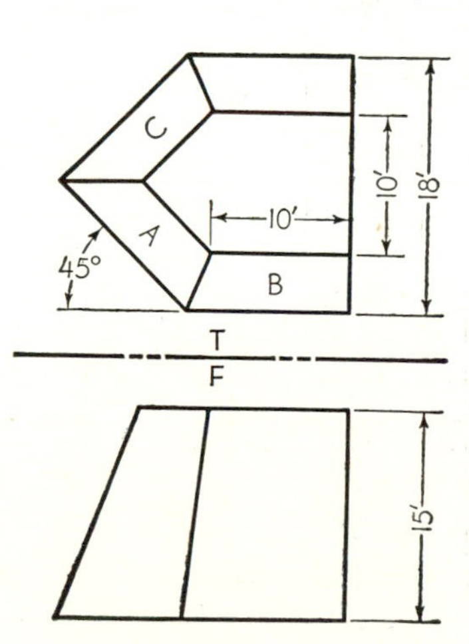

Fig. 78. Problem 44·18.

44·20 FIG. 80 B. Scale: ½ in. = 1 ft. The hopper shown feeds coal from an overhead storage bin into a boiler stoker. Determine the dihedral angle between sides *A* and *B*. ANS. 93½°.

44·21 FIG. 80 A. Same as Prob. 44·20, but determine the angle between sides *A* and *C*. ANS. 98°.

44·22 FIG. 81 B. Scale: ⅛ in. = 1 ft. The structural members of one leg of a tower are shown. Find the dihedral angle between planes *ABC* and *ABE*. ANS. 94°.

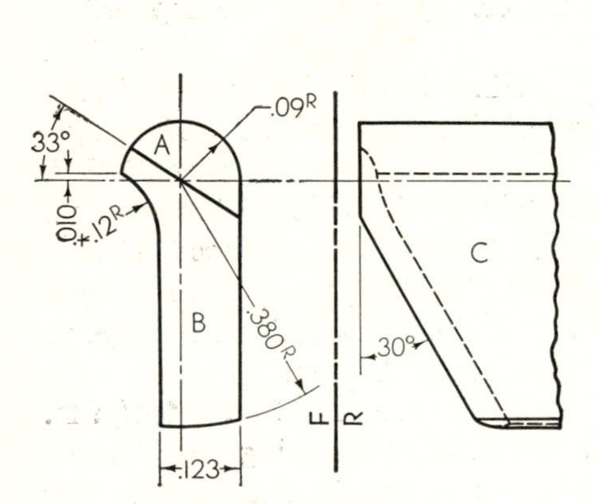

Fig. 79. Problem 44·19.

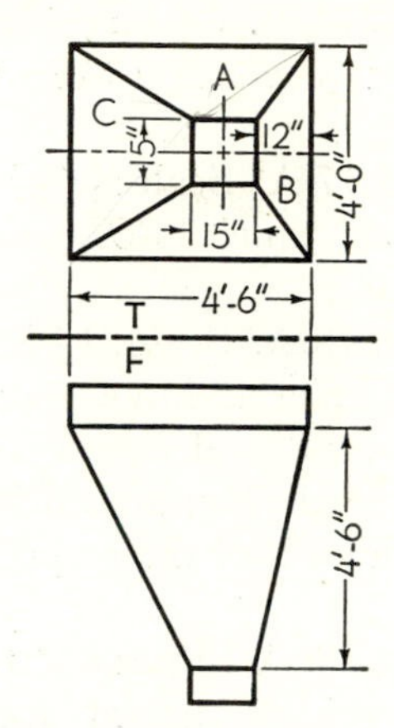

Fig. 80. Problems 44·20 and 44·21.

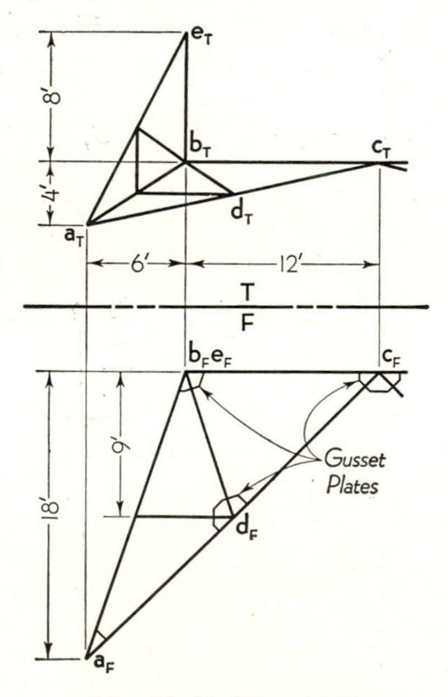

Fig. 81. Problem 44·22.

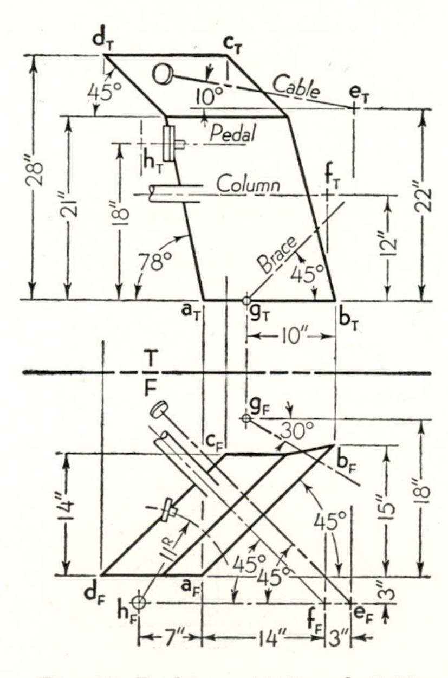

Fig. 82. Problems 44·23 and 45·13.

44·23 FIG. 82 A. Scale: 1½ in. = 1 ft. Through how many degrees must the floor plate *ABCD* be bent to assume the shape shown in the figure? (Disregard the cable, column, brace, and pedal.) ANS. 26°.

Group 45. Angle between a line and a plane (Arts. 4·40, 4·41)

In Probs. 45·1 to 45·12 determine the angle in degrees between the line *MN* and the given plane. Use the method of either Art. 4·40 or 4·41 or that specified by the instructor. Use Layout B.

45·1 FIG. 71(1). ANS. 50°.
45·2 FIG. 71(2). ANS. 45°.
45·3 FIG. 71(3). ANS. 53½°.
45·4 FIG. 71(4). ANS. 18°.
45·5 FIG. 71(5). ANS. 40°.
45·6 FIG. 71(6). ANS. 14½°.
45·7 FIG. 71(7). ANS. 23½°.
45·8 FIG. 71(8). ANS. 57½°.
45·9 FIG. 71(9). ANS. 36°.
45·10 FIG. 71(10). ANS. 67°.
45·11 FIG. 71(11). ANS. 23½°.
45·12 FIG. 71(12). ANS. 61½°.

45·13 FIG. 82 C. Scale: 1½ in. = 1 ft (B, 1 in. = 1 ft). Find the angles that the column, cable, and brace make with the part of the floor plate *ABCD* that each passes through. (Disregard the pedal.)
ANS. Column, 81½°. Cable, 48°. Brace, 54½°.

45·14 FIG. 83 C. Scale: ¼ in. = 1 ft (B, 3/16 in. = 1 ft). A mast is to be erected on a roof and is supported by three 2-by-2-by-¼ angle-iron braces *AB*, *AC*, and *AD*, anchored at points *B*, *C*, and *D* on the roof. The braces will be attached to the roof surface with bent angle plates.

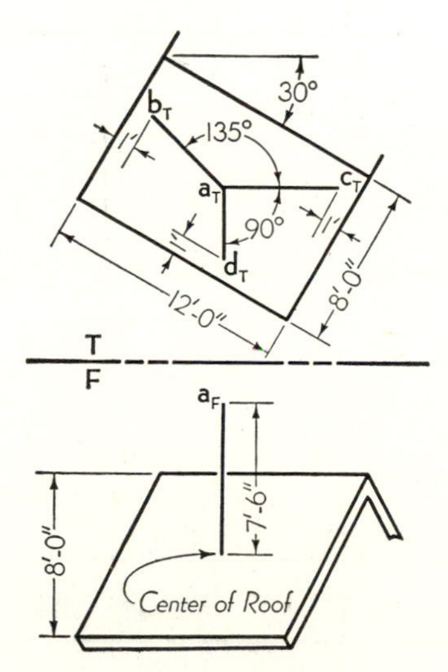

Fig. 83. Problem 45·14.

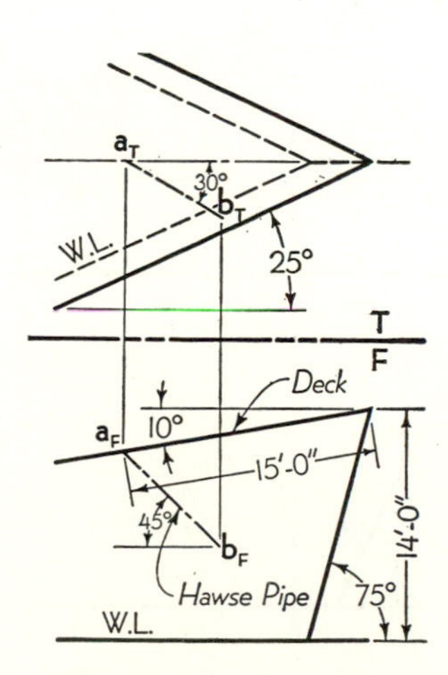

Fig. 84. Problem 45·15.

Determine the true length of each brace and the bend angle for each angle plate.

ANS. *B*, 8 ft 1 in., 41½°. *C*, 7 ft 6 in., 46°. *D*, 11 ft 2 in., 28½°.

45·15 FIG. 84 C. Scale: $\frac{3}{16}$ in. = 1 ft. Line *AB* is the center line of a hawsepipe (a pipe through which the anchor chain passes) connecting the deck and hull of a ship (compare Fig. 9·22). For simplicity the deck and side of the bow have been drawn as plane surfaces. Determine the true length of center line *AB* and the angle it makes with the deck and with the hull.

ANS. 8 ft 8 in.; deck, 48½°; hull, 44°.

45·16 FIG. 85 B. Scale: ½ in. = 1 ft. Line *MN* is the center line of a pipe that intersects the side of a hopper at *M* and the floor at *N*. Find the angles that the pipe makes with the hopper side and with the floor. ANS. Hopper, 54½°. Floor, 33°.

45·17 FIG. 86 B. Scale: ½ in. = 1 ft. Line *MN* is the center line of a pipe that intersects two vertical walls at *M* and *N*, respectively. Find the angles that the pipe makes with each wall.

ANS. At *M*, 36°. At *N*, 20°.

45·18 FIG. 42 B. Scale: 3 in. = 1 ft. What are the angles of warp of the ends *AB* and *CD* of the bent tube? (Angle of warp is the angle *CD* makes with plane *ABC* or *AB* with plane *BCD*.)

ANS. *CD*, 28°. *AB*, 58°.

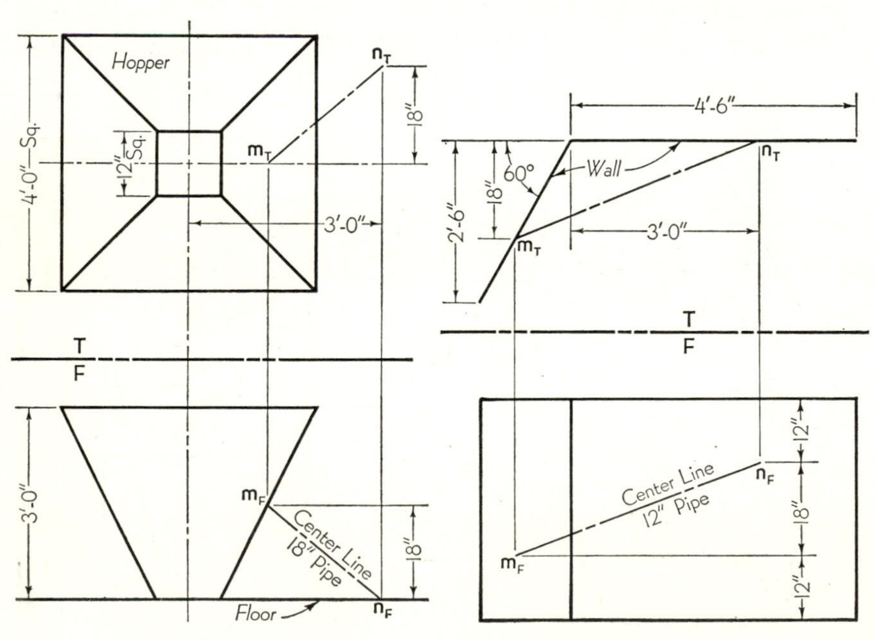

Fig. 85. Problem 45·16.

Fig. 86. Problem 45·17.

45·19 FIG. 16 C. Scale: 1 in. = 1 ft (B, ¾ in. = 1 ft). Change the engine mount shown (see statement of Prob. 6·22) to fit on the canted bulkhead. Do not change the front view. Find the angle that each straight tube makes with the canted bulkhead.
ANS. *AB*, 51½°. *AC*, 46°. *CD*, 48½°. *DE*, 43°.

Group 46. Solids on a plane surface (Art. 4·42)

In Probs. 46·1 to 46·9 locate the given object in its correct position on the given plane. The object should be drawn completely in the auxiliary views but may also be drawn, partly or completely, in the top and front views as required by the instructor. Use Layout C.

46·1 FIGS. 87 and 88. Show the tie-rod bracket of Fig. 87(1) attached to the upper side of surface *RSTV* of Fig. 88(1). The center line *XY* on the bracket must coincide with the tie-rod center line *AB*. In the true-size view line *AB* appears at 45° to line *RS*, and the bracket angle *A* is also 45°.

46·2 Same as Prob. 46·1, but use the bracket of Fig. 87(2).

46·3 Same as Prob. 46·1, but use the bracket of Fig. 87(3).

46·4 FIGS. 87 and 88. Show the tie-rod bracket of Fig. 87(1) attached to the right side of surface *RSTV* of Fig. 88(2). The center line *XY* on the bracket must coincide with the tie-rod center line *AB*. Determine angle *A* for the bracket and the angle *B* that *XY* makes with *RV* in the true-size view. ANS. *A*, 38°. *B*, 51½°.

46·5 Same as Prob. 46·4, but use the bracket of Fig. 87(2).

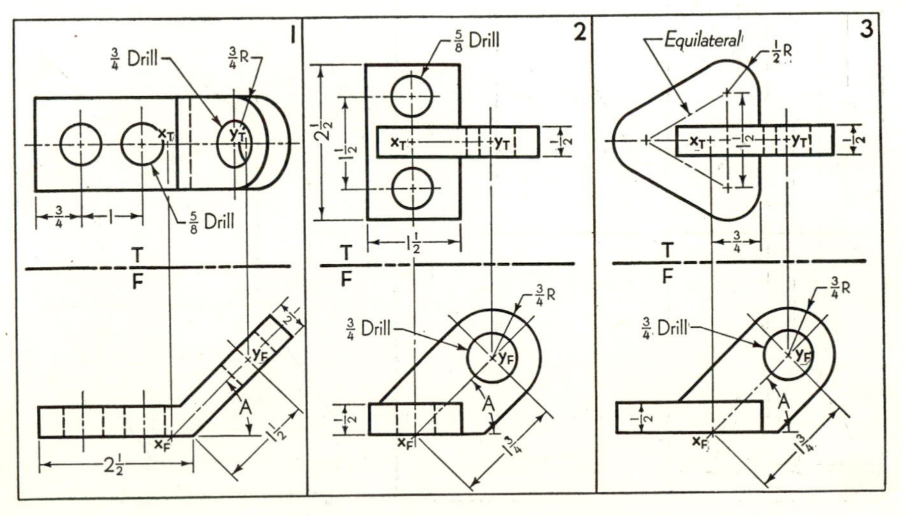

Fig. 87. Problems 46·1 to 46·9.

46·6 Same as Prob. 46·4, but use the bracket of Fig. 87(3).

46·7 FIGS. 87 and 88. Show the tie-rod bracket of Fig. 87(1) attached to the *upper* side of surface *RSTV* of Fig. 88(3). The center line *XY* on the bracket must coincide with the tie-rod center line *AB*. Determine angle *A* for the bracket and the angle *B* that *XY* makes with *RS* in the true-size view. ANS. *A*, 50°. *B*, 24°.

46·8 Same as Prob. 46·7, but use the bracket of Fig. 87(2).

46·9 Same as Prob. 46·7, but use the bracket of Fig. 87(3).

46·10 FIG. 89 C. Scale: half size. Line *ABC* is the center line of a ¼-in. airplane control cable that passes around a pulley at *B* and continues to *C* through a 1½-in. hole in the fuselage bulkhead. The pulley bracket shown at the left in Fig. 89 must be attached to the bulkhead surface so that the pulley lies in the plane *ABC* tangent to lines *AB* and *BC*. Determine angle *A* for the bracket and the angle *B* that the bracket base makes with the bent edge of the bulkhead. ANS. *A*, 15°. *B*, 37°.

46·11 FIG. 90 C. Scale: full size. Across the center of the sloping face of the casting a dovetail groove is to be cut parallel to side *AB*. Dimensions of the dovetail are shown at the right in the figure. Draw the top and front views of the grooved casting.

46·12 Same as Prob. 46·11, but cut the dovetail parallel to *BC*.

46·13 Same as Prob. 46·11, but cut the dovetail from *A* to *C*.

46·14 Same as Prob. 46·11, but use the T slot shown in Fig. 90.

46·15 Same as Prob. 46·14, but cut the T slot parallel to *BC*.

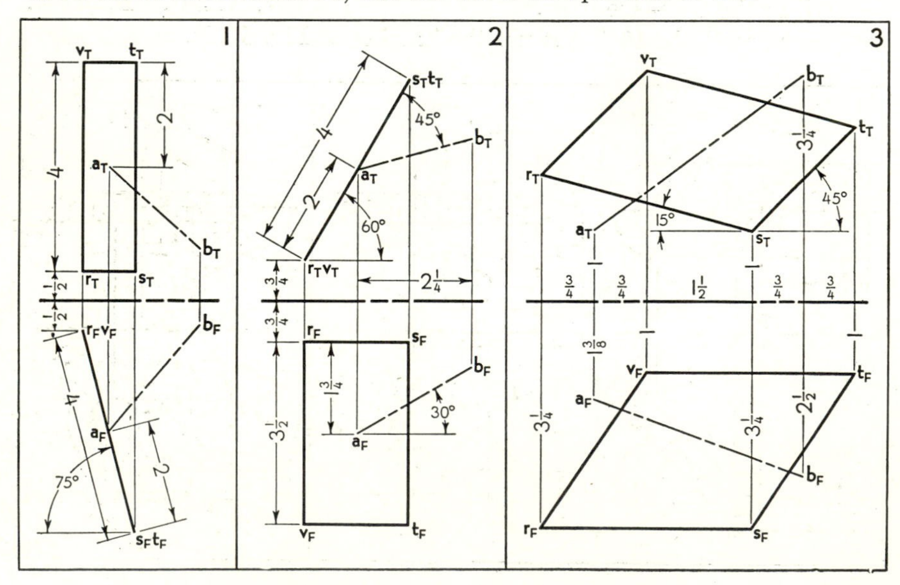

Fig. 88. Problems 46·1 to 46·9.

46·16 Same as Prob. 46·14, but cut the T slot from *A* to *C*.

46·17 FIGS. 90 and 87 C. Scale: full size. On the sloping face of the casting shown in Fig. 90 the bracket of Fig. 87(2) is to be centered with the 2½-in. side of the bracket parallel to the 2½-in. edge of the face. Bracket angle *A* is 45° and center line *XY* is vertical. Draw the top and front views of the assembled casting and bracket. Allow 2½ in. between the given views of the casting.

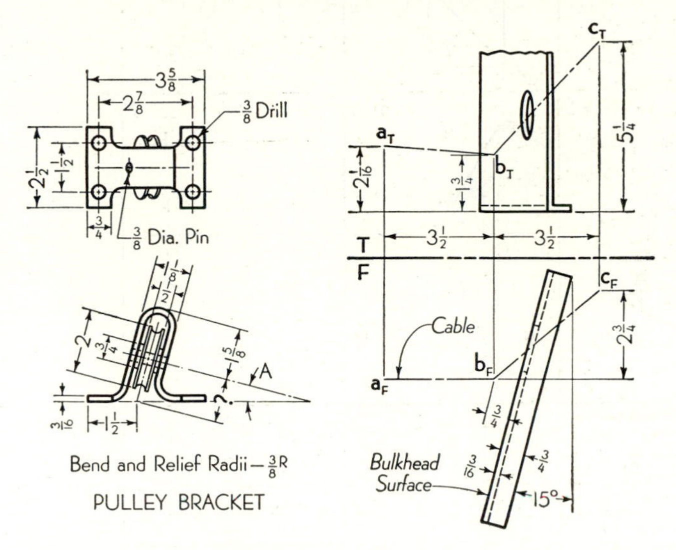

Fig. 89. Problem 46·10.

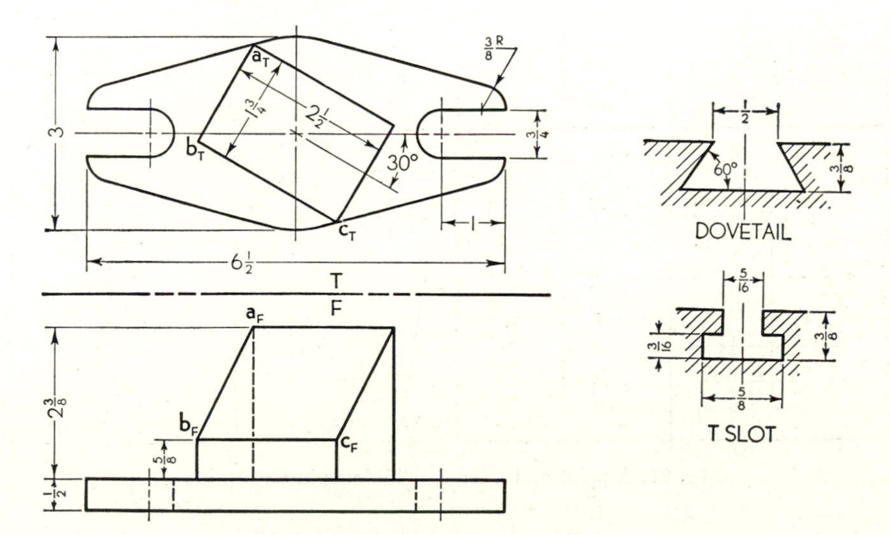

Fig. 90. Problems 46·11 to 46·17.

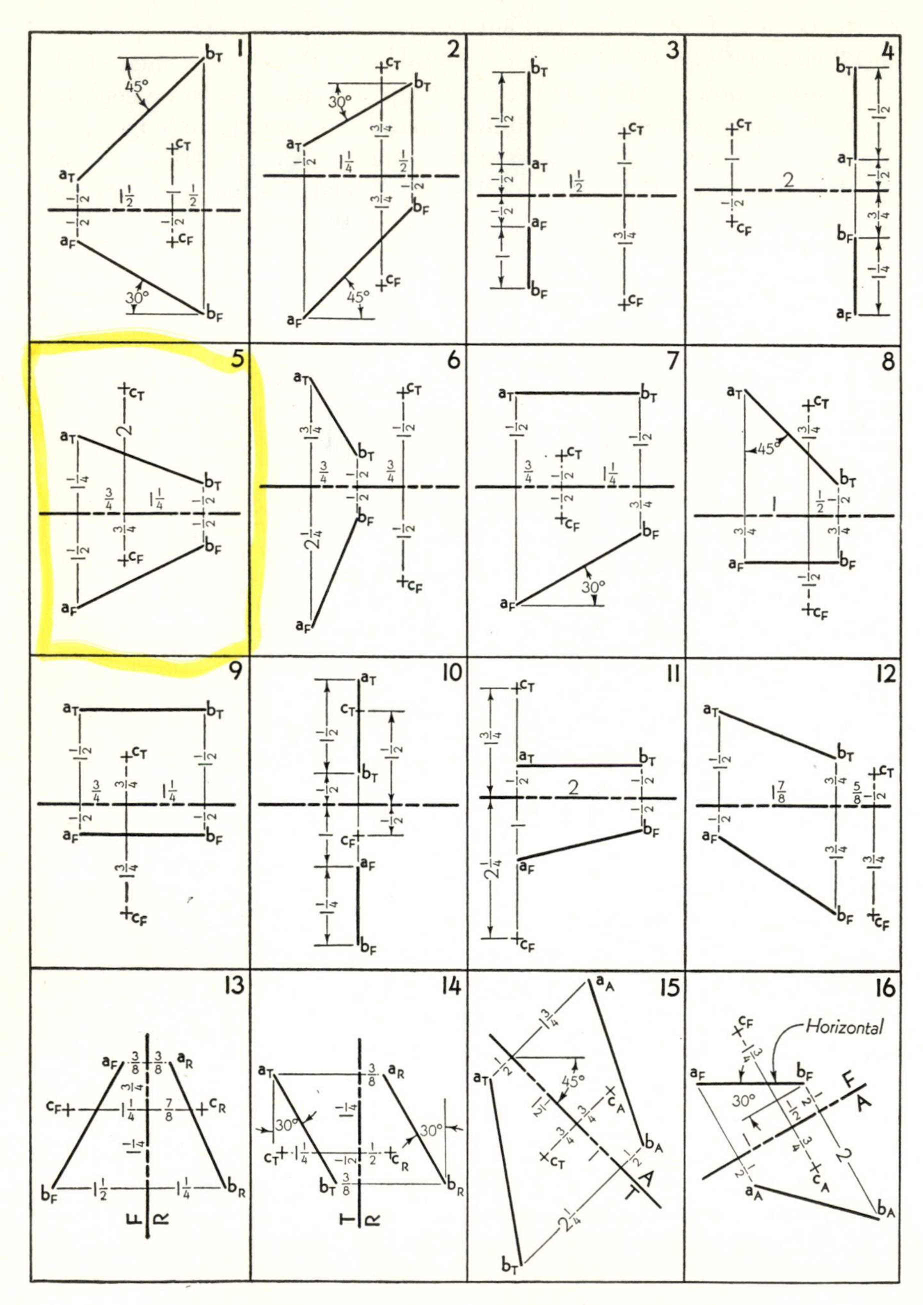

Fig. 91. A point and a line. Problems in Group 47.

Group 47. Revolution of a point (Arts. 5·1, 5·2)

In Probs. 47·1 to 47·14 revolve point C about the line AB as an axis into the specified position. Direction of rotation, when specified, assumes that the axis AB is being viewed from A to B. Show the revolved point in all views. Use Layout A.

47·1 FIG. 91(1). Revolve C 90° clockwise.
47·2 FIG. 91(2). Revolve C 60° counterclockwise.
47·3 FIG. 91(3). Revolve C 45° clockwise.
47·4 FIG. 91(4). Revolve C 30° clockwise.
47·5 FIG. 91(5). Revolve C into its lowest position.
47·6 FIG. 91(6). Revolve C into its most forward position.
47·7 FIG. 91(7). Revolve C into the higher position, ½ in. behind line AB.
47·8 FIG. 91(8). Revolve C into the rearmost position on the same level as line AB.
47·9 FIG. 91(9). Revolve C 90° clockwise.
47·10 FIG. 91(10). Revolve C into the position farthest to the left.
47·11 FIG. 91(11). Revolve C into its lowest position.
47·12 FIG. 91(12). Revolve C into its most forward position.
47·13 FIG. 91(13). Revolve C until it is directly in front of line AB.
47·14 FIG. 91(14). Revolve C until it is directly behind AB.
47·15 FIG. 54(1). Revolve X about line BC as an axis until it lies in plane ABC.
47·16 FIG. 54(5). Revolve X about line AB as an axis until it lies in plane ABC.
47·17 FIG. 54(6). Revolve X about line CD as an axis until it lies in plane $ABCD$.
47·18 FIG. 92 A. Scale: 1½ in. = 1 ft. Line AB is the axis of a large bolt whose hexagonal head is at A. A wrench 9 in. long is used to tighten the bolt. Locate the point X where the tip of the wrench handle will strike the vertical side of the engine and the point Y where it strikes the drip pan. Show each wrench position as a heavy line in all views. Through how many degrees can the wrench be turned between points X and Y? What length wrench will just clear the drip pan?

ANS. 191°. 8¼ in.

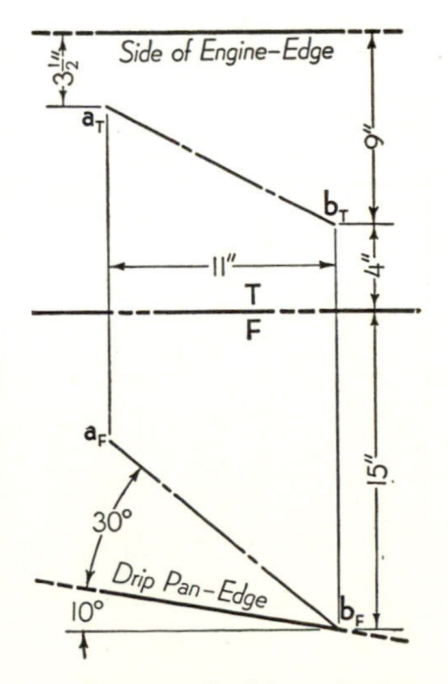

Fig. 92. Problem 47·18.

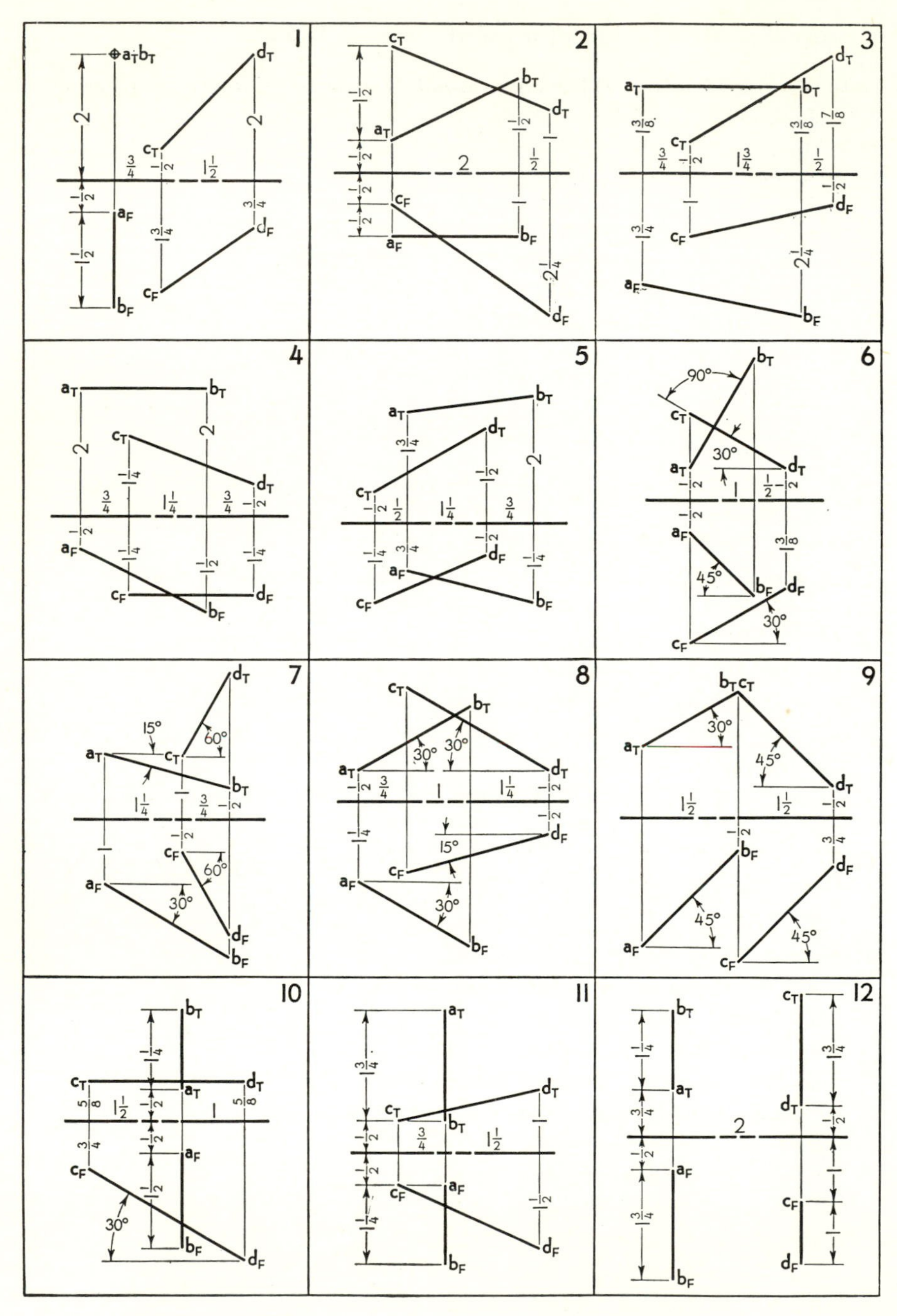

Fig. 93. Two skew lines. Problems in Groups 48 and 56.

Group 48. Revolution of a line (Art. 5·3)

In Probs. 48·1 to 48·12 revolve the given line about the other line as an axis into the specified position. Direction of rotation, when specified, assumes that the axis is being viewed from A to B (or from C to D). Show the revolved position as a dash line in all views. Use Layout A, except as indicated.

48·1 FIG. 93(1). Revolve CD 135° clockwise about AB.
48·2 FIG. 93(2). Revolve CD about AB until CD is level and below AB.
48·3 FIG. 93(3). Revolve CD about AB until CD is frontal and behind AB.
48·4 FIG. 93(4). Revolve AB about CD until AB is level and above CD.
48·5 FIG. 93(5). Revolve AB 90° counterclockwise about CD.
48·6 FIG. 93(6). Revolve AB 30° clockwise about CD.
48·7 FIG. 93(7). Revolve CD 45° clockwise about AB.
48·8 FIG. 93(8). Revolve CD 60° counterclockwise about AB.
48·9 FIG. 93(9). Revolve CD 90° counterclockwise about AB.
48·10 FIG. 93(10). Revolve AB about CD until AB is level. Show both solutions. (HINT: Use the true length of AB and the known length of $a_F^R b_F^R$ of the revolved position to establish the *direction* of the revolved line in the top view.)
48·11 FIG. 93(11) B. Revolve CD about AB until CD is a profile line to the right of AB.
48·12 FIG. 93(12) B. Same as Prob. 48·10, but reduce given distance between lines from 2 to 1 in.
48·13 FIG. 94 C. Scale: ½ in. = 1 ft. The retractable landing gear swings upward to the left, pivoting about the hinge center line AB. Show the oleo AC and wheel axle CD in the retracted position after rotating 90°. Complete the top view, and record the angles that the

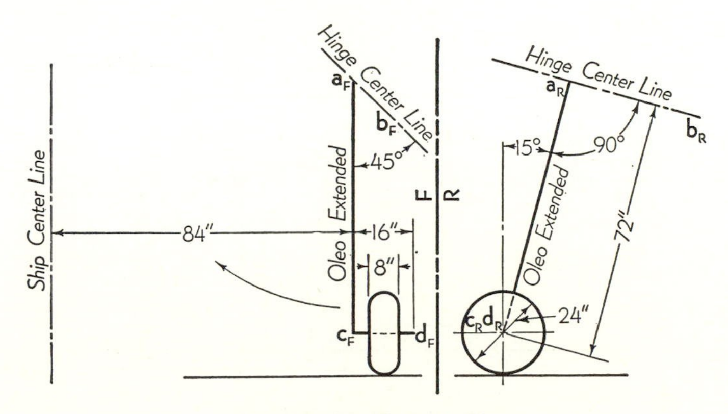

Fig. 94. Problem 48·13.

retracted oleo makes with reference line T–F in top and front views.
ANS. Top view, 14°12′. Front view, 3°43′.

48·14 FIG. 95 C. Scale: ½ in. = 1 ft. The retractable landing gear swings upward to the left until the wheel axle CD occupies the retracted position $C^R D^R$ inside the wing. Locate the hinge center line, or axis, about which rotation must take place. (HINT: Points C and D move in parallel planes during rotation.) Complete the top view, and record the angles that the hinge center line makes with the ship center line in top and front views. Through how many degrees was the landing gear revolved? What angle A must the extended oleo make with the vertical in order to intersect the hinge axis? ANS. Axis, top view, 8°30′; front view, 28°30′. Rotation, 86½°. Angle A, 16°.

48·15 FIG. 93(5) A. Locate the axis about which line CD may be revolved until it coincides with line AB, point C coinciding with point A. Show the axis in all views, and record its bearing and slope angle. (HINT: A segment CE of line CD will coincide with line AB, and the paths of points C and E, moving to A and B, respectively, will lie in parallel planes.) ANS. N 57° W; 27° slope.

48·16 FIG. 93(8) A. Same as Prob. 48·15. ANS. N 64° W; 35° slope.

Group 49. True length and slope of a line by revolution (Arts. 5·4 to 5·7)

49·1–49·12 Same as Probs. 6·1 to 6·12, except that the true length and slope angle of the line are to be found by the method of revolution. Check the true length by revolving the line about a different axis. Draw and label each axis used. Show the revolved positions as dash lines.

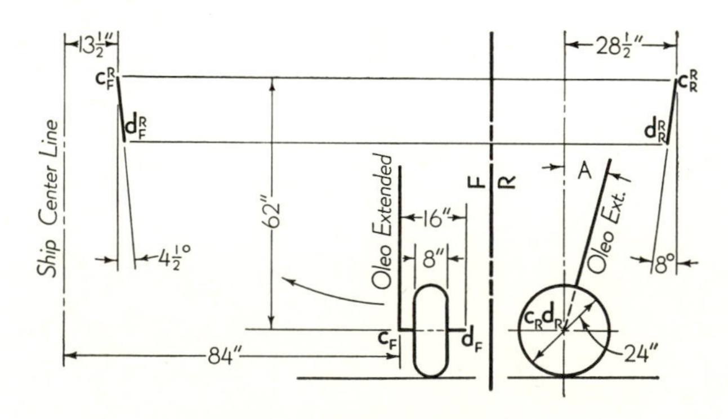

Fig. 95. Problem 48·14.

49·13–49·23 Same as Probs. 6·13 to 6·23, except that the solution is to be made by the method of revolution.

In Probs. 49·24 to 49·29 construct the top and front views of the given line, using only the given views.

49·24 Line *AB* of Prob. 7·1.

49·25 Line *AB* of Prob. 7·2.

49·26 Line *AB* of Prob. 7·3.

49·27 Line *AB* of Prob. 7·4.

49·28 Line *AB* of Prob. 7·5.

49·29 Line *AB* of Prob. 7·6.

49·30 FIG. 96 A. Scale: ¼ in. = 1 ft. A mast on a roof is supported by three guy wires *AB*, *AC*, and *AD*, anchored at points *B*, *C*, and *D* on the roof. Complete the front view, and determine the true length of each guy wire. Solve, using only the given views.

ANS. *AB*, 8 ft 1 in. *AC*, 7 ft 6 in. *AD*, 11 ft 2 in.

49·31 FIG. 97 A. Scale: ⅛ in. = 1 ft. A derrick erected on a hillside of 15 per cent grade has a mast *AB* and boom *CD*. The four guy wires attached to the top of the mast are shown in the top view. They must just clear the boom as it revolves at any angle of elevation. Locate the guy-wire-anchor points *W*, *X*, *Y*, and *Z* in both views, and determine the true length of each guy wire. Solve, using only the given view. ANS. *AW*, 32 ft 0 in. *AX*, 36 ft 11 in. *AY*, 38 ft 7 in. *AZ*, 33 ft 3 in.

49·32 A. Scale: 1 in. = 1 ft. A trapeze bar is 3 ft long and is supported by two ropes each 3½ ft long. How far will the bar be raised if it is

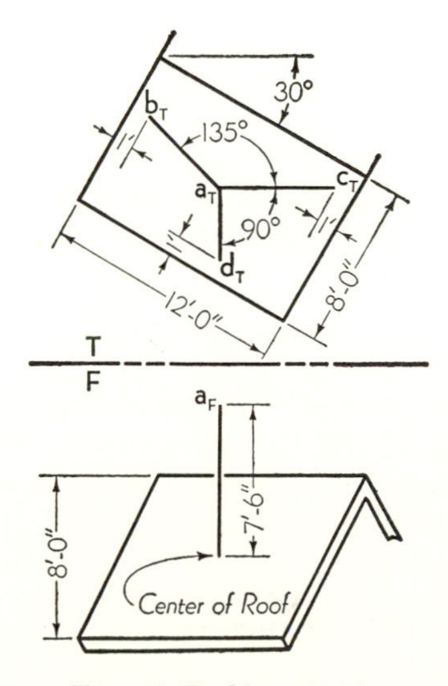

Fig. 96. Problem 49·30.

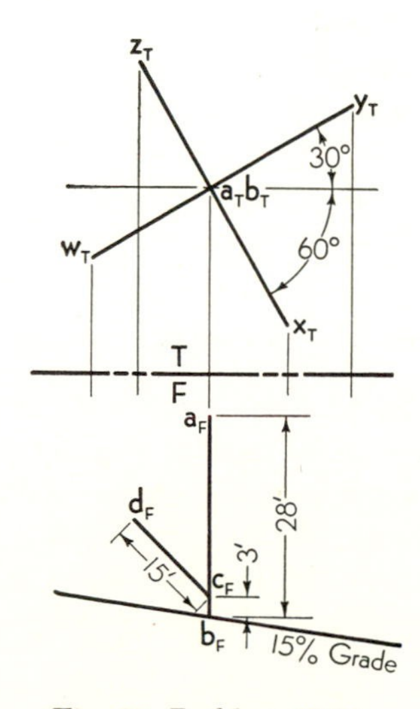

Fig. 97. Problem 49·31.

rotated through an angle of 60°? Solve also for rotations of 90° and 180°. ANS. 60°, 4 in. 90°, 8⅝ in. 180°, 20⅝ in.

49·33 A. Scale: ¼ in. = 1 ft. A brace 18 ft long is attached to the floor and a vertical wall. The upper end is 12 ft above the floor, and the lower end is 10 ft in front of the wall. Draw top and front views of the brace, and determine the angles it makes with the floor and wall. ANS. Floor, 42°. Wall, 34°.

49·34 FIG. 15 C. Scale: ¾ in. = 1 ft (B, ½ in. = 1 ft). Member *CD* of the tripod-type airplane landing gear is a tension-type shock absorber. Under the impact of landing, *CD* elongates, and point *D* rotates about the hinge line *AB*. Locate the position of *D* when *CD* elongates 4 in. from the position shown. How far does point *D* rise vertically? ANS. 7 in.

Group 50. Plane as an edge by revolution (Art. 5·8)

In Probs. 50·1 to 50·9 revolve the given plane until it appears as an edge in either the top or the front view as specified. Draw and label the axis. Show the revolved position with dash lines. Solve, using only the given views. Use Layout A.

50·1 FIG. 98(1). Edge in front view.
50·2 FIG. 98(2). Edge in top view.
50·3 FIG. 98(3). Edge in front view.
50·4 FIG. 98(4). Edge in top view.
50·5 FIG. 98(5). Edge in front view.
50·6 FIG. 98(6). Edge in front view.
50·7 FIG. 98(7). Edge in top view.
50·8 FIG. 98(8). Edge in front view.
50·9 FIG. 98(9). Edge in top view.
50·10 FIG. 76(4). Determine the dihedral angle between planes *ABC* and *BCD* by revolving them until both appear edgewise in the front view.

Group 51. True size of a plane by revolution (Art. 5·9)

51·1–51·12 Same as Probs. 22·1 to 22·12, except that the plane is to be revolved until it appears true size in one of the given views. When space permits, choose the axis so that the revolved position will fall clear of the given view (compare Fig. 5·11). Label the axis, and show the revolved position with dash lines.

51·13 Same as Prob. 22·19, but solve by revolution.
51·14 Same as Prob. 22·20, but solve by revolution.

51·15 Same as Prob. 22·21, but solve by revolution.
51·16 Same as Prob. 23·3, but solve by revolution.
51·17 Same as Prob. 23·4, but solve by revolution.
51·18 Same as Prob. 23·5, but solve by revolution.
51·19 Same as Prob. 23·6, but solve by revolution.
51·20 Same as Prob. 23·7, but solve by revolution.
51·21 Same as Prob. 23·8, but solve by revolution.
51·22 Same as Prob. 23·9, but solve by revolution.

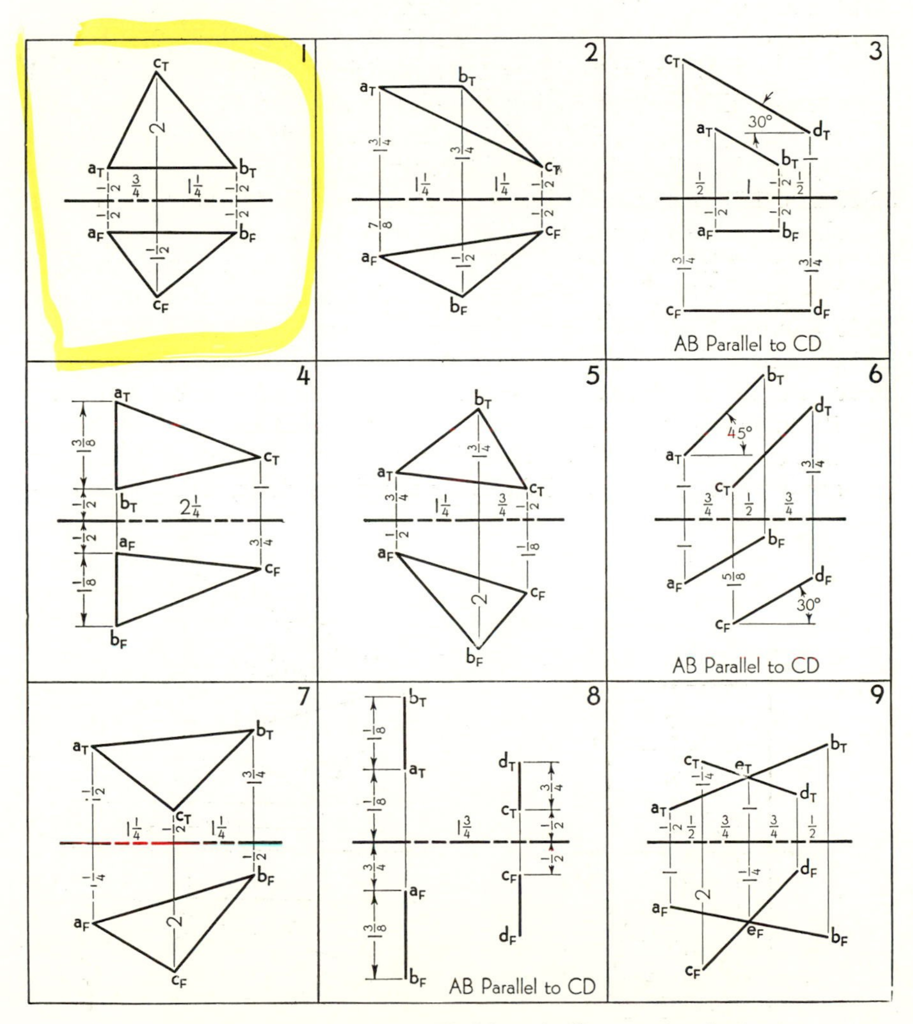

Fig. 98. Plane surfaces. Problems in Groups 50 and 55.

Group 52. Counterrevolution (Art. 5·10)

52·1–52·9 Same as Probs. 26·1 to 26·9, except that the given plane figure is to be located in the plane by the method of counterrevolution. Label the axis, and show the revolved lines of the plane as dash lines.

In Probs. 52·10 to 52·15 locate the required circle in the plane by counterrevolving 12 points on the circle. Use Layout A, except in Prob. 52·15.

52·10 Same as Prob. 27·3.
52·11 Same as Prob. 27·4.
52·12 Same as Prob. 27·5.
52·13 Same as Prob. 27·6.
52·14 Same as Prob. 27·7.
52·15 B. Same as Prob. 27·12.

52·16 FIG. 99 A. Scale: ½ in. = 1 ft. A tripod stands on an inclined surface with the legs resting on points *A*, *B*, and *C*. The legs *AD*, *BD*, and *CD* are 6, 5½, and 5 ft long, respectively. Locate vertex *D*, and show the tripod in the given views.

52·17 FIG. 100 B. Scale: ½ in. = 1 ft. The cable whose center line is shown passes through a slot in the trap door. Plot the center line of the slot in the door that will permit the fixed cable to pass through the door while the door swings upward to a horizontal position.

52·18 Same as Prob. 23·1, but solve by counterrevolution.

52·19 Same as Prob. 23·2, but solve by counterrevolution.

52·20 FIG. 101 A. Scale: 1 in. = 1 ft. The ends of a wooden timber, 12 in. square, must be cut to fit against a wall and floor as shown. The completed top view shows that two faces, or sides, of the timber are in vertical planes. Using only the given views, revolve the tim-

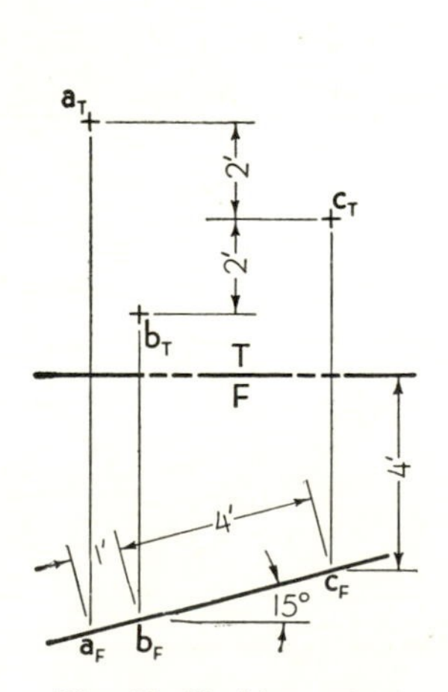

Fig. 99. Problem 52·16.

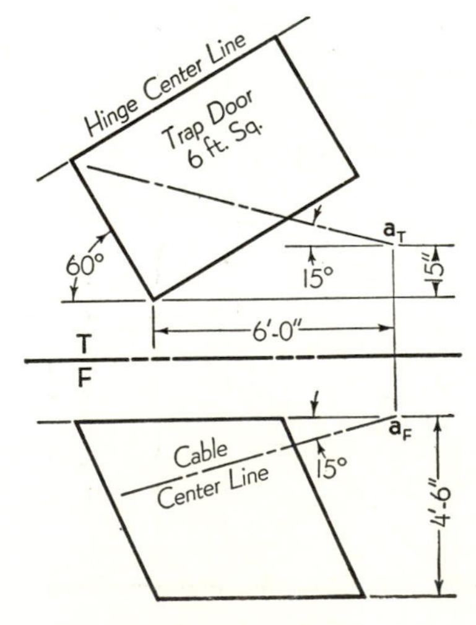

Fig. 100. Problem 52·17.

ber to show the true size of its vertical faces, and then complete the front view.

Group 53. Dihedral angle by revolution (Art. 5·11)

53·1–53·23 Same as Probs. 44·1 to 44·23, except that the solution is to be made by the method of revolution. Use the same layouts suggested in Group 44, except that Prob. 53·19 can be done on Layout B. In Probs. 53·10 to 53·15 the line of intersection of the two planes must be determined first by one of the methods described in Arts. 4·24, 4·26, or 4·27.

53·24 FIG. 102 B. Scale: 3 in. = 1 ft. A triangular cover plate *ABC* is riveted in a corner between two walls and a floor as shown in the figure. A 1¼-in. strip along each edge of the plate must be bent to fit against the adjacent surfaces. Using only the given views and a right-side view, determine the bend angle along each edge. In the upper right part of the sheet make a detail drawing of the flattened plate (note typical detail in the figure).

ANS. Bend angles: *AB*, 63°; *AC*, 47°; *BC*, 55°.

53·25 A. A pyramid has a horizontal base that is an equilateral triangle, 2½ in. on a side. The dihedral angle between each pair of adjacent lateral faces is 75°. Construct top and front views of the pyramid. What is its altitude? ANS. 1.65 in.

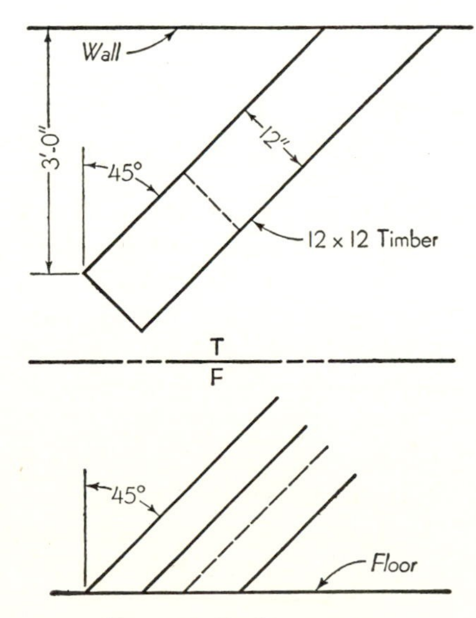

Fig. 101. Problem 52·20.

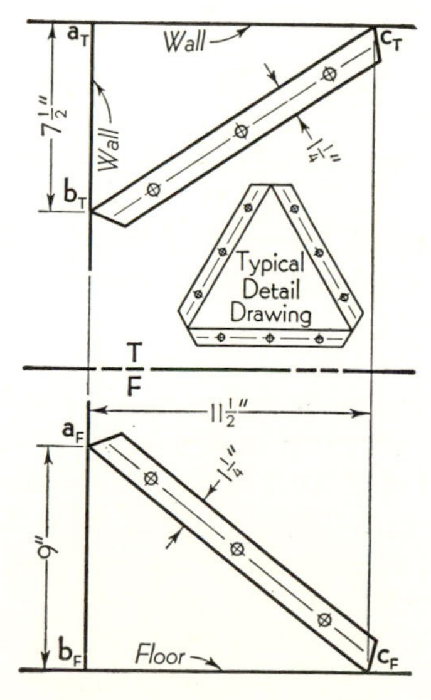

Fig. 102. Problem 53·24.

53·26 A. A pyramid has a horizontal base that is a 2-in. square. The dihedral angle between each pair of adjacent lateral faces is 120°. Construct top and front views of the pyramid. What is its altitude?
Ans. 1.00 in.

Group 54 Angle between a line and a plane by revolution (Art. 5·12)

54·1–54·19 Same as Probs. 45·1 to 45·19, except that the solution is to be made by the method of revolution. Use the same layouts suggested in Group 45, except that Prob. 54·15 can be done on Layout A and Prob. 54·19 on Layout B.

Group 55. Line making given angles with two intersecting lines (Arts. 5·13 to 5·15)

In Probs. 55·1 to 55·5 locate a line (or lines) through the intersection point of the given lines, making the specified angle with each line. Show all possible solutions in the given views, and state the bearing of each line. Use Layout B.

55·1 FIG. 98(1). 30° with *AB* and 45° with *AC*.
Ans. N 60° E, N 78° E.

55·2 FIG. 98(2). 60° with *AB* and 45° with *AC*.
Ans. N 76° E, S 25° E.

55·3 FIG. 98(4). 45° with *AB* and 75° with *AC*.
Ans. N 38° E, N 66° E, S 53° E, S 6° W.

55·4 FIG. 98(7). 75° with *AB* and 45° with *AC*.
Ans. N 84° E, S 20° E, S 2° E.

55·5 FIG. 98(9). 45° with *AB* and 30° with *CD*.
Ans. N 73° W, S 72° W.

55·6 A. A line passing through the corner of a cube makes 60° with one edge and 75° with another. What angle does it make with the third edge?
Ans. 34½°.

55·7 FIG. 103 C. Scale: 1 in. = 1 ft. Two braces *AB* and *CD* are connected to a wall and intersect at point *E*. Locate additional braces at point *E* that make 60° with *AB* and 75° with *CD*. State the bearing of each brace. Ans. S 87° W, S 45° W, S 12° E, S 13° E.

55·8 FIG. 104 C. Scale: 1 in. = 1 ft. Redesign the tripod-type airplane landing gear so that member *BD* will make 45° with both members *AD* and *CD*. Members *AD* and *CD* remain as given. If *B* is kept at the same elevation as point *A*, what are the new values of the given 16- and 18-in. dimensions?
Ans. 16 in. unchanged; 18 in. becomes 23¼ in.

Group 56. Line making given angles with two skew lines (Art. 5·16)

56·1 B. Scale: 1 in. = 1 ft. A horizontal pipe *AB* runs due east, and a vertical pipe *CD* is located 18 in. north of *AB*. Connect the two pipes with a third pipe that makes 45° with pipe *AB* and 60° with the vertical pipe *CD*. Connect so that water flowing eastward in *AB* can flow downward to *CD*. How far west of the vertical pipe, and how far below the horizontal pipe, must the connections be made? How long is the connecting pipe?

ANS. 25 in. west, 17¾ in. below. 35½ in.

56·2 B. Similar to Prob. 54·1, except that pipe *AB* slopes 15° downward to the east. How far west of the vertical pipe must the connection be made? How long is the connecting pipe?

ANS. 17½ in. west. 29 in.

56·3 B. Scale: 1 in. = 1 ft. A horizontal pipe *AB* runs due west; a second horizontal pipe *CD* runs N 30° W but is 2 ft 6 in. lower than *AB*. Connect the two pipes with a third pipe using 45° elbows at the connections. What are the length, slope, and bearing of the connecting pipe? ANS. 4 ft 4¼ in., 35°, N 60° W.

56·4 FIG. 93(1) B. Locate a line (or lines) connecting skew lines *AB* and *CD* and making 45° with *AB* and 60° with *CD*. State the true length, slope, and bearing of each line.

ANS. 2.38 in. 45°. S 64° E, S 25° E.

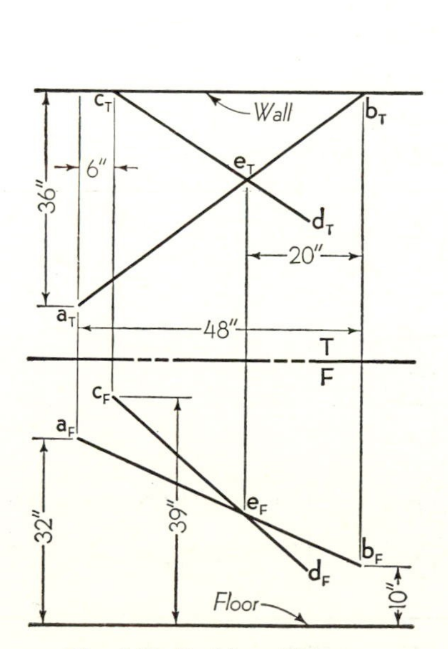

Fig. 103. Problem 55·7.

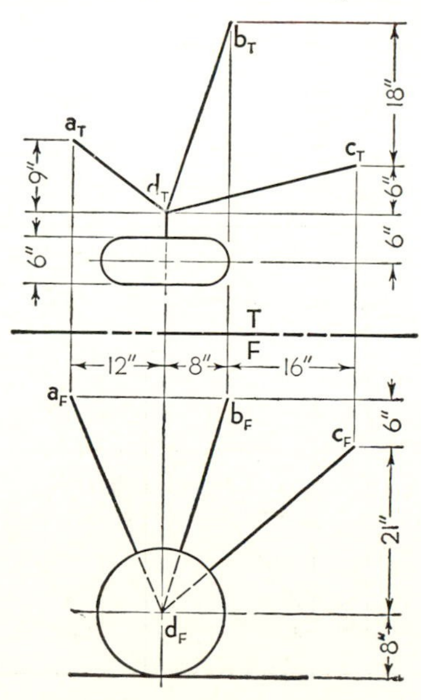

Fig. 104. Problem 55·8.

56·5 FIG. 93(3) B. Same as Prob. 56·4.
ANS. 1.72 in. 49½°, N 60° E. 22½°, S 59° E.

56·6 FIG. 93(4) B. Same as Prob. 56·4.
ANS. 1.15 in. 1°, N 50° E. 56°, S 43° E.

56·7 FIG. 93(10) B. Same as Prob. 56·4, but making 45° with both *AB* and *CD*. ANS. 1.09 in. 19°, S 42° W. 74°, S 79° E.

56·8 FIG. 93(12) B. Same as Prob. 56·4, but making 75° with *AB* and 45° with *CD*. ANS. 2.96 in. 11°, S 44° E and N 44° E. 3.18 in. 39°, S 55° E and N 55° E.

Group 57. Line making given angles with two planes (Arts. 5·17, 5·18)

In Probs. 57·1 to 57·6 locate a line (or lines) through the given point, making the specified angle with each plane. Show all possible solutions in the given views, and state the bearing of each line.

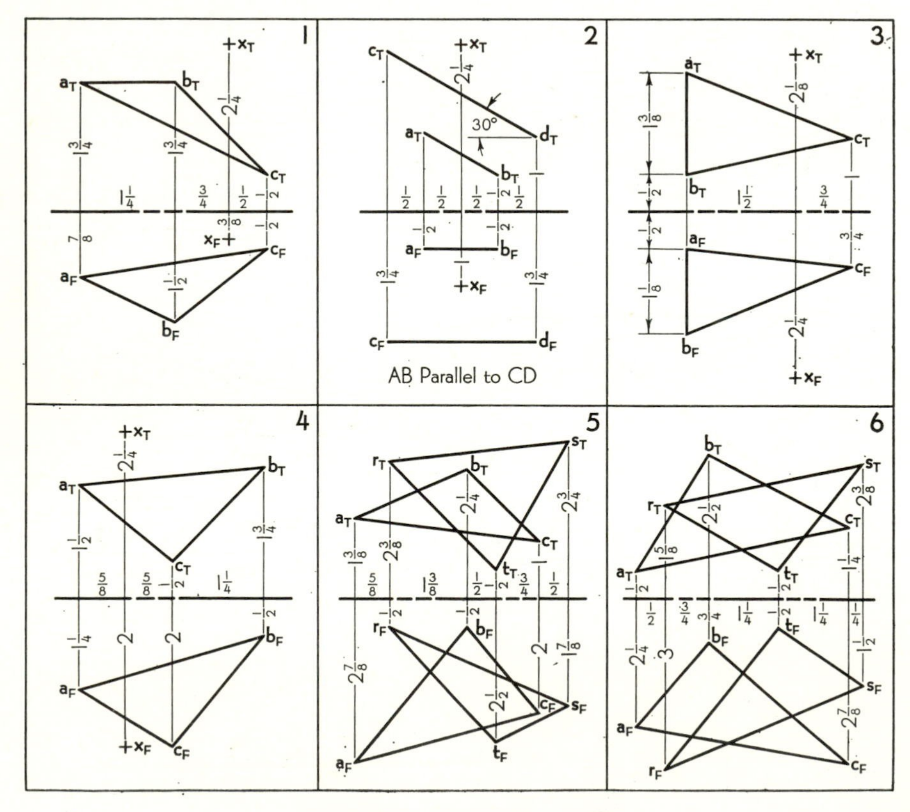

Fig. 105. Problems 57·1 to 57·6.

57·1 FIG. 105(1) B. Through point X, making 60° with plane ABC and 45° with the horizontal plane. ANS. N $17\frac{1}{2}$° W, N $59\frac{1}{2}$° E.

57·2 FIG. 105(2) B. Through point X, making 30° with plane $ABCD$ and 15° with the horizontal plane.
ANS. N $12\frac{1}{2}$° W, N $72\frac{1}{2}$° E, S 34° E, N 86° W.

57·3 FIG. 105(3) B. Through point X, making 45° with plane ABC and 30° with the horizontal plane. ANS. S 40° E, S 68° W.

57·4 FIG. 105(4) B. Through point X, making 30° with plane ABC and 60° with the horizontal plane. ANS. N 85° E, N $62\frac{1}{2}$° W.

57·5 FIG. 105(5) C. Through point B, making 60° with plane ABC and 30° with plane RST. State bearing and slope of each line.
ANS. N, $5\frac{1}{2}$°. N 50° E, 41°.

57·6 FIG. 105(6) C. Through point B, making 45° with plane ABC and 15° with plane RST. State bearing and slope of each line.
ANS. E, $53\frac{1}{2}$°. S 72° E, $26\frac{1}{2}$°. S 23° W, 17°. S 43° W, 42°.

57·7 FIG. 106 C. Scale: $\frac{1}{4}$ in. = 1 ft. Shown is the plan (top view) and elevation of a portion of a factory. (*a*) From the hatch A in the wall a chute runs to the floor, making an angle of 45° with the floor and 30° with the wall. What will be the length of the parallel bottom stringers, and will they clear the bin? (*b*) From hatch B a second chute makes an angle of 45° with the floor and just clears the

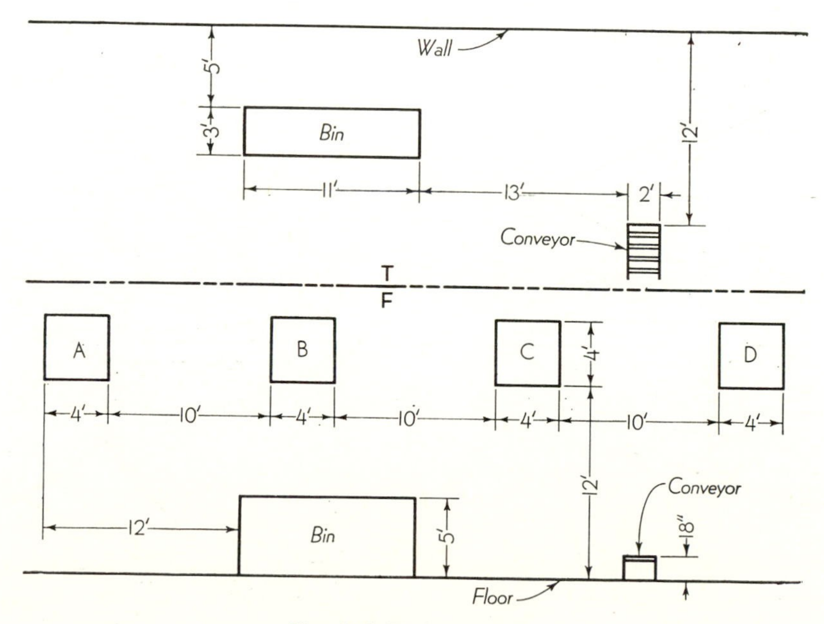

Fig. 106. Problem 57·7.

right end of the bin. What will be the length of the parallel bottom stringers, and what angle must they make with the wall? (*c*) From hatches *C* and *D* two chutes deliver to the top of the roller conveyor. What will be the length of the converging bottom stringers, and what angles will they make with floor and wall? ANS. (*a*) 16 ft 11 in. (*b*) 16 ft 11 in., 20°. (*c*) 17 ft 1 in., 38°, 45°. 17 ft 10 in., 36°, 42°.

57·8 FIG. 107 C. Scale: $\frac{3}{16}$ in. = 1 ft. Line *AB* is the center line of a hawsepipe connecting the deck and hull of a ship (compare Fig. 9·22). This center line must make an angle of 45° with the deck and 30° with the side of the ship. Locate point *B* on the hull, and determine angles *X* and *Y*. What is the true length of *AB*?
ANS. *X*, 6½°. *Y*, 35½°. 12 ft 2 in.

57·9 FIG. 101 A. Scale: 1 in. = 1 ft. Redraw the 12-by-12 timber, starting at the located point, so that the timber makes an angle of 30° with the floor and 40° with the wall. (Disregard the given 45° angles.) Do not copy the timber shown, but note that two sides of the timber must be vertical. Complete the top and front views of the new timber. Solve, using only the given views. What is the over-all length of the new timber? ANS. 5 ft 9¼ in.

57·10 FIG. 108 A. Scale: 1½ in. = 1 ft. A 4-by-8 rafter must be cut to fit against the plate and ridge as shown. Note that the 4-in. sides of the rafter appear edgewise in the front view (the 8-in. sides will

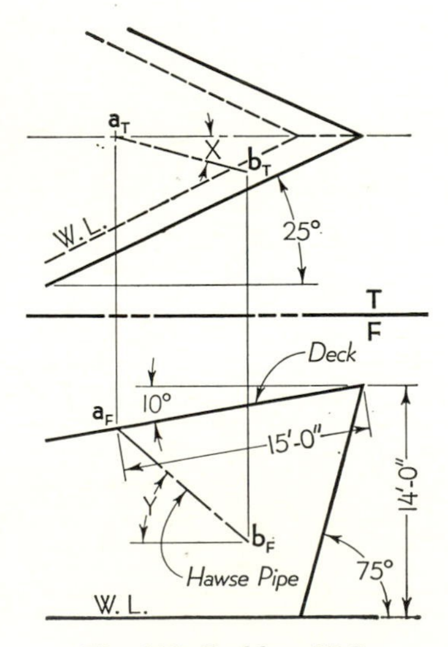

Fig. 107. Problem 57·8.

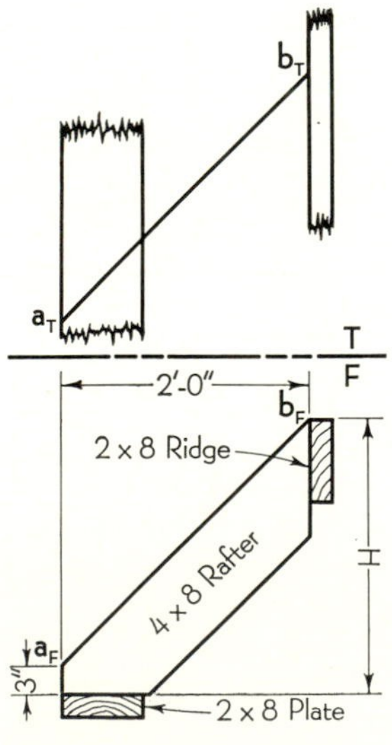

Fig. 108. Problem 57·10.

not appear edgewise in the top view). The rafter makes an angle of 40° with the horizontal plane of the plate and 30° with the vertical plane of the ridge. Using only the given views, complete the top and front views of the rafter. What is the over-all length of the rafter, and how high is the ridge above the plate (dimension *H*)?

Ans. 4 ft 5¾ in. *H*, 2 ft 10 in.

Group 58. Line making given angles with a line and a plane (Art. 5·19)

In Probs. 58·1 to 58·6 locate a line (or lines) through the mid-point of the given line and making the specified angle with the line and plane. Show all possible solutions in the given views, and state the bearing of each line.

58·1 FIG. 109(1) A. 45° with line *AB*, and 60° with the horizontal plane.

Ans. N 11° E, N 80° E.

58·2 FIG. 109(2) A. 60° with line *AB*, and 30° with the horizontal plane.

Ans. N 15° W, S 45° E.

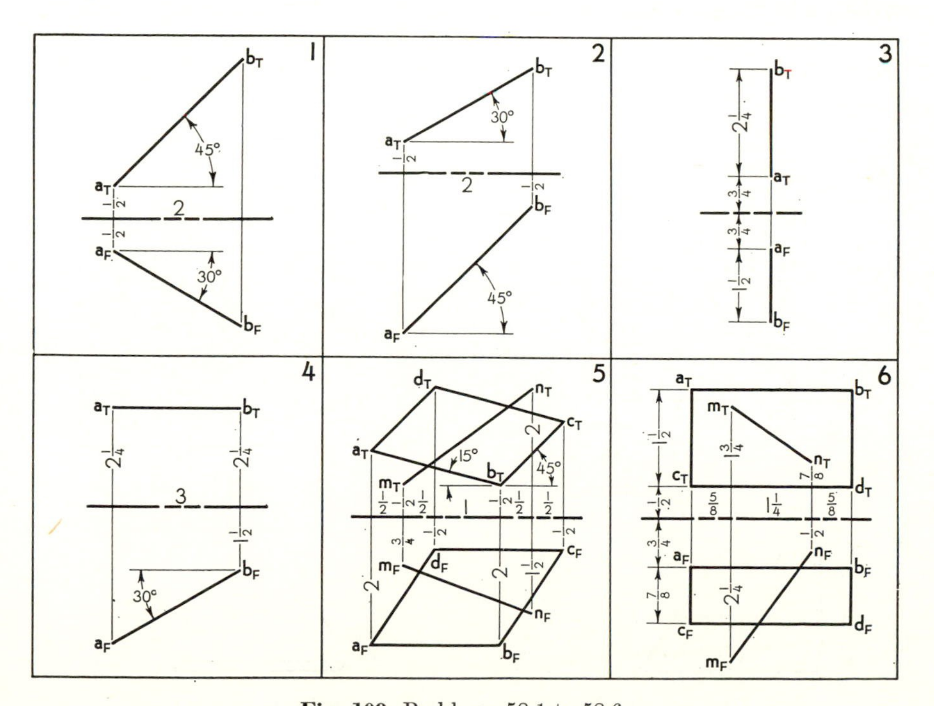

Fig. 109. Problems 58·1 to 58·6.

58·3 FIG. 109(3) B. 60° with line AB, and 15° with the horizontal plane. ANS. N 37° E, N 64° E, S 37° E, S 64° E.

58·4 FIG. 109(4) A. 30° with line AB, and 45° with the horizontal plane. ANS. N 56½° E, N 56½° W.

58·5 FIG. 109(5) B. 45° with line MN, and 30° with plane $ABCD$. State bearing and slope of each line. ANS. N 26° E, 20°. S 82° E, 41°.

58·6 FIG. 109(6) B. 30° with line MN, and 60° with plane $ABCD$. State bearing and slope of each line.
ANS. N 21° W, 33°. N 79° W, 77½°.

58·7 FIG. 110 A. Scale: ¾ in. = 1 ft. The hopper must be connected to the bin by a sheet-metal duct that makes 60° with the bin cover and 45° with the vertical axis of the hopper. Draw the connecting duct in the top and front views.

58·8 FIG. 111 C. Scale: ¼ in. = 1 ft (B, 3/16 in. = 1 ft). The vertical mast on the roof is to be supported by three guy wires AB, AC, and AD, spaced roughly as shown. To use standard fittings guys AB and AC must make 45° with both roof and mast and AD must make 30° with the roof and 15° with the mast. Locate the three guys in the top and front views.

Group 59. Location of a point on a cone (Arts. 6·4 to 6·6)

59·1–59·6 FIG. 112(1) to (6). In each problem points X and Y lie on the surface of the cone, but only one view of each point is given. Locate

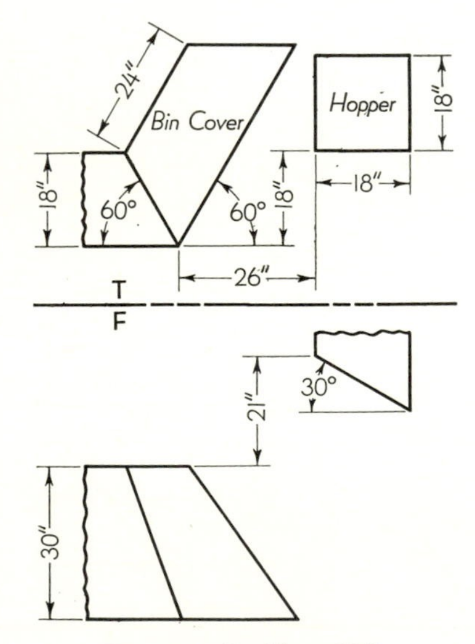

Fig. 110. Problem 58·7.

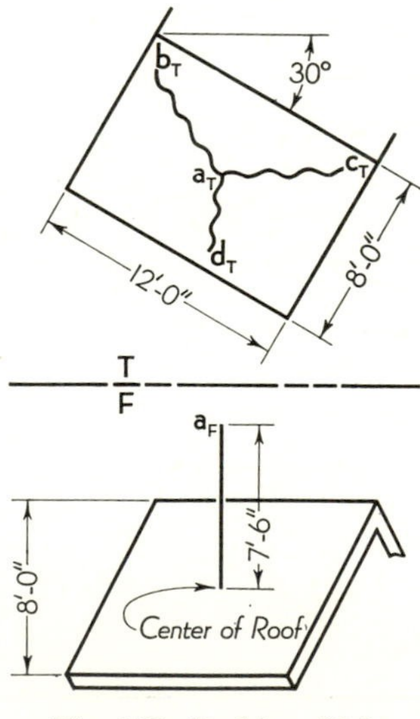

Fig. 111. Problem 58·8.

each point in the other view, and show the cone element through the point. The notation x_T^V means that point X is *visible* in the top view; y_F^H means that Y is *hidden* in the front view. Use Layout A.

59·7 FIG. 113(1) A. Line AB is the axis of a right circular cone with vertex at A and a 2-in.-diameter base at B. Show the cone in the top and front views. Point X is on the front side of the cone.

59·8 FIG. 113(2) A. Same as Prob. 59·7, but locate X on the underside of the cone.

59·9 FIG. 113(3) B. Same as Prob. 59·7, but locate X on the right side of the cone.

59·10 FIG. 113(4) A. Same as Prob. 59·7, but locate X on the rear side of the cone.

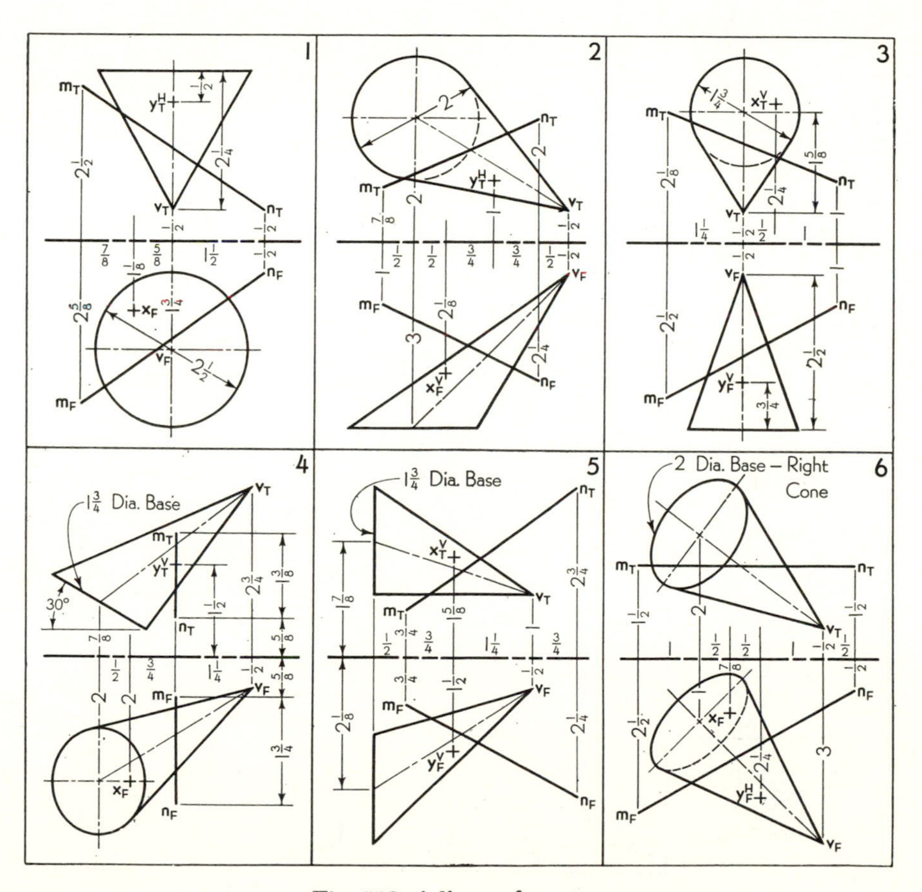

Fig. 112. A line and a cone.

Group 60. Intersection of a plane and a cone (Arts. 6·7, 6·8)

60·1–60·6 FIG. 112(1) to (6). In each problem find the line of intersection between the cone and a vertical plane that contains line *MN*. Use at least 12 elements, being sure to locate accurately those points which lie on extreme elements. Draw a view showing the true size of the plane intersection. Use Layout A.

Problems 60·7 to 60·11 are similar to Probs. 60·1 to 60·6, but use an edge-front plane that contains line *MN*, instead of a vertical plane. Use Layout A.

60·7 FIG. 112(1).
60·8 FIG. 112(2).
60·9 FIG. 112(3).
60·10 FIG. 112(5).
60·11 FIG. 112(6).

In Probs. 60·12 to 60·17 find the line of intersection between the cone and the plane *ABC*, assuming that the plane is indefinite in extent and cuts completely through the given cone. Show the curve in both the top and front views, locating accurately those points which lie on extreme elements in each view. Draw a view showing the true size of the plane intersection.

60·12 FIG. 114(1) B.
60·13 FIG. 114(2) B.
60·14 FIG. 114(3) B.
60·15 FIG. 114(4) B.
60·16 FIG. 114(5) B.
60·17 FIG. 114(6) C.

60·18 FIG. 114(4) A. Extend the cone, and find its intersection with a horizontal plane passing through point *B*. (Disregard plane *ABC*.)

60·19 FIG. 114(5) A. Same as Prob. 60·18, but take the horizontal plane through point *O*.

60·20 FIG. 6·8(*a*) B. Using 24 equally spaced elements, construct an elliptical conic section by the method shown in the figure (see Art. 6·8).

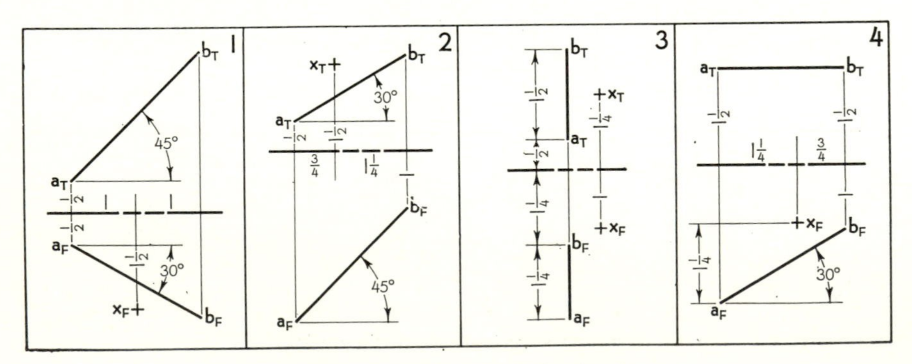

Fig. 113. Problems 59·7 to 59·10.

Show the ellipse in both the top view and view A. The base diameter and altitude are each $2\frac{3}{4}$ in., angle A is 45°, and the cutting plane intersects the cone axis $1\frac{1}{4}$ in. below the vertex. Locate the foci of the ellipse (see Appendix, Art. A·12).

60·21 FIG. 6·8(b) B. Same as Prob. 60·20, but construct a parabolic conic section. Use the same dimensions, except that angle A must now equal angle B. Locate the focus and directrix of the parabola (see Appendix, Art. A·16).

60·22 FIG. 6·8(c) A. Same as Prob. 60·20, but construct a hyperbolic conic section. Use the same cone proportions, but draw both nappes of the cone as shown. Angle A is 75°, and the cutting plane intersects the cone axis $1\frac{1}{4}$ in. above the vertex. Locate the foci and asymptotes of the hyperbola (see Appendix, Art. A·19).

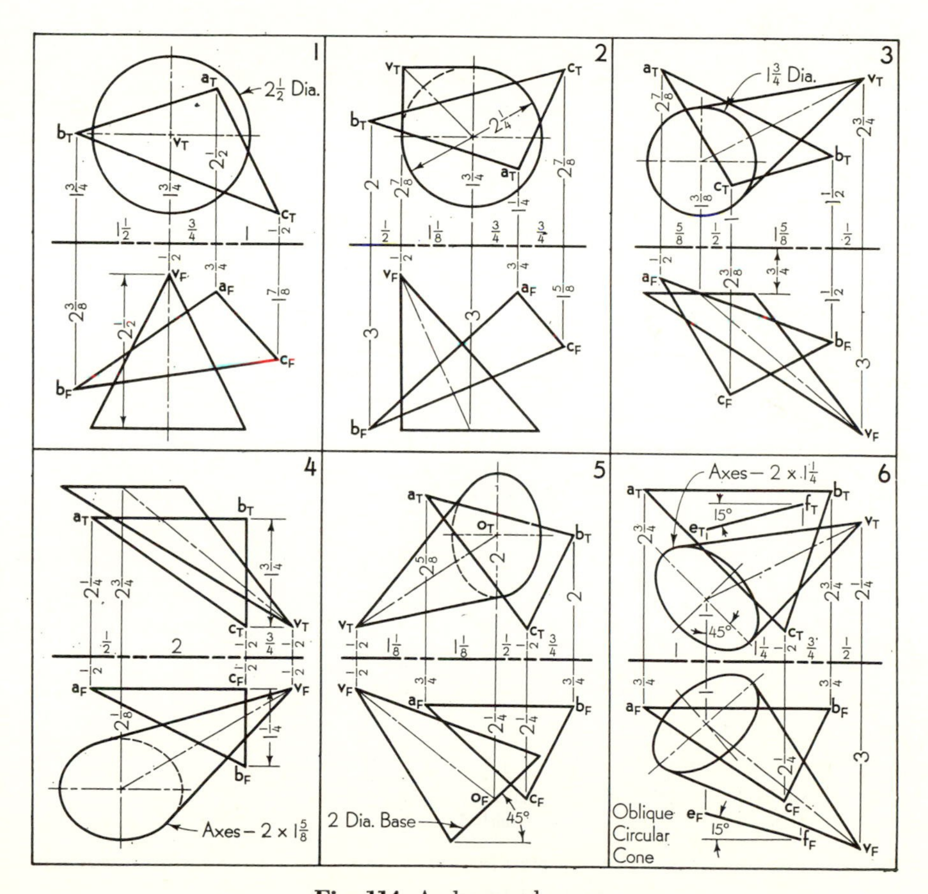

Fig. 114. A plane and a cone.

Group 61. Intersection of a line and a cone (Arts. 6·9, 6·10)

61·1–61·6 FIG. 112(1) to (6). In each problem locate the points P and Q where line MN intersects the cone, and show line MN in correct visibility. Points very close to an extreme element should be labeled (p_T^V or p_T^H, for example) to indicate whether they are *visible* or *hidden*. Use Layout A.

In Probs. 61·7 to 61·12 determine which of the three lines AB, AC, and BC intersect the cone. Locate and label all intersection points, and then show each of the three lines in correct visibility.

61·7 FIG. 114(1) A.
61·8 FIG. 114(2) A.
61·9 FIG. 114(3) A.
61·10 FIG. 114(4) A.
61·11 FIG. 114(5) A.
61·12 FIG. 114(6) C.

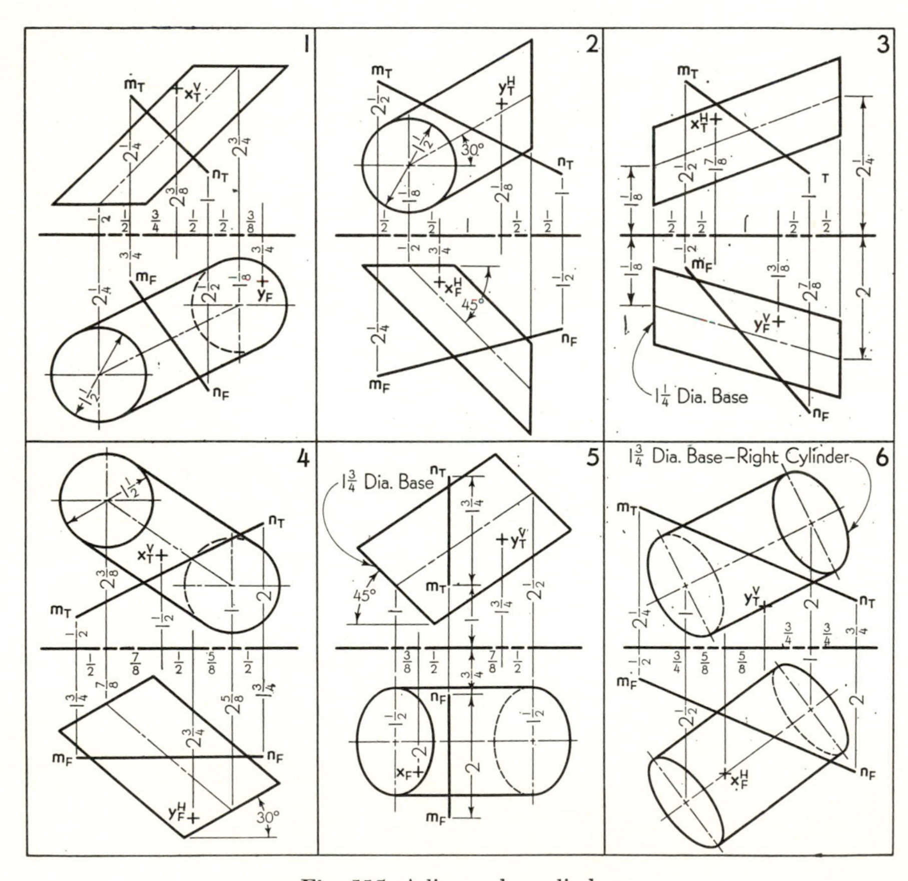

Fig. 115. A line and a cylinder.

Group 62. Location of a point on a cylinder (Arts. 6·11, 6·12)

62·1–62·6 FIG. 115(1) to (6). In each problem points X and Y lie on the surface of the cylinder, but only one view of each point is given. Locate each point in the other view, and show the cylinder element through the point. The notation x_T^V means that point X is *visible* in the top view; y_F^H means that Y is *hidden* in the front view. Use Layout A.

62·7 FIG. 116(1) A. Line AB is the axis of a right circular cylinder, 2 in. in diameter, with bases at A and B. Show the cylinder in the top and front views. Locate point X on the front side of the cylinder.

62·8 FIG. 116(2) A. Same as Prob. 62·7, but locate X on the upper side of the cylinder.

62·9 FIG. 116(3) B. Same as Prob. 62·7, but locate X on the left side of the cylinder.

62·10 FIG. 116(4) A. Same as Prob. 62·7, but locate X on the rear side of the cylinder.

Group 63. Intersection of a plane and a cylinder (Art. 6·13)

63·1–63·6 FIG. 115(1) to (6). In each problem find the line of intersection between the cylinder and a vertical plane that contains line MN. Use at least 12 elements, being sure to locate accurately those points which lie on extreme elements. Use Layout A, except for Prob. 63·5, which requires Layout B to show the intersection curve in true size.

Problems 63·7 to 63·11 are similar to Probs. 63·1 to 63·6, but use an edge-front plane that contains line MN, instead of a vertical plane. Use Layout A.

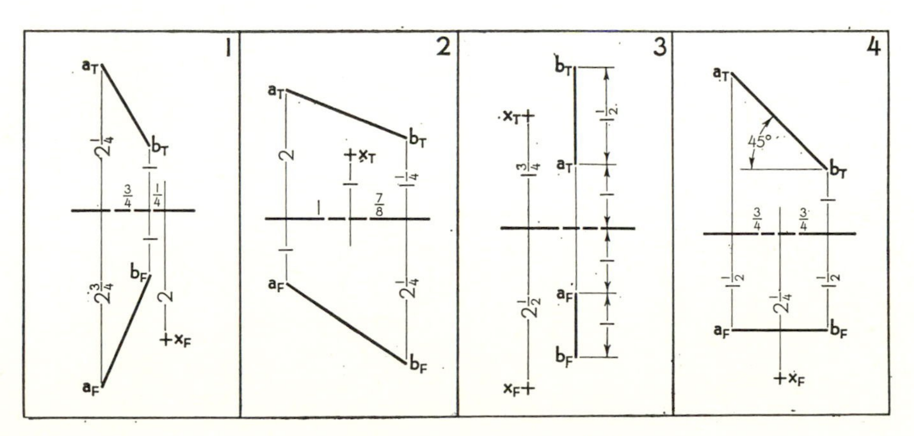

Fig. 116. Problems 62·7 to 62·10.

63·7 FIG. 115(1).
63·8 FIG. 115(2).
63·9 FIG. 115(3).
63·10 FIG. 115(4).
63·11 FIG. 115(6).

In Probs. 63·12 to 63·17 find the line of intersection between the cylinder and the plane *ABC*, assuming that the plane is indefinite in extent and cuts completely through the given cylinder. Show the curve in both the top and front views, locating accurately those points which lie on extreme elements in each view. Use either of the methods suggested in Art. 6·13 or that specified by the instructor.

63·12 FIG. 117(1) A.
63·13 FIG. 117(2) B.
63·14 FIG. 117(3) B.
63·15 FIG. 117(4) A.
63·16 FIG. 117(5) B.
63·17 FIG. 117(6) B.
63·18 FIG. 117(4) B. Extend the cylinder, and find its intersection with a horizontal plane passing through point *A*. (Disregard plane *ABC*.)

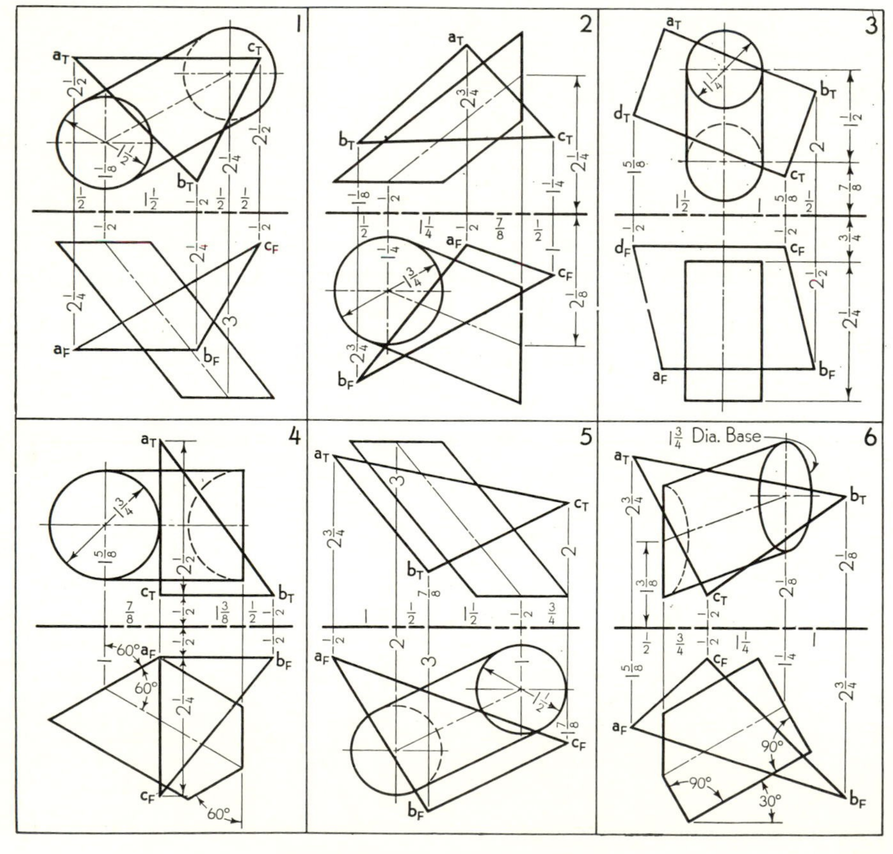

Fig. 117. A plane and a cylinder.

Group 64. Intersection of a plane and a right circular cylinder (Art. 6·14)

64·1–64·6 FIG. 118(1) to (6). In each problem a 1-in.-diameter cylinder, having *AB* as its axis, is cut off at *B* by plane *X*. Show the cylinder in both given views, and represent the end at *A* as a conventional break. Use Layout A, except on Probs. 64·5 and 64·6, where Layout B should be used.

In Probs. 64·7 to 64·9 find the line of intersection between a 1-in.-diameter cylinder, having *MN* as its axis, and the given plane. Show the cylinder and its intersection, each in correct visibility, in both given views. Represent the ends at *M* and *N* as conventional breaks. Use Layout B, full size, or Layout C, $1\frac{1}{2} \times$ full size.

64·7 FIG. 119(1). Plane *ABC*.
64·8 FIG. 119(2). Plane *ABC*.
64·9 FIG. 119(3). Plane *ABCD*.

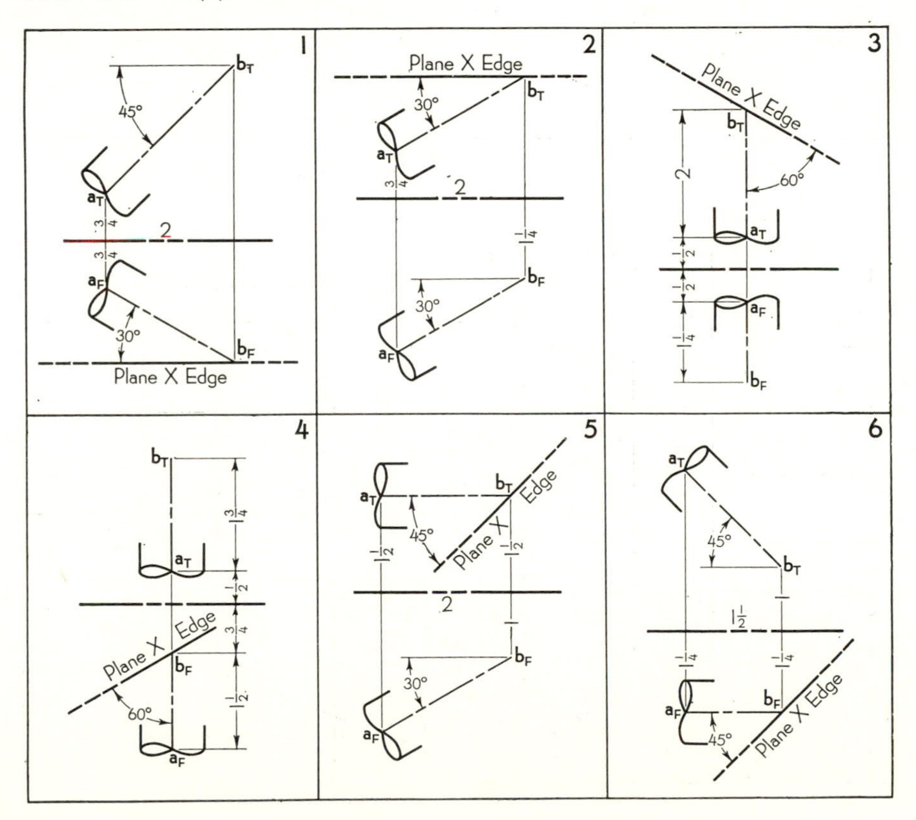

Fig. 118. Problems 64·1 to 64·6.

64·10 FIG. 120 C. Scale: ¾ in. = 1 ft. Line *MN* is the center line of a pipe, 18 in. in diameter, that intersects the side of the hopper at *M* and the floor at *N*. Find the true size of the openings that must be cut in hopper and floor, and show in the given views the section of pipe between the two openings.

64·11 FIG. 121 C. Scale: 1 in. = 1 ft. Line *MN* is the center line of a pipe, 12 in. in diameter, that intersects two vertical walls at *M* and *N*, respectively. Find the true size of the opening that must be cut in each wall, and show in the given views the section of pipe between the walls.

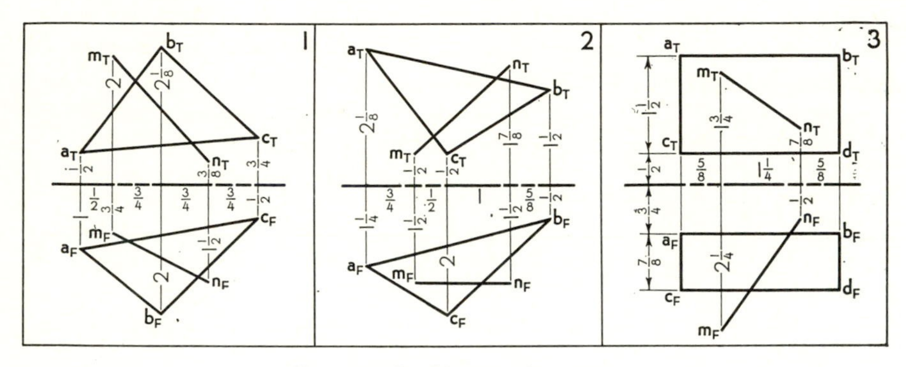

Fig. 119. Problems 64·7 to 64·9.

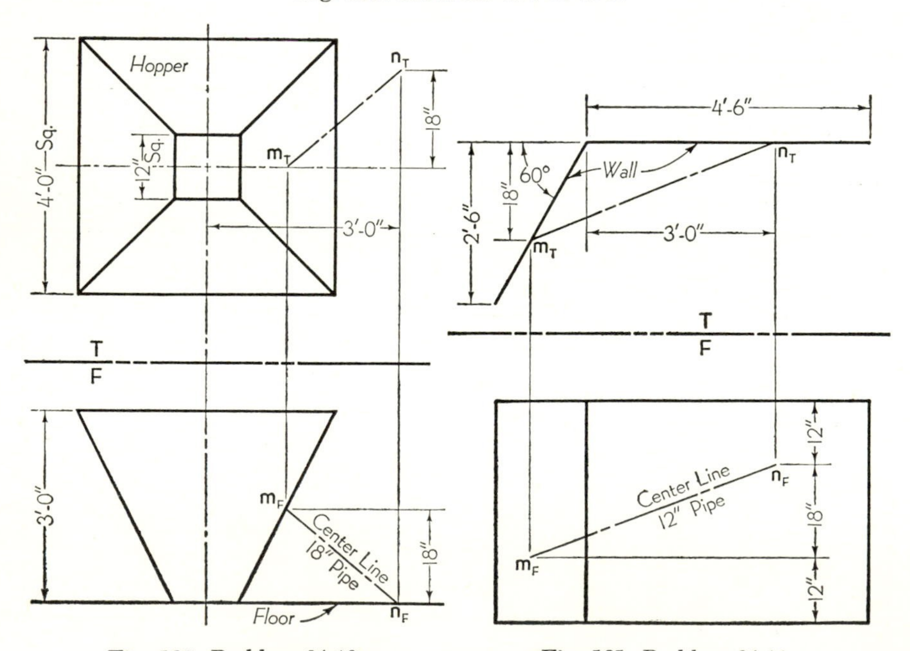

Fig. 120. Problem 64·10.

Fig. 121. Problem 64·11.

64·12 FIG. 122 C. Scale: full size. Two inclined rods support a ball as shown in the figure. Plot the path of the center of the ball as it rolls from a position directly over point A to the point where it drops between the rods. At what distance x from point A will the ball be in stable equilibrium, that is, having no tendency to roll either to the right or to the left? At what distance from point A will the ball drop through the rods? Place given views in upper right corner of sheet. (HINT: See third paragraph of Art. 6·11.)

ANS. 3⅜ in. 6⅜ in.

Group 65. Intersection of a line and a cylinder (Art. 6·15)

65·1–65·6 FIG. 115(1) to (6). In each problem locate the points P and Q where line MN intersects the cylinder, and show line MN in correct visibility. Points very close to an extreme element should be labeled (p_T^V or p_T^H, for example) to indicate whether they are *visible* or *hidden*. Use Layout A.

In Probs. 65·7 to 65·12 locate the points P and Q where the specified line intersects the cylinder, and show this line in correct visibility. (Disregard other lines.) See also statement of Probs. 65·1 to 65·6 above. Use Layout A.

65·7 FIG. 117(1). Line AB.
65·8 FIG. 117(2). Line BC.
65·9 FIG. 117(3). Line AC.
65·10 FIG. 117(4). Line AC.
65·11 FIG. 117(5). Line AC.
65·12 FIG. 117(6). Line AB.

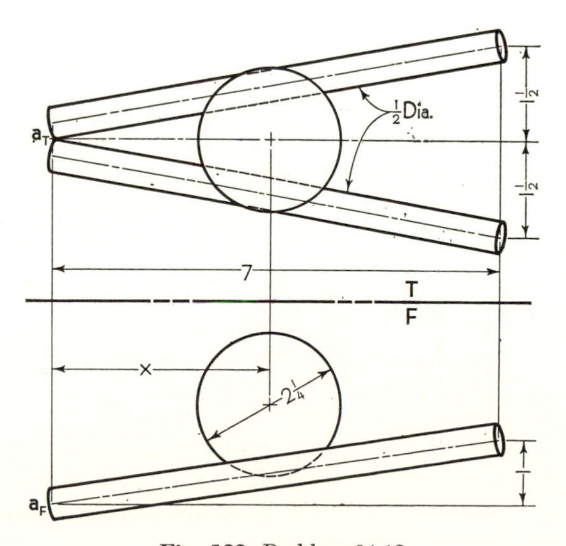

Fig. 122. Problem 64·12.

Group 66. Plane tangent to a cone through a given point on the cone (Art. 6·19)

66·1–66·6 FIG. 112(1) to (6). In each problem construct a plane tangent to the cone at point X on its surface. Construct a second plane tangent to the cone at point Y on its surface. State what lines define each plane. Use Layout A.

Group 67. Plane tangent to a cone through a given external point (Art. 6·20)

In each problem construct a plane tangent to the cone through the given point. Draw the element of tangency, and state what lines define the plane. If there are two solutions, show both planes. Use Layout A.

67·1 FIG. 112(1). Point N.
67·2 FIG. 112(2). Point N.
67·3 FIG. 112(3). Point N.
67·4 FIG. 112(4). Point M.
67·5 FIG. 112(5). Point N.
67·6 FIG. 112(6). Point M.
67·7 FIG. 114(1). Point C.
67·8 FIG. 114(2). Point C.
67·9 FIG. 114(3). Point B.
67·10 FIG. 114(4). Point C.
67·11 FIG. 114(5). Point C.
67·12 FIG. 114(6). Point C.

Group 68. Plane tangent to a cone and parallel to a given line (Art. 6·21)

In each problem construct a plane tangent to the cone and parallel to the given line. Draw the element of tangency, and state what lines define the plane. If there are two solutions, show both planes. Use Layout A.

68·1 FIG. 112(1). Line MN.
68·2 FIG. 112(2). Line MN.
68·3 FIG. 112(3). Line MN.
68·4 FIG. 112(4). Line MN.
68·5 FIG. 112(5). Line MN.
68·6 FIG. 112(6). Line MN.
68·7 FIG. 114(1). Line AC.
68·8 FIG. 114(2). Line AB.
68·9 FIG. 114(3). Line AC.
68·10 FIG. 114(4). Line AB.
68·11 FIG. 114(5). Line AB.
68·12 FIG. 114(6). Line EF.

Group 69. Plane tangent to a cylinder through a given point on the cylinder (Art. 6·22)

69·1–69·6 FIG. 115(1) to (6). In each problem construct a plane tangent to the cylinder at point X on its surface. Construct a second plane tangent to the cylinder at point Y on its surface. State what lines define each plane. Use Layout A.

Group 70. Plane tangent to a cylinder through a given external point (Art. 6·23)

In each problem construct a plane tangent to the cylinder through the given point. Draw the element of tangency, and state what lines define the plane. If there are two solutions, show both planes. Use Layout A.

70·1 FIG. 115(1). Point M.
70·2 FIG. 115(2). Point N.
70·3 FIG. 115(3). Point N.
70·4 FIG. 115(4). Point M.
70·5 FIG. 115(5). Point M.
70·6 FIG. 115(6). Point M.
70·7 FIG. 117(1). Point A.
70·8 FIG. 117(2). Point C.
70·9 FIG. 117(3). Point B.
70·10 FIG. 117(4). Point C.
70·11 FIG. 117(5). Point B.
70·12 FIG. 117(6). Point C.

Group 71. Plane tangent to a cylinder and parallel to a given line (Art. 6·24)

In each problem construct a plane tangent to the cylinder and parallel to the given line. Draw the element of tangency, and state what lines define the plane. If there are two solutions, show both planes. Use Layout A.

71·1 FIG. 115(1). Line MN.
71·2 FIG. 115(2). Line MN.
71·3 FIG. 115(3). Line MN.
71·4 FIG. 115(4). Line MN.
71·5 FIG. 115(5). Line MN.
71·6 FIG. 115(6). Line MN.
71·7 FIG. 117(1). Line BC.
71·8 FIG. 117(2). Line BC.
71·9 FIG. 117(3). Line BC.
71·10 FIG. 117(4). Line AC.
71·11 FIG. 117(5). Line AB.
71·12 FIG. 117(6). Line AB.

Group 72. Plane through a line making a given angle with a given plane (Art. 6·25)

72·1 FIG. 123(1) A. Line AB is one inclined edge of a pyramid with a horizontal triangular base (a tetrahedron). The vertex is at B, and A is one corner of the base. The two faces that intersect at edge AB both have 60° slope. The third face is perpendicular to AB. Draw the top and front views of the pyramid, and record the length of the sides of the base ACD. Solve, using only the given views.
ANS. $AC = AD$, 4.10 in. CD, 4.05 in.

72·2 FIG. 123(2) A. Line AB is one inclined edge of a pyramid whose horizontal base is a parallelogram. The vertex is at B, and A is one corner of the base. One of the faces intersecting at edge AB slopes 60°; the other face intersecting AB slopes 45°. Opposite faces have the same slope. Draw the top and front views of the pyramid, and record the length of the sides of the base. Solve, using only the given views. ANS. 1.45 in. and 2.50 in.

72·3 FIG. 123(3) A. Line AB is one inclined edge of a pyramid having a horizontal four-sided base $ADEF$. The vertex is at B, and A is one corner of the base. Point C lies on a second inclined edge BD adjacent to edge AB. One face adjacent to face ABD slopes 60°; the other face adjacent to ABD slopes 45°. The fourth face—that opposite face ABD—slopes 75° due south. Draw the top and front views of the pyramid, and record the length of the sides of the base. Solve, using only the given views.

ANS. AD, 2.12 in. DE, 2.32 in. EF, 3.10 in. FA, 1.76 in.

72·4 FIG. 123(4) A. The hexagonal inclined plane is the upper surface of a polyhedron whose six side faces all slope outward and downward at 60° slope. The base of the polyhedron is horizontal as shown. Draw the top and front views of the polyhedron. Solve, using only the given views. Is the polyhedron thus formed a truncated pyramid?

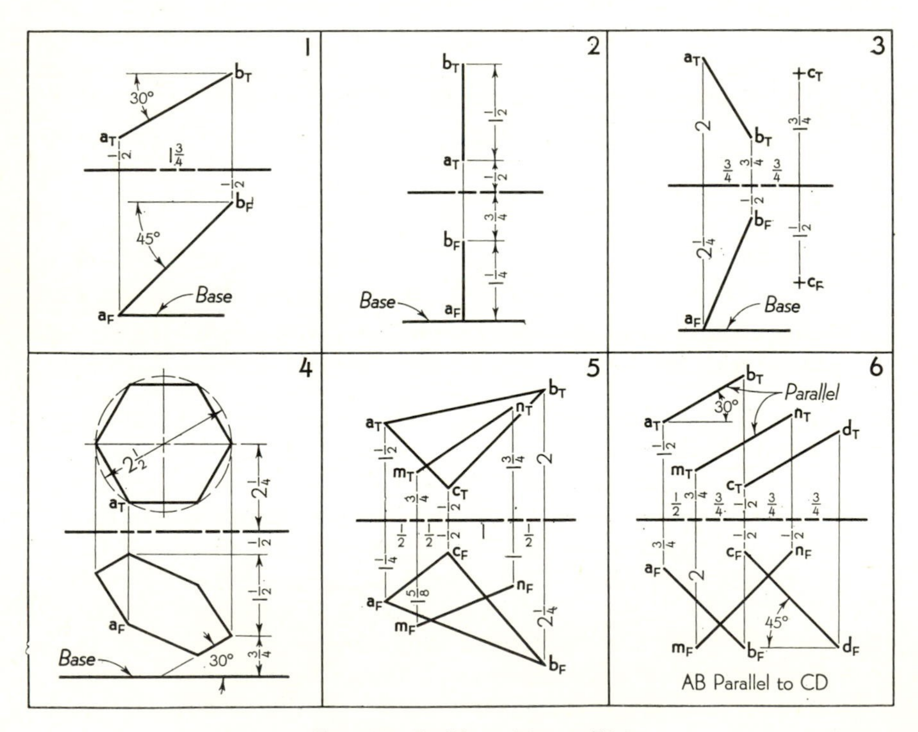

Fig. 123. Problems 72·1 to 72·6.

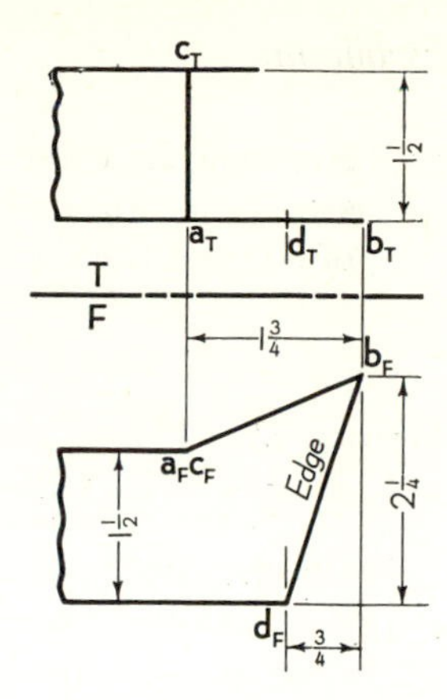

Fig. 124. Problem 72·7.

72·5 FIG. 123(5) C. Scale: 1½ × full size (B, full size). Construct two planes through line *MN* making an angle of 60° with plane *ABC*. Show in the given views the intersection of each plane with plane *ABC*. State the strike of each plane.

ANS. N 56° E and N 38° W.

72·6 FIG. 123(6). Same as Prob. 72·5, but making 75° with plane *ABCD*.

ANS. Due north and N 79° E.

72·7 FIG. 124 B. Scale: full size. The tip of a cutting tool is shown, but the views are incomplete. The top surface to the left of line *AC* is horizontal. The right-end surface appears as an edge in the front view as shown. Complete the views by constructing through line *AB* a plane *ABE* that slopes downward to the rear and makes 60° with the right-end surface. Draw the right-side view of the tool, and find the dihedral angle between plane *ABE* and the vertical front surface. ANS. 48°.

72·8 FIG. 125 B. Scale: full size. The tip of a cutting tool is shown, but the views are incomplete. The top surface to the left of line *AC* is horizontal. To the right of line *AC* the upper surface passes through line *AB*, slopes downward to the rear, and makes 60° with the front surface. The right-end surface passes through line *BD* and makes 75° with the front surface. Complete the given views, draw the right-side view, and find the dihedral angle between the two constructed planes. ANS. 73°.

72·9 FIG. 126 A. Scale: ½ in. = 1 ft. The base of a bay window is shown. The triangular face *ABC* has a slope of 45°, and the true

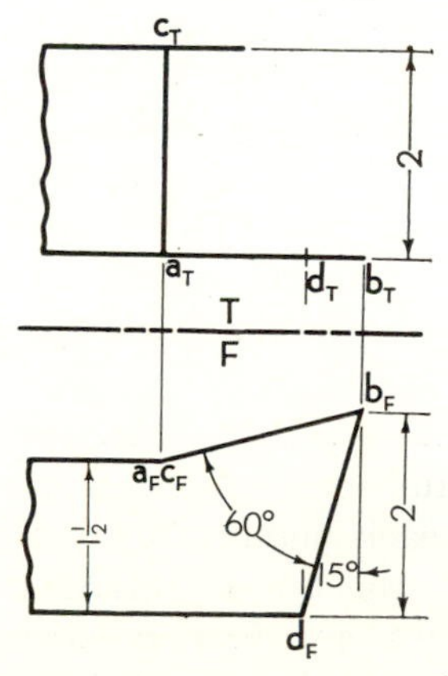

Fig. 125. Problem 72.8.

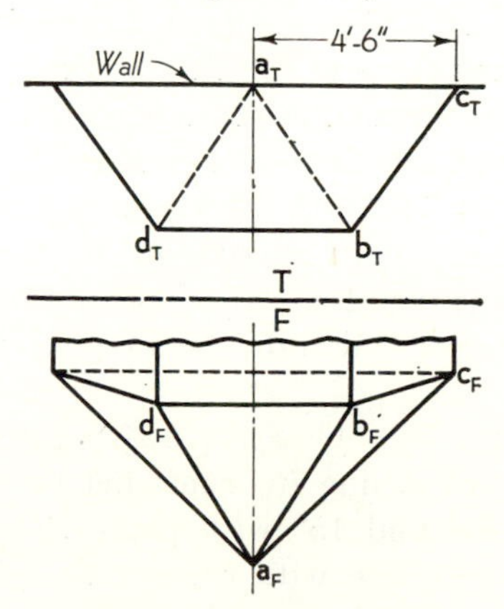

Fig. 126. Problem 72·9.

length of its three sides are *AB*, 5 ft; *AC*, 6 ft; and *BC*, 4 ft. Construct the top and front views as shown, and determine the length of *BD* and the slope of face *ABD*. Solve, using only the given views.
ANS. *BD*, 4 ft 2½ in. Slope, 46°.

Group 73. Plane through a point making given angles with two planes (Art. 6·26)

In Probs. 73·1 to 73·4 draw the top and front views of a point *X* that is 1½ in. above a horizontal plane and 1½ in. in front of a vertical plane. Construct a plane through point *X* making the specified angles with the horizontal and vertical planes, and show its intersection with each of these planes. Use Layout B, placing point *X* in the left half of the sheet and point *O*, the center of the cone-loci, in the right half.

73·1 Plane making 75° with the horizontal, 45° with the vertical, and sloping forward and downward to the right.

73·2 Plane making 60° with the horizontal, 50° with the vertical, and sloping backward and downward to the right.

73·3 Plane making 60° with the horizontal, 75° with the vertical, and sloping forward and downward to the left.

73·4 Plane making 45° with the horizontal, 60° with the vertical, and sloping backward and downward to the left.

73·5 B. A square stick is cut off obliquely so that the cut section makes angles of 75° and 45° with the adjacent faces of the stick. The longer sides of the section are 1½ in. long. Draw top and front views of the stick, and determine its dimensions. ANS. 1 in. square.

73·6 B. The upper right-front corner of a 2-in. cube is cut off by a plane that passes through the mid-point of the upper front edge. The plane makes angles of 120° and 105°, respectively, with the top and front faces of the cube. Draw top and front views of the cut cube, and determine the angle that the oblique surface makes with the right-side face of the cube. ANS. 145½°.

73·7 FIG. 76(4) B. Through the mid-point of line *BC* construct two planes, each making 45° with plane *ABC* and 30° with plane *BCD*. Show the line of intersection of each plane with each of the given planes, and state the strike of each plane.
ANS. N 10½° E, N 49½° E.

73·8 FIG. 76(1) C. Scale: 1½ × full size (B, full size). Through the mid-point of line *BC* construct two planes, each making 30° with plane *ABC* and 15° with plane *BCD*. Show the line of intersection of each plane with each of the given planes, and find the strike and dip of each plane. ANS. N 72° W, 27°. S 59° E, 51°.

Group 74. The helical convolute (Arts. 6·27 to 6·29)

In Probs. 74·1 to 74·4 draw the top and front views of one turn of the helical convolute generated by line *AB* moving tangent to the given helix. The helix axis is vertical. Use 16 elements, and show the convolute surface in correct visibility, assuming the helix cylinder to be opaque. Use Layout A.

74·1 Right-hand helix, 1½ in. diameter, 3 in. lead, starting at the front center of the lower base. Line *AB*, 1½ in. long, generates the lower nappe only.

74·2 Left-hand helix, 2 in. diameter, 3 in. lead, starting at the front center of the lower base. Line *AB*, 2 in. long, generates the lower nappe only.

74·3 Right-hand helix, 1½ in. diameter, 4 in. lead, starting at the left side of the lower base. Line *AB*, 1½ in. long, generates the lower nappe only.

74·4 Left-hand helix, 1 in. diameter, 3 in. lead, starting at the right side of the lower base. Line *AB*, 2 in. long, generates the upper nappe only.

In Probs. 74·5 to 74·8 draw the top and front views of one turn of the helical convolute generated by a line moving tangent to the given helix. The helix axis is vertical, and the generating line extends downward to the horizontal plane of the helix base. Use 16 elements, and show the convolute surface in correct visibility, assuming the helix cylinder to be opaque. Show the line of intersection of the convolute and a horizontal plane 1½ in. above the base plane. Use Layout A.

74·5 Right-hand helix, 1¼ in. diameter, 3 in. lead, starting at the front center of the lower base. (Place helix axis 2¾ in. from right border of sheet.)

74·6 Left-hand helix, 1 in. diameter, 4 in. lead, starting at the front center of the lower base. (Place helix axis 3 in. from left border of sheet.)

74·7 Right-hand helix, 1½ in. diameter, 3 in. lead, starting at the rear center of the lower base. (Place helix axis 2⅝ in. from left border of sheet.)

74·8 Left-hand helix, 1¼ in. diameter, 4 in. lead, starting at the rear center of the lower base. (Place helix axis 3¾ in. from right border of sheet.)

Group 75. Tangent-plane convolutes (Arts. 6·30, 6·31)

In each problem show in the given top and front views a series of straight-line elements on the convolute surface that joins the two curves. Use 24 elements on the closed surfaces and 12 or 16 on those of lesser extent (such as that of Prob. 75·2). Elements should be equally spaced along one of the two curves and arranged symmetrically wherever possible. Use Layout A, except as noted.

75·1 FIG. 127(1). Stoker distributing hood.
75·2 FIG. 127(2). Automobile hood.
75·3 FIG. 127(3). Pipe connection.
75·4 FIG. 127(4). Offset pipe connection.
75·5 FIG. 127(5). Elliptical pipe connection.
75·6 FIG. 127(6). Elliptical elbow.
75·7 FIG. 127(7). Oil-drum hinged cover (surface connects circular arcs *ABC* and *ADC*).
75·8 FIG. 127(8). Parabolic air scoop (surface connects parabola *ABC* and semicircle *ADC*).
75·9 FIG. 127(9). Flared connection from a circular pipe to a cylindrical drum (the curve that appears as an ellipse in the top view is a double-curved line lying on the surface of the 6-in.-diameter drum).
75·10 FIG. 129(1). Ventilating hood. (Place given views in upper left corner of the sheet.)
75·11 FIG. 129(2). Eccentric pipe connection.
75·12 FIG. 129(3) C. Eccentric pipe connection. Draw top, front, and right-side views. (Allow 2½ in. between given views.)

Group 76. General warped surfaces—three line directrices (Arts. 7·1 to 7·4)

76·1–76·6 FIG. 128(1) to (6). In each problem draw top, front, and right-side views of the warped surface having the three given lines as directrices. Show 13 straight-line elements, spaced equally along one of the directrices. The hidden part of elements should be drawn with very light dash lines. Determine the intersection line of the given cutting plane and the warped surface. Use Layout B.

Group 77. Warped cones (Arts. 7·5, 7·6)

In Probs. 77·1 to 77·4 draw the top, front, and right-side views of the warped cone having the specified line as its straight-line directrix. Show 24 straight-line elements, spaced equally and symmetrically along one of

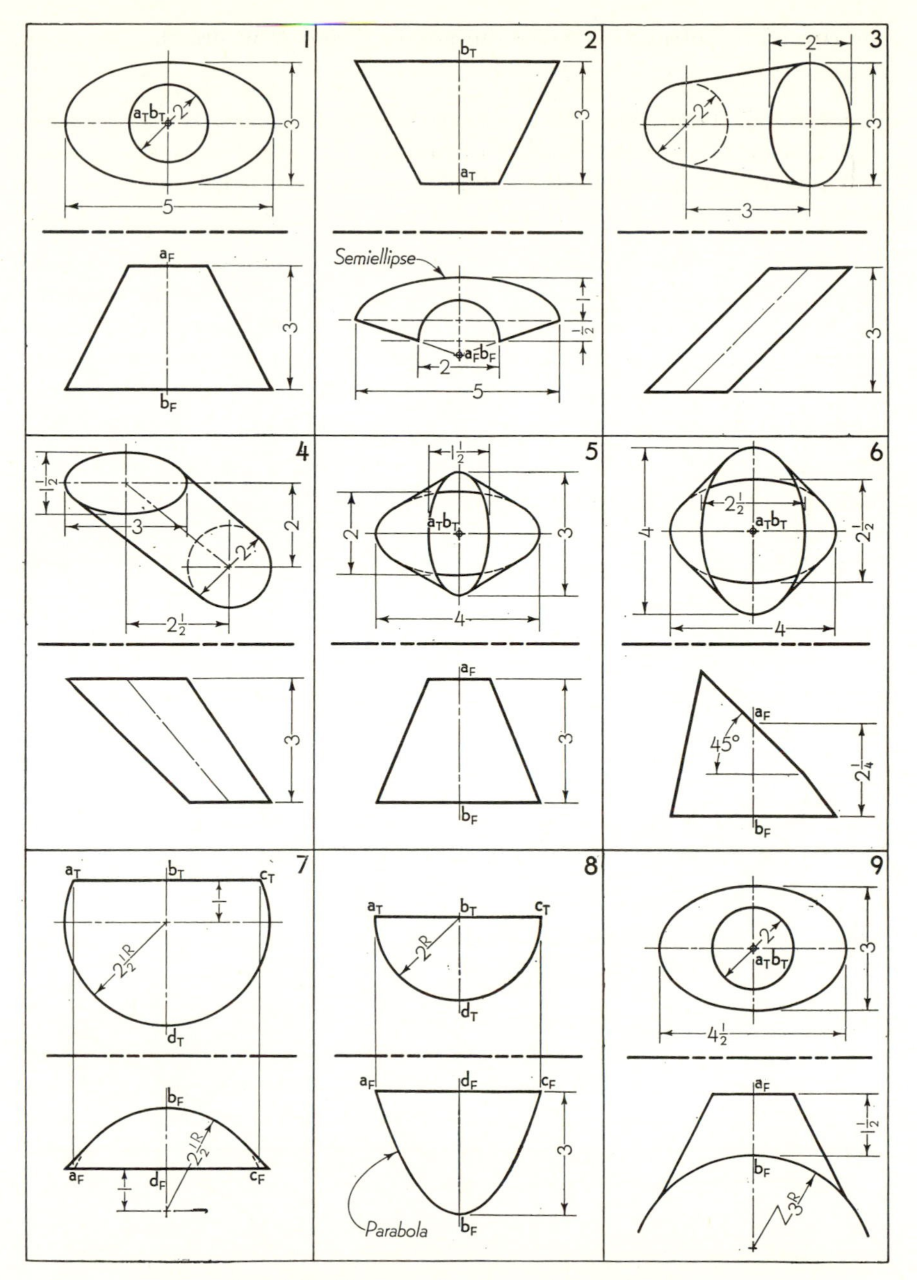

Fig. 127. Tangent-plane convolute surfaces.

the curved directrices. Locate the points P and Q where the line that begins at M intersects the surface.

77·1 FIG. 129(1) B. Axis AB as directrix.
77·2 FIG. 129(2) B. Element CD as directrix.
77·3 FIG. 129(2) A. Axis AB as directrix. Omit side view.
77·4 FIG. 129(3) B. Axis AB as directrix.

In Probs. 77·5 to 77·9 draw the top and front views of the surface shown. Treat the surface as a warped cone, using axis AB as the straight-line directrix. Show 24 straight-line elements spaced equally and symmetrically along one of the curved directrices. Intersect the surface with two cutting planes perpendicular to AB and spaced to divide AB into three equal parts, and draw the curves of intersection. Use Layout A.

77·5 FIG. 127(1). **77·7** FIG. 127(5). **77·9** FIG. 127(9).
77·6 FIG. 127(2). **77·8** FIG. 127(6).

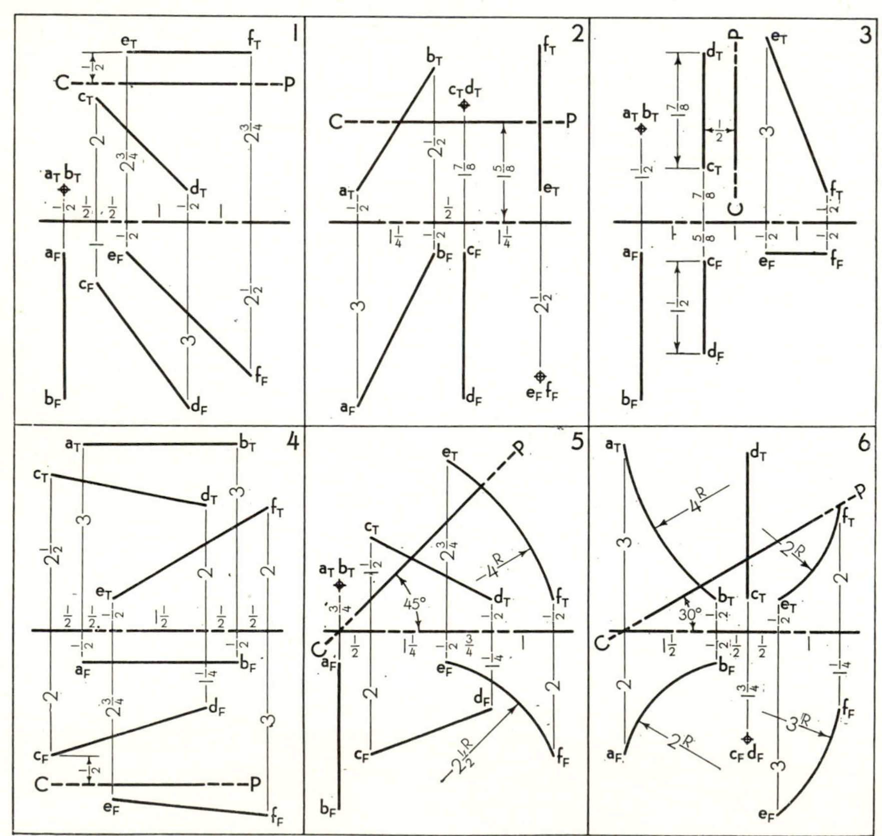

Fig. 128. Problems in Group 76.

Group 78. Cow's horns (Arts. 7·7, 7·8)

78·1–78·3 FIG. 130(1) to (3). Draw top, front, and right-side views of the cow's-horn surface having the two given curves as its directrices. Show 24 straight-line elements uniformly spaced. Hidden elements may be omitted in the side view. Points X and Y lie on the surface; locate each point in the other two views. Use Layout B.

78·4 B. Scale: $\frac{1}{16}$ in. = 1 ft. A railroad running north and south passes over a level highway running N 60° W on a cow's-horn arch. The openings in the arch are semicircles of 24 ft radius in planes parallel to the railroad, and 40 ft apart. The pavement is 30 ft wide, and

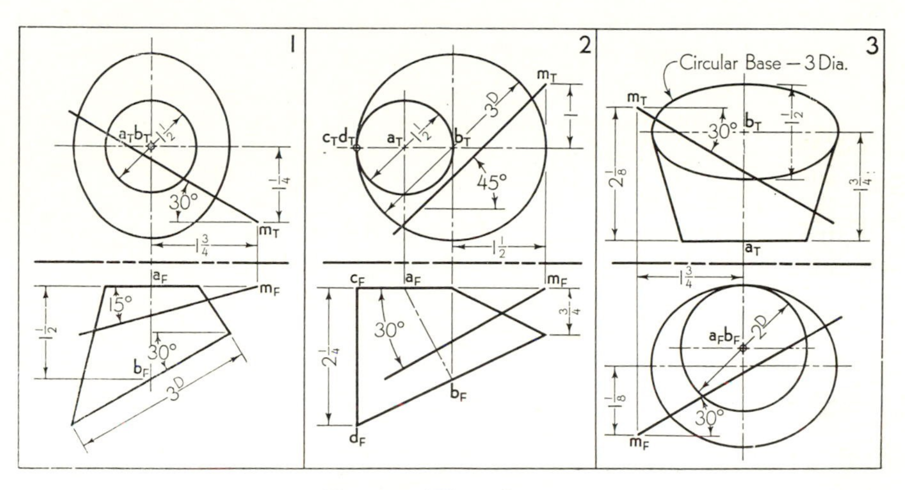

Fig. 129. Warped cones.

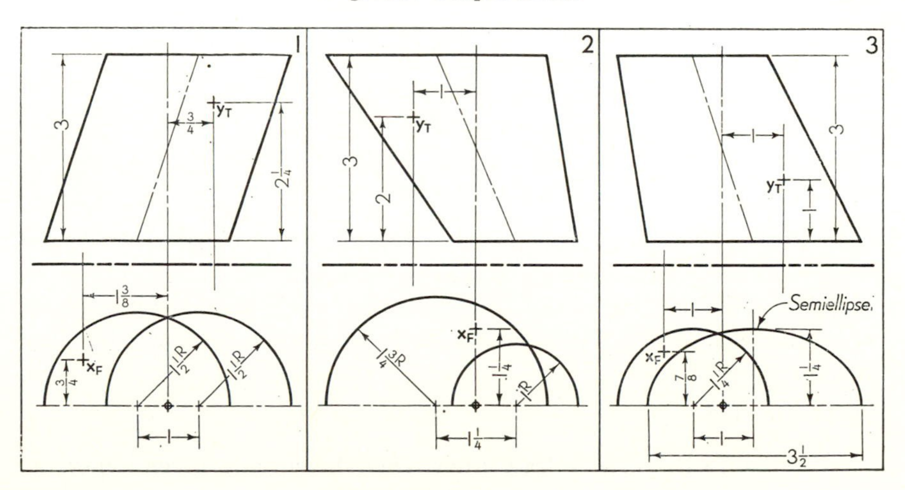

Fig. 130. Cow's horns.

its center line (the "white line") passes through the centers of the two openings. Sidewalks on each side of the pavement are at the same level as the center of the highway, but the pavement is cambered (arched) to make curbings 1 ft high at the sides. Draw a top view and three elevation views of the arch, one elevation looking north, one looking west, and one looking N 60° W. Use 24 elements. What is the minimum height clearance under the arch for vehicles?
ANS. 15 ft 6 in.

Group 79. Hyperbolic paraboloids (Arts. 7·9 to 7·13)

79·1–79·4 FIG. 131(1) to (4). Draw top, front, and right-side views of the hyperbolic paraboloid having the plane director shown and lines *AB* and *CD* as the line directrices. Show 13 equally spaced elements; draw the hidden part of elements with very light dash lines. Locate the point (or points) where line *MN* intersects the surface. Use Layout B.

In Probs. 79·5 to 79·13 draw the top and front views of the hyperbolic paraboloid having lines *AB* and *CD* as line directrices and having the specified lines as the first and last of 13 equally spaced elements. Draw an auxiliary view in which the elements appear parallel. When line *MN* is given, locate its intersection with the surface; when points *X* and *Y* are given, locate these points on the surface. Use Layout B, except as noted.

79·5 FIG. 131(1). Elements *AD* and *BC*.
79·6 FIG. 131(2). Elements *AC* and *BD*.
79·7 FIG. 131(3). Elements *AD* and *BC*.
79·8 FIG. 131(4). Elements *AD* and *BC*.
79·9 FIG. 131(5). Elements *AD* and *BC*.
79·10 FIG. 131(5) A. Elements *AC* and *BD*.
79·11 FIG. 131(6). Elements *AD* and *BC*.
79·12 FIG. 131(6) A. Elements *AC* and *BD*.
79·13 FIG. 131(7). Elements *AD* and *BC*.

In Probs. 79·14 and 79·15 draw the top and front views of the hyperbolic paraboloid having lines *AB* and *CD* as line directrices and lines *AC* and *BD* as the first and last of 13 equally spaced elements. Find the line of intersection of each cutting plane with the warped surface. Use Layout A.

79·14 FIG. 131(8). **79·15** FIG. 131(9).

In Probs. 79·16 to 79·18 draw a new view showing the given hyperbolic paraboloid in *standard position*, that is, with line directrices *AB* and *CD*

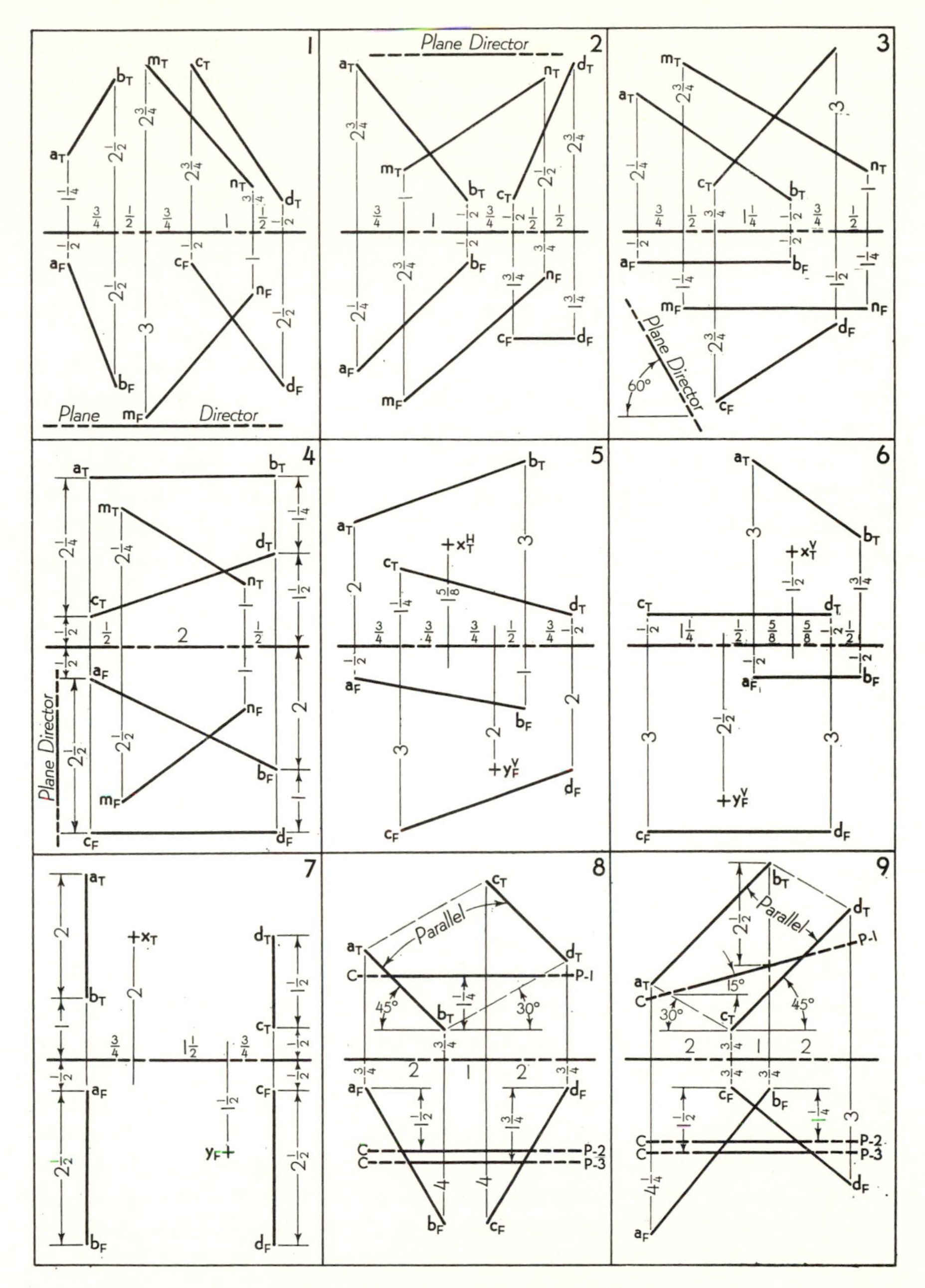

Fig. 131. Hyperbolic paraboloids.

appearing parallel and of equal length (as in the top view of Fig. 7·8). Use seven equally spaced elements, and show them in all views. Disregard line *MN* and the given plane director. Use Layout C.

79·16 FIG. 131(1). First and last elements are *AD* and *BC*.
79·17 FIG. 131(2). First and last elements are *AC* and *BD*.
79·18 FIG. 131(3). First and last elements are *AD* and *BC*.

Group 80. Conoids (Arts. 7·14, 7·16)

80·1–80·6 FIG. 132(1) to (6). Draw top, front, and right-side views of the conoid having the line *AB*, the curve, and the given plane director as the three directrices. Show 24 straight-line elements spaced equally and symmetrically along the curved directrix. Determine the intersection line of the given cutting plane and the conoid. Use Layout B.

80·7 FIG. 133(2) A. Substitute for the semicircle of 2 in. radius a rectangle 2 in. high by 4 in. wide. A surface connecting the curve and rectangle can be designed by using one plane triangle and two conoids. Draw the top and front views of this surface, using 12 elements in each conoid.

80·8 FIG. 133(5) B. Substitute for the lower circle a square whose sides are equal to the diameter of the replaced circle. A transition piece connecting the square and the upper circle can be designed by using two plane triangles and two conoids. Draw top, front, and right-side views of this surface, using 12 elements in each conoid.

80·9 A. Scale: 1 in. = 1 ft. A transition piece must be designed to connect a 4-ft-diameter opening to a 2-ft-square opening. The square and circle both lie in horizontal planes, and the square is 3½ ft directly above the center of the circle. Draw top and front views of the transition, designing it as a surface composed of four conoids and four cones. Show eight elements in each conoid and four elements in each cone.

80·10 B. Scale: ¼ in. = 1 ft. The sheet-metal roof of a subway entrance slopes downward from the arched entrance opening to the pavement. The entrance is 12 ft wide and 12 ft high, the upper half of the opening being a semicircular arch of 6 ft radius. The roof slopes downward for a horizontal distance of 20 ft to join the flat pavement at the lower end. Design the roof as a conoid, using about 12 elements. Draw top, front, and right-side views, and determine the shape of a vertical cross section 10 ft from the entrance.

Group 81. Cylindroids (Arts. 7·15, 7·16)

In each problem draw the top and front views of the cylindroid having the two curves and the given plane director as the three directrices. Show 24 straight-line elements spaced equally and symmetrically along one of the curved directrices. In Probs. 81·1 to 81·3 the line of intersection of the given cutting plane is also required; in Probs. 81·4 to 81·6 a right-side view is also required.

81·1 FIG. 133(1) A.
81·2 FIG. 133(2) A.
81·3 FIG. 133(3) A.
81·4 FIG. 133(4) B.
81·5 FIG. 133(5) B.
81·6 FIG. 133(6) B. (See Appendix, Art. A·9.)

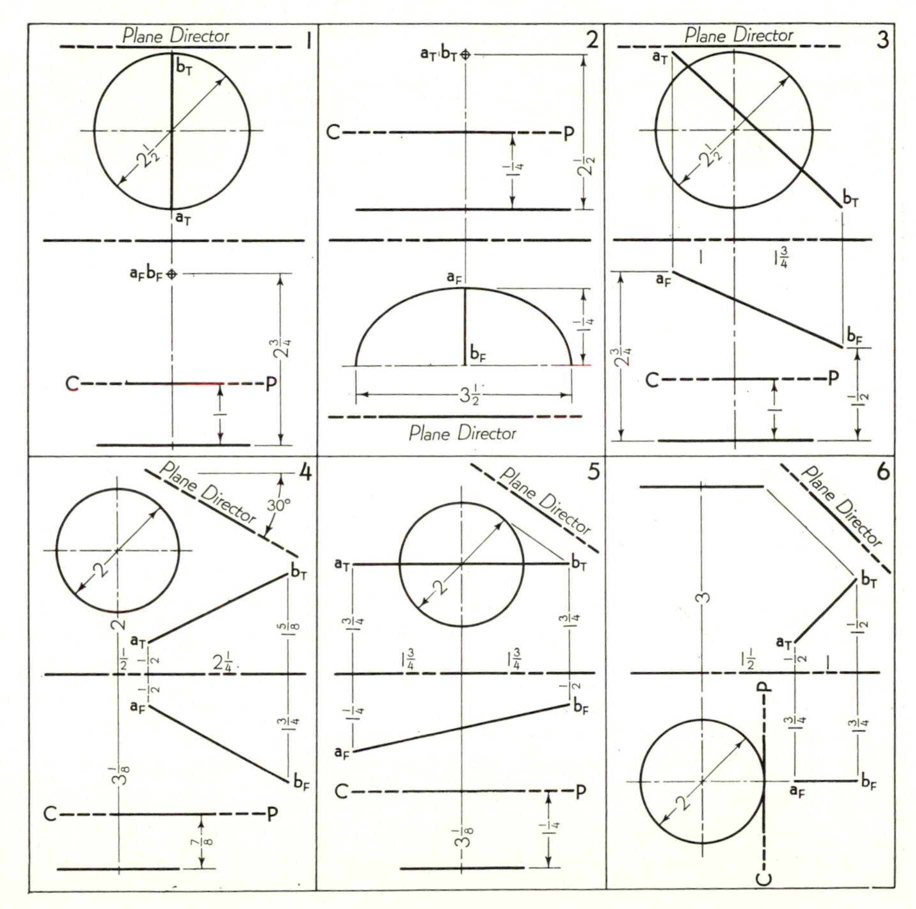

Fig. 132. Conoids.

Group 82. Helicoids (Arts. 7·17 to 7·19)

82·1 A. A right helicoid has a vertical axis as its straight-line directrix. The helical directrix is right-hand, 3½ in. in diameter, and has a lead of 4½ in. The helicoidal surface extends from the given helix to an inner solid cylinder 1 in. in diameter. Draw top and front views of one turn of the helicoid, starting on the lower left side. Use 24 elements.

82·2 A. A right helicoid has a vertical axis as its straight-line directrix. The helical directrix is left-hand, 4½ in. in diameter, and has a lead of 3½ in. The helicoidal surface extends from the given helix to an inner solid cylinder 1½ in. in diameter. Draw top and front views of one turn of the helicoid, starting at the lower front center. Use 24 elements.

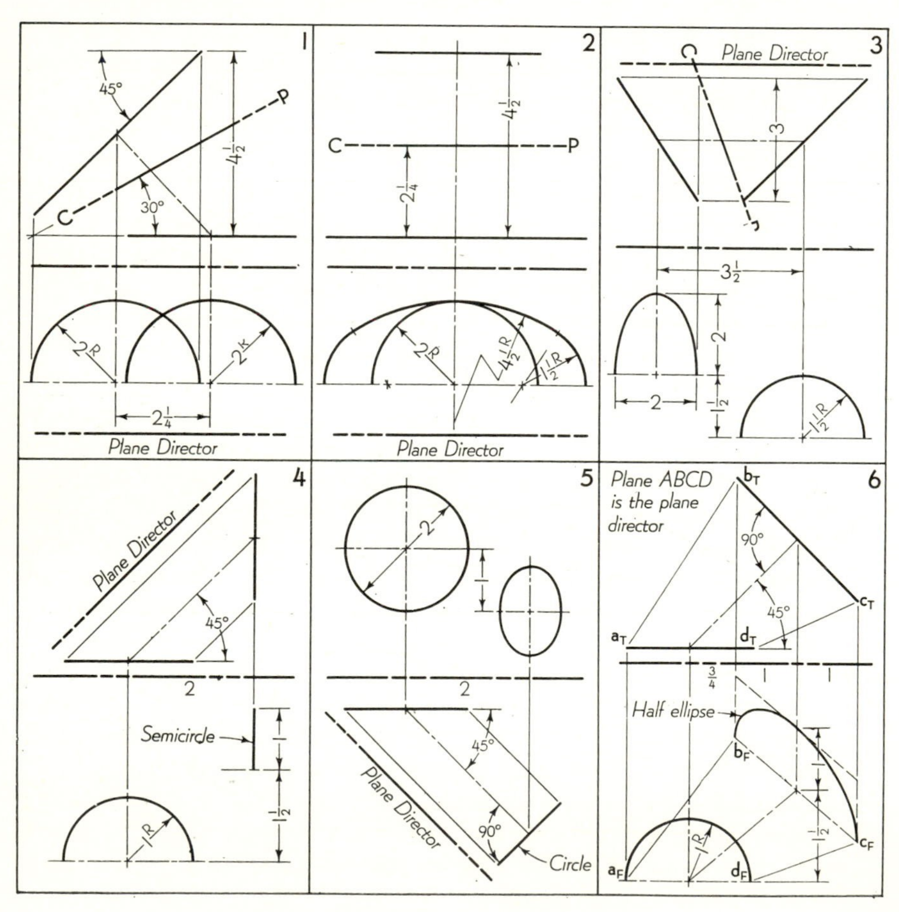

Fig. 133. Cylindroids.

82·3 B. A screw 3½ in. in diameter has a right-hand single square thread of 1½ in. lead [see Fig. 7·15(*a*)]. Draw front and side (end) views of a 4½-in. length of the screw. Omit hidden lines.

82·4 B. Using the data of Prob. 82·3 draw in section a cylindrical nut 4½ in. long and 5 in. in diameter. To save space only half the end view need be drawn. Omit hidden lines, and note that the inner helices should be drawn first.

82·5 B. A coil spring, 4 in. outside diameter, has a lead of 2 in. and is left-hand-wound from 1-in.-square stock [see Fig. 7·15(*c*)]. Draw front and side (end) views of two turns of the spring. Omit hidden lines.

82·6 B. Scale: 3 in. = 1 ft. A horizontal screw conveyor is to be designed capable of delivering 6,000 cu ft of grain per hour at 100 rpm. Outside diameter of the screw is 12 in., and the shaft diameter is 2 in. Calculate, to the nearest ¼ in., the lead of the screw, and neatly letter your calculations on the drawing. Draw front and end views of a 2-ft length of the conveyor screw. Omit hidden lines.

Ans. Lead, 1 ft 3¾ in.

82·7 A. Scale: ½ in. = 1 ft. A right-hand spiral chute is to be designed to carry packages from a second to a first floor, a distance of 18 ft, in three complete turns [see Fig. 7·15(*e*)]. The outside diameter is 6 ft, the core diameter 2 ft, and the guard sheet at the outer edge is 12 in. high. On the first floor, delivery of packages is to be toward the front. Draw top and front views of the lower 1½ turns of the chute. What is the slope of the chute at the inner and outer edges?

Ans. 43°40′, 17°40′.

82·8 A. Scale: ½ in. = 1 ft. An open semicircular stairway has an outside radius of 6 ft, an inside radius of 2 ft 6 in., and ascends clockwise. The height from floor to floor is 8 ft 9 in. There are 14 steps, hence 15 risers of 7 in. height. The soffit (ceiling under the steps) is a helicoidal surface 8 in. below the risers. Draw top and front views of the stairway, omitting hidden lines. The front view should be taken looking into the open stair well. [There is no central column as in Fig. 7·15(*f*).]

82·9 A. A right helicoid has a vertical axis, and its helical directrix is 3½ in. in diameter with a lead of 4 in. The generating line remains tangent to a 1½-in.-diameter inner cylinder, extends to the outer helix cylinder on both sides of the point of tangency, and rotates counterclockwise in moving downward. Draw top and front views of one turn of the helicoid, starting with the generating line at the top front-center position. Use 24 elements, and show hidden lines in both top and front views.

82·10 A. A right helicoid has a vertical axis, and its helical directrix is 4 in. in diameter with a lead of 4 in. The generating line remains tangent to a $1\frac{1}{4}$-in.-diameter inner cylinder, extends to the outer helix on only one side of the point of tangency, and rotates clockwise in moving downward. Draw top and front views of one turn of the helicoid, starting with the generating line at the top rear position and extending to the left. Use 24 elements, and omit the inner cylinder in the front view.

82·11 A. A right-hand twist drill 2 in. in diameter is made of flat bar stock $\frac{1}{2}$ in. thick [see Fig. 7·15(*d*)]. The *spiral* angle (angle of helix with *axis*) is 50°, and the *lip* angle (angle of cutting edge with axis) is 60°. By graphical construction determine the lead, and then draw front and top (end) views of the drill showing the point and about 6 in. below the point. Disregard clearance on the cutting edge, and omit hidden lines.

82·12 A. An oblique helicoid has a vertical axis as its straight-line directrix. The helical directrix is right-hand, 3 in. in diameter, and has a lead of 4 in. The generating line slopes downward at 30° from the given helix to an inner cylinder 1 in. in diameter. Draw top and front views of one turn of the helicoid, starting on the lower left side. Use 24 elements.

82·13 A. An oblique helicoid has a vertical axis as its straight-line directrix. The helical directrix is left-hand, 4 in. in diameter, and has a lead of 3 in. The generating line slopes upward at 45° from the given helix to an inner cylinder $1\frac{1}{2}$ in. in diameter. Draw top and front views of one turn of the helicoid, starting on the lower right side. Use 24 elements, and omit the inner cylinder in the front view.

82·14 B. A screw 5 in. in diameter has a right-hand single 60° sharp V thread of $1\frac{1}{2}$ in. lead [see Fig. 7·15(*b*)]. Draw front and side (end) views of a $4\frac{1}{2}$-in. length of the screw. Only half the end view need be drawn, and hidden lines should be omitted. Show the intersection of the thread surface with the end plane of the screw.

82·15 B. Same as Prob. 82·14, except that the screw has a double thread. Such a screw has two separate V threads each $1\frac{1}{2}$ in. wide, but the lead of each thread is 3 in.

82·16 FIG. 134 C. Scale: 1 in. = 1 ft. The hub and one blade of a right-hand screw propeller are shown. The face of the blade is a right helicoid of constant pitch from root to tip. The propeller diameter is 16 ft, and the pitch/diameter ratio is 1.25; hence the pitch (lead) is 20 ft. Disregarding the thickness of the blade, complete the front view and draw a right-side view of the hub and face of the blade. Draw elements of the helicoid at $1\frac{1}{2}$° intervals, and note that the blade is symmetrical about element AB.

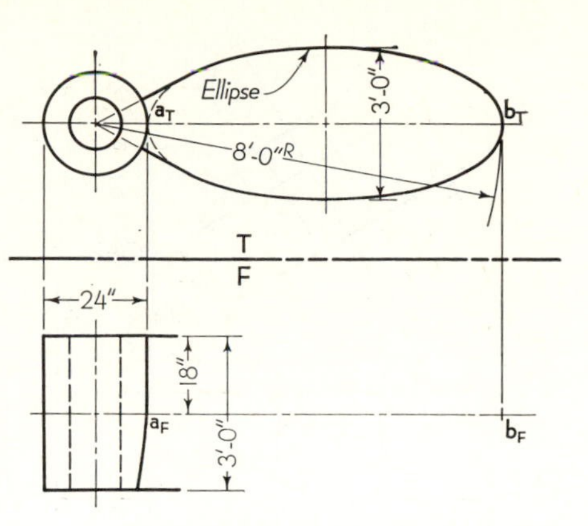

Fig. 134. Problem 82·16.

Group 83. Hyperboloids of revolution (Arts. 7·20, 7·21)

83·1–83·3 FIG. 135(1) to (3). In each problem line AB is the axis, and line CD is the generatrix of a hyperboloid of revolution. Show 24 equally spaced elements in the top and front views. Show hidden elements in both views, but draw them lightly. Point X is on the surface; locate it in the other view. Use Layout A.

83·4 A. A hyperboloid of revolution with axis vertical is 3½ in. high. Its diameter at the top is 2½ in. and at the bottom 4½ in. Its smallest diameter is 1½ in. Draw top and front views of the surface, using 24 elements of either generation. Omit hidden lines in the front view, but show them in the top view.

83·5 A. A hyperboloid of revolution with axis vertical is 4 in. high. Its diameter at the top is 4 in. and at the bottom 3 in. At 1½ in. below the top circle its diameter is 2 in. Using these three circles as directrices, locate one straight-line element that touches all three circles. Find the diameter and location of the gorge circle. Construct the hyperbolic contour curves in the front view by locating a series of points like point M in Fig. 7·16(a).

ANS. Gorge circle, 1.48 in. diameter; 2.34 in. below top.

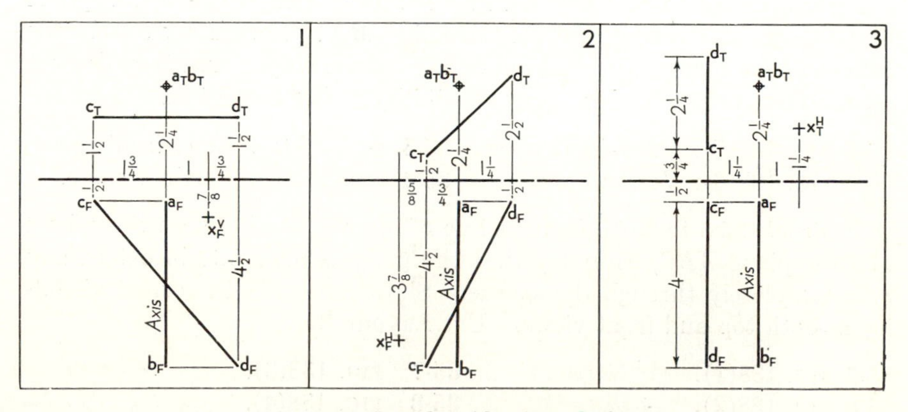

Fig. 135. Hyperboloids of revolution.

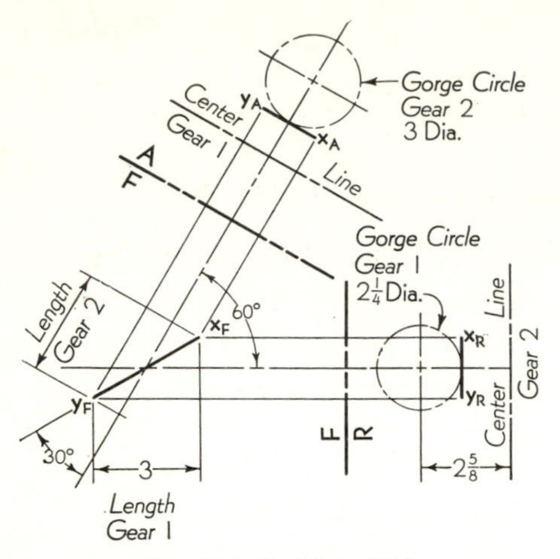

Fig. 136. Problem 83·6.

83·6 FIG. 136 C. Scale: full size (B, ⅔ size). The pitch surfaces of two mating skew gears are generated by the common element XY revolving about each gear center line. Show 12 equally spaced elements on gear 1 and 16 on gear 2. Draw both gears completely in the front and right-side views, but use view A only for construction. Show all hidden lines and elements, assuming that the hyperboloids are solids closed on the ends.

Group 84. Location of a point on a sphere (Arts. 8·3 to 8·5)

84·1–84·6 FIG. 137(1) to (6). In each problem points X, Y, and Z lie on the surface of the sphere, but only one view of each point is given. Locate each point in the other view, and indicate its visibility. Use Layout A.

84·7 FIG. 137(2) A. Using only the given views, find the shortest great-circle distance between points X and Y on the surface of the sphere. ANS. 1.99 in.

84·8 FIG. 137(5) A. Same as Prob. 84·7. ANS. 3.04 in.

Group 85. Intersection of a plane and a sphere (Art. 8·6)

In Probs. 85·1 to 85·4 find the line of intersection between the sphere and the specified plane that contains line MN. Locate accurately the tangent points and major and minor axes of the ellipse, and then draw it by the trammel method in correct visibility. Use Layout A.

85·1 FIG. 137(2). Edge-front plane.
85·2 FIG. 137(3). Vertical plane.
85·3 FIG. 137(5). Edge-front plane.
85·4 FIG. 137(6). Vertical plane.

In Probs. 85·5 to 85·10 find the line of intersection between the sphere and the plane ABC, assuming that the plane is indefinite in extent and cuts completely through the sphere. Show the curve in correct visibility in both top and front views. Use Layout B.

85·5 FIG. 138(1).
85·6 FIG. 138(2).
85·7 FIG. 138(3).
85·8 FIG. 138(4).

85·9 FIG. 138(5).

85·10 FIG. 138(6).

Group 86. Intersection of a line and a sphere (Art. 8·7)

86·1–86·6 FIG. 137(1) to (6). In each problem locate the points P and Q where line MN intersects the sphere, and show line MN in correct visibility. Label points to indicate visibility. Use Layout A.

In Probs. 86·7 to 86·12 determine which of the three lines AB, AC, and BC intersect the sphere. Locate and label all intersection points, and then show each of the three lines in correct visibility. Use Layout A.

86·7 FIG. 138(1).
86·8 FIG. 138(2).
86·9 FIG. 138(3).
86·10 FIG. 138(4).
86·11 FIG. 138(5).
86·12 FIG. 138(6).

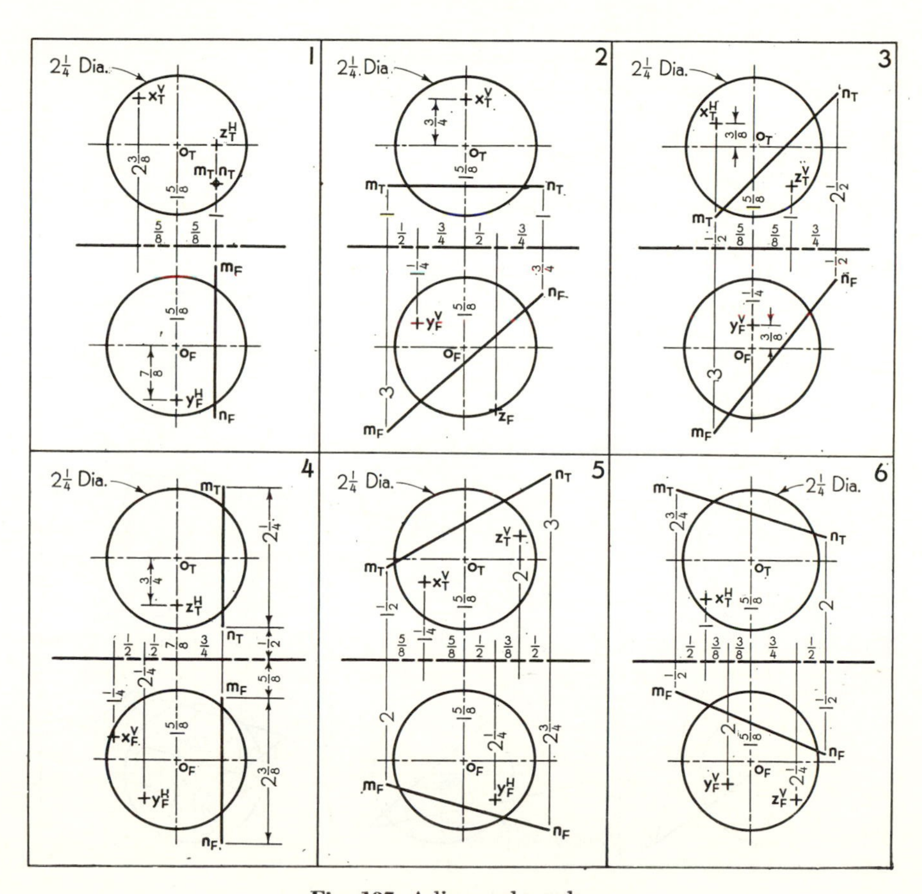

Fig. 137. A line and a sphere.

Group 87. Planes tangent to a sphere (Arts. 8·8, 8·9)

87·1–87·6 FIG. 137(1) to (6). In each problem construct a plane tangent to the sphere at point X on its surface. Construct a second plane tangent to the sphere at point Y on its surface. Solve, using only the given views. State what lines define each plane. Use Layout A.

In Probs. 87·7 to 87·14 construct the two planes that contain the specified line and are tangent to the sphere. Show the planes in all views, representing each as a triangle having one vertex at the point of tangency. Show sphere and planes in correct visibility. Use Layout B.

87·7 FIG. 138(1). Line AB.
87·8 FIG. 138(2). Line AB.
87·9 FIG. 138(2). Line AC.
87·10 FIG. 138(3). Line BC.
87·11 FIG. 138(4). Line AC.
87·12 FIG. 138(4). Line BC.

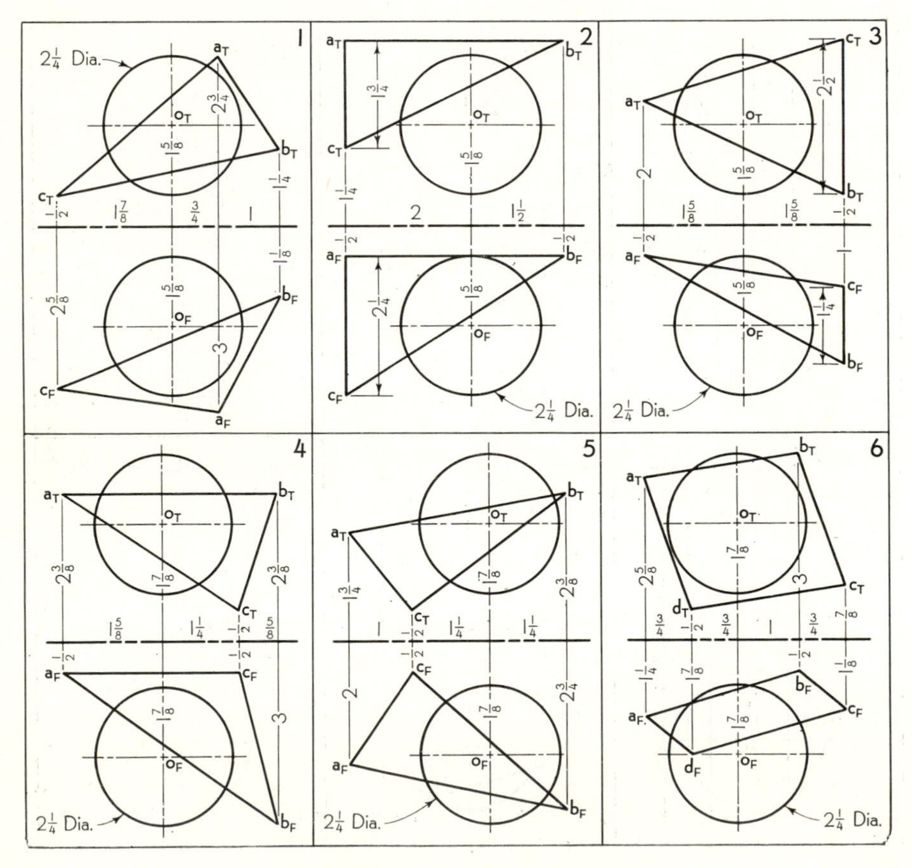

Fig. 138. A plane and a sphere.

87·13 FIG. 138(5). Line *AC*. **87·14** FIG. 138(6). Line *BC*.

Group 88. Cone-envelope and miscellaneous sphere problems (Art. 8·10)

88·1 C. Points *A* and *B* are the centers of two spheres 2 in. and 1 in. in diameter, respectively. Through a third point *X* construct two planes, each tangent to, and passing between, the two spheres. Point *B* is 2½ in. to the right of and ⅜ in. higher than *A*. Point *X* is 3½ in. to the right of, 2 in. behind, and 1⅜ in. lower than *A*. Label the four tangent points 1, 2, 3, and 4, and represent the planes as triangles *XV*-1, *XV*-2, etc., in the top and front views. (*V* is the vertex of the cone envelope.)

88·2 C. Same as Prob. 88·1, except that the two spheres both lie on the same side of each tangent plane.

88·3 C. Same as Prob. 88·1, except that points *A*, *B*, and *X* are located as follows: *B* is 1½ in. to the left of, 1¾ in. before, and 1¼ in. higher than *A*; *X* is 3 in. to the left of, 1 in. behind, and ½ in. lower than *A*. Separate the top and front views generously.

88·4 C. Same as Prob. 88·3, except that the two spheres both lie on the same side of each tangent plane.

88·5 C. Points *A*, *B*, and *C* are the centers of three spheres 2½ in., 1¼ in., and ¾ in. in diameter, respectively. The three centers lie in the same horizontal plane; *B* is 2 in. to the right of and 2¼ in. behind *A*; *C* is 1 in. to the left of and 3½ in. behind *A*. Draw the top view of the three spheres, placing point a_T 6 in. from the left border and 4½ in. above the lower border. Omit the front view. Draw four top-adjacent views, each showing as an edge two of the eight planes that are tangent to all three spheres. Number the planes 1 to 8, and then locate and label in the top view the eight tangent points on each sphere.

88·6 B. Point *A* is the center of a 2½-in.-diameter sphere. Point *X* is 2 in. to the right of, ¼ in. before, and 1¼ in. higher than *A*. Through point *X* construct a plane tangent to the sphere and having a slope of 45°. Show all possible solutions, and state what lines define each tangent plane. Solve, using only the given views.

88·7 B. Points *A* and *B* are the centers of two spheres 2 in. and 1 in. in diameter, respectively. *B* is 2½ in. to the right of, ½ in. behind, and ¾ in. lower than *A*. Construct a plane having a slope of 60°, tangent to both spheres but not passing between them. Show all possible solutions, and state what lines define each tangent plane. Solve, using only the given views.

88·8 C. Points A, B, C, and D are four points on the surface of a sphere. B is $\frac{3}{4}$ in. to the left of, 2 in. behind, and $1\frac{1}{2}$ in. lower than A. C is $1\frac{1}{2}$ in. to the right of, 1 in. behind, and $\frac{1}{4}$ in. lower than A. D is 1 in. to the right of, $\frac{1}{4}$ in. behind, and 2 in. lower than A. Locate the center, and determine the diameter of the sphere. (HINT: Any three of the points lie on a small circle of the required sphere.)

ANS. 2.94 in. diameter.

88·9 C. Points A, B, C, and D are the vertices of a tetrahedron. B is $\frac{3}{4}$ in. to the right of, $2\frac{1}{2}$ in. behind, and $2\frac{1}{2}$ in. higher than A. C is 2 in. to the right of, $2\frac{1}{2}$ in. behind, and $1\frac{1}{4}$ in. higher than A. D is 3 in. to the right of, $\frac{1}{2}$ in. behind, and $1\frac{1}{4}$ in. higher than A. Locate the center, and determine the diameter of the inscribed sphere. Show the sphere and its points of tangency with each face of the tetrahedron in the top and front views.

ANS. 0.84 in. diameter.

88·10 C. Three spheres, 3 in., 2 in. and $1\frac{1}{2}$ in. in diameter, respectively, are tangent to each other and stand on a horizontal plane. The 2-in. sphere is directly to the left of the 3-in. sphere, and the $1\frac{1}{2}$-in. sphere is in front of these two. Draw top and front views of the three spheres, and then show a fourth sphere, $1\frac{1}{2}$ in. in diameter, on top of and tangent to them. Show all four spheres in correct visibility.

88·11 A. Three spheres, each $1\frac{1}{2}$ in. in diameter, are tangent to each other and rest on the bottom of a hemispherical bowl 6 in. in diameter. The three spheres are in equilibrium; that is, they are each equidistant from the vertical axis of the bowl. Locate the center and determine the diameter of a fourth sphere that rests on top of the first three and just touches the center of the hemisphere. Show the four spheres in correct visibility in the top and front views. Solve, using only the given views.

ANS. 1.59 in. diameter.

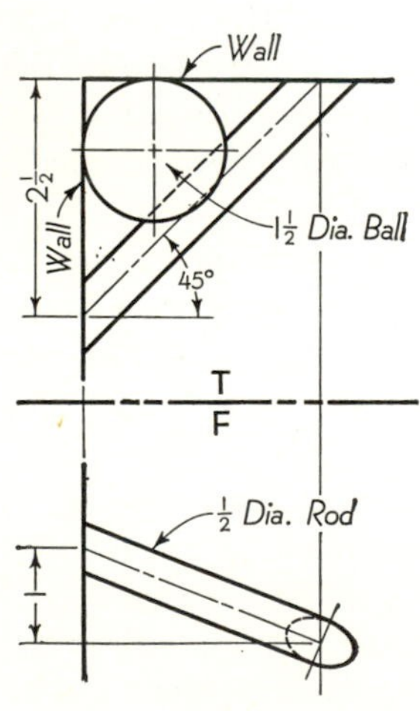

Fig. 139. Problem 88·12.

88·12 FIG. 139 B. A $\frac{1}{2}$-in.-diameter rod extends across the corner formed by two vertical walls as shown in the figure. A $1\frac{1}{2}$-in.-diameter sphere rests on the rod and touches each wall. Show in the top and front views the position of the sphere, and locate the point where it touches the rod.

Group 89. Surfaces of revolution (Arts. 8·11, 8·12)

In Probs. 89·1 to 89·6 points *X* and *Y* lie on the surface of revolution, but only one view of each point is given. Locate each point in the other view, and indicate its visibility. Draw a third view *A* in the direction of the given arrow, establishing the contours by the method of inscribed spheres. Show points *X* and *Y* in view *A* also.

89·1 FIG. 140(1) B.
89·2 FIG. 140(2) B. (See Appendix, Art. A·14.)
89·3 FIG. 140(3) B.
89·4 FIG. 140(4) A.
89·5 FIG. 140(5) A.
89·6 FIG. 140(6) A.

In Probs. 89·7 to 89·11 draw a third view *A* in the direction of the given arrow, establishing the contours by the method of inscribed spheres. Disregard plane *ABC*.

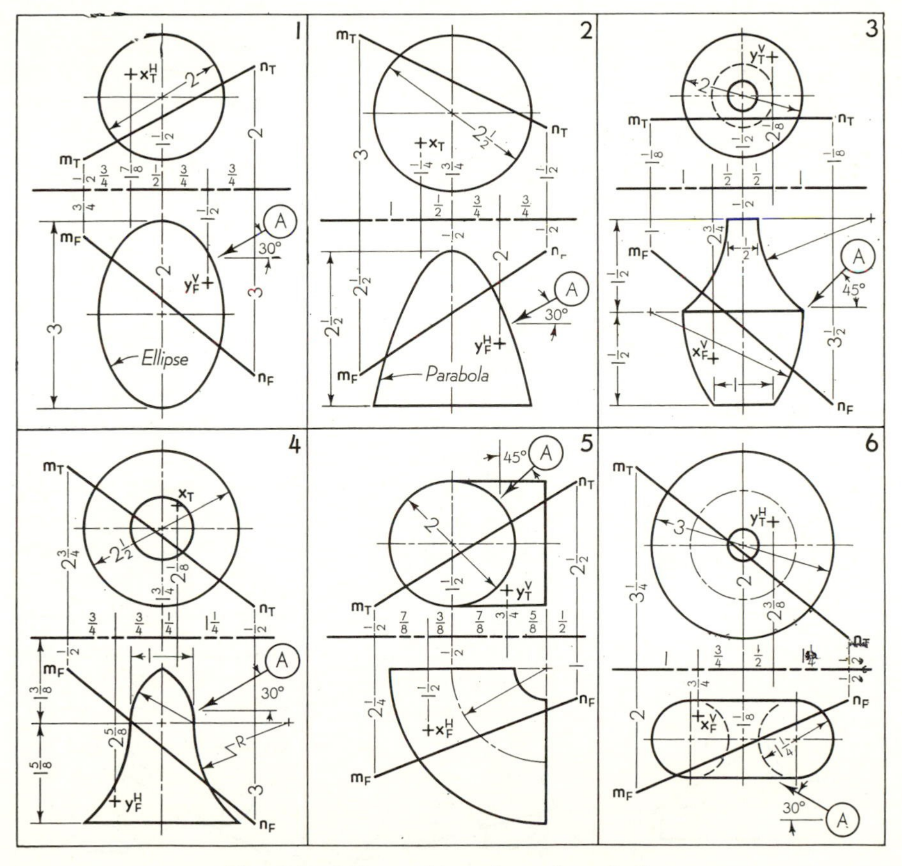

Fig. 140. A line and a surface of revolution.

89·7 FIG. 141(1) B. (HINT: Use three or four rings of various-sized spheres.)

89·8 FIG. 141(2) B. (See Appendix, Art. A·18.)

89·9 FIG. 141(3) A. (See Appendix, Art. A·15.) Sheet-metal cowling, open top and bottom.

89·10 FIG. 141(4) B. **89·11** FIG. 141(5) A.

Group 90. Intersection of a plane and a surface of revolution (Art. 8·13)

In Probs. 90·1 to 90·6 find the line of intersection between the surface of revolution and the plane *ABC*, assuming that the plane is indefinite in extent and cuts completely through the given surface. Show the curve in both the top and front views, locating accurately those points which lie on the contour curves. Use Layout B, except as noted.

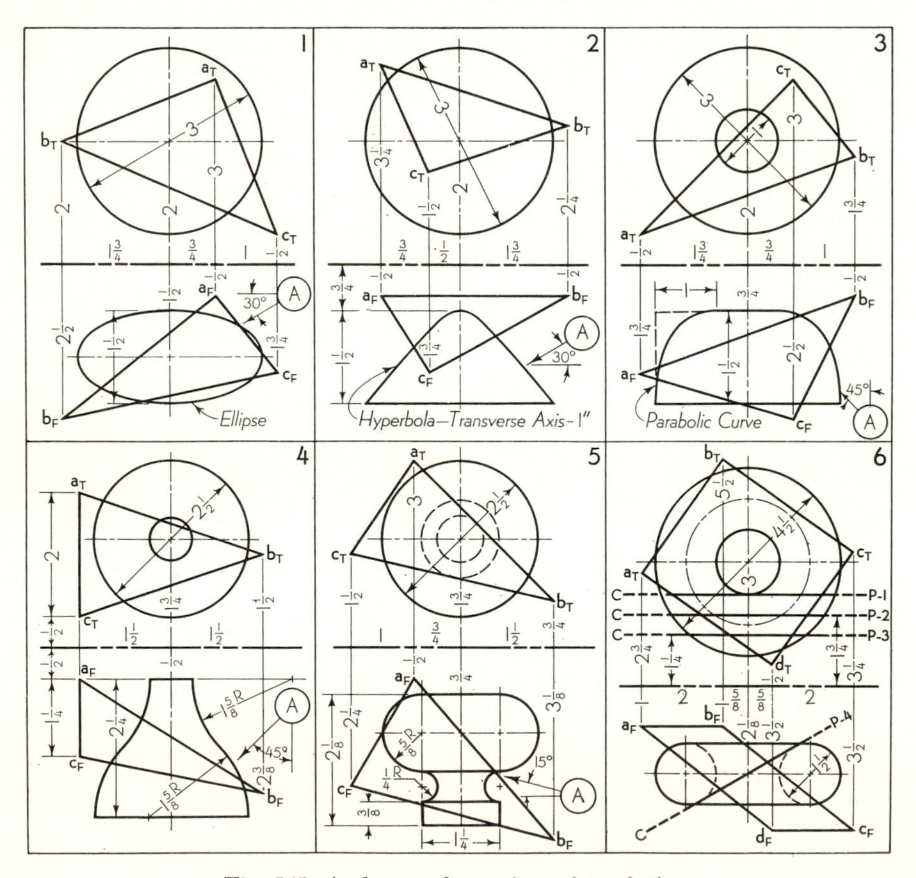

Fig. 141. A plane and a surface of revolution.

90·1 FIG. 141(1).
90·2 FIG. 141(2). (See Appendix, Art. A·18.)
90·3 FIG. 141(3). (See also Prob. 89·9.)
90·4 FIG. 141(4).
90·5 FIG. 141(5).
90·6 FIG. 141(6) A. Plane *ABCD* only. Solve without auxiliary view.
90·7 FIG. 141(6) A. Find the lines of intersection of the three vertical cutting planes and the torus. All curves obtained thus, by planes parallel to the axis of an annular torus, are called *Cassinian ovals,* of which the *lemniscate* (plane *C-P*-1) is one special form. Disregard planes *ABCD* and *C-P*-4.
90·8 FIG. 141(6) A. Find the line of intersection of the *bitangent* plane *C-P*-4 and the torus. (Plane is tangent to the surface at *two* points.) Examination and measurement of the intersection curve will reveal that it is two overlapping *circles.*

Group 91. Intersection of a line and a surface of revolution (Art. 8·14)

91·1–91·6 FIG. 140(1) to (6). In each problem locate the points *P* and *Q* where line *MN* intersects the surface of revolution, and show line *MN* in correct visibility. Use either a vertical or an edge-front cutting plane through line *MN*, and draw the complete curve of intersection in correct visibility. Use Layout A.

In Probs. 91·7 to 91·12 locate the points *P* and *Q* where the specified line intersects the surface of revolution, and show this line in correct visibility. (Disregard other lines.) See also statement of Probs. 91·1 to 91·6 above. Use Layout A.

91·7 FIG. 141(1). Line *AB*.
91·8 FIG. 141(2). Line *BC*.
91·9 FIG. 141(3). Line *AB*.
91·10 FIG. 141(4). Line *AB*.
91·11 FIG. 141(5). Line *AB*.
91·12 FIG. 141(6). Line *AD*.

Group 92. Plane tangent to a surface of revolution (Art. 8·15)

92·1–92·6 FIG. 140(1) to (6). In each problem construct a plane tangent to the surface of revolution at point *X* on its surface. State what lines define the plane. Use Layout A.

In Probs. 92·7 to 92·10 a plane tangent to the surface of revolution at point *Y* on its surface would also intersect the surface. Construct a line perpendicular (normal) to the surface at point *Y*. Use Layout A.

92·7 FIG. 140(3).
92·8 FIG. 140(4).
92·9 FIG. 140(5).
92·10 FIG. 140(6).

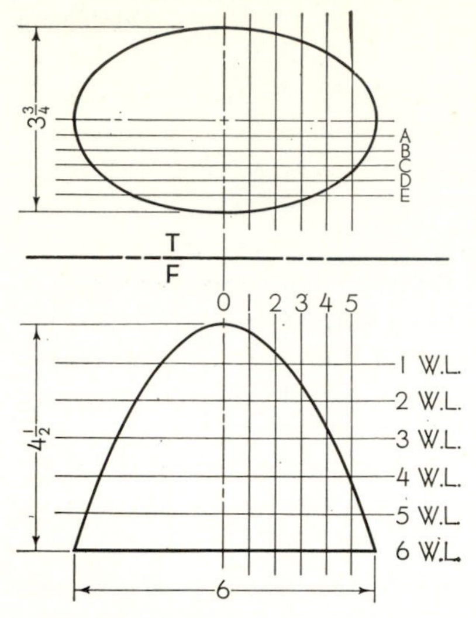

Fig. 142. Problem 93·1.

Group 93. Contoured surfaces (Arts. 8·16 to 8·19)

In each problem the contour outlines of a double-curved surface are given. Draw the required views, constructing the given curves very accurately and smoothly. Keep your pencils very sharp. Then construct the required frame lines, water lines, and buttock lines, testing and adjusting each curve until all curves are fair and all points check by alignment and measurement with the other two views. Unless otherwise indicated, all section planes are uniformly spaced. Problems may be shortened by fairing only a portion of the given surface.

93·1 FIG. 142 C. Scale: full size (B, ⅔ size). The surface shown is an elliptical paraboloid; hence all water lines will be ellipses, and all vertical sections will be parabolas. Draw the top, front, and right-side views of the surface. Construct the water lines first as a series of concentric ellipses, and then use these curves to construct the buttock lines and frame lines. Test for fairness by checking the alignment of points between the front and side views. Curves may be drawn only in the right front quarter of the surface if desired.

93·2 FIG. 143 C. Scale: full size (B, ⅔ size). The surface shown is a portion of a hyperbolic paraboloid in standard position (see Art. 7·11 and Fig. 7·8). All contour curves shown in the given views are parabolas, and all frame lines will be identical parabolas, differing only in vertical position. Draw the top, front, and right-side views of the surface. Construct the parabolic frame lines in the side view

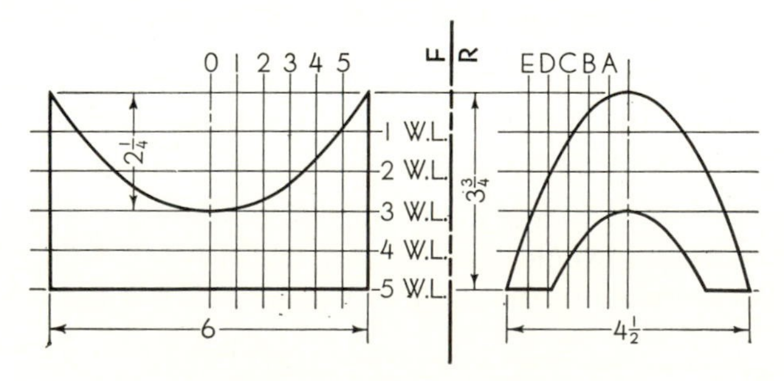

Fig. 143. Problem 93·2.

first, and then use these curves to construct the buttock and water lines. Test for fairness by checking the alignment of points between the top and front views. Curves may be drawn only in the right front quarter if desired.

93·3 FIG. 144 C. Scale: ¾ in. = 1 ft. The surface shown is the hull of a 14-ft boat. The parabolic curves shown in the sheer plan and in the half-breadth plan are tangent to the dash lines (see Appendix, Art. A·15). The 14 frame lines (of which a few have been drawn in the figure) are described in the table of offsets, which gives the displacement from the vertical center line of four points on each frame line. (Distances in the table are *not to scale* but are in *actual inches on the drawing* for ease of plotting with a scale divided in fiftieths or hundredths.) Plot and draw the frame lines first, and then construct the indicated water lines and buttock lines. Check and adjust curves until all are fair. For a shorter problem fair only the forward or aft portion of the hull.

93·4 FIG. 145 C. Scale: 1½ in. = 1 ft. The surface shown is an automobile fender. The parabolic curves in the side view are tangent to the dash lines (see Appendix, Art. A·15). Undimensioned curves in the front view must be constructed by alignment and measurement. Since no complete set of lines is given, it is suggested that the buttock lines be assumed first in the side view. These lines should be circular arcs in the forward half of the fender and approximately parabolic in the rear half. From these assumed buttock lines con-

TABLE OF OFFSETS FOR PROB. 93·3
(Distances in actual inches)

Frame	A.W.L.	L.W.L.	2 W.L.	3 W.L.
1	0.34	0.24	0.11	0.01
2	0.74	0.65	0.50	0.23
3	1.07	1.01	0.88	0.59
4	1.30	1.27	1.19	0.93
5	1.43	1.42	1.37	1.17
6	1.50	1.50	1.45	1.27
7	1.49	1.49	1.44	1.25
8	1.47	1.47	1.42	1.22
9	1.44	1.42	1.35	1.12
10	1.39	1.35	1.25	0.92
11	1.33	1.26	1.10	0.67
12	1.25	1.16	0.89	0.42
13	1.16	1.04	0.64	0.21
14	1.08	0.88	0.46	0.10

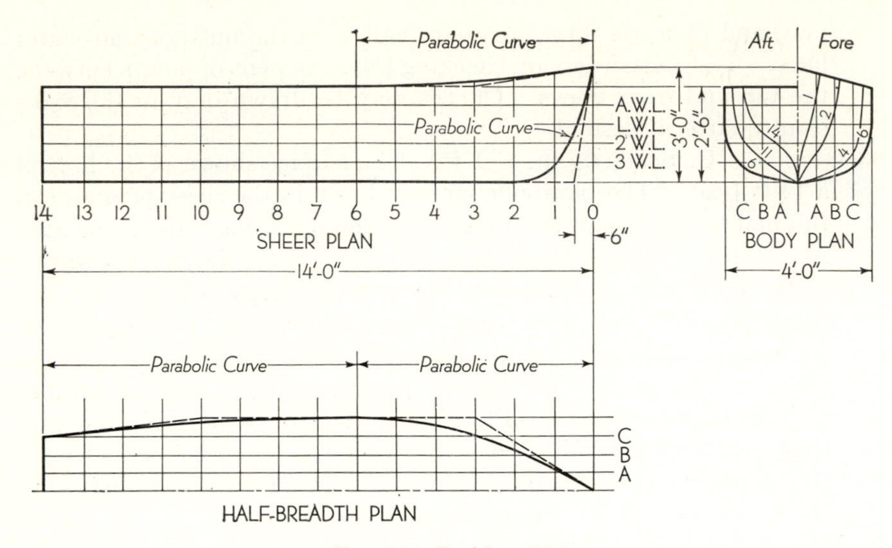

Fig. 144. Problem 93·3.

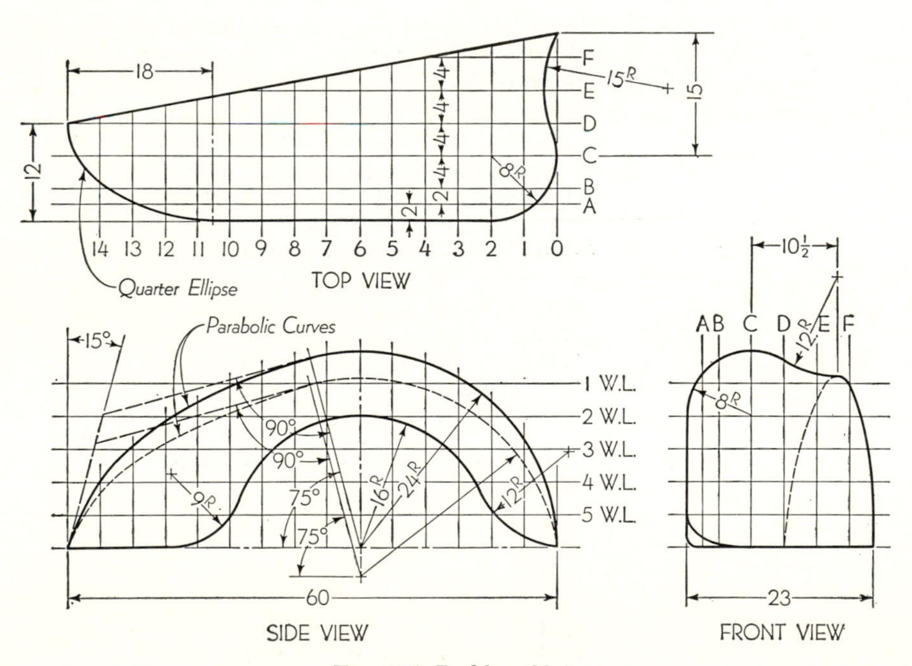

Fig. 145. Problem 93·4.

struct tentative frame and water lines. Change, adjust, and correct all lines until satisfactory agreement is achieved and all lines are fair. If the whole fender is faired, it is suggested that the front view be drawn twice, once for frame lines 0 to 6, once for frame lines 6 to 14. For a shorter problem fair only the forward or rear half of the fender.

93·5 FIG. 146 C. Scale: 1 in. = 1 ft. The surface shown is a portion of an airplane fuselage. The given curves are all conics, and they define the surface exactly. Draw the three given views, constructing the curves by the method of Art. A·23 in the Appendix. Note that, except for the top view of the shoulder line, each curve is tangent to a pair of intersecting dash lines at its extremities and passes through a given control point. Construct one or more of the frame lines at stations *B*, *C*, and *D*.

Group 94. Intersection of two prisms (Arts. 9·3, 9·4)

In each problem determine the line of intersection of the two prisms, and complete each edge in correct visibility. Omit that part of each edge which is inside the other prism. Use either the individual-line method or the cutting-plane method, whichever seems better for the par-

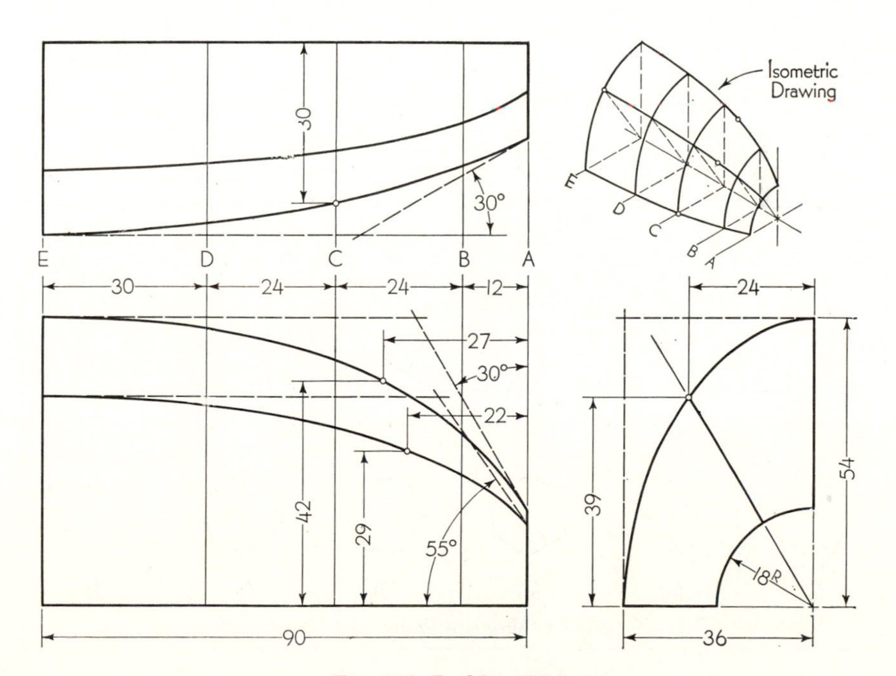

Fig. 146. Problem 93·5.

ticular problem. Any of the geometric figures shown at the bottom of Fig. 147 may be substituted for the 1½-in. square to provide alternate problems. In each substitution, match the center lines marked with a small arrow. Designate such alternate problems as 94·1*a*, 94·1*b*, etc. (For combined intersection and development problems see Probs. 108·10 to 108·15.)

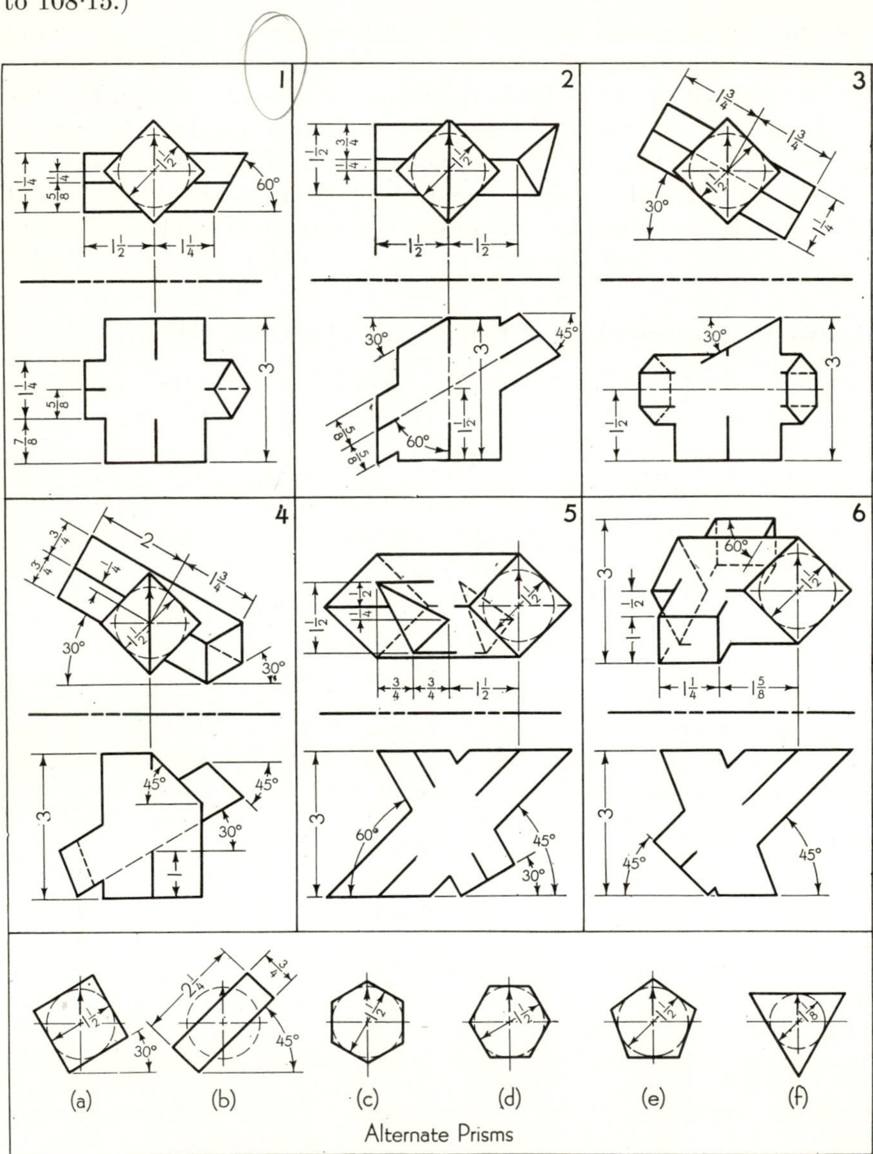

Fig. 147. Intersecting prisms.

94·1 FIG. 147(1) B.
94·2 FIG. 147(2) B.
94·3 FIG. 147(3) B.
94·4 FIG. 147(4) A.
94·5 FIG. 147(5) B.
94·6 FIG. 147(6) B.

Group 95. Intersection of two cylinders (Arts. 9·5, 9·6)

In each problem determine the line of intersection of the two cylinders, and complete the extreme elements in correct visibility. Choose *first* those cutting planes which locate points on the extreme elements. The problems may be varied by substituting one of the alternate dimensions for the one in the figure that is marked with a star. Designate such alternate problems as 95·1*a*, 95·1*b*, etc. (For combined intersection and development problems see Probs. 110·10 to 110·15.)

95·1 FIG. 148(1) B. Alternate dimensions: (*a*) 0, (*b*) 3/8, (*c*) 5/8, (*d*) 7/8.
95·2 FIG. 148(2) B. Alternate dimensions: (*a*) 0, (*b*) 3/8, (*c*) 5/8, (*d*) 7/8.
95·3 FIG. 148(3) B. Alternate dimensions: (*a*) 0, (*b*) 3/8, (*c*) 5/8, (*d*) 7/8.
95·4 FIG. 148(4) A. Alternate dimensions: (*a*) 0, (*b*) 3/8, (*c*) 5/8, (*d*) 7/8.
95·5 FIG. 148(5) B. Alternate dimensions: (*a*) 0, (*b*) 3/8, (*c*) 1/2.
95·6 FIG. 148(6) B. Alternate dimensions: (*a*) 0, (*b*) 3/8, (*c*) 5/8, (*d*) 7/8.
95·7 FIG. 148(7) A. Offset outlet pipe.
95·8 FIG. 148(8) B. Collar with offset hole perpendicular to axis.
95·9 FIG. 148(9) B. Bolthead boss and recess in machine base. Add right-side view.

Group 96. Intersection of a cylinder and a prism (Art. 9·7)

In Probs. 96·1 to 96·6 substitute a 1½-in.-diameter circle for the 1½-in. square shown in the figure, and determine the line of intersection of the cylinder and prism. Choose *first* those cutting planes which locate points of special significance.

96·1 FIG. 147(1) B.
96·2 FIG. 147(2) B.
96·3 FIG. 147(3) B.
96·4 FIG. 147(4) A.
96·5 FIG. 147(5) B.
96·6 FIG. 147(6) B.

In Probs. 96·7 to 96·12 substitute one of the geometric figures shown at the bottom of Fig. 147 for the 2-in.-diameter circle shown in the given figure, and determine the line of intersection of the prism and cylinder. In each substitution, match the center lines marked with a small arrow.

96·7 FIG. 148(1) B.
96·8 FIG. 148(2) B.
96·9 FIG. 148(3) B.
96·10 FIG. 148(4) A.
96·11 FIG. 148(5) B.
96·12 FIG. 148(6) B.

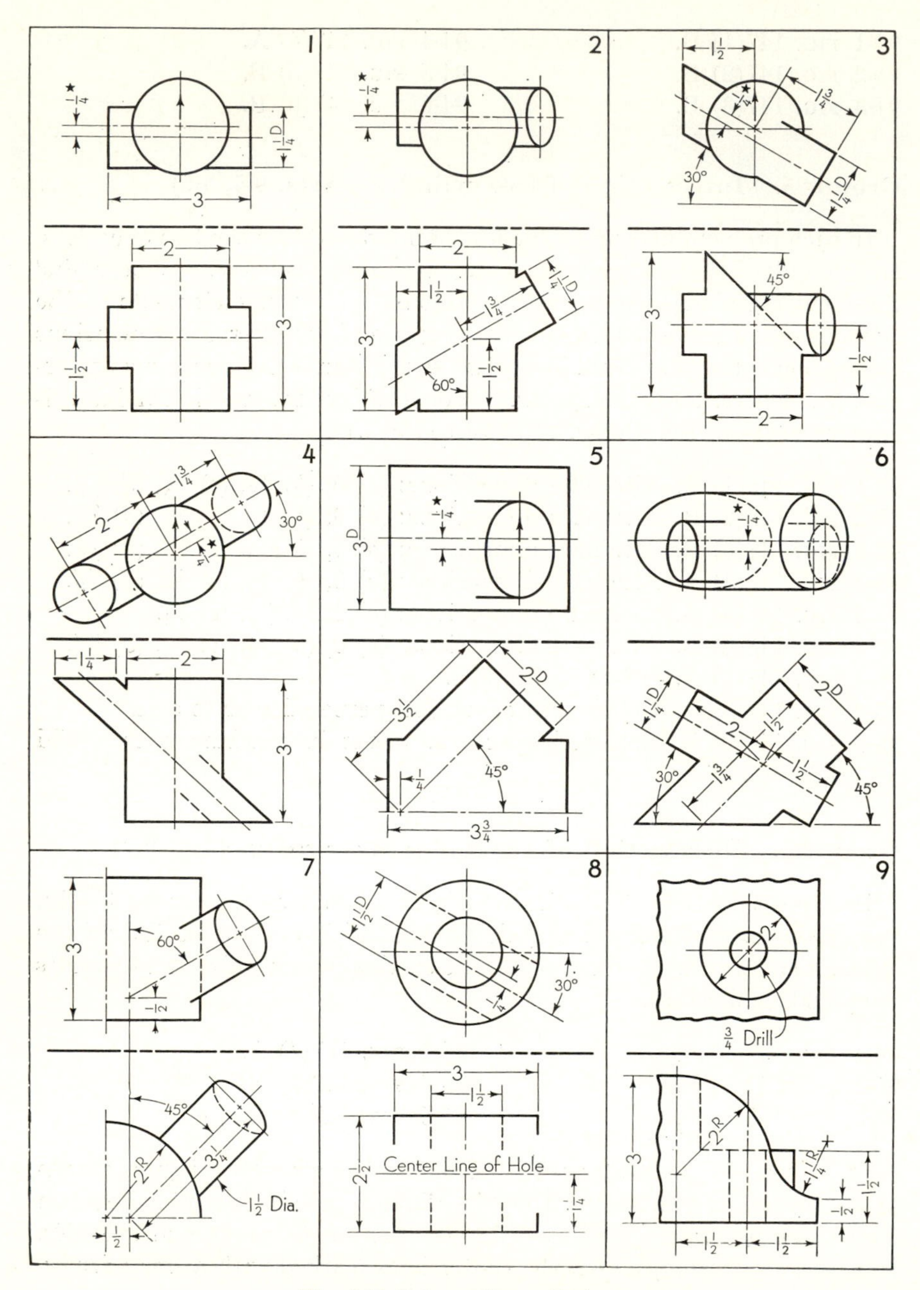

Fig. 148. Intersecting cylinders.

Group 97. Intersection of cones, pyramids, and cylinders (**Arts. 9·8 to 9·10**)

In each problem determine the line of intersection of the two surfaces, and complete the extreme elements in correct visibility. Use that method or combination of methods that seems best for the particular problem. The problems may be varied by substituting one of the alternate dimensions for the one in the figure that is marked with a star. (For combined intersection and development problems see Groups 111, 113, and 114.)

97·1 FIG. 149(1) A. Alternate dimensions: (*a*) 0, (*b*) 1.
97·2 FIG. 149(2) A. Alternate dimensions: (*a*) 0, (*b*) 1.
97·3 FIG. 149(3) A. Alternate dimensions: (*a*) 0, (*b*) 7/8.
97·4 FIG. 149(4) A. Alternate dimensions: (*a*) 0, (*b*) 1/2, (*c*) 3/4.
97·5 FIG. 149(5) B. Alternate dimensions: (*a*) 0, (*b*) 1/4, (*c*) 5/8, (*d*) 1.
97·6 FIG. 149(6) B. Alternate dimensions: (*a*) 0, (*b*) 1/2, (*c*) 3/4.
97·7 FIG. 149(7) B. Alternate dimensions: (*a*) 0, (*b*) 1/2.
97·8 FIG. 149(8) A. Distributing hopper.
97·9 FIG. 149(9) A. Exhaust-pipe branch connection.

Group 98. Intersection of cones, pyramids, and prisms (**Arts. 9·8 to 9·11**)

In each problem substitute one of the geometric figures shown at the bottom of Fig. 147 for the 2- or 1½-in. circle shown in the given figure, and determine the line of intersection of the two surfaces. In each substitution, match the center lines marked with a small arrow.

98·1 FIG. 149(1) A.
98·2 FIG. 149(2) A.
98·3 FIG. 149(3) A.
98·4 FIG. 149(4) A.
98·5 FIG. 149(5) B.
98·6 FIG. 149(6) B.
98·7 FIG. 149(7) B.

Group 99. Intersection of two oblique cylinders (**Arts. 9·12 to 9·14**)

99·1–99·6 FIG. 150(1) to (6). Determine the line of intersection of the two oblique cylinders, using the method of Art. 9·13, and complete the extreme elements in correct visibility. Choose *first* those cutting planes which locate points on the extreme elements. An accurate solution will require 8 to 12 cutting planes. Use Layout A.

99·7–99·9 FIG. 150(7) to (9). Determine the line of intersection of the two oblique circular cylinders, using the method of Art. 9·14, and complete the extreme elements. Use Layout A.

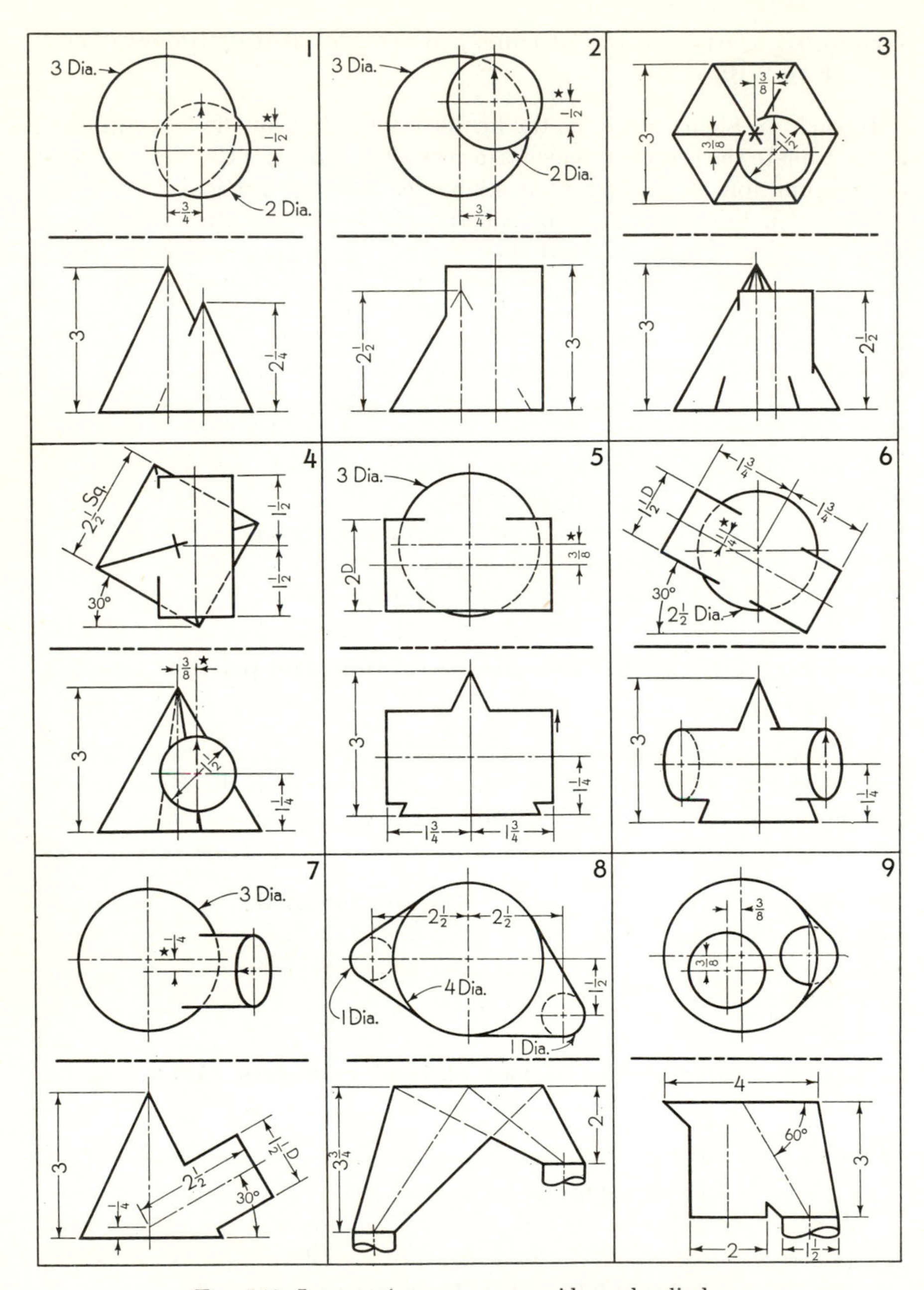

Fig. 149. Intersecting cones, pyramids, and cylinders.

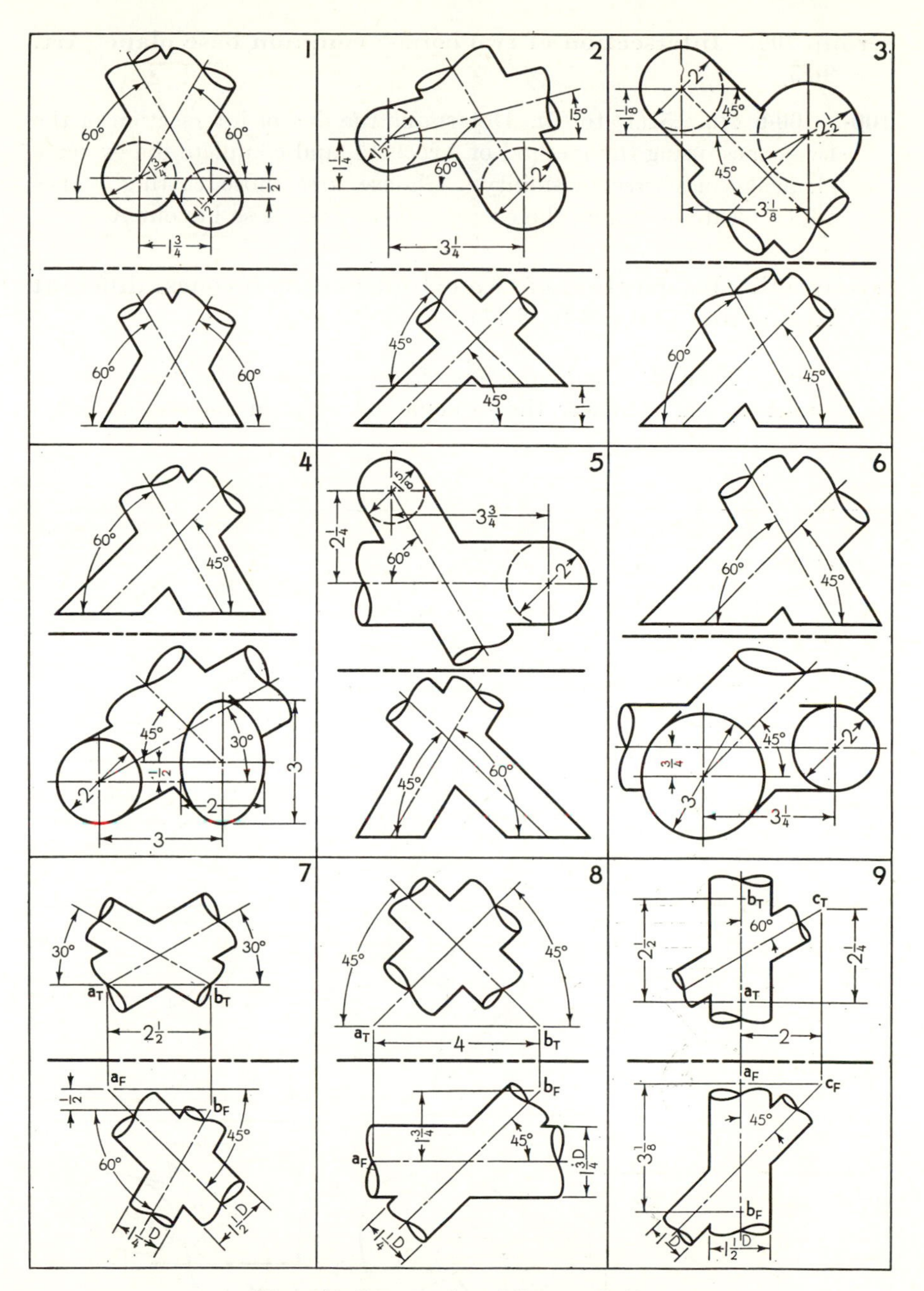

Fig. 150. Intersecting oblique cylinders.

Group 100. Intersection of two cones: common base plane (Art. 9·15)

100·1–100·6 FIG. 151(1) to (6). Determine the line of intersection of the two cones, using the method of Art. 9·15, and complete the extreme elements in correct visibility. Choose *first* those cutting planes which locate points on the extreme elements. Use Layout A.

Group 101. Intersection of two cylinders or two cones: different base planes (Arts. 9·16, 9·17)

In each problem determine the line of intersection of the two cylinders or two cones, and complete the extreme elements in correct visibility.

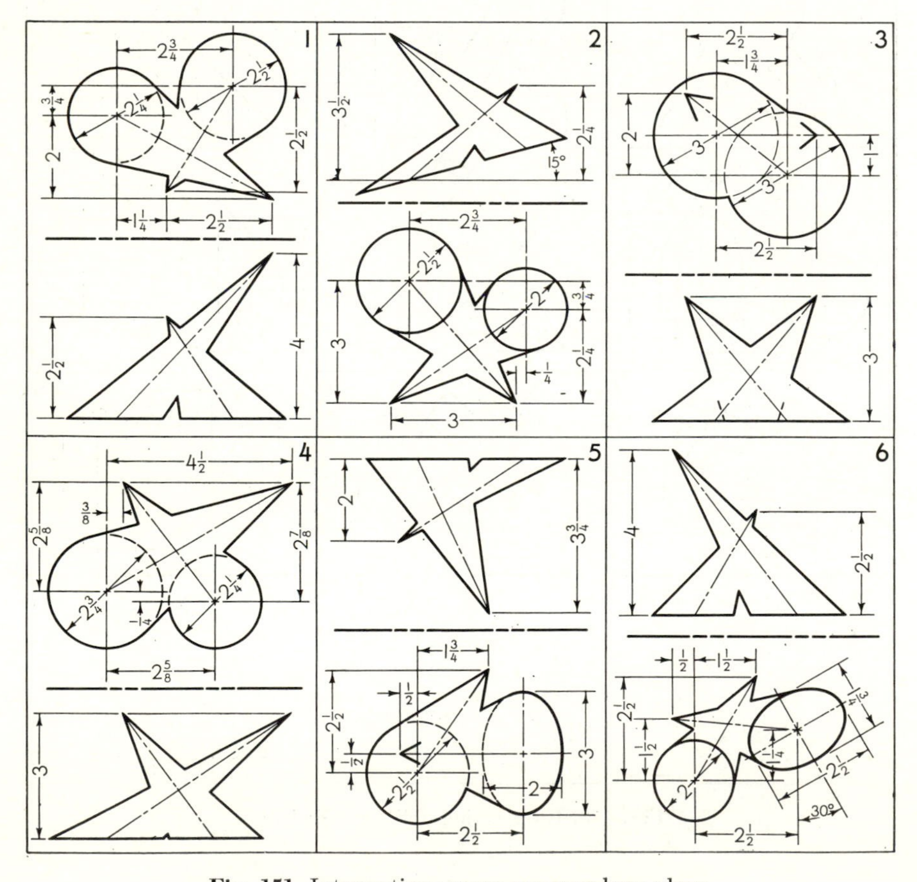

Fig. 151. Intersecting cones: common base plane.

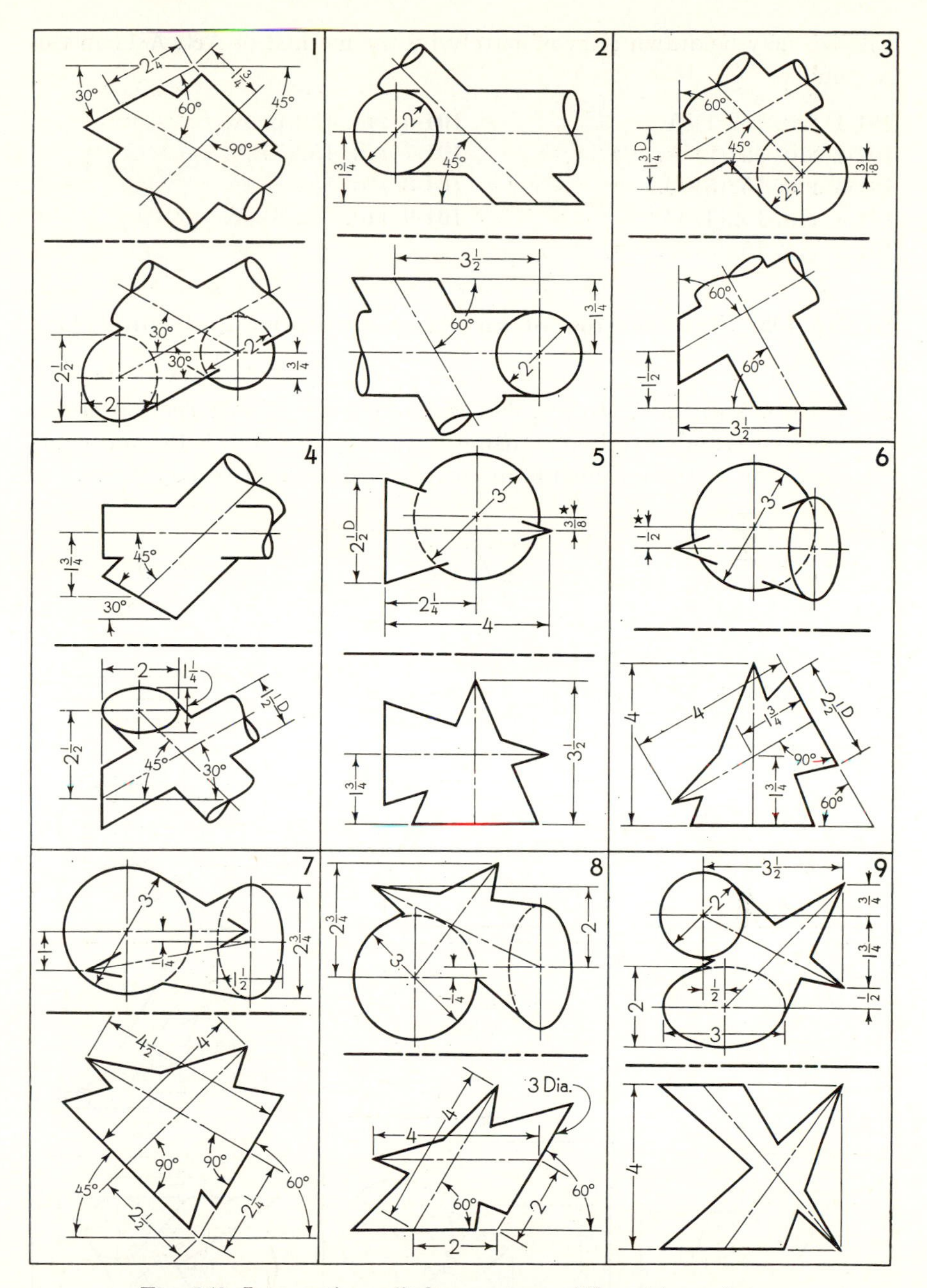

Fig. 152. Intersecting cylinders or cones: different base planes.

Ellipses may be drawn approximately by the method of Art. A·11 in the Appendix.

101·1 FIG. 152(1) A.
101·2 FIG. 152(2) A.
101·3 FIG. 152(3) C.
101·4 FIG. 152(4) C.
101·5 FIG. 152(5) C.
101·6 FIG. 152(6) A.
101·7 FIG. 152(7) A.
101·8 FIG. 152(8) C.
101·9 FIG. 152(9) A.

Group 102. Intersection of an oblique cylinder and cone (**Art. 9·18**)

102·1–102·6 FIG. 153(1) to (6). Determine the line of intersection of the oblique cylinder and cone, using the method of Art. 9·18, and complete the extreme elements in correct visibility. Use Layout A.

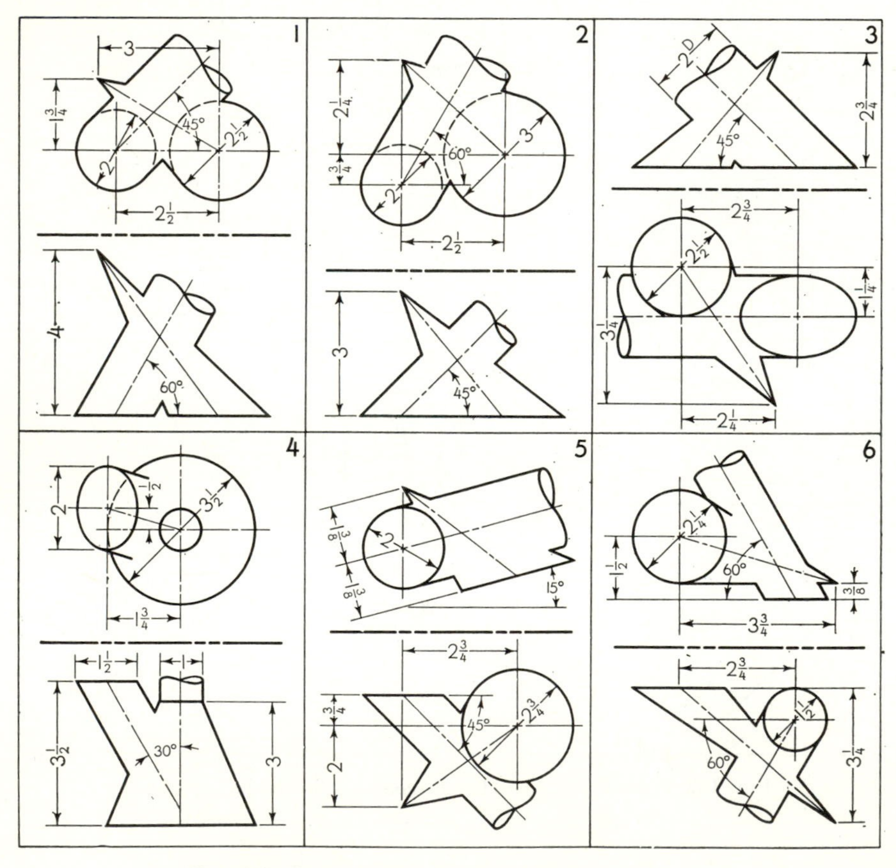

Fig. 153. Intersecting oblique cylinders and cones.

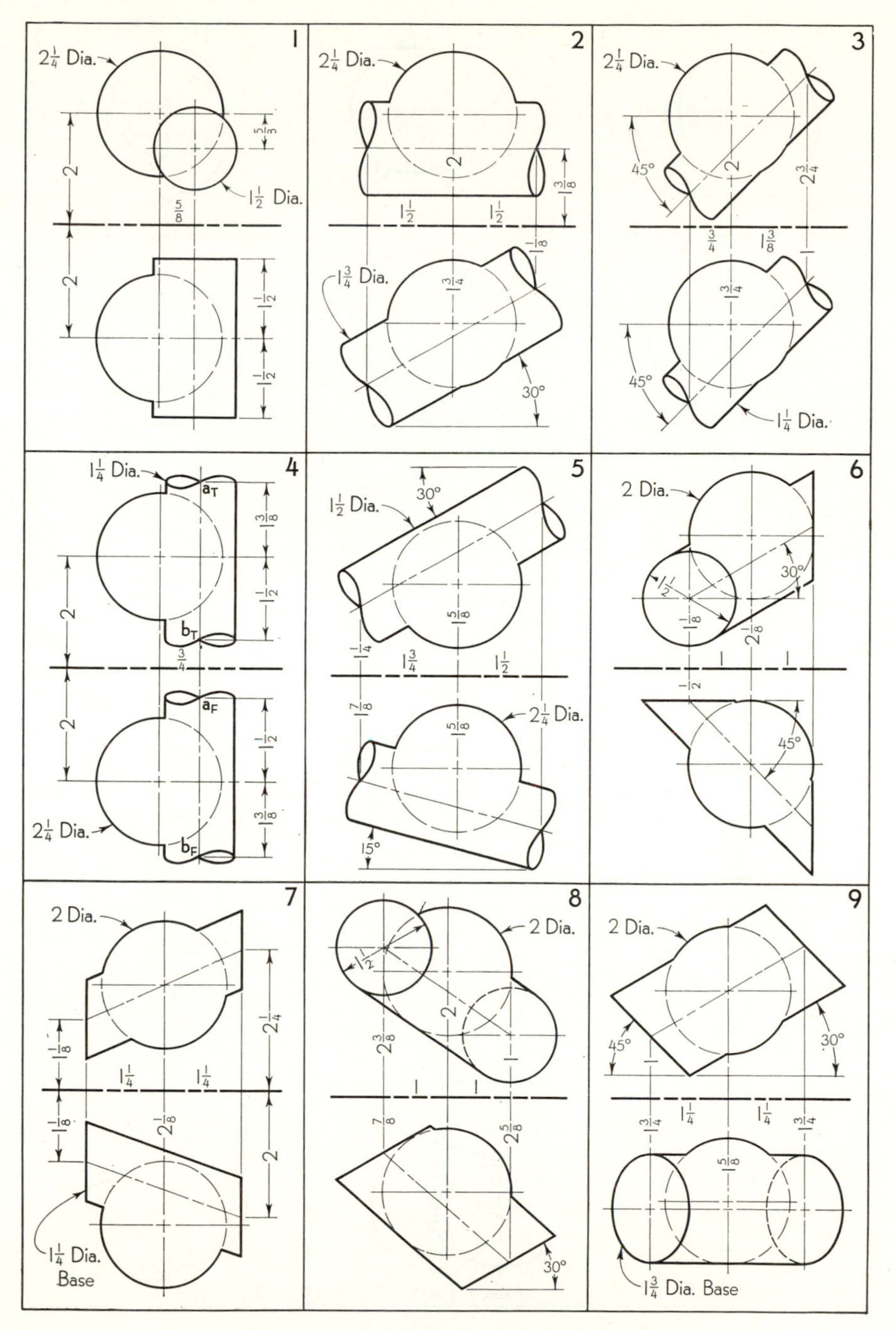

Fig. 154. Intersecting spheres and cylinders.

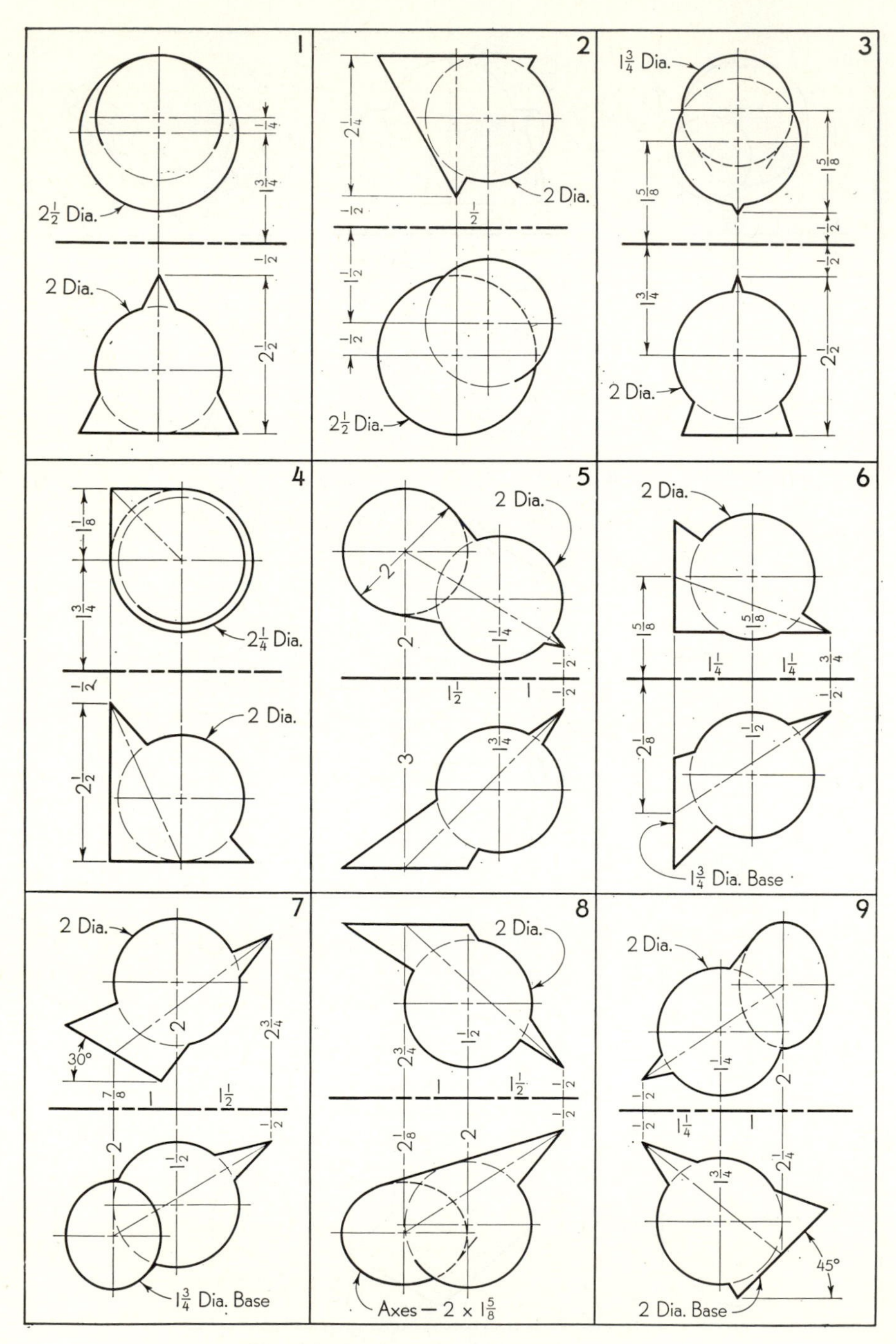

Fig. 155. Intersecting spheres and cones.

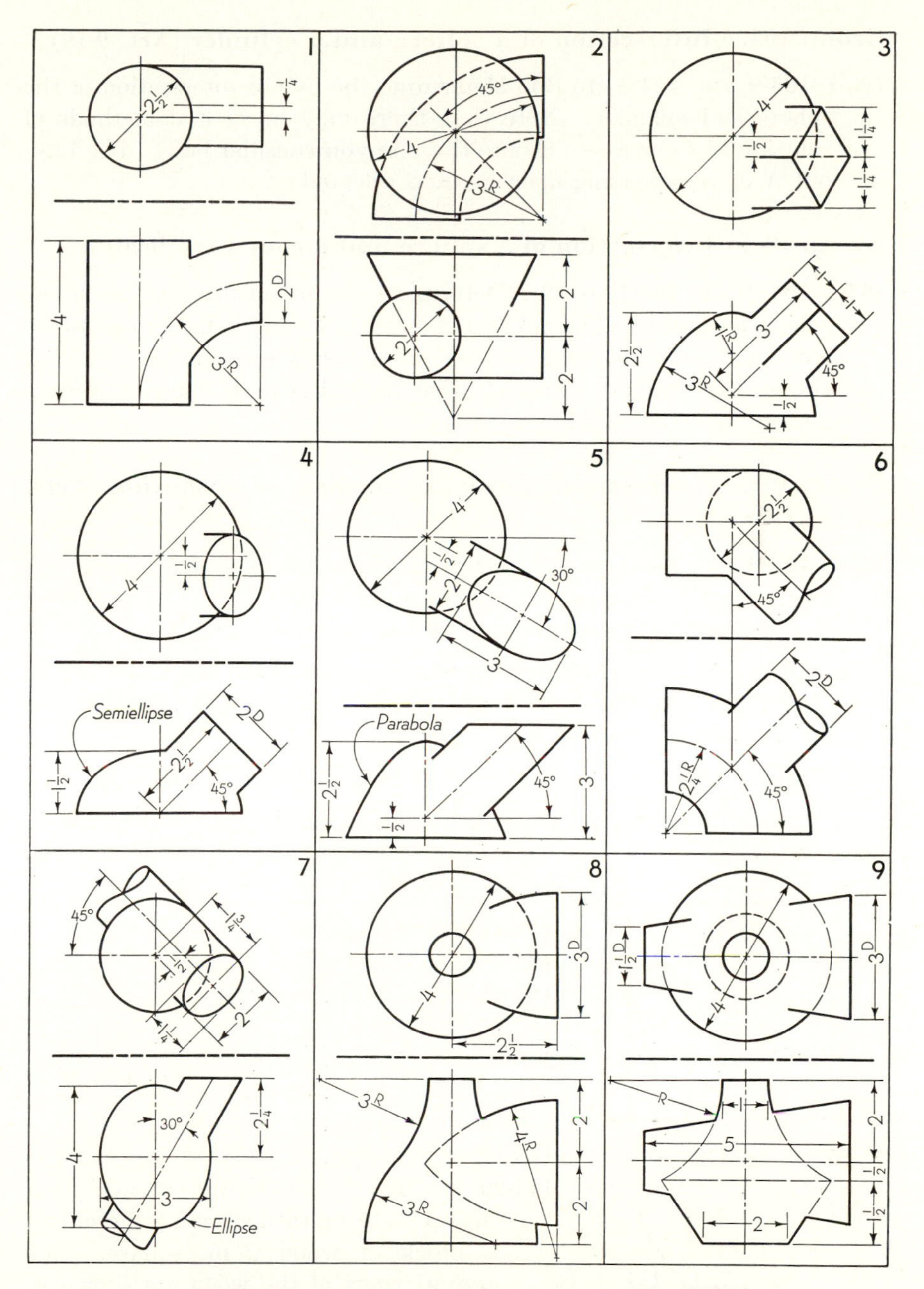

Fig. 156. Intersections involving surfaces of revolution.

Group 103. Intersection of a sphere and a cylinder (Art. 9·19)

103·1–103·9 FIG. 154(1) to (9). Determine the line of intersection of the sphere and cylinder. Note that there may be several methods of solution in some cases; choose the one you consider best. Use Layout A or B depending upon method selected.

Group 104. Intersection of a sphere and a cone (Art. 9·20)

104·1–104·9 FIG. 155(1) to (9). Determine the line of intersection of the sphere and cone. The method of Art. 9·20 may be used in all problems, but in some cases simpler methods of solution are feasible; choose the simplest solution. Use Layout A or B depending upon method selected.

Group 105. Intersections involving surfaces of revolution (Arts. 9·21 to 9·23)

105·1–105·9 FIG. 156(1) to (9). Determine the line of intersection of the two surfaces. Study the assigned problem carefully in order to select the best method of solution. Use Layout A.

In Probs. 105·10 and 105·11 determine the line of intersection of the two surfaces, using the method of cutting cylinders (Art. 9·21). Use Layout A.

105·10 FIG. 149(7).

105·11 FIG. 153(4).

In Probs. 105·12 to 105·19 let the starred dimension be *zero*, and determine the line of intersection by the method of cutting spheres (Art. 9·23). Use Layout A.

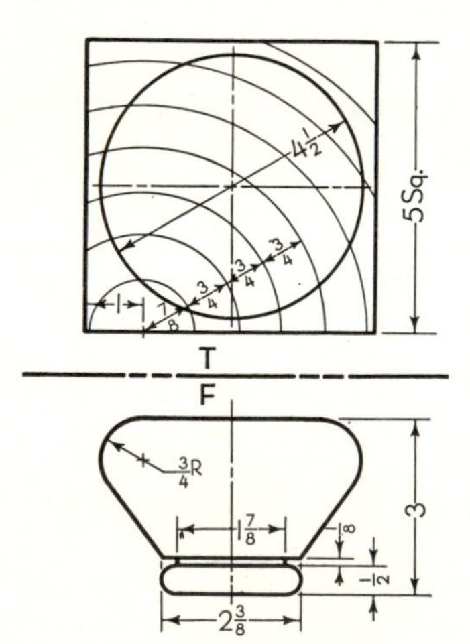

Fig. 157. Problem 105·20.

105·12 FIG. 148(1).
105·13 FIG. 148(2).
105·14 FIG. 148(5).
105·15 FIG. 148(6).
105·16 FIG. 149(5).
105·17 FIG. 149(7).
105·18 FIG. 152(5).
105·19 FIG. 152(6).
105·20 FIG. 157 A. The surface of revolution shown is to be turned on a lathe out of a block of wood 5 in. square. The annual rings of the wood are shown in the top view. Trace the emergence of these rings on the surface of the finished piece.

105·21 FIG. 158 A. Scale: 1½ in. = 1 ft. The figure shows the top and front views of a doorway in a semicircular arch. Determine how far the door can be opened before it interferes with the arch over the door. Design a curve for the arch soffit that will allow the door to open 90°. ANS. 70°.

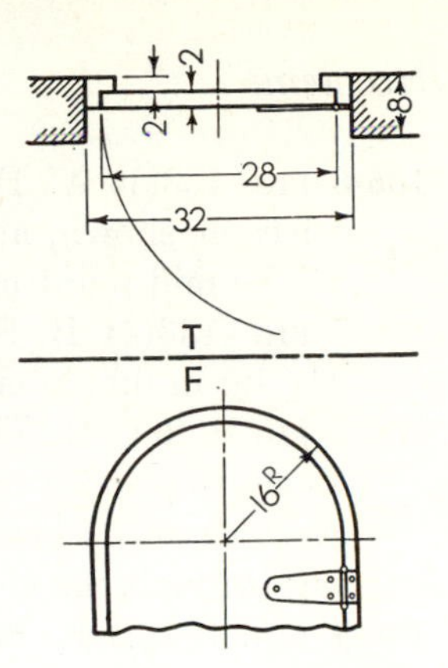

Fig. 158. Problem 105·21.

Group 106. Intersections involving contoured and warped surfaces (Art. 9·24)

106·1–106·3 FIG. 159(1) to (3). Determine the line of intersection of the cylinder or cone with the double-curved surface that has been represented by its outline and five equally spaced contour lines. In each case the contour lines are circular arcs with centers equally spaced along a given arc or line *AB*. Note that regardless of the number of contour lines employed—and more than the five shown are desirable—all distances marked *AB* must be divided into the same number of equal divisions. Use Layout A.

106·4 FIG. 131(5) A. Find the intersection of the hyperbolic paraboloid, whose first and last elements are *AC* and *BD*, and a vertical cylinder, 1 in. in diameter, center at x_T.

106·5 FIG. 131(6) A. Find the intersection of the hyperbolic paraboloid, whose first and last elements are *AC* and *BD*, and a vertical cylinder, 1½ in. in diameter, center at x_T.

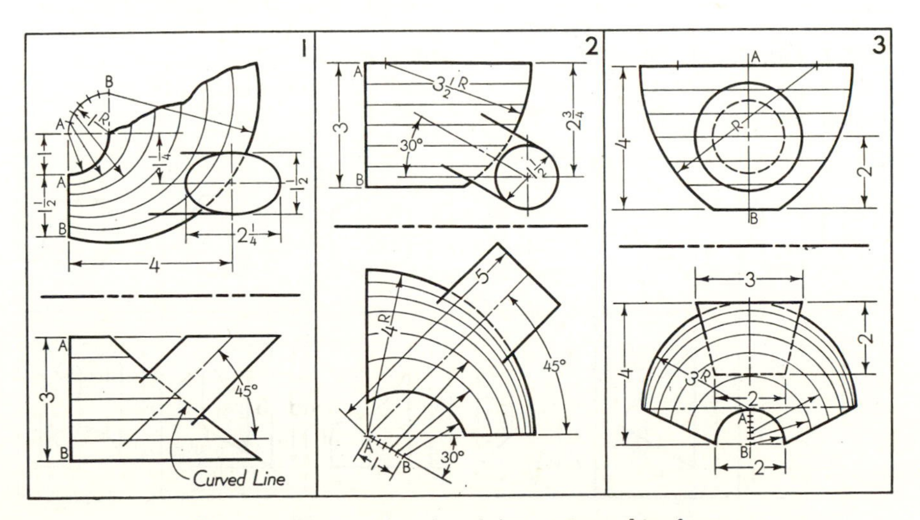

Fig. 159. Intersections involving contoured surfaces.

106·6 FIG. 133(1) A. Find the intersection of the cylindroid, with directrices as shown, and a vertical cylinder, 2¼ in. in diameter, center at the mid-point of the cylindroid axis.

106·7 FIG. 133(4) B. Same as Prob. 106·6, except that the cylinder is 1½ in. in diameter.

106·8 FIG. 132(1) B. Find the intersection of the conoid, with directrices as shown, and a horizontal cylinder, 2 in. in diameter, with its axis parallel to the plane director, 1 in. above the conoid base, and passing through the center of the conoid.

106·9 FIG. 132(3) B. Same as Prob. 106·8.

Group 107. Development of right prisms (Art. 10·4)

107·1–107·6 FIG. 160(1) to (6). Develop the lateral surfaces of the right prism. Use Layout B, and place the given views ½ in. from the left margin.

In Probs. 107·7 to 107·10 complete the front view of the truncated prism, and then develop all the surfaces except the bottom. Use Layout C, and place the given views on the left side.

107·7 FIG. 39(1).
107·8 FIG. 39(2).
107·9 FIG. 39(3). See also Art. 10·7.
107·10 FIG. 39(4).

Group 108. Development of oblique prisms (Arts. 10·5, 10·6)

In Probs. 108·1 to 108·5 develop the lateral surfaces of the oblique prism.

108·1 FIG. 161(1) B.
108·2 FIG. 161(2) A.
108·3 FIG. 161(3) B.
108·4 FIG. 161(4) C.
108·5 FIG. 161(5) C.

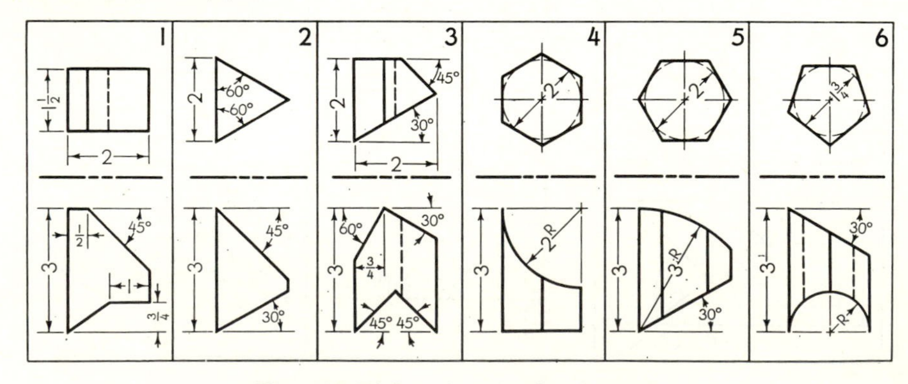

Fig. 160. Right prisms for development.

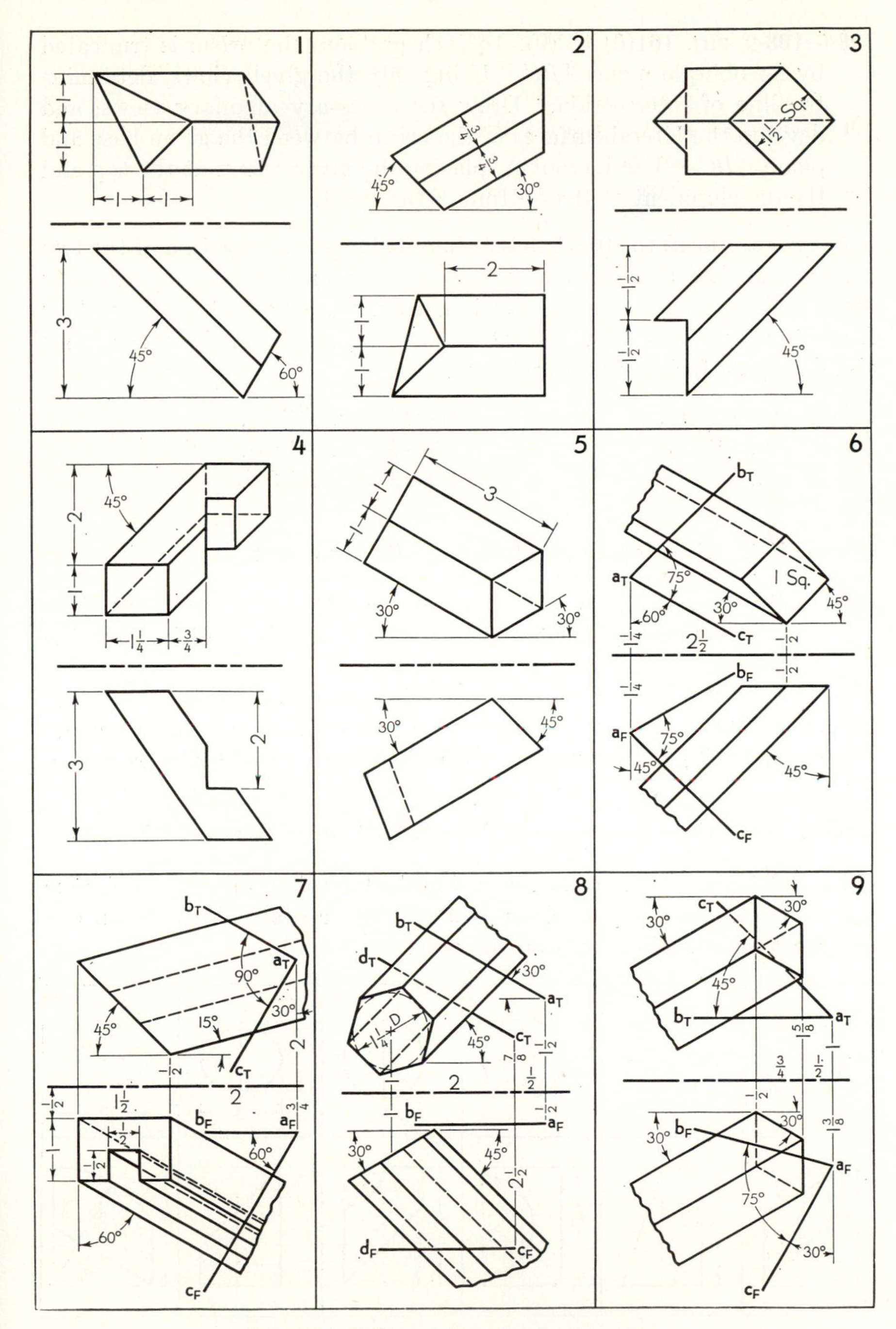

Fig. 161. Oblique prisms for development.

108·6–108·9 FIG. 161(6) to (9). In each problem the prism is truncated by an oblique plane *ABC*. Using only the given views, determine the line of intersection. Draw the necessary auxiliary views, and develop the lateral surfaces of the prism between the given base and plane *ABC*. Use Layout A, placing the given views at the top and the development at the bottom of the sheet.

In Probs. 108·10 to 108·15 determine the line of intersection of the two prisms, and then develop one or both of the prisms as assigned, showing the line of intersection on the developments. Use Layout C or two A or B sheets with views on one sheet and developments on the other. See also Group 94 for alternate problems.

108·10 FIG. 147(1).
108·11 FIG. 147(2).
108·12 FIG. 147(3).
108·13 FIG. 147(4).
108·14 FIG. 147(5).
108·15 FIG. 147(6).

Group 109. Development of right cylinders (Art. 10·7)

109·1–109·6 FIG. 162(1) to (6). Develop the lateral surface of the right cylinder. Use Layout B, and place the given views ½ in. from the left margin.
109·7 FIG. 160(4) B. Substitute a 2-in.-diameter circle for the hexagon.
109·8 FIG. 160(5) B. Substitute a 2-in.-diameter circle for the hexagon.
109·9 FIG. 160(6) B. Substitute a 2-in.-diameter circle for the pentagon.

Group 110. Development of oblique cylinders (Arts. 10·8, 10·9)

110·1–110·4 FIG. 163(1) to (4). Develop the lateral surface of the oblique cylinder. Use Layout C or two A or B sheets with views on one sheet and developments on the other.

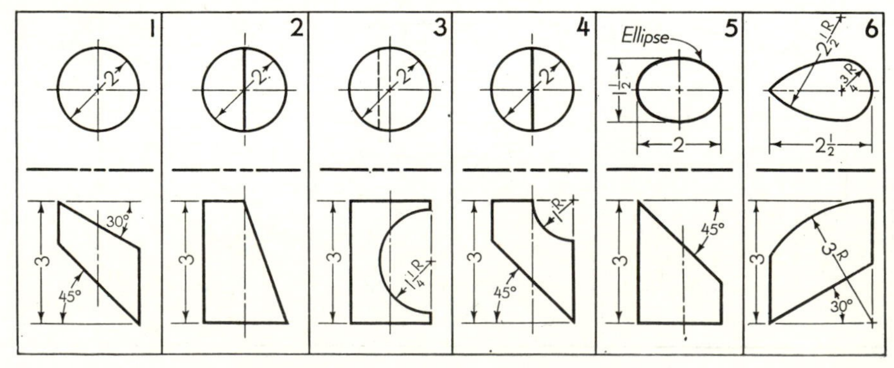

Fig. 162. Right cylinders for development.

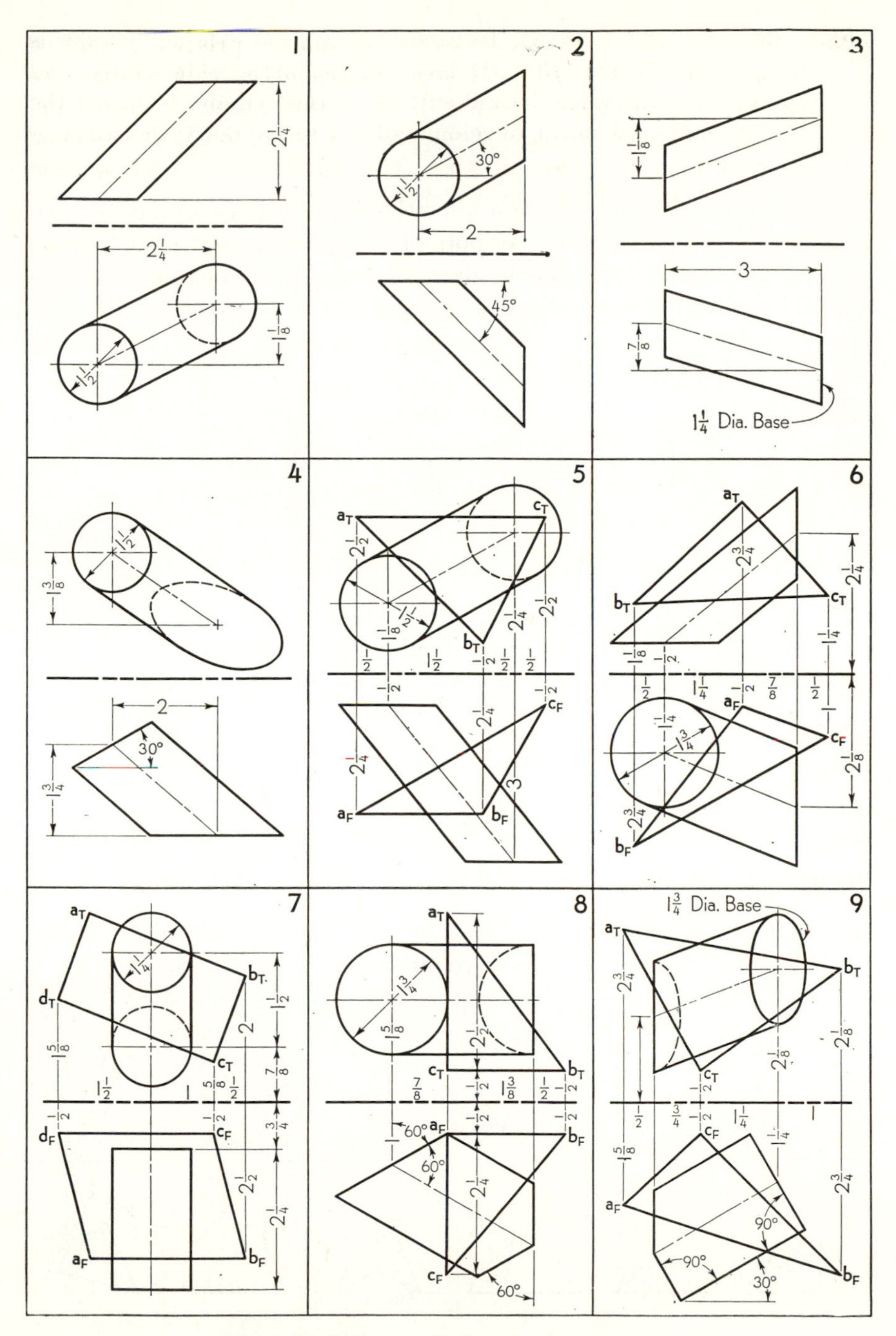

Fig. 163. Oblique cylinders for development.

110·5–110·9 FIG. 163(5) to (9). In each problem the oblique cylinder is truncated by plane *ABC*. Determine the line of intersection and then develop the whole lateral surface of the cylinder, showing the intersection curve on the development. Use Layout C or two A or B sheets.

In Probs. 110·10 to 110·15 determine the line of intersection of the two cylinders and then develop one or both of the cylinders as assigned, showing the intersection curve on the developments. Use Layout C or two A or B sheets. See also Group 95 for alternate dimensions.

110·10 FIG. 148(1).
110·11 FIG. 148(2).
110·12 FIG. 148(3).
110·13 FIG. 148(4).
110·14 FIG. 148(5).
110·15 FIG. 148(6).

Additional problems on the development of intersected cylinders can be selected from Figs. 149, 150, 152, 153, 154, 156, and 159.

Group 111. Development of right pyramids (Art. 10·10)

111·1–111·6 FIG. 164(1) to (6). Complete the top view, and develop the lateral surfaces of the right pyramid. Use Layout B.
111·7 FIG. 65(1) B. Determine the intersection of the right pyramid and plane *ABC*, and develop the portion of the pyramid below the plane.
111·8 FIG. 149(3) B. Determine the intersection of the right pyramid and the cylinder, and develop the pyramid.
111·9 FIG. 149(4) B. Same as Prob. 111·8.

Group 112. Development of oblique pyramids (Art. 10·11)

112·1–112·6 FIG. 165(1) to (6). Complete the top view, and develop the lateral surfaces of the oblique pyramid. Use Layout B.

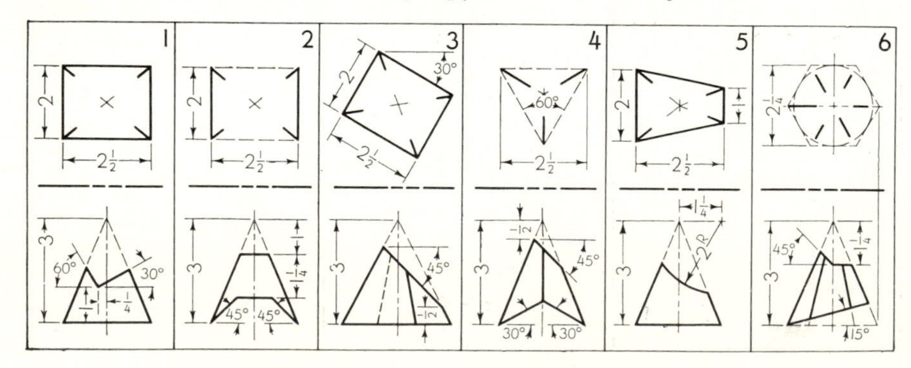

Fig. 164. Right pyramids for development.

112·7 FIG. 65(3) B. Determine the intersection of the oblique pyramid and plane *ABC*, and develop the whole pyramid, showing the intersection line on the development.

112·8 FIG. 65(5) B. Same as Prob. 112·7.

112·9 FIG. 65(8) B. Same as Prob. 112·7.

Group 113. Development of right circular cones (Art. 10·12)

113·1–113·6 FIG. 166(1) to (6). Complete the top view, and develop the lateral surface of the right circular cone. Use Layout B.

113·7 FIG. 114(1) B. Determine the intersection of the cone and plane *ABC*, and develop the portion of the cone below the plane.

In Probs. 113·8 to 113·13 determine the line of intersection of the two given surfaces, and then develop the lateral surface of the cone, showing the intersection curve on the development. See also Group 97 for alternate dimensions.

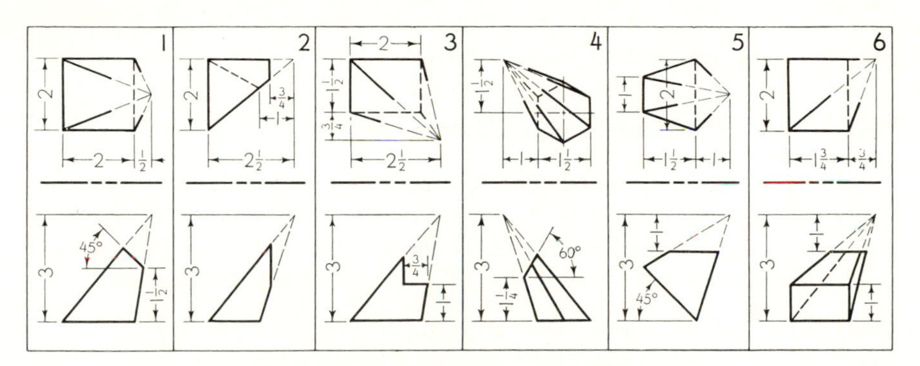

Fig. 165. Oblique pyramids for development.

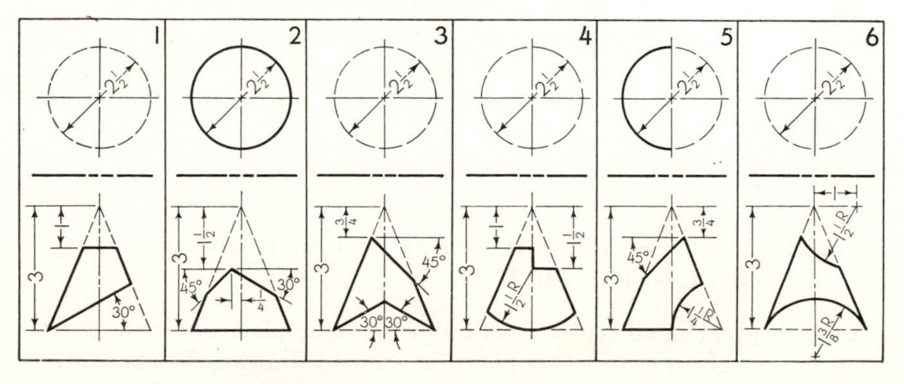

Fig. 166. Right circular cones for development.

113·8 FIG. 149(1) B.
113·9 FIG. 149(2) B.
113·10 FIG. 149(5) C.
113·11 FIG. 149(6) C.
113·12 FIG. 149(7) C.
113·13 FIG. 153(4) C.

Group 114. Development of oblique cones (Arts. 10·13, 10·14)

114·1–114·4 FIG. 167(1) to (4). Develop the lateral surface of the oblique cone. Use Layout B.

114·5–114·9 FIG. 167(5) to (9). In each problem the oblique cone is truncated by plane *ABC*. Determine the line of intersection, and then develop the whole lateral surface of the cone, showing the intersection curve on the development. Use Layout B, except for Prob. 114·9, which requires Layout C.

114·10 FIG. 149(8) C. Determine the line of intersection of the two cones, and then develop one or both of the cones as assigned.

114·11 FIG. 149(9) C. Determine the line of intersection of the cone and cylinder, and then develop the cone.

Additional problems on the development of intersected cones can be selected from Figs. 151 to 153 and 155.

Group 115. Cylindrical pipe elbows (Art. 10·15)

In each problem draw the cylindrical pipe elbow and its development, using the specified values of N, D, R, and θ. Use Layout B, showing elbow and development as in Figs. 10·14(*d*) and 10·14(*e*). (If desired, only the end section and one adjacent section need be developed.)

115·1 $N = 4$, $D = 1\frac{1}{2}$ in., $R = 2\frac{1}{2}$ in., $\theta = 75°$.
115·2 $N = 6$, $D = 1\frac{1}{2}$ in., $R = 3\frac{1}{2}$ in., $\theta = 90°$.
115·3 $N = 3$, $D = 1\frac{3}{4}$ in., $R = 2\frac{1}{2}$ in., $\theta = 60°$.
115·4 $N = 4$, $D = 1\frac{3}{4}$ in., $R = 2$ in., $\theta = 75°$.
115·5 $N = 5$, $D = 1\frac{3}{4}$ in., $R = 2$ in., $\theta = 90°$.
115·6 $N = 6$, $D = 1\frac{3}{4}$ in., $R = 2\frac{1}{2}$ in., $\theta = 90°$.
115·7 $N = 3$, $D = 2$ in., $R = 1\frac{1}{2}$ in., $\theta = 45°$.
115·8 $N = 4$, $D = 2$ in., $R = 1\frac{1}{2}$ in., $\theta = 75°$.
115·9 $N = 5$, $D = 2$ in., $R = 1\frac{1}{2}$ in., $\theta = 90°$.

Group 116. Development of cones and cylinders with plane intersections (Arts. 10·16, 10·17)

In each problem all center lines lie in the same plane and appear true length in the given view. Complete the view by drawing the plane intersections between sections, and then develop one symmetrical half of each

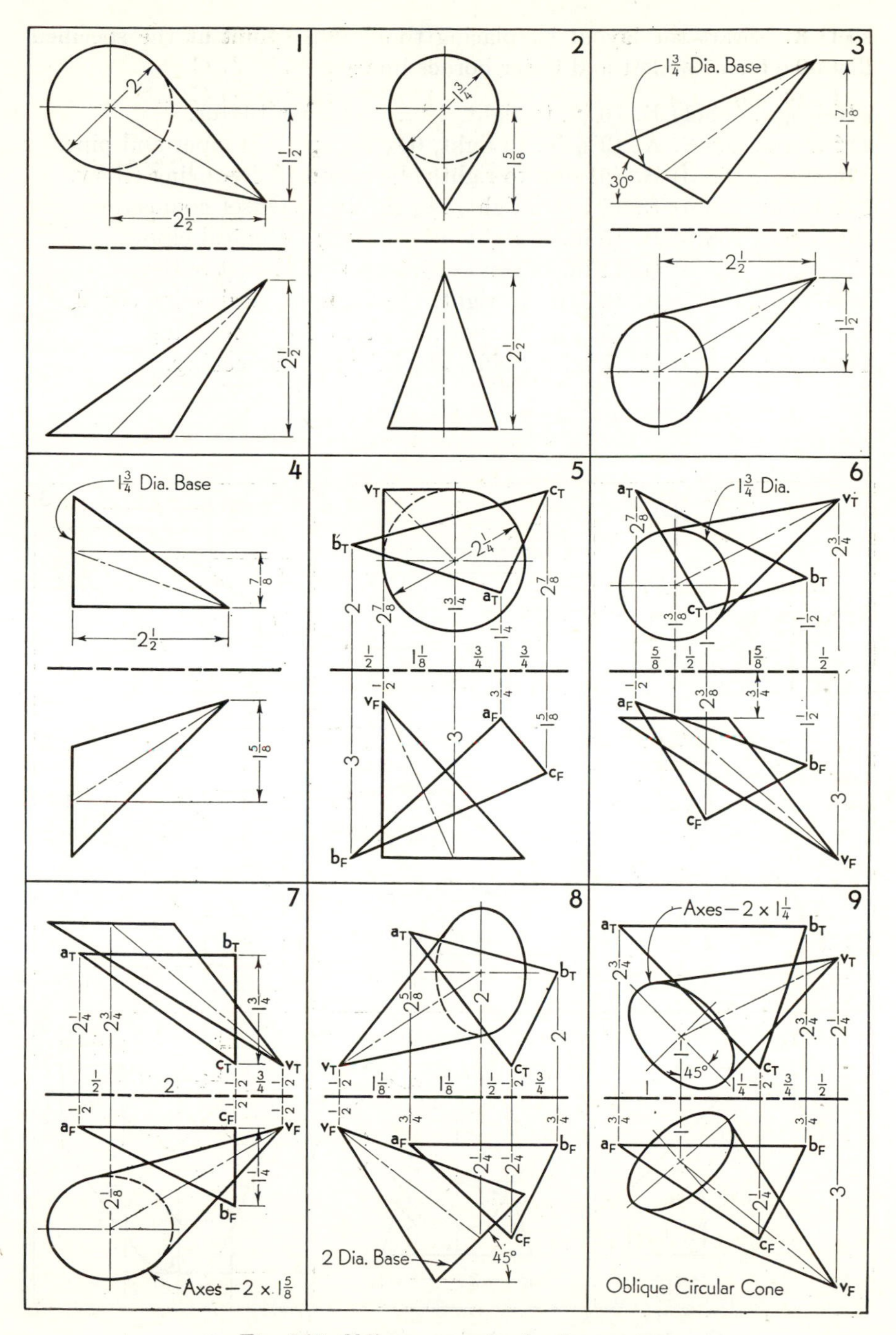

Fig. 167. Oblique cones for development.

section. Start the layout by placing the starred point at the specified distance from the left and lower border line.

116·1 FIG. 168(1) B. (6 in. to right, 1 in. up.) Ventilator.
116·2 FIG. 168(2) A. ($2\frac{3}{4}$ in. to right, $6\frac{3}{4}$ in. up.) Hopper and pipe.
116·3 FIG. 168(3) B. ($5\frac{1}{2}$ in. to right, $1\frac{1}{4}$ in. up.) Reducing elbow.
116·4 FIG. 168(4) B. (5 in. to right, $1\frac{1}{4}$ in. up.) Offset connection.
116·5 FIG. 168(5) B. (5 in. to right, $1\frac{3}{4}$ in. up.) Conical elbow.
116·6 FIG. 168(6) B. (4 in. to right, 1 in. up.) Exhaust outlet.
116·7 FIG. 168(7) B. ($6\frac{1}{4}$ in. to right, $1\frac{1}{4}$ in. up.) Rain-pipe cutoff.
116·8 FIG. 168(8) B. ($5\frac{1}{2}$ in. to right, $1\frac{3}{4}$ in. up.) Breeching.
116·9 FIG. 168(9) C. (7 in. to right, $2\frac{1}{4}$ in. up.) Breeching.

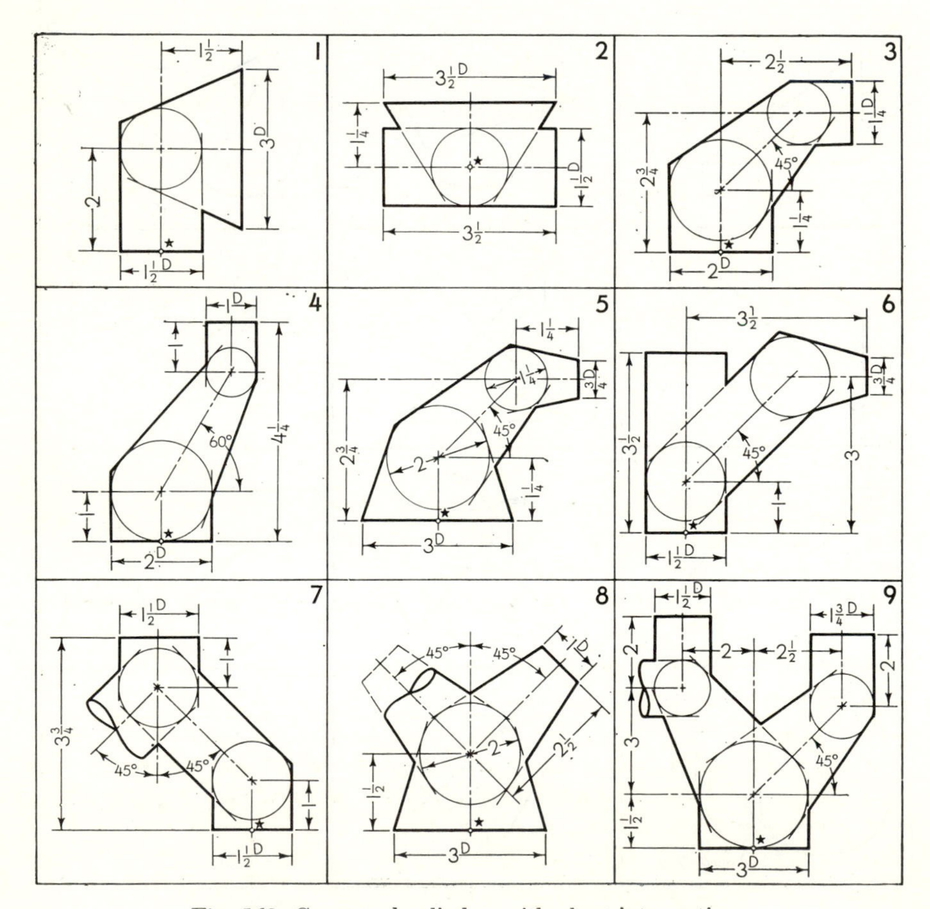

Fig. 168. Cones and cylinders with plane intersections.

Group 117. Development by triangulation (Arts. **10·18, 10·19**)

In each problem divide the transition piece into developable surfaces, and then develop it. If the piece is symmetrical, develop only one symmetrical half. Use Layout B, and place the given views on the left and the development on the right.

117·1 FIG. 169(1).
117·2 FIG. 169(2).
117·3 FIG. 169(3).
117·4 FIG. 169(4).
117·5 FIG. 169(5).
117·6 FIG. 169(6).
117·7 FIG. 169(7).
117·8 FIG. 169(8).
117·9 FIG. 169(9).
117·10 FIG. 169(10).
117·11 FIG. 169(11). Use plane surfaces wherever possible.
117·12 FIG. 169(12). Show also the developed shape of the hole in the cylinder.

Group 118. Development of convolute and warped surfaces (**Arts. 10·20, 10·21**)

In each problem the given surface may be developed as a convolute (Art. 10·20) or as a warped surface (Art. 10·21). Use either method or that assigned by the instructor. Develop only one symmetrical half of the surface (except in Prob. 118·4). Problems marked for Layout C or B can be solved on Layout B by drawing and developing only the symmetrical right half of the surface.

118·1 FIG. 170(1) C or B. Stoker distributing hood.
118·2 FIG. 170(2) C or B. Automobile hood.
118·3 FIG. 170(3) C. Pipe connection.
118·4 FIG. 170(4) C. Offset pipe connection.
118·5 FIG. 170(5) C or B. Elliptical pipe connection.
118·6 FIG. 170(6) C. Elliptical elbow.
118·7 FIG. 170(7) C or B. Oil-drum hinged cover (surface connects circular arcs *ABC* and *ADC*).
118·8 FIG. 170(8) C or B. Parabolic air scoop (surface connects parabola *ABC* and semicircle *ADC*).
118·9 FIG. 170(9) C or B. Flared connection from a circular pipe to a cylindrical drum (the curve that appears as an ellipse in the top view is a double-curved line lying on the surface of the 6-in.-diameter drum).
118·10 FIG. 171(1) B. Ventilating hood.
118·11 FIG. 171(2) B. Eccentric pipe connection.
118·12 FIG. 171(3) C. Eccentric pipe connection.

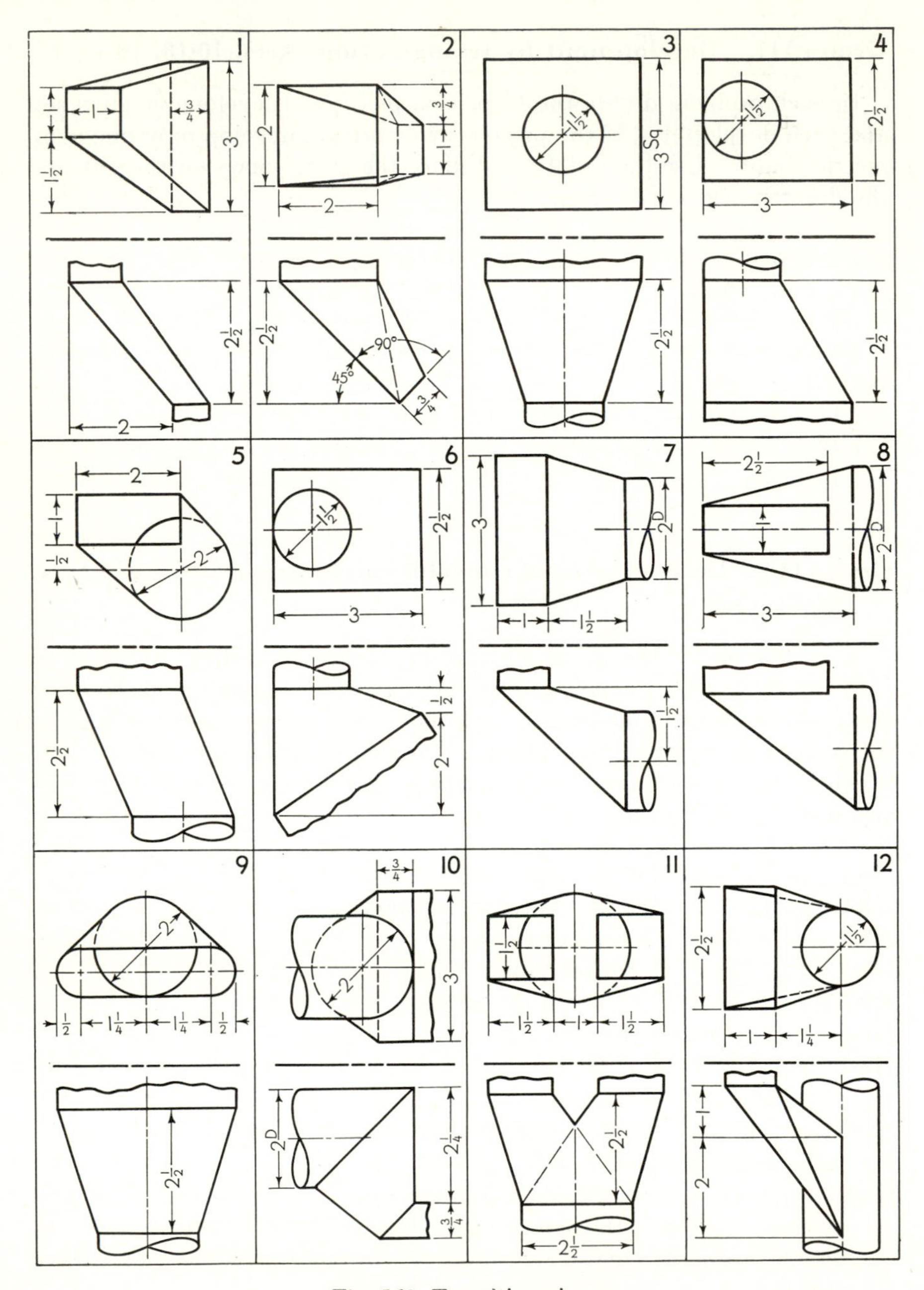

Fig. 169. Transition pieces.

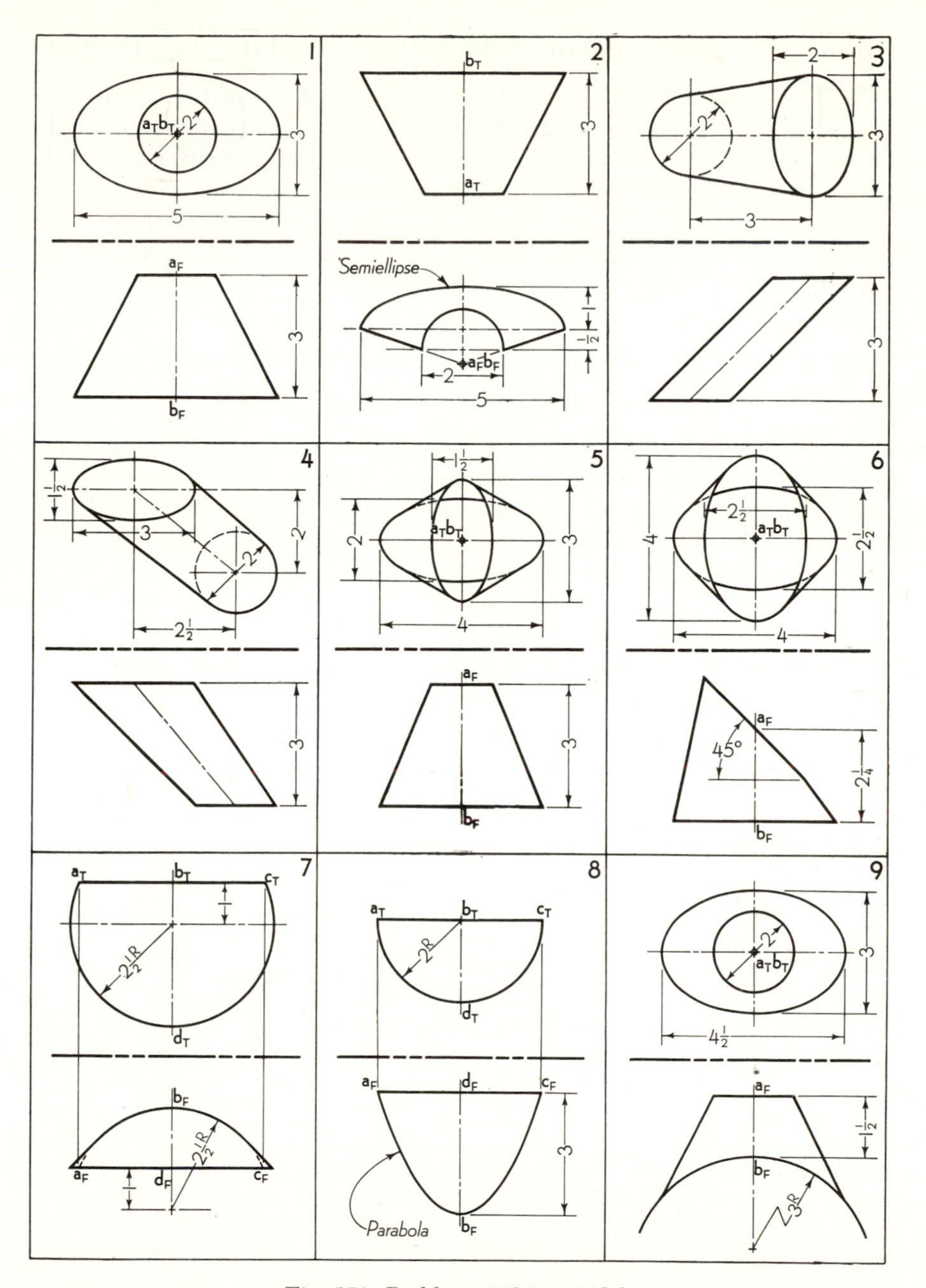

Fig. 170. Problems 118·1 to 118·9.

Group 119. Development of the helical convolute (Art. 10·22)

In Probs. 119·1 to 119·8 construct the development of the helical convolute whose dimensions are given in the correspondingly numbered problem of Group 74. Determine all values graphically (helix angle as in Fig. 6·28), but check each by calculation. Do *not* draw views of the surface; draw only the development and necessary constructions. For Probs. 119·1 to 119·4 use Layout A; for Probs. 119·5 to 119·8 use Layout B.

Group 120. Approximate development of double-curved surfaces (Art. 10·23)

120·1 B. Make an approximate development of one 30° gore (one-twelfth of the surface) of a 4-in.-diameter hemisphere.

120·2 B. Make an approximate development of a 2½-in.-diameter hemisphere by the zone method. Use four equal width zones.

120·3 B. Same as Prob. 120·2, but use three zones.

In Probs. 120·4 to 120·11 make an approximate development of one 45° gore (one-eighth of the total surface) of the given double-curved surface. Use Layout B.

120·4 FIG. 140(1).
120·5 FIG. 140(2).
120·6 FIG. 140(3).
120·7 FIG. 140(4).
120·8 FIG. 141(1).
120·9 FIG. 141(2).
120·10 FIG. 141(3).
120·11 FIG. 141(4).

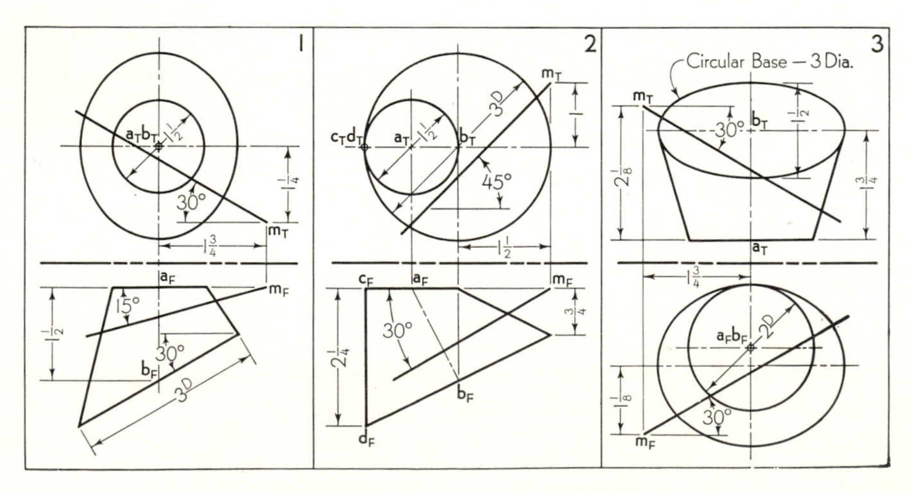

Fig. 171. Problems 118·10 to 118·12.

In Probs. 120·12 to 120·20 make an approximate development of one symmetrical half of the given double-curved surface by the zone method. Use three to six zones depending upon the curvature of the surface. Make zones of equal width when the meridian curve is an arc of a circle; make them unequal when the meridian curve is parabolic or hyperbolic. Use Layout B.

120·12 FIG. 140(1).
120·13 FIG. 140(2).
120·14 FIG. 140(3).
120·15 FIG. 140(4).
120·16 FIG. 140(5).
120·17 FIG. 141(1).
120·18 FIG. 141(2).
120·19 FIG. 141(3).
120·20 FIG. 141(4).

Group 121. Bend allowance (Art. 10·26)

In Probs. 121·1 to 121·5 calculate the bend allowance for the specified bend. The given angle is the angle of bend, and the radius is inside. Thickness of stock is U.S. standard gauge stated in inches. Show complete calculations.

121·1 90°, ½ in. radius, 0.125 in. thick. ANS. 0.874 in.
121·2 45°, 1 in. radius, 0.188 in. thick. ANS. 0.850 in.
121·3 135°, ⅜ in. radius, 0.062 in. thick. ANS. 0.950 in.
121·4 60°, ⅝ in. radius, 0.250 in. thick. ANS. 0.770 in.
121·5 180°, ¾ in. radius, 0.125 in. thick. ANS. 2.533 in.

In Probs. 121·6 to 121·9 draw the given views of the sheet-metal object, and then make a development showing the location of all bends, holes, etc. Dimension the development completely and accurately. All bend radii given in the figures are inside. Answers given are total over-all length of the development. Use Layout B.

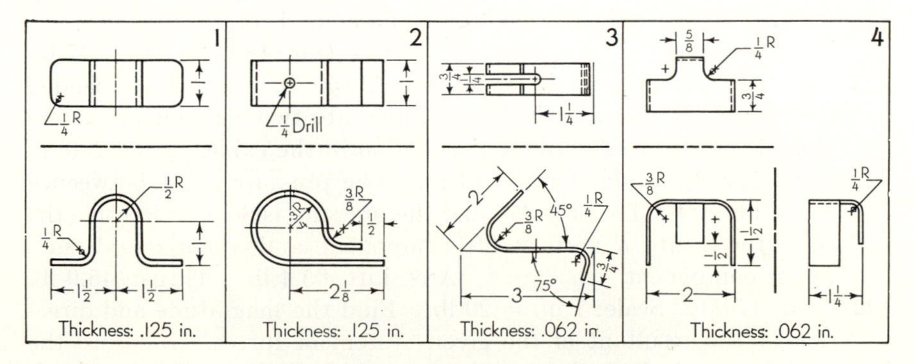

Fig. 172. Bend allowance applications.

121·6 FIG. 172(1). ANS. 5.21 in.
121·7 FIG. 172(2). ANS. 6.98 in.
121·8 FIG. 172(3). ANS. 4.29 in.
121·9 FIG. 172(4). ANS. 4.52 in.
121·10 FIG. 89 C. Scale: full size. Draw the given views of the pulley bracket, and make a fully dimensioned development of the sheet-metal part. Make angle A 30°, and the questioned dimension $1\frac{3}{8}$ in. All bend and relief radii are $\frac{3}{8}$ in. inside. (Relief radii are those in the inside corners of the development.)
ANS. Over-all length = 7.92 in.

Group 122. Coplanar-vector problems (Arts. 11·1 to 11·4)

In Probs. 122·1 to 122·15 use Layout B, and place the space diagram on the left side of the sheet, the vector diagram on the right side. Space diagrams can be small, and need not be drawn to scale. Scales given are for the vector diagram.

122·1 FIG. 173(1). Scale: 1 in. = 100 in.-lb. Find the magnitude and direction of the resultant of the two given vectors. Measure the direction angle from the right horizontal. ANS. 500 in.-lb, $36\frac{3}{4}$°.
122·2 FIG. 173(2). Scale: 1 in. = 25 fps. Same as Prob. 122·1.
ANS. 86.5 fps, 45°.
122·3 FIG. 173(3). Scale: 1 in. = 100 lb. Same as Prob. 122·1.
ANS. 397 lb, −19°.
122·4 FIG. 173(4). Scale: 1 in. = 200 lb. Same as Prob. 122·1.
ANS. 905 lb, −96°.
122·5 FIG. 173(5). Scale: 1 in. = 20 fps. Resolve the given vector into two components acting along lines OM and ON. Show the sense and magnitude of each component.
ANS. OM, 111.5 fps. ON, 30.0 fps.
122·6 FIG. 173(6). Scale: 1 in. = 50 lb. Same as Prob. 122·5.
ANS. OM, 180 lb. ON, 147 lb.
122·7 FIG. 173(7). Scale: 1 in. = 50 lb. A 150-lb weight is suspended from a cable. Resolve the 150-lb force into two components acting along the cable to determine the tension in the cable. ANS. 290 lb.
122·8 FIG. 173(8). Scale: 1 in. = 10 lb. The pressure angle between a cam and a roll follower is 35°, and the pressure is 80 lb. Resolve the 80-lb force into a vertical lift component and a horizontal side-thrust component. ANS. Lift, 65.4 lb. Thrust, 45.9 lb.
122·9 FIG. 174(1). Scale: 1 in. = 20 lb. Find the magnitude and direction of the resultant of the given system of forces. Measure the direction angle from the right horizontal. ANS. 26.5 lb, $76\frac{1}{2}$°.

122·10 FIG. 174(2). Scale: 1 in. = 100 lb. Same as Prob. 122·9.
ANS. 400 lb, $-40\frac{1}{2}°$.

122·11 FIG. 174(3). Scale: 1 in. = 1 lb. Same as Prob. 122·9.
ANS. 1.67 lb, $-141°$.

122·12 FIG. 174(4). Scale: 1 in. = 100 lb. Same as Prob. 122·9.
ANS. 0 lb.

122·13 FIG. 174(5). Scale: 1 in. = 20 fps. An automobile is traveling at 60 mph (88 fps). What is the absolute velocity of point *B* on the tire? Does any point on the wheel have zero absolute velocity?
ANS. 163 fps.

122·14 FIG. 174(6). Scale: 1 in. = 1 fps. The 6-in. crank rotates at 150 rpm. At the instant shown what is the velocity component along the center line of the rod? ANS. 5.55 fps.

122·15 FIG. 174(7). Scale: 1 in. = 200 fps. Steam at 1,500 fps velocity is directed through a nozzle against the blades of a steam turbine at an angle of 20°. The blade angle is 30°, and the peripheral velocity of the blades is 500 fps. What is the velocity of the entering steam relative to the blades? If the steam loses 20 per cent of its velocity in passing through the blades, what is the exit velocity of the steam?
ANS. 1,028 fps, 454 fps.

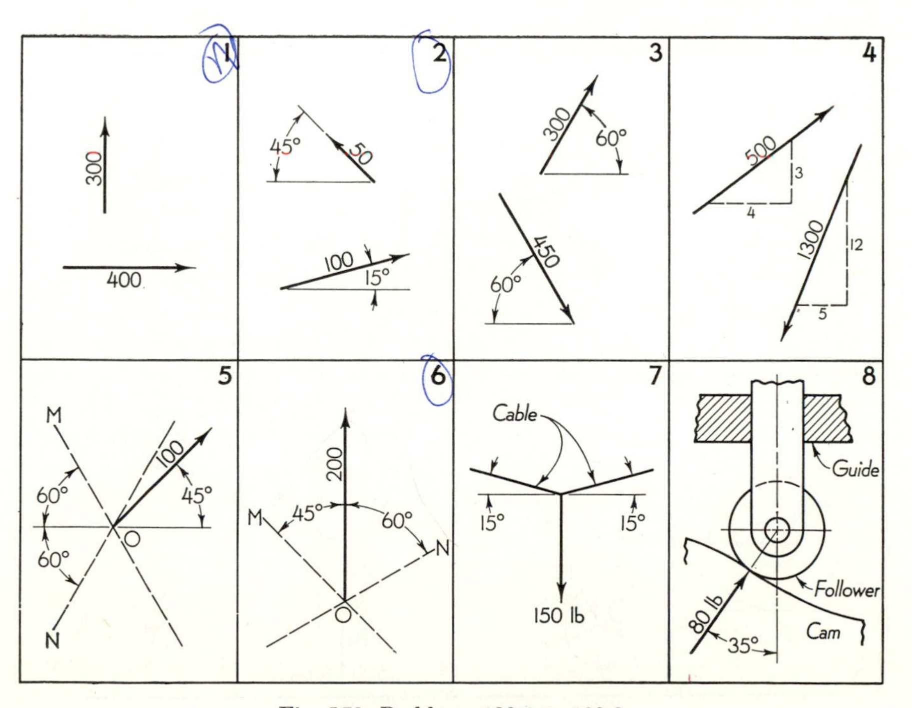

Fig. 173. Problems 122·1 to 122·8.

122·16 FIG. 174(8) B. Space scale: full size. Vector scale: 1 in. = 10 fps. The crankshaft of a small gas engine turns at 2,000 rpm. Determine the velocity of the piston at 15° crank intervals, and plot a graph of piston velocity versus piston position. Place the given view at the left with crankshaft center $2\frac{1}{4}$ in. above the lower border. Plot the graph at the right, and make the 2-in. stroke distance double size.
ANS. At 90° piston velocity is 17.5 fps.

122·17 B. Space scale: 1 in. = 2 nautical miles. Vector scale: 1 in. = 5 knots. Ship A is running S 60° E at 15 knots. Ship B, 10 miles due south of ship A, is running N 45° E at 12 knots. How many minutes will it be before the ships are closest to each other? How far apart will they then be?
ANS. 35 min, 2.7 miles.

122·18 B. Space scale: 1 in. = 1 nautical mile. Vector scale: 1 in. = 5 knots. A freighter is running due east at 14 knots. A submarine, 6 miles due southeast of the freighter, alters course and runs for a position $\frac{1}{2}$ mile off the starboard side of the freighter. If the submarine makes 10 knots, what should be its course? If the submarine makes only 8 knots, how close can it get to the freighter for a broadside shot?
ANS. N 19° E, 1.25 miles.

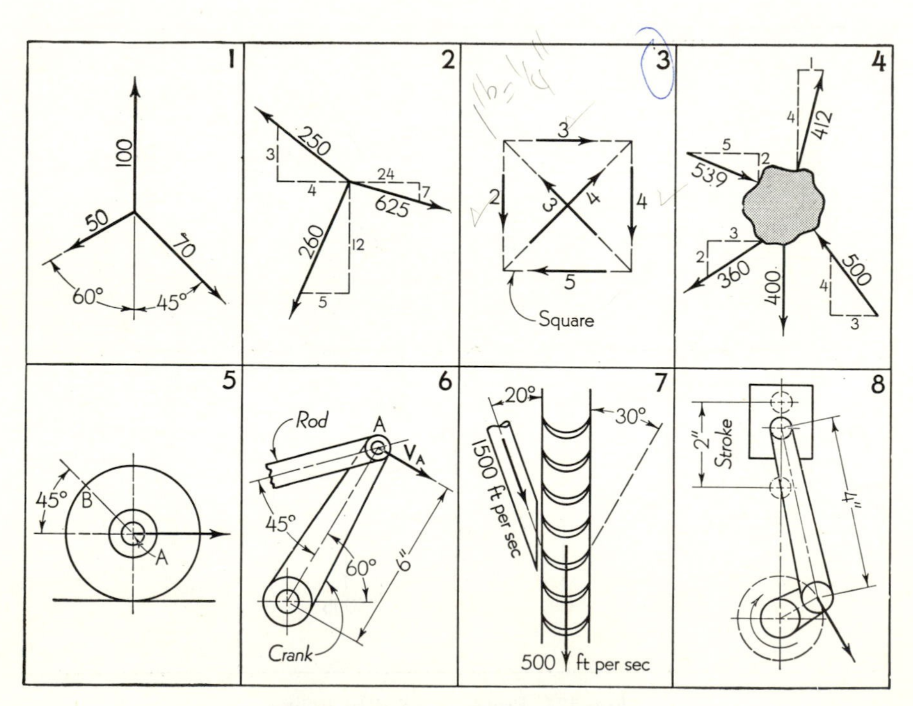

Fig. 174. Problems 122·9 to 122·16.

Group 123. Resultant of concurrent forces (Arts. 11·5 to 11·9)

In the following problems find the magnitude of the resultant of the given force system. Show the resultant on the space diagram. Use Layout B, and place the space diagram on the left, the vector diagram on the right. Space scale is full size, vector scale as given. Vectors shown in the figures are not to scale; the given points establish direction only. Make a quick freehand sketch of the vector polygon to determine general shape before selecting the starting point.

123·1 FIG. 175(1). Scale: 1 in. = 10 lb. ANS. 27.0 lb.
123·2 FIG. 175(2). Scale: 1 in. = 20 lb. ANS. 69.6 lb.
123·3 FIG. 175(3). Scale: 1 in. = 50 lb. ANS. 92.0 lb.
123·4 FIG. 175(4). Scale: 1 in. = 100 lb. ANS. 274 lb.
123·5 FIG. 175(5). Scale: 1 in. = 20 lb. ANS. 52.0 lb.
123·6 FIG. 175(6). Scale: 1 in. = 100 lb. ANS. 0 lb.

Group 124. Resultant of parallel forces (Arts. 11·10 to 11·12)

124·1 FIG. 176(1) A. Scales: ½ in. = 1 ft, and 1 in. = 4 tons. Find the magnitude and location of the resultant of the four beam loads.
ANS. 18 tons, 7 ft 3 in. from left end.

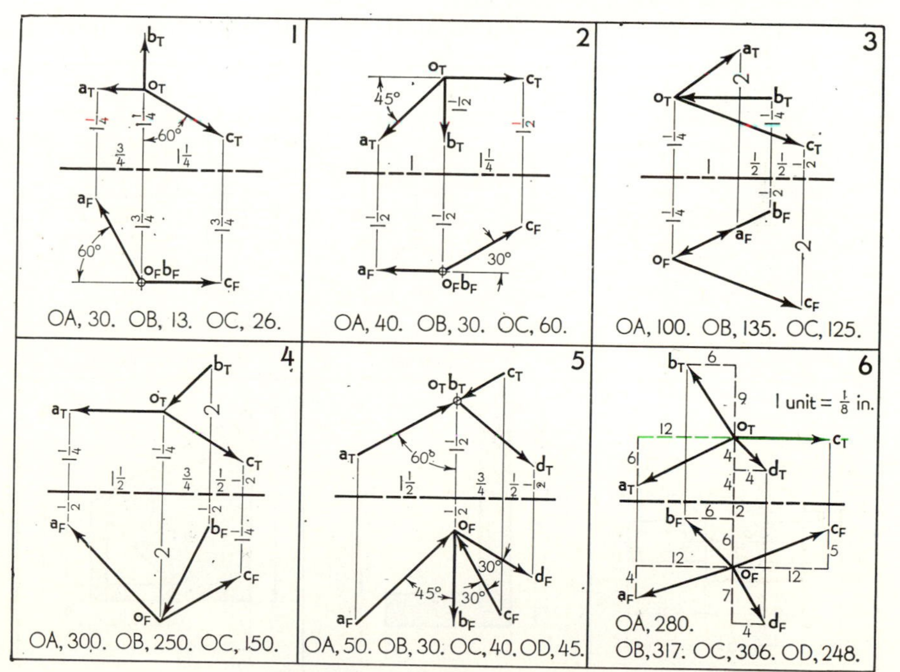

Fig. 175. Problems in Group 123.

124·2 FIG. 176(2) A. Scales: half size, and 1 in. = 20 lb. To remain in horizontal equilibrium the resultant R of the four given forces must act through the center of the pivot pin. What must be the value of distance X? ANS. $4\frac{3}{4}$ in.

124·3 FIG. 176(3) A. Scales: half size, and 1 in. = 50 lb. Find the resultant of the five forces that act on the lever.
ANS. Clockwise couple of 285 in.-lb.

124·4 FIG. 176(4) A. Scales: full size; vector scale, to suit computed values. Locate the center of gravity of the shaft.
ANS. 2.43 in. from left end.

124·5 FIG. 176(5) B. Scales: full size; vector scale, to suit computed values. Locate the center of gravity of the area from the left edge and bottom edge. ANS. 0.58 in., 1.08 in.

124·6 FIG. 176(6) B. Same as Prob. 124·5. ANS. 1.03 in., 1.22 in.

124·7 FIG. 177(1) B. Scales: $\frac{1}{4}$ in. = 1 ft, and 1 in. = 2,000 lb. Find the magnitude and location of the resultant of the four floor loads.
ANS. 4,750 lb, 9 ft 4 in. from left, 7 ft 3 in. from front.

124·8 FIG. 177(2) B. Scales: $\frac{1}{2}$ in. = 1 ft, and 1 in. = 2,000 lb. The floor reactions have been weighed at three points under a 7,300-lb compressor. From this data determine the horizontal location of the center of gravity of the machine.
ANS. 4 ft $9\frac{1}{2}$ in. from left support, 2 ft 4 in. from front support.

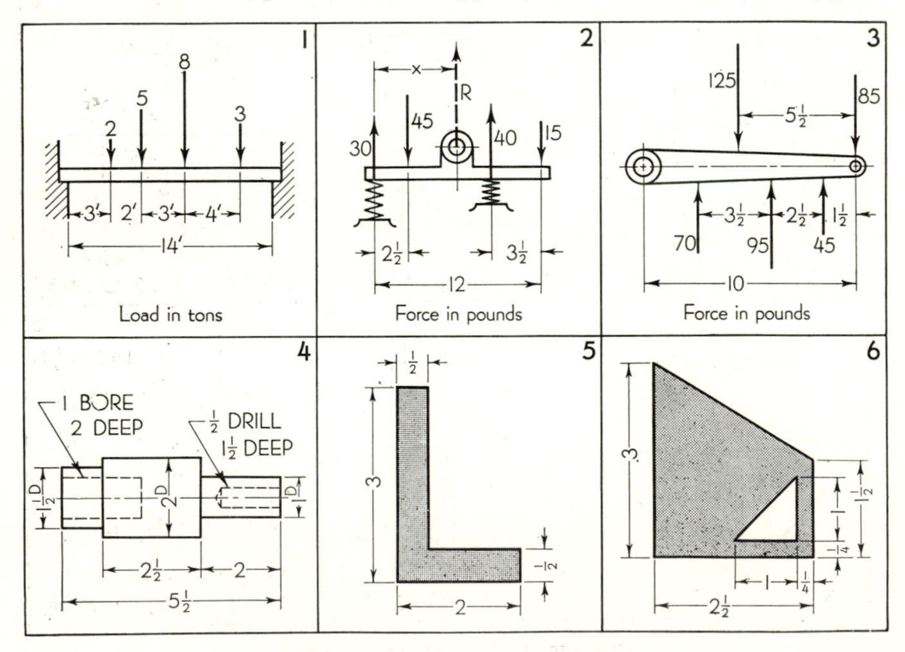

Fig. 176. Problems 124·1 to 124·6.

124·9 FIG. 177(3) B. Scales: half size, and 1 in. = 50 lb. The five forces acting on the pressure plate are supposed to be in equilibrium to keep the plate horizontal, but the location of the 135-lb force has been questioned. Check for equilibrium, and relocate the 135-lb force if necessary.

ANS. 135-lb force must be moved 0.22 in. to rear.

124·10 FIG. 177(4) B. Scales: half size; vector scale, to suit computed values. Locate the center of gravity of the solid from the left side, bottom, and rear. ANS. 1.90 in., 1.34 in., 1.14 in.

124·11 FIG. 177(5) B. Same as Prob. 124·10.

ANS. 2.36 in., 1.32 in., 1.34 in.

124·12 FIG. 177(6) B. Scales: full size; vector scale, to suit computed values. The adjusting disk was designed to keep the center of gravity as close as possible to the exact center of the 3-in. disk. Check for maximum deviation in a coordinate direction. ANS. 0.05 in.

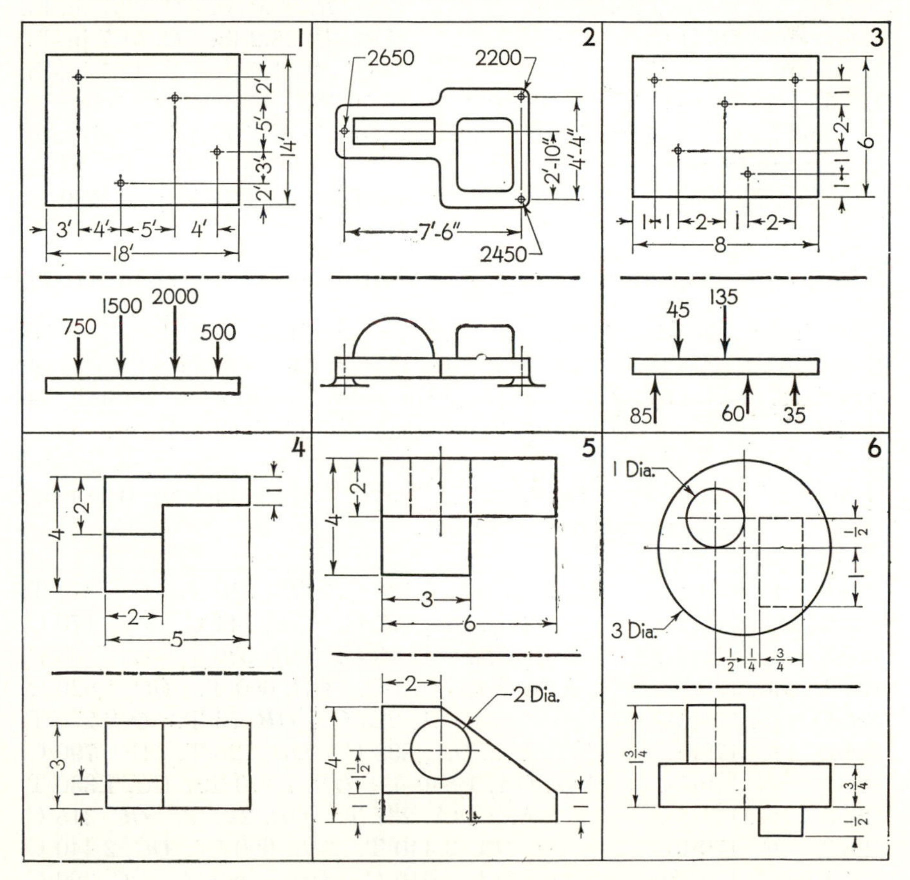

Fig. 177. Problems 124·7 to 124·12.

Group 125. Resultant of noncoplanar nonconcurrent forces (Arts. 11·13 to 11·15)

In the following problems reduce the given force system to a single force acting through the given point *O* and a couple. Determine the magnitudes of the force and couple, and represent them as in stage 6 of Fig. 11·20. Center the given top and front views in the left half of the sheet, and add a right-side view. Vectors shown in the figures are not to scale; the given points only establish position and direction. For all problems the space scale is full size, force scale is 1 in. = 10 lb (except Prob. 125·6), and couple vector scale is 1 in. = 20 in.-lb.

125·1 FIG. 178(1) B. ANS. *R*, 19.3 lb. *C*, 20.6 in.-lb.
125·2 FIG. 178(2) B. ANS. *R*, 35.4 lb. *C*, 43.9 in.-lb.
125·3 FIG. 178(3) B. Also locate point o_T so that the resultant couple will lie in a vertical plane. ANS. *R*, 30.0 lb. *C*, 26.7 in.-lb.
125·4 FIG. 178(4) C. ANS. *R*, 28.2 lb. *C*, 51.7 in.-lb.
125·5 FIG. 178(5) C. ANS. *R*, 31.0 lb. *C*, 43.1 in.-lb.
125·6 FIG. 178(6) C. Force scale: 1 in. = 50 lb. Also locate points o_T and o_F so that there is no component couple in either the horizontal or frontal plane.
ANS. *R*, 300 lb. *C*, 105 in.-lb. From *T–F*: o_T, 1.77 in.; o_F, 2.30 in.

Group 126. Equilibrium of concurrent forces (Arts. 11·16, 11·17)

In Probs. 126·1 to 126·12 determine the magnitude and sense (*tension* or *compression*) of the force acting along each of the members of the given structure. Use Layout B, and place the space diagram on the left, the vector diagram on the right. Space scale is full size. Make a quick freehand sketch of the vector polygon to determine general shape and then select a suitable vector scale. All given force values are true magnitudes. Answers are given in pounds.

126·1 FIG. 179(1). ANS. *OA*, 2,120 C. *OB*, 810 T. *OC*, 810 T.
126·2 FIG. 179(2). ANS. *OA*, 227 C. *OB*, 144 C. *OC*, 170 C.
126·3 FIG. 179(3). ANS. *OA*, 66 C. *OB*, 226 T. *OC*, 153 C.
126·4 FIG. 179(4). ANS. *OA*, 1,070 C. *OB*, 660 T. *OC*, 1,920 C.
126·5 FIG. 179(5). ANS. *OA*, 223 C. *OB*, 73 T. *OC*, 276 T.
126·6 FIG. 179(6). ANS. *OA*, 808 C. *OB*, 226 T. *OC*, 790 C.
126·7 FIG. 179(7). ANS. *OA*, 1,640 T. *OB*, 2,110 T. *OC*, 1,330 T.
126·8 FIG. 179(8). ANS. *OA*, 237 T. *OB*, 212 T. *OC*, 315 C.
126·9 FIG. 179(9). ANS. *OA*, 3,440 T. *OB*, 990 C. *OC*, 2,440 C.
126·10 FIG. 179(10). ANS. *OA*, 2,040 C. *OB*, 1,980 T. *OC*, 990 C.

126·11 FIG. 179(11). ANS. *AC*, 925 C. *BC*, 4,610 T. *AD*, 3,030 C. *CE*, 3,400 T. *DE*, 4,710 C.

126·12 FIG. 179(12). ANS. *AB*, 7,210 T. *BC*, 2,130 C. *BD*, 5,880 C. *BE*, 6,500 T. *CE*, 4,070 C. *DE*, 4,800 C.

126·13 FIG. 10 B. Scales: ⅛ in. = 1 ft, and 1 in. = 10,000 lb. Assume that the shear-leg derrick supports a vertical load of 40,000 lb suspended from joint *C*. Find the magnitude and sense of the force in leg *BC* and cable *AC*. From the symmetry of the members it may be assumed that the stresses in the cables are equal.

ANS. *AC*, 15,600 T. *BC*, 32,800 C.

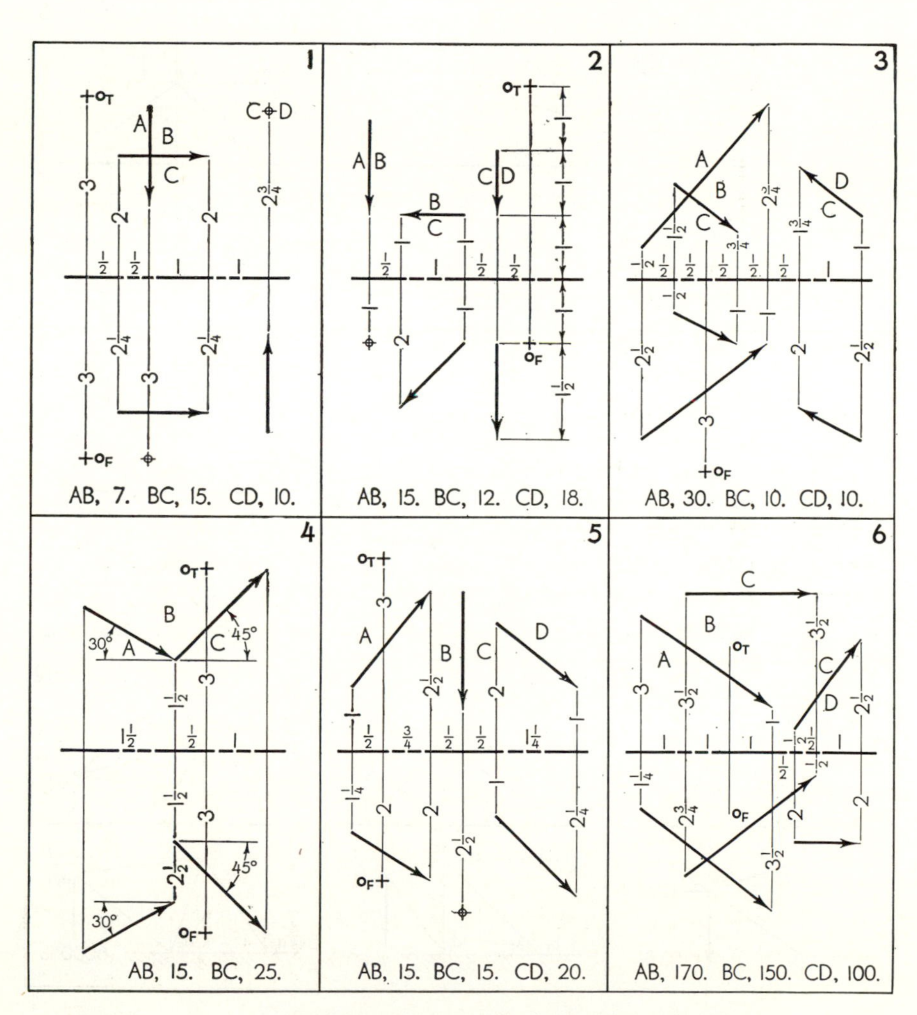

Fig. 178. Problems in Group 125.

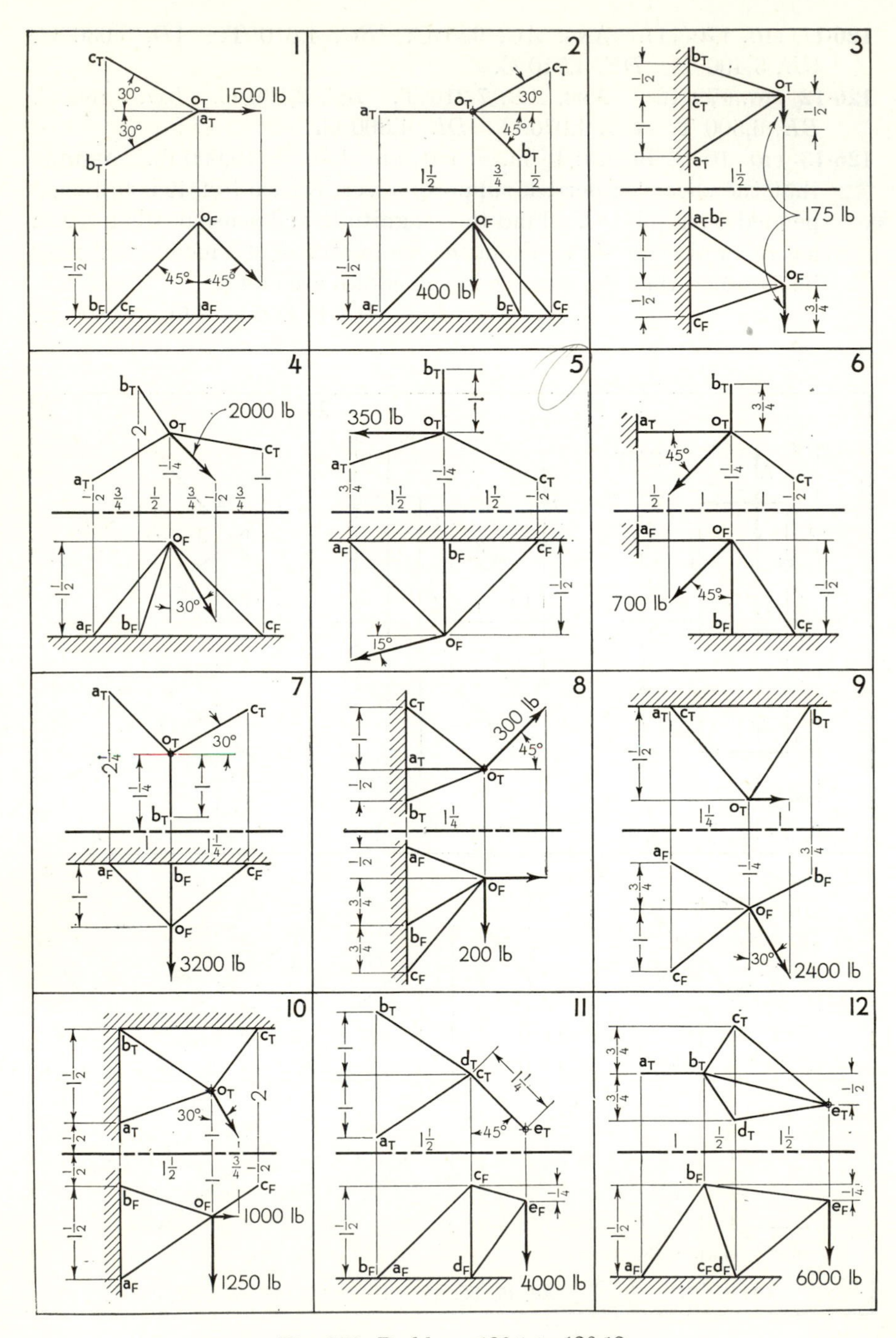

Fig. 179. Problems 126·1 to 126·12.

126·14 FIG. 14 A. Scales: 1 in. = 1 ft, and 1 in. = 500 lb. Assume that a vertical wheel load of 1,800 lb acts upward at joint *D* (neglect bending of members caused by offset wheel). Find the magnitude and sense of the axial force in each strut of the landing gear.
ANS. *AD*, 2,080 C. *BD*, 1,380 C. *CD*, 1,250 T.

126·15 FIG. 15 A. Scales: ¾ in. = 1 ft, and 1 in. = 500 lb. Same as Prob. 126·14. ANS. *AD*, 1,595 C. *BD*, 2,355 C. *CD*, 2,005 T.

126·16 FIG. 180(1) C. Scales: 1 in. = 1 ft, and 1 in. = 200 lb. A tripod rests on a smooth floor and supports a load of 1,000 lb. At the base the legs are connected by three tie rods. Find the compressive force in each leg, and the tension in each tie rod. ANS. *OA*, 500 C. *OB*, 325 C. *OC*, 395 C. *AB*, 145 T. *AC*, 165 T. *BC*, 87 T.

Group 127. Equilibrium of noncoplanar parallel forces (Art. 11·18)

127·1 FIG. 180(1) B. Scales: ¾ in. = 1 ft, and 1 in. = 500 lb. A tripod rests on a smooth floor, and supports a load of 1,000 lb. At the base the legs are connected by three tie rods. Determine the vertical force that the floor exerts on each leg.
ANS. F_A, 410 lb. F_B, 250 lb. F_C, 340 lb.

127·2 FIG. 180(2) B. Scales: ½ in. = 1 ft, and 1 in. = 500 lb. A triangular platform is supported by three vertical cables—one at each vertex—and carries three loads as shown. Determine the tension in each cable. ANS. F_A, 280 lb. F_B, 300 lb. F_C, 420 lb.

127·3 FIG. 180(3) B. Scales: half size, and 1 in. = 50 lb. A 5- by 6-in. pressure plate rests on five springs. The springs are adjusted so that the forces exerted by the springs at *A* and *B* are equal, and the force exerted by the center spring at *E* is twice that of spring *A*. What compressive force is produced in each spring by the 100-lb load? ANS. F_A, 10 lb. F_B, 10 lb. F_C, 47 lb. F_D, 13 lb. F_E, 20 lb.

127·4 FIG. 180(4) B. Scales: half size, and 1 in. = 50 lb. The rocker arm, supported on bearings at *A* and *B*, is operated by a 100-lb force as shown. Determine force *T* and the vertical bearing forces.
ANS. *T*, 125 lb. F_A, 172 lb. F_B, 53 lb.

127·5 FIG. 180(5) B. Scales: 3 in. = 1 ft, and 1 in. = 50 lb. A treadle, spring arm, and rocker arm are keyed to a shaft which is supported on bearings at *A* and *B*. The return spring exerts a force of 70 lb when 100 lb is applied on the treadle. Determine force *T* and the vertical bearing forces. ANS. *T*, 109 lb. F_A, 73 lb. F_B, 66 lb.

127·6 FIG. 180(6) B. Scales: 3 in. = 1 ft, and 1 in. = 20 lb. A ½-hp motor weighing 50 lb is pivoted on bearings at *A* and *B*. The belt tensions T_1 and T_2 partly support the motor, and under running load $T_1 = 4T_2$. Determine the belt tensions and the vertical bearing forces.

ANS. T_1, 32.2 lb. T_2, 8.05 lb. F_A, 13.2 lb down, F_B, 22.9 lb up.

Group 128. Equilibrium of general noncoplanar force systems (Arts. 11·19, 11·20)

In the following problems determine the magnitude and sense of all unknown forces acting on the given body. Lay out the given views to scale, and add a right-side view. Detail size dimensions omitted from

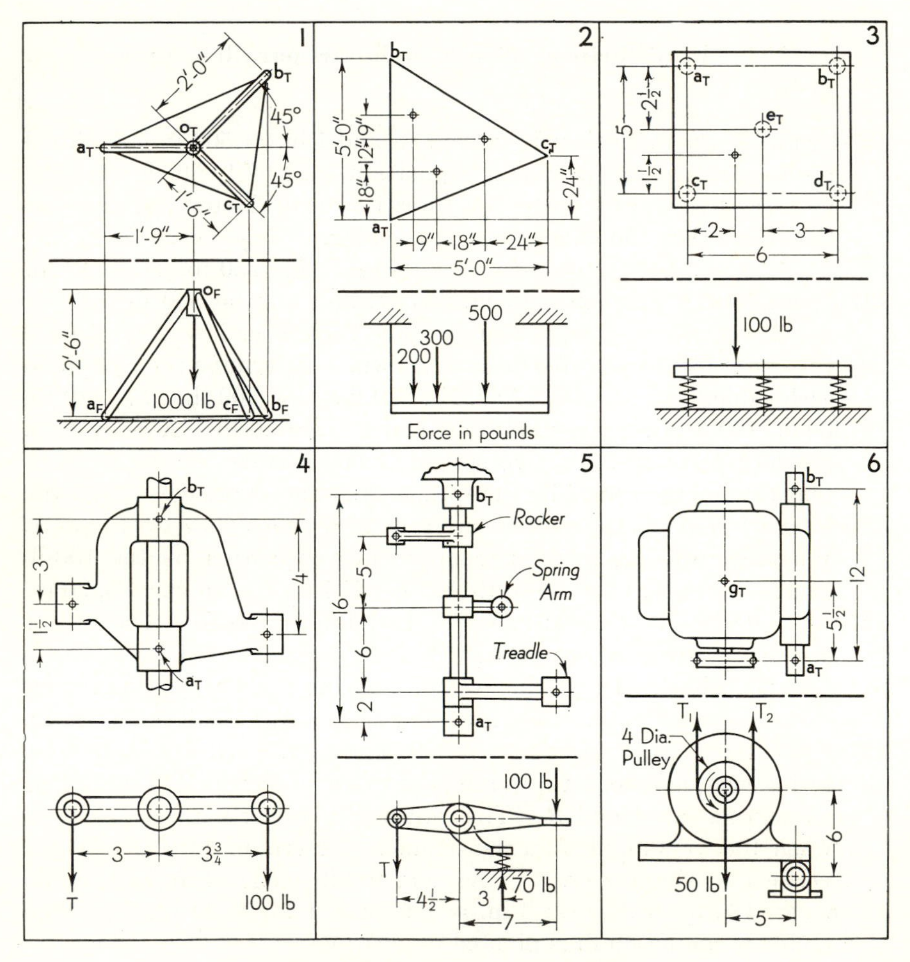

Fig. 180. Problems in Group 127.

the figures have no effect on the solution, and may be assumed in any reasonable proportions. Forces of unknown direction should be represented by x, y, and z components, taking the axis directions conventionally, as in Fig. 11·11. When known forces intersect off the paper, combine them by the method of Fig. 11·13. When the force triangle is narrow and elongated, the method of Fig. 11·13 can also be employed to improve accuracy. When the point of concurrency falls off the paper, use the construction given in Art. A·2 of the Appendix. Scales given are for Layout C, which is recommended for more accurate results, but Layout B can be used by halving the given scales. All answers are given in pounds.

128·1 FIG. 181(1). Scales: half size, and 1 in. = 100 lb. The offset bell crank is mounted on a vertical shaft supported on bearings at A and B. A 200-lb force pulling on the ball joint is opposed by the force P. Determine force P, the axial thrust T, and the bearing force components at A and B.

ANS. P, 280. T, 90. A_X, 84. A_Y, 42. B_X, 107. B_Y, 113.

128·2 FIG. 181(2). Scales: 3 in. = 1 ft, and 1 in. = 50 lb. The inner pin of the take-up rocker arm supports two loads—150 lb vertically and 80 lb horizontally, as shown. Determine the adjusting force P, the axial thrust T, and the bearing force components at A and B.

ANS. P, 128. T, 80. A_X, 141. A_Z, 3. B_X, 30. B_Z, 89.

128·3 FIG. 181(3). Scales: ⅜ in. = 1 ft, and 1 in. = 200 lb. The upper and lower ends of the windlass are set in sockets capable of exerting compression only. Determine the force P, and the components of the socket forces at A and B.

ANS. P, 425. A_X, 115. A_Y, 365. A_Z, 225. B_X, 335. B_Y, 385.

128·4 FIG. 181(4). Scales: ⅛ in. = 1 ft, and 1 in. = 500 lb. The boom, weighing 640 lb, rests on a hinge-and-swivel joint at A, and is supported by cables CE and DE. The 1,200-lb load acts at an angle of 45° with the horizontal. Find the tension in each cable and the components of the force at A.

ANS. CE, 330 T. DE, 1,930 T. A_X, 960. A_Y, 440. A_Z, 840.

128·5 FIG. 181(5). Scales: half size, and 1 in. = 200 lb. A 45-tooth worm gear of ½-in. pitch is supported between ball bearings at A and B, and drives a 5-in. chain sprocket on the extended shaft. The worm (not shown) applies a force of 670 lb on the teeth of the gear. This force acts at 15°30′ (the *lead angle*), as shown, and at 20° slope angle (the *pressure angle*). Determine the chain tension P (slack side tension is zero), the axial thrust T, and the radial load components on the bearings at A and B.

ANS. P, 870. T, 170. A_X, 305. A_Z, 1,265. B_X, 305. B_Z, 170.

128·6 FIG. 181(6). Scales: half size, and 1 in. = 200 lb. A pair of helical gears is keyed to a shaft supported in bearings at A and B. The $17\frac{1}{2}°$ *normal pressure angle* of the gears appears as 20° (70° with the gear radius) in the front view. Both gears have a 30° *helix angle*, but this angle appears as 30° only for the smaller gear. Determine the gear pressure P, the axial thrust T, and the bearing-force components at A and B.

ANS. P, 515. T, 105. A_X, 740. A_Z, 75. B_X, 60. B_Z, 125.

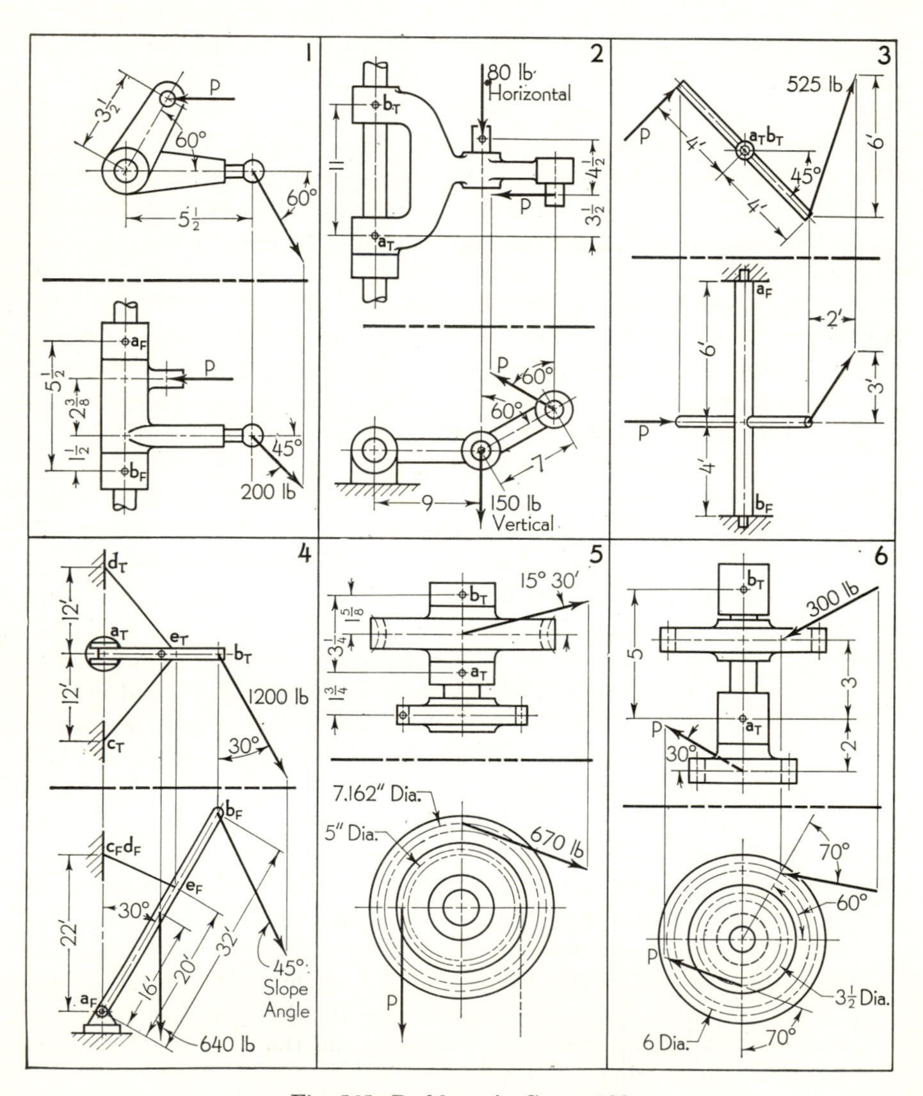

Fig. 181. Problems in Group 128.

Group 129. Strike and dip (Arts. 12·1 to 12·6)

In the following problems find the strike and dip of the stratum determined by points *A*, *B*, and *C*. Use Layout B, and solve by the method of either Art. 12·5 or 12·6 or as specified by the instructor. In Fig. 182 north is toward the top of the figure in all problems. In Probs. 129·4 to 129·6, distances along the given bearings are *map distances* (Art. 3·2) from the bench mark (B.M.).

129·1 FIG. 182(1). Scale: 1 in. = 100 ft. ANS. N 66½° E, 34° SE.
129·2 FIG. 182(2). Scale: 1 in. = 200 ft. ANS. N 41° W, 38½° SW.
129·3 FIG. 182(3). Scale: 1 in. = 500 ft. ANS. N 83½° W, 53° S.
129·4 FIG. 182(4). Scale: 1 in. = 100 ft. ANS. N 83° W, 40° N.
129·5 FIG. 182(5). Scale: 1 in. = 200 ft. ANS. N 31½° E, 23½° NW.
129·6 FIG. 182(6). Scale: 1 in. = 500 ft. ANS. N 49° W, 48° NE.

Group 130. Apparent dip (Art. 12·7)

In the following problems the dip and bearing of two apparent dip lines are given. Find the strike and dip of the stratum. No specific scale is necessary. A common point *A* in the stratum may be assumed in any convenient location on the paper. Use either the conventional or the geologist's method or that specified by the instructor.

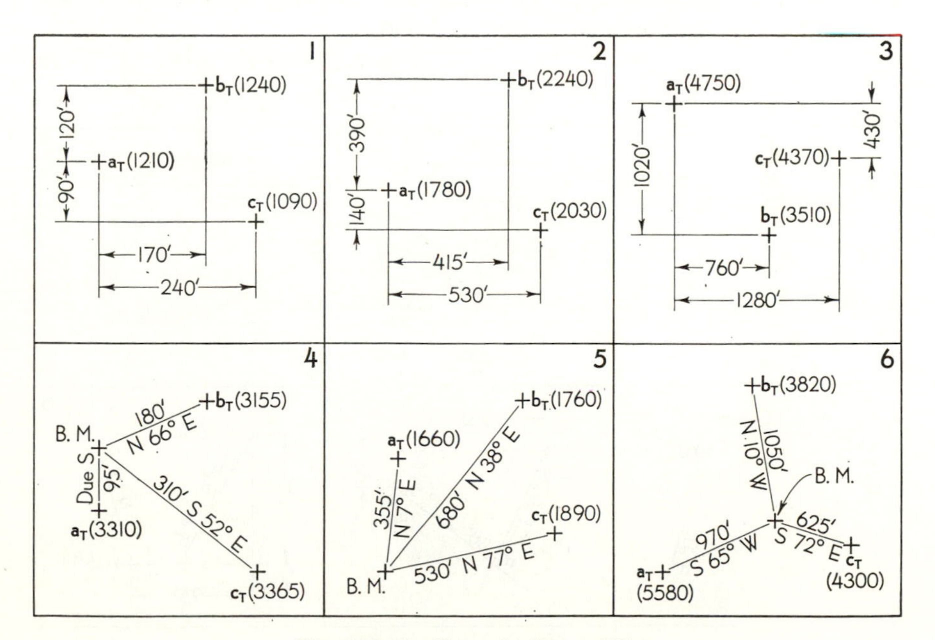

Fig. 182. Problems in Group 129.

130·1 B. 21° N 40° E, and 24° N 65° W. Ans. N 74° E, 34½° NW.
130·2 A. 42° N 70° E, and 33° S 35° E. Ans. N 5½° E, 45° E.
130·3 A. 19° S 70° W, and 14° S 15° W. Ans. N 30½° W, 19½° SW.
130·4 A. 37° N 70° E, and 15° S 60° E. Ans. N 40° W, 39° NE.
130·5 A. 27° N 10° E, and 49° N 30° W. Ans. N 33½° E, 51½° NW.
130·6 B. 28° S 65° E, and 33° S 40° W. Ans. N 82° E, 44° S.

Group 131. Strike, dip, and thickness from skew boreholes (Art. 12·8)

In the following problems points X and Y (elevations in parentheses) mark the surface ends of two skew boreholes that penetrate a stratum. The slope angle and bearing of each hole is given on the figure. The depths (measured along the hole) at which the hanging wall and then the footwall were struck are given in the problem statement. Find the strike, dip, and thickness of the stratum.

131·1 FIG. 183(1) B. Scale: 1 in. = 50 ft. Depths: in borehole X, 25 and 120 ft; in borehole Y, 20 and 40 ft. Ans. N 69° E, 42½° SE, 15 ft.
131·2 FIG. 183(2) B. Scale: 1 in. = 100 ft. Depths: in borehole X, 70 and 310 ft; in borehole Y, 50 and 120 ft.
Ans. N 78° W, 27° NE, 62 ft.
131·3 FIG. 183(3) A. Scale: 1 in. = 200 ft. Depths: in borehole X, 300 and 420 ft; in borehole Y, 200 and 540 ft.
Ans. N 30½° E, 36½° SE, 78 ft.

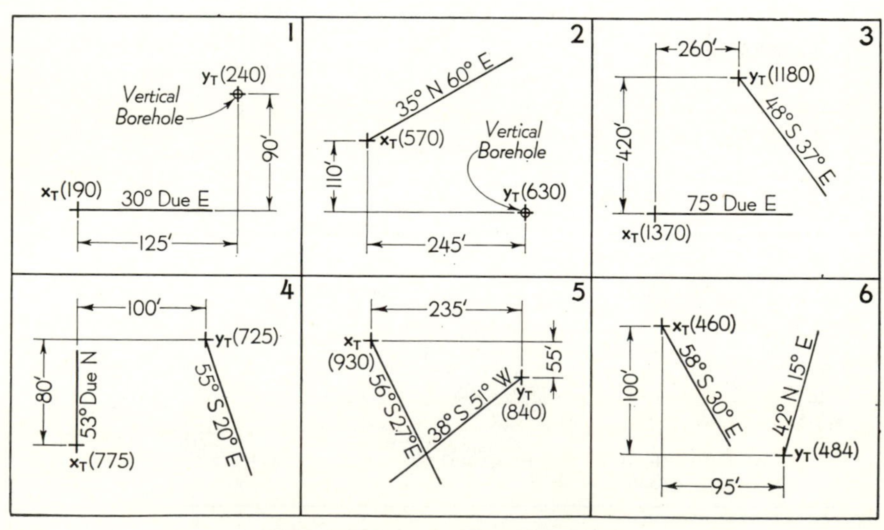

Fig. 183. Problems in Group 131.

131·4 FIG. 183(4) B. Scale: 1 in. = 50 ft. Depths: in borehole *X*, 30 and 95 ft; in borehole *Y*, 20 and 55 ft. ANS. N 64° W, 25° NE, 32 ft.

131·5 FIG. 183(5) B. Scale: 1 in. = 100 ft. Depths: in borehole *X*, 78 and 308 ft; in borehole *Y*, 85 and 356 ft. ANS. N 78½° E, 42° SE, 59 ft.

131·6 FIG. 183(6) A. Scale: 1 in. = 50 ft. Depths: in borehole *X*, 57 and 120 ft; in borehole *Y*, 33 and 101 ft. ANS. N 11° W, 34° SW, 50 ft.

Group 132. Location of workings (Art. 12·9)

132·1 A. A vein has a strike of N 60° E and a dip of 50° SE. What is the slope of an incline in the vein driven due east? ANS. 31° down.

132·2 B. A vein has a strike of N 70° W and a dip of 25° NE. What is the grade of an incline in the vein driven N 55° W? ANS. 12½ per cent down.

132·3 A. A vein has a strike of N 20° E and a dip of 60° SE. What is the bearing of an incline in the vein driven northerly on a 15° downward slope? ANS. N 29° E.

132·4 B. A vein has a strike of N 65° W and a dip of 35° SW. What is the bearing of an incline in the vein driven westerly on a 20 per cent downward grade? ANS. N 81½° W.

132·5 A. Scale: 1 in. = 50 ft. A vein having a strike of N 48° E and a dip of 57° SE outcrops at point *A* (elevation, 690 ft). At *B*, 125 ft due south of *A*, a vertical shaft (collar elevation, 655 ft) is sunk to the vein. From the bottom of the shaft a drift is driven N 48° E to meet a slope driven down on the dip from *A*. What is the depth of the vertical shaft? What is the length of the drift? What is the depth of the slope from *A*? ANS. 108 ft, 84 ft, 171 ft.

132·6 A. Scale: 1 in. = 100 ft. A vein outcrops at point *A* (elevation, 2,210 ft), and has a strike of N 28° W and a dip of 42° NE. On a hillside, 330 ft due east of *A*, is point *B* at an elevation of 2,120 ft. What is the bearing and length of the shortest tunnel that can be driven from *B* to the vein on an upward grade of 8 per cent? If the workings begin at *A* with a slope on the dip followed by a drift, what depth slope and what length drift will be required to meet the vein end of the tunnel? ANS. S 62° W, 212 ft. Slope, 54 ft; drift, 76 ft.

132·7 FIG. 183(4) B. Scale: 1 in. = 50 ft. Shafts have been sunk from surface points *X* and *Y* (elevations in parentheses) in the slope and direction of the given boreholes. What is the bearing, elevation, and length of the shortest level crosscut connecting the two shafts? ANS. N 80½° E. Elevation, 707 ft. Length, 106 ft.

132·8 FIG. 183(5) B. Scale: 1 in. = 100 ft. Same as Prob. 132·7, but also determine the depth of the shortest vertical shaft connecting the given shafts. How far down the higher shaft should digging begin on the vertical shaft? ANS. S 9° E. Elevation, 682 ft. Length, 32 ft. Vertical shaft, 50 ft deep; 240 feet from *Y*.

132·9 FIG. 183(6) A. Scale: 1 in. = 50 ft. Same as Prob. 132·7.
ANS. N 89° E. Elevation, 413 ft. Length, 101 ft.

132·10 B. Scale: 1 in. = 100 ft. A vein has a strike of N 56° W and a dip of 50° NE. From the mine entrance a slope is driven 320 ft down on the dip of the vein. From the bottom of the slope a drift is driven 145 ft N 56° W, and here encounters a second vein having a strike of N 66° E and a dip of 37° SE. How far must the original slope be extended to meet the second vein? ANS. 90 ft.

132·11 B. Scale: 1 in. = 50 ft. A vein has a strike of N 72° W and a dip of 42° SW. From the mine entrance (elevation, 6,295 ft) a slope is driven down on the dip to the 6,220-ft level. At this level a drift is driven 90 ft S 72° E, and here encounters a second vein having a strike of N 79° E and a dip of 31° NW. In this second vein a raise is driven 120 ft upward on the dip to exit on the surface. What is the map distance, bearing, and elevation of this exit with respect to the original mine entrance?
ANS. 223 ft S 21° E. Elevation, 6,282 ft.

132·12 A. Scale: 1 in. = 50 ft. A vein having a strike of N 38° E and a dip of 62° SE outcrops on a level plateau. From point *A* on the surface a slope is driven 115 ft down on the dip of the vein, and from this point a drift is driven 80 ft N 38° E. At the end of the drift a winze is sunk 40 ft on the dip, and here encounters a second vein having a strike of N 24° W and a dip of 47° SW. What would be the length of the original slope if it were extended to meet the second vein? If a raise is started at the foot of the winze and is continued upward on the dip of the second vein, where would it surface on the plateau? ANS. 226 ft. 235 ft N 73° E of point *A*.

Group 133. Intersection of veins and faults (Arts. 12·10 to 12·13)

133·1 A. Scale: 1 in. = 50 ft. A vein having a strike of N 22° E and a dip of 64° SE outcrops at point *A* (elevation, 2,760 ft). A second vein outcrops at point *B* (elevation, 2,715 ft) 148 ft S 73° E of *A* and has a strike of N 36° W and a dip of 52° SW. Locate the intersection line of the two veins, and determine its bearing and slope.
ANS. S 2½° W, 34½° S.

133·2 B. Scale: 1 in. = 50 ft. A vein having a strike of N 70° E and a dip of 38° SE outcrops at point *A* (elevation, 1,775 ft). A second vein outcrops at point *B* (elevation, 1,840 ft) 170 ft S 32° E of *A* and has a strike of N 55° W and a dip of 70° NE. Locate the intersection line of the two veins, and determine its bearing and slope.

ANS. S 66½° E, 28½° E.

133·3 A. Scale: 1 in. = 100 ft. A vein having a strike of N 40° W and a dip of 39° NE outcrops at point *A* (elevation, 3,550 ft). A second vein outcrops at point *B* (elevation, 3,400 ft) 270 ft due east of *A* and has a strike of N 28° E and a dip of 77° SE. Locate the intersection line of the two veins, and determine its bearing and slope.

ANS. N 38½° E, 38½° NE.

133·4 FIG. 185. Scale: 1 in. = 100 ft. Draw a 4-in. square in the upper center part of Layout B. Trace, or transfer with dividers, the position of the six points *A*, *B*, *C*, and *E*, *F*, *G* (do not trace the contour lines). Points *A* and *B* are outcrop points on a vein; a borehole at *C* strikes the vein at a depth of 20 ft. Points *E* and *F* are outcrop points on a second vein; a borehole at *G* strikes this second vein at a depth of 73 ft. All points are on the upper bedding planes; hence thickness of the veins may be neglected. Use the geologist's method (Art. 12·6) to find the strike of each vein. Then locate the intersection line of the two veins, and determine its bearing and slope.

ANS. S 10° E, 18° S.

133·5 FIG. 186. Same as Prob. 133·4, except that the borehole at *C* is 46 ft deep, and the borehole at *G* is 42 ft deep.

ANS. S 86° E, 21½° E.

133·6 FIG. 187. Same as Prob. 133·4, except that the borehole at *C* is 50 ft deep, and the borehole at *G* is 40 ft deep.

ANS. N 12° W, 28° N.

133·7 B. Scale: 1 in. = 100 ft. From a shaft at *S* a drift is driven 175 ft S 30° E at the 1,660-ft level in a vein that dips 54° SW. The drift ends at a fault having a strike of due east and a dip of 48° S. The striations on the slickensides are pitched 50° downward in a southwesterly direction. A borehole at *C*, 280 ft due south of *S*, relocates the vein at the 1,575-ft level. Find the net slip, strike slip, dip slip, bearing, and plunge of the fault displacement.

ANS. 197 ft, 126 ft, 151 ft, S 51° W, 34½°.

133·8 B. Scale: 1 in. = 100 ft. A vein of ore has a strike of N 15° E and a dip of 45° SE and outcrops at point *A* (elevation, 2,750 ft). At *B* (elevation, 2,700 ft), 260 ft S 35° W of *A*, a faulted portion of the same vein outcrops. The fault plane is located and found to have a strike of N 75° E, a dip of 60° NW, and striations that pitch 60° downward in a northwesterly direction. Find the net slip, bearing, and plunge of the fault displacement.

ANS. 98 ft, N 64° W, 49°.

133·9 B. Scale: 1 in. = 50 ft. A vein of ore outcrops on a level plateau and has a strike of N 50° E and a dip of 40° NW. To the southeast on the same plateau a second outcrop of the same vein is found to be offset 55 ft from the first outcrop. The fault plane has a strike of N 60° W, a dip of 65° NE, and striations that pitch 70° downward in a northwesterly direction. Find the net slip, bearing, and plunge of the fault displacement. ANS. 99 ft, N 11½° W, 58°.

133·10 FIG. 184(1) B. Scale 1 in. = 100 ft. The map shows the location, strike, and dip of a fault plane and two veins that have been displaced by a translatory fault movement. All strike lines have been adjusted to the same elevation. Find the net slip, strike slip, dip slip, bearing, and plunge of the fault displacement.

ANS. 172 ft, 57 ft, 162 ft, S 29° W, 46½°.

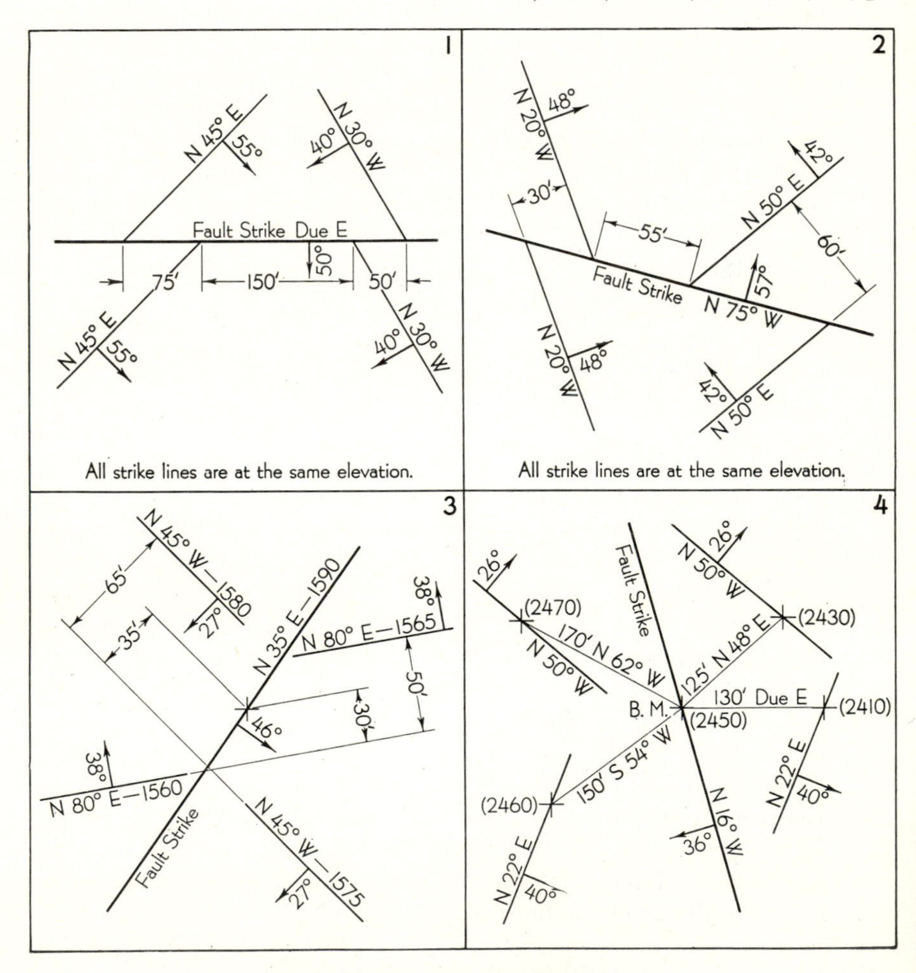

Fig. 184. Problems 133·10 to 133·13.

133·11 FIG. 184(2) B. Scale: 1 in. = 50 ft. Same as Prob. 133·10.
ANS. 84 ft, 10 ft, 83 ft, N 27° E, 56°.

133·12 FIG. 184(3) B. Scale: 1 in. = 50 ft. Same as Prob. 133·10, except that the strike lines are at the indicated elevations.
ANS. 39 ft, 4 ft, 38 ft, S 46° E, 45°.

133·13 FIG. 184(4) A. Scale: 1 in. = 100 ft. Same as Prob. 133·10, except that the strike lines are at the elevations of the indicated points. ANS. 106 ft, 82 ft, 66 ft, N 50° W, 22°.

Group 134. Line of outcrop; profiles; cuts and fills (**Arts. 12·14 to 12·17**)

These problems should be solved on thin bond paper or tracing paper. Draw a 4-in. square on the paper, and then trace the given contour map within its outline, marking carefully all points or lines pertinent to the problem. Areas of outcrop, strata in edge views, and cuts and fills should be distinctively shaded. All problems are to a scale of 1 in. = 100 ft, corresponding to the scale of the maps. Use Layout A.

134·1 FIG. 185. (Place map in the upper right corner of sheet.) Points *A* and *B* are two points on the upper outcrop line of a bed of coal. A vertical borehole drilled at *C* strikes the upper bedding plane at a depth of 20 ft. Continued drilling strikes the underlying rock at a depth of 40 ft. Find the strike, dip, and thickness of the bed. Plot the probable upper and lower outcrop lines.
ANS. N 67° E, 18° SE, 19 ft.

134·2 FIG. 186. Same as Prob. 134·1, except that the borehole at *C* strikes the upper and lower bedding planes at depths of 46 and 88 ft, respectively. ANS. N 50° E, 29½° SE, 36 ft.

134·3 FIG. 187. Same as Prob. 134·1, except that the borehole at *C* strikes the upper and lower bedding planes at depths of 50 and 70 ft, respectively. ANS. N 39° E, 35° NW, 16 ft.

134·4 FIG. 185. (Place map in upper left corner of sheet.) Points *E* and *F* are two points on the upper outcrop line of a stratum of limestone. A vertical borehole drilled at *G* strikes the stratum at a depth of 73 ft. Continued drilling strikes the underlying rock at a depth of 106 ft. Find the strike, dip, and thickness of the stratum. Plot the probable upper and lower outcrop lines.
ANS. N 39° W, 34½° SW, 27 ft.

134·5 FIG. 186. Same as Prob. 134·4, except that the borehole at *G* strikes the stratum and underlying rock at depths of 42 and 64 ft, respectively. ANS. N 24° W, 24° NE, 20 ft.

134·6 FIG. 187. Same as Prob. 134·4, except that the borehole at *G* strikes the stratum and underlying rock at depths of 40 and 70 ft, respectively. ANS. N 71° W, 32° NE, 25 ft.

134·7 FIG. 185. Trace the map in the upper center of the sheet, and then construct below the map profiles taken along lines *XX*, *YY*, and *ZZ*. Vertical scale: 1 in. = 100 ft.

134·8 FIG. 186. Same as Prob. 134·7.

134·9 FIG. 187. Same as Prob. 134·7.

134·10 FIG. 185. Line *PQ* is the center line of a level 40-ft highway at an elevation of 60 ft. Construct a profile taken along the highway center line, and draw sections at stations 15 + 60 and 17 + 80. Using a slope of 1 to 1 for cuts and 1½ to 1 for fills, draw the lines of cut and fill.

134·11 FIG. 186. Same as Prob. 134·10, except that the 40-ft highway is at a constant elevation of 130 ft, and sections are to be taken at stations 7 + 70, 9 + 40, and 10 + 60.

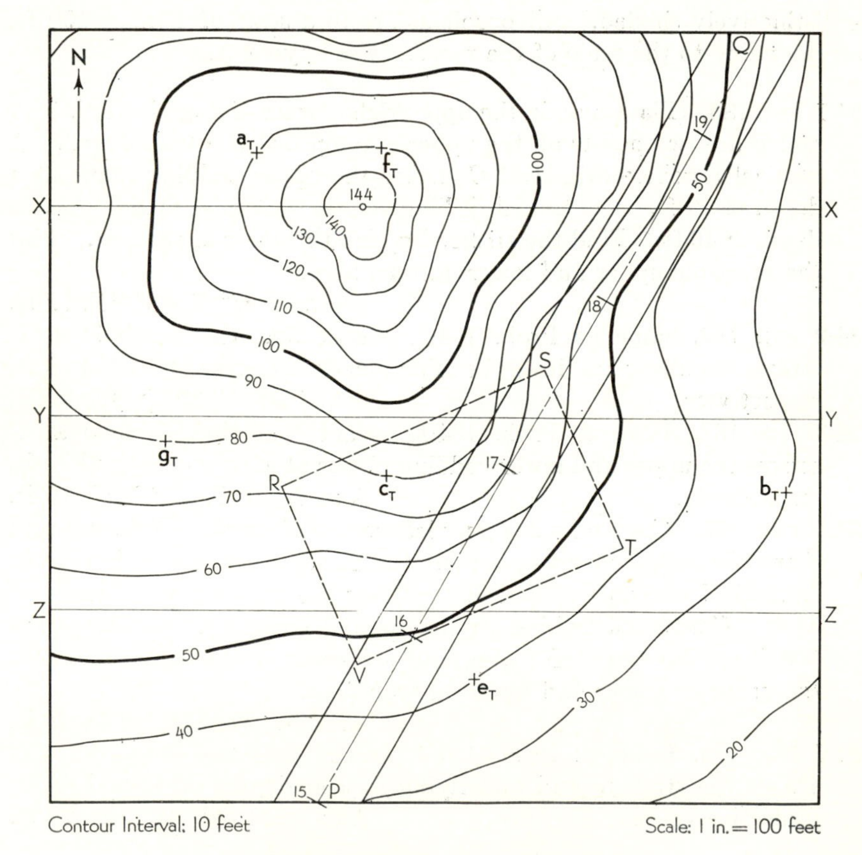

Fig. 185. Problems in Group 134.

134·12 FIG. 187. Area *PQRS* is the level top surface of an earth dam. The top of the dam is at an elevation of 340 ft and is 30 ft wide. Construct a profile along line *YY* and a section at the highest part of the dam. Using a slope of 2 to 1 on both sides of the dam, draw the lines of fill.

134·13 FIG. 185. The rectangle *RSTV* is a level area, 100 by 150 ft, cut into the hillside at an elevation of 60 ft. Construct a profile through the center of the area perpendicular to line *RS*. Using a slope of 1 to 1 for the cut and 2 to 1 for the fill, draw the lines of cut and fill. Note that at the corners of the area the embankments will be conical surfaces.

134·14 FIG. 186. The center line of a curved highway (not shown on the map) is an arc of 300-ft radius with center at *K*. The highway is 40 ft wide and is at a constant elevation of 90 ft. Construct a profile taken along the highway center line. Using a slope of 1 to 1 for cuts and $1\frac{1}{2}$ to 1 for fills, draw the lines of cut and fill.

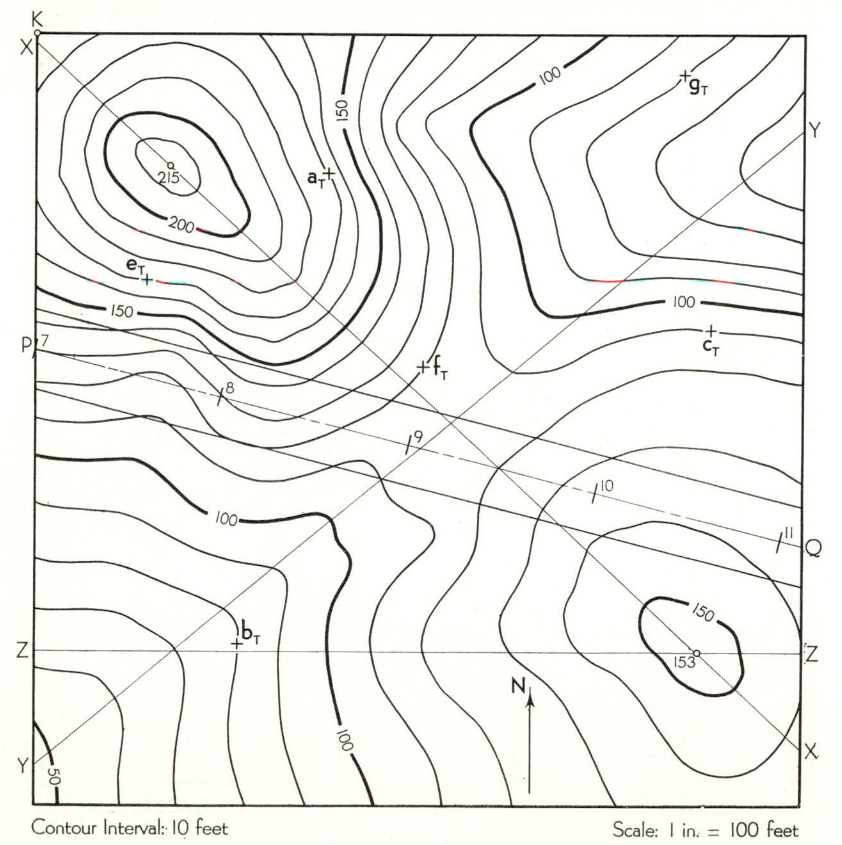

Fig. 186. Problems in Group 134.

134·15 FIG. 187. Same as Prob. 134·14, except that the highway is at a constant elevation of 300 ft and the slope of fill is 2 to 1.

134·16 FIG. 185. Line *PQ* is the center line of a 40-ft highway having a 10 per cent upward grade from *P* to *Q*. Highway elevation at station 15 is 20 ft. Using a slope of 1 to 1 for cuts and 1½ to 1 for fills, draw the lines of cut and fill.

134·17 FIG. 186. Line *PQ* is the center line of a 40-ft highway having a 10 per cent downward grade from *P* to *Q*. Highway elevation at station 7 is 140 ft. Using a slope of 1 to 1 for cuts and 1½ to 1 for fills, draw the lines of cut and fill.

134·18 FIG. 187. Line *YY* is the center line of a 40-ft highway that crosses the valley on a 10 per cent downward grade to the northeast. Highway elevation at *Y* in the southwest corner of the map is 340 ft. Using a slope of 1 to 1 for cuts and 2 to 1 for fills, draw the lines of cut and fill.

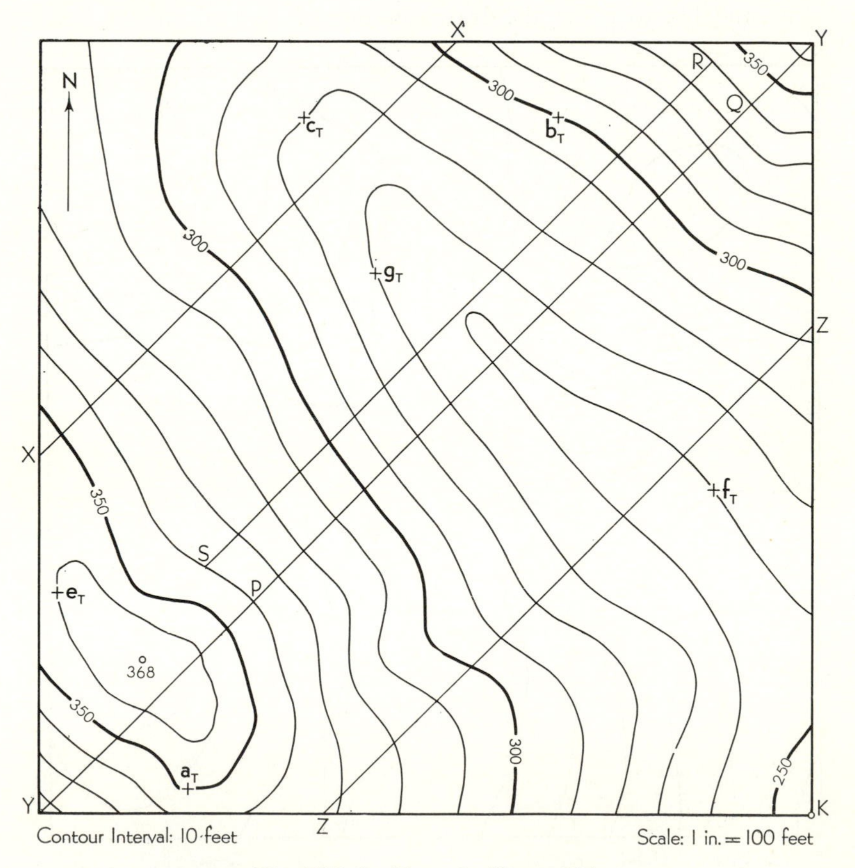

Fig. 187. Problems in Group 134.

134·19 FIG. 186. The center line of a curved highway (not shown on the map) is an arc of 300-ft radius with center at *K*. The highway is 40 ft wide and has a 10 per cent downward grade from the north to the southwest. Highway elevation at the top of the map (due east of point *K*) is 120 ft. Using a slope of 1 to 1 for cuts and 2 to 1 for fills, draw the lines of cut and fill.

134·20 FIG. 187. Same as Prob. 134·19, except that the highway has a 10 per cent downward grade from the south to the northeast, and the highway elevation at the bottom of the map (due west of point *K*) is 320 ft.

APPENDIX

The plane constructions given in this appendix are those which the author has found to be useful accessories in the solution of descriptive geometry problems. Most of the elementary constructions dealing with triangles and polygons have been omitted; others, such as those on conic sections, have been included because they are frequently needed or cannot readily be found in other texts. Where several methods of construction are possible, the best one for general practical use has been given. Proofs are beyond the scope of this appendix and hence have been omitted.

A·1. Measurement and layout of angles (Fig. A·1)

Angles can be measured or constructed by using a table of tangents. Let A be an unknown angle formed by the intersection of lines BC and BD. At any point D on BD draw the line CD perpendicular to BD, thus forming the right triangle BCD. Measure BD and CD, using any desired scale (the fiftieth scale is convenient). Then the tangent of angle A equals CD/BD, or 47.5/74.0 equals 0.645, and angle A therefore equals 32°50′. The accuracy of the result depends upon the size of the triangle; hence side BD should be made as large as possible. The computation

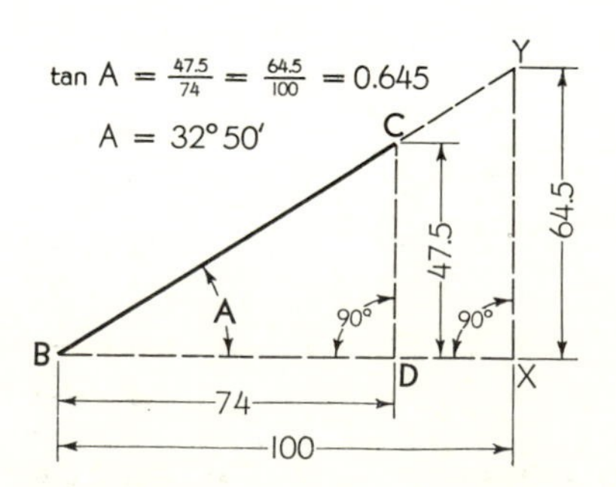

Fig. A·1. Measurement of angles.

can be simplified by making the base of the triangle exactly 100 units long, as at BX, in which case side XY will be 64.5 units long. Then the tangent of angle A equals 64.5/100, or 0.645, as before.

To construct angle A, begin by laying off the base BX 100 units long. On the perpendicular XY lay off a distance equal to 100 times the tangent of angle A to locate point Y. Connect points B and Y to complete the triangle, and establish the required angle.

A·2. To draw a line through a point and through the inaccessible intersection of two lines (Fig. A·2)

Let P be the given point and AB and CD the given lines. Draw any triangle PEF so that two of its vertices lie on the given lines and the third is at the point P. At any convenient distance away draw GH parallel to EF, then GQ parallel to EP and HQ parallel to FP, thus locating point Q. Line PQ is the required line. Note that the resulting similar triangles are analogous to parallel plane sections of a triangular pyramid.

A·3. To draw a line tangent to a circle and through the inaccessible intersection of two given lines (Fig. A·3)

Let the circle with center at O be the given circle and AB and CD the given lines. Draw any triangle EFG so that two of its vertices lie on the given lines, its third vertex lies anywhere on the circle, and one side of the triangle passes through the center of the given circle at O. At any convenient distance away draw HK parallel to EF, then HM parallel to EG and KM parallel to FG, thus locating point M. Draw MN parallel to GO to locate N on HK. With N as center and MN as radius draw a circle. Line PQ, tangent to this circle and the given circle, is the required line.

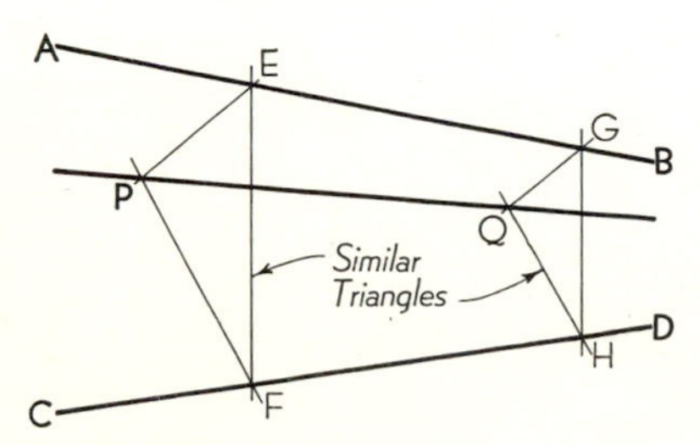

Fig. A·2. Line through an inaccessible point.

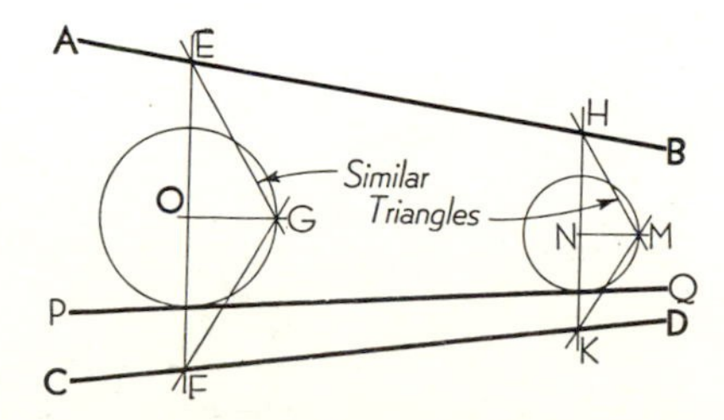

Fig. A·3. Tangent through an inaccessible point.

A·4. To rectify any curved line (Fig. A·4)

To rectify a curved line means to construct a straight line whose length will equal the length of the curved line. Let AB be the given curve. Draw a tangent to the curve at A. With the bow dividers set to a small distance, start at B, and step off equal chordal distances along the curve until less than one whole division remains near A. Without lifting the dividers step back along the tangent the same number of divisions counted on the curve to reach point C. The tangent AC is very nearly equal to the length of curve AB. Note that the divider setting must be taken small enough so that there is no appreciable difference between the curve segment and its chord. This method is applicable to all curved lines and is the one commonly used by draftsmen.

A·5. To rectify a circular arc of less than 60° (Fig. A·5)

a. Given the circular arc AB with center at O. Draw a tangent to the arc at A. Draw the chord AB, and extend it beyond A. Bisect chord AB, and lay off AD equal to one-half of AB. Then, with D as center and DB as radius, draw an arc to intersect the tangent at E. Line AE is slightly shorter than the arc AB. For an arc of 30° the error is only about 1/14,000, but for an arc of 60° the error is about 1/900. Arcs longer than 60° should be divided into several equal arcs of less than 60° and the construction then applied to one of the smaller arcs.

b. Given the tangent length AB, let it be required to determine the equivalent length on arc AC. Divide the line AB into *four* equal parts. With the first division point (number 1) as center and radius 1–B draw an arc to intersect the given arc AC at D. Arc AD is slightly longer than tangent AB, the error being approximately the same as that given in part *a* above.

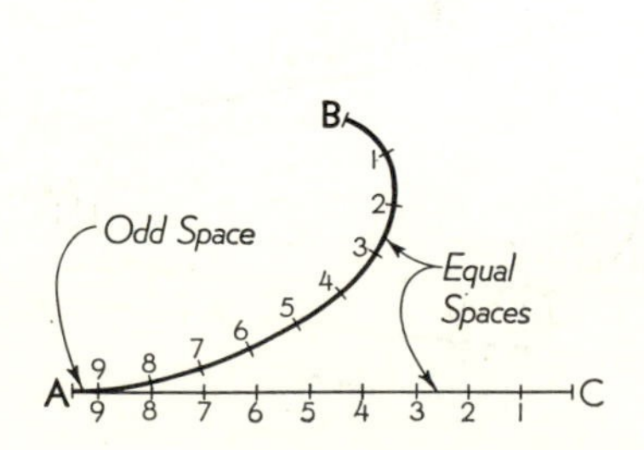

Fig. A·4. To rectify a curved line.

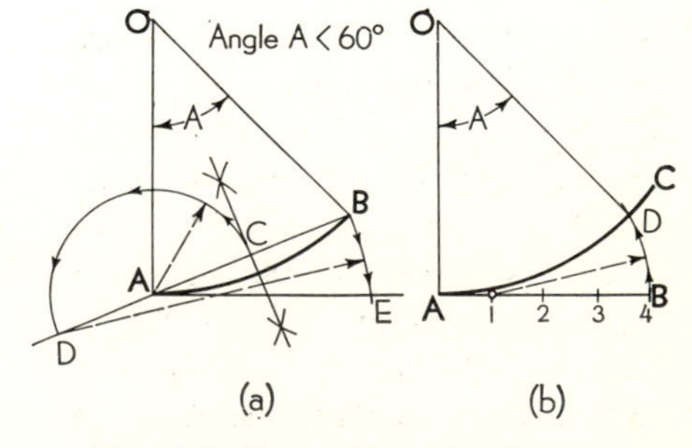

Fig. A·5. To rectify a circular arc.

A·6. To rectify the circumference of a circle (Fig. A·6)

Draw line AC tangent to the circle, and lay off on this line the distance AC equal to three times the diameter of the circle. Draw the radius OE at an angle of 30° with the diameter AB as shown. From E draw EF perpendicular to AB. The line CF is very slightly longer than the circumference of the circle, the error being about 1/21,800.

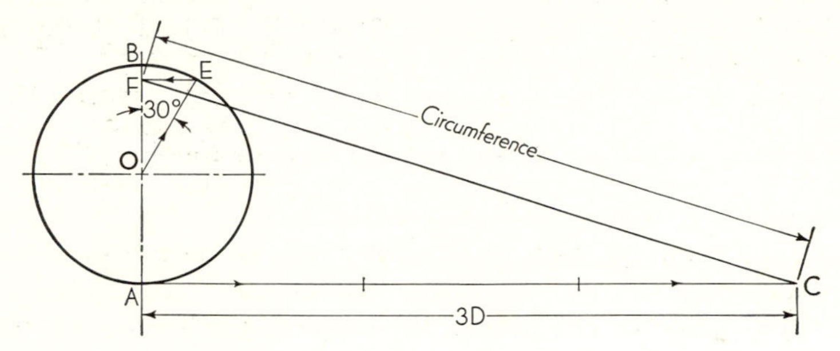

Fig. A·6. To rectify the circumference of a circle.

A·7. To draw an ellipse by the trammel method (Fig. A·7)

Given the major and minor axes, AB and CD, either of two trammels may be used.

a. On the straight edge of a strip of paper mark off the distance PM equal to AO, one-half of the major axis, and on the same side of P mark off PN equal to CO, one-half of the minor axis. Then, if this trammel is laid across the axes so that point N is on the major axis and point M is on the minor axis, the point P marks a point on the ellipse. By repeatedly shifting the trammel, but always keeping N on the major axis and M on the minor axis, the marking point P will locate a series of points on the ellipse. With the irregular curve draw a smooth curve through the points.

b. When the axes AB and CD are nearly equal in length, more accurate results can be obtained by marking distances PM and PN on opposite sides of the marking point P. The trammel is then moved as described in part a, N being always kept on the major axis and M on the minor axis and points marked at P. The trammel method produces an

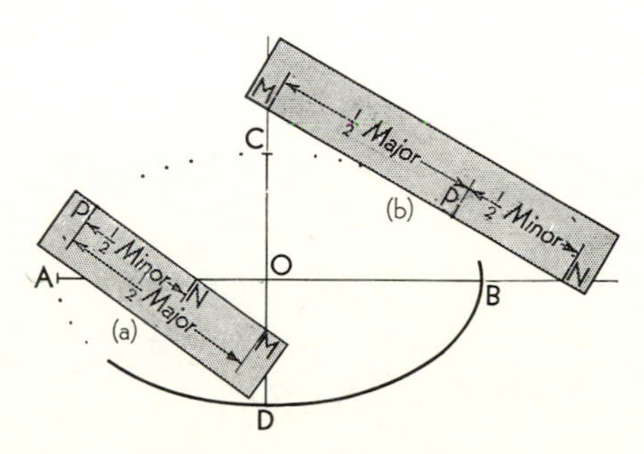

Fig. A·7. Ellipse by the trammel method.

accurate ellipse without construction lines and is the method preferred by draftsmen.

A·8. To determine the minor axis of an ellipse (Fig. A·8)

Given the major axis AB and any other diameter CD. Through O draw a perpendicular to the major axis AB. With $\frac{1}{2}AB$ as radius and center at C (or D) strike an arc cutting the perpendicular at E. Draw line CE to intersect the major axis at G. Then CG is the length of the semiminor axes OH and OK.

A·9. To draw an ellipse by the parallelogram method (Fig. A·9)

Given any pair of conjugate axes AB and CD. Through the ends of each axis draw lines parallel to the other axis to form a parallelogram that circumscribes the required ellipse. Divide AO and AE into the same number of equal parts, and number the division points from A. From D extend lines through points 1, 2, and 3 on AO, and from C draw lines to points 1, 2, and 3 on AE. The intersections of correspondingly numbered lines will be points on one quarter of the required ellipse. Points

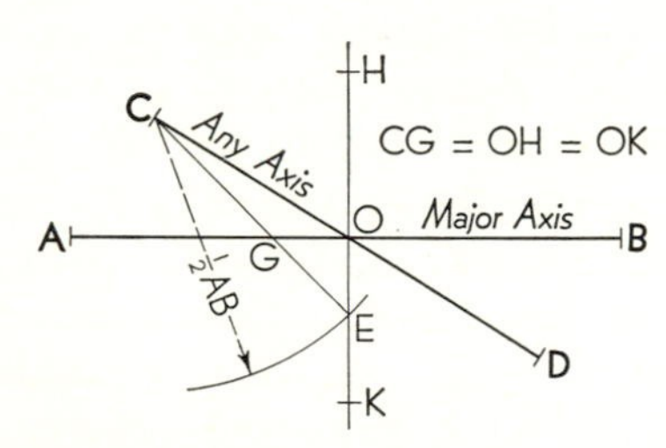

Fig. A·8. To find the minor axis of an ellipse.

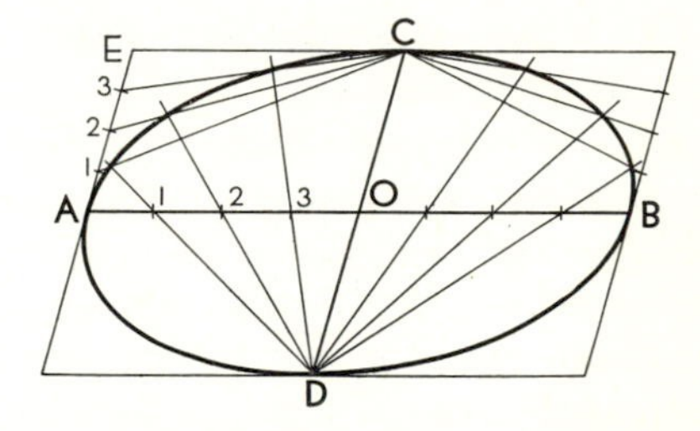

Fig. A·9. Ellipse by the parallelogram method.

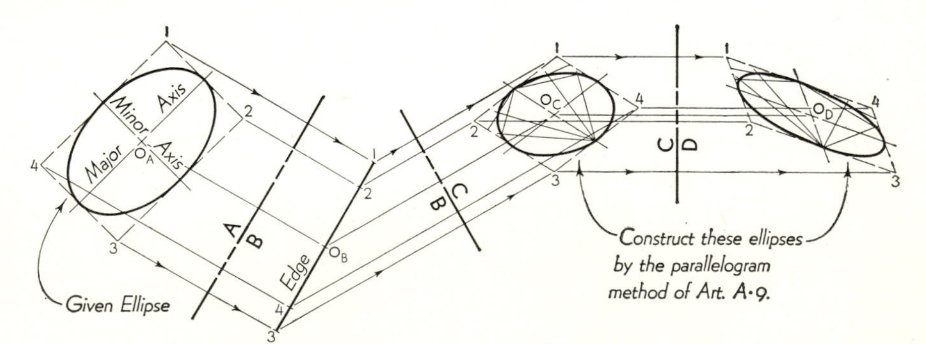

Fig. A·10. To draw additional views of a given ellipse.

on the other three quarters of the ellipse may be located in similar manner.

A·10. To draw additional views of a given ellipse (Fig. A·10)

Given an ellipse with center at O which appears true size in view A and as an edge in view B. The ellipse is required in views C and D. Enclose the given ellipse in a rectangle 1–2–3–4. Locate points 1, 2, 3, and 4 in views B, C, and D by alignment and measurement. Draw the parallelograms 1–2–3–4 in views C and D, and through o_C and o_D draw conjugate axes parallel to the sides of the parallelograms. The ellipses in views C and D may now be constructed by the parallelogram method of Art. A·9.

A·11. To draw an approximate ellipse (Fig. A·11)

Given the major and minor axes AB and CD. Draw line AC. With O as center and OA as radius draw an arc to intersect the minor axis extended at E. With C as center and CE as radius draw a second arc to intersect AC at F. Bisect AF, and extend the perpendicular bisector GH to cross the major axis at P and the minor axis at Q. Locate symmetrically the points P' and Q' by making OP' equal to OP and OQ' equal to OQ. Draw lines PQ', $P'Q'$, and $P'Q$, and extend them. Then, P, Q, P', and Q' are the centers, PA, $P'B$, QC, and $Q'D$ the radii, and T_1, T_2, T_3, and T_4 the points of tangency, for four arcs that may be drawn to approximate the required ellipse.

Although this method may be used for any ellipse, the approximation is poor for long, slender ellipses but improves as the ratio of the axes approaches unity. When the minor axis is greater than three-fourths of the major axis, the approximate ellipse is for most practical purposes an adequate substitute for the exact ellipse.

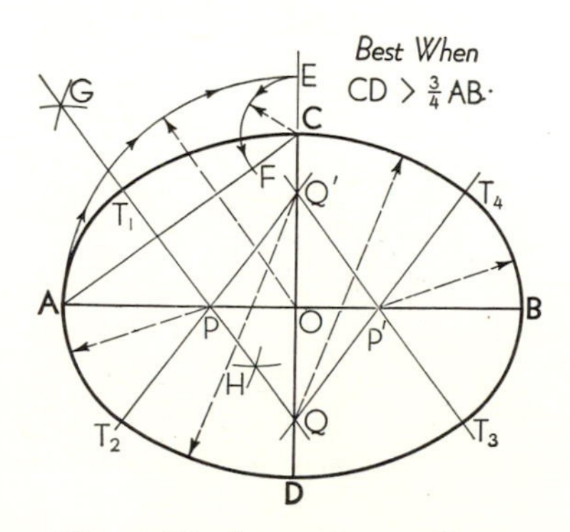

Fig. A·11. Approximate ellipse.

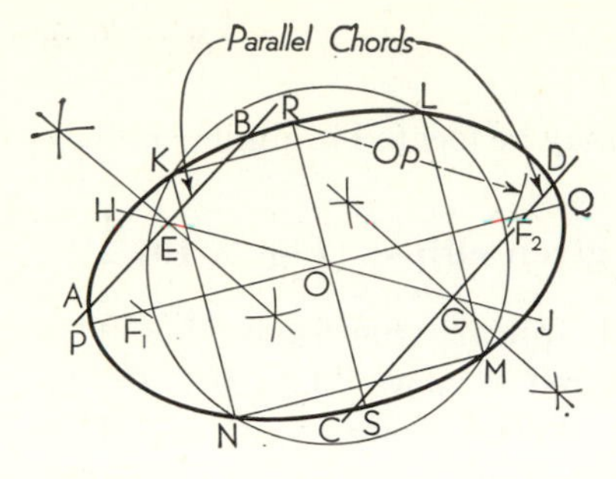

Fig. A·12. Axes and foci of an ellipse.

A·12. To locate the axes and foci of a given ellipse (Fig. A·12)

Given only the curve of the ellipse. Draw any pair of parallel chords intersecting the curve at A and B, C and D. Bisect chords AB and CD to locate their mid-points E and G. Draw line EG, and extend it to intersect the ellipse at H and J. Then HJ is one diameter of the ellipse, and point O, its mid-point, is the center of the ellipse. With O as a center and any convenient radius draw a circle to intersect the ellipse at the four points K, L, M, and N, thus forming the rectangle $KLMN$. Through O draw PQ parallel to KL, and draw RS parallel to KN. Lines PQ and RS are the required major and minor axes.

With R (or S) as center and radius equal to OP (half the major axis) draw an arc to intersect the major axis at points F_1 and F_2. These points are the required foci of the ellipse.

A·13. To draw a tangent to an ellipse (Fig. A·13)

Given the ellipse and its major and minor axes. Locate the foci as described in Art. A·12. Three cases will be considered.

a. Tangent at a given point P on the curve [***Fig. A·13***(***a***)]. Draw the focal radii PF_1 and PF_2, extending one of them outside the ellipse. Bisect the exterior angle formed by the focal radii. The bisector is the required tangent.

b. Tangent from a given external point Q [***Fig. A·13***(***a***)]. With Q as center and QF_2 as radius (F_1 and F_2 may be interchanged) draw an

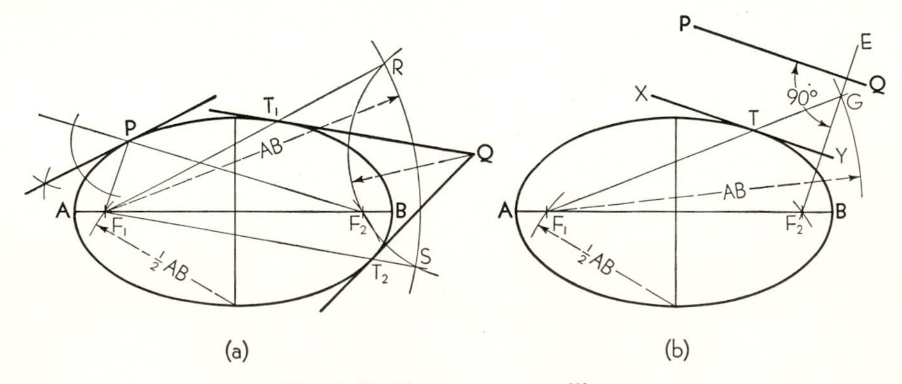

Fig. A·13. Tangents to an ellipse.

arc of indefinite length. With F_1 as center and AB, the major axis, as radius draw a second arc to intersect the first arc at points R and S. Draw lines RF_1 and SF_1 to intersect the ellipse at T_1 and T_2, the required points of tangency. Lines QT_1 and QT_2 are the required tangents.

***c. Tangent parallel to a given line PQ* [*Fig. A·13*(*b*)].** From F_2 (F_1 and F_2 may be interchanged) draw F_2E perpendicular to PQ. With F_1 as center and AB as radius draw an arc to intersect F_2E at G. Draw line F_1G to intersect the ellipse at T, the required point of tangency. Draw the required tangent XY through T parallel to PQ.

A·14. To draw a parabola by the parallelogram method (Fig. A·14)

Given the axis AB and one point C on the required parabola (or given the *rise* AB and *span* CD). Draw the parallelogram $CDFE$. Divide CB and CE into the same number of equal parts, and number the division points from C. From points 1, 2, 3, and 4 on CE draw lines to point A, and from points 1, 2, 3, and 4 on CB draw lines parallel to axis AB. The intersections of correspondingly numbered lines will be points on one half of the parabola. Points on the other half may be located in similar manner. This method is also applicable when CD is not perpendicular to AB, but AB will then be a diameter of the parabola, not its axis.

A·15. To draw a parabola by the envelope method (Fig. A·15)

Given the intersecting lines AB and AC; required, the parabolic curve tangent to these lines at points B and C. Divide AB and AC into the same number of equal parts, and number the division points on each line in opposite sequence as shown. Lines connecting correspondingly numbered points will be tangent to and thus envelop the required curve. Draw the curve from B to C tangent to the connecting lines.

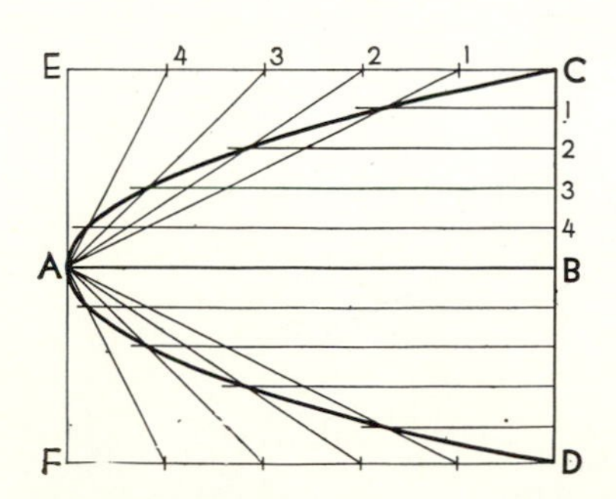

Fig. A·14. Parabola by the parallelogram method.

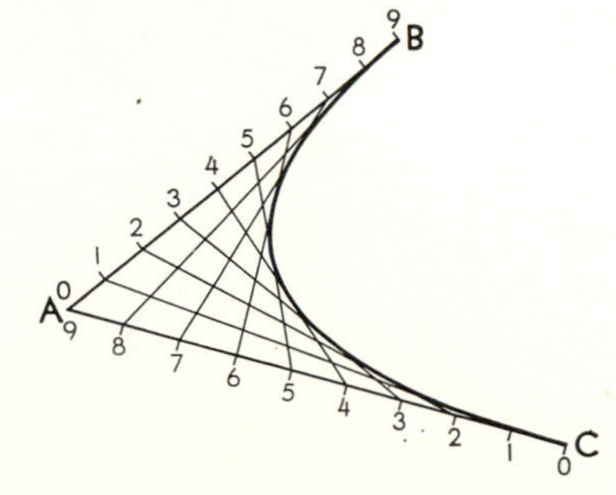

Fig. A·15. Parabola by the envelope method.

A·16. To locate the axis, focus, and directrix of a given parabola (Fig. A·16)

Given only the curve of a parabola. Draw any pair of parallel chords intersecting the curve at *A* and *B*, *C* and *D*. Bisect chords *AB* and *CD* to locate their mid-points *E* and *G*. Then line *EG* is one diameter of the parabola. Draw any line perpendicular to *EG*, and extend it to intersect the curve at *H* and *J*. The perpendicular bisector of *HJ* is line *KL*, the required axis of the parabola.

To locate the focus, extend diameter *EG* to intersect the curve at point *T*. Through *T* draw *MN* parallel to chords *AB* and *CD*. Then *MN* is tangent to the parabola at point *T*. With *T* as center and any convenient radius draw an arc to intersect *EG* extended at *P* and *MN* at *O*. With *O* as center and radius *OP* draw a second arc to intersect the first arc at *Q*. Draw *TQ*. Then tangent *MN* is the bisector of angle *PTQ*, and point *F* where line *TQ* crosses axis *KL* is the required focus.

To locate the directrix lay off on *KL* from *S* (the parabola vertex) the distance *RS* equal to *SF*. Draw the required directrix *XY* through point *R* perpendicular to axis *KL*.

A·17. To draw a tangent to a parabola (Fig. A·17)

Given the parabola and its axis. Locate the focus, and the directrix when needed, as described in Art. A·16. Three cases will be considered.

***a. Tangent at a given point P on the curve* [*Fig. A·17*(*a*)].** *First method.* From the given point P_1 draw P_1B perpendicular to the axis of the parabola. With *A* (the vertex) as center and *AB* as radius draw an arc to intersect the axis extended at *C*. Line P_1C is the required tangent.

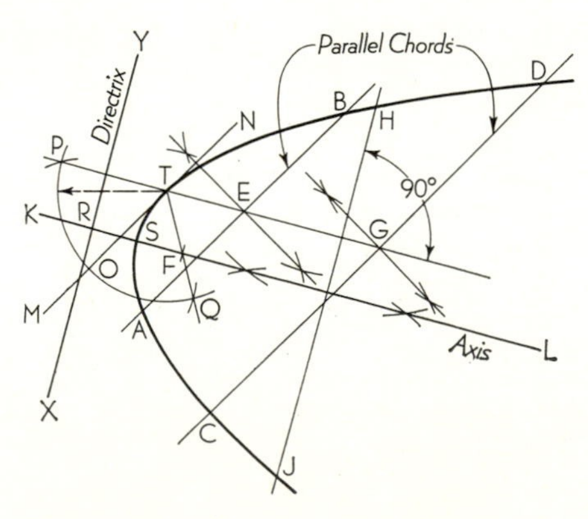

Fig. A·16. Axis, focus, and directrix of a parabola.

Second method. From the given point P_2 draw the focal radius P_2F and line P_2D parallel to the axis. Bisect angle FP_2D. The bisector P_2E is the required tangent.

***b. Tangent from a given external point Q* [*Fig. A·17*(*b*)].** With point Q as center and QF as radius draw an arc to intersect the directrix at points B and C. From B and C draw lines parallel to the axis to intersect the curve at points T_1 and T_2, the required points of tangency. Lines QT_1 and QT_2 are the required tangents.

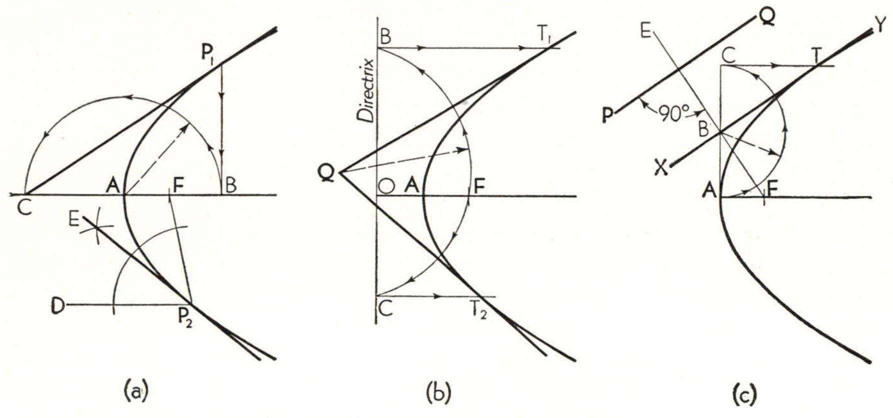

Fig. A·17. Tangents to a parabola.

***c. Tangent parallel to a given line PQ* [*Fig. A·17*(*c*)].** From the focus F draw line FE perpendicular to PQ. From A draw line AC perpendicular to the axis, thus locating point B on line FE. Locate C by making BC equal to AB. From C draw a line parallel to the axis to intersect the curve at point T, the required point of tangency. Line XY, passing through points B and T, is the required tangent.

A·18. To draw a hyperbola by the parallelogram method (Fig. A·18)

Given the transverse axis AB and one point D on the required hyperbola (or given the axis AB, the rise BC, and the span DE). Draw the parallelogram $DEGF$. Divide DC and DF into the same number of equal parts, and number the division points from D. From A draw lines to points 1, 2, and 3 on DC, and from B draw lines to points 1, 2, and

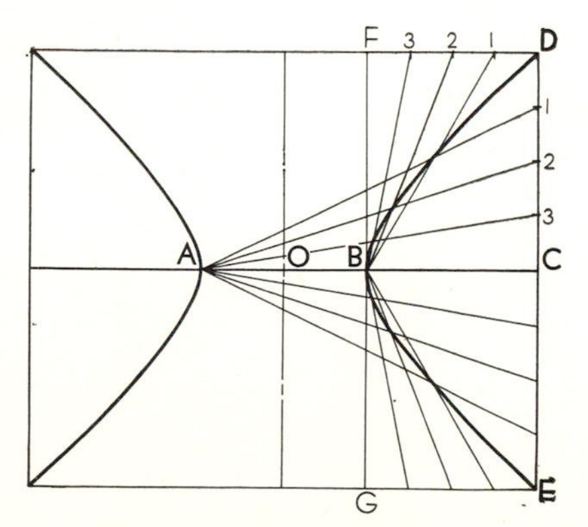

Fig. A·18. Hyperbola by the parallelogram method.

3 on *DF*. The intersections of correspondingly numbered lines will be points on one quarter of the complete hyperbola. Points on the other three quarters of the curve may be located in similar manner. This method is also applicable when *DE* is not perpendicular to *AB*, but *AB* will then be a diameter of the hyperbola, not its axis.

A·19. To locate the axes, asymptotes, and foci of a given hyperbola (Fig. A·19)

Given only the curve of a hyperbola. Locate the transverse axis *AB* and the conjugate axis *PQ* by the same method as that described for an ellipse in Art. A·12 (construction not shown in Fig. A·19).

To locate the asymptotes draw from *D* (any point on the curve) the ordinate *DC*, thus locating *C* on the transverse axis extended. With *O*, the center of the hyperbola, as center and *OC* as radius draw the arc *CG*. From *D* draw a line *DH* parallel to axis *AB*. With *O* as center again and radius *OB* draw an arc to intersect the conjugate axis at *J*. From *J* draw a line parallel to axis *AB* to intersect arc *CG* at *K*. From *K* draw a line parallel to the conjugate axis to intersect *DH* at *M*. Then *OM* is one of the required asymptotes, and *ON*, located by symmetry, is the other.

To locate the foci, draw at *B* a perpendicular to the transverse axis. Note points *R* and *S* where the perpendicular intersects the asymptotes. With *O* as center and radius *OR* draw a circle to intersect axis *AB* extended on both sides at F_1 and F_2, the required foci.

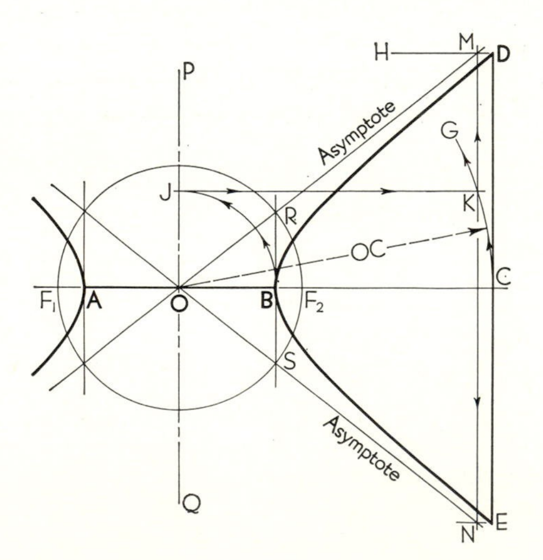

Fig. A·19. Foci and asymptotes of a hyperbola.

A·20. To draw a tangent to a hyperbola (Fig. A·20)

Given the hyperbola and its axis AB. Locate the foci as described in Art. A·19. Three cases will be considered.

a. Tangent at a given point P on the curve [*Fig. A·20(a)*]. From P draw the focal radii PF_1 and PF_2. The bisector of angle F_1PF_2 is the required tangent.

b. Tangent from a given external point Q [*Fig. A·20(b)*]. With Q as center and QF_1 as radius (F_1 and F_2 may be interchanged) draw an arc of indefinite length. With F_2 as center and AB, the transverse axis, as radius, draw a second arc to intersect the first arc at points C and D. Draw lines CF_2 and DF_2, extending each as necessary, to intersect the hyperbola at T_1 and T_2, the required points of tangency. Lines QT_1 and QT_2 are the required tangents.

c. Tangent parallel to a given line MN [*Fig. A·20(a)*]. From F_1 (F_1 and F_2 may be interchanged) draw F_1E perpendicular to MN. With F_2 as center and AB as radius draw an arc to intersect F_1E at G. Draw line F_2G, and extend it to intersect the hyperbola at T, the required point of tangency. Draw the required tangent XY through T parallel to MN. Verify the accuracy of the solution by checking point H on the tangent (GH should equal HF_1).

A·21. Pascal's theorem for hexagons inscribed in conics (Fig. A·21)

Generally defined, a hexagon is any plane polygon having six angles and hence six sides. Therefore, any six points, no three of which lie on the same straight line, may be connected in any sequence to form a hexagon. Pascal's theorem states that *the three pairs of opposite sides of a hexagon inscribed in a conic intersect in three points which lie on a straight*

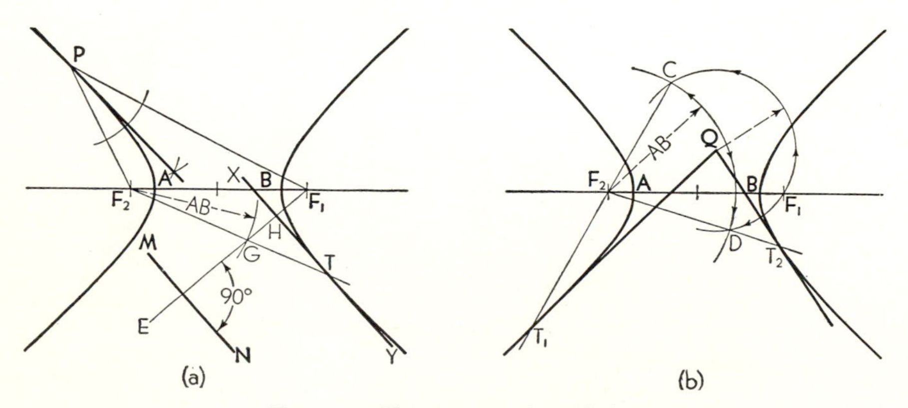

Fig. A·20. Tangents to a hyperbola.

line. Two examples are shown to illustrate the validity of the theorem.

a. Choose any six points on the ellipse, and number these points from 1 to 6 in any desired sequence. Connect the six points in numerical sequence with six straight lines to form a hexagon. Reference to a similarly numbered regular hexagon shows that the three pairs of opposite sides are 1–2 and 4–5, 2–3 and 5–6, and 3–4 and 6–1. On the ellipse note that sides 1–2 and 4–5 intersect at point P, sides 2–3 and 5–6 at point Q, and sides 3–4 and 6–1 at point R. The straight line passing through P, Q, and R is called *Pascal's line.*

b. Six points on the parabola have been numbered and connected as described in (*a*) above. Note again that sides 1–2 and 4–5 (each extended) intersect at P, sides 2–3 and 5–6 (also extended) intersect at Q, etc. Points P, Q, and R lie on the same straight line.

As an aid in remembering the pairs of opposite sides, it is convenient to sketch a numbered regular hexagon, or the numbers designating the sides may be arranged in tabular form as shown. A simple test of Pascal's theorem can be made by using any six points on a circle and completing the construction described above.

A·22. To draw a conic through five points (Fig. A·22)

Through any five points, no three of which lie on the same straight line, one, and only one, conic curve can be drawn. If five points on the required curve are given, Pascal's theorem can be employed to locate any number of sixth points.

Given the five points 1, 2, 3, 4, 5 (the numbers may be interchanged if desired). Let x be any sixth point on the required conic curve. Then the polygon 1, 2, 3, 4, 5, x will be a hexagon inscribed in a conic. Draw lines 1–2 and 4–5, and extend them to intersect at point P, thus establishing what we shall call the *pivot point.* Draw line 2–3, and call this

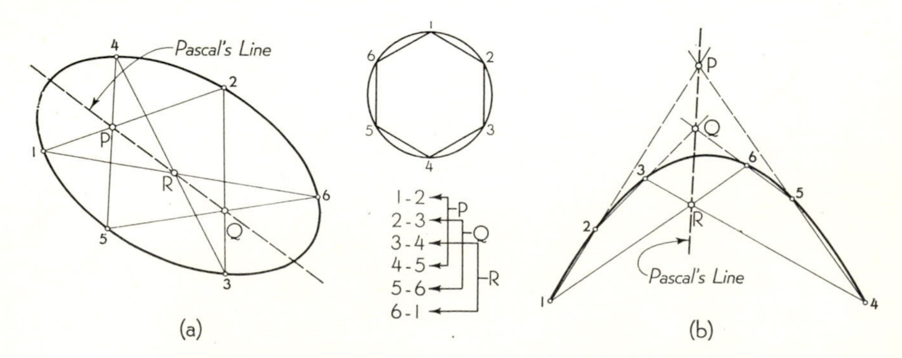

Fig. A·21. Pascal's line for hexagons inscribed in conics.

line the *Q control;* draw line 3–4, and call this line the *R control.* Through P draw any line (a Pascal line) to intersect the Q control at Q and the R control at R. Draw line 5–Q and 1–R (extended); the intersection of these two lines is one point x on the required curve. By repeatedly altering the position of the Pascal line through point P any number of additional x points can be located to establish the curve fully. Pascal line $PQ'R'$, for example, determines point x'.

A·23. To draw a conic through three points and tangent to given lines at two of the points (Fig. A·23)

If, in Fig. A·22, point 2 is moved along the curve until it coincides with point 1, then the chord 1–2 will become a tangent to the curve at point 1. Thus, a conic may be determined by four points, and the tangent at one of these points. In this case the construction for additional x points on the curve is identical with that described in Art. A·22, except that points 1 and 2 coincide at the tangent point. If point 4 also moves into coincidence with point 5, then the chord 4–5 becomes a tangent to the curve at point 5 and the conic is determined by three points and the tangents at two of these points. This second case is illustrated in Fig. A·23.

Given the three points 1,2 and 3 and 4,5 and the tangents at points 1,2 and 4,5 (double numbering of the tangent points makes the construction analogous to that of Fig. A·22 and hence easier to remember). Extend the tangents at 1,2 and 4,5 to intersect at point P, the pivot point. Draw line 2–3, the Q control; draw line 3–4, the R control. Through P draw a Pascal line to intersect the control lines at Q and R, respectively. Draw lines 5–Q and 1–R; the intersection of these two lines is one point x on the

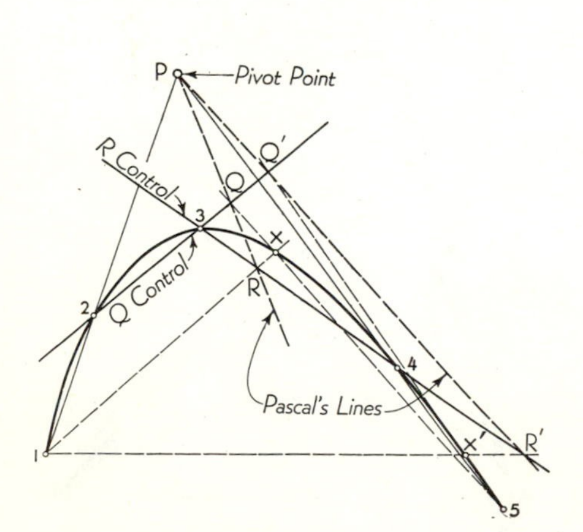

Fig. A·22. Conic through five points.

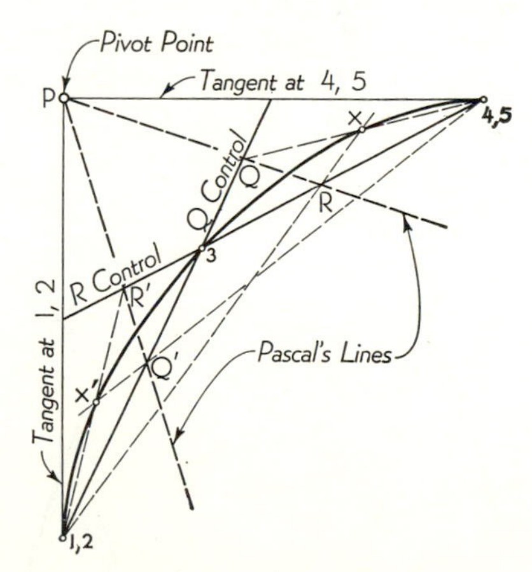

Fig. A·23. Conic through three points with tangents at two points.

required curve. Alter the position of the Pascal line through P to locate additional points such as x'.

A·24. To draw a tangent to a conic (Fig. A·24)

Given only the conic curve; required to draw a tangent to the curve at point T. Choose any four points on the curve, two on either side of point T. Number the points on one side of T, 1 and 2; on the other side, 3 and 4. Let T be point 5,6. Then in the inscribed hexagon 1, 2, 3, 4, 5, 6, side 5,6 will be tangent to the conic at point T. Draw lines 1–2 and 4–5, and extend them to intersect at P. Draw lines 3–4 and 6–1, and extend them to intersect at R. Draw the Pascal line through P and R. Draw line 2–3, and extend it to intersect the Pascal line at Q. Line QT is the required tangent.

A·25. To test a given conic to determine its type (Fig. A·25)

Given the three conics AEB, AMB, and AHB. Each may be tested in the following manner to determine whether it is a portion of an ellipse, a parabola, or a hyperbola. Draw the chord AB. Draw tangents to the curve at A and B, and extend them to intersect at D. Locate C at the mid-point of AB, and draw line CD. Locate M at the mid-point of CD. Then, if the given curve intersects line CD at any point E between M and C, the curve is an ellipse; if it intersects CD at any point H between M and D, it is a hyperbola; if it intersects CD exactly at M, it is a parabola.

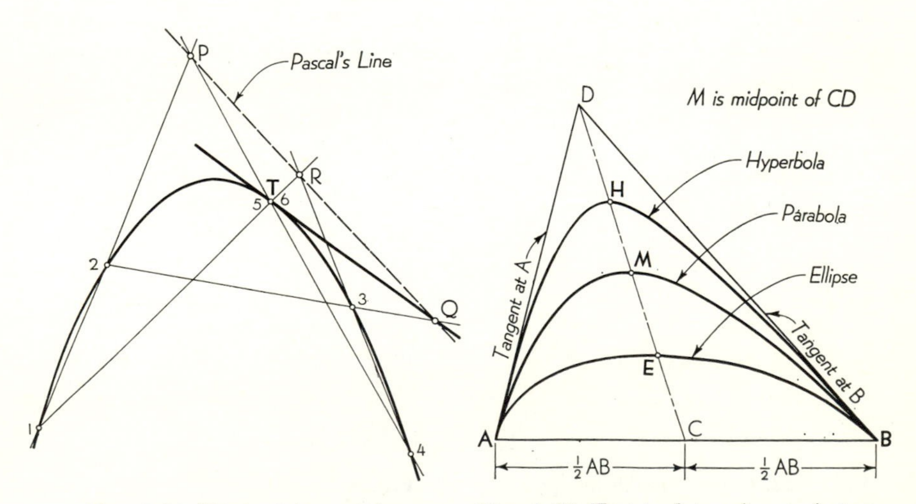

Fig. A·24. Tangent to a conic.

Fig. A·25. Test to determine conic type.

A·26. Brianchon's theorem for hexagons circumscribed about conics (Fig. A·26)

Brianchon's theorem states that *the three diagonals joining the opposite vertices of a hexagon circumscribed about a conic pass through the same point.* Two examples are shown to illustrate the validity of the theorem.

a. Choose any six tangents to the ellipse, and number these tangents from 1 to 6 in any desired sequence. Extend the tangents until 1 intersects 2, 2 intersects 3, 3 intersects 4, etc., to form a hexagon. Reference to a similarly numbered regular hexagon shows that the opposite vertices are the points of intersection of tangents 1,2 and 4,5; 2,3 and 5,6; and 3,4 and 6,1. On the ellipse note that the diagonal from vertex 1,2 to vertex 4,5, the diagonal from 2,3 to 5,6, and the diagonal from 3,4 to 6,1 all pass through the same point *B*. Point *B* is called *Brianchon's point.*

b. Six tangents to the parabola have been numbered and extended to their consecutive intersections as described in (*a*) above. Note again that the three diagonals joining opposite vertices all pass through the same point *B*.

A regular hexagon whose sides have been numbered in sequence is a convenient device for remembering the opposite vertices; the tabular form shown may also be used. Note that for Pascal's theorem the six points are numbered, whereas for Brianchon's theorem the tangent lines are numbered.

A·27. To draw a conic tangent to five lines (Fig. A·27)

Tangent to any five lines, no three of which intersect at the same point, one, and only one, conic curve can be drawn. Since a sixth tangent would form a circumscribed hexagon, Brianchon's theorem may be used to locate any number of such sixth tangents, thus enveloping the required curve.

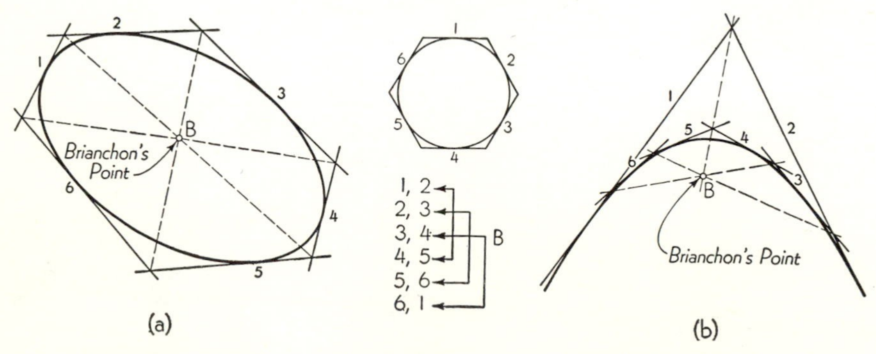

Fig. A·26. Brianchon's point for hexagons circumscribed about conics.

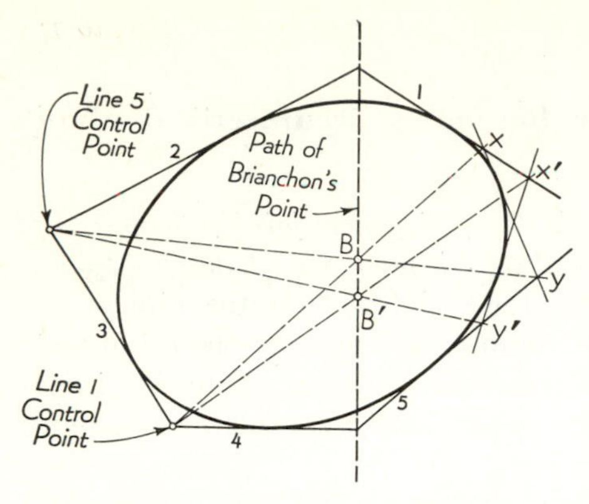

Fig. A·27. Conic tangent to five lines.

Given the five tangents 1, 2, 3, 4, 5 (the numbers may be interchanged if desired). Extend tangent 1 to its intersection with tangent 2 (vertex 1,2), tangent 2 to its intersection with tangent 3 (vertex 2,3), etc. Vertex 1,5 is not required. Draw the diagonal from vertex 1,2 to vertex 4,5, thus establishing the *path of Brianchon's point*. Now choose any point B on this path. From vertex 2,3 draw a line through B, and extend it to intersect tangent 5 at y. From vertex 3,4 draw a line through B, and extend it to intersect tangent 1 at x. Then line xy is a sixth tangent to the required curve. By repeatedly altering the position of point B along its path, any number of additional tangents can be located to completely envelop the required curve. Point B', for example, establishes tangent $x'y'$. Note that diagonal 1,2 to 4,5 (the path of Brianchon's point) may be extended beyond the vertices and that tangents 1 and 5 may also be extended indefinitely as needed.

If tangent 1 is moved along the curve until it coincides with tangent 2, then vertex 1,2 will become a point on one continuous tangent and the required curve will be tangent to this line at point 1,2. Thus, a conic may be determined by four tangents and the point of tangency on one of these tangents. If tangent 5 also moves into coincidence with tangent 4, then vertex 4,5 will similarly become a point of tangency and the conic is then determined by three tangents and the points of tangency on two of the

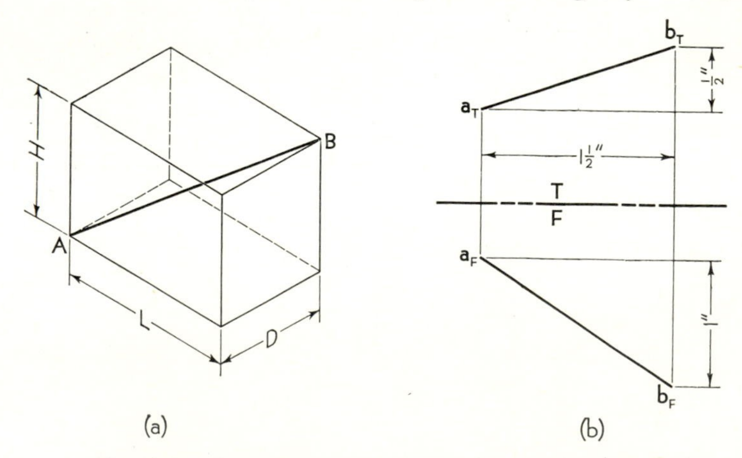

Fig. A·28. Calculation of true length of an oblique line.

tangents. In either case the construction is similar to that for five tangents, except that one or two of the vertices have become points of tangency.

A·28. To calculate the true length of an oblique line (Fig. A·28)

It is sometimes necessary to determine the true length of an oblique line more accurately than is possible by graphical methods. In such cases, the length can readily be calculated to any desired accuracy consistent with the accuracy of the given data.

Let AB be a given oblique line. Then, as shown in Fig. A·28(a), line AB is the diagonal of an imaginary rectangular box whose linear dimensions may be designated as L, H, and D. The true length of AB is equal to the square root of the sum of the squares of L, H, and D, or

$$AB = \sqrt{L^2 + H^2 + D^2}$$

For a numerical example, let us calculate the true length of line AB in Fig. A·28(b). Here $L = 1\frac{1}{2}$ in., $H = 1$ in., and $D = \frac{1}{2}$ in. Then

$$AB = \sqrt{1.5^2 + 1^2 + 0.5^2} = 1.8708 \text{ in.}$$

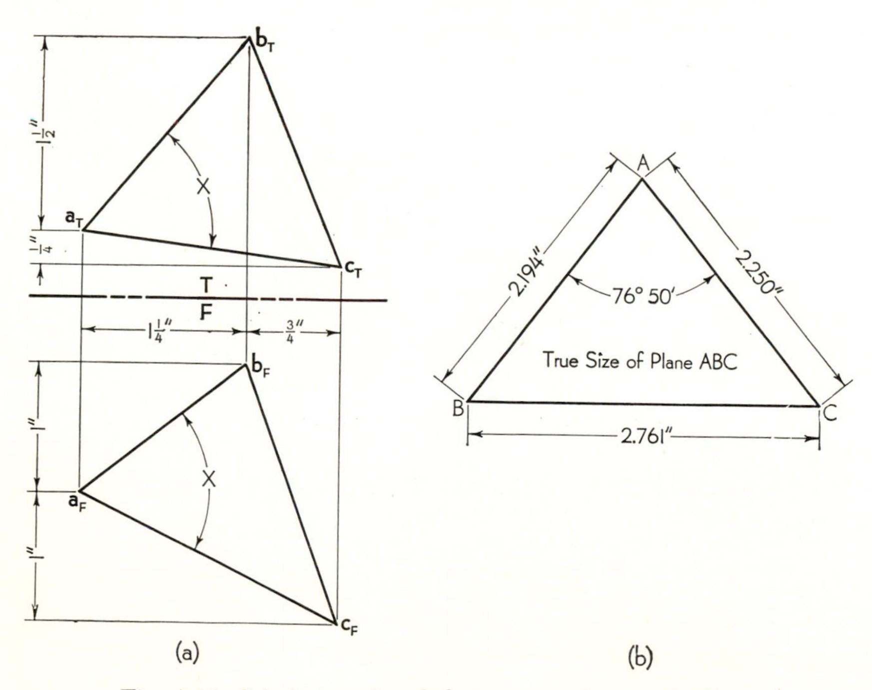

Fig. A·29. Calculation of angle between two intersecting lines.

A·29. To calculate the angle between two intersecting lines (Fig. A·29)

For the purpose of illustration, assume that it is required to calculate the true angle X between lines AB and AC of the triangle ABC shown in Fig. A·29(*a*). By the method of Art. A·28, the true lengths of lines AB, AC, and BC can each be calculated from the given coordinate dimensions. Then, as shown in Fig. A·29(*b*), the true size of plane ABC is a triangle whose sides are equal to the calculated true lengths.

The required angle X can now be calculated from the formula

$$\cos X = \frac{(AB)^2 + (AC)^2 - (BC)^2}{2(AB)(AC)}$$

Substituting numerical values,

$$\cos X = \frac{2.194^2 + 2.250^2 - 2.761^2}{2(2.194)(2.250)} = 0.2279$$

$$X = 76°50'$$

Many other problems that involve only the true length of lines or the true size of a plane can easily be solved trigonometrically as described here. For the more complex problems, however, accurately calculated results can best be obtained by the methods of analytic geometry.

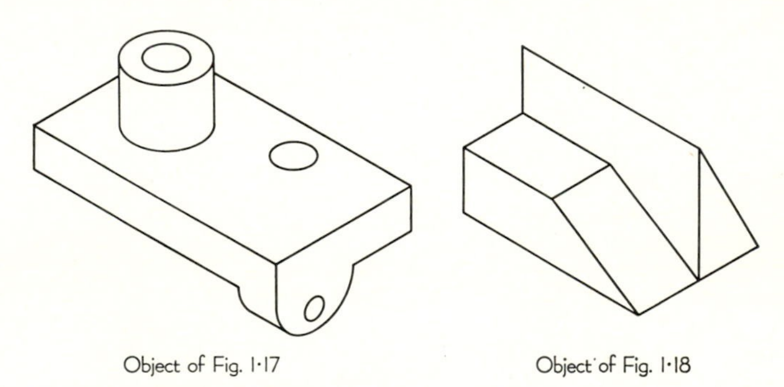

Fig. A·30. Pictorial drawings of objects in Chapter 1.

Answers to Fig. 3·21. Triangles (*b*), (*c*), and (*d*) are right triangles; (*a*) and (*e*) are not.

INDEX